# COSMIC RAYS FOR PARTICLE AND ASTROPARTICLE PHYSICS

Proceedings of the 12th ICATPP Conference

## Astroparticle, Particle, Space Physics, Radiation Interaction, Detectors and Medical Physics Applications

**Series Editors:** Claude Leroy *(Université de Montréal, Canada)*
Pier-Giorgio Rancoita *(Instituto Nazionale di Fisica Nucleare (INFN), Italy)*

The book series is dedicated to an up-to-date coverage of investigations, physics requirements, survey of technologies and performance of detectors employed — or to be employed — in fundamental and particle physics experiments at accelerators, underground laboratories, submarine facilities, and in space environment, including Earth magnetosphere and heliosphere, for astroparticle and cosmic-ray physics experiments and astrophysics research, special applications like medical imaging, radiotherapy, simulation codes for dose estimates in radiotherapy, biological effects, radiation effects on devices, synchrotron radiation and new advanced detectors.

---

Vol. 1 Advanced Technology and Particle Physics
Proceedings of the 7th International Conference on ICATPP-7
*edited by M. Barone, E. Borchi, J. Huston, C. Leroy, P.-G. Rancoita P. Riboni & R. Ruchti*

Vol. 2 Astroparticle, Particle and Space Physics, Detectors and Medical Physics Applications
Proceedings of the 8th Conference
*edited by M. Barone, E. Borchi, C. Leroy, P.-G. Rancoita, P.-L. Riboni & R. Ruchti*

Vol. 3 Astroparticle, Particle and Space Physics, Detectors and Medical Physics Applications
Proceedings of the 9th Conference
*edited by M. Barone, E. Borchi, A. Gaddi, C. Leroy, L. Price, P.-G. Rancoita & R. Ruchti*

Vol. 4 Astroparticle, Particle and Space Physics, Detectors and Medical Physics Applications
Proceedings of the 10th Conference
*edited by M. Barone, A. Gaddi, C. Leroy, L. Price, P.-G. Rancoita & R. Ruchti*

Vol. 5 Astroparticle, Particle and Space Physics, Detectors and Medical Physics Applications
Proceedings of the 11th Conference
*edited by C. Leroy, P.-G. Rancoita, M. Barone, A. Gaddi, L. Price & R. Ruchti*

Vol. 6 Cosmic Rays for Particle and Astroparticle Physics
Proceedings of the 12th ICATPP Conference
*edited by S. Giani, C. Leroy & P.-G. Rancoita*

# COSMIC RAYS FOR PARTICLE AND ASTROPARTICLE PHYSICS

Proceedings of the 12th ICATPP Conference

Villa Olmo, Como, Italy     7 – 8 October 2010


Editors

**S Giani**
CERN, Switzerland

**C Leroy**
University of Montreal, Canada

**P G Rancoita**
INFN Milano-Bicocca, Italy



World Scientific

NEW JERSEY • LONDON • SINGAPORE • BEIJING • SHANGHAI • HONG KONG • TAIPEI • CHENNAI

*Published by*

World Scientific Publishing Co. Pte. Ltd.
5 Toh Tuck Link, Singapore 596224
*USA office:* 27 Warren Street, Suite 401-402, Hackensack, NJ 07601
*UK office:* 57 Shelton Street, Covent Garden, London WC2H 9HE

**British Library Cataloguing-in-Publication Data**
A catalogue record for this book is available from the British Library.

**Astroparticle, Particle, Space Physics, Radiation Interaction, Detectors and Medical Physics Applications — Vol. 6**
**COSMIC RAYS FOR PARTICLE AND ASTROPARTICLE PHYSICS**
**Proceedings of the 12th ICATPP Conference**



ISBN-13 978-981-4329-02-6
ISBN-10 981-4329-02-9

Printed in Singapore by Mainland Press Pte Ltd.

**Organizing Committee:**

| | |
|---|---|
| S. Giani | CERN |
| C. Leroy | University of Montreal |
| L. Price | ANL |
| P.G. Rancoita | INFN Milano-Bicocca (chairman) |
| R. Ruchti | University of Notre Dame |

**Local organizers**

| | |
|---|---|
| A. Cazzaniga | Centro Volta, Conference Secretariat |
| F. Mercalli | Director of Centro Volta |

**Session Oraganizers and International Advisory Committee**

| | |
|---|---|
| A. Barrau | LPSC Grenoble |
| A. Chilingarian | Yerevan Institute of Physics |
| M. Cirelli | CNRS IPhT Saclay and CERN |
| F. Donato | INFN and University of Torino |
| J. Edsjoe | Stockholm University |
| M. Gervasi | INFN and University of Milano-Bicocca |
| A. Ibarra | TUM Munich |
| F. Jansen | DLR, Bremen |
| J. Kota | University of Arizona, Tucson |
| K. Kudela | Slovak Academy of Sciences, Kosice |
| L. Lazutin | Moscow State University |
| J. Pinfold | University of Alberta |
| M. S. Potgieter | North-WestUniversity, Potchefstroom |
| V.S. Ptuskin | IZMIRAN,Troitsk |
| S. Sarkar | Oxford University |
| P. Serpico | CERN and CNRS LAPTH Annecy |
| G. Servant | IPhT Saclay and CERN |
| Y. Stozhkov | Lebedev Physics Institute, Moscow |
| A.W. Strong | Max Planck Institute, Garching |
| F. Takahashi | IPMU, Tokyo |
| K. Zurek | University of Michigan, Ann Arbor |

**Scientific Organization Assistence:**

| | |
|---|---|
| S. Della Torre | INFN Milano-Bicocca and University of Insubria |
| D. Grandi | INFN Milano-Bicocca |

**Plenary Session Organizers:**

| | |
|---|---|
| M. Cirelli and A. Ibarra | Production of Cosmic Rays from Exotic Matter |
| F. Donato | Production and Propagation of Matter in the Galaxy |
| J. Pinfold | Experimental Observations of Cosmic Rays |
| S. Sarkar | Production of Cosmic Rays from Astrophysical Sources |

**Parallel Session Organizers:**

| | |
|---|---|
| M. Cirelli and A. Ibarra | Production of Cosmic Rays from Exotic Matter |
| F. Donato | Production and Propagation of Matter in the Galaxy |
| S. Giani and C. Leroy | Poster Session |
| K. Kudela and M. S. Potgieter | Sun, Heliosphere and Cosmic Rays |
| K. Kudela and L. Lazutin | Energetic Particles in the Magnetosphere and Atmosphere of the Earth |
| J. Pinfold | Experimental (and Future Experimental) Observations of Cosmic Rays |
| J. Pinfold | Double Beta Decay and Dark Matter Search for Astroparticle Physics |
| J. Pinfold and R. Ruchti | Broader Impacts Activities in Cosmic Ray Science |
| S. Sarkar and P. Serpico | Production of Cosmic Rays from Astrophysical Sources |

**The Article Committee:**

| | |
|---|---|
| S. Giani | CERN |
| C. Leroy | University of Montreal |
| P.G. Rancoita | INFN Milano-Bicocca |

**Conference Secretariat:**

| | |
|---|---|
| A. Cazzaniga | Centro di Cultura Scientifica A. Volta, Como |
| N. Tansini | Centro di Cultura Scientifica A. Volta, Como |
| G. Fontanive | Centro di Cultura Scientifica A. Volta, Como |

# PREFACE

The 12th conference of the ICATPP (International Conference on Advanced Technology and Particle Physics) was dedicated to Cosmic Rays for Particle and Astroparticle Physics - Cosmic rays as a tool for probing fundamental particle physics, astrophysical sources and contents of the Universe - held during 7-8 October 2010 at the "Centro di Cultura Scientifica A. Volta". There were more than 120 participants representing more than 100 institutions from laboratories located in East and West Europe, USA, Canada, Africa and Asia. The Participants were experienced researchers but also graduate students and recent postdoctoral fellows, students receiving financial support from the Conference organization.

This conference was dedicated to presentations and discussions on observations and results obtained in cosmic ray physics in the energy range from sub-GeV to TeV and beyond, and their implications for fundamental physics, i.e., particle physics, astroparticle physics and cosmology. Among the topics covered by the conference were new results from space-based and terrestrial astroparticle experiments, including searches for dark matter and double beta decay, a better understanding of the production of ordinary and/or exotic matter, propagation of cosmic rays in the Galaxy and within the solar system, processes inducing trapped particles inside the Earth magnetosphere and their effects, space environment, Solar-Terrestrial physics.

The conference was aimed at promoting contacts among scientists involved in Solar-Terrestrial physics, space physics, particle physics, astroparticle physics and cosmology both from the theoretical and the experimental approach. The conference was devoted to physics and physics requirements, survey of theoretical models and performance of detectors employed (or to be employed) in experiments for fundamental physics, astroparticle physics, astrophysics research and space environment including Earth magnetosphere and Heliosphere and Solar-Terrestrial physics. Presentations included the following subjects: advances in physics from present and next generation ground and space experiments, dark matter, double beta decay, high-energy astrophysics, space environment, trapped particles, propagation of cosmic rays in the Earth magnetosphere, Heliosphere and Galaxy.

Plenary and parallel sessions covered:

**1** Production of Cosmic Rays from Exotic Matter (organized by M. Cirelli and A. Ibarra),
**2** Production and Propagation of Matter in the Galaxy (organized by F. Donato),
**3** Experimental Observations of Cosmic Rays (organized by J. Pinfold),
**4** Production of Cosmic Rays from Astrophysical Sources (organized by S. Sarkar and P. Serpico),
**5** Sun, Heliosphere and Cosmic Rays (organized by K.Kudela and M. S. Potgieter),
**6** Energetic Particles in the Magnetosphere and Atmosphere of the Earth (organized by K. Kudela and L. Lazutin),
**7** Double Beta Decay and Dark Matter Search for Astroparticle Physics (organized by J. Pinfold),
**8** Broader Impacts Activities in Cosmic Ray Science(organized by J. Pinfold and R. Ruchti).

Several rooms were dedicated to the poster exhibition organized by C. Leroy and S. Giani. From the quality of the material of the posters, it is clear that a lot of effort went into their preparation making the poster exhibition a truly important part of this Conference.

The Article Committee - with Simone Giani, Claude Leroy and Pier-Giorgio Rancoita - was set to follow the article submission, review and publication in the Conference Proceedings.

We would like to thank the staff of the Centro A. Volta for the excellent support provided to the Conference organization at Villa Olmo. In particular, we would like to extend our appreciation and thanks to the Secretariat of Centro di Cultura Scientifica A. Volta for their help and efficiency with the organization of the Conference and its running. The sponsorships of the University of Milano-Bicocca, the Direction de la Science de la Matière (DSM) - Commissariat à l'Énergie Atomique (CEA) - centre de Saclay and the Town of Como (for the free usage of the Villa Olmo) are gratefully acknowledged. The organizers would like to thank the strong support of INFN which made the conference possible.

Finally, we would like to thank the speakers for the high quality of their contributions and the participants for their enthusiasm and contributions to the discussions.

Article and Organizing Committee

Simone Giani
Claude Leroy
Pier-Giorgio Rancoita

Organizing Committee

Larry Price
Randal Ruchti

Local Organizers

Alessandra Cazzaniga
Franco Mercalli
December 2010

CONTENTS

# Broader Impacts Activities and Treaments

| | |
|---|---|
| A.S. Boldyrev | VHE spectral energy distribution of Crab Nebula compared with the prediction of a synchrotron self-Compton emission model |
| C. Consolandi | Nuclear and Non-Ionizing Energy-Loss for Coulomb Scattered Particles from Low Energy up to Relativistic Regime in Space Radiation Environment |
| E. Giroletti | Study of the natural radioactivity influence on ARGO-YBJ detector |
| S. Maltezos | High-accuracy determination of Fabry-Perot effective mirror spacing used for the receivers of atmospheric monitoring in VHE gamma ray astronomy |
| E. Memola | AMS-02 photon data reduction approach |
| K. Smolek | CZELTA: An overview of the Czech large-area time coincidence array |
| V. Sola | Calibration of the CMS Electromagnetic Calorimeter with first LHC data |
| A. Tartari | On the detectability of cosmic ray electron spectral featuresin the microwave/mm-wave range |
| M.C.S. Williams | Science in the schools - the Extreme Energy Events project |
| S.A. Wotton | A cosmic ray detector array for schools in the Cambridge region |
| O. Zaymidoroga | Observation of electroscalar radiation during a solar eclipse |
| J.L. Pinfold | Young researchers focus on the extreme energy universe |

# VHE SPECTRAL ENERGY DISTRIBUTION OF CRAB NEBULA COMPARED WITH THE PREDICTION OF A SYNCHROTRON SELF-COMPTON EMISSION MODEL

V.G. SINITSYNA, A.Y. ALAVERDIAN, A.S. BOLDYREV, S.S. BORISOV, R.M. MIRSAFATIKHOV, V.Y. SINITSYNA

*P.N.Lebedev Physical Institute,*
*Leninsky prospect 53, Moscow, 119991, Russia*
*E-mail: sinits@sci.lebedev.ru*

Crab Nebula has an extraordinary broad spectrum, attributed to synchrotron radiation of electrons with energies from GeV to PeV. This continuous spectrum appears to terminate near $10^8$ eV and photons, produced by relativistic electrons and positrons via Inverse Compton, form a new component of spectrum in GeV TeV energy range. Crab Nebula has been regularly observed by high mountain SHALON telescope. The integral spectra, spectral energy distribution and detailed images at $E > 0.8$TeV of Crab by SHALON are presented. Also, in order to find relation between TeV and X-ray emission and source characteristics, the combination of SHALON and Chandra images were analyzed. The TeV gamma-ray spectrum of Crab by SHALON is generated via Inverse Compton of soft, mainly optical, photons which are produced by relativistic electrons and positrons, in the nebula region around 1.5ʹfrom the pulsar with specific average magnetic field of about 67nT.

*Keywords*: Crab Nebula; TeV gamma-rays; Synchrotron Self-Compton Emission Model.

## Introduction

The observations on Tien-Shan high-mountain station with SHALON Atmospheric Cherenkov Telescopic System had been carried out since 1992 year.[1–4] The mirror telescopic system of SHALON consists of a composed mirror with the area of $11.2m^2$. It is equipped with a 144-photomultipher light-receiver that has the field of view $> 8°$ (the largest in the world). It enables one to continuously control the background of cosmic-ray particle emission and the atmosphere transparency, thus increasing the observation efficiency. During the period since 1992 12 metagalactic and galactic sources have been observed. Among them are galactic sources Crab Nebula, Cyg

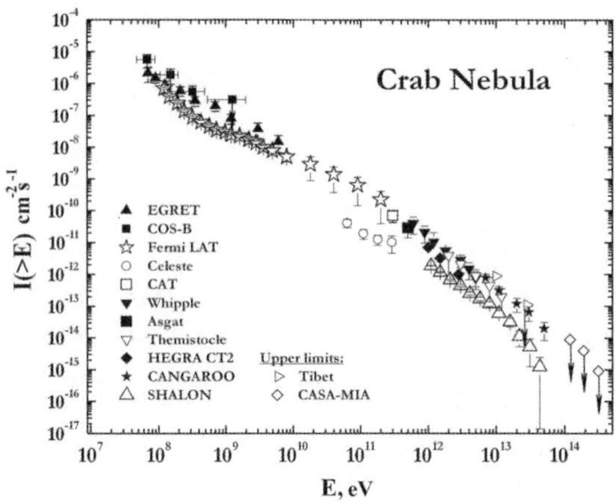


Fig. 1. The Crab Nebula gamma-quantum integral spectrum by SHALON[1–7] in comparison with other experiments: Fermi LAT, EGRET, COS-B, CELESTE, CAT, Asgat, Whipple, Themistocle, HEGRA CT2, CANGAROO, Tibet, CASA-MIA[8–16].

Table 1. The flux from Crab Nebula[4–17]

| Group | VHE Spectrum ($10^{-11}$ $photons\ cm^{-2}s^{-1}TeV^{-1}$) | $E_{th}$ (TeV) |
|---|---|---|
| Whipple (1991) | $25 \times (E/0.4TeV)^{-2.4\pm0.3}$ | 0.4 |
| Whipple (1998) | $(3.2 \pm 0.7) \times (E/TeV)^{-2.49\pm0.06_{stat}\pm0.04_{syst}}$ | 0.3 |
| SHALON (2009) | $(1.72 \pm 0.12) \times 10^{-1} \times (E/TeV)^{-2.40\pm0.07}$ | 0.8 |
| CANGAROO (1998) | $(2.01 \pm 0.36) \times 10^{-2} \times (E/7TeV)^{-2.53\pm0.18}$ | 7.0 |
| CAT (1999) | $(2.7 \pm 0.17 \pm 0.40) \times (E/TeV)^{-2.57\pm0.14_{stat}\pm0.08_{syst}}$ | 0.25 |
| HEGRA (1999) | $(2.7 \pm 0.2 \pm 0.8) \times (E/TeV)^{-2.60\pm0.05_{stat}\pm0.05_{syst}}$ | 0.5 |
| Magic (2005) | $(1.5 \pm 0.18) \times 10^{-3} \times (E/GeV)^{-2.58\pm0.16}$ | 0.3 |
| HESS (2005) | $(2.86 \pm 0.06 \pm 0.57) \times (E/TeV)^{-2.67\pm0.01_{stat}\pm0.1_{syst}}$ | 0.44 |
| | $(3.76 \pm 0.07) \times (E/TeV)^{-2.39\pm0.03} \times exp(E/(14.3 \pm 2.1))$ | |
| Tibet HD (1999) | $(4.61 \pm 0.1) \times 10^{-1} \times (E/3, TeV)^{-2.62\pm0.17}$ | 3.0 |

X-3, Tycho's SNR, Geminga and 2129+47)[1–7]. The results of observation data analysis for the each source are integral spectra of events coming from source - $k_{ON}$, and background events, coming simultaneously with source observation - $k_{OFF}$, temporal analysis of these two kind events and the source images. At Figs. 1, 2, and Table 1, the observation results Crab Nebula are showed.

## Crab Nebula

The Crab Nebula, most famous supernova remnant, pays an important role in the modern astrophysics. No other space object has such impact on the progress and development of the modern experimental and theoretical as-

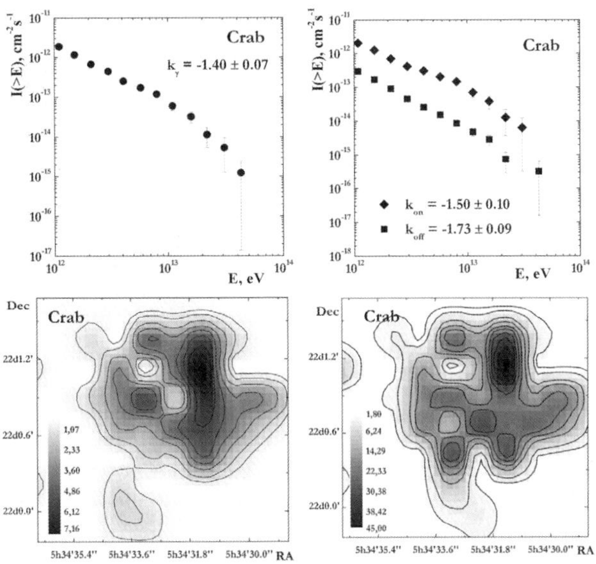


Fig. 2. **Top. left** – The Crab Nebula $\gamma$-quantum integral spectrum with power index of $k_\gamma = -1.40 \pm 0.07$; **right** – The event spectrum from Crab Nebula with background with index of $k_{ON} = -1.50 \pm 0.10$ and spectrum of background events observed simultaneously with Crab Nebula with index $k_{OFF} = -1.73 \pm 0.09$. **Bottom. left** – The image of $\gamma$-ray emission from Crab; **right** – The energy image (TeV units) of Crab by SHALON.

trophysics methods. Pulsar located at the centre of the Crab Nebula is the first whos optical, X-ray and $\gamma$-emission has been detected. As in many other bands of electromagnetic spectrum, the Crab Nebula has become the standard candle for TeV gamma-ray astronomy. It is available as steady source to test and calibrate the telescope and can be seen from both hemispheres. Since the first detection with ground based telescope the Crab has been observed by the number of independent groups using different methods of registration of gamma-initiated showers. Some of these detections are presented below and shown on Fig. 1. The SHALON observation results of well-known gamma-source Crab Nebula are presented at Figure 1 in comparison with other experiments EGRET, COS-B, CELESTE, CAT, Asgat, Whipple, Themistocle, HEGRA CT2, CANGAROO, Tibet, CASA-MIA. The spectrum of gamma rays from the Crab Nebula has been measured in the energy range 0.8 TeV to 30 TeV at the SHALON Alatoo Observatory by the atmospheric Cerenkov technique. The integral energy spectrum is well described by the single power law $I(>E_\gamma) \propto E_\gamma^{-1.40 \pm 0.07}$(Fig. 2, Table 1). The spectrum indices for Crab Nebula obtained by Whipple, SHALON,

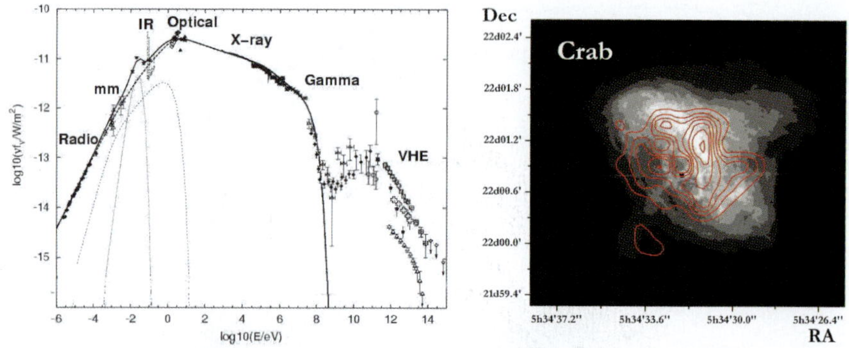


Fig. 3. **left** – The Crab Nebula $\gamma$-quantum integral spectrum by SHALON[1–7] in comparison with other experiments: Fermi LAT, EGRET, COS-B, CELESTE, CAT, Asgat, Whipple, Themistocle, HEGRA CT2, CANGAROO, Tibet, CASA-MIA ;[8–16] **right** – A Chandra X-ray image of Crab Nebula. The central part 200" × 200" of Crab Pulsar Wind Nebula (PWN) in the energy range 0.2-4 keV. The contour lines show the TeV - structure by SHALON observations.

Table 2. Magnetic Fields and Lifetimes[18]

| Region | Field ($10^4$ Gauss) | Lifetime (years) | Extent of region (light years) |
|---|---|---|---|
| PWN average | 5.8 | 6.7 | 1.9 (60" radius) |
| bright Torus | 7.7 | 4.3 | 1.7 (radius) |
| NW Loops | 9.1 | 2.9 | 0.6 |
| Jet center, region 7 | 9.1 | 2.9 | 1.8 (length of jet) |
| bright inner ring | 10.8 | 2.6 | 0.08 (thickness of ring) |
| knot in bright inner ring | 15.3 | 1.5 | 0.07 (size of knot) |
| S Finger, region 5 | 6.2 | 5.9 | 1.3 |
| SW Finger, region 7 | 6.2 | 5.9 | 1.6 |

CANGAROO, CAT, HEGRA atmospheric Cherenkov telescopes, Magic, HESS and Tibet are presented in Table 1 and .[4–17] Also, the results of observation data analysis are the images of Crab. A detailed image of gamma-ray emission from Crab Nebula by SHALON telescope is shown in Fig. 2.

Crab Nebula has an extraordinary broad spectrum, attributed to synchrotron radiation of electrons with energies from GeV to PeV. This continuous spectrum appears to terminate near $10^8$ eV and photons, produced by relativistic electrons and positrons ($\sim 10^{15}$ eV) via Inverse Compton, form a new component of spectrum in GeV TeV energy range. To made a description of the intensity and spectral shape in the TeV region of ¿0.8 TeV, the model of inverse Compton scattering of the ambient photons in

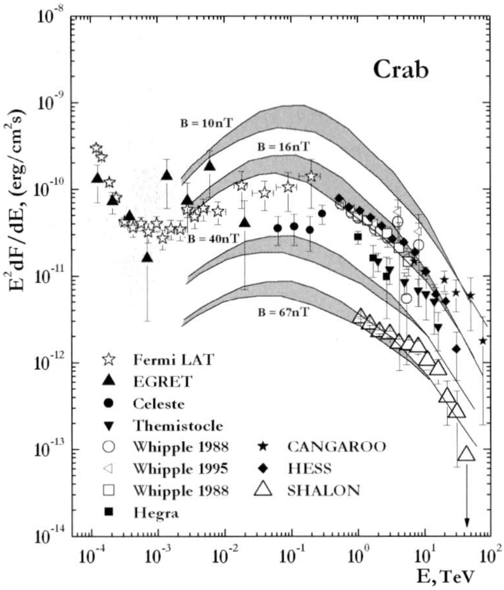


Fig. 4. The Crab Nebula $\gamma$-quantum spectrum by SHALON together with other experiments:[4–17] Fermi, Celeste, Whipple, EGRET, CANGAROO, HEGRA, Magic, HESS and with the predicted inverse Compton spectrum for the different field strengths [11] and for the 67 nT.

the nebula in the Ref.[11] is used. For the purpose of calculating the Inverse Compton scattered radiation, the complete spectrum of Crab Nebula need to be taken into account to deduce the spectrum of relativistic electrons.[11]

Additionally, we need the assuming about magnetic field strength in the region of emission (Figs. 2, 3). The average magnetic field in the region of VHE gamma-ray emission is extracted from the comparison of $0.8 - 30$ TeV (SHALON data) and X-ray (Chandra data) emission regions. In order to find relation between TeV and X-ray emission and characteristics which are necessary to calculate a VHE gamma-ray spectrum, the combination of SHALON and Chandra images were analyzed.

Figure 3 presents a Chandra X-ray image of the central part 200" $\times$ 200" of Crab Nebula in the energy range $0.2 - 4$ keV. In this energy band most of the PWN X-rays come from a torus surrounding the pulsar. The contour lines show the TeV - structure by SHALON observations. The most part of TeV energy gamma-quanta come from the region of bright torus (see Fig. 3, Table 2) whereas the contribution of energy gives the region of the southern jet (see Fig. 2). Magnetic fields and lifetimes of representative regions in

Chandra image of Crab have been derived[18] and are listed in Table 2.

Then using the complete spectrum of the Crab Nebula and extracted average magnetic field strength, the spectrum of relativistic electrons in the nebula is deduced. And finally, the spectrum of emitted Inverse Compton scattered photons (see. Fig. 4) can be deduced from the electron and low energy photon spectra .[11] The spectrum of the Inverse Compton emission that they would generate is in good agreement with the observed by SHALON gamma-ray flux, if the average magnetic field in the region where these scattered photons originate is 67nT (Table 2 and Figs. 3, 4).

## Conclusion

The TeV gamma-quantum spectrum of Crab by SHALON is generated via Inverse Compton of soft, mainly optical, photons which are produced by relativistic electrons and positrons, in the nebula region around 1.5' (see Figs. 3 and 4) from the pulsar with specific average magnetic field of about 67nT (Table 2).

## References

1. V. G. Sinitsyna, AIP Conf. Proc., 515, (1999) pp. 205 and 293.
2. V. G. Sinitsyna, S. I. Nikolsky, et al.,Izv. Ross. Akad. Nauk Ser. Fiz., 66(11),(2002) pp. 1667 and 1660.
3. V. G. Sinitsyna, et al.,in Proc.27th ICRC, Hamburg, 3, (2001) p. 2665; in Proc. 29th ICRC, Puna, 4, (2005) p. 231.
4. V. G. Sinitsyna et al.,Nucl. Phys. B (Proc.Suppl.), 151, (2006), p. 112;
5. V. G. Sinitsyna et al., Nucl. Phys. B (Proc.Suppl.), 122,(2003) p. 247;
6. S. I. Nikolsky and V. G. Sinitsyna, Nucl. Phys. B (Proc.Suppl.), 122, (2003) p. 409;
7. V. G. Sinitsyna et al., Nucl. Phys. B (Proc.Suppl.), 97,(2001) pp. 215,219.
8. T. C. Weekes, AIP Conf. Proc., 515, (1999) p. 3.
9. M. Catanese and T. C.Weekes, Preprint Series, no. 4811, (1999).
10. J. Prahl and C.Prosch (for the HEGRA Collaboration), in Proc. 25th Int. Cosmic Ray Conf., Durban, 3, (1997) p. 217.
11. A. M. Hillas, et al., ApJ, 503, (1998) p. 744.
12. T. Tanimori, et al., ApJ, 492, (1998) p. 133.
13. F. Piron, et al., in Proc. 28th ICRC, Tsukuba, (2003) p. 2607.
14. F. Aharonian, A.G. Akhperjanian, et al. ApJ, 539, (2000) p. 317.
15. C. Masterson, et al., in Proc. 29th ICRC, Puna, 4 (2005) p. 143.
16. A. A. Abdo, et al., ApJ, 708 (2010) p. 1254.
17. R. M. Wagner, et al., in Proc. 29th ICRC, Puna, 4 (2005) p. 163.
18. F. D. Seward, W. H. Tucker and R. A. Fesen, ApJ. 652 (2006) p. 1277.

# Nuclear and Non-Ionizing Energy-Loss for Coulomb Scattered Particles from Low Energy up to Relativistic Regime in Space Radiation Environment

M.J. Boschini[1,2], C. Consolandi[*,1], M. Gervasi[1,3], S. Giani[4], D. Grandi[1], V. Ivanchenko[4], S. Pensotti[3], P.G. Rancoita[**,1], M. Tacconi[1]

[1] *INFN-Milano Bicocca, P.zza Scienza,3 Milano, Italy*
[2] *CILEA Via R. Sanzio, 4 Segrate, MI-Italy*
[3] *Milano Bicocca University, Piazza della Scienza, 3 Milano, Italy*
[4] *CERN, Geneva, 23, CH-1211, Switzerland*
[*] *E-mail: cristina.consolandi@mib.infn.it*
[**] *E-mail: piergiorgio.rancoita@mib.infn.it*

In the space environment, instruments onboard of spacecrafts can be affected by displacement damage due to radiation. The differential scattering cross section for screened nucleus–nucleus interactions - i.e., including the effects due to screened Coulomb nuclear fields -, nuclear stopping powers and non-ionization energy losses are treated from about 50 keV/nucleon up to relativistic energies.

## 1. Introduction

In the space environment near Earth, low energy particles are, for instance, found trapped within the radiation belts and are partially coming from the Sun. On the other hand, the energies of Galactic Cosmic Rays (GCR) extend up to relativistic range. Protons are the most abundant, but alpha particles and heavier nuclei are also present (e.g., see Sections 4.1.2–4.1.2.5 of Ref. [1]). Abundances and energy spectra of GCRs depend on the position inside the solar cavity and are affected by the solar activity. Above (30–50) MeV/nucleon, the dominant radiation consists of GCRs. At lower energies, from 1 MeV/nucleon up to about 30 MeV/nucleon, one also finds the so-called Anomalous Cosmic Rays (ACRs). GCRs can reach Earth's magnetosphere and interact with upper layers of the atmosphere. These interactions produce secondary particles, like for example protons with (10–100) MeV energies, which may - in turn - become trapped within the radiation belts. In addition, during transient phenomena like solar flares and coronal mass ejections, Solar Energetic Particles (SEP) are produced in the energy range from few keV's to GeV's.

All these energetic particles can inflict permanent damages to onboard electronic devices employed in space missions. While passing through matter, they can lose energy by Coulomb interactions with electrons (*electronic energy-loss*) and nuclei (*nuclear energy-loss*) of the material. In particular, the nuclear energy-loss - due to screened Coulomb scattering on nuclei of the medium - is relevant for the creation of permanent defects inside the lattice of the material; thus, for instance, it is mostly responsible for the displacement damage which is a cause of degradation of silicon devices.

The developed model - presented in this article - for screened Coulomb elastic scattering up to relativistic energies is included into Geant4 distribution [2] and is available with Geant4 version 9.4 (December 2010).

## 2. Nucleus–Nucleus Interactions and Screened Coulomb Potentials

At small distances from the nucleus, the potential energy is a Coulomb potential, while - at distances larger than the Bohr radius - the nuclear field is screened by the fields of atomic electrons. The interaction between two nuclei is usually described in terms of an interatomic Coulomb potential (e.g., see Section 2.1.4.1 of Ref. [1] and Section 4.1 of Ref. [3]), which is a function of the radial distance $r$ between the two nuclei

$$V(r) = \frac{zZe^2}{r}\,\Psi_{\rm I}(r_{\rm r}), \tag{1}$$

where $ez$ (projectile) and $eZ$ (target) are the charges of the bare nuclei and $\Psi_{\rm I}$ is the *interatomic screening function.* This latter function depends on the *reduced radius* $r_{\rm r}$ given by

$$r_{\rm r} = \frac{r}{\rm a_I}, \tag{2}$$

where $\rm a_I$ is the so-called *screening length* (also termed *screening radius*). In the framework of the Thomas–Fermi model of the atom (e.g., see Chapters 1 and 2 of Ref. [4]) - thus, following the approach of ICRU Report 49 (1993) -, a commonly used screening length for $z = 1$ incoming particles is that from Thomas–Fermi (e.g., see Refs. [5, 6])

$$\rm a_{TF} = \frac{C_{TF}\,a_0}{Z^{1/3}}, \tag{3}$$

and - for incoming particles with $z \geq 2$ - that introduced by Ziegler, Biersack and Littmark (1985) (and termed *universal screening length*[a])

$$\mathrm{a_U} = \frac{C_{\mathrm{TF}}\,\mathrm{a_0}}{z^{0.23} + Z^{0.23}}, \tag{4}$$

where

$$\mathrm{a_0} = \frac{\hbar^2}{me^2}$$

is the Bohr radius, $m$ is the electron rest mass and

$$C_{\mathrm{TF}} = \frac{1}{2}\left(\frac{3\,\pi}{4}\right)^{2/3} \simeq 0.88534$$

is a constant introduced in the Thomas–Fermi model.

The simple scattering model due to Wentzel [10] - with a single exponential screening-function $\Psi_{\mathrm{I}}(r_{\mathrm{r}})$ {e.g., see Ref. [10] and Equation (21) in Ref. [11]} - was repeatedly employed in treating single and multiple Coulomb-scattering with screened potentials (e.g., see Ref. [11] - and references therein - for a survey of such a topic and also Refs. [12–15]). The resulting elastic differential cross section differs from the Rutherford differential cross section by an additional term - the so-called *screening parameter* - which prevents the divergence of the cross section when the angle $\theta$ of scattered particles approaches 0°. The screening parameter $A_{\mathrm{s,M}}$ [e.g., see Equation (21) of Bethe (1953)] - as derived by Molière (1947, 1948) for the single Coulomb scattering using a Thomas–Fermi potential - is expressed[b] as

$$A_{\mathrm{s,M}} = \left(\frac{\hbar}{2\,p\,a_{\mathrm{I}}}\right)^2 \left[1.13 + 3.76 \times \left(\frac{\alpha z Z}{\beta}\right)^2\right] \tag{5}$$

where $a_{\mathrm{I}}$ is the screening length - from Eqs. (3, 4) for particles with $z = 1$ and $z \geq 2$, respectively; $\alpha$ is the fine-structure constant; $p$ ($\beta c$) is the momentum (velocity) of the incoming particle undergoing the scattering onto a target supposed to be initially at rest; $c$ and $\hbar$ are the speed of light and the reduced Planck constant, respectively. When the (relativistic) mass

---

[a] Another screening length commonly used is that from Lindhard and Sharff (1961) (e.g., see Ref. [8] ; see also Ref. [9] and references therein):

$$\mathrm{a_L} = \frac{C_{\mathrm{TF}}\,\mathrm{a_0}}{\left(z^{2/3} + Z^{2/3}\right)^{1/2}}.$$

[b] It has to be remarked that the screening radius originally used in Refs. [12, 13] was that from Eq. (3).

- with corresponding rest mass $m$ - of the incoming particle is much lower than the rest mass ($M$) of the target nucleus, the differential cross section - obtained from the Wentzel–Molière treatment of the single scattering - is:

$$\frac{d\sigma^{\rm WM}(\theta)}{d\Omega} = \left(\frac{zZe^2}{p\,\beta c}\right)^2 \frac{1}{(2A_{\rm s,M}+1-\cos\theta)^2} \tag{6}$$

$$= \left(\frac{zZe^2}{2\,p\,\beta c}\right)^2 \frac{1}{\left[A_{\rm s,M}+\sin^2(\theta/2)\right]^2} \tag{7}$$

(e.g., see Section 2.3 in Ref. [11] and references therein). Equation (7) differs from Rutherford's formula - as already mentioned - for the additional term $A_{\rm s,M}$ to $\sin^2(\theta/2)$. The corresponding total cross section {e.g., see Equation (25) in Ref. [11]} per nucleus is

$$\sigma^{\rm WM} = \left(\frac{zZe^2}{p\,\beta c}\right)^2 \frac{\pi}{A_{\rm s,M}(1+A_{\rm s,M})}. \tag{8}$$

Thus, for $\beta \simeq 1$ (i.e., at very large $p$) and with $A_{\rm s,M} \ll 1$, from Eqs. (5, 8) one finds that the cross section approaches a constant:

$$\sigma_{\rm c}^{\rm WM} \simeq \left(\frac{2\,zZe^2 a_{\rm I}}{\hbar c}\right)^2 \frac{\pi}{1.13+3.76\times(\alpha z Z)^2}. \tag{9}$$

In case of a scattering under the action of a central potential (for instance that due to a screened Coulomb field), when the rest mass of the target particle is no longer much larger than the relativistic mass of the incoming particle, the expression of the differential cross section must properly be re-written - in the center of mass system - in terms of an "effective particle" with momentum ($p'_{\rm r}$) equal to that of the incoming particle ($p'_{in}$) and rest mass equal to the relativistic reduced mass

$$\mu_{\rm rel} = \frac{mM}{M_{1,2}},$$

where $M_{1,2}$ is the invariant mass; $m$ and $M$ are the rest masses of the incoming and target particles, respectively (e.g., see Refs. [15–17] and references therein). The "effective particle" velocity is given by:

$$\beta_{\rm r} c = c\sqrt{\left[1+\left(\frac{\mu_{\rm rel}c}{p'_{in}}\right)^2\right]^{-1}}.$$

Thus, the differential cross section[f] per unit solid angle of the incoming particle results to be given by

$$\frac{d\sigma^{\mathrm{WM}}(\theta')}{d\Omega'} = \left(\frac{zZe^2}{2\,p'_{in}\,\beta_{\mathrm{r}}c}\right)^2 \frac{1}{\left[A_{\mathrm{s}} + \sin^2(\theta'/2)\right]^2}, \tag{10}$$

with

$$A_{\mathrm{s}} = \left(\frac{\hbar}{2\,p'_{in}\,a_{\mathrm{I}}}\right)^2 \left[1.13 + 3.76 \times \left(\frac{\alpha z Z}{\beta_{\mathrm{r}}}\right)^2\right] \tag{11}$$

and $\theta'$ the scattering angle in the center of mass system.

The energy[c] $T$ transferred to the recoil target is related to the scattering angle as $T = T_{max}\sin^2(\theta'/2)$ - where $T_{max}$ is the maximum energy which can be transferred in the scattering (e.g., see Section 1.5 of Ref. [1]) -, thus, assuming an isotropic azimuthal distribution one can re-write Eq. (10) in terms of the kinetic energy transferred from the projectile - , i.e., $[-T]$, where the negative sign indicates that energy is lost by the projectile - to the recoil target as

$$\begin{aligned} d\sigma^{\mathrm{WM}} &= \left(\frac{zZe^2}{2\,p'_{in}\,\beta_{\mathrm{r}}c}\right)^2 \frac{1}{\left[A_{\mathrm{s}} + \sin^2(\theta'/2)\right]^2} \sin(\theta')\,d\theta' \int_0^{2\pi} d\phi \\ &= \pi \left(\frac{zZe^2}{p'_{in}\,\beta_{\mathrm{r}}c}\right)^2 \frac{T_{max}}{\left[T_{max}\,A_{\mathrm{s}} + T\right]^2}\, d[-T]. \end{aligned} \tag{12}$$

Finally, from Eq. (12), the differential cross section with respect to the kinetic recoil energy ($T$) of the target is given by:

$$\frac{d\sigma^{\mathrm{WM}}(T)}{dT} = \pi \left(\frac{zZe^2}{p'_{in}\,\beta_{\mathrm{r}}c}\right)^2 \frac{T_{max}}{\left[T_{max}\,A_{\mathrm{s}} + T\right]^2}. \tag{13}$$

---

[f]By inspection of Eqs. (5, 7, 10, 11), one finds that for $\beta_{\mathrm{r}} \approxeq 1$ the cross section is given by Eq. (9).

[c]One can show - e.g., see Section 1.5 of Ref. [1] - that the four momentum transfer is given by

$$t = -2MT.$$

Since $t$ is invariant, then the kinetic energy transferred is also invariant. Furthermore, since $T = T_{max}\sin^2(\theta'/2)$, then one finds that

$$d[-T] = T_{max}\,d[-\sin^2(\theta'/2)] = \frac{T_{max}}{2}\sin(\theta')\,d\theta'$$

(e.g., see Section 1.5 of Ref. [1]).

Furthermore, since

$$\begin{aligned}\beta_{\rm r} c &= \frac{p\,c^2}{E} \\ p'_{in} &= \frac{pM}{M_{1,2}} \\ T_{max} &= \frac{2p^2 M}{M_{1,2}^2}\end{aligned} \tag{14}$$

with $p$ and $E$ the momentum and total energy of the incoming particle in the laboratory, then one finds

$$\frac{T_{max}}{(p'_{in}\,\beta_{\rm r} c)^2} = \frac{2E^2}{p^2 M c^4}.$$

Therefore, Eq. (13) can be re-written as

$$\frac{d\sigma^{\rm WM}(T)}{dT} = 2\,\pi\left(zZe^2\right)^2 \frac{E^2}{p^2 M c^4}\,\frac{1}{\left[T_{max}\,A_{\rm s} + T\right]^2}. \tag{15}$$

Equation (15) expresses - as already mentioned - the differential cross section as a function of the (kinetic) energy $T$ achieved by the recoil target.

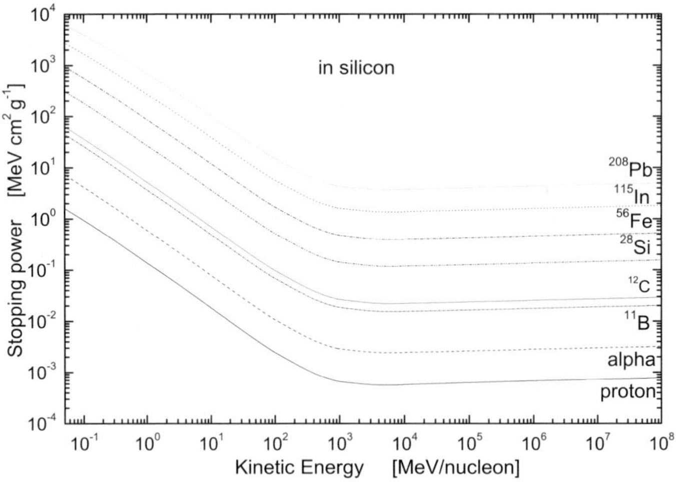


Fig. 1. Nuclear stopping power - in $\mathrm{MeV\,cm^2\,g^{-1}}$ - calculated using Eq. (17) in silicon is shown as a function of the kinetic energy per nucleon - from 50 keV/nucleon up 100 TeV/nucleon - for protons, $\alpha$-particle and $^{11}$B-, $^{12}$C-, $^{28}$Si-, $^{56}$Fe-, $^{115}$In-, $^{208}$Pb-nuclei.

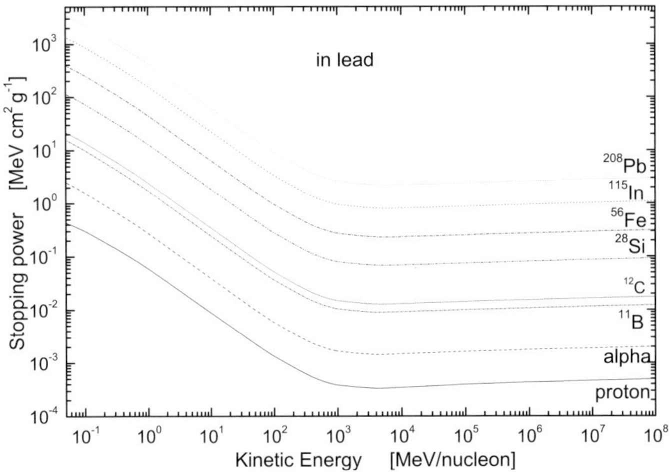


Fig. 2. Nuclear stopping power - in MeV cm$^2$ g$^{-1}$ - calculated using Eq. (17) in lead is shown as a function of the kinetic energy per nucleon - from 50 keV/nucleon up 100 TeV/nucleon - for protons, $\alpha$-particles and $^{11}$B-, $^{12}$C-, $^{28}$Si-, $^{56}$Fe-, $^{115}$In-, $^{208}$Pb-nuclei.

## 3. Nuclear Stopping Power

Using Eq. (15) the nuclear stopping power - in MeV cm$^{-1}$ - is obtained as

$$-\left(\frac{dE}{dx}\right)_{\rm nucl} = n_A \int_0^{T_{max}} \frac{d\sigma^{\rm WM}(T)}{dT}\, T\, dT \tag{16}$$

$$= 2\, n_A \pi \left(zZe^2\right)^2 \frac{E^2}{p^2 M c^4} \int_0^{T_{max}} \frac{T}{\left[A_{\rm s} T_{max} + T\right]^2}\, dT$$

$$= 2\, n_A \pi \left(zZe^2\right)^2 \frac{E^2}{p^2 M c^4} \left[\frac{A_{\rm s}}{A_{\rm s}+1} - 1 + \ln\left(\frac{A_{\rm s}+1}{A_{\rm s}}\right)\right] \tag{17}$$

with $n_A$ the number of nuclei (atoms) per unit of volume and, finally, the negative sign indicates that the energy is lost by the incoming particle (thus, achieved by recoil targets). For energies higher than a few tens of keVs, because $A_{\rm s} \ll 1$, Eq. (17) can be re-written as

$$-\left(\frac{dE}{dx}\right)_{\rm nucl} = 2\,\pi n_A \left(zZe^2\right)^2 \frac{E^2}{p^2 M c^4} \left[\ln\left(\frac{1}{A_{\rm s}}\right) - 1\right]. \tag{18}$$

It has to be noted that, as the incoming momentum increases to a value for which $p \simeq E$, the set of terms - in front of those included in brackets - decreases and approaches a constant; while the term $\ln(1/A_{\rm s})$ increases as $\ln(p)$ for $E \gg mc^2, Mc^2$ [e.g., see Eqs. (11, 14)]. Thus, a slight increase of the nuclear stopping power with energy is expected because of the decrease of the screening parameter with energy.

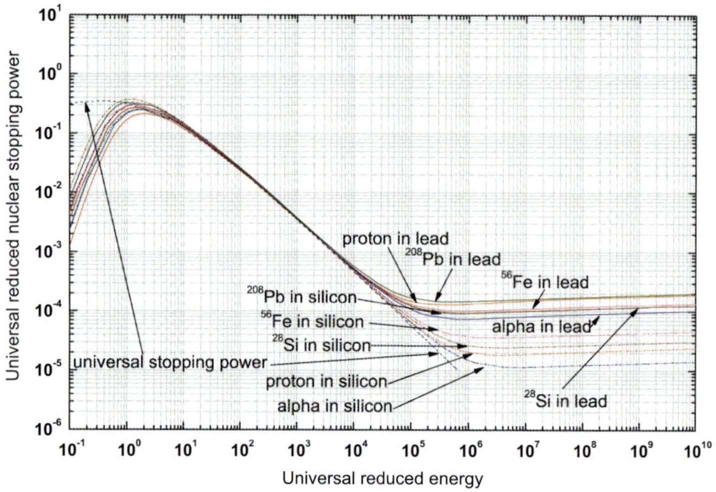


Fig. 3. Universal stopping power (dashed line) as a function of the universal reduced energy [Eq. (19)]. The other curves correspond to the dimensionless nuclear stopping power obtained from Eq. (17) - for protons, $\alpha$-particles, $^{28}$Si-, $^{56}$Fe-, $^{208}$Pb-nuclei in silicon (Fig. 1) and lead (Fig. 2) absorbers - and divided by the parameter $\mathbb{K}$ [Eq. (21)].

For instance, in Fig. 1 (Fig. 2) the nuclear stopping power in silicon (in lead) - in MeV cm$^2$ g$^{-1}$ - is shown as a function of the kinetic energy per nucleon - from 50 keV/nucleon up 100 TeV/nucleon - for protons, $\alpha$-particles and $^{11}$B-, $^{12}$C-, $^{28}$Si-, $^{56}$Fe-, $^{115}$In-, $^{208}$Pb-nuclei.

It has to be remarked that - at very low energies - the Wentzel–Molière nuclear stopping power [Eq. (17)] differs from that obtained by Ziegler, Biersack and Littmark (1985) using the so-called *universal screening potential* (see also Ref. [18]). However, they have shown (e.g., see Figure 2-18 in Ref. [7] or, equivalently, Figure 2-18 in Ref. [18]) that different screening potentials - including the Bohr potential in which $\Psi_{\rm I}(r_{\rm r})$ is assumed to be an exponential function similarly to Wentzel's assumption - result in nuclear stopping powers which exhibit marginal differences for $\epsilon_{\rm r,U}$ above 10 (see also Fig. 3). $\epsilon_{\rm r,U}$ is the so-called *universal reduced energy* expressed as:

$$\epsilon_{\rm r,U} = \frac{\mathbb{R}}{zZ\left(z^{0.23}+Z^{0.23}\right)}\left(\frac{M}{m+M}\right)E_k, \tag{19}$$

where $E_k$ is in MeV and the numerical constant is $\mathbb{R} = 32.536\times10^3\,{\rm MeV}^{-1}$ {e.g., see Equation (2-73) of Ref. [7] or Equation (2-88) of Ref. [18], see also Section 2.1.4.1 of Ref. [1]}. For instance, in silicon $\epsilon_{\rm r,U} \simeq 10$ corresponds to $E_k \simeq 13$ keV [67 keV/nucleon] for protons [lead nuclei]. Ziegler, Biersack and Littmark (1985) provided a general expression for the nuclear stopping

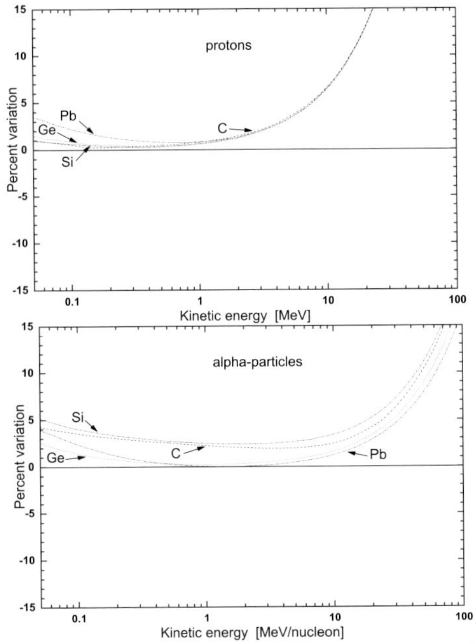


Fig. 4. Variation (in percentage) of nuclear stopping powers - calculated with Eq. (17) for energies from 50 keV/nucleon up to 100 MeV/nucleon - with respect to ICRU tabulated values [3] as a function of the kinetic energy in MeV/nucleon, for protons (top) and $\alpha$-particles (bottom) traversing amorphous carbon, silicon, germanium and lead media.

power (e.g., see Section 2.1.4.1 of Ref. [1]), i.e.,

$$-\left(\frac{dE}{dx}\right)^{\mathrm{U}}_{\mathrm{nucl}} = \mathbb{K}\,\mathfrak{R}(\epsilon_{\mathrm{r,U}}) \quad [\mathrm{MeV/cm}], \tag{20}$$

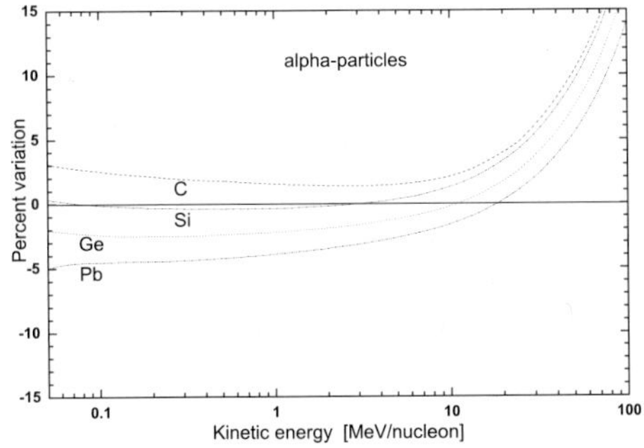


Fig. 5. Variation (in percentage) of nuclear stopping powers - calculated with Eq. (17) with the expressions (22, 23) for the screening parameter - with respect to ICRU tabulated values [3] as a function of the kinetic energy in MeV/nucleon, for $\alpha$-particles traversing amorphous carbon, silicon, germanium and lead media for energies from 50 keV/nucleon up to 100 MeV/nucleon.

with

$$\mathbb{K} \simeq 5.1053 \times 10^{3} \frac{\rho z Z}{A\,(1 + M/m)\,(z^{0.23} + Z^{0.23})} \quad [\mathrm{MeV/cm}] \tag{21}$$

where $\rho$ and $A$ are the density and atomic weight of the target medium, respectively; $\mathfrak{R}(\epsilon_{\mathrm{r,U}})$ - the so-called (*universal*) *reduced nuclear stopping power* [termed, also, (*universal*) *scaled nuclear stopping power*] given in Equations (2-89)–(2-90) in Ref. [18] (see also page 80 in Ref. [1]) - is dimensionless. Additionally, the present calculations can be compared with values of stopping powers - obtained using the universal screened potential - available in SRIM (2008) [19]. Usually an agreement - to better than a few percents - is achieved down to about 150 keV/nucleon, where - for instance - one finds $\approx 5.5\ (9.9)\,\%$ for $\alpha$-particles (lead ions) in silicon. At large energies, the non-relativistic approach due to Ziegler, Biersack and Littmark (1985) becomes less appropriate and deviations from stopping powers calculated by means of the universal screening potential are expected and observed for $\epsilon_{\mathrm{r,U}} \gtrsim (1.5\text{–}2.5) \times 10^{4}$ (e.g., see Fig. 3).

The non-relativistic approach - based on the universal screening potential - of Ziegler, Biersack and Littmark (1985) was also used by ICRU (1993) to calculate nuclear stopping powers - currently available on the web (e.g., see Ref. [20]) - due to protons and $\alpha$-particles in materials. ICRU (1993) used as screening lengths those from Eqs. (3, 4) for protons and $\alpha$-particles, respectively. In Fig. 4, the variation (in percentage) of nuclear stopping powers - calculated with Eq. (17) - with respect to ICRU tabulated values [3] is shown as a function of the kinetic energy per nucleon (in MeV/nucleon) - for energies from 50 keV/nucleon up to 100 MeV/nucleon - for protons and $\alpha$-particles traversing amorphous carbon, silicon, germanium and lead media. The stopping powers for protons ($\alpha$-particles) from Eq. (17) are less than $\approx 5\%$ larger than those reported by ICRU (1993) from 50 keV/nucleon up to $\approx 8$ MeV (19 MeV/nucleon) - the upper energy corresponds to $\epsilon_{\mathrm{r,U}} \approx 1.6 \times 10^{4}$ $(2.6 \times 10^{4})$ for protons in an amorphous carbon ($\alpha$-particles in a silicon) medium -. At larger energies the stopping powers from Eq. (17) largely differ from those from ICRU - as expected - due to the complete relativistic treatment of the present approach.

The simple screening parameter used so far [Eq. (11)] - derived by Molière (1947) - can be modified by means of a *practical correction*, i.e.,

$$A'_{\mathrm{s}} = \left(\frac{\hbar}{2\,p'_{in}\,a_{\mathrm{I}}}\right)^{2} \left[1.13 + 3.76 \times \mathbb{C} \left(\frac{\alpha z Z}{\beta_{\mathrm{r}}}\right)^{2}\right], \tag{22}$$

to achieve a better agreement with low energy calculations of Ziegler, Biersack and Littmark (1985). For instance, for $\alpha$-particles and heavier ions, with

$$\mathbb{C} = (10\pi z Z\alpha)^{0.12} \tag{23}$$

the stopping powers obtained from Eq. (17) - in which $A'_\mathrm{s}$ replaces $A_\mathrm{s}$ - differ from the values of SRIM (2008) by less than $\approx 4.7\ (3.6)\,\%$ for $\alpha$-particles (lead ions) in silicon down to about 50 keV/nucleon. With respect to the tabulated values of ICRU (1993), the agreement for $\alpha$-particles is usually better than 4% at low energy down to 50 keV/nucleon (Fig. 5) - a 5% agreement is achieved at about 50 keV/nucleon in case of a lead medium. At very high energy, the stopping power is slightly affected when $A'_\mathrm{s}$ replaces $A_\mathrm{s}$: for example, a) in silicon at 100 TeV/nucleon the nuclear stopping power of $\alpha$-particles (lead-ions) is decreased by about 0.5 (1.9) % and b) in lead it is decreased by about 1.0 (2.7) %. It has to be remarked that a more appropriate expression for the screening parameter and a practical correction factor may require a further understanding.

## 4. Non-Ionizing Energy Loss due to Coulomb Scattering

A relevant process - which causes permanent damage to the silicon bulk structure - is the so-called *displacement damage* (e.g., see Chapter 4 of Ref. [1], Refs. [21–23] and references therein). Displacement damage may be inflicted when a *primary knocked-on atom* (PKA) is generated. The interstitial atom and relative vacancy are termed Frenkel-pair (FP). In turn, the displaced atom may have sufficient energy to migrate inside the lattice and - by further collisions - can displace other atoms as in a collision cascade. This displacement process modifies the bulk characteristics of the device and causes its degradation. The total number of FPs can be estimated calculating the energy density deposited from displacement processes. In turn, this energy density is related to the *Non-Ionizing Energy Loss* (NIEL), i.e., the energy per unit path lost by the incident particle due to displacement processes.

In case of Coulomb scattering on nuclei, the non-ionizing energy-loss can be calculated using the Wentzel–Molière differential cross section [Eq. (15)] discussed in Sect. 2, i.e.,

$$-\left(\frac{dE}{dx}\right)^{\mathrm{NIEL}}_{\mathrm{nucl}} = n_A \int_{T_d}^{T_{max}} T\,L(T)\,\frac{d\sigma^{\mathrm{WM}}(T)}{dT}\,dT\ , \tag{24}$$

where $E$ is the kinetic energy of the incoming particle, $T$ is the kinetic energy transferred to the target atom, $L(T)$ is the fraction of $T$ deposited

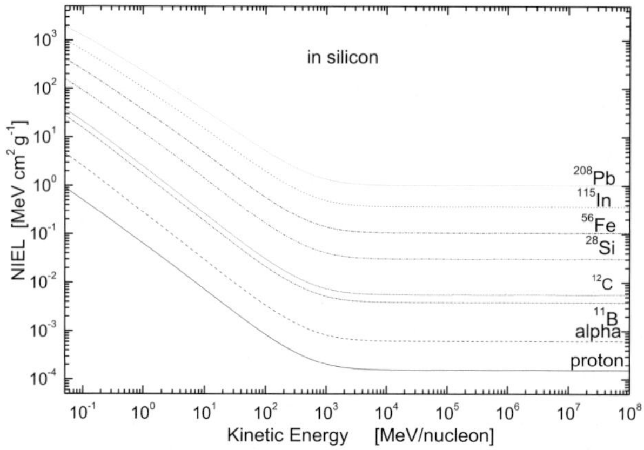


Fig. 6. Non-ionizing stopping power - in MeV cm$^2$ g$^{-1}$ - calculated using Eq. (24) in silicon is shown as a function of the kinetic energy per nucleon - from 50 keV/nucleon up 100 TeV/nucleon - for protons, $\alpha$-particles and $^{11}$B-, $^{12}$C-, $^{28}$Si-, $^{56}$Fe-, $^{115}$In-, $^{208}$Pb-nuclei. The threshold energy for displacement is 21 eV in silicon.

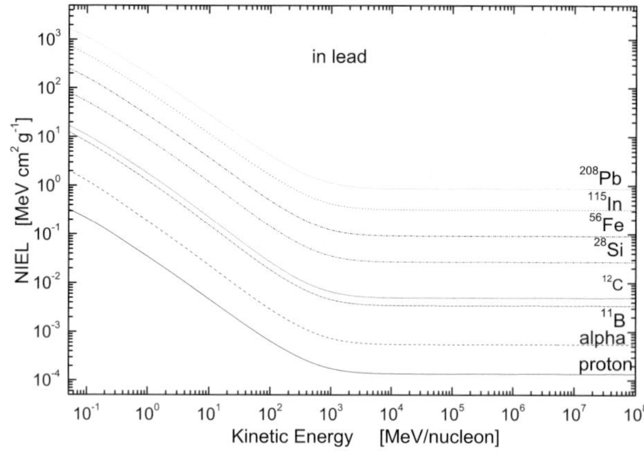


Fig. 7. Non-ionizing stopping power - in MeV cm$^2$ g$^{-1}$ - calculated using Eq. (24) in lead is shown as a function of the kinetic energy per nucleon - from 50 keV/nucleon up 100 TeV/nucleon - for protons, $\alpha$-particles and $^{11}$B-, $^{12}$C-, $^{28}$Si-, $^{56}$Fe-, $^{115}$In-, $^{208}$Pb-nuclei. The threshold energy for displacement is 25 eV in lead

by means of displacement processes. The expression of $L(T)$ - the so-called *Lindhard partition function* - can be found, for instance, in Refs. [24, 25] and in Equations (4.94, 4.96) of Section 4.2.1.1 in Ref. [1] (see also references therein). $T_{\rm de} = T\,L(T)$ is the so-called *damage energy*, i.e., the energy deposited by a recoil nucleus with kinetic energy $T$ via displacement damages inside the medium. The integral in Eq. (24) is computed from the minimum energy $T_d$ - the so-called *threshold energy for displacement*, i.e., that energy necessary to displace the atom from its lattice position - up to

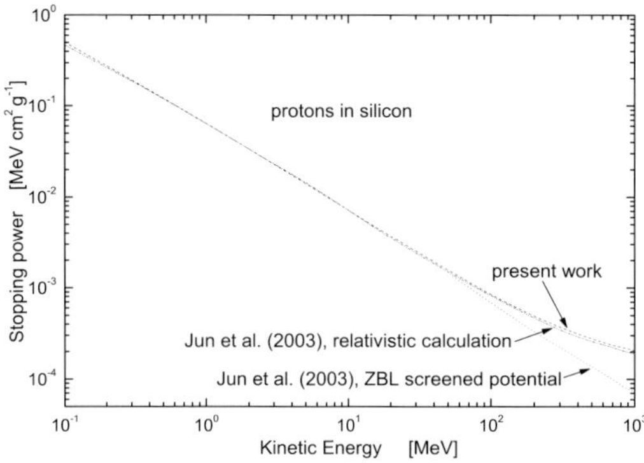


Fig. 8. Non-ionizing stopping powers of protons in silicon with energies from 100 keV up to 1 GeV: data for the dashed-dotted and dotted lines are the relativistic and Ziegler-Biersack-Littmark screened potential calculations from Jun and collaborators (2003), respectively (e.g., see Ref. [26]); data for the dashed line are obtained from Eq. (17) (e.g, see Fig. 1).

the maximum energy $T_{max}$ that can be transferred during a single collision process. $T_d$ is about 21 eV in silicon (e.g., see Table 1 in Ref. [26] and references therein) and 25 eV in lead (e.g., see Table 22 at page 83 in Ref. [27] and references therein). For instance, in Fig. 6 (Fig. 7) the non-ionizing energy loss - in $\mathrm{MeV\,cm^2\,g^{-1}}$ - in silicon (lead) is shown as a function of the kinetic energy per nucleon - from 50 keV/nucleon up 100 TeV/nucleon - for protons, $\alpha$-particles and $^{11}$B-, $^{12}$C-, $^{28}$Si-, $^{56}$Fe-, $^{115}$In-, $^{208}$Pb-nuclei. As already remarked by Boschini and collaborators (2010), at high energy the Coulomb NIEL - similarly to the nuclear stopping power - does not decrease with energy as it is found by Ziegler, Biersack and Littmark (1985) or in other calculations based on their universal screening potential derived in the framework of a non-relativistic treatment of the screened Coulomb scattering.

Furthermore, Jun and collaborators (2003) have already demonstrated that a relativistic treatment [24] of Coulomb scattering of protons - with kinetic energies above 50 MeV - upon silicon results into a non-ionizing energy loss which is larger than that expected from calculations using the Ziegler–Biersack–Littmark screened potential with a universal screening length (e.g., see Refs. [7, 25, 26]). The relativistic cross section used for treating the Coulomb scattering is the one derived by McKinley and Feshbach (1948) to describe the scattering of electrons on nuclei (e.g., see Section 4.2.1.4 of Ref. [1] and references therein). Seitz and Koehler (1956) suggested that - when the mass of the projectile is much lower than the target rest-mass (e.g., see Section 13 of Ref. [29] and references therein)

- this cross section can also describe - although screening effects are neglected - the scattering of protons and light nuclei, thus, providing - at high energy - a damage cross section which does not decrease with increasing energy. The data from Jun and collaborators (2003) - for protons with energies from 100 keV up to 1 GeV - are shown in Fig.8: the Ziegler–Biersack–Littmark (1985) screened potential was used to treat the Coulomb scattering of protons with energies lower than 50 MeV. In the same figure, the data obtained using Eq. (17) - e.g., see Fig. 1 - are also shown. There is an agreement to better than $\approx 6.5\%$ - achieved at $\approx 1$ GeV - between the results obtained by Jun and collaborators (2003) and the present calculations.

## 5. Conclusions

The treatment of nucleus–nucleus interactions due to relativistic Coulomb scatterings with screened potentials - like the present approach based on Wentzel–Molière scattering - allows one to determine both the total and differential cross sections, thus, to calculate the resulting nuclear and non-ionizing stopping powers. At high energies, the nuclear stopping powers exhibit a very slight logarithmic increase with energy. At low energies - i.e., above $\approx 50$ keV/nucleon and up to $\approx 8$ and 19 MeV/nucleon, for protons and $\alpha$-particles, respectively -, the present results are in agreement to better than $\approx 5\%$ with ICRU tabulated values of stopping powers obtained using the universal screening potential for the scattering of protons and $\alpha$-particles in matter. An agreement to better than to (5.5–9.9) % - for instance, for a silicon medium - is found with SRIM (2008) values down to 150 keV/nucleon. Furthermore, these calculations are also in agreement to better than $\approx 6.5\%$ with those obtained by Jun and collaborators (2003) for the non-ionizing stopping powers of protons - with energies from 50 MeV up to 1 GeV - in silicon.

Finally, it has to be remarked that - with a simple correction factor applied to the screening factor - an agreement to a few percents can be achieved with SRIM (2008) stopping powers down to about 50 keV/nucleon. However, a more appropriate expression for the screening parameter and a practical correction factor may require a further understanding.

## References

1. C. Leroy and P.G. Rancoita, *Principles of Radiation Interaction in Matter and Detection*, 2nd Edition, World Scientific (Singapore) 2009.

2. S. Agostinelli et al., Geant4 a simulation toolkit, *Nucl. Instr. and Meth. in Phys. Res. A* 506 (2003), 250-303; see also the website: *http://geant4.cern.ch/*
3. ICRU, ICRU Report 49, Stopping Powers and Ranges for Protons and Alpha Particles (1993).
4. I.M. Torrens, *Interatomic Potentials*, Academic Press (New York) 1972.
5. L.H. Thomas, *Proc. Cambridge Phil. Soc.* 23 (1927), 542.
6. E. Fermi, *Z. Phys.* 48 (1928), 73–79.
7. J.F. Ziegler, J.P. Biersack and U. Littmark, The *Stopping Range of Ions in Solids*, Vol. 1, Pergamon Press (New York) 1985.
8. J. Lindhard and M. Sharff, *Phys. Rev.* 124 (1961), 128–130.
9. S. Kalbitzer and H. Oetzmann, *Phys. Lett. A* 59 (1976), 197–198.
10. G. Wentzel, *Z. Phys.* 40 (1926), 590–593.
11. J.M. Fernandez-Vera et al., *Nucl. Instr. and Meth. in Phys. Res. B* 73 (1993), 447–473.
12. von G. Molière, *Z. Naturforsh.* A2 (1947), 133–145; A3 (1948), 78.
13. H. A. Bethe, *Phys. Rev.* 98 (1953), 1256–1266.
14. A. V. Butkevick et al., *Nucl. Instr. and Meth. in Phys. Res. A* 488 (2002), 282-194.
15. M. Boschini et al., *Geant4-based application development for NIEL calculation in the Space Radiation Environment*, Proc. of the 11th ICATPP Conference, October 5–9 2009, Villa Olmo, Como, Italy, World Scientific, Singapore (2010), 698–708, IBSN: 10-981-4307-51-3.
16. A. Staruszkiewicz and K. Zalewski *Acta Phys. Pol. B* **8 (no. 10)** (1977), 815–817.
17. P.P. Fiziev and I.T. Todorov, *Phys. Rev. D* **63** (2001), 104007-1–104007-9.
18. J.F. Ziegler, M.D. Ziegler and J.P. Biersack, *The Stopping and Range of Ions in Matter*, SRIM Co. (Chester.) 2008.
19. SRIM: J.F. Ziegler, M.D. Ziegler and J.P. Biersack, The Stopping and Range of Ions in Matter, version 2008.03 (2008), available at: *http://www.srim.org/*
20. M.J. Berger, J.S. Coursey, M.A. Zucker and J. Chang, data from ESTAR, PSTAR and ASTAR (version 1.2.3, August 2005); available on the web site (2010): *http://physics.nist.gov/Star*; originally published as: M.J. Berger, NISTIR 4999, NIST, Gaithersburg, MD (1993).
21. C. Leroy and P.G. Rancoita, *Reports on Progress in Physics* 70, 4 (2007) 493–625.
22. E. Borchi et al., *Nucl. Instr. and Meth. in Phys. Res. A* 279 (1989), 277–280.
23. C. Consolandi et al., *Nucl. Instr. and Meth. in Phys. Res. B* 252 (2006), 276.
24. I. Jun, *IEEE Trans. on Nucl. Sci.* 48 (2001), 162–175
25. S.R. Messenger et al., *IEEE Trans. on Nucl. Sci.* 50 (2003), 1919–1923.
26. I. Jun, M.A Xapsos, S.R. Messenger, E.A. Burke, R.J. Walters, G.P. Summers and T. Jordan, *IEEE Trans. on Nucl. Sci.* 50 (2003), 1924–1928.
27. G.S. Was, *Fundamentals of radiation materials science: metals and alloys* Springer (Berlin), 2007.
28. A.Jr. McKinley and H. Feshbach *Phys. Rev.* 74 (1948), 1759.
29. F. Seitz and J.S. Koehler, *Solid State Physics* vol. 2, edited by F. Seitz and D. Turnbull, Academic Press Inc. (New York), 1956.

# STUDY OF THE NATURAL RADIOACTIVITY INFLUENCE ON ARGO-YBJ DETECTOR

I. Bolognino[1,2] , C. Cattaneo[2] , E. Giroletti*[1,2] , G. Liguori[1,2] , P. Salvini[2] , P. Vallania[3,4] , C. Vigorito[4,5]

[1] *Dipartimento di Fisica Nucleare e Teorica, Univerità di Pavia*
[2] *Istituto Nazionale di Fisica Nucleare, Sezione di Pavia*
[3] *INAF-IFSI Torino*
[4] *Istituto Nazionale di Fisica Nucleare, Sezione di Torino*
[5] *Dipartimento di Fisica Generale, Università degli studi di Torino*
*Italy*

* E-mail: elio.giroletti@unipv.it
on behalf of the ARGO-YBJ Collaboration

ARGO-YBJ is an extensive air shower detector located at the Yangbajing Cosmic Ray Laboratory (4300 m a.s.l., Tibet, P.R. China). It is made by a single layer of Resistive Plate Chambers organized into 153 units called "clusters"(7.6 x 5.7 $m^2$ of area). The lowest energy threshold of the experiment (about 1 GeV) is obtained in the "scaler operation mode".

In this paper the preliminary results of a study on the correlation between Radon concentration in the experimental hall and counting rates are presented.

*Keywords*: Low energy cosmic rays, Instrumentation, Extensive air showers, Natural radioactivity, Radon indoor, MC simulations

## 1. Introduction

ARGO-YBJ (Astrophysical Radiation Ground-based Observatory at YangBaJing) is an extensive air shower detector located at the Yangbajing Cosmic Ray Laboratory (4300 m a.s.l., Tibet, P.R. China).

ARGO-YBJ is made by a single layer of Resistive Plate Chambers (RPCs), housed in a large building (100x110 m$^2$). The detector has a modular structure: the basic module is a cluster (5.7 x 7.6 m$^2$), made of 12 RPCs (2.850 x 1.225 m$^2$ each). 130 clusters are organized in a full coverage carpet of 5600 m$^2$ with active area of $\sim$ 93%, surrounded by 23 additional clusters, extending the detector surface to $\sim$ 6700 $m^2$, up to (110 x 100) $m^2$ with a coverage of $\sim$ 20%.

Two main operation modes have been designed: shower mode, with an energy threshold of a few hundreds of GeVs, and scaler mode, with a threshold of about 1 GeV. In the scaler operation mode, for each cluster the signals generated by shower particles are put in coincidence in a narrow time window (150 ns) and measured by four independent scaler channels, giving the counting rates of $n \geq 1, \geq 2, \geq 3$,and $\geq 4$ hits, that are read by four indipendent scaler channels[1] . The corresponding measured counting rates are $\sim 40\ kHz$, $\sim 2\ kHz$, $\sim 300\ Hz$ and $\sim 120\ Hz$.

The study of the scaler mode counting rates has given unexpected results: our Montecarlo simulations are in agreement for $\geq 2$ hits meanwhile for channel $n \geq 1$ is about half of the measured value and counting rates are badly correlated with atmospheric pressure.[2] Since this lack of correlation is at the same time all over the carpet it was supposed that our counters are detecting physical phenomena also variable in time.

Following studies showed that Radon concentration in air can give an highly variable contribution to the single particle countings since some Radon daughters are gamma emitters and time variability of indoor Rn gas is expected to be significative (greater than 100%). On the other side contribution from natural emitters in soil is expected to be less variable in time so that this contribution is not as much relevant in transient signal analysis.

In this paper we focus on the radioactivity in the air, showing also preliminary analysis about correlation between counting rates and Radon concentration.

## 2. Influence of environmental variables on counting rates

Counting rate due to cosmic rays can be influenced by several effects. First of all, variations in the atmospheric pressure affect secondary particle flux: an increase (decrease) of the ground level atmospheric pressure results in a reduction (enhancement) of the background rate, because of the larger (smaller) absorption of the electromagnetic component. By monitoring the pressure at the detector position it is possible to correct the data for this effect.

Moreover, the experiment hall temperature modifies the detector performances, changing the gas density inside the RPCs and the bakelite electrodes resistivity.

Besides these well known effects we study here the small contribution ($\sim 1\%$) of the natural radioactivity due to Radon gas on scaler 1.

## 3. Simulations & Measures of ambient radioactivity

$^{222}Rn$ is an $^{238}U$ daughter and it comes out from the soil because it is an inert gas; its variations seem to be a chaotic phenomena, too.[3]

The presence of doors and fans implies that Radon concentration is not constant inside the experiment hall.[4] Radon daughters are gamma-rays emitters with emission energies up to 3 MeV and their concentration could influence the cluster counting rate. Since $^{222}Rn$ daughters have an half-life much shorter than their father, they reach the Rn activity in about two hours, even if other local conditions (i.e. air particulate concentration) can decrease their activity.[5]

Radon gas has been directly measured in place with two lucas cells, at the center and in the north side of the hall, since 2009.

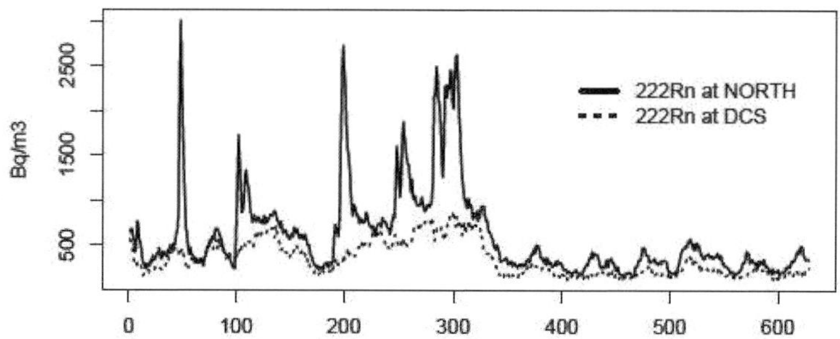


Figure 1. 13 days (one unit is equal to 30 minutes) Rn gas variation (Bq/m$^3$) inside ARGO-YBJ building in two places: at the carpet center (DCS position, dashed line) and at the hall north side.

As you can see in the Figure 1 Radon concentration in the north side is greater than in the center one: the reason is that Rn enters in the hall from this side. Rn at the center is taken as reference in this preliminary analysis, because this time serie represents the mean concentration inside the hall. Three campaignes of measurements made with passive detectors (CR-39) confirmed this result.

## 4. The analysis

We considered the temporal series of Radon concentration, scaler 1, pressure and gas temperature with the same span (every 30 minutes) and lapse of time (from the 2nd to 15th of June).

The factors influencing scaler 1 ($C_1$) can be seen in the following formula:

$$C_1(t) \sim \gamma(P,t) + B_{ck}(T,t) + Rn(t) \quad (1)$$

The first term represents the cosmic rays contributions depending on pressure (P) and time (t), the second one is the intrinsic noise that can change with temperature (T) and time and the last one is Radon (Rn) concentration. Other physical phenomena could influence scaler 1 in addition to these parameters.

In a first approximation $C_1$ can be written as the sum of the result of its linear fit depending on pressure and temperature and a term ($C_{1RESIDUE}(t)$ ), which is the residual after the fit includeing the Rn influence.

$$C_1(t) = a - bP(t) + cT(t) + C_{1RESIDUE}(t) \quad (2)$$

The next step was to calculate the correlation factor between $^{222}Rn$ and $C_{1RESIDUE}(t)$ that was expected to be good, when the influence of other physical parameters is negligible.

We studied different clusters placed in several part of the carpet. Figure 2 shows pressure, temperature, Scaler1 counting of cluster 104 (in the middle of the experiment hall) and Radon time series. Figure 3 shows the difference between $n \geq 1$ and its linear fit $(a - bP + cT)$ . Since the two series are not superimposed, the influence of other physical phenomena on counting rates is evident.

If we compare $C_{1RESIDUE}(t)$ and the Radon concentration (Fig. 4), it is clear how the two series are similar, being the correlation coefficient 0.94.

This data analysis was repeated for other clusters giving similar results; the best correlation coefficient belongs to cluster 104 . The worst correlation we found is refered to a cluster in the north side of the carpet: the correlation coefficient is 0.7. These results confirm the Fluka simulations[6] we did to assess Rn influence in different conditions. The difference between the correlation coefficient of cluster 104 (center) and the 91 (north) one, could be due to an influence of Radon at north of the carpet that is greater than at center. Analysis to check it are in progress.

This kind of analysis was repeated using scaler 2. This last should not be correlated with Radon concentration, because the probability that two gammas, producted by Rn daughters, hit the same cluster within a window of 150 ns is negligile. The results confirm our convictions because the value of the correlation coefficient is 0.2.

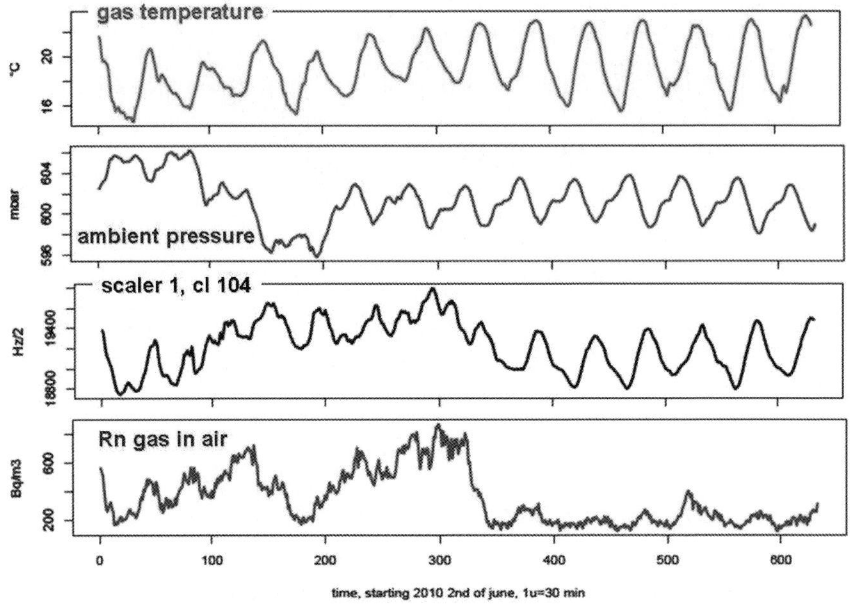


Figure 2. From the top: pressure, temperature, scaler 1 of cluster 104, and Radon vs time. The span is 30 minutes from 2nd June to 15th June.

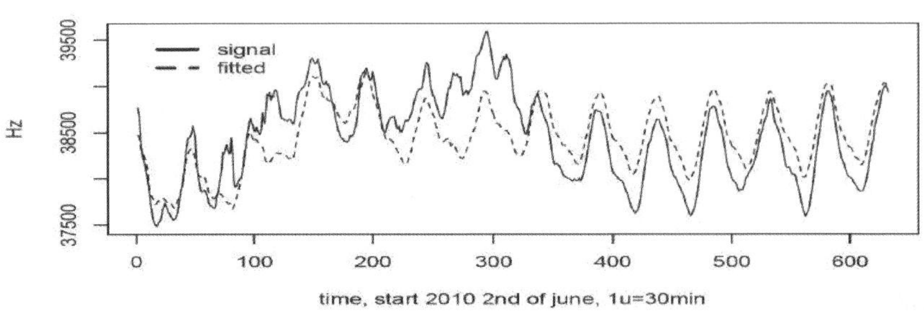


Figure 3. The dash line represents $C_1$ fit vs time while the other one scaler 1 cluster 104 vs time.

Moreover if we fit $n \geq 1$ also taking into account Radon contribution the $C_{1RESIDUE}$ obtained is very similar to scaler 2: both are representing the same signal.

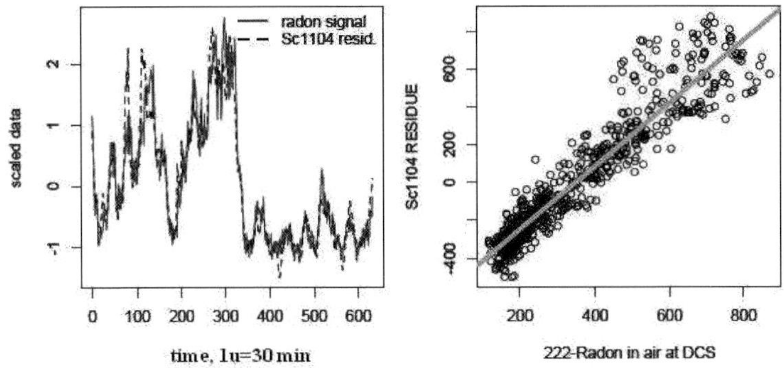


Figure 4. Left side: comparison between Radon concentration and the residue on scaler 1, cluster 104 (dashed line). Right side: correlation between the two series. The value of the correlation coefficient is 0.94.

## 5. Conclusions

Scaler 1 data without P and T corrections haven't a good correlation with the Radon concentration, but Radon influence doesn't allow a good correlation between scaler1 and pressure and temperature as expected.

Our analysis show good correlation between Rn and Scaler 1 if at the last one is subtracted the influence of pressure and of the temperature that is a predominant effect. We stress that the high correlation with Radon is a very good result since Radon influences counting rates only for about 1% according to Fluka simulations. This kind of test was repeated with other clusters all over the carpet: each cluster is an independent counter but similar values of correlation coefficient were obtained. This method confirms also that scaler 2 is not correlated with $^{222}Rn$ concentration, as expected, and it is well correlated with scaler 1 when the last one is "cleaned" by the pressure, temperature and Radon influence. We are repeating our analysis using other periods in order to generalize our results. Finally we are looking for other physical phenomena which are able to influence scaler 1.

## References


1. G. Aielli *et al.*, *Nucl. Instr. Meth. A* **562**, 92 (2006).
2. G. Aielli *et al.*, *Astrop. Phys.* **30**, 85 (2008).
3. J. Bejar, U. Facchini, E. Giroletti and S. Magnoni, *J. Environ. Radioactivity* **28**, 73 (1995).
4. NCRP, *Report 103, Chilton* (1989).
5. A. Nero and W. Nazaroff, *Whiley intersc, New York* (1988).
6. G. Battiston *et al.*, *Proc. of the Hadronic Shower Simulation Workshop, Fermilab* **896**, 31 (2007).

# HIGH – ACCURACY DETERMINATION OF FABRY – PEROT EFFECTIVE MIRROR SPACING USED FOR THE RECEIVERS OF ATMOSPHERIC MONITORING IN VHE GAMMA RAY ASTRONOMY

S. MALTEZOS*, E. FOKITIS, N. MARAGOS, V. GIKA, A. GEORGAKOPOULOU, E. KOUBLI AND G. KOUTSOURAKIS

*Physics Department, National Technical University of Athens, 9, Heroon Polytechniou Zografos, Athens, ZIP 15780, Greece*

In the framework of the design of a High Spectral Resolution Lidar (HSRL) for the atmospheric monitoring in High Energy Cosmic Ray Observatories, we demonstrate an accurate and non-invasive method for determination of the effective mirror spacing of the Fabry-Perot of one of the etalons being used for the receiver and having a nominal value of 5 mm. We used as a reference three very well known spectral lines of natural isotope mixture of Hg I at 435 and 546 and 579 nm. Furthermore, we present the investigation of the robustness of the obtained solution for the effective mirror spacing, by means of conclusions concerning the validity of the experimentally measured "excess fractions". Because as the light source of the HSRL prototype we intent to use an SLM Nd:$YVO_4$ laser it is very crucial to study and verify its spectral stability. This could be done using three Fabry–Perot etalons with precisely known mirror spacing.

## 1. Introduction

In the Imaging Atmospheric Cherenkov Technique (IACT) the atmosphere is itself part of the detector. It is where the particle shower is initiated by the incident gamma-ray and the medium the Cerenkov photons are propagated. The energy estimation of the individual gamma-ray is based on the calorimetric energy deposit in the atmosphere, which in turn can be measured recording the Cherenkov radiation. Any change in atmosphere quality and its constituents can affect the signal detected. Many current Cherenkov telescope arrays have in-situ single-scattering Elastic Lidars. Eventually, more accurate Lidars have to be installed at the sites of future Cherenkov telescopes in order to significantly reduce the systematic error in energy measurements and in deriving source fluxes.

The motivation of the present work is associated with the development of a prototype system for atmospheric monitoring dedicated for the aforementioned

* Corresponding author, e-mail: maltezos@central.ntua.gr

Cherenkov telescopes in gamma ray astronomy field. We are aiming to apply the technique of High Spectral Resolution Lidar using multi-wavelength emission at 1064 nm, 532 nm and down to near UV at 355 nm [1,2]. The high accuracy spectral resolution can be accomplished using Fabry-Perot etalons, one for the aerosol scattering signal (aerosol channel) and another for the molecular scattering signal (molecular channel). This feature can allow higher levels of accuracy in measuring the optical effects of the atmospheric constituents (i.e. aerosols). The characterization of the etalons being used, by means of accurate determination of the effective mirror spacing, is very important because we can use them to evaluate the stability of the light source. The accuracy of wavelength determination is quantified by the order of magnitude of wavelength broadening due to aerosol scattering which is of the order of 50 MHz.

In section 2, we describe the application topic which is the HSRL prototype, in section 3, we present the experimental demonstration of the method and its successive steps, in section 4, the result of effective mirror spacing of the etalon characterized and an analysis of the robustness of the solution is given. Finally, in section 5, the conclusions and some prospects are discussed.

## 2. Application topic: HSRL Prototype Design

For atmospheric monitoring, the HSRL and the associated technique takes advantage of the spectral distinction of the returned signal from aerosols and molecules and thereby to measure the aerosol extinction and backscatter coefficients at the same time. According to our design of a prototype HSRL, the spectral analysis is based on two Fabry-Perot etalons of different Free Spectral Ranges corresponding to mirror spacing, 5 mm and 100 mm, corresponding to molecular and aerosol channels respectively. As a light source, we plan to use an available DPSS CW SLM laser at 532 nm with power 100 mW and coherence length of the order of 50 m.

## 3. Experimental demonstration of the method

As effective mirror spacing we consider the final spacing after mounting the etalon using the required spacers between the mirrors. This effective mirror spacing is mostly affected by the thickness of the multi-layers deposited on the inner surfaces of the two mirrors and is of the order of 1 μm in each surface. In order to determine the effective mirror spacing we follow four separate steps. The first step includes the experimental data, that is, the capturing of interferograms. The choice of three wavelengths can provide highly robustness in the solution. Because the light source has to be very coherent, we used a low pressure discharge lamp of Hg I emitting lines of natural-isotope-mixture in air.

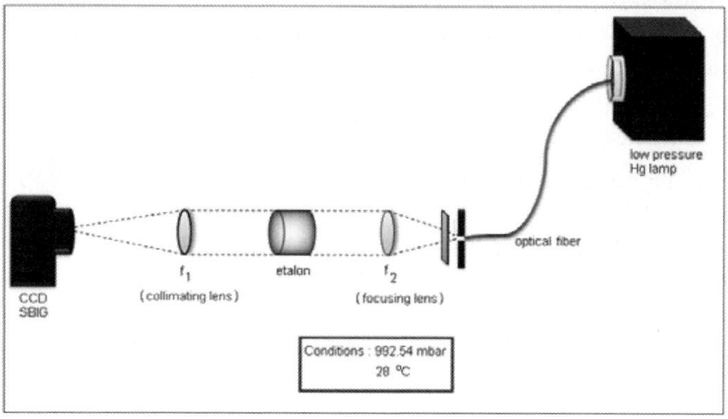


Figure 1. The configuration of the experimental setup for capturing the required interferograms.

The configuration of the experimental setup is illustrated in Figure 1. Below we describe the optical components used in more details.

1. CCD sensor: SBIG† 1600x1200 pixels (7.4 ×7.4 $mm^2$) controlled by computer software.
2. Collimating lens: $f_1$=50 mm.
3. Focusing lens: $f_2$=100 mm.
4. Etalon mirror spacing (nominal): *d*=5 mm.
5. Optical filter: bandpass, FWHM=10 nm.
6. Optical filter: bandpass, FWHM=50 nm.
7. Diaphragm: 2 mm wide.
8. Light source: low pressure discharge lamp of natural-isotope-mixture of Hg I in air.

The second step concerns the selection and further processing of the fringe patterns. By appropriate new homemade software we are able to select the fringe system belonging to an identified spectral line for further analysis and afterwards to produce the corresponding contours taking into account the intensity profile in the radial direction.

In the third step we have to determine the excess fractions of each fringe pattern corresponding to the wavelength selected and identified. The spectral

† Santa Barbara Instrument Group, 147-A Castilian Drive, Santa Barbara, CA 93117.

lines of Hg I which we identified and used were that having the following wavelengths: $\lambda_1 = 435.83277$ nm, $\lambda_2 = 546.07348$ nm and $\lambda_3 = 579.06630$ nm using eight significant digits. The method for determining the excess fraction with high accuracy has been presented recently in [3]. The wavelengths of the spectral lines have been taken from [4] and are consistent with the results of a more recent article [5]. In Figure 2 we give an indicative example of the fringe pattern obtained with the relatively broad optical filter where three fringe systems are produced at the same time.

## 4. Data analysis and results

In the fourth step we are investigating the solution to the problem. The central idea of the method is to find the optimum solution of the modular equation system among infinite possible ones (see [6] for more details). In a vicinity of the optimal solution point in Excess Fraction Space (EFS) the optimal value of thickness, $d_{opt}$, is locked-in due to properties of the modulo operator. We applied this method in the etalon of 5 mm spacing and also we went further: That is, we

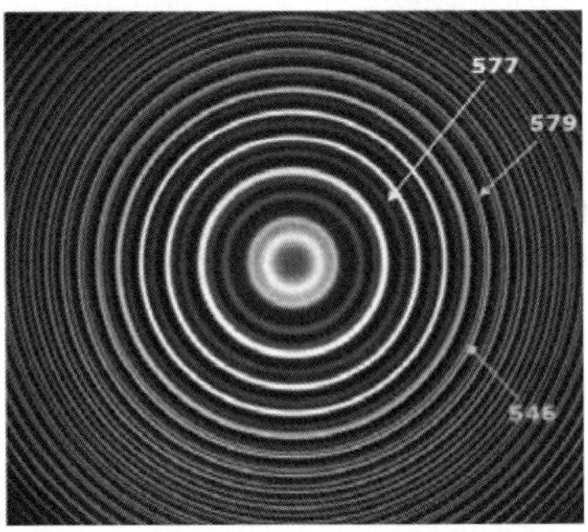


Figure 2. The obtained picture-interferogram using a 50 nm FWHM optical filter. The fringe systems produced by the lines 546, 577 and 579 nm are shown by the arrows.

investigated the criteria of robustness and, as well as, the uncertainty limits in the parameters used (see Figure 3). The three-equations system using 3 wavelengths, $\lambda_i$, $\lambda_j$, $\lambda_k$, can be expressed by a single equation as seen below:

$$f_m(d;\lambda_m,\varepsilon_m) = \left(\frac{2d}{\lambda_m} \bmod 1\right) - (\varepsilon_m + \Delta\varepsilon_m) \text{ , where } m = (i,j,k) \tag{1}$$

The wavelengths $\lambda_m$ have to be adjusted due to barometric pressure and temperature conditions with respect to standard ones used in the literature. Nevertheless, we have to apply a reliable model for the correction of the refractive index, by means of normalizing for slightly different atmospheric conditions (pressure, temperature and humidity). The function to be minimized is:

$$f(d,\Delta\varepsilon_i,\Delta\varepsilon_j,\Delta\varepsilon_k;\lambda_i,\lambda_j,\lambda_k,\varepsilon_i,\varepsilon_j,\varepsilon_k)=f_i^2+f_j^2+f_k^2 \qquad (2)$$

The data analysis is based on the method given in [4]. From the modeling with ellipses we found: $\varepsilon_1=0.74958$, $\varepsilon_2=0.25668$ and $\varepsilon_3=0.095654$.

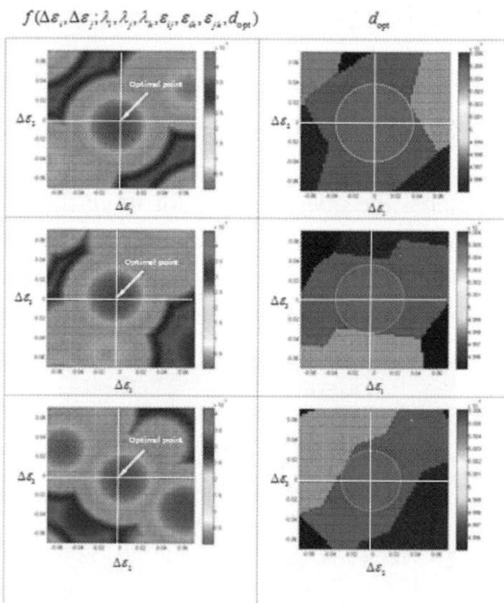


Figure 3. Projections of the variation of the function $f$ (plots in the left column) and the determined effective mirror spacing, $d_{opt}$ (plots in the right column) in a vicinity of the origin in EFS. The optimal solution point (0,0) in each case lies effectively in to the valleys of the function while the $d_{opt}$ is fixed in a large area of the space. The circles indicate the permitted margin in $\varepsilon_m$.

The accuracy of these values, according to our method, is estimated to be of the order of $10^{-5}$, but due to overlapping of the fringe patterns it should reduced to $10^{-4}$. In the optimal point the value of effective mirror spacing is $d_{opt}=5.00377(6)$ mm. This value is corrected by referring to the conditions of pressure and temperature held in our Laboratory (correction affecting the wavelengths). The error is affected mostly by the accuracy of the wavelengths used. According to the manufacturer, who measured the etalon spacers before mounting the result is, $d=5.001\pm0.00D_x$ mm, where $D_x$ is an unknown uncertain digit. The result we derived is not only consistent, but the precision we accomplished is much beyond by a factor of the order of 100. The free parameters of the model, $\Delta\varepsilon_m$, are used for more accurate minimization and are also useful for studying the robustness of the optimal solution found. For this study the parameters $\Delta\varepsilon_m$ are varied in pairs $\Delta\varepsilon_{mn}$, where $mn$=($i,j$), ($i,k$), ($j,k$), in a vicinity of the origin in the EFS ( $\Delta\varepsilon_{mn}$=0, $m\neq n$). In this study the function $f$ is modified, from the point of view of variables we are using, as follows:

$$f_{ijk,opt}(\Delta\varepsilon_m,\Delta\varepsilon_n,0;d_{\text{opt}},\lambda_i,\lambda_j,\lambda_k,\varepsilon_i,\varepsilon_j,\varepsilon_k)=f_i^2+f_j^2+f_k^2 \qquad (3)$$

From the diagrams shown in Figure 3 it is evident that we could come up to the equally accurate values of the $d$ (the regions with identical color).

## 5. Conclusions and Prospects

Our research is motivated by the need to characterize the laser used for atmospheric monitoring by the HSRL method. Towards this aim, a complete, robust and non-invasive method for determination of the effective mirror spacing of Fabry-Perot etalons with a precision of few nanometers has been developed, further extended and applied. This precision, in addition, allows measurement of absolute frequency of the laser and as well as quantification of its spectral fluctuations.

## Acknowledgments

We would like to thank Prof. I. Raptis for providing us, for the required period, his low pressure discharge lamp of Hg.

## References

1. S.T. Shipley, D. H. Tracy, E.W. Eloranta, J. T. Trauger, J. T. Sroga, F. L. Roesler, and J. A. Weinman, Appl. Opt., Vol. 22, No. 23 (1983).
2. E. Fokitis, P. Fetfatzis, A. Georgakopoulou, S. Maltezos and A. Aravantinos Nuclear Physics B (Proc. Suppl.) 190 (2009) 261–265.
3. S. Maltezos, E. Fokitis, P. Fetfatzis, G. Koutelieris and V. Gika, Proc. of 12th ICATPP09, Como, Italy (2009).
4. K. Burns, K. B. Adams and J. Longwell, Journal of the Opt. Soc. Of America, Vol. 40, No. 6 (1950).
5. E. B. Saloman, J. Phys. Chem. Ref. Data, Vol. 35, No. 4, (2006).
6. S. Maltezos, E. Fokitis, N. Maragos, V. Gika, E. Koubli, G. Koutsourakis ans A. Aravantinos, Presented at IPRD10 Conference, Siena, Italy (2010).

# AMS-02 photon data reduction approach

G. Boella[1,2], M.J. Boschini[1,4], C. Consolandi[1], S. Della Torre[1,3], M. Gervasi[1,2], D. Grandi[1], E. Memola[1], S. Pensotti[1,2], P.G. Rancoita[1], and M. Tacconi[1]

[1] *Istituto Nazionale di Fisica Nucleare, INFN Milano-Bicocca, Milano (Italy)*
[2] *Department of Physics, University of Milano Bicocca, Milano (Italy)*
[3] *Department of Physics, University of Insubria, Como (Italy)*
[4] *CILEA, Segrate (MI) (Italy)*
*E-mail: cristina.consolandi@mib.infn.it*
*E-mail: elisabetta.memola@mib.infn.it*

The Alpha Magnetic Spectrometer is going to be launched in April 2011 from the Kennedy Space Center. It will be located by the Space Shuttle on the International Space Station and will have its same lifetime: 10 years or even more. The experiment will observe high energy gamma-ray photons from several Astrophysical sources. We consider here the AMS-02 *single-photon mode*, i.e. gamma ray photons revealed by the ECAL through electromagnetic shower production. By means of Monte Carlo simulations and Test Beam data, we present a preliminary approach to the gamma-ray photon selection procedure and discuss the potentiality of the detector.

*Keywords*: gamma-ray photons; data reduction.

## 1. AMS-02

The Alpha Magnetic Spectrometer (AMS-02)[1,2] is a particle detector that is going to be launched in April 2011 from the Kennedy Space Center, onboard the shuttle Endeavour, within the mission STS-134. AMS-02 will operate on the International Space Station (ISS) for, at least, 10 years. This experiment is going to measure composition and intensity of cosmic rays with a precision never reached from other observatories, in space and on Earth. In addition to charged particles, the physics goals achivable by AMS-02 are greatly broadened by the detection of gamma-ray photons emitted from Galactic (Pulsars, Microquasars) and extragalactic (Gamma-Ray Bursts, Active Galactic Nuclei) sources in the Universe. The Universe, in fact, is largely transparent to the energies of the gamma-ray photons detectable by AMS-02 (from GeV to TeV). Unlikely high-energy cosmic

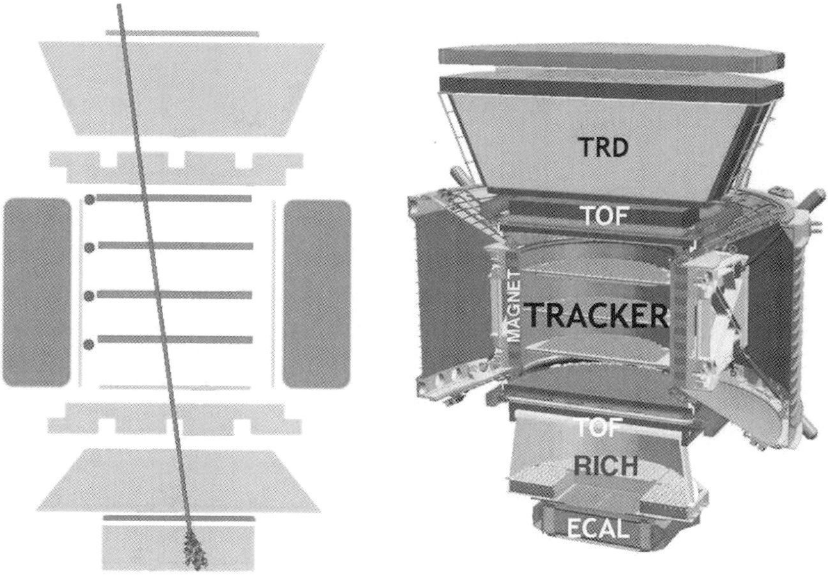


Fig. 1. AMS-02 single photon mode.

rays, which are deflected by magnetic fields, gamma-ray photons point back to their sources, and are ideal to provide a direct view into Nature's highest-energy acceleration processes.

## 2. ECAL Single-Photon Mode and Hadron Discrimination

AMS-02 will reveal photons by means of two gamma-ray detection modes: the *conversion mode*, and the *single-photon mode* (Fig. 1). The probability that a high energy photon converts into a pair before the inner silicon tracker layers is $\sim 20\%$, while $\sim 80\%$ is the remaining probability that such a photon produces an electromagnetic (e.m.) shower directly reaching the calorimeter (ECAL). Our goal is to present here how to approach the *single-photon mode* for the gamma-ray selection, in order to extract this signal from the cosmic-ray background (mainly protons).

Thanks to the ECAL *stand alone trigger*, it is possible to know that a particle has produced a shower inside the electromagnetic calorimeter. In addition, to discriminate an e.m shower gamma signal from an hadronic one produced by a proton, a shower shape analysis is necessary. For each reconstructed event, four estimators are generally used: *1)* the shower lat-

eral size ($E_{2MR}/E_{tot}$); *2)* the shower collimation (*Thrust* value); *3)* the shower *footprint*, and *4)* the depth at which the longitudinal shower has its maximum value ($T_{max}$).
In the next sections we will describe these estimators and show that to apply the correct cut value to each of them may provide a powerful tool to discriminate e.m. from hadronic events.

### 2.1. *Lateral Size:* $E_{2MR}/E_{tot}$

The transverse development of an e.m. shower, in different materials, scales fairly accurately with the Molière's radius ($R_M$), which, for the AMS-02 ECAL, is $R_M \sim 2$ cm. On average, for an e.m. shower only 10% of the total energy deposit ($E_{tot}$) is expected to lie outside a cylinder with radius $R_M$ centered around the reconstructed axis shower. For each reconstructed event, the ratio $E_{2MR}/E_{tot}$ represents the fraction of energy deposited in a cylinder of radius $2R_M$, which is expected to be larger than 90% for an e.m. shower. For an hadron shower, instead, a considerable fraction of $E_{tot}$ lies outside this cylinder. The distribution of $E_{2MR}/E_{tot}$ is therefore expected to be broader. The cut of events with $E_{2MR}/E_{tot} \lesssim 0.96$ contributes to suppress the spurious hadron signal.

### 2.2. *Shower Collimation: Thrust*

The *Thrust* is a widely used shape estimator to determine jets direction and collimation in accelerator physics. It is defined as:

$$\mathrm{T} = \mathrm{Max}\frac{\sum_i |\hat{\mathbf{n}} \cdot \mathbf{p}_i|}{\sum_i |\mathbf{p}_i|} \tag{1}$$

where the sum is performed over all the $i$-particles with momentum $\mathbf{p}_i$ that are generated inside the shower. The maximum refers to the direction of the trust axis indicated by the $\hat{\mathbf{n}}$ versor. Highly collimated events have $T \simeq 1$. A cut at $0.97 - 0.98$ is suggested to reject an hadronic signal.

### 2.3. *Shower Footprint*

The *footprint* value ($S_{FP}$) represents the shower transversal profile surface. It is possible to obtain a measure of the area of the *footprint* by computing the determinant of the tensor $\mathbf{I}_j$:

$$S_{FP} = \sum_j \sqrt{det(\mathbf{I}_j)}\ ; \qquad \mathbf{I}_j = \begin{bmatrix} \sigma_j^2 & \sigma_{jz} \\ \sigma_{jz} & \sigma_z^2 \end{bmatrix} \tag{2}$$

where the sum is performed over $j = x, y$ directions. The coefficients of the tensor are:

$$\sigma_j^2 = \frac{\sum_i^{ncell}(c_j^i - b_j)^2 E_i}{\sum_i E_i}$$
$$\sigma_{jz} = \frac{\sum_i^{ncell}(c_j^i - b_j)(c_z^i - b_z)E_i}{\sum_i E_i}$$
$$\sigma_z^2 = \frac{\sum_i^{ncell}(c_z^i - b_z)^2 E_i}{\sum_i E_i}$$

where $c_i$ are the coordinates of the $i_{th}$ cell in the AMS-02 reference frame, $b_i$ are the shower center of gravity ones, and $E_i$ is the energy deposited in the $i_{th}$ cell. The sum is performed over all the cells associated to the shower. A cut in the range $1.2 - 5$ can be used to discriminate hadron showers.

### 2.4. *Shower-Max Estimator:* $T_{Max}$

The maximum of the longitudinal profile of e.m. showers occurs at a depth proportional to the logarithm of the shower energy. It can be described by the following parametrization[3,4]:

$$\langle T_{Max}\rangle = log(E_{tot}[\text{MeV}]/9.6) - 0.3 \tag{3}$$

where $\langle T_{Max}\rangle$ represents the mean value of the e.m. shower maximum. The effective value of the e.m. shower maximum $T_{Max}$ instead is calculated event by event from the fit of the longitudinal profile of the shower. By comparing effective and mean values a selection cut may be applied. An equivalent method considers the following parametrization[5]:

$$T_{MaxP} = 6.5 + 1.2log(E_{tot}[\text{GeV}]). \tag{4}$$

In this case events with $T_{Max}/T_{MaxP} > 1$ are rejected. We use this method in the present paper for e.m. events selection cuts.

## 3. Using Monte Carlo and Test Beam data to Unveil a Gamma-Ray Signal

We performed an analysis of Monte Carlo data simulated via GEANT3 (MC 2005) and Beam Test data (BT 4-9 February 2010). The simulated data consist of proton and gamma runs. Fluxes of particles are generated in random directions and are distributed according to a power-low in the energy range 10-200 GeV. BT data were taken at CERN from an highly collimated

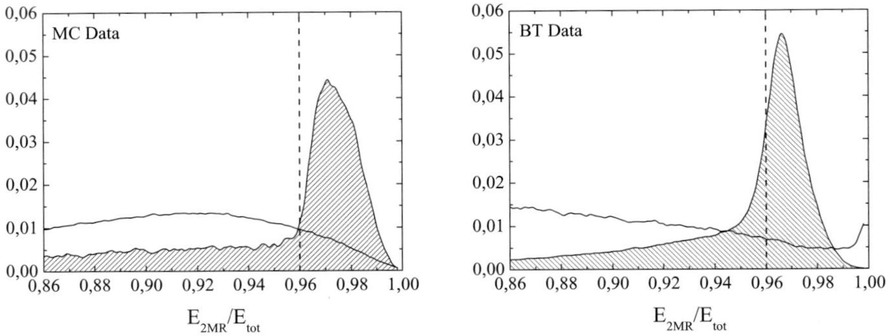


Fig. 2. Comparison of the $E_{2MR}/E_{tot}$ distribution for gamma (hatched area) and proton counts. Histograms are normalized to the total number of events. *Left*: Monte Carlo data for gammas and protons in the range 10–200 GeV. *Right*: Beam Test February 2010 data for gammas ($\sim$ 100 GeV) and protons (400 GeV). Vertical dashed lines denote the cut value.

flux of protons (@ 400 GeV) and photons (@ $\sim$100 GeV). We developed an analysis code in the ROOT framework based on the AMS-02 standard analysis tool (AMS-ROOT). In our code, we considered only events giving a signal in the ECAL by means of the ECAL *stand alone trigger*. For each of them the following *pre-selection cuts* were applied: *1)* the total energy deposited in ECAL had to be $>$ 8 GeV to guarantee an ECAL trigger efficiency $>$ 99%; *2)* events that produced more then one shower were rejected. In addition, we applied specific *direction cuts*: *1)* the reconstructed shower axis direction is within the time of flight (TOF) acceptance; *2)* the reconstructed shower impact point is within the ECAL fiducial area, i.e. $(60 \times 60)$ cm$^2$.

To perform the *shower shape analysis* we created the histograms related to the four estimators described in §2. The $E_{2MR}/E_{tot}$ distribution for both sets of data is shown in Fig. 2. To compute this estimator we used the already existing function available in the AMS-ROOT libraries that provide the $E_{2MR}$ and $E_{tot}$ values. The *Thrust* distribution instead was newly implemented through a modified algorithm as described in Ref. 4 and Ref. 6. In order to get rid of the *Thrust* dependence from the total shower energy, the following parametrization was applied[6] :

$$\hat{T} = \frac{Thrust}{0.99 - 0.017e^{-0.18E_{tot}}} \tag{5}$$

whose distribution is shown in Fig. 3. The shower *footprint* has been also newly implemented as described in § 2.3. In this case the energy dependence

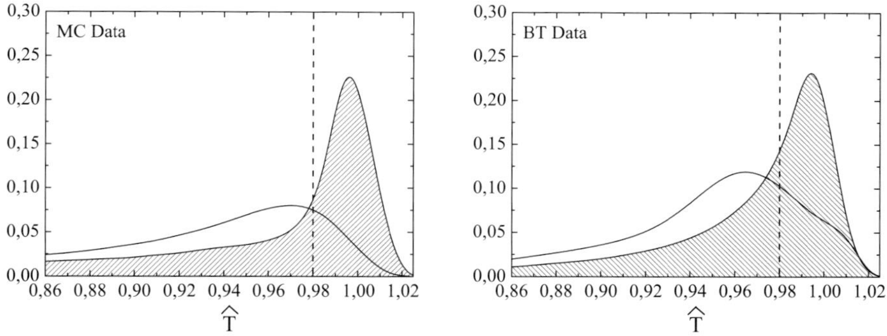


Fig. 3. Comparison of the *Trust* distribution for gamma (hatched area) and proton counts. Histograms are normalized to the total number of events. *Left*: Monte Carlo data for gammas and protons in the range 10–200 GeV. *Right*: Beam Test February 2010 data for gammas ($\sim$ 100 GeV) and protons (400 GeV). Vertical dashed lines denote the cut value.

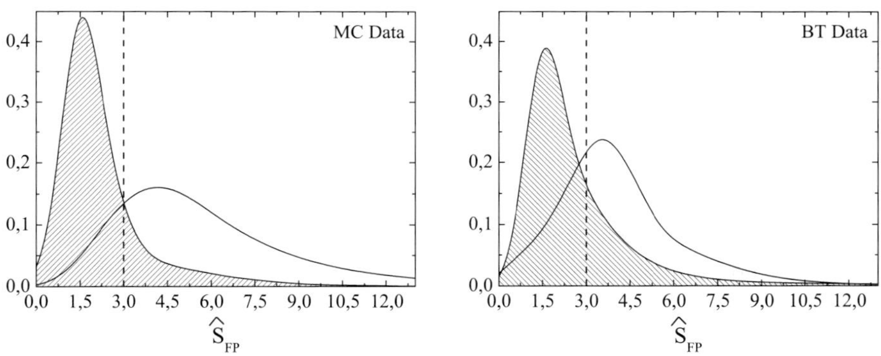


Fig. 4. Comparison of the *footprint* distribution for gamma (hatched area) and proton counts. Histograms are normalized to the total number of events. *Left*: Monte Carlo data for gammas and protons in the range 10–200 GeV. *Right*: Beam Test February 2010 data for gammas ($\sim$ 100 GeV) and protons (400 GeV). Vertical dashed lines denote the cut value.

was removed by applying the following parametrization[6]:

$$\hat{S}_{FP} = \frac{S_{FP}}{4.9 - 0.31 log(E_{tot})} \tag{6}$$

whose distribution is shown in Fig. 4. The shower maximum estimator $T_{Max}$, instead, was directly obtained by means of the AMS-ROOT libraries and the ratio $T_{Max}/T_{MaxP}$, with $T_{MaxP}$ as in Eq. (4), is shown in Fig. 5.

As expected, for both sets of data, the shower estimator distributions

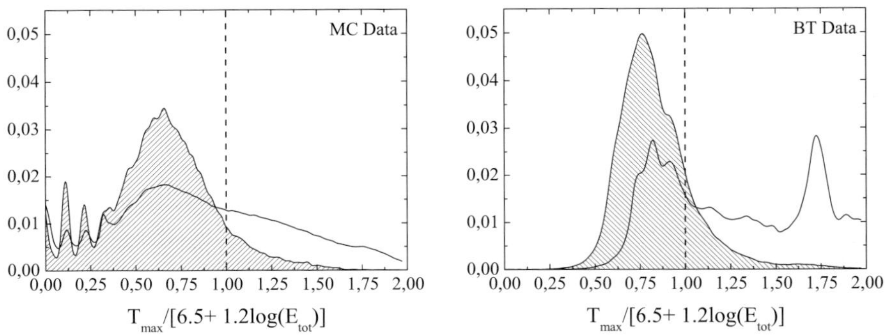


Fig. 5. Comparison of the $T_{Max}/T_{MaxP}$ distribution for gamma (hatched area) and proton counts. Histograms are normalized to the total number of events. *Left*: Monte Carlo data for gammas and protons in the range 10–200 GeV. *Right*: Beam Test February 2010 data for gammas ($\sim$ 100 GeV) and protons (400 GeV). Vertical dashed lines denote the cut value.

of gammas are narrower than the proton ones. Note that an abundant electromagnetic component is visible in the proton distributions due to the muon secondary production. Muons, in fact, give most of the spurious signal and are hardly removed.

### 3.1. *Additional Cuts and Rejection Power*

To get rid of the spurious signals, *additional cuts* involving AMS-02 subdetectors are needed. In particular, for the *single photon detection mode* we have that :*1)* no track has to be reconstructed in both the transition radiation detector (TRD) and the silicon tracker layers; *2)* the time of flight (TOF) measurement must be zero; *3)* the number of hits in the ring imaging Cherenkov detector (RICH) must be $< 5$.

In order to provide an estimation of the proton rejection power we took into account all the criteria mentioned so far. The *rejection power* is defined as:

$$R = \frac{A_C^\gamma}{A_C^p} \tag{7}$$

where $A_C^\gamma$ and $A_C^p$ are the selection criteria *acceptances* for gammas and protons respectively, and are defined as:

$$A_C^\gamma = \frac{N_C^\gamma}{N^\gamma} A_{gen} \tag{8}$$

$$A_C^P = \frac{N_C^p}{N^p} A_{gen} \tag{9}$$

where $N_C^\gamma$ ($N_C^p$) is the number of gammas (protons) survived after a set of selection cuts $C$; $N^\gamma$ ($N^p$) is the number of generated gammas (protons), which represents the signal (background) to be studied; $A_{gen}$ is the generation acceptance (286.7 $m^2$sr for MC data). We applied all the selection cuts listed above and obtained $R \sim 10^3$ for both MC and BT data. This value has the same order of magnitude of the one given in Ref. 5. Other authors[4], instead, obtained an higher value ($R \sim 3.8 \times 10^5$). This discrepancy may be due to the strict constraints adopted for TOF and TRD. A time of flight equal to zero or no reconstructed track, in fact, may decrease the gamma statistics by suppressing the events that produce particles back scattered from the e.m. shower. In addition, further and very accurate simulations are necessary in order to establish the most suitable shower shape analysis cuts that will be able to increase the rejection power.

## 4. Conclusions

We developed an analysis tool to extract the gamma-ray signal from AMS-02 data for the *single photon mode*. We introduced and implemented some estimators for the *shower shape* analysis, taking into account pre-selection cuts, direction cuts, and additional sub-detector cuts. By applying all of them, a rejection power of $\sim 10^3$ was achieved. Further investigations and analysis are ongoing to improve this value in order to reach a higher statistics together with a even less contaminated signal in order to optimize the AMS-02 detection capabilities.

## References

1. R. Battiston, and AMS collaboration, Proc. of the 11th ICATPP, Como 5-9/10/2009, World Scientific, Singapore, **741** (2009).
2. J. Casaus, *J. Phys. Conf. Ser.*, **171**, 012045 (2009).
3. Experimental Techniques in High Energy Physics, ed. by T. Ferbel (Addison-Wesley, Menlo Park CA 1987).
4. P. Maestro, AMS Note 2003-01-01.
5. V. Choutko http://ams.cern.ch/AMS/Analysis/hpl3itp1/ams02_sg.ps.
6. F. Pilo, PhD Thesis, 2005.

# CZELTA: AN OVERVIEW OF THE CZECH LARGE-AREA TIME COINCIDENCE ARRAY

K. Smolek*, J. Čermák, J. Hubík, S. Pospíšil, P. Přidal, J. Smejkal, and I. Štekl

*Institute of Experimental and Applied Physics, Czech Technical University in Prague, Horská 3a/22, 128 00 Praha 2, Czech Republic*
**E-mail: karel.smolek@utef.cvut.cz*

F. Blaschke, P. Lichard

*Faculty of Philosophy and Science, Silesian University in Opava, Bezručovo náměstí 13, 746 01 Opava, Czech Republic*

V. Vícha

*Gymnázium Pardubice, Dašická 1083, 530 03 Pardubice, Czech Republic*

CZELTA (CZEch Large-area Time coincidence Array) is a sparse network of stations for the detection of cosmic rays with energy higher than $10^{14}$ eV. The detector system covers large area in the Czech Republic and is related to the similar experiment ALTA in Canada. The network is designed to study non-random component of cosmic rays over large distances. We present the design of the detection system, obtained results, and our outreach activities connected to the project.

*Keywords*: CZELTA; Large-area array

## 1. Introduction

Nowdays, most of the ground based experiments for the detection of extensive air showers (EAS) are designed as a relatively dense array of detectors (see e.g. Ref. 1). In such experiments, each individual EAS hits several detectors and from the measured size of a detected shower, the energy of a primary particle can be estimated. However, such kind of experiments is not able to study large scale cosmic ray phenomena.

The effect of Gerasimova-Zatsepin[2] is an example of a process that can produce EAS correlated on large distance. The Solar radiation field can break apart a high energy cosmic ray nuclei. The remnants of the nuclei will be deflected in the interplanetary magnetic field and, after coming to

Fig. 1. Three boxes containing detection part of the station.

the Earth's atmosphere, they can produce correlated EAS. The distance of both EAS depends on the primary particle energy and composition and can exceed thousands of kilometers[3]. For study of such large scale phenomena, a very large sparse network of individual detectors is needed.

## 2. CZELTA

The Institute of Experimental and Applied Physics of the Czech Technical University in Prague and the Faculty of Philosophy and Science of the Silesian University in Opava realize a project for the detection of high energy cosmic rays, called CZELTA (CZEch Large-area Time coincidence Array)[4]. In the experiment, a relatively sparse network of detection stations is being built in the Czech Republic. The project CZELTA is related to the older project ALTA (Alberta Large-area Time coincidence Array)[5] of the University of Alberta in Canada. Both experiments use the same design and hardware of the detection stations.

The primary objective of the project CZELTA is to find correlations in the arrival times of air showers over large distances and to find non-random component of air showers arriving at a single site. The detection stations are placed mainly on the roofs of high school buildings in Canada and the Czech Republic. Students from those schools who are interested in research participate in the project. Therefore, the experiment has not only scientific, but pedagogical impact, too.

The same hardware, which is used in the CZELTA and ALTA detection stations in the Czech Republic and Canada, is used also in the parallel starting projects in Great Britain, Slovakia, and Romania. For the purpose of easier manipulation with measured data from the mentioned experiments, we developed a special web-based interface. Regularly every day, the system downloads new measured data from all connected stations to our server and provides them in a simple way for further analysis. The web-interface provides on-line check of running stations and information about meteorological conditions in the place of the detection site, too. The system is broadly used by the students of collaborating high schools and researchers involved in the project.

Similar projects, however with different design and hardware, are in operation in other countries, too. The network NALTA (North America Large-scale Time coincidence Array)[6] was established within a Canada–USA cooperation. It covers local detection networks in Canada and the USA (ALTA, SALTA[7], WALTA[8], CHICOS[9], CROP[10], etc.). A similar project is running in Japan (LAAS[11]) and detection stations are being built on the roofs of European high schools, too — in Sweden (SEASA[12]), Germany (SkyView[13]), the Netherlands (HiSPARC[14]), Great Britain (Cosmic School[15]) among other countries.

## 3. Design of the Project

The hardware of the CZELTA station is identical with the hardware of the ALTA network (see Refs. 16 and 17 for more details). Each station is composed of three scintillation detectors, which are placed into an equilateral triangle configuration with 10 m long sides (Fig. 1) and use GPS signals for time-labeling of events. One scintillation detector has dimensions of 60 cm $\times$ 60 cm $\times$ 1 cm and is connected to a photomultiplier, which detects photons originating from the passage of an extensive air shower through the scintillator. The detectors work in a coincidence. A shower of secondary particles must hit all three scintillation detectors in order to be recorded. These cases, the energy of the primary particle is at least $\sim 10^{14}$ eV.

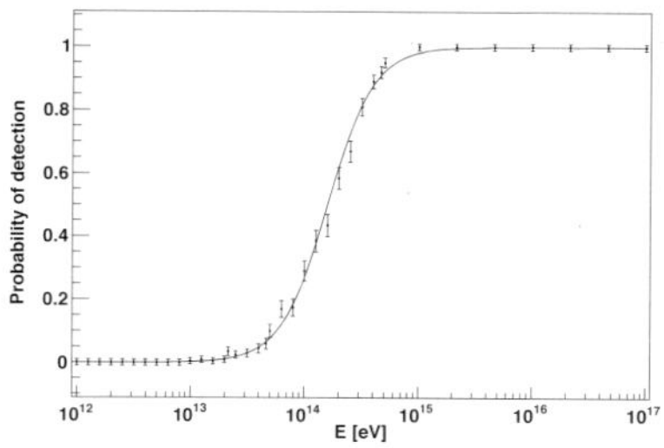


Fig. 2. The probability of detection of a shower produced by a primary proton with energy $E$. The direction of the primary particle is perpendicular to the surface and is oriented to the center of the detector triangle.

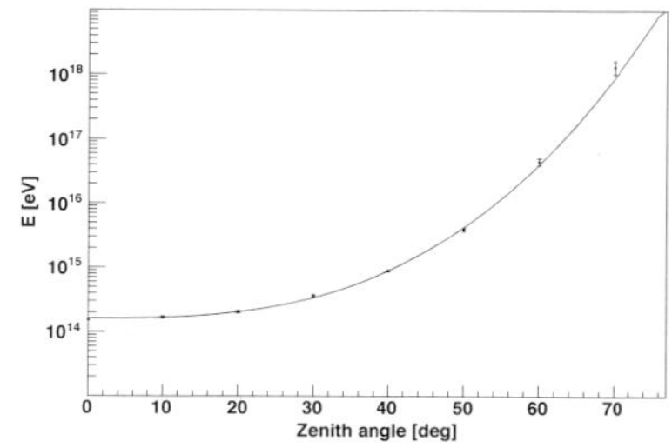


Fig. 3. Minimal energy of a primary proton which can produce a detectable shower in an ALTA/CZELTA station as a function of a zenith angle.

Table 1. Geographical locations of detection stations

| ID | Name | Latitude | Longitude | $\tau$ [day] | $\lambda$ [s] |
|---|---|---|---|---|---|
| 1 | IEAP CTU | 50° 4′ 2″ | 14° 25′ 22″ | 387 | 39.91 |
| 2 | SU | 49° 56′ 16″ | 17° 53′ 52″ | 1178 | 42.26 |
| 3 | Pardubice | 50° 2′ 9″ | 15° 47′ 22″ | 731 | 45.50 |
| 4 | Opava Mendel | 49° 56′ 18″ | 17° 54′ 26″ | 660 | 35.95 |
| 5 | Kladno | 50° 8′ 26″ | 14° 6′ 3″ | 426 | 39.53 |
| 6 | Alta BC | 48° 27′ 43″ | -123° 18′ 37″ | 98 | 83.87 |
| 7 | O'Leary | 53° 35′ 34″ | -113° 28′ 30″ | 699 | 25.07 |
| 8 | O'Brien | 53° 32′ 8″ | -113° 25′ 37″ | 998 | 28.34 |
| 9 | Trinity | 53° 27′ 27″ | -113° 26′ 35″ | 1006 | 25.79 |
| 10 | MacDonald | 53° 33′ 10″ | -113° 34′ 1″ | 1144 | 22.64 |
| 11 | Maddock | 53° 13′ 6″ | -114° 57′ 50″ | 719 | 19.82 |
| 12 | Thorhild | 54° 9′ 11″ | -113° 6′ 43″ | 716 | 26.78 |
| 13 | G. Prairie | 55° 10′ 56″ | -118° 48′ 36″ | 1133 | 30.66 |
| 14 | Norbuck | 53° 31′ 41″ | -113° 31′ 33″ | 913 | 6.19 |
| 15 | Vegreville | 53° 29′ 56″ | -112° 4′ 26″ | 977 | 27.29 |
| 16 | Med. Hat | 50° 3′ 17″ | -110° 41′ 6″ | 494 | 26.34 |
| 17 | Phys. Roof | 53° 31′ 41″ | -113° 31′ 34″ | 284 | 24.21 |
| 18 | Laurent | 53° 28′ 46″ | -113° 31′ 23″ | 585 | 24.46 |
| 19 | Page | 53° 27′ 24″ | -113° 26′ 46″ | 345 | 25.91 |
| 20 | McNally | 53° 32′ 54″ | -113° 27′ 32″ | 455 | 24.17 |
| 21 | Beaumont | 53° 20′ 56″ | -113° 25′ 23″ | 713 | 23.99 |

For our purposes, extensive Monte-Carlo simulations have been provided using the CORSIKA package[18]. Figure 2 shows the dependance of the probability of detection of an extended air shower produced by a primary proton with an energy $E$. The showers were simulated with the perpendicular direction aimed to the center of the detection triangle. Primary particle with energy higher than $\sim 2 \times 10^{14}$ eV can produce a shower whose probability of detection is higher than 50%.

Probability of a shower detection depends on the length and density of a penetrated atmospheric column. Therefore, the minimal energy of a primary particle which can produce a detectable shower strongly increases with the zenith angle (see Fig. 3). Measured flux of showers strongly depends on the atmospheric pressure. At the beginning of the project, the meteorological data from a national meteorological service network were used. However, for some detection sites a distance of these stations is too large and the meteorological data are not fully valid. Based on this fact the decision to add a local meteorological station was accepted. Stored data are used for corrections of observed flux of cosmic rays on the influence of atmospheric pressure.

From the time difference among the measured signals, the direction of the extensive air shower as well as primary particle flight can be determined. For the estimation of the error of the direction measurement, data from two stations (rotated by 60°) located at the same place were analysed. The mean of the error of the reconstructed direction of the primary particle (for the showers with the altitude angle $> 50°$) can be estimated as 5°, which conforms with our CORSIKA simulations.

Recognition of a shower arrival time is done by GPS and an additional internal electronic board (with accuracy $\sim 10$ ns). This method enables the study of time correlations of data measured with distant stations.

## 4. Very large area coincidencies

The main research aim of the ALTA/CZELTA project is to search for very large area coincidencies due to cosmic ray phenomena. Some highly energetical process in the Universe could produce at one moment several particles or even a wave of particles coming towards the Earth. After interaction in the atmosphere, such particles could produce extensive air showers detectable at large distances during a short time interval.

The data analysis provided in the article is connected with the phenomena of large area coincidencies. The analysis takes into account only times of registration of showers and not directions of primary particles. Let us suppose the detection of events by a single station is a Poisson process. If the total length of overlapping measurement by two stations with mean detected frequencies $1/\lambda_1$ and $1/\lambda_2$ ($\lambda_1$, $\lambda_2$ are mean time intervals between two consecutive events) is $\tau$, the probability of observing of $k$ pairs of randomly correlated events by both the stations in a time window with length $2t_{12}$ is

$$P_2(k) = \frac{x_2^k}{k!} \mathrm{e}^{-x_2} \equiv S_k(x_2), \quad x_2 = \frac{2t_{12}\tau}{\lambda_1 \lambda_2}, \tag{1}$$

where an expected number of observed pairs is $x_2$. For triple-coincidencies, we can derive a similar formula:

$$P_3(k) = \frac{x_3^k}{k!} \mathrm{e}^{-x_3}, \quad x_3 = \frac{4\tau(t_{12}t_{13} + t_{12}t_{23} + t_{13}t_{23})}{3\lambda_1 \lambda_2 \lambda_3}. \tag{2}$$

For two selected stations $i$, $j$, we use a maximal half-width of a time window $t_{ij}$ computed as their distance divided by the speed of light. In the Tab. 1, geographical locations, total duration of measurement and mean time intervals between two consecutive events for all ALTA/CZELTA stations are depicted.

In the data, we found many double-coincidencies. With considered a combinatorial factor of selection pairs of stations, our results show no statistically significant doublets and deeper analysis using shorter selection windows is necessary.

Let us suppose the mean time interval between two consecutive physical (caused by some phenomena in the Universe) double-coincidence events measured with a pair of stations is $\lambda^{(s)}$. If we measure $k$ events, we can use Bayes' theorem and estimate a probability density of $\lambda^{(s)}$ as

$$f(\lambda^{(s)}|k) = \frac{\tau}{S_{k+1}(x_2)} \frac{1}{k!} \left( \frac{\tau}{\lambda^{(s)}} + x_2 \right)^k \mathrm{e}^{-x_2 - \tau/\lambda^{(s)}}. \tag{3}$$

We can suppose the data from all pairs are almost statistically independent. If we serialize the time series from all pairs of stations, we can use Eq. 3 and estimate the lower limit of a mean time interval between two consecutive physical (non-statistical) double-coincidence events:

$$\lambda^{(s)} > 0.38 \text{ yr}, \quad 95\% \text{ C.L.} \tag{4}$$

In other words, we can expect less than 2.6 physical coincidencies detected with all 21 ALTA/CZELTA stations per year.

Whilst we can expect many double-coincidencies in the data, triple-coincidencies are expected to occur with much smaller probability. Indeed, we found only two triple-coincidencies. We present them in Tab. 2. In the second column, there are distances between pairs of stations. In the fourth column, the observed time differences between detected events are presented. Last two columns show expected numbers of triple-coincidencies (with maximal time windows presented in the fourth column) and the probability of a detection at least one such coincidence (p-value). From both found triplets, the second one is most promising. However, after considering a combinatorial factor of selection of triplets of stations, we do not observe a statistically significant excess of triple-coincidencies.

Table 2. Triple-coincidencies

| ID | $\text{Dist}_{ij}$ [km] | $\tau$ [day] | $t_{ij}^{obs}$ [ms] | $n_{exp}$ | p-value |
|---|---|---|---|---|---|
| 1 & 9 & 15 | 7110 / 7050 / 90 | 171 | 15.7 / 15.5 / 0.2 | 0.17 | 0.16 |
| 11 & 14 & 16 | 90 / 450 / 420 | 366 | 0.1 / 0.4 / 0.3 | 0.0028 | 0.0028 |

Another way of analysis can consider sudden increases of event rate (so called bursts or clusters of events) measured with an individual detection station. We have observed several clusters, however we need to make a profound analysis of their statistical significance. With our detection system, we plan to analyse time coincidencies of clusters detected with distant stations.

## 5. Conclusion

Within the common project ALTA/CZELTA, we built up a very large area network for the detection of extended air showers. Our global network is exceptional with its dimension — it covers the area with size more than 7 000 km. All detection stations use the same hardware and the same data format. This enables to study non-random properties of cosmic rays at large distances. In the paper, we demonstrated the ability of the network to detect such phenomena and presented first preliminary results. We believe, after using other measured properties of detected showers (reconstructed direction of the shower) and more complex statistical methods, we will obtain more rewarding results concerning large-area properties of cosmic rays.

## Acknowledgment

We would like to thank all our colleagues from the collaborating Canadian and Czech high schools for their help with installation of the detection stations and running of the project. This work was supported by the Ministry of Education, Youth and Sports of the Czech Republic under Grants No. MSM 6840770029 and No. LC 07050.

## References

1. The Auger Collaboration, *The Pierre Auger Observatory Design Report*, *FERMILAB-PUB-96-024* (1996).
2. N. Gerasimova and G. Zatsepin, *Soviet Phys. JETP* **11**, 899 (1960).
3. S. Lafèbre, H. Falcke, J. Hörandel, and J. Kuijpers, *A&A* **485**, 1 (2008).
4. http://www.utef.cvut.cz/czelta/czelta-en
5. http://csr.phys.ualberta.ca/~alta/
6. http://csr.phys.ualberta.ca/nalta/
7. http://faculty.washington.edu/wilkes/salta/
8. http://neutrino.phys.washington.edu/~walta/
9. http://www.chicos.caltech.edu
10. http://crop.unl.edu
11. N. Ochi et al., *J. Phys. G: Nucl. Part. Phys.* **29**, 1169 (2003).
12. http://www.particle.kth.se/SEASA/
13. http://skyview.uni-wuppertal.de
14. http://www.hisparc.nl
15. http://hep.ph.liv.ac.uk/~green/cosmic/
16. W. Brouwer et al., *Nucl. Instr. and Meth. A* **493**, 79 (2002).
17. W. Brouwer et al., *Nucl. Instr. and Meth. A* **539**, 595 (2005).
18. D. Heck et al., *CORSIKA Physics Description*, report FZKA 6019 (Forschungszentrum Karlsruhe, 1998); http://www-ik.fzk.de/corsika/physics_description/corsika_phys.html.

# Calibration of the CMS Electromagnetic Calorimeter with first LHC data

V. Sola

vsola@to.infn.it

*on behalf of the CMS Collaboration*

The first 7 TeV LHC collisions recorded with the CMS detector have been used to perform a channel-by-channel calibration of the electromagnetic calorimeter (ECAL). Decays of $\pi^0$ and $\eta$ into two photons as well as the azimuthal symmetry of the average energy deposition at a given pseudorapidity are utilized to equalize the response of the individual channels. The ECAL comprises a central *barrel* section and two *endcaps*. Based on an integrated luminosity of up to 250 $nb^{-1}$, a channel-by-channel *in-situ* calibration precision of 0.6% has been achieved in the barrel ECAL in the pseudorapidity region $|\eta| < 0.8$. The energy scale of the ECAL has been investigated and found to agree with the simulation to within 1% in the barrel and 3% in the endcaps.

## 1. Introduction

The electromagnetic calorimeter (ECAL) of the Compact Muon Solenoid (CMS) experiment consists of a barrel section (EB) and two endcaps (EE). The EB contains 61 200 lead tungstate crystals arranged in 170 $\eta-$rings of 360 $PbWO_4$ crystals each. It is subdivided into 36 supermodules each containing 1700 crystals and provides coverage in pseudorapidity up to $|\eta| < 1.5$. The two ECAL Endcap (EE) subsystems consist of four half-disk Dees, each containing 3662 crystals. The EE provides a coverage of $1.5 < |\eta| < 3.0$. The scintillation light is detected by avalanche photodiodes (APDs) in the EB and vacuum phototriodes (VPTs) in the EE.

The ECAL energy resolution has been measured in test beams to be:

$$\frac{\sigma(\mathrm{E})}{\mathrm{E}} = \frac{2.8\%}{\sqrt{\mathrm{E}(GeV)}} \oplus \frac{12\%}{\mathrm{E}(GeV)} \oplus 0.3\%\,, \qquad (1)$$

where the three contributions correspond to the stochastic, noise, and constant terms, respectively. This result was obtained in the absence of magnetic field, with almost no inert material in front of the calorimeter and with the beam aligned on the centres of crystals.

The high resolution of the ECAL is particularly important for a discovery of the Higgs boson in the decay channel $H \to \gamma\gamma$, where the constant term dominates.

This article describes results of the first crystal-by-crystal calibrations and global energy response measurements of the CMS ECAL at the LHC, performed during the first months of operation.[1]

## 2. The pre-calibration

The relative channel response spread affects the constant term of the energy resolution of the electromagnetic calorimeter. The main sources of chanel-by-channel response variation are the crystal light yield differences of the barrel, about 13% at construction, and the gain spread of the photodetectors in the endcaps, about 25%.

All the channels have been pre-calibrated prior to installation with laboratory measurements of crystal light yield and photo-detector gain during the construction phase (all EB and EE channels), with test beam electrons (9 EB supermodules and about 500 EE crystals) and with cosmic ray muons (all EB channels). After installation in the LHC, circulating beams were stopped in collimators 150 m away from CMS in September 2008 and November 2009. The resulting *beam dump* events have been used to improve pre-calibration precision.

Inter-calibration constants coming from different sources are combined as a weighted average. The precision of a given calibration method is determined empirically by comparing the new inter-calibration constants to those derived at the test beams. The comparison for the EE constants (derived from the combination of laboratory measurements and beam dump calibration) is shown in Figure 1 (left) for a subset of 162 of those crystals exposed to the test beam.

The channel-by-channel calibration precision at the start-up at 7 TeV is estimated to be:

- EB: about 0.5% for the 9 supermodules calibrated with test beams and 1.4% to 1.8%, depending on pseudorapidity, for the other 27 supermodules;
- EE: below 1% in the $\sim$ 500 crystals calibrated with beam and about 5% elsewhere;

In the rest of the article, a crystal-by-crystal calibration constant is defined as a multiplicative factor $c_i$ relative to the precalibration constant derived for this crystal: $C_i = C_i^{precalib} \cdot c_i$ .

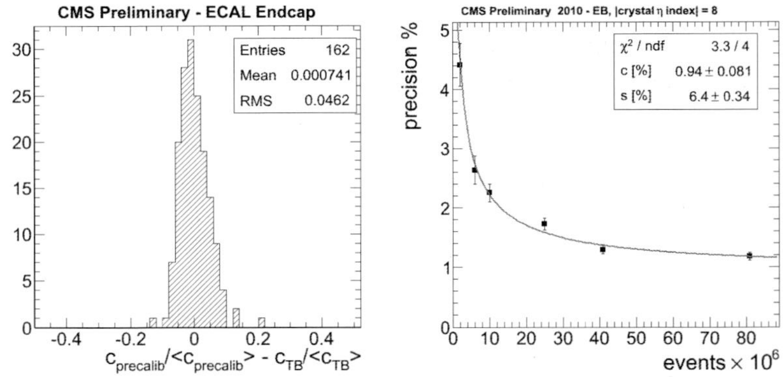


Fig. 1. Left: Relative difference between the inter-calibration coefficients measured at the test beam and the coefficients derived from the combination of all other available measurements (laboratory, 2008 and 2009 beam dump data) in the EE. Right: Precision of $\phi$-symmetry method as a function of number of events for ring at |crystal $\eta$ index| = 8.

## 3. Calibration with the $\phi$-symmetry method

The $\phi$-symmetry method is based on the expectation that for a large sample of minimum bias events the total deposited transverse energy ($E_T$) should be the same for all crystals in a ring at fixed pseudorapidity ($\eta$). Inter-calibration in $\phi$ is performed by comparing the total transverse energy ($\Sigma E_T$) deposited in one crystal with the mean of the total $\Sigma E_T$ collected by crystals at the same absolute value of $\eta$. In the EB, crystals are arranged in 85 pairs of rings symmetric in pseudorapidity with 360 crystals along $\phi$ in each ring. In the determination of the transverse energy sum, only deposits with energies between lower and upper thresholds are considered. The former is applied to remove the noise contribution, the latter is applied to avoid a possible bias from very high $E_T$ deposits. The $\phi$ inhomogeneities of the detector are taken into account introducing a data driven correction.

### 3.1. *Crystal-by-crystal calibration*

The $\phi$-symmetry inter-calibration procedure has been applied on $1.6{\cdot}10^8$ minimum bias events and inter-calibration constants $c_i$ are derived for all crystals in the EB. The calibration precision is estimated as the distribution of the $c_i$ for each ring of constant $\eta$. The width of the distribution gives the convolution of the precision of the pre-calibration with the precision of this method. For the nine test-beam calibrated supermodules, the expected pre-calibration precision is about 0.5%, several times better than expected for the $\phi$-symmetry method. Therefore, the width of distribution of the $c_i$ limited to those supermodules gives an upper bound of the inter-calibration precision achieved with the $\phi$-symmetry method. The width of the $c_i$ distribution is estimated by performing a Gaussian fit and extracting the sigma. Figure 1 (right) shows the inter-calibration precision as a function of the number of events for $\eta$−ring 8. The measured points are fitted with the function $\sqrt{c^2 + s^2/N}$. The fitted value of $c$ (the constant term) suggests that the systematic limit of the method is nearly reached.

## 4. Calibration with $\pi^0$ and $\eta$ decays

An inter-calibration technique based on reconstructing $\pi^0 \to \gamma\gamma$ and $\eta \to \gamma\gamma$ decays has been studied using both the simulations and a dedicated test-beam study.

The photon candidates are reconstructed using a simple 3×3 window clustering algorithm that requires that the energy deposited in a seed crystal is above 0.5 GeV. The cluster energy is then computed as the sum of nine crystal energies. The shape of the cluster energy deposition should be consistent with that of an electromagnetic shower produced by an unconverted photon. To be considered in the combinatorial selection of di-photon pairs, a photon candidate is required to have a transverse energy above 0.3 GeV. The $\pi^0$ candidates are then selected by requiring their transverse energy to be above 0.9 GeV and an invariant mass below 0.25 GeV/c$^2$. Figure 2 shows the invariant mass distributions of the selected $\pi^0$ candidates after all selection cuts both for data and simulated minimum bias events, for the EB. In total, close to $4.3 \cdot 10^7$ signal $\pi^0$ decays are selected from the 250 nb$^{-1}$ of data collected at 7 TeV.

For $\eta$ candidates, the same selection and reconstruction methods are used except that the transverse energy cuts are tightened to $E_T^\gamma > 0.5$ GeV and $E_T^{\gamma\gamma} > 2.5$ GeV. In addition, a veto cut is applied to reject photons coming from di-photon candidates within the $3\sigma$ window around the fitted $\pi^0$ peak.

### 4.1. *Inter-calibration Results*

To investigate the potential of the $\pi^0 \to \gamma\gamma$ inter-calibration, two independent calibration algorithms are used. Both algorithms are based on an iterative procedure where the inter-calibration constants are updated after each iteration step.

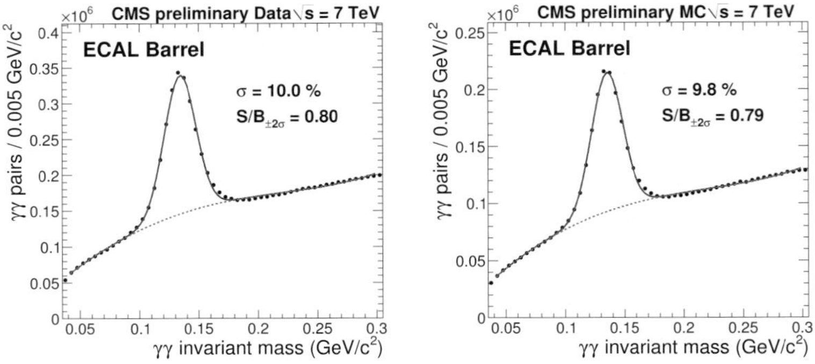


Fig. 2. $\pi^0$ invariant mass reconstructed from photon pairs for the data (left) and simulation (right).

Nine supermodules in this region have been pre-calibrated to a precision of about 0.5% in the test-beams. Crystals in those supermodules are used to estimate the in-situ inter-calibration precision in the same way as for the $\phi$−symmetry method. In the region |crystal $\eta$ index| $\leq 45$, the crystal-by-crystal calibration precision is found to be 1.2%. The systematic limit on the precision of this method has not yet been reached and the current inter-calibration accuracy is dominated by the statistical uncertainty.

## 4.2. *The ECAL energy scale with $\pi^0$ and $\eta$ decays*

The absolute scale of the ECAL, the ADC to GeV conversion factor, has been measured during the test beam campaign for EB and EE separately. In-situ, the ECAL energy scale will be determined by reconstructing di-electron and di-photon invariant mass peaks. A first measurement can be done using $\pi^0$ and $\eta$ reconstructed decays.

The energy scale is derived by measuring the ratio of the reconstructed invariant mass peak position between data and Monte Carlo. Assuming perfect simulation of the material in front of ECAL and alignment of the detector, this ratio provides the correction to be applied to the ECAL scale. More conservatively, we consider this measurement as an indication of the current ECAL scale accuracy.

The EB scale is found to agree with the simulation at a level of 1%, with the measured mass shift − 0.7% ± 0.02% (stat.) ± 0.9% (syst.). In the EE, the scale is found to agree with the simulation to about 3%. The average mass shift is measured to be + 2.5% ± 0.2% (stat.) ± 2.2% (syst.).

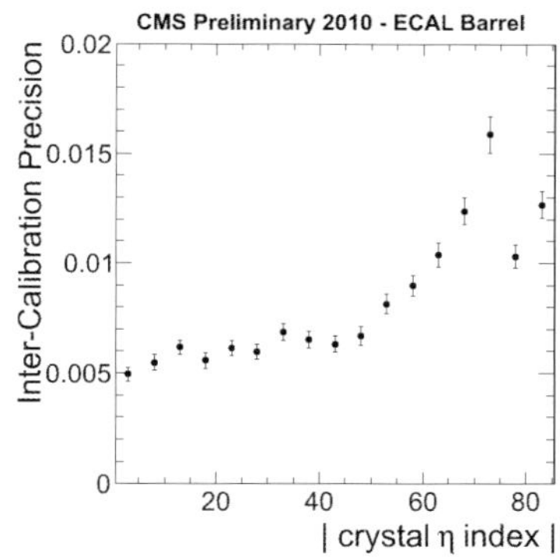


Fig. 3. Precision of combined inter-calibration constants as a function of pseudorapidity.

## 5. Combination of the in-situ inter-calibration results

The crystal-by-crystal calibration constants obtained with the $\phi$-symmetry and $\pi^0$ methods are combined together with the beam dump constants. The precision of the resulting constants is shown in Figure 3 as a function of crystal pseudorapidity. The assumed 0.5% test beam precision is not negligible with respect to the combination precision and therefore subtracted in quadrature. For the central barrel ($|\text{crystal } \eta \text{ index}| \leq 45$) the combined inter-calibration precision is found to be 0.6%.

## 6. Summary

Using the first 250 $\text{nb}^{-1}$ collected with the CMS detector in 2010 at a center of mass energy of 7 TeV, studies of the CMS ECAL calibration methods and procedures are carried out. The achieved channel-by-channel calibration precision in the central barrel with the sample size currently available is about 0.6%.

The global energy response scale of the ECAL is also studied, and found to agree with the expectation to within about 1% in the barrel and 3% in the endcaps. The calibration of the CMS ECAL will continue to improve with the upcoming LHC data, with collection of large samples of neutral pion decays as well as W and Z bosons decaying into energetic and isolated electrons.

## References

1. CMS Collaboration, "Electromagnetic calorimeter calibration with 7 TeV data", CMS PAS **EGM-10-003** (2010).

# On the detectability of cosmic ray electron spectral features in the microwave/mm-wave range

A. TARTARI[1*], M. GERVASI[1,2], G. SIRONI[1], M. ZANNONI[1] and S. SPINELLI[1]

[1] *Physics Department "G.Occhialini", University of Milano-Bicocca,*
[2] *INFN, Milano-Bicocca,*
*Piazza della Scienza 3, Milano, 20126, Italy*
*[*]E-mail: andrea.tartari@mib.infn.it*
*www.unimib.it*

Recent measurements of cosmic ray electron energy spectra suggest that above 10 GeV there may be deviations from a single power law spectrum. There are hints (ATIC) for a bump occurring between 100 GeV and 1TeV, meaning that there might be more high energy electrons than expected. Whether these electrons are produced within pulsar magnetospheres, or due to Dark Matter annihilation or decay, this is still matter of debate. Understanding the nature of these ultra high energy particles is a difficult task that can be fulfilled using all the available astrophysical observables. We investigate how different energy spectra produce different observable manifestations in the radio/microwave/mm-wave domain, where corresponding deviations from a synchrotron power law could appear. We raise the question around the detectability of these possible radio spectral features, which may be interesting for a wide scientific community including astrophysicists and scientists working on foregrounds removal for CMB experiments.

*Keywords*: Cosmic rays; synchrotron radiation; diffuse radio continuum.

## 1. Science case

Recently several experiments have pointed out interesting features in the measured energy spectrum of the $e^+$ and $e^-$ components of cosmic rays. In particular, we recall here the positron excess detected by Pamela[1], the excess of electron counts ($e^+ + e^-$) between 100 GeV and 1 TeV detected by Fermi[2], and the bump in the cosmic ray electron (CRE) spectrum, centered around $\sim$ 500 GeV, detected by ATIC[3] (not confirmed by H.E.S.S. low energy analysis[4]). The observational picture will be probably improved after the observations of AMS-02[5] (which will be operating in February 2011), due to the large acceptance and accuracy of the apparatus and to the long operation time (at least ten years). In this work we do not look for possible interpretations of these data, rather we study the possibility of using radio observations to confirm the above CREs spectral features. In particular, we concentrate on the evaluation of CRE synchrotron emission in a galactic environment, with typical values of the magnetic field strength of $1-5\ \mu$G.

Due to the explorative aims of this study, we limit our attention to the intensity and degree of polarization of radiation produced by an ensemble of electrons, in the hypothesis of homogeneity of the galactic magnetic field on scales greater than the gyro-radius, and discuss the properties of radiation at the source, not taking into account radiative transfer through the interstellar medium (ISM), which affects both intensity and degree of polarization. This issue can be the object of a subsequent paper. Here we point out the relevant frequency bands for possible observations and signatures that may allow one to disentangle their signal from contaminants like thermal dust emission, peaking in the sub-mm range. We show also that expected signals are not a limiting foreground for Cosmic Microwave Background experiments, being expected at $\sim$ 1 THz and above. Finally we discuss a few observational aspects of some astrophysical sources which could be used as targets for this investigation.

## 2. Method

The connection between diffuse radio emission and cosmic ray propagation within the Galaxy was firmly established in the Sixties[6]. Here we recall the basic results concerning the single particle synchrotron emission, and the emission of an ensemble of particles, with a known energy spectrum, propagating in a uniform magnetic field[7]. A charged particle (an electron, from now on) moving in a region pervaded by a uniform magnetic field describes an helical trajectory, with a pitch angle $\alpha$ between the velocity vector and the magnetic field lines. Being accelerated, it looses energy by radiation. If the velocity of the charge is relativistic, the radiation is strongly beamed, and is almost completely emitted within a cone of aperture $\sim 1/\gamma$, where $\gamma$ is the Lorentz factor, typically $\geq 10^3$ for the electrons associated to the observed radio continuum of the Galaxy. This cone is centered around the instantaneous velocity of the particle, sweeping periodically the line of sight of the observer. Only in a short ($\sim 1/\gamma$) fraction of the gyration period, when the line of sight lies within the radiation cone, the observer registers a pulse of radiation. The Fourier transform of this pulse, in the observer's frame, is a continuous spectrum and most of the radiation is concentrated in a narrow peak at frequency $\nu \sim \gamma^2 \nu_g$, with gyro-frequency

$\nu_g = eB/2\pi m_e$, where $B$ is the magnetic field, $e$ the charge of the electron, and $m_e$ its mass.

Following Longair, we introduce a cartesian frame suitable for the description of synchrotron radiation, including its polarization. To this purpose, let's consider a plane orthogonal to the wave vector $\mathbf{k}$: we can project the magnetic field on this plane. Let's call $B_{\perp\mathsf{k}}$ the projected component. We can form a triad with a versor parallel to $B_{\perp\mathsf{k}}$ (components in this direction will be labeled with $\|$), one orthogonal to both $B_{\perp\mathsf{k}}$ and $\mathbf{k}$ (components in this direction will be labeled with $\perp$), and $\mathbf{k}$ itself. In this frame, synchrotron emissivities of a single electron in the two polarizations are:

$$j_\perp(\nu) = \Gamma\ B_\perp \sin\alpha\,[F(\nu/\nu_c) + G(\nu/\nu_c)]$$
$$j_\|(\nu) = \Gamma\ B_\perp \sin\alpha\,[F(\nu/\nu_c) - G(\nu/\nu_c)]$$

where: (1) $\Gamma \cong 1.86 \times 10^{-26}$ W/THz; (2) $\nu_c = 3\gamma^2\nu_g \sin(\alpha/2)$ is the critical frequency ; (3) $F(x)$ and $G(x)$ are spectral functions defined through modified Bessel functions as $F(x) = x\int_x^\infty K_{5/3}(z)dz$ and $G(x) = xK_{2/3}(x)$. Now, if we have $N(E)dE$ electrons per energy interval $dE$, per unit volume, the spectral intensity per unit volume radiated by the cloud of electrons is $I(\nu) = \int_0^\infty j(x)N(E)dE$, where $x = \nu/\nu_c = x(E,B)$ and $j = j_\perp + j_\|$. Therefore we arrive at a useful integral:

$$I(\nu) = \frac{\kappa m_e c^2}{\sqrt{6\nu_g \sin\alpha}}\sqrt{\nu}\int_0^\infty j(x)N[E(x)]x^{-3/2}dx \qquad (1)$$

that can be computed analytically in some remarkable cases (e.g. $N(E) \propto E^{-\beta}$), but that we compute numerically in order to deal with arbitrary energy distributions. Here $\kappa$ is normalization factor. An average on the pitch angle $\alpha$ in the range $(0,\pi)$ will lead to the results presented in the next section. If we deal separately with the intensity emitted in two orthogonal polarizations, we can compute its degree of linear polarization, that is:

$$\Pi(\nu) = \frac{I_\perp(\nu) - I_\|(\nu)}{I_\perp(\nu) + I_\|(\nu)}. \qquad (2)$$

In terms of Stokes parameters, $\Pi = \sqrt{Q^2+U^2}/I$. Its value may be up to $\sim 75\%$ for a power law distribution of electrons with slope around $-3$. Nevertheless this value is strictly an upper limit, being obtained without considering radiative transfer effects occurring in the ISM, leading to depolarization. Conversely, the degree of circular polarization of an ultra-relativistic electron is vanishingly small, scaling as $1/\gamma$: therefore, we do not consider $\Pi_c$ in the following discussion. Like the intensity, the degree of polarization is computed numerically.

## 3. Results

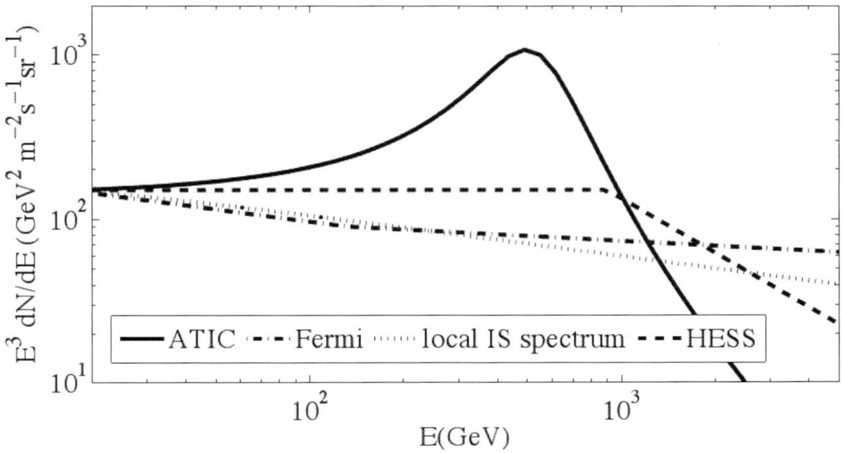


Fig. 1. Different analytical description of the electron spectra used for calculating synchrotron spectra. All these spectra are normalized at the value of 150 GeV$^2$m$^{-2}$s$^{-1}$sr$^{-1}$ at 10 GeV energy.

The calculations described in the previous sections (equation 1 and 2), have been implemented using different electron spectra. In particular we used: (1) the description of the local interstellar spectrum by Zhang & Cheng[8]; (2) a broken power law to describe H.E.S.S. high energy knee[4]; (3) a synthetic representation of a Fermi-like spectrum exhibiting a hardening (from a $-3.2$ to a $-3.0$ slope) around $\sim$ 100 GeV[2]; (4) a synthetic representation of an ATIC-like spectrum[3] with a lorentzian-shaped bump on a constant slope ($\beta = -3$) background. This oversimplified analytical description is enough to capture the main features arising in synchrotron emission. All these spectra have been normalized at a common value of $E^3J(E)$ of 150 GeV$^2$m$^{-2}$s$^{-1}$sr$^{-1}$ at 10 GeV, and are shown in Fig.1 .

In turn, the synchrotron spectra obtained from these sources are ultimately normalized to a measured value of the minimum synchrotron sky brightness at $\delta = +42°$ and $RA \simeq 10h$ (Galactic halo), using the results of the TRIS experiment presented in Tartari et al.[9]: our reference brightness at 1 GHz is $3.8 \times 10^4$ Jy/sr. The obtained synchrotron spectra are shown in Fig.2, while in Fig.3 we show the degree of linear polarization.

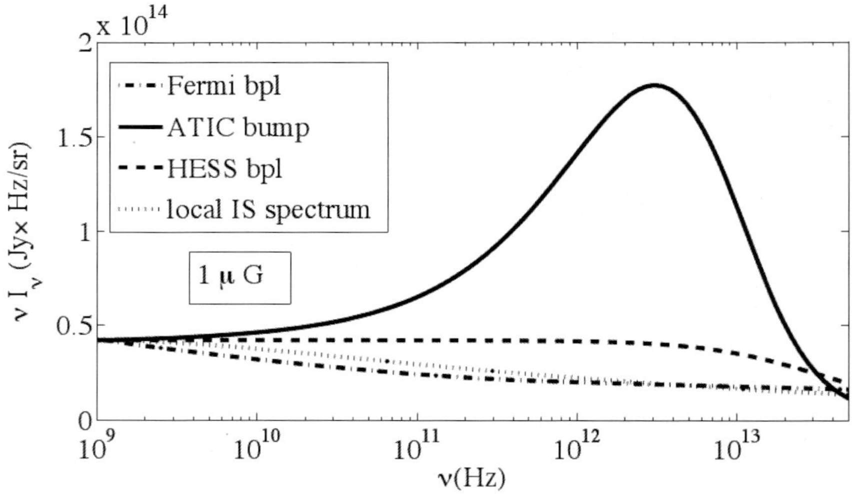


Fig. 2. Spectra of the synchrotron emission forecasts for the different CRE populations. These intensities are normalized to a radio brightness of $3.8 \times 10^4$ Jy/sr measured by the TRIS experiment in a purely synchrotron emitting region on the Galactic halo at $\delta = +42°$ and $RA \simeq 10h$. A Galactic magnetic field of $1\mu$G is assumed.

In Fig.2 and 3 we show only the values obtained for a magnetic field of $1\mu$G. Results at 3 and 5 $\mu$G are similar except that all the features moves at higher frequency as expected.

## 4. Discussion and Conclusions

We have shown that the features in synchrotron intensity and polarization produced by $> 100$ GeV electrons fall into the sub-mm wave/far IR regime (therefore, they are not an issue when removing foregrounds from CMB maps). Moreover a bump in the electron spectrum (ATIC) modulates significantly (more than 10%) the degree of synchrotron linear polarization at frequencies around 1 THz. Unfortunately, the Galactic dust, through its grey-body emission, $I_d(\nu) = k_d(\nu/\nu_0)^{\beta_d} BB(T_d)$, completely dominates the sky brightness in these bands. In Fig.4 we show the expected brightness of a clean region of the sky normalized (through $k_d$ emission coefficient) at the DIRBE[10] 100 $\mu$m channel. We assumed a dust temperature $T_d = 20$ K,

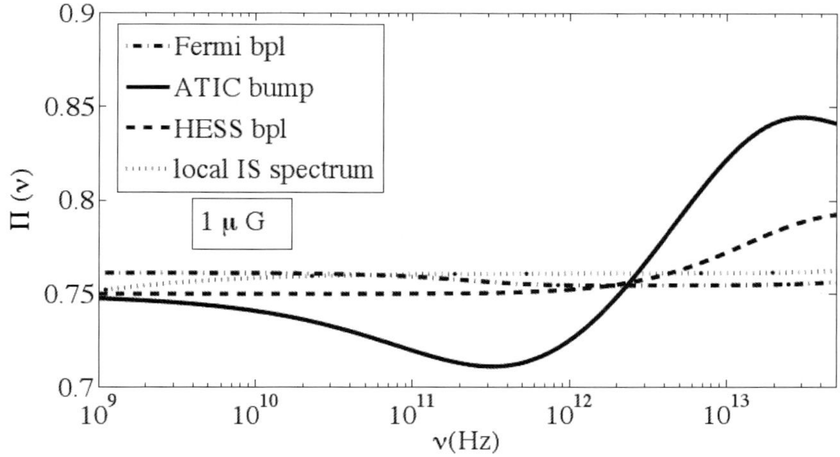


Fig. 3. Degree of linear polarization (Π) *vs* frequency. A Galactic magnetic field of 1$\mu$G and the maximum degree of linear polarization are assumed.

and grey-body emissivity scaling as $\nu^{\beta_d}$ ($\beta_d = 1.7$). We see that dust signal overcomes the synchrotron one by several orders of magnitude.

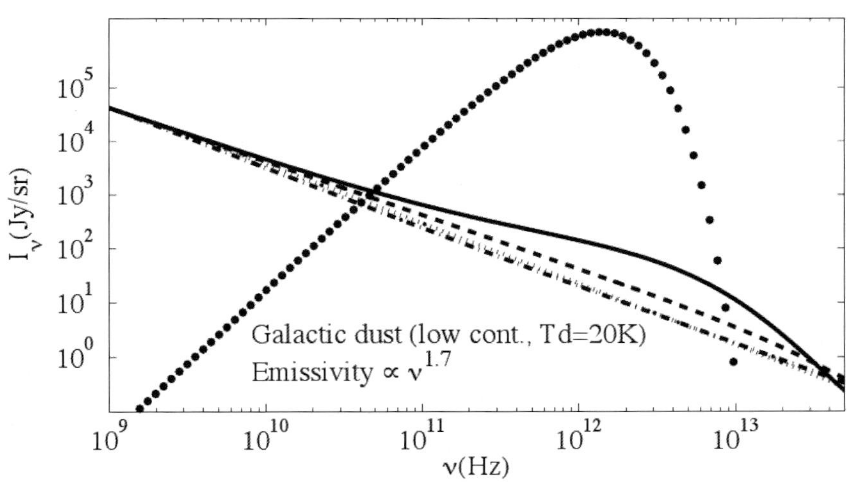


Fig. 4. Synchrotron brightness corresponding to different electron spectra (same line style as in fig.2 and 3.). The filled circles correspond to the emission of cold (20 K) dust in a low contamination region of the sky, normalized to the measurements of the 100 $\mu$m DIRBE channel.

This is just an example, since in general dust contamination depends on the region observed (different temperature, density and composition of dust). Let's now consider different sources of the observed CRE excess.

*A) Dark Matter decay or annihilation in our galaxy.* The distribution of synchrotron radiation is probably similar to the expected Dark Matter distribution in the galactic halo. We expect a smooth distribution with a maximum of intensity in direction of the galactic center decreasing away from it. The best directions of observations are far from the galactic disk, where the synchrotron emission of the background electrons and the thermal emission of the interstellar dust are faint and more uniform. This is the situation we have considered to estimate the signals shown in Fig.4: above 100 GHz thermal dust emission is the dominating signal also far from the Galactic disk. Therefore it is difficult to imagine a detection of a diffuse, smoothly varying synchrotron signal coming from our own Galaxy, such as what could be associated to a Dark Matter halo. In this case, also considering the typical polarization signatures of synchrotron radiation, the removal of the thermal background would be extremely difficult, despite the small degree of linear polarization of thermal dust emission.

*B) Electrons from Pulsar or SNR.* A galactic source like a pulsar or a SNR can be found (much more easily) on the galactic disk. Here both the synchrotron emission of the background electrons and the thermal emission of the interstellar dust are stronger respect to the signals coming from the halo and shown in Fig.4. Dust emission is expected to increase more than synchrotron. In addition the angular distribution is also more anisotropic. This situation is compensated by a possible *boost factor* enhancing the synchrotron signal coming from these sources, if CRE come out along the direction of observation. In fact electrons radiate only towards the speed vector, and we detect their radiation only if this direction is aligned with our line of sight. Besides electrons originated by these sources come to the Earth position through a diffusion process, because the gyro-radius, in the interstellar magnetic field even at energy up to 1 TeV, is much smaller than the distance of the closest candidate sources. In addition around these sources the magnetic field is far from uniform. This *boost factor* could be large up to the Lorentz factor $\gamma$, but we must take into account also that the detected CRE could come from sources not radiating in the direction of our line of sight. In conclusion one could map a region surrounding a galactic source, like a pulsar or a SNR, looking for an anisotropic signal at small angular scales. In this case the background subtraction could be easier, in particular if we look at the spectral feature in the polarized signal.

*C) Electrons in extragalactic sources.* An alternative approach is to observe extragalactic radio sources: spiral or elliptical galaxies. In fact we expect the same phenomenology, regarding cosmic rays and synchrotron radiation, in external galaxies as in our own. Also regions surrounding AGNs could be used as target to observe features in the synchrotron radiation (which here is much more intense), but our understanding of these sources is still uncomplete and CRE producing synchrotron radiation are not observable. *(1)* Spiral galaxies should have synchrotron and dust emissions similar to the Milky Way, and should have a similar population of pulsars and SNR in the disk. We can look for an anisotropic signal against the background at small angular scale, but the presence of the thermal dust emission does not facilitate the background removal. In this case we can not decouple the effect generated by local sources from a Dark Matter signature. *(2)* Elliptical galaxies show a low dust and gas content. This means that both thermal emission and SNR generated by supernova explosions in a low density ISM are no longer important contaminants. Therefore these galaxies can be used as favorite targets for looking at a signature of Dark Matter annihilation or decay from the galactic halo.

In order to investigate the features we pointed out in this paper, spectral and polarimetric information, together with good angular resolution, would help. A spectro-polarimeter, or a polarimeter operating in different photometric bands, installed in the focal plane of a large far-IR telescope would be necessary. Because of the frequencies under investigation a space or a balloon-borne experiment could be preferred.

To have a more realistic estimate of the signals, in all the situations considered above, detailed calculations have to be done, including radiative transfer effects, in particular concerning polarized signals.

## References

1. O. Adriani, et al., *Nature*, **458**, 607 (2009).
2. A.A. Abdo, et al., *Phys. Rev. Lett.*, **102**, 181101 (2009).
3. J. Chang, et al., *Nature*, **456**, 362 (2008).
4. F. Aharonian, et al., *Astron. & Astrophys.*, **508**, 561 (2009).
5. J. Casaus, *J. Physics: Conf. Ser.*, **171**, 012045 (2009).
6. V.L. Ginzburg & S.I. Syrovatskii, *Ann.Rev.Astron.& Astrophys.*, **3**, 297 (1965).
7. M.S. Longair, *High energy astrophysics*, Cambridge University Press, Cambridge, United Kingdom, 2nd edition (1994).
8. L. Zhang & K.S. Cheng, *Astron. & Astrophys.*, **368**, 1063 (2001).
9. A. Tartari, et al., *Astrophys. J.*, **688**, 32 (2008).
10. D.J. Schlegel, D.P. Finkbeiner & M. Davis, *Astrophys. J.*, **500**, 525 (1998).

## Science in the schools - the Extreme Energy Events project

M.Abbrescia[1], R. Antolini[11], R. Baldini Ferroli[6,7], G. Bencivenni[7], E. Bressan[2], A.Chiavassa[8], C. Cical[10], L. Cifarelli[2], F. Coccetti[11,6], D. De Gruttola[12], S.DePasquale[12], M. DIncecco[11], F. L. Fabbri[7], V. Frolov[8], M. Garbini[2], ?C. Gustavino[11], D.Hatzifotiadou[2], P. La Rocca[4,6], F. Librizzi[4], A. Maggiora[8], H.Menghetti[2], S.Miozzi[7], R. Moro[11,6], M. Panareo[9], G. Piragino[8], F. Riggi[4], F.Romano[1], G. Sartorelli[2], E.Scapparone[2], M. Selvi[2], S. Serci[10], E. Siddi[10], M.C.S.Williams[2], A. Zichichi[2,6,5] and R. Zuyeuski[3]

Presented by: M.C.S. Williams crispin.williams@cern.ch

1. INFN and Dipartimento di Fisica, Universià di Bari - Bari, Italy
2. INFN and Dipartimento di Fisica, Università di Bologna - Bologna, Italy
3. World Laboratory - Geneva, Switzerland
4. INFN and Dipartimento di Fisica e Astronomia, Università di Catania, Italy
5. CERN - Geneva, Switzerland
6. Museo Storico della Fisica, Centro Studi e Ricerche E.Fermi, Rome, Italy
7. INFN, Laboratori Nazionali di Frascati - Frascati (RM), Italy
8. INFN and Dipartimento di Fisica, Università di Torino - Torino, Italy
9. INFN and Dipartimento di Fisica, Università di Lecce - Lecce, Italy
10. INFN and Dipartimento di Fisica, Università di Cagliari - Cagliari, Italy
11. INFN, Laboratori Nazionali del Gran Sasso - Assergy (AQ), Italy
12. INFN and Dipartimento di Fisica, Universià di Salerno - Salerno, Italy

The Extreme Energy Events (EEE) project aims to study extended air showers from high energy cosmic rays and extreme energy events by detecting the muon component of the shower. To achieve this goal, a network of muon telescopes has been installed in high schools distributed all over Italy. Each muon telescope consists of three large area (80 x 160 $cm^2$) Multigap Resistive Plate Chambers (MRPCs). Each MRPC has 24 pickup strips read out at both ends; the hit position along the strip is thus deduced from the time difference. This design offers pointing capability, so that the muon direction can be reconstructed. The project has been conceived by Prof. A. Zichichi in order to rekindle the interest of young people in science and give them a first-hand experience of scientific research.

*Keywords*: Outreach, cosmic rays, extreme energy events, multigap resistive plate chambers, muon telescope.

## 1. Introduction

The Extreme Energy Events (EEE) project aims to study extended air showers from high energy cosmic rays and extreme energy events by detecting the muon component of the shower. To achieve this goal, a network of muon telescopes has been installed in high schools distributed all over Italy. The project has been conceived by Professor A. Zichichi in order to rekindle the interest of young people in science and give them a first-hand experience of scientific research. To ensure their full involvement in all stages of the project, they participate in the construction of the muon detectors, setting up and commissioning of the telescopes, installation in the schools and the data-taking and analysis. The project is financed by the Italian Ministry for Education, Universities and Research (MUIR), the National Institute for Nuclear Physics (INFN) and the Centro Studi e Ricerche e Museo Storico della Fisica Enrico Fermi; CERN is also a partner in the project.

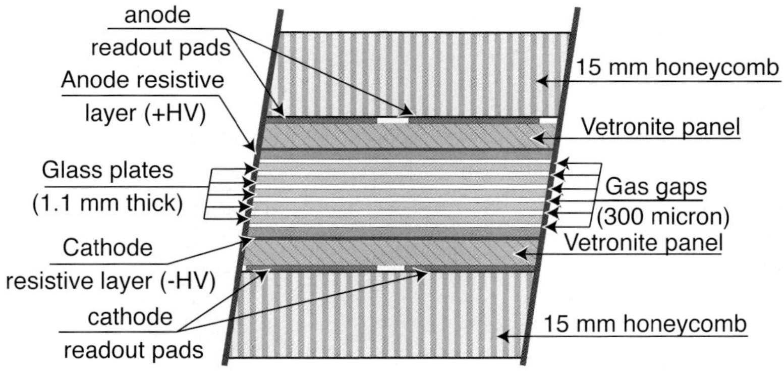


Fig. 1. Cross section of the MRPC for the EEE project.

## 2. The detector

The requirements that led to the design of the muon telescopes and the choice of the detectors for the EEE project were : large area coverage, ease of construction and pointing capability, so that the direction of incoming muons can be reconstructed. Each muon telescope consists of three large area (80x160 $cm^2$) Multigap Resistive Plate Chambers (MRPCs). These are gaseous detectors of the parallel plate type and consist of a stack of resistive

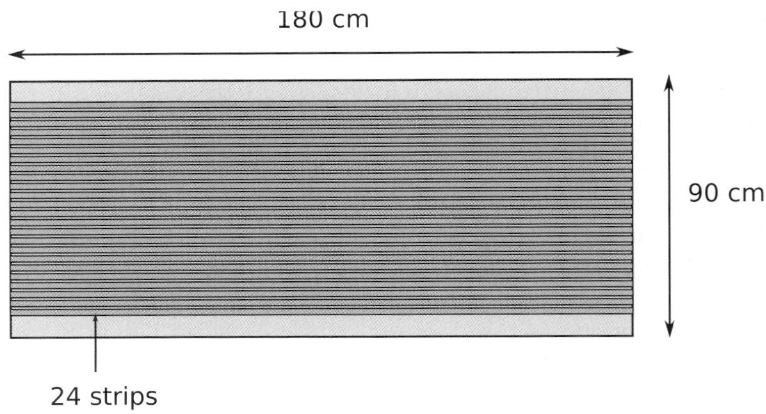


Fig. 2. Readout strips made of copper tape glued on vetronite.

Fig. 3. Making a readout plane: Glueing copper strips on a vetronite sheet.

plates made of commercial glass. The distance between the glass plates is kept fixed by means of nylon fishing line spacers in a zig-zag layout. This MRPC has 6 gas gaps with a gap width of 300 microns. The external glass plates, 1.9 mm thick, have their outside surface painted with a conductive layer; this is to apply the high voltage needed for the chamber operation; the intermediate plates, with a thickness of 1.1 mm, are electrically floating. A cross section of this MRPC is shown in Fig. 1. This design is based on the 10 gap MRPCs used for the Time of Flight (TOF) system of the ALICE experiment at LHC.[1]

Charged particles going through the MRPC ionise the gas; the electrons produced from the ionisation are multiplied due to the high electric field and avalanches are created inside the gas volume; their movement towards the anode induces electrical signals on pick-up strips. The intermediate glass plates physically stop the development of the avalanches, but they are transparent to the induced signal; thus the signal induced on the readout electrodes is the sum of the signals from all avalanches created in all gas gaps. The advantage of this multigap design is that the time precision of the signal is correlated with the gap width; thus the result is an excellent time resolution.

The readout is done with 24 pickup strips, made of adhesive copper tape glued on a vetronite plate, as shown in Fig 2. The width of the strips is 2.5 cm and the distance between them is 7 mm. There are both anode and cathode pickup strips; in this way a differential signal is sent to the readout electronics.

The MRPC is placed in an aluminum, gas tight box, with dimensions $200 \times 100 \times 5$ cm$^2$.

## 3. Electronics and readout

The front end electronics is based on the ultrafast NINO amplifier and discriminator developed for the ALICE TOF.[2,3] This is a low-power, 8-channel chip, which accepts differential input and produces LVDS output. The leading edge of the signal provides the time measurement; the trailing edge is used to calculate the time width. The width of this LVDS depends on the amount of input charge and thus can be used for slewing corrections. The front end cards have 3 NINO chips, with a total of 24 channels, which matches the 24 strips per readout plane. Each 24-channel card provides also the OR of the 24 inputs. Two such cards are used for each MRPC to read out the signals from both ends of the pickup strips. The hit position along the strip is obtained from the time difference.

The trigger is created by a six-fold coincidence of the OR signals from both sides of all three MRPCs. A home-made (Catania-CERN) trigger card is used in most of the EEE stations, a variant of the ALICE LTU is used in others; the trigger can alternatively be formed by standard NIM electronics.

A VME based system is used for the readout. High resolution commercial TDCs are used to measure the time of the hits from the MRPCs: CAEN V1190A with 128 channels and CAEN V1190B with 64 channels. These are based on the High Performance Time to Digital Converter (HPTDC),[4] developed by the CERN microelectronics group, in the 100 ps/bin mode. A GPS card (HYTEC) is included in the readout, providing the absolute time stamp needed to correlate events from different stations.

The Data Acquisition is done by running a LabView program on a Windows PC. The communication between the PC and the VME crate is with the CAEN V1718 USB bridge. In addition to data readout and recording, the program used for data acquisition provides online monitoring of the raw data.

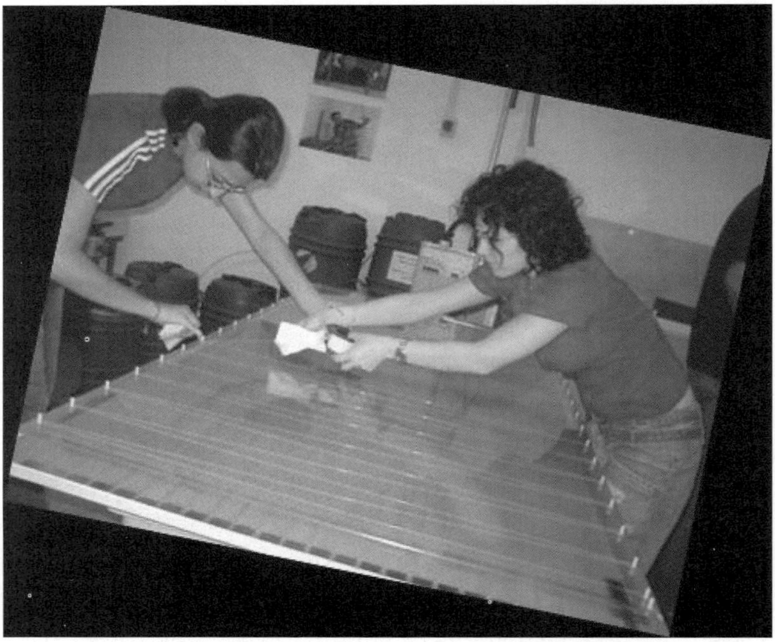

Fig. 4. Stretching the fishing line spacer in a zig-zag pattern between glass plates.

## 4. The MRPC construction

The construction of the MRPCs for the muon telescopes took place at CERN, in the laboratory and assembly hall of the ALICE TOF. Italian

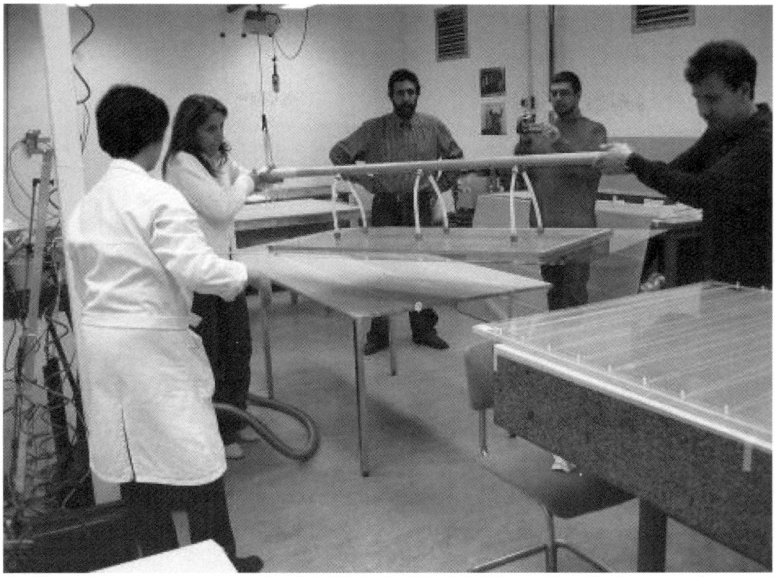

Fig. 5. Lifting a glass plate with a vacuum system to position it on top of the previous one.

Fig. 6. The 3 MRPCs of a muon telescope, inside their aluminium boxes, packaged with boat fenders for safe transport.

Fig. 7. A muon telescope on its support frame in the Gran Sasso laboratory.

High Schools participating in the project have been sending groups of 5-10 persons, consisting of pupils and teachers, accompanied by researchers from the University or INFN section from the same town or a neighbouring one. Each group stayed a period of a week and constructed, under the supervision and guidance of INFN/CERN physicists, the chambers for their muon telescope. There were construction periods of 2-3 months in 2005, 2006 and 2009.

A total of more than 100 MRPCs have been built from the beginning of the project until now. Details on the chamber construction procedure, including preparation of the different components, are given in ref. 5. Figures 3,4,5 show photographs of some of the construction steps (preparation of the readout plane; lifting of the glass plate; stretching the fishing-line spacers).

This was a valuable experience for the pupils, from various aspects. Making with their own hands a detector that works was very important, especially as many of them were using tools (to drill holes, to solder etc) for the first time.

After the end of the construction the pupils had the opportunity to get some experience with operating gaseous detectors. They flowed gas in

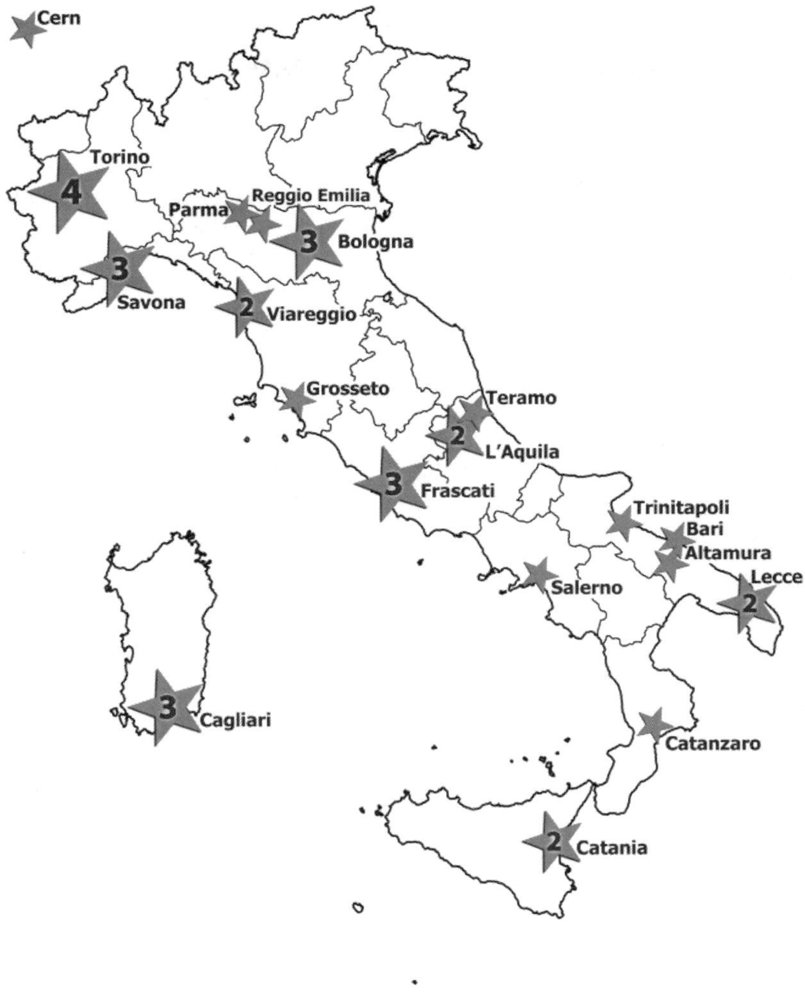


Fig. 8. Map of Italy showing the towns where EEE telescopes have been installed; the numbers inside the stars show the number of telescopes.

the chambers, checked the gas-tightness and hunted for and repaired leaks. They slowly increased the high voltage and monitored the dark current. Eventually they measured the efficiency plateau.

## 5. Chamber performance

The performance of the MRPCs for the EEE muon telescopes has been studied extensively, both with cosmic rays[6] and in a test beam at the CERN East Hall.[5] The MRPCs are filled with a gas mixture of 98% Freon $C_2F_4H_2$ and 2% $SF_6$ and are operated in avalanche mode. For a voltage difference between anode and cathode in the 18 kV range their efficiency is better than 95% and the time resolution is of the order of 100 ps. This results in a position resolution of the order of 1 cm along the readout strip; thus the muon direction is reconstructed with an angular resolution of $0.3°$(RMS).

## 6. Installation of the muon telescopes

After their construction and first tests at CERN, the muon chambers were transported to Italy and distributed to the schools participating in the project. Figure 6 shows one of the methods used for packaging in order to ensure that the glass plates do not break during transport. Each school was equipped with a mechanical support structure, offering the possibility to vary the distance between the MRPC planes (typical distances are between 50 cm and 100 cm); a telescope on its support is shown in Fig. 7. The schools were also equipped with a gas mixing system (Bronkhorst), a special cupboard to contain the gas bottles, and the front end and readout electronics, described in section 3.

The High Voltage necessary for the MRPC operation is provided by DC-DC converters, as the use of High Voltage cables is not allowed in the schools for security reasons. Two such converters per chamber are used, providing positive and negative output of 0-10 kV for an input of 0-5 V.

## 7. Status of the project

At the moment there are 33 muon telescopes installed in Italian high schools and two at CERN; their geographical distribution is shown in Fig. 8. Many of them are taking data; others are in various stages of being equipped with front end and readout electronics, commissioned and debugged. There is also a long list of schools interested in joining the project.

The stations that take data analyse them locally and, in addition, transfer them to the Ettore Majorana Centre for Scientific Culture (EMCSC) where they are stored.

## 8. First coincidence results

Extensive air showers have been detected with the EEE project telescopes.[7] Using two telescopes installed in two High Schools in L' Aquila, at a distance of 180 m from each other, the first time coincidences have been observed. The track reconstruction achieved with the MRPCs allows one to reduce the background due to accidental coincidences and determine the shower axis.

## References

1. A.V. Akindinov et al., Nucl. Instr. Meth. Phys. Res. A 56 (2000) 16-122.
2. F. Anghinolfi, P. Jarron, A.N. Martemiyanov, E. Usenko, H. Wenninger, M.C.S. Williams and A. Zichichi, Nucl. Instr. Meth. Phys. Res. A 533(2004) 183-187.
3. F. Anghinolfi, P. Jarron, F. Krummenacher, E. Usenko and M.C.S. Williams, IEEE Transactions on Nuclear Science, Vol. 51, No 5, Oct. 2004, 1974-1978.
4. J. Christiansen, a write-up can be found at the following web address: http://tdc.web.cern.ch/tdc/hptdc/docs/hptdc_manual_ver2.2.pdf
5. M. Abbrescia et al.,. Nucl. Instr. Meth. Phys. Res. A 593 (2008) 263-268.
6. Shaohui An, D. Hatzifotiadou, Jinsook Kim, M.C.S. Williams, A. Zichichi and R. Zuyeuski, Nucl. Instr. Meth. Phys. Res. A 578 (2007) 139-145.
7. M. Abbrescia et al., Il Nuovo Cimento, Vol. 125B, N.2, 243-254.

# A COSMIC RAY DETECTOR ARRAY FOR SCHOOLS IN THE CAMBRIDGE REGION

S.A. WOTTON, M.J. GOODRICK, B. HOMMELS, M.A. PARKER
e-mail: wotton@hep.phy.cam.ac.uk

*Cavendish Laboratory, University of Cambridge, JJ Thomson Avenue, Cambridge CB3 0HE, UK*

Particle physics, astrophysics and cosmology are areas of research that have captured the imagination of the general public in recent years. By giving school students first-hand experience of building and operating a particle detector and the analysis of the data in a collaborative environment we anticipate that they will gain a deeper insight into the many and diverse facets of experimental particle physics. Cosmic rays provide a readily available source of high energy particles and other projects have already exploited this in building arrays of cosmic ray detectors located in schools and linked together via the internet. We aim to extend this concept by creating our own network of detectors in our region with a particular emphasis on hands-on involvement by school students in the partner schools. This talk outlines our plans towards the implementation of this project and our wider goals of integrating our local network with other projects both nationally and internationally.

## 1. Introduction

The Cavendish Laboratory High Energy Physics (HEP) group at the University of Cambridge has a very active outreach program to engage with the general public. In particular we host a number of events targeted at school students with the aim of inspiring young people to pursue the study of science at school and beyond. Our group have found in particular that the use of real particle detectors is effective not only at catching the attention of the students but also to provide a catalyst for the discussion of many aspects of experimental and theoretical particle physics. These aspects include not only particle physics, astrophysics and cosmology, but also experimental techniques, diverse topics in classical and quantum physics, engineering, information technology and many others. Although many of these topics touch on existing areas of the school curriculum, our aim is not to teach the curriculum but rather to try to show how scientific research is performed outside of the classroom. We are fortunate that particle physics has benefited in recent years from exposure through the media and this has generated a lot of interest in the Large Hadron Collider and its detectors amongst the general public. However, it is important to go beyond the passive absorption of pre-digested information presented by the media in order to promote a deeper and longer-lasting appreciation of the science. In the following sections I will briefly review the detectors that we have developed in the HEP group for our outreach activities and present the status of our new cosmic ray detector array for schools.

## 2. Demonstration particle detectors

The HEP group has developed several particle detectors for the purposes of outreach. These include a diffusion cloud chamber, a table-top spark chamber and a cosmic ray counter. The first two provide a direct and immediate visualisation of the paths of charged particles. In the case of the cloud chamber the traces are mainly due to the decay of naturally occurring radioactive elements in the environment while the spark chamber reveals the paths of muons produced in cosmic ray air showers. The cosmic ray counter is effective when used as an illustration of the techniques used in the detectors used for research.

### 2.1. *The cloud chamber*

Our diffusion cloud chamber is very popular amongst all age groups. The reasons for its popularity are the simplicity of construction, the direct and immediate visualisation of particle traces and the fact that its basic operation can be readily explained using simple physical principles and everyday analogies. Even very young children are frequently enthralled by the ever-changing landscape inside the chamber.

Figure 1 is a photograph of the fully assembled chamber. The chamber is made from an inverted aquarium purchased from a pet shop. The open side is closed with a flat metal plate and placed in contact with a layer of dry ice in an insulated container made from a block of expanded polystyrene or similar. A tissue moistened with isopropanol is supported from the inside top face of the chamber. The heavy alcohol vapour falls towards the very cold plate so that, after a few minutes, when the conditions inside the chamber have stabilised, particle traces can be discerned in the supersaturated layer of alcohol vapour that forms a few millimetres above the plate. Some experimentation with lighting may be needed for the best contrast. Our apparatus uses an approximately parallel beam of light produced using a strip of bright white light-emitting diodes and a hemi-cylindrical lens although a desktop strip lamp can also be effective. More details about its construction and operation can be found on our web site [1].

Figure 1. The fully assembled cloud chamber.

### 2.2. *The spark chamber*

Like the cloud chamber, a spark chamber provides a very immediate visualisation of the path of charged particles. In particular, it provides a very dramatic display of muons produced in cosmic ray air showers. It is very good at attracting the attention of visitors during crowded open days and is also effective for public lectures as it can be appreciated even from the back of a lecture theatre. Although the construction is more complex than the cloud chamber, the basic principle of operation is nevertheless also easily explained using familiar analogies such as lightning and lightning conductors.

The HEP group, in association with the Cambridge Hands-On Science (CHaOS) organisation [2], has recently obtained funding to build a spark chamber that will be used as a touring exhibit to promote science amongst young people. Figure 2 is a photograph of our prototype chamber (about 20cm high) with which we have already had considerable success at various events. The final version is now under construction and this version has been designed to require minimum maintenance for the duration of a typical exhibition as it will be operated by volunteers without direct expertise in the details of its construction. The chamber is also intended to be sufficiently portable and self-contained so that it can be transported easily in a family car. More details about the project can be found on our web site [3].

Figure 2. The prototype spark chamber.

### 2.3. *The cosmic ray counter*

A more quantitative measurement of the cosmic ray flux can be achieved in the classroom using one or more scintillators with suitable photo-detectors operating in coincidence. Although not as dramatic in its simplest form as the cloud or spark chamber it is perhaps more similar to the detectors used in real research projects and provides more opportunities for an extended programme of measurements in the school environment. Measurements of the flux of muons can be readily performed and investigations of the dependence of the flux on various parameters are an excellent introduction to the analysis of data, treatment of errors and scientific methods in general.

## 3. The detector array project

The remaining sections outline our plans to create an array of cosmic ray detectors distributed amongst a number of schools in the south-east of England and the current status of the project. The array will be similar in principle to large area research arrays and will complement existing educational arrays.

Several such research and educational arrays have been reported at this conference.

Our aim is to achieve as wide participation as practical and this imposes a few important constraints on the project. The financial cost of the project to the schools must be kept low and, since we have no guaranteed source of long term funding for the project, this implies that the equipment must be inexpensive. It is also necessary that the equipment should be sufficiently robust and simple to construct that it does not require frequent expert intervention. This is particularly important as we anticipate that the project will run for an extended duration and our long-term support may be limited to a small number of trained volunteers. Ideally, most troubleshooting would be done by the schools themselves but it is crucially important that there is sufficient expert back-up when required otherwise it is likely that the equipment will quickly fall into disuse. For similar reason it is also important to fully engage the teachers in the partner schools and maintain regular contact with them.

### 3.1. *Status of the prototype*

As part of a pilot study we are currently building a prototype detector that we will use to evaluate our project aims in the school environment and adapt them if necessary after feedback from the schools. Our prototype uses scintillator reclaimed from the CERN NA48 experiment. The scintillator is in the form of 1cm thick strips, 25cm wide and up to about 150cm long. A photograph of a sample of the scintillator is shown in Figure 3 where a slot to accept a wavelength shifting (WLS) fibre can be seen running across the middle of the tile and a WLS fibre lying on the surface.

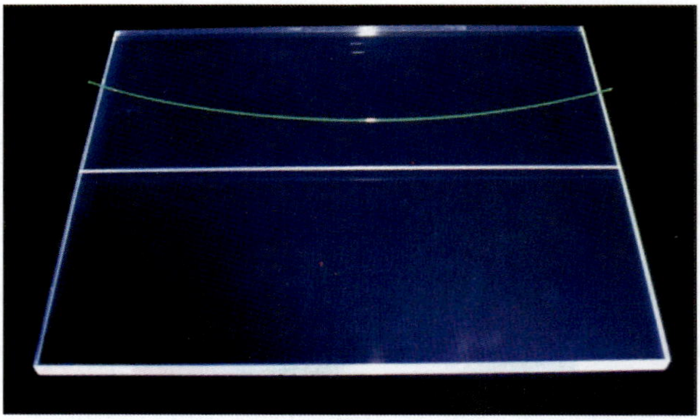

Figure 3. The scintillator sample and wavelength shifting fibre.

We plan to use silicon photomultipliers as the photo-detector element. Because of the small size of the photo-detector (1mm square) it is necessary to use WLS fibre to collect the scintillation light while at the same time matching the light output to the photo-detector. Unlike conventional photomultipliers, silicon photomultipliers have the significant advantage of not requiring high voltage to operate. This is an important consideration for equipment that is required to operate in a school. The device we have selected, from the company Photonique SA [4], operates at a voltage less than 20V and is shown in Figure 4 with its preamplifier.

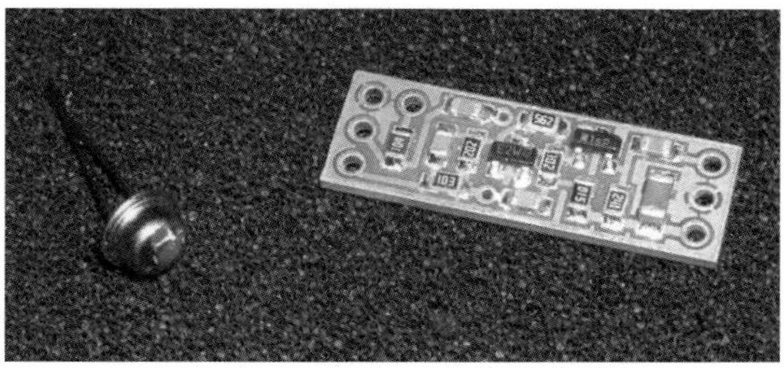

Figure 4. The silicon photomultiplier and its preamplifier.

During the course of this year we have measured some of the properties of the silicon photomultiplier and the light collection efficiency of the scintillator, WLS fibre and silicon photomultiplier assembly. This work has been largely performed by physics undergraduates as part of their assessed project work. Particularly notable, however, have been some measurements performed by a high school student from The Netherlands. A typical example of his measurements of the silicon photomultiplier and detector assembly is shown in Figure 5. It provides encouraging evidence of what we can expect to be achieved by the participating school students.

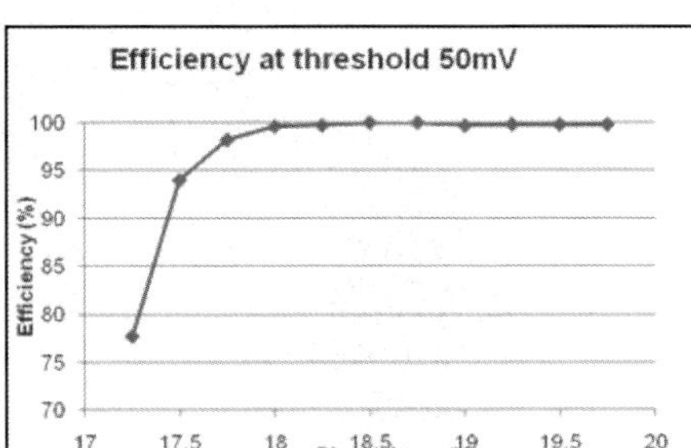


Figure 5. Detector efficiency vs. bias voltage at fixed threshold.

### 3.2. *The school pilot study*

Before finalising the detector design, the apparatus will be taken to one of our partner schools in order to gain experience working in the school environment. This will be an important phase of the project as we will get valuable feedback from the teachers and students on the practical aspects of the project. Our first sessions in the school will be in an after-school science club where the students will progressively build and test the detector assembly from a kit of parts. In doing so they will be encouraged to make a series of measurements that will give them a deeper understanding of how the detector works and its limitations.

Having successfully constructed one of the detector array elements the students will then duplicate it and make some exploratory measurements that combine the data from two or more elements. At this stage they will be able to start making some quantitative measurements of the detector efficiency and eventually of the cosmic ray muon flux in various experimental configurations. Later, when more schools are equipped with detectors, the students will be able to extend their measurements to look for correlated activity over a wide area by combining the data with those from the other schools.

### 3.3. *Next steps*

We anticipate that the pilot school studies will be done over a number of weeks during which time we will continue to develop the experimental set-up in preparation for distributing detector kits to more school sites. The scintillator pieces will be cut to size and the WLS fibre slots machined in the Cavendish Laboratory. We will also provide pre-built parts to facilitate the coupling of the WLS fibre, silicon photomultiplier and pre-amplifier. We expect most other mechanical work to be within the capabilities of the students themselves or the school workshop.

It is unlikely that the schools have the resources and expertise to design and build the electronic circuits that will be required so these will be provided by us. However, the initial stages of assembly and testing of the detectors in the schools will provide many opportunities for interested students to gain some practical electronics experience.

In order to combine data from the distributed array of detectors we will need to develop an interface module that incorporates the necessary timing information. This new module will be an updated version of a low-cost, FPGA based module that we developed for a previous project with the addition of a GPS receiver interface. The module will also be designed to allow auxiliary modules to be connected using a standard connector so that its functionality can be extended easily as may be required.

In parallel with our programme of hardware development and with the feedback from the teachers and students, we will also develop resources that will help to fully exploit the potential educational benefit of the project. These could include worksheets with ideas on possible measurements, links to related sources of information, tools for the analysis of the data both standalone in the school and also in collaboration with other schools.

In the future we plan to explore the possibility of integrating our detector array with other similar arrays nationally and internationally in order to realise a truly collaborative project that gives school students the experience of working in a global research endeavour.

## 4. Conclusion

The use of cosmic ray detector arrays in schools is now widely exploited as a way of promoting a deeper understanding of particle physics and cosmology. Our array is in the early stages of commissioning but has already provided a rewarding project for a number of young scientists. We are confident that our partner schools will also find the project stimulating and exciting and we are looking forward to rolling out the detectors during the coming year.

## Acknowledgements

The HEP group are grateful for the sponsorship provided by The Ogden Trust for pilot studies related to the detector array project.

The spark chamber project is funded in part by the UK Science and Technology Facilities Council.

The projects described here have benefited from the enthusiastic participation of University of Cambridge undergraduates Philip de Grouchy, Jack Collins, Zoe Penfold-Fitch and Ben Harrison and, from The Netherlands, high school student Floris Keizer.

## References

1. http://www.hep.phy.cam.ac.uk/outreach/cloud.php
2. http://www.chaosscience.org.uk/
3. http://www.hep.phy.cam.ac.uk/outreach/spark.php, Project leaders C. Lester (University of Cambridge) and D. Ansell (CHaOS).
4. Now ceased trading. Device portfolio taken over by Centre of Prospective Technology & Apparatus, 1-ya Vladimirskaya Street 22-2, 111141 Moscow, Russian Federation.

# OBSERVATION OF ELECTROSCALAR RADIATION DURING A SOLAR ECLIPSE

O.A. ZAYMIDOROGA and D.V. PODGAINY

*Joint Institute for Nuclear Research,*
*Dubna, 141980, Moscow Region, Russia*
*E-mail: zaimidor@jinr.ru*

In this work, a non-relativistic theory of the electroscalar field being an expansion of the classical Maxwell's electrodynamics is briefly discussed and results of observation of a solar eclipse which took place on 1 August 2008 are presented.

*Keywords*: Longitudinal Electroscalar Wave, Transport of the Coulomb Field, Sun Eclipse.

## 1. Introduction

As is known, classical Maxwell equations contain wave solutions only for the transverse waves, which can propagate in vacuum, however, there are no solutions corresponding to longitudinal waves. The recent years have seen growing interest in the theoretic research of the possibility of introducing both longitudinal electroscalar [1,2] and magnetoscalar waves [3] and ongoing attempts to discover them experimentally [4]. One of the motives for expanding classical Maxwell electrodynamics by way of introducing an additional longitudinal mode into this theory is the absence of wave transport of the Coulomb field in this theory, the latter factor leading to the appearance of long-range action in this theory. In the framework of quantum electrodynamics, an approach is developed for describing the Coulomb interaction of charged particles by way of longitudinal photons [5], although the latter are to be considered non-physical (virtual), otherwise, the quantum theory faces a number of unsurmountable difficulties [6,7]. However, the methods for excluding non-physical photons elaborated in the framework of quantum electrodynamics (see review [7]) exclude also the wave transport of the Coulomb field in this theory. Indicative of one of the possibilities of introducing the longitudinal wave mode into Maxwell electrodynamics is the existing analogy between the electrodynamics and linear elasticity theory, which allows one to consider electromagnetic vacuum as a compressible medium [8]. Radial oscillations of the electron gas of a spherically symmetric metallic particle, which are sketched out in Fig. 1, can serve as a possible macroscopic source for such a longitudinal wave.

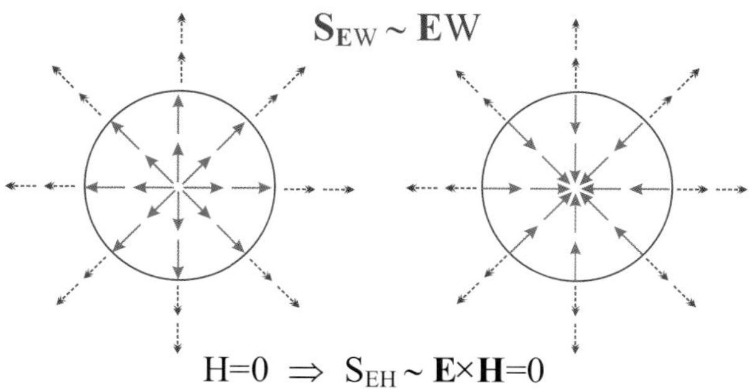


Fig. 1. Possible source for a longitudinal electroscalar wave. The electron current (green arrows) and the wave energy flux (blue arrows) flow strictly along the radius.

Owing to the spherical symmetry, the magnetic field of the fluctuating radial transfer current equals to zero and, thus, radiation losses of this system can be realized only at the expense of radiation of a longitudinal wave in which the electric-field vector is collinear to the wave energy flux vector.

In the event of existence of such longitudinal waves in nature, the Sun will be one of the most powerful sources for them as a plasmoid object. The most favourable conditions for the detection of longitudinal solar waves are realized during a solar eclipse. In this case, the Moon screens part of the flux of transverse electromagnetic solar waves and longitudinal waves, presumably possessing greater penetration power [9], can reach the surface of the Earth. The present paper outlines a theoretical scheme describing transport of the Coulomb field via a longitudinal electroscalar wave and presents results of observation of a solar eclipse which took place on 1 August 2008.

## 2. Electroscalar Field Equations

The considered expansion of Maxwell electrodynamics which allows one to describe transport of the Coulomb field using a longitudinal wave is based on the assumption that the Coulomb field, from the viewpoint of the quantum-field concept, is a superposition of scalar photons or superposition of massless scalar particles with a zero spin, rather than a superposition of vector photons with a zero spin projection [8]. At the same time, it is assumed that the introduced scalar particles realize physical conditions of the field, in other words, they appear as observables. Constructively, this assumption is expressed in that along with the 4-vector potential $A^\mu$, there exists a 4-scalar massless field $\lambda$. In the non-relativistic case, the equations for potentials take the form [8]:

$$-\frac{1}{c^2}\frac{\partial^2 \mathbf{A}}{\partial t^2} + \Delta\mathbf{A} = -\frac{4\pi}{c}\mathbf{j}, \quad -\frac{1}{c^2}\frac{\partial^2 \lambda}{\partial t^2} + \Delta\lambda = -4\pi\rho. \tag{1}$$

Here, $c$ is the velocity of light in vacuum, $\rho$ and $\mathbf{j}$ are, respectively, the density of charges and density of transfer current, and $\mathbf{A}$ is the spatial component of the 4-vector potential $A^\mu$ (the time component $\varphi$ of the 4-potential can be excluded via gauge transformations [10]), with $\mathbf{A}$ satisfying the Coulomb condition: $\mathrm{div}\mathbf{A} = 0$. By introducing the following field definitions:

$$\mathbf{E}_\perp = -\nabla\varphi - \frac{1}{c}\frac{\partial \mathbf{A}}{\partial t}, \qquad \mathbf{H} = \mathbf{rot}\ \mathbf{A}, \tag{2}$$

$$\mathbf{E}_{||} = \nabla\lambda, \qquad W = -\frac{1}{c}\frac{\partial \lambda}{\partial t},$$

one obtains two systems of equations from (1). The first system is for the fields $\mathbf{E}_\perp$ and $\mathbf{H}$:

$$-\frac{1}{c}\frac{\partial \mathbf{E}_\perp}{\partial t} + \mathbf{rot}\ \mathbf{H} = \frac{4\pi}{c}\mathbf{j}, \tag{3}$$

$$\frac{1}{c}\frac{\partial \mathbf{H}}{\partial t} + \mathbf{rot}\ \mathbf{E}_\perp = 0,$$

while the second one is for the fields $\mathbf{E}_{||}$ and $W$, being henceforth referred to as a system of electroscalar fields:

$$\frac{1}{c}\frac{\partial \mathbf{E}_{||}}{\partial t} + \nabla W = 0, \tag{4}$$

$$\frac{1}{c}\frac{\partial W}{\partial t} + \mathrm{div}\mathbf{E}_{||} = -4\pi\rho.$$

It is seen from these systems that the electrostatics equation $\mathrm{div}\mathbf{E}_{||} = -4\pi\rho$ follows from the system (4) and magnetostatics one $\mathbf{rot}\ \mathbf{H} = 4\pi\mathbf{j}/c$ follows from (3). System (3) describes propagation of strictly transverse waves (proof for this can be found, for example, in [10]), and system (4), in contrast, describes propagation of longitudinal waves. Now, let us consider the vacuum solution of system (4) in the form of plane waves $\mathbf{E}_{||} = \mathbf{E}_0 \exp[i(\omega t + \mathbf{kr})], W = W_0 \exp[i(\omega t + \mathbf{kr})]$, where $\mathbf{E}_0$ and $W_0$ are the amplitudes, $\mathbf{k}$ is the wave vector defining the direction of propagation of a wave with the frequency $\omega$. Then, we obtain from (4):

$$\frac{\omega}{c}\mathbf{E}_0 + \mathbf{k}W_0 = 0, \qquad \frac{\omega}{c}W_0 + (\mathbf{kE}_0) = 0, \qquad \mathbf{k} \times \mathbf{E}_0 = 0,$$

in other words, the vector $\mathbf{E}_{||}$ oscillates along the direction of wave propagation. The dispersion relationship for the given wave takes the normal form $\omega = kc$.

In order to derive the equations of energy balance for the fields $\mathbf{E}_{||}$ and $W$, let us multiply the first equation (4) scalarwise by $\mathbf{E}_{||}$ and second equation by $W$, and obtain after volume integration:

$$\frac{\partial}{\partial t}\int \epsilon_{EW} dV + \oint \mathbf{s}_{EW} \mathbf{d\sigma} = -c\int (\rho W) dV, \tag{5}$$

where $\mathbf{d\sigma}$ is the surface area element for the volume $V$ and the following definitions are introduced for the energy density $\epsilon_{EW}$ and for the energy flux vector $\mathbf{s}_{EW}$ of the electroscalar field:

$$\epsilon_{EW} = \frac{\mathbf{E}_{||}^2 + W^2}{8\pi}, \qquad \mathbf{s}_{EW} = \frac{c}{4\pi}\mathbf{E}_{||}W. \tag{6}$$

From the definition of the vector $\mathbf{s}_{EW}$ it follows that the energy flux in the above plane wave is collinear to the vector of the field $\mathbf{E}_{||}$. The system of equations (4) can be presented in the form of a system of wave equations for the fields $\mathbf{E}_{||}$ and $W$:

$$-\frac{1}{c^2}\frac{\partial^2 \mathbf{E}_{||}}{\partial t^2} + \Delta \mathbf{E}_{||} = -4\pi \nabla \rho, \qquad \mathbf{rot}\ \mathbf{E}_{||} = 0, \tag{7}$$

$$-\frac{1}{c^2}\frac{\partial^2 W}{\partial t^2} + \Delta W = \frac{4\pi}{c}\frac{\partial \rho}{\partial t}.$$

It is seen from these equations that time-dependent non-homogeneous charge density serves as a source for the longitudinal waves, while the radial oscillations of the electrons of a spherically symmetric metal particle, mentioned in the introduction, can serve as an example of a source for longitudinal electroscalar waves. Thus, the worked out formalism allows one to describe the wave transport of the Coulomb field using a longitudinal electroscalar wave.

## 3. Results of an Observation of a Solar Eclipse

An observation of an eclipse of the Sun was held on 01 August 2008. Registration of radiation during an eclipse of the Sun covered by the Moon by 58% was provided by four copper spheres placed in a metallic screen. The diameter of the spheres was $28mm$, $36mm$, and two spheres had a diameter of $43mm$; the total surface of the spheres amounted to $43.3sm^2$. The centers of the spheres were connected to each other with a cooper wire. The sum signal from the spheres was fed to the diodes plug input, and the diode plug output was connected to the Fluke289 multimeter. Directly before the metallic screen, two tourmaline plates were fixed. The detection layout is shown in Fig. 2.

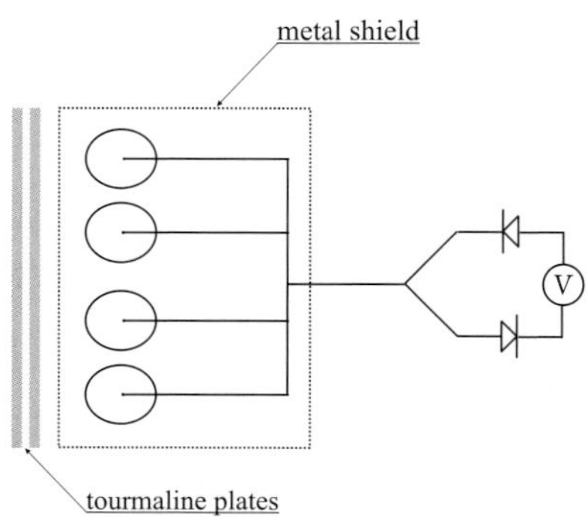


Fig. 2. Schematic illustration of radiation detector.

The measurement results are presented in Fig. 3. The horizontal axis is the registration time axis with the start at 1 pm on 1 August 2008. The maximal signal was registered at 2.38 pm. Along the vertical axis, the volt value of the detected signal is plotted. The maximal value of the signal was 1.04 DC volt. The direct current amounted to a few microamperes, being not dependent on the load resistance.

The maximal signal corresponds with a great accuracy to the highest value of the eclipse of the Sun. (Conspicuous is the surge to the left of the peak that arose 10 minutes before the peak of the eclipse, which can represent acts of back-scattering of the Earths radio-frequency radiation on the Moon). Let us note also that the radiation frequency 47 MHz in the solar radiation spectrum, for a wavelength of 4 cm, apparently, can point to the Stokes line of excitation of the spheres electron gas by the

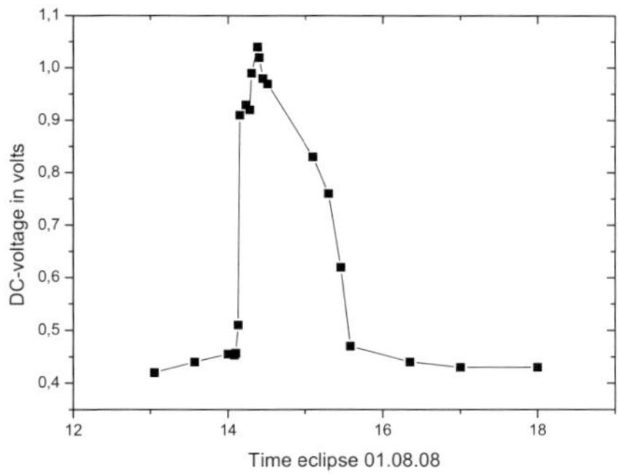


Fig. 3. The measurement results. The volt value of the direct current as function of registration time.

high-frequency part of the Suns spectrum.

As for the nature of the detected radiation, it should be noted that it possesses great penetration power (anyway, this radiation penetrated through the metallic screen surrounding the antennae) and the value of the registered signal does not depend on the mutual orientation of the tourmaline plates, which corresponds to the absence of transverse polarization for the registered radiation. These properties ensure explicit evidence that the character of the detected radiation is that of a longitudinal electroscalar wave.

## References


1. K.J. van Vlaenderen and A. Waser, *Hadronic Journal* **24**, 609 (2001).
2. K.J. van Vlaenderen, *A generalisation of classical electrodynamics for the prediction of scalar field effects.*, arXiv:physics/0305098v1 [physics.class-ph] .
3. N.P. Khvorostenko, *Izvestiya vuzov, Fizika* **3**, 24 (1992).
4. C. Monstein and J. P. Wesley, *Europhys. Lett.* **59**, 514 (2002).
5. P. A. M. Dirac, *Proc.Roy.Soc. A* **136**, 453 (1932).
6. P. A. M. Dirac, *Scientific Papers Collection. V. II. Quantum Theory (scientific articles 1924-1947)*, ed. A.D. Sukhanov (Moscow, Fizmatlit, 2002).
7. Polubarinov I. V., *Physics of Elementary Particles and Atomic Nuclei* **34**, 738 (2003).
8. D.V. Podgainy, O.A. Zaimidoroga, *Nonrelativistic theory of electroscalar field and Maxwell electrodynamics*, arXiv:1005.3130.
9. Gertsenshtein M. E., *JETP Letters.* **14**, 427 (1971).
10. L.D. Landau, E.M. Lifshitz, *Theory of Elasticity, Vol. 7* (Butterworth-Heinemann, 1986).

# YOUNG RESEARCHERS FOCUS ON THE EXTREME ENERGY UNIVERSE

JAMES L. PINFOLD

*Centre for Particle Physics, University of Alberta, Edmonton Alberta, T6G 0V1, Canada*

A new dimension to cosmic-ray research opened in 1998 in Alberta when the first node of a new kind of sparse very-large-area network of cosmic-ray detectors began to take data. The innovative aspect of the Alberta Large-area Time-coincidence Array (ALTA) is that it is deployed in high schools, involving school children in the excitement of fundamental research. Independently, in 2000 the Cosmic Ray Observatory Project (CROP), centred at the University of Nebraska, set up five schools with cosmic ray detectors. Around the same time the Washington Large-area Time-coincidence Array (WALTA) installed its first detectors. I shall describe the development of "educational cosmic ray arrays" from 1998 and briefly discuss the physics issues that such arrays could address. Also, I will introduce the idea of using experimental cosmic ray physics as a means to involve developing countries in the global scientific culture. In conclusion I will attempt to elucidate future directions.

## 1. Introduction

On 15 October 1991 the highest-energy cosmic-ray particle ever measured struck Earth's atmosphere tens of kilometres above the Utah Desert. Colliding with a nucleus, it lit up the night for an instant and then was gone. The Fly's Eye detector at the Dugway Proving Grounds in Utah captured the trail of light emitted as the cascade of secondary particles created in the collision made the atmosphere fluoresce. The Fly's Eye researchers measured the energy of the unusual ultra-high-energy cosmic-ray event - dubbed the "Oh-My-God (OMG) event" - at 320 exa-electron-volts (EeV), or $320 \times 10^{18}$ eV. In SI units, the particle, probably a proton, hit the atmosphere with a total kinetic energy of about 5 J. For a microscopic particle this is a truly macroscopic energy - enough to lift a mass of 1 kg half a metre against gravity. On 3 December 1993, on the opposite side of the world, the Akeno Giant Air Shower Array (AGASA) in Japan recorded another OMG event with an energy of 200 EeV. In this case the cosmic ray was recorded using a large array of detectors on the ground to measure the extended air shower (EAS) resulting from the primary cosmic ray interacting with the atmosphere.

However signs of the extreme-energy universe may also come in a different guise - not as a single OMG event but rather as bursts of events of more-modest energy. On 20 January 1981, near Winnipeg, a cluster of 32 EASs - with an estimated mean energy of 3000 tera-electron-volts - was observed within 5 min [1]. Only one such event would have been expected. This observation was the only one of its kind during an experiment that recorded 150,000 showers in 18 months. In the same year an Irish group reported an unusual simultaneous increase in the cosmic-ray shower rate at two recording stations 250 km apart [2]. The event, recorded in 1975, lasted 20s and was the only one of its kind detected in three years of observation.

One mechanism that could generate correlated showers over hundreds of kilometres is the photodisintegration of high-energy cosmic-ray nuclei passing through the vicinity of the Sun, first proposed by N M Gerasimova and Georgy Zatsepin back in the 1950s. Other more recent and more exotic examples of phenomena that could give rise to large-area non-random cosmic-ray correlations include relativistic dust grains, neutrino bursts from collapsing stellar systems, primordial black-hole evaporation and even mechanisms arising from the presence of extra dimensions.

## 2. Involving Young Researchers

Now, however a new dimension to cosmic-ray research has opened up. In 1998 in Alberta, building on a proposal first presented in 1995, the first node of a new kind of sparse very-large-area network of cosmic-ray detectors began to take data. The innovative aspect of the Alberta Large-area Time-coincidence Array [3] is that it is deployed in high schools. Figure 1 shows the first high school based cosmic ray detectors being installed on the roof of Archbishop O'Leary High School in Edmonton, Alberta, on 23 February 1998. The idea came from my search for a practicable way to set up a large area sparse array to search for non-random cosmic phenomena using air showers and from my wife Melinda – a gifted and experienced educator. Melinda's brilliant suggestion was to involve high-schools and high school students and teachers. By the end of 1999 three high-school sites were operating, each communicating with the central site at the University of Alberta. In 2000 the Cosmic Ray Observatory Project (CROP) [4], centred at the University of Nebraska, set up five schools with detectors from the decommissioned Chicago Air Shower Array. Around the same time the Washington Large-area Time-coincidence Array (WALTA) installed its first detectors.

Figure 1 Installing the first educational cosmic-ray array on the roof of Archbishop O'Leary High School in Edmonton, Alberta, on 23 February 1998. The array is part of the ALTA experiment.

## 2.1. *The Startup in North America*

The ALTA, CROP and WALTA [4] projects have a distinct purpose - to forge a connection between two seemingly unrelated but equally important aims. The first is to study the extreme-energy universe by searching for large-area cosmic-ray coincidences and their sources; the second is to involve high-school students and teachers in the excitement of fundamental research. These "educational arrays", with their serious research purpose, provide a unique educational experience, and the paradigm has spread to many other sites in North America. The detector systems are simple but effective. Following the ALTA/CROP model they use a small local array of plastic scintillators, which are read by custom-made electronics and which use GPS for precise coincidence timing with other nodes in a network of local arrays over a large area. Most of the local systems forming an array use three or more detectors, which, with a separation of the order of 10 m and a hard-wired coincidence, allow accurate pointing at each local site. Today the ALTA/CROP/WALTA arrays involve more than 40 high schools and there are three further North American educational arrays in operation: the California High School Cosmic Ray Observatory (CHICOS) and the Snowmass Area Large-scale Time-coincidence Array (SALTA) [4] in the US, and the Victoria Time-coincidence Array (VICTA) in Canada. At least seven more North American projects are planned.

The CHICOS array [5] is the largest ground-based array in the Northern Hemisphere. Its detectors, donated by the CYGNUS collaboration, are deployed on more than 70 high-school rooftops across 400 $km^2$ in the Los Angeles area. Each site has two 1 $m^2$ plastic scintillator detectors separated by a few metres. Local pointing at each site is not possible, nor is it required as CHICOS uses GPS pointing across multiple sites to concentrate on the search for single ultra-high-energy cosmic-ray air showers.

Innovative detection techniques have also been employed in this burgeoning collaboration between researchers and high-school students and teachers in North America. A prime example is the project for Mixed Apparatus for Radar Investigation of Cosmic Rays of High Ionization [6], based at Brookhaven National Laboratory, New York. The plan is for the experiment to detect ultra-high-energy cosmic rays using the passive bistatic radar technique, where stations continuously listen to a radio frequency that illuminates the sky above it. The ionization trails of ultra-high-energy cosmic-ray showers - as well as meteors, micro-meteors and even aeroplanes - in the field of the radio beam will reflect radio waves into the high-school-based detectors. These schools will also be equipped with conventional cosmic-ray air-shower detectors. The technique, if successful, will speed the construction of ultra-large-area cosmic-ray detectors.

## 2.2. *Young European Researchers Get Involved*

Across the Atlantic, schools in many European countries are also getting involved in studying the extreme-energy universe. In 2001 physicists from the University of Wuppertal proposed SkyView - the first European project to suggest using high-school-based cosmic-ray detectors. This ambitious project proposed an immense 5000 $km^2$ array, the size of the PAO, using thousands of universities, colleges, schools and other public buildings in the North Rhine-Westphalia area. Roughly a year later CERN entered the field with a collaborative effort to distribute cosmic-ray detectors from the terminated High Energy Gamma Ray Astronomy project in schools around Dusseldorf. A test array of 20 counters was set up at Point 4 on the tunnel for the Large Electron-

Positron (LEP) collider, with the aim of studying coincidences with counters installed about 5 km away at Point 3 as part of cosmic-ray studies by the L3 experiment on the LEP.

Also in 2002 the High School Project on Astrophysics Research with Cosmics (HISPARC) [7], initiated by physicists from the University of Nijmegen in the Netherlands, joined the European effort. HiSPARC now has five regional clusters of detectors being developed in the areas of Amsterdam, Groningen, Leiden, Nijmegan and Utrecht. Around 40 high schools are participating so far and more are joining. In March 2005 the HiSPARC array registered an event of energy $8 \times 10^{19}$ eV, in the ultra-high-energy "ankle" region of the cosmic-ray energy spectrum, which was reported at the international conference in Pune [7].

Various interested European universities have organized themselves in the EuroCosmics [8] consortium which is a network of research institutes and high-schools across Europe to study the physics, properties and origin of cosmic rays: very energetic subatomic particles reaching the Earth from deep space. Currently the Europhysics consortium includes the following institutes and their accompanying partner schools: FOM-Nikhef (the Netherlands); Universite Libre de Bruxelles (Belgium); the Czech Technical University (Czech Republic); University of Arhus (Denmark); University of Jyvaskyla (Finland) Bergische Universitat Wuppertal (Germany); the Hellenic Open University, (Greece); University of Catania (Italy); A. Soltan Institute for Nuclear Studies (Poland); Instituto Superior Tecnico, LIP (Portugal); the Institute for Space Science (Romania); Moscow Engineering Physics Institute (Russia); Alcala University (Spain); Kungliga Tekniska Hogskolan (Sweden); and, the University of Sheffield (UK). In the EuroCosmics network, cosmic rays are studied with particle detectors placed at schools and research laboratories.

Detectors are being designed which are advanced enough to satisfy present-day scientific requirements and simple enough that they can be constructed by high-school students under supervision of scientists and/or technical staff at universities and research institutes. The large network of detectors will allow the fluxes of particles at different places to be correlated in space and time. With the help of specific software tools much of the data analysis can be performed by high-school students guided by science teachers or university staff with only a moderate amount of preparation. The EuroCosmics network is committed to sharing educational, scientific and technical expertise, as well as data and material.

The original pioneering ALTA Collaboration is directly involved in a few European projects in addition to the networks they have a direct hand in setting up in Canada. A few examples are as follows. The CZELTA network in the Czech Republic [9] currently has 5 ALTA-like systems in operation with 3 in preparation. We are also working with King's College London in collaboration that has already deployed two detector stations in the London area and is preparing to deploy two more in the near future. In addition, we have supplied a detector system to the institute of Space Sciences in Romania, that is planning a larger array for the future. Last, but not least, we are working with the Cavendish laboratory at the University of Cambridge that is planning on joining joining the "Eurocosmics" club. Cambridge will probably adopt a HISPARC-like technology for their detector systems.

While the majority of the European projects are based on plastic scintillators, the Italian Extreme Energy Events (EEE) project [10] has opted instead for multigap resistive plate chambers (MRPCs) as their basic detector element. These allow a precise measurement of the direction and time of arrival of a cosmic ray. The aim of this project, the roots of which date back to 1996, is to have a system of MRPC telescopes distributed over a surface of 106 $km^2$, for precise detection of extreme-energy events (Zichichi 1996). These chambers are similar to those that will be used in the time-of-flight detector for the ALICE experiment at CERN's Large Hadron Collider. Three MRPC chambers form a detector "telescope" that can reconstruct the trajectories of cosmic muons in a shower. At present 23 schools from across Italy are involved in the pilot project, with around 100 others on a waiting list from the length and breadth of the Italian peninsula.

## 3. A Worldwide Network

Most of the major groups in Canada and the US have formed a loose collaboration - the North American Large-area Time Coincidence Arrays [4] - with more than 100 detector stations spread across North America. The aim is to share educational resources and information. However, it is also planned to have one central access point where students and researchers can use data from all of the NALTA sites, creating in effect a single giant array. Such a combined network across North America could eventually consist of thousands of cosmic-ray detectors, with the primary research aim of studying ultra-high-energy cosmic-ray showers and correlated cosmic-ray phenomena over a very large area.

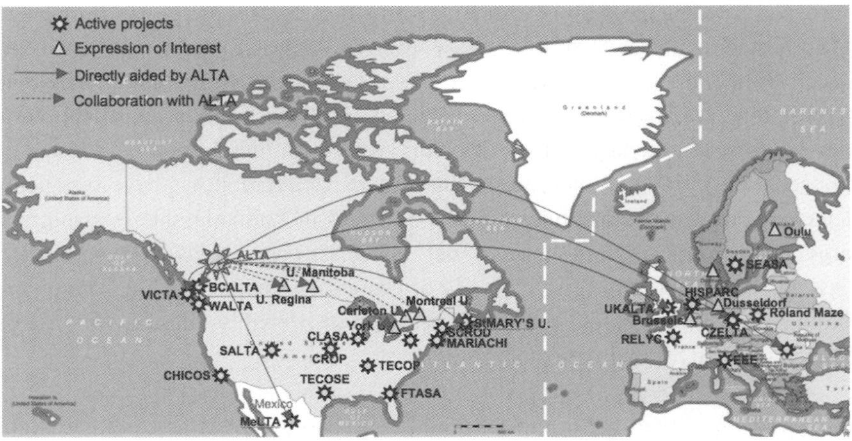

Figure 2 The geographical distribution of the main educational cosmic ray array centres

The European groups are also developing a similar collaboration, called Eurocosmics. It is clear that a natural next step is to combine the North American and European networks into a worldwide network that could contribute significantly to elucidating the extreme-energy universe. Such a network could aid and encompass other efforts throughout the world, including in developing countries where it could provide a natural bridgehead into the global scientific culture [11]. A picture of the geographical distribution of educational cosmic ray arrays is given in Figure 2.

## 4. Developing Countries and the Global Scientific Culture

Many developed countries face the challenge of encouraging more young people to take up science to ensure future innovation to benefit society. However, there is a related and equally important challenge - to promote a scientific infrastructure to aid the academic and career ambitions of members of under-represented and economically disadvantaged groups, as well as scientists from developing countries, to increase their participation in scientific and technical fields worldwide.

Severe constraints on resources, which are a common feature in developing countries, mean that research there does not usually consist of designing and making equipment for a new experiment at the forefront of the field. In many schools, colleges and universities laboratories either do not exist or are poorly equipped. Consequently, the brain drain of bright young scientists from developing to developed countries seems to be the norm, and further intellectually impoverishes the developing world. Collaborative programmes between scientists from developed and developing countries are urgently needed.

In many areas of research, leading-edge science is expensive and there are few support networks for disadvantaged groups. However, cost-effective projects to investigate the nature of ultra-high-energy cosmic rays (UHECR) are already being developed for high schools and could provide an ideal vehicle for such an effort. These projects demonstrate the basic elements of research and technology, with modern detectors, fast electronics, GPS timing, computerized data acquisition and data analysis. Perhaps just as importantly, they also teach social skills such as collaborative effort, organization, long-term planning and teamwork.

## 5. Conclusion

We believe that one of the keys to developing a world-wide network of school based cosmic rays detectors is the reduction of the cost of the detector system. The ALTA group at Alberta is presently concentrating on reducing the cost of their detector system by utilizing water Cerenkov detectors rather than plastic scintillator detectors and using silicon PMTs rather than photomultiplier tubes with their concomitant requirement for high voltage. The aim is to produce a detector system for around 1500 Euros.

The educational paradigm created by the networks of cosmic-ray arrays in high schools is one that can be employed in many areas. The "cosmic connection" between research and education in both the developed and developing world is not the only way of fostering fundamental research in developing countries. One can also develop similar projects topics in, for example, biology (e.g. the effects of natural radiation), mathematics, computer science and programming, space physics, chemistry, and environmental and Earth sciences. Such large-area, national and international school-based detector networks could aid and encompass other efforts throughout the world including developing countries, where it could provide entry to the global scientific community.

**References**


1. Gary R Smith *et al.* 1983 *Phys. Rev. Letts.* **50** 2110.
2. D J Fegan *et al.* 1983 *Phys. Rev. Letts.* **51** 2341.
3. W. Brouwer et al., Nucl.Instrum.Meth. **A539** 595-605 (2005).
4. J. L. Pinfold, for theALTA/NALTA Collaborations. Prepared for 8th International Conference ICATPP 2003: Como, Italy. Published in *Como 2003, Astroparticle, particles and space physics, detectors and medical physics applications* 108-120 (2003)
5. R D McKeown *et al.*, CHICOS Collaboration 2005 *Proceedings of the 29th International Cosmic Ray Conference, Pune, India* 10.
6. See the talk by Helio Takai at this conference.
7. C Timmermans, HiSPARC Collaboration, The 29th Proceedings of the *29th* International Cosmic Ray Conference, Pune, India 104 (2005).
8. http://www.nevod.mephi.ru/English/eurocosm.htm
9. (CZELTA) K. Smolek et al., IEEE Nuclear Science Symposium, **N02-11** 720 (2008); also, see talk by K. Smolek at this conference.
10. http://www.centrofermi.it/eee/index3.html
11. J. L. Pinfold, “Viewpoint a cosmic vision for world science”, CERN Courier May 3rd, (2006).

# Cosmic Rays Experimental Observations and Searches

| | |
|---|---|
| A.Y. Alaverdian | Galactic cosmic ray production in Tychos SNR and Geminga |
| T.I. Banks | The CUORICINO and CUORE neutrinoless double beta decay experiments |
| P. Belli | Results from DAMA/LIBRA |
| E. Bonamente | Recent Results from the Fermi Large Area Space Telescope |
| S.S. Borisov | Gamma-ray activity of Cygnus X-3 at energy range of 1-100 TeV during 15 year observations of SHALON |
| I. Buesching | Signatures of middle aged, nearby pulsars in the cosmic ray lepton spectrum? |
| P. Camarri | Highlights from the ARGO-YBJ Experiment |
| A. De Angelis | Status of MAGIC and recent results |
| B. Degrange | Recent H.E.S.S. results |
| C. Fruck | Atmospheric evaluation with LIDAR for MAGIC |
| S. Haino | The AMS-02 silicon tracker |
| A. Haungs | From the Knee to the Ankle: From Galactic to Extragalactic Origin of Cosmic Rays? |
| A. Karelin | High energy cosmic-ray photons and Helium |
| P. Klimov | Status of UHE CR orbital fluorescence detector TUS |
| P. Montini | The observation of the light component spectrum in the 5250 TeV region by the ARGO-YBJ experiment |
| F. Orio | Status and plans of the LUCIFER Experiment |
| M. Pesce-Rollins | In-flight measurement of the absolute energy scale of the Fermi Large Area Telescope |
| J.L. Pinfold | The synergy between astroparticle and collider physics in the search for dark matter |
| M.-C. Piro | PICASSO: Search for dark matter in the spin-dependent sector |
| J. Quinn | Recent Results from VERITAS |

| | |
|---|---|
| S.B. Ricciarini | Recent results from the PAMELA experiment |
| T. Sako | First results of LHCF; very forward particles at LHC collision |
| E.S. Seo | Status and recent results from the CREAM experiment |
| O.P. Shustova | On the possibility of registering UHE EAS Cherenkov light by the TUS detector |
| V.G. Sinitsyna | TeV gamma-rays from NGC 1275 detected in 15 year observation of SHALON telescope |
| V.Y. Sinitsyna | Constraints on extragalactic background light from distant quasars 3C454.3 (z = 0.859) and 1739+522 (z = 1.375) detected by SHALON |
| W. Springer | Status of the high altitude water Cherenkov (HAWC) gamma ray observatory |
| N. Tomassetti | Light Nuclei and Isotope Abundances in Cosmic Rays. Results from AMS-01 |

# GALACTIC COSMIC RAY PRODUCTION IN TYCHO'S SNR AND GEMINGA

V.G. SINITSYNA, A.Y. ALAVERDIAN, S.S. BORISOV,
S.I. NIKOLSKY, V.Y. SINITSYNA
*P.N.Lebedev Physical Institute,*
*Leninsky prospect 53, Moscow, 119991, Russia*
*E-mail: sinits@sci.lebedev.ru*

A nonlinear kinetic model of cosmic ray acceleration in supernova remnants is used for Tycho's SNR. The expected $\pi^o$-decay gamma-ray flux from Tycho's SNR extends up to > 30 TeV, whereas the inverse Compton gamma-ray flux has a cutoff above a few TeV. Hence, the detection of gamma-rays at energies 10 - 80 TeV by SHALON provides an evidence of their hadronic origin. Thus, the shell-type SNRs should be detected at the TeV energy range in gamma-rays of predominantly, hadronic origin and the expected flux of gamma-quanta from $\pi^o$-decay $(2-5)\times 10^{-13} erg/(cm^2 s)$ extends up to 100TeV if the distance is within the range 3.1-3.3 kpc. The SHALON results for Geminga are presented. Its integral gamma-ray flux is found to be $(0.48\pm0.17)\times 10^{-12} cm^{-2}s^{-1}$ at energies of > 0.8 TeV. Within the range 0.8 -6 TeV, the integral energy spectrum is well described by the power law $I(>E)E^{-0.59\pm0.10}$.

*Keywords*: TeV gamma-rays; Tycho's SNR; Geminga; electronic and hadronic Cosmic Ray components.

## Introduction

TeV energies gamma-rays, measurable by the imaging Cherenkov technique, are the most interesting for searching hadronic CRs in SNRs because they provide the information about CRs of highest possible energies $10^{13}-10^{14}$ eV. Direct information about high-energy CR population in SNRs can be obtained from gamma-ray observation. The gamma-quantum spectra produced by the electronic and hadronic components of cosmic rays have similar shapes at the energies from 1GeV to 1 TeV due to the synchrotron losses of the electrons. So, the only observational possibility to discriminate between leptonic and hadronic contributions is to measure the gamma-quantum spectrum at energies higher than 1 TeV, where these two spectra are expected to be essentially different. High-energy gamma-rays are pro-

Table 1. The SHALON catalogue of Galactic gamma-quantum sources with energy > 0.8 TeV

| Sources | Observable flux $(cm^{-2}s^{-1})$ | Distance (kpc) |
|---|---|---|
| Crab Nebula (Plerion) | $(1.72 \pm 0.11) \times 10^{-12}$ | 2 |
| Cygnus X-3 (binary) | $(0.68 \pm 0.07) \times 10^{-12}$ | 10 |
| Geminga (radioweak pulsar) | $(0.48 \pm 0.12) \times 10^{-12}$ | 0.25 |
| Tycho' SNR (Shell type SNR) | $(0.52 \pm 0.09) \times 10^{-12}$ | 2.3 - 3.5 |
| 2129+47XR (low-mass X-ray binary) | $(0.19 \pm 0.09) \times 10^{-12}$ | 6 |

duced by electronic and hadronic CR components in the inverse Compton (IC) scattering and in the hadronic collisions leading to pion production and subsequent decay respectively.

The gamma-quantum emitting objects in our Galaxy are the supernova remnants and binary. SNe of type Ib and II are more numerous in our Galaxy. According to the theoretical prediction about 20 SNRs should be visible in the TeV gamma-rays whereas only two were detected up to now by SHALON, namely Tycho's SNR and Geminga.

The observations on Tien-Shan high-mountain station with SHALON had been carried out since 1992 year.[1–4] During this period 12 metagalactic and galactic sources have been observed. Among them are galactic sources Crab Nebula (supernova remnant), Cygnus X-3 (binary), Tycho's SNR (supernova remnant), Geminga (radioweak pulsar) and 2129+47 (binary)[1–4] and Table 1. The results of observation data analysis for the each source are integral spectra of events coming from source - $k_{ON}$, and background events, coming simultaneously with source observation - $k_{OFF}$, temporal analysis of these two kind events and the source images. At Figs. 1, 2, 3 the observation results of galactic gamma-sources are showed.

## Geminga

A neutron star in the constellation Gemini is the second brightest source of high-energy gamma-rays in the sky, discovered in 1972, by the SAS-2 satellite. For nearly 20 years, the nature of Geminga was unknown, since it didn't seem to show up at any other wavelengths. In 1991, an regular periodicity of 0.237 second was detected by the ROSAT satellite in soft X-ray emission, indicating that Geminga is almost certainly a pulsar. Geminga is the closest known pulsar to Earth.

Geminga is one of the brightest source of MeV - GeV gamma-ray, but the only known pulsar that is radio-quiet. Geminga has been the object for study at TeV energies with upper limits being reported by three ex-

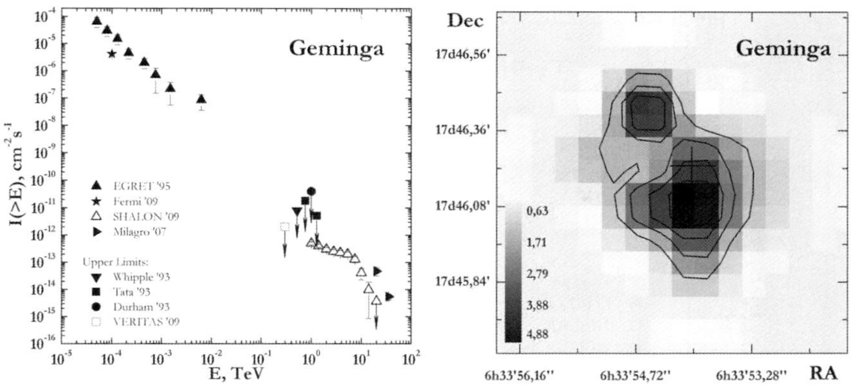


Fig. 1. **left:** The Geminga gamma - quantum ($E > 0.8$ TeV) integral spectrum by SHALON in comparison with other experiments EGRET, Fermi LAT,[10] Milagro[9] and upper limits: Whipple'93,[5] Tata'93,[6] Durham'93[7] and Veritas;[8] **right:** The image of gamma-ray emission from Geminga by SHALON.

periments Whipple'93,[5] Tata'93,[6] Durham'93[7] and very recently by VERITAS.[8] Also Geminga has observed with Milagro[9] at energies of 20TeV and 35 TeV and Fermi LAT at energies 30MeV - 200GeV.[10] The spectrum by Fermi is fitted with a power law with exponential cutoff in the form: $dN/dE = N_0 \times (E^{-\gamma}) exp(-E/E_0)$, where $N_0 = (1.19 \pm 0.08) \times 10^{-6} GeV^{-1} cm^{-2} s^{-1}$, $\gamma = 1.3 \pm 0.05$, $E_0 = (2.47 \pm 0.19) GeV$.

Figure 1 shows the SHALON results for this gamma-source. An image of gamma-ray emission from Geminga by SHALON telescope is shown in Fig. 1. As is seen from this figure the value Geminga flux obtained by SHALON is lower than the upper limits published before. Its integral gamma-ray flux is found to be $(0.48 \pm 0.12) \times 10^{-12} cm^{-2} s^{-1}$ at energies of $> 0.8$ TeV.[1] Within the range 0.8 - 6 TeV, the integral energy spectrum is well described by the single power law $I(> E_\gamma) \propto E_\gamma^{-0.59 \pm 0.10}$(Fig. 1). The energy spectrum of supernova remnant Geminga $F(E_O > 0.8 TeV) \propto E^k$ is harder than Crab spectrum.

## Tycho's SNR

Tycho Brage supernova remnant has been observed by SHALON atmospheric Cherenkov telescope of Tien-Shan high-mountain observatory. This object has long been considered as a candidate to cosmic ray hadrons source in Northern Hemisphere, although it seemed that the sensitivity of the present generation of Imaging Atmospheric Cherenkov System's too small

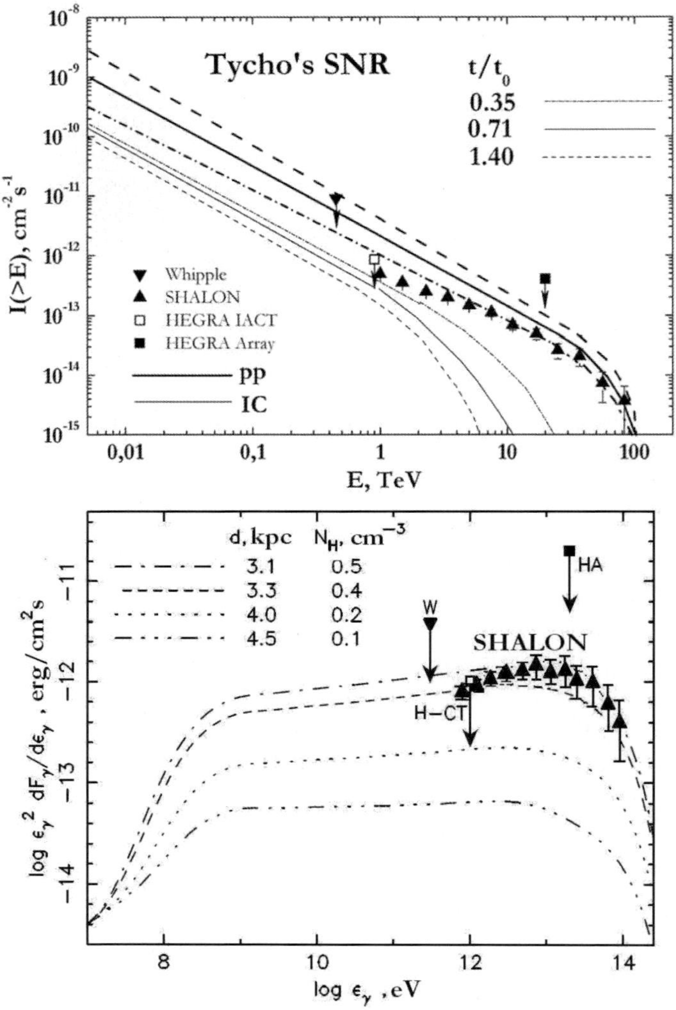


Fig. 2. **top:** The Tycho's SNR gamma-quantum integral spectrum by SHALON in comparison with other experiments: the observed upper limits Whipple, HEGRA IACT system, HEGRA AIROBICC and calculations: IC emission (thin lines), $\pi^{\circ}$ - decay (thick lines).[11] **bottom:** Spectral energy distribution of the gamma-ray emission from Tycho's SNR.[16,17] All cases have dominant hadronic gamma-ray flux.

for Tycho's detection. Tycho's SNR has been detected by SHALON at TeV energies.[1] The integral gamma-ray flux above 0.8 TeV was estimated as $(0.52 \pm 0.09) \times 10^{-12}$ (Fig. 2).

Figures 2 and 3 show the observational results for the Tycho's SNR. An

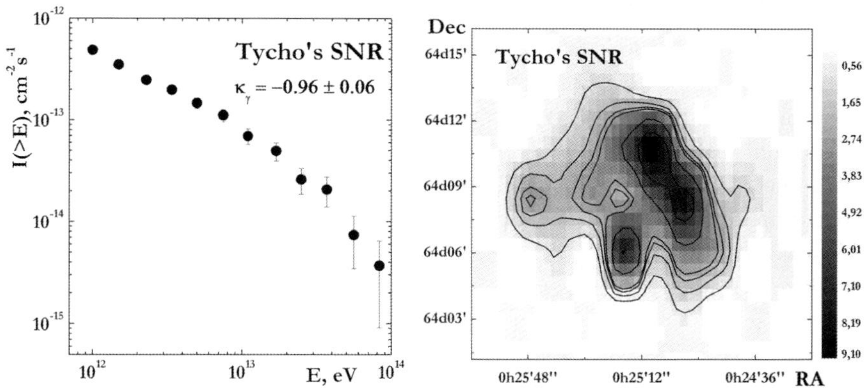


Fig. 3. **left:** The Tycho's SNR gamma-quantum integral spectrum with power index of $k_\gamma = -0.96 \pm 0.06$ by SHALON experiment; **right:** The SHALON image of gamma-ray emission from Tycho's SNR.

image of gamma-ray emission from Tycho's SNR by SHALON telescope is shown in Fig. 3. The energy spectrum of Tycho's SNR at 0.8 – 20 TeV can be approximated by the power law $F(>E_O) \propto E^{k_\gamma}$ , with $k_\gamma = -0.96 \pm 0.06$. The energy spectrum of supernova remnant Tycho's SNR $F(E_O > 0.8TeV) \propto E^k$ is also harder than Crab spectrum.

A nonlinear kinetic model of cosmic ray acceleration in supernova remnants is used in [11] (Fig. 2), to describe the properties of Tycho's SNR. The kinetic nonlinear model for cosmic ray acceleration in SNR has been applied to Tycho's SNR in order to compare model results with recently found very low observational upper limits on TeV energy range. In fact, HEGRA didn't detect Tycho's SNR, but established a very low upper limit at energies > 1 TeV. This value is consistent with that previously published by Whipple collaboration, being a factor of 4 lower (the spectral index of −1.1 for this comparison [11]). The $\pi^\circ$-decay gamma-quantum flux turns out to be some greater than inverse Compton flux at 1 TeV becomes strongly dominating at 10 TeV. The predicted gamma-quanta flux is in consistent with upper limits published by Whipple[12–14] and HEGRA.[15] The expected flux of gamma-quanta from $\pi^\circ$-decay, $F_\gamma \propto E_\gamma^{-1}$, extends up to ∼ 30 TeV, while the flux of gamma-rays originated from the Inverse Compton scattering has a sharp cutoff above the few TeV , so the detection of gamma-rays with energies of ∼ 10 to 80 TeV by SHALON is an evidence of their hadronic origin.[11,16,17]

Figure 2 presents spectral energy distribution of the gamma-ray emission from Tychos SNR, as a function of gamma-ray energy $\epsilon_\gamma$ , for a mechanical SN explosion energy of $E_{SN} = 1.2 \times 10^{51}$ erg and four different distances d and corresponding values of the interstellar medium number densities $N_H$. All cases have dominant hadronic gamma-ray flux.[17] The additional information about parameters of Tycho's SNR can be predicted in frame of nonlinear kinetic model [11,16,17] if the TeV gamma- quantum spectrum of SHALON telescope is taken into account: a source distance 3.1 - 3.3 kpc and an ambient density $N_H$ $0.5 - 0.4$ $cm^{-3}$ and the expected $\pi^\circ$-decay gamma-ray energy spectrum extends up to about 100 TeV.

## References


1. V. G. Sinitsyna, AIP Conf. Proc., 515, (1999) pp. 205 and 293.
2. V. G. Sinitsyna, S. I. Nikolsky, et al.,Izv. Ross. Akad. Nauk Ser. Fiz., 66(11),(2002) pp. 1667 and 1660.
3. V. G. Sinitsyna, et al.,in Proc. 27th ICRC, Hamburg, 3, (2001) p. 2665; in Proc. 29th ICRC, Puna, 4, (2005) p. 231.
4. V. G. Sinitsyna et al.,Nucl. Phys. B (Proc.Suppl.), 151, (2006), p. 112; V. G. Sinitsyna et al., *ibid.* 122,(2003) 247; S. I. Nikolsky and V. G. Sinitsyna, *ibid.*, 122, (2003) 409; V. G. Sinitsyna et al., *ibid.*, 97,(2001) pp. 215 and 219.
5. C. W. Akerlof et al. in Proc. 23rd ICRC, Calgary, 1, (1993) p. 305.
6. P. R. Vishwanath et al. Astron.&Astrophys., 267, (1993) L5.
7. C. C. G. Bowden et al. J. Phys. G: Nucl. Part. Phys. 19, (1993) L29-L31
8. G. Finnegan; "Search for TeV emission from Geminga by VERITAS" Proceedings of the 31st ICRC, Lodz 2009;
9. A. A. Ado et al. "Milagro observations of TeV emission from galactic sources in the Fermi bright source list", arXiv:0904.1018v1; "TeV gamma-ray sources from a survey of the Galactic plane with Milagro", *arXiv:0705.0707v1*;
10. O. Celik; "Fermi LAT observations of the Geminga Pulsar" Proc. of the 31st ICRC, Lodz 2009;
11. H. J. Völk, E. G. Berezhko, et al., in Proc. 27th ICRC, Hamburg, 2, (2001) p. 2469.
12. J. H. Buckley, C. W. Akerlof, D. A. Carter-Lwis, et al., Astron. Astrophys., 329, (1998) p. 639.
13. M. Catanese and T. C. Weekes, Preprint Series, no. 4811, (1999).
14. T. C. Weekes, AIP Conf. Proc., 515, (1999) p. 3.
15. J. Prahl and C. Prosch (for the HEGRA Collaboration), in Proc. 25th ICRC, Durban, 3, (1997) p. 217.
16. H. J. Völk, E. G. Berezhko and L. T. Ksenofontov, in Proc. 29th ICRC, Pune, 3, (2005) p. 235.
17. H. J. Völk, E. G. Berezhko, L. T. Ksenofontov Astron.&Astrophys., 483(2), (2008) p.529.

# The CUORICINO and CUORE neutrinoless double beta decay experiments

T. I. BANKS, on behalf of the CUORICINO and CUORE Collaborations

*Department of Physics, University of California*
*366 LeConte Hall MC 7300*
*Berkeley, CA 94720-7300, USA*

*INFN – Laboratori Nazionali del Gran Sasso*
*S.S. 17 bis km 18.910*
*67010 Assergi (AQ), Italy*

*E-mail: tbanks@berkeley.edu*

CUORICINO and CUORE are closely affiliated cryogenic bolometer experiments designed to search for neutrinoless double beta ($0\nu\beta\beta$) decay in $^{130}$Te. Both experiments are carried out underground at the Gran Sasso National Laboratory, Italy, in order to achieve the extremely low backgrounds necessary for high-sensitivity searches of their kind. CUORICINO ran from 2003–2008 and served as a prototype for the upcoming CUORE detector, which will be 20 times larger. This article describes: the basics of $0\nu\beta\beta$ decay; CUORICINO and its recent final result, which has set the most stringent limit to date on $0\nu\beta\beta$ decay in $^{130}$Te; and the physics goals and status of CUORE, which is currently under construction.

*Keywords*: Neutrinoless double beta decay, cryogenic bolometers, $^{130}$Te.

## 1. Introduction

Neutrinoless double beta ($0\nu\beta\beta$) decay is a spontaneous process in which a nucleus $(A, Z)$ undergoes two simultaneous $\beta^-$ decays yet emits no neutrinos:

$$(A, Z) \rightarrow (A, Z{+}2) + 2e^- \ . \tag{1}$$

This lepton-number-violating reaction occurs extremely rarely, if at all, with a half-life $T_{1/2} \geq 10^{24}$ years. To date there has been only one claim[1] of observation of $0\nu\beta\beta$ decay, and the result is a subject of some controversy. Confirmation by other experiments is therefore considered essential.

The signature of $0\nu\beta\beta$ decay is the simultaneous emission of two electrons of combined energy $Q_{\beta\beta}$, the $Q$ value of the decay in the particular isotope under study. $0\nu\beta\beta$ decay would thus appear as a small peak at the endpoint of the two-neutrino double beta ($2\nu\beta\beta$) decay spectrum. $2\nu\beta\beta$ decay has been observed in several elements (e.g., $^{76}$Ge, $^{130}$Te, and $^{136}$Xe) for which single beta decay is energetically forbidden, and in principle any element that undergoes $2\nu\beta\beta$ decay is a candidate for $0\nu\beta\beta$ decay. $^{130}$Te is attractive because it has a relatively high natural isotopic abundance (hence enrichment is not essential) and its $Q$ value of 2527.5 keV (see Ref. 2) is above the energy range of most naturally occurring gamma-ray backgrounds.

There is currently great interest in searching for $0\nu\beta\beta$ decay because of its unique potential to provide insight into the nature of the neutrino. First, $0\nu\beta\beta$ decay can occur only if the neutrino is a massive Majorana particle—i.e., its own antiparticle. In the past decade, multiple oscillation experiments have convincingly established that neutrinos are massive particles, while observation of $0\nu\beta\beta$ decay is the only feasible way known to determine if neutrinos are Majorana particles. If they are, it would have profound implications for our understanding of how neutrinos acquire mass. Second, the transition width of $0\nu\beta\beta$ decay is proportional to the square of the effective Majorana mass, $|m_\nu|^2$, which is a mixture of the three neutrino mass eigenstates. A measurement of the $0\nu\beta\beta$ decay rate could therefore provide information about the neutrino mass hierarchy and absolute mass scale, two open questions at the forefront of neutrino physics.

The Cryogenic Underground Observatory for Rare Events (CUORE)[3] is an upcoming experiment that will search for $0\nu\beta\beta$ decay in $^{130}$Te. The project represents the culmination of nearly 30 years of research in the use of cryogenic bolometers for studying $0\nu\beta\beta$ decay, a technique first proposed in 1984.[4] CUORE's most recent predecessor, the CUORICINO experiment, ran from 2003–2008 using the same basic techniques as CUORE and recently announced its final result. CUORICINO and CUORE are collaborative projects involving Italian, U.S., and Chinese institutions and sharing many of the same facilities and personnel. Both experiments are located in the underground Laboratori Nazionali del Gran Sasso (LNGS), Italy, where the cosmic-ray muon flux is attenuated by a factor of $10^6$ to $\sim 1\ \mu/\mathrm{hr/m}^2$ (see Refs. 5 and 6), thus providing an extremely low-background environment for searches for rare processes.

## 2. Cryogenic bolometers

The general approach in searching for $0\nu\beta\beta$ decay is to detect the two final-state electrons and measure their energy. Two complementary experimental techniques have historically been employed: "source=detector," and "source≠detector." In the "source=detector" technique, the detector is made from material containing the isotope of interest, and the signature of $0\nu\beta\beta$ decay is an excess of events at the $Q$ value of the decay. The advantages of this technique are that it offers excellent energy resolution, a large source mass, and high efficiency; the primary disadvantage is that it does not offer the capability for identifying particles. In the "source≠detector" approach, material containing the $0\nu\beta\beta$ decay isotope is surrounded by tracking detectors. In this case the detectors provide particle discrimination, which is a powerful means of suppressing backgrounds, but the energy resolution, rates, and efficiencies are relatively low. It should be emphasized that good energy resolution is critical to observing any $0\nu\beta\beta$ peak in the detected energy spectrum.

In CUORICINO and CUORE we utilize the "source=detector" approach by cooling crystals of $TeO_2$ to $\sim 10$ mK inside a dilution-refrigerator cryostat. At cryogenic temperatures a $TeO_2$ crystal has such a small heat capacity that an individual particle interaction inside it produces a measurable rise in its temperature—i.e., the crystal functions as a bolometer. The temperature pulses are detected using special neutron-transmutation-doped (NTD) thermistors which are glued to each crystal's surface and whose resistance varies with the system's temperature. The amplitude of a temperature pulse is proportional to the energy deposited in the crystal, and we look for an excess of pulses (i.e., above background) at 2527.5 keV, the $Q$ value for $0\nu\beta\beta$ in $^{130}$Te. The response of each crystal—namely, its energy scale and gain—is monitored via monthly calibrations with a $^{232}$Th gamma source and by injecting controlled amounts of energy into the crystals at 5-minute intervals using Joule heaters glued to the crystal surface.

## 3. CUORICINO

The CUORICINO detector consisted of 62 $TeO_2$ crystal bolometers arranged into a single 13-floor tower (Fig. 1). The crystals had a combined mass of 40.7 kg, corresponding to 11.6 kg of $^{130}$Te. The experiment ran from April 2003 to May 2008, resulting in a total exposure of 19.6 kg-yr $^{130}$Te. Down time was due primarily to monthly energy calibrations, cryostat maintenance, and a single extended period of detector repairs.

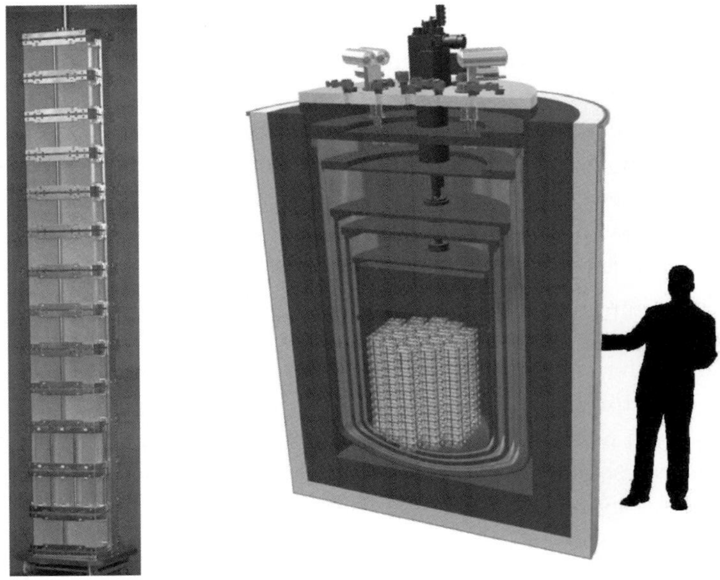

Fig. 1. (Left) Photograph of the CUORICINO crystal tower. (Right) Drawing of the future CUORE cryostat and its array of 19 CUORICINO-like crystal towers.

The first CUORICINO result[7] was published in 2008 based on analysis of data from 2003–2006. Two semi-independent data analyses of the full 2003–2008 data set were conducted by the U.S. and Italian sides and completed in 2010. The analyses shared a common framework for low-level data processing but diverged in the higher-level analysis and fitting. Nevertheless, they shared many of the same essential features—e.g., events involving multiple crystals were excluded, because the electrons from $0\nu\beta\beta$ decay are expected to usually deposit all of their energy within a single crystal.

Despite the differences in their details, the two analyses yielded near-identical results—namely, no evidence of $0\nu\beta\beta$ decay in $^{130}$Te was found, as is reflected in the absence of a visible peak at $Q_{\beta\beta}(^{130}\text{Te}) = 2527.5$ keV in the measured energy spectrum (Fig. 2). As a result, the experiment was able to set a lower limit on the half-life of $0\nu\beta\beta$ decay in $^{130}$Te,

$$T_{1/2}^{0\nu\beta\beta}(^{130}\text{Te}) > 2.8 \times 10^{24}\ \text{y} \quad (90\%\text{C.L.}), \tag{2}$$

which corresponds to an upper limit of $m_\nu < 300$–$710$ meV on the Majorana mass.[8] The background level in the region of interest was $0.169 \pm 0.006$ c/keV/kg/y, roughly half of which comes from Compton events from the 2615 keV $\gamma$ peak of $^{208}$Tl (itself coming from $^{232}$Th contamination

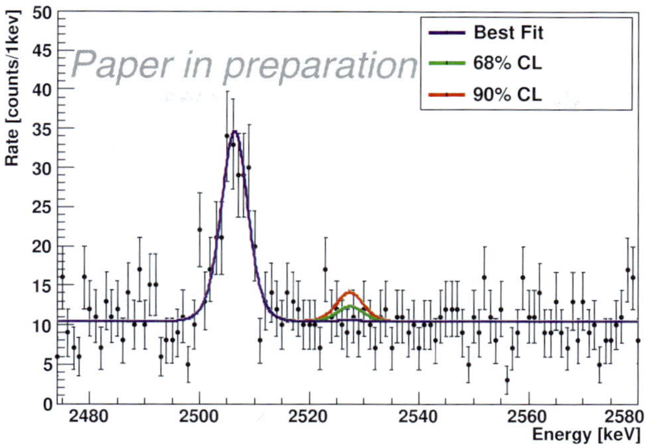


Fig. 2. Superposition of the summed energy spectra from the CUORICINO crystals and the likelihood-fit results in the region of interest. No peak is visible at the $^{130}$Te $0\nu\beta\beta$ decay $Q$ value of 2527.5 keV. The peak at 2506 keV is from the decay of $^{60}$Co, likely produced by cosmic-ray activation of the copper in the cryostat.

in the cryostat), while the other half is attributed to degraded alphas from copper surfaces, with a small fraction coming from the crystal surfaces.

## 4. CUORE

The CUORE experiment will use the same basic techniques as CUORICINO but it will be roughly 20 times larger and incorporate numerous improvements. The detector will consist of 988 $TeO_2$ crystals arranged in 19 CUORICINO-like towers (Fig. 1). Each crystal will measure $5 \times 5 \times 5$ cm$^3$ and weigh 750 g, resulting in a total mass of 740 kg of $TeO_2$, or 210 kg of $^{130}$Te. The background level is expected to be reduced to 0.01 c/keV/kg/y (see Ref. 9) due to: improvements in the radiopurity of the copper and crystal surfaces and in the assembly environment; the use of better frames which hold the crystals more firmly and thus transmit less vibrational noise; the use of thicker shields of low-activity ancient Roman lead; and the fact that the 19-tower array affords superior self-shielding and better anti-coincidence coverage. The experiment should be capable of probing the Majorana neutrino mass down to the range $m_\nu \approx$ 30–90 meV.

The challenges in constructing CUORE are due largely to its size and the complexity of the engineering. Namely, it is difficult to prepare and maintain the cleanliness of such a large amount of radiopure material, and the apparatus involves many interconnected systems occupying the same space under unique conditions. Moreover, it is anticipated that the loaded cryostat will require a relatively long cooldown time ($\sim$ 1 month), so careful planning and robust systems are essential.

The first tower from the CUORE assembly line, CUORE-0, will be cooled down in the former CUORICINO cryostat from 2011–2014 while the remaining 19 CUORE towers are assembled. CUORE-0 is intended to serve primarily as a test of the assembly line protocols, but it will also represent a valuable measurement in its own right since it will be comparable in size to CUORICINO but exhibit lower backgrounds due to the use of improved copper frames. The assembly of the 19 CUORE towers is expected to be completed in 2013, and data taking is anticipated for 2013–2018.

## 5. Conclusions

The nature of neutrino mass is one of the frontier problems in contemporary physics, and $0\nu\beta\beta$ decay offers a unique and powerful tool for investigating the neutrino mass hierarchy and absolute mass scale, as well as possible extensions to the standard model. CUORE is one of several major efforts underway searching for evidence of $0\nu\beta\beta$ decay. Observation of the process in multiple isotopes and via different techniques would provide robust confirmation of its existence.

## 6. Acknowledgments

CUORICINO and CUORE are funded by the Italian Instituto Nazionale di Fisica Nucleare (INFN), the Commission of the European Community, and the US DOE and NSF.

## References

1. H. V. Klapdor-Kleingrothaus and I. V. Krivosheina, *Mod. Phys. Lett.* **A21**, 1547 (2006).
2. M. Redshaw, B. J. Mount, E. G. Myers and F. T. Avignone, *Phys. Rev. Lett.* **102**, p. 212502 (2009).
3. C. Arnaboldi *et al.*, *Nucl. Instrum. Meth.* **A518**, 775 (2004).
4. E. Fiorini and T. O. Niinikoski, *Nucl. Instr. Meth.* **A224**, p. 83 (1984).
5. F. Bellini *et al.*, *Astropart. Phys.* **33**, 169 (2010).
6. E. Andreotti *et al.*, *Astropart. Phys.* **34**, 18 (2010).
7. C. Arnaboldi *et al.*, *Phys. Rev.* **C78**, p. 035502 (2008).
8. Announced by M. Pavan at the Nu2010 Conference; to be published in *Astropart. Phys.*
9. M. Pavan *et al.*, *Eur. Phys. J.* **A36**, 159 (2008).

# RESULTS FROM DAMA/LIBRA

R. BERNABEI, P. BELLI*, F. MONTECCHIA* and F. NOZZOLI

*Dip. di Fisica, Università di Roma "Tor Vergata" and*
*INFN, sez. Roma "Tor Vergata", I-00133 Rome, Italy*
**E-mail: pierluigi.belli@roma2.infn.it*
*www.roma2.infn.it*

F. CAPPELLA, A. d'ANGELO, A. INCICCHITTI and D. PROSPERI†

*Dip. di Fisica, Università di Roma "La Sapienza" and*
*INFN, sez. Roma, I-00185 Rome, Italy*

R. CERULLI

*Laboratori Nazionali del Gran Sasso, I.N.F.N., Assergi, Italy*

C.J. DAI, H.L. HE, X.H. MA, X.D. SHENG, Z.P. YE‡ and R.G. WANG

*IHEP, Chinese Academy, P.O. Box 918/3, Beijing 100039, China*

The DAMA/LIBRA experiment has a sensitive mass of about 250 kg highly radiopure NaI(Tl). It is running at the Gran Sasso National Laboratory of the I.N.F.N. in Italy, and is mainly devoted to the investigation of Dark Matter (DM) particles in the Galactic halo by exploiting the model independent DM annual modulation signature. The present DAMA/LIBRA experiment and the former DAMA/NaI one (the first generation experiment having an exposed mass of about 100 kg) have cumulatively released so far the results obtained with the data collected over 13 annual cycles (total exposure: 1.17 ton $\times$ yr). They give a model independent evidence of the presence of DM particles in the galactic halo on the basis of the investigated DM signature at 8.9 $\sigma$ C.L. for the cumulative exposure. Few aspects of the obtained results are summarized and some comments addressed.

*Keywords*: Scintillation detectors; elementary particle processes, Dark Matter

*also: Lab. Sperim. Policentrico di Ingegneria Medica, Università di Roma Tor Vergata
†Deceased
‡also: University of Jing Gangshan, Jiangxi, China

## 1. Introduction

DAMA is an observatory for rare processes located deep underground at the Gran Sasso National Laboratory of the I.N.F.N.. It is based on the development and use of low background scintillators. Several set-ups have been realized and are in operation to investigate many rare processes.[1–16]

In particular, DAMA/LIBRA is investigating the presence of Dark Matter (DM) particles in the galactic halo by exploiting the model independent DM annual modulation signature.[17] In fact, as a consequence of its annual revolution around the Sun, which is moving in the Galaxy traveling with respect to the Local Standard of Rest towards the star Vega near the constellation of Hercules, the Earth should be crossed by a larger flux of Dark Matter particles around $\sim$ 2 June (when the Earth orbital velocity is summed to the one of the solar system with respect to the Galaxy) and by a smaller one around $\sim$ 2 December (when the two velocities are subtracted). This signature offers an efficient model independent signature, able to test a large number of DM candidates, a large interval of cross sections and of halo densities. The DM annual modulation signature is very distinctive since the corresponding signal must simultaneously satisfy all the following requirements: the rate must contain a component modulated according to a cosine function (1) with one year period (2) and a phase that peaks roughly around $\simeq 2^{nd}$ June (3); this modulation must only be found in a well-defined low energy range, where DM particle induced events can be present (4); it must apply only to those events in which just one detector of many actually "fires" (*single-hit events*), since the DM particle multi-interaction probability is negligible (5); the modulation amplitude in the region of maximal sensitivity must be $\lesssim$7% for usually adopted halo distributions (6), but it can be larger in case of some possible scenarios such as e.g. those in Refs. 18,19. Only systematic effects or side reactions able to simultaneously fulfil all these requirements and to account for the whole observed modulation amplitude could mimic this signature; thus, no other effect investigated so far in the field of rare processes offers a so stringent and unambiguous signature.

In the following, we will just briefly summarize the results on the DM particle investigation obtained by DAMA/LIBRA, exploiting over six annual cycles the model independent DM annual modulation signature (exposure: 0.87 ton$\times$yr).[15,20] The data have also been combined with those previously collected by DAMA/NaI over 7 annual cycles (0.29 ton$\times$yr). Thus, the total exposure collected over 13 annual cycles is 1.17 ton$\times$yr, this is orders of magnitude larger than the exposures typically collected in the field.

The DAMA/NaI set up and its performances are reported in Refs. 1,3–5, while the DAMA/LIBRA set-up and its performances are described in Ref. 14. Here we just summarized the main features: i) the sensitive part of the set-up is made of 25 highly radiopure NaI(Tl) crystal scintillators placed in a 5-rows by 5-columns matrix; ii) the detectors' responses range from 5.5 to 7.5 photoelectrons/keV; iii) the hardware threshold of each PMT is at single photoelectron (each detector is equipped with two low background photomultipliers working in coincidence); iv) energy calibration with X-rays/$\gamma$ sources are regularly carried out down to few keV in the same conditions as the production runs; v) the software energy threshold of the experiment is 2 keV.

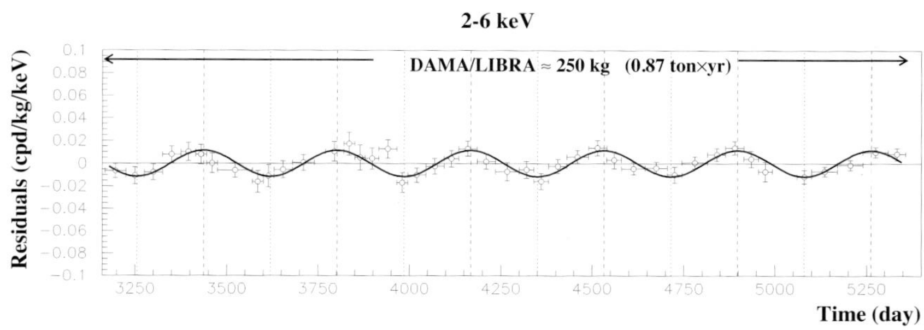


Fig. 1. Experimental model-independent residual rate of the *single-hit* scintillation events, measured by DAMA/LIBRA,1,2,3,4,5,6 in the (2 – 6) keV energy intervals as a function of the time. The experimental points present the errors as vertical bars and the associated time bin width as horizontal bars. The superimposed curves are the cosinusoidal functions behaviors $A\cos\omega(t-t_0)$ with a period $T=\frac{2\pi}{\omega}=1$ yr, with a phase $t_0=152.5$ day (June $2^{nd}$) and with modulation amplitudes, $A$, equal to the central values obtained by best fit over the whole data including also the exposure previously collected by the former DAMA/NaI experiment (1.17 ton × yr) The dashed vertical lines correspond to the maximum expected for the DM signal (June $2^{nd}$), while the dotted vertical lines correspond to the minimum.[20]

Several analyses on the model-independent DM annual modulation signature have been performed (see Refs. 15,20 and references therein); here just few arguments are mentioned. In particular, Fig. 1 shows the time behaviour of the experimental residual rates of the *single-hit* events collected by DAMA/LIBRA in the (2–6) keV energy interval.[15,20] The superimposed curve is the cosinusoidal function: $A\cos\omega(t-t_0)$ with a period $T=\frac{2\pi}{\omega}=1$ yr and with a phase $t_0=152.5$ day (June $2^{nd}$), and modulation amplitude,

$A$, obtained by best fit over the seven cycles of DAMA/NaI[4,5] and the six of DAMA/LIBRA.[15,20] When the period and the phase parameters are also released in the fit, values well compatible with those expected for a DM particle induced effect are obtained for the cumulative exposure[20] in the cumulative (2–6) keV energy interval: $A = (0.0116 \pm 0.0013)$ cpd/kg/keV, $T = (0.999 \pm 0.002)$ yr and $t_0 = (146 \pm 7)$ day. Summarizing, the analysis of the *single-hit* residual rate favours the presence of a modulated cosine-like behaviour with proper features at 8.9 $\sigma$ C.L..[20]

The same data of Fig. 1 have also been investigated by a Fourier analysis, obtaining a clear peak corresponding to a period of 1 year;[20] this analysis in other energy region shows instead only aliasing peaks. Moreover, while in the (2–6) keV *single-hit* residuals a clear modulation is present, it is absent at energies just above.[20] The measured energy distribution has been investigated in other energy regions not of interest for Dark Matter, also verifying the absence of any significant background modulation. In fact, the background in the lowest energy region is essentially due to "Compton" electrons, X-rays and/or Auger electrons, muon induced events, etc., which are strictly correlated with the events in the higher energy part of the spectrum; thus, if a modulation detected in the lowest energy region would be due to a modulation of the background (rather than to a signal), an equal or larger modulation in the higher energy regions should be present. The data analyses have allowed one to exclude the presence of a background modulation in the whole energy spectrum at a level much lower than the effect found in the lowest energy region for the *single-hit* events.[20]

A further relevant investigation has been done by applying the same hardware and software procedures, used to acquire and to analyse the *single-hit* residual rate, to the *multiple-hits* one. In fact, since the probability that a DM particle interacts in more than one detector is negligible, a DM signal can be present just in the *single-hit* residual rate. Thus, this allows the test of the background behaviour in the same energy interval of the observed positive effect. A clear modulation is present in the *single-hit* events, while the fitted modulation amplitude of the *multiple-hits* residual rate is well compatible with zero.[20] Similar results were previously obtained also for the DAMA/NaI case.[5] Thus, again evidence of annual modulation with proper features, as required by the DM annual modulation signature, is present in the *single-hit* residuals (events class to which the DM particle induced events belong), while it is absent in the *multiple-hits* residual rate (event class to which only background events belong). Since the same identical hardware and the same identical software procedures have been used to analyse the two classes of events, the obtained result offers an additional strong support for the presence of DM particles in the galactic halo further excluding any side effect either from hardware or from software procedures or from background.

The annual modulation present at low energy has also been analyzed by depicting the differential modulation amplitudes, $S_{m,k}$, as a function of the energy (the $k$ index identifies the energy interval); the $S_{m,k}$ is the modulation amplitude of the modulated part of the signal obtained by maximum likelihood method over the data, considering $T = 1$ yr and $t_0 = 152.5$ day. The $S_{m,k}$ values are reported as function of the energy in Fig. 2 and in Ref. 20. It has been also verified that the measured mod-

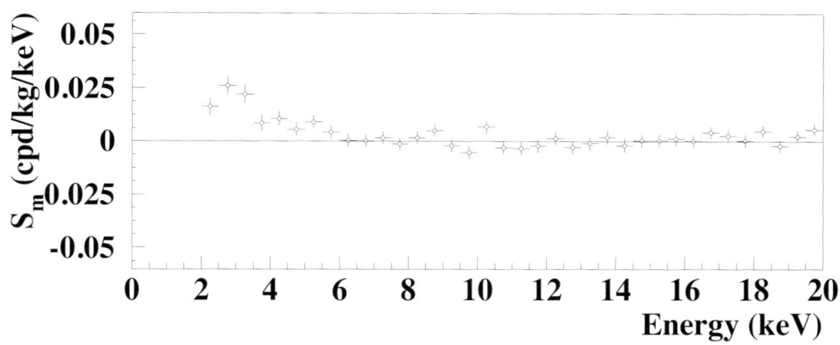


Fig. 2. Energy distribution of the $S_m$ variable for the total cumulative exposure 1.17 ton×yr. The energy bin is 0.5 keV. A clear modulation is present in the lowest energy region, while $S_m$ values compatible with zero are present just above. In fact, the $S_m$ values in the (6–20) keV energy interval have random fluctuations around zero with $\chi^2$ equal to 27.5 for 28 degrees of freedom. See Refs. 15,20.

ulation amplitudes are statistically well distributed in all the crystals, in all the annual cycles and energy bins; these and other discussions can be found in Ref. 20. It is also interesting the results of the analysis performed by releasing the assumption of a phase $t_0 = 152.5$ day in the procedure of maximum likelihood to evaluate the modulation amplitudes from the data of the seven annual cycles of DAMA/NaI and the six annual cycles of DAMA/LIBRA. In this case alternatively the signal is written as: $S_{0,k} + S_{m,k} \cos\omega(t - t_0) + Z_{m,k} \sin\omega(t - t_0) = S_{0,k} + Y_{m,k} \cos\omega(t - t^*)$, where $S_{0,k}$ is the constant part of the signal in $k$-th energy interval. Obviously, for signals induced by DM particles one would expect: i) $Z_{m,k} \sim 0$ (because of the orthogonality between the cosine and the sine functions); ii)

$S_{m,k} \simeq Y_{m,k}$; iii) $t^* \simeq t_0 = 152.5$ day. In fact, these conditions hold for most of the dark halo models; however, it is worth noting that slight differences in the phase can be expected in case of possible contributions from non-thermalized DM components, such as e.g. the SagDEG stream[7] and the caustics.[21] The $2\sigma$ contours in the plane $(S_m, Z_m)$ for the (2–6) keV and (6–14) keV energy intervals and those in the plane $(Y_m, t^*)$ are reported in Fig. 3. The best fit values for the (2–6) keV energy interval are ($1\sigma$ errors): $S_m = (0.0111 \pm 0.0013)$ cpd/kg/keV; $Z_m = -(0.0004 \pm 0.0014)$ cpd/kg/keV; $Y_m = (0.0111 \pm 0.0013)$ cpd/kg/keV; $t^* = (150.5 \pm 7.0)$ day; while for the (6–14) keV energy interval (also shown there) are: $S_m = -(0.0001 \pm 0.0008)$ cpd/kg/keV; $Z_m = (0.0002 \pm 0.0005)$ cpd/kg/keV; $Y_m = -(0.0001 \pm 0.0008)$ cpd/kg/keV and $t^*$ obviously not determined. These results confirm those achieved by other kinds of analyses. In particular, a modulation amplitude is present in the lower energy intervals for single hit events and the period and the phase agree with those expected for DM induced signals. For more detailed discussions see Ref. 20.

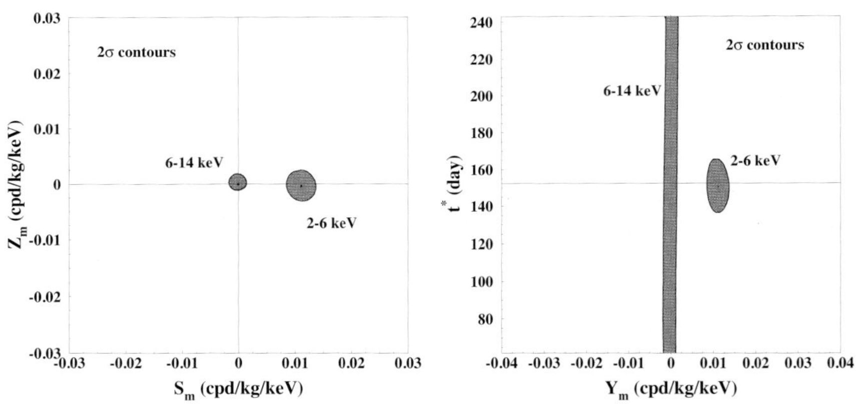


Fig. 3. $2\sigma$ contours in the plane $(S_m, Z_m)$ (*left*) and in the plane $(Y_m, t^*)$ (*right*) for the (2–6) keV and (6–14) keV energy intervals. The contours have been obtained by the maximum likelihood method, considering the cumulative exposure of 1.17 ton $\times$ yr. A modulation amplitude is present in the lower energy intervals and the phase agrees with that expected for DM induced signals. See Refs. 15,20.

Both the data of DAMA/LIBRA and of DAMA/NaI fulfil all the requirements of the DM annual modulation signature.

As previously done for DAMA/NaI,[4,5] careful investigations on absence of any significant systematics or side reaction effect in DAMA/LIBRA have been quantitatively carried out and reported in details in Ref. 15 and some other arguments have also been addressed in Refs. 22–24. No systematics or side reactions able to mimic the signature (that is, able to account for the measured modulation amplitude and simultaneously satisfy all the requirements of the signature) has been found or suggested by anyone over more than a decade. For detailed quantitative discussions on all the related topics and for results see Refs. 15,20 and Refs. therein.

In conclusion, DAMA/LIBRA has confirmed the presence of an annual modulation satisfying all the requirements of the DM annual modulation signature, as previously pointed out by DAMA/NaI; in particular, the evidence for the presence of DM particles in the galactic halo is cumulatively supported at 8.9 $\sigma$ C.L..

As regards the corollary investigation on the nature of the DM candidate particle(s) and related astrophysical, nuclear and particle physics scenarios, it has been shown that the obtained model independent evidence can be compatible with a wide set of possibilities; see for example Refs. 2,4–11,15. Many other interpretations of the annual modulation results are available in literature (as e.g. Refs. 18,25–31, etc.); others are open.

It is worth recalling that no other experiment exists, whose result can be directly compared in a model-independent way with those by DAMA/NaI and DAMA/LIBRA, and that – more in general – results obtained with different target materials and/or different approaches cannot be directly compared among them in a model-independent way. This is in particular due to the existing experimental and theoretical uncertainties, not last e.g. how many kinds of dark matter particles can exist in the Universe[a], the nature, the interaction types, the different nuclear and/or atomic correlated aspects, the unknown right halo model, the right DM density, etc. as well as the uncertainties on the values of each one of the many involved experimental and theoretical parameter/assumption/approximation used in the calculations. Moreover, some experimental aspects of some techniques used in the field have also to be addressed.[4,23,32,33] Another relevant argument is the methodological robustness.[34] In particular, the general considerations on comparisons reported in Appendix A of Ref. 15 still hold. Hence, claims for contradiction have no scientific basis. On the other hand, whatever possible positive result has to be interpreted and a large room of compatibility with DAMA annual modulation evidence is present.

Similar considerations can also be done for the indirect detection

[a]In fact, it is worth noting that, considering the richness in particles of the visible matter which is less than 1% of the Universe density, one could also expect that the particle part of the Dark Matter in the Universe may also be multicomponent.

searches, since it does not exist a biunivocal correspondence between the observables in the direct and indirect experiments. However, if possible excesses in the positron to electron flux ratio and in the $\gamma$ rays flux with respect to some modeling of the background contribution, which is expected from the considered sources, might be interpreted – under some assumptions – in terms of Dark Matter, this would also be not in conflict with the effect observed by DAMA experiments. However, it is worth noting that different possibilities either considering different background modeling or accounting for other kinds of sources can also explain the indirect observations.[35]

Finally, considering the relevance to lower the software energy threshold of the experiment, in order to improve the performance and the sensitivity of the experiment and to allow a highly precise determination of all the DM modulation parameters and deeper corollary information on the nature of the DM particle(s) and on the various related astrophysical, nuclear and particle Physics scenarios, the replacement of all the PMTs with new ones with higher quantum efficiency has been planned and work is in progress. DAMA/LIBRA will also study second order effects and several other rare processes as done by the former DAMA/NaI apparatus in the past[12] and by itself so far.[16]

## References

1. R. Bernabei et al., *Il Nuovo Cim.* **A112**, 545–576 (1999).
2. R. Bernabei et al., *Phys. Lett.* **B389**, 757–766 (1996); R. Bernabei et al., *Phys. Lett.* **B424**, 195–201 (1998); R. Bernabei et al., *Phys. Lett.* **B450**, 448–455 (1999); P. Belli et al., *Phys. Rev.* **D61**, 023512–9 (2000); R. Bernabei et al., *Phys. Lett.* **B480**, 23–31 (2000); R. Bernabei et al., *Phys. Lett.* **B509**, 197-203 (2001); R. Bernabei et al., *Eur. Phys. J.* **C23**, 61–64 (2002); P. Belli et al., *Phys. Rev.* **D66**, 043503–17 (2002).
3. R. Bernabei et al., *Eur. Phys. J.* **C18**, 283–292 (2000).
4. R. Bernabei el al., *La Rivista del Nuovo Cimento* **26** n.1, 1-73 (2003).
5. R. Bernabei et al., *Int. J. Mod. Phys.* **D13**, 2127–2159 (2004).
6. R. Bernabei et al., *Int. J. Mod. Phys.* **A21**, 1445–1469 (2006).
7. R. Bernabei et al., *Eur. Phys. J.* **C47**, 263–271(2006).
8. R. Bernabei et al., *Int. J. Mod. Phys.* **A22**, 3155–3168 (2007).
9. R. Bernabei et al., *Eur. Phys. J.* **C53**, 205–213 (2008).
10. R. Bernabei et al., *Phys. Rev.* **D77**, 023506–9 (2008).
11. R. Bernabei et al., *Mod. Phys. Lett.* **A23**, 2125–2140 (2008).
12. R. Bernabei et al., *Phys. Lett.* **B408**, 439–444 (1997); P. Belli et al., *Phys. Lett.* **B460**, 236–241 (1999); R. Bernabei et al., *Phys. Rev. Lett.* **83**, 4918–4921 (1999); P. Belli et al., *Phys. Rev.* **C60**, 065501–7 (1999); R. Bernabei et al., *Il Nuovo Cimento* **A112**, 1541–1552 (1999); R. Bernabei et al., *Phys.*

*Lett.* **B515**, 6–12 (2001); F. Cappella et al., *Eur. Phys. J.-direct* **C14**, 1-6 (2002); R. Bernabei et al., *Eur. Phys. J.* **A23**, 7–10 (2005); R. Bernabei et al., *Eur. Phys. J.* **A24**, 51-56 (2005); R. Bernabei et al., *Astrop. Phys.* **4**, 45-54 (1995); R. Bernabei, in *The identification of Dark Matter*, World Sc. Pub., Singapore, 1997, pp. 574.

13. DAMA web page: http://people.roma2.infn.it/dama
14. R. Bernabei et al., *Nucl. Instr. & Meth.* **A592**, 297–315 (2008).
15. R. Bernabei et al., *Eur. Phys. J.* **C56**, 333–355 (2008).
16. R. Bernabei et al., *Eur. Phys. J.* **C62**, 327–332 (2009).
17. K. A. Drukier et al., Phys. Rev. **D33**, 3495–3508 (1986); K. Freese et al., Phys. Rev. **D37**, 3388–3405 (1988).
18. D. Smith and N. Weiner, Phys. Rev. D 64 (2001) 043502; D. Tucker-Smith and N. Weiner, Phys. Rev. D 72 (2005) 063509; D. P. Finkbeiner et al, Phys. Rev. D 80 (2009) 115008.
19. K. Freese et al., Phys. Rev. **D71**, 043516–15 (2005); *Phys. Rev. Lett.* **92**, 111301–4 (2004).
20. R. Bernabei et al., *Eur. Phys. J.* **C67** (2010) 39.
21. F. S. Ling, P. Sikivie and S. Wick, *Phys. Rev.* **D70**, 123503–19 (2004).
22. R. Bernabei et al., AIP Conf. Proceed. **1223** (2010) 50, arXiv:0912.0660[astro-ph.GA].
23. R. Bernabei et. al, *J. Phys.: Conf. Ser.* **203**, 012040 (2010) (arXiv:0912.4200); http://taup2009.lngs.infn.it/slides/jul3/nozzoli.pdf, talk given by F. Nozzoli.
24. R. Bernabei et. al, arXiv:1007.0595
25. A. Bottino, N. Fornengo, and S. Scopel, Phys. Rev. D 67 (2003) 063519; A. Bottino, F. Donato, N. Fornengo, and S. Scopel, Phys. Rev. D 69 (2003) 037302; Phys. Rev. D 78 (2008) 083520; A. Bottino, F. Donato, N. Fornengo, S. Scopel, arXiv:0912.4025.
26. R. Foot, Phys. Rev. D 78 (2008) 043529.
27. Y. Bai and P.J. Fox, arXiv:0909.2900
28. K. Belotsky, D. Fargion, M. Khlopov and R.V. Konoplich, Phys. Atom. Nucl.71 (2008) 147.
29. E.M. Drobyshevski et al., Astrohys. & Astronom. Trans. 26:4 (2007) 289; Mod. Phys. Lett. A 23 (2008) 3077..
30. Nima Arkani-Hamed et al., Phys. Rev. D79 (2009) 015014.
31. Daniele S.M. Alves et al., arXiv:0903.3945.
32. R. Bernabei et al., *Liquid Noble gases for Dark Matter searches: a synoptic survey*, Exorma Ed., Roma, ISBN 978-88-95688-12-1, 2009, pp. 1–53 (arXiv:0806.0011v2).
33. J.I. Collar and D.N. McKinsey, arXiv:1005.0838; J.I. Collar and D.N. McKinsey, arXiv:1005.3723; J.I. Collar, arXiv:1006.2031.
34. R. Hudson, *Found. Phys.* **39**, 174–193 (2009).
35. F. Donato et al., Phys. rev. Lett. 102 (2009) 071301; T. Delahaye et al., Astron. Astrophys. 501 (2009) 821; S. Profumo, arXiv:0812.4457; P. Blasi, Phys. Rev. Lett. 103 (2009) 051104; M. Ahlers et al., arXiv:0909.4060.

## Recent Results from the Fermi Large Area Space Telescope

ICATPP Conference on Cosmic Rays for Particle and Astroparticle Physics
Villa Olmo, Como 7-8 October 2010

Emanuele BONAMENTE

*Istituto Nazionale di Fisica Nucleare, Sezione di Perugia, I-06123 Perugia, Italy*
*Dipartimento di Fisica, Università degli Studi di Perugia, I-06123 Perugia, Italy*
*Representing the Fermi LAT Collaboration*
** E-mail: emanuele.bonamente@pg.infn.it*

The Fermi mission was launched in summer of 2008, carrying aboard the gamma-ray Large Area Telescope, an instrument with top performances in sensitivity, effective area and energy resolution in the range between 20 MeV and 300 GeV. Latest results obtained with first two years of data by the Fermi collaboration are presented, providing an overview of current analyses on active galactic nuclei, diffuse gamma-ray emission, galactic and extragalactic sources, gamma-ray bursts, source catalogs and dark matter searches.

### 1. The Fermi Space Telescope

The Fermi Space Telescope is an international mission with approximately 400 members from 5 different Countries, France, Italy, Japan, Sweden and United States. The satellite has been launched on June 11, 2008, and now it is in its second year of data taking. It is performing very well, with low degradation rate for silicon strips, and passed the record of 100 billion triggers.

It is composed by two instruments, the Gamma-ray Burst Monitor[1] (GBM), which is an array of crystals, disposed around the spacecraft, designed to measure the energy emission of bursts occurring in the energy range from 8 $keV$ up to 40 $MeV$, and the Large Area Telescope[2] (LAT), made by an array of 4x4 towers, each one having a converting-tracker and an electromagnetic calorimeter, surrounded by tiles of scintillators. The energy range of the LAT goes from 30 $MeV$ up to $> 300$ $GeV$, producing an overall energy coverage of more than 7 energy decades for the Fermi Telescope. It is able to observe the gamma-ray emission with unprecedented performances in terms of spatial resolution (0.1 $deg$, $E > 10$ $GeV$) and sensitivity ($3 \cdot 10^{-9} ph \cdot cm^{-2} s^{-1}$, $E > 100\ MeV$), covering the unexplored region above 10 $GeV$. The satellite is orbiting at an altitude of $\sim 550\ km$, with a period of 90 minutes. The LAT has been designed with a very compact aspect ratio, resulting in a broad field of view (2.4 $sr$), corresponding approximately to the 20% of the sky. This two characteristics allow one to have a uniform exposure coverage of the sky every 2 orbits.

## 2. Source Catalogs

Data taken in the first 11 months have been used to produce the so called first-year source catalog[3] (1FGL, FIG.1). In the catalog there are 1451 gamma-ray sources, the 43% of them not yet associated. This means that such sources have no counterpart at other wavelengths. This number is very high if compared with the previous gamma-ray catalog, the $3^{rd}$ EGRET catalog, which has 271 sources.

Among these sources there are 709 AGNs (1LAC[4] catalog), and it appears that the AGNs make up almost one half of the associated 1FGL source. Within the 1FG there is the same fraction of FSRQs and BL Lacs.

Up to now Fermi has seen more than 60 gamma-ray pulsars, and discovered a population of milli-second pulsars and 24 radio-quiet pulsars[5–7] (FIG.2).

## 3. Particle acdeleration sites

One of the scientific objectives for the mission is that of exploring the production and acceleration mechanisms of Cosmic Rays (CRs). Four Supernova Remnants (SNRs) have been spatially resolved by the LAT (Cas A, W51C, W44 and IC 443), and this analysis shows very strong evidence for CRs to be produced in SNRs by the decay of $\pi^0$, confirming, with the joint analysis of radio data, that the adronic model should be preferred[8] (FIG.3).

Very surprisingly gamma-rays have been observed for the fist time, and yet the only one, from a Nova in the Cygnus region[9] (FIG.4). The measured spectrum appears to favor the adronic model, but can not rule out the leptonic model (Inverse Compton and Bremsstrahlung).

## 4. Gamma-ray Bursts

The GBM is detecting Gamma-ray Bursts (GRBs) with an almost constant rate of $\sim 250$ per year (FIG.5). One half of them were in the field of view of the LAT, but only 20 showed emission in its energy range, some with very high energy photons, up to 33 $GeV$.

A very particular short GRB[10] (GRB 090510) produced a high energy photon (31 $GeV$) less than 1 *sec* after the X-ray trigger. In the hypothesis that this photon was not produced before the first X-ray photon, it is possible to put very constraining limits on the scale of a possible Lorentz Invariance Violation ($M_{QG} > 1.2 M_{Planck}$).

## 5. Cosmic-ray electrons and positrons

The LAT produced also very high statistic spectrum of cosmic-ray $e^+e^-$ in the range from 7 $GeV$ up to 1 $TeV$[11] (FIG.6). The CR electron spectrum appears to be substantially flat, with no evidence for a flattening at low energies, but with a possible contribute, in the 100 $GeV$ – 1 $TeV$ range, which exhibited an exponential cutoff behavior. Analyses are in progress to determine if such excess can be attributed to some known or unknown sources.

The spatial distribution of CR electrons[12] does not show any appreciable anisotropy. The dipole anisotropy analysis sets limits on the CR electrons production sites, disfavoring a bright nearby source (the Vela pulsar), but not constraining on any interesting DM model.

## 6. Diffuse Emission

The analysis of the medium latitude galactic Diffuse Emission does not confirm the GeV EGRET excess, and it is consistent with the local CR spectrum[13] (FIG.7).

This makes also any contribute due to Dark Matter (DM) annihilation negligible. It is then possible to put some limits for the DM annihilation cross-section, especially for the $\mu^+/\mu^-$ channel. Once the Inverse Compton Scattering off the Cosmic Microwave Background is taken into account, one gets very constraining limits, that rule out the possible interpretation of the CR electron bump, both from Fermi and from PAMELA, as given by DM.

In the same way the Isotropic Diffuse is consistent with a power law (index= $2.41 \pm 0.05$), without any evidence for an excess[14] (FIG.8). This rules out also a possible DM component, because of its unique spectral signature, and gives interesting limits[15] on different models.

## 7. Dark Matter Search

DM annihilation is searched also in very high mass-to-light ratio objects, hence DM dominated, like dwarf spheroidal galaxies within the Milky Way, and galaxy clusters. In both cases[16,17] no gamma ray detection is observed at all (FIG.9). The non detection sets limits on the flux and hence on the self-annihilation cross-section ($\sigma v$). In both cases thermal values of $\sigma v$ are not reached, but the results in the $\mu^+/\mu^-$ channel rule out the DM interpretation of the $e^+/e^-$ CR bump.

The smoking gun for DM detection could be that of seeing a line in the gamma-ray spectrum produced by the direct annihilation or decay into a $\gamma\gamma$ or $\gamma X$ final state, where $X$ represents a generic boson. The search in the galactic center and at high latitudes[18] produced no evidences for lines and begins to set some interesting limits, especially for for decaying models.

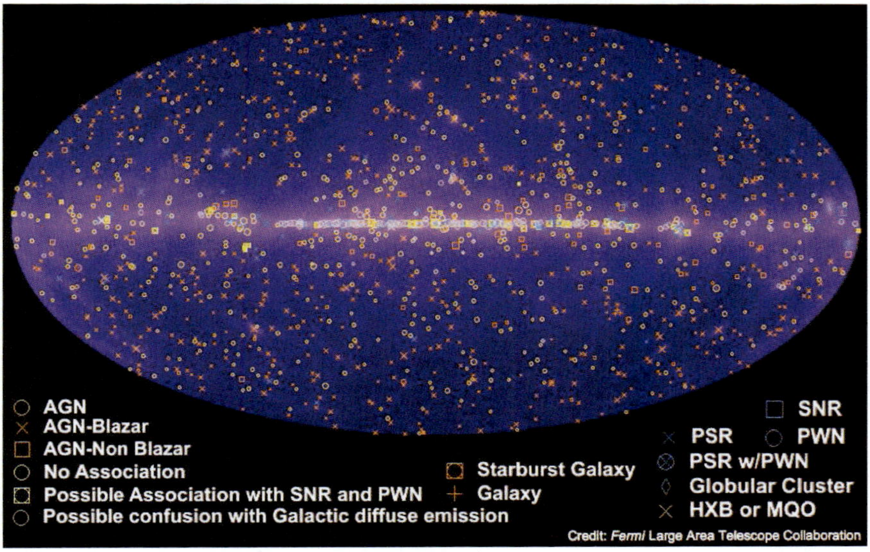


Fig. 1. — 1FGL gamma-ray source catalog with associations.

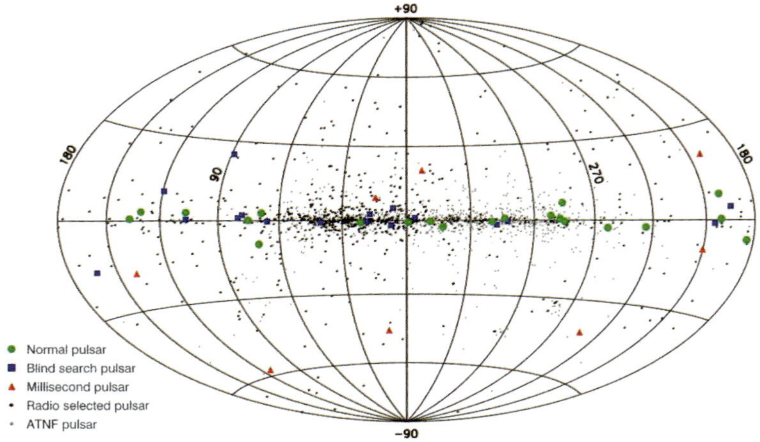


Fig. 2. — Gamma-ray Pulsar catalog.

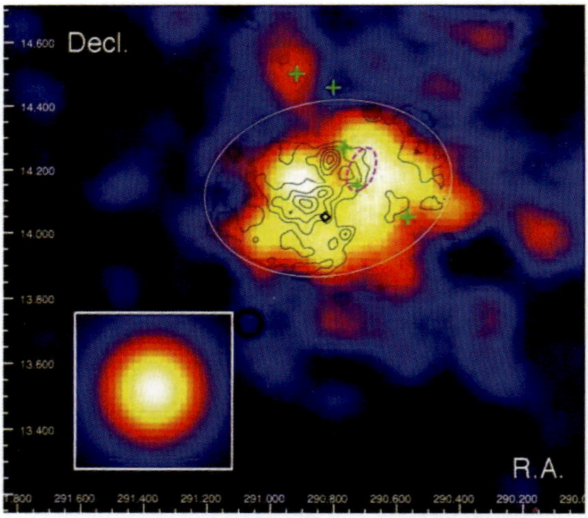


Fig. 3. — Close up view around SNR W51C. The simulated point source image (smoothed by the same gaussian) is shown in the inset. The outer boundary of W51C is indicated by a white ellipse. Superposed is the ROSAT X-ray map (contours) from Koo et al. (1995). The region where shocked CO clumps (Koo & Moon 1997b) were found is represented by a dashed magenta ellipse. A diamond near the SNR centroid indicates CXO J192318.5+143035 (see the text). The positions of five HII regions are indicated by green crosses (Carpenter & Sanders 1998).

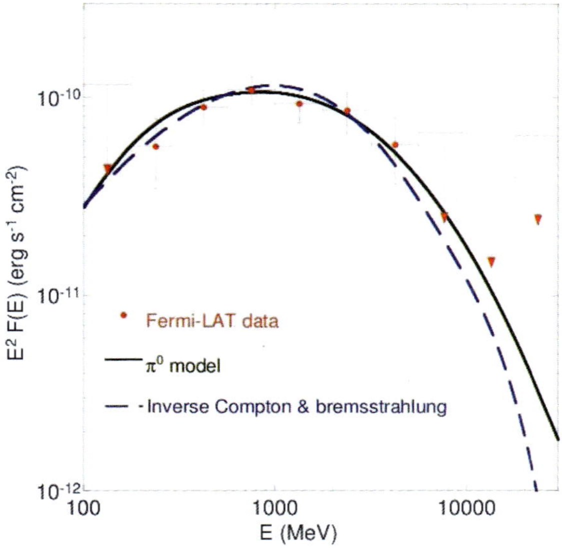


Fig. 4. — SED of V407 Cyg in MeV/GeV gamma-rays measured by the Fermi-LAT over the period 10 March 18:00 – 29 March 00:00 2010. Vertical bars indicate $1\sigma$ statistical errors, arrows indicate $2\sigma$ upper limits, and horizontal bars indicate energy ranges. The best-fit $\pi^0$ (black solid line) and leptonic (blue dashed line) models are indicated.

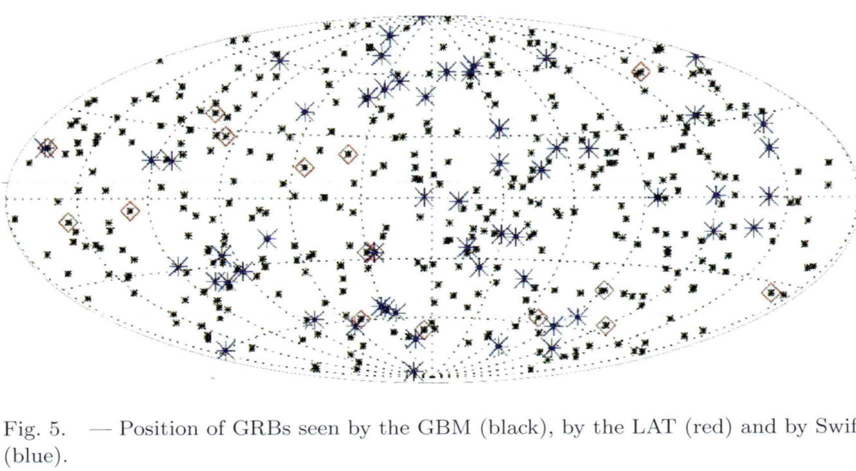

Fig. 5. — Position of GRBs seen by the GBM (black), by the LAT (red) and by Swift (blue).

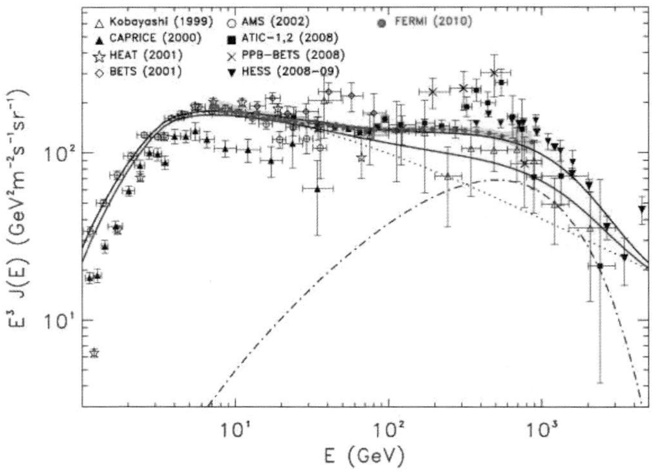


Fig. 6. — The $e^{+}+e^{-}$ spectrum (solid line) computed with the conventional GALPROP model but with a different in-jection spectrum: an injection index $\Gamma = 1.6/2.7$ below/above 4 *GeV* (dotted line). An additional component with an injection index $\Gamma = 1.5$ and exponential cut-off is shown by the dashed line. Blue line shows $e^{-}$ spectrum only. Fermi-LAT data points are shown by red filled circles.

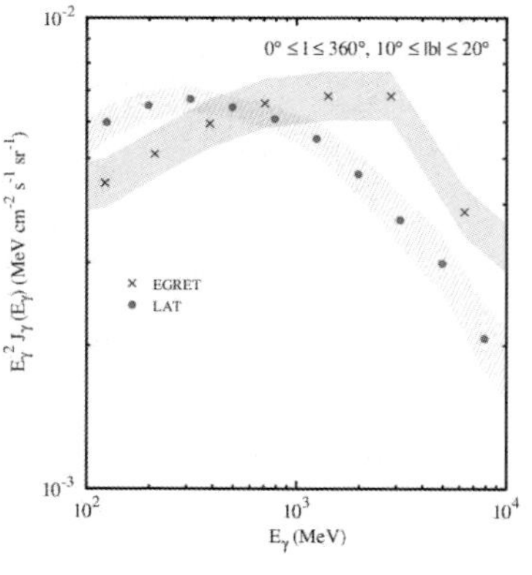


Fig. 7. — Diffuse emission intensity averaged over all Galactic longitudes for latitude range $10° \leq |b| \leq 20°$ . Data points: LAT, red dots; EGRET, blue crosses. Systematic uncertain- ties: LAT, red; EGRET, blue.

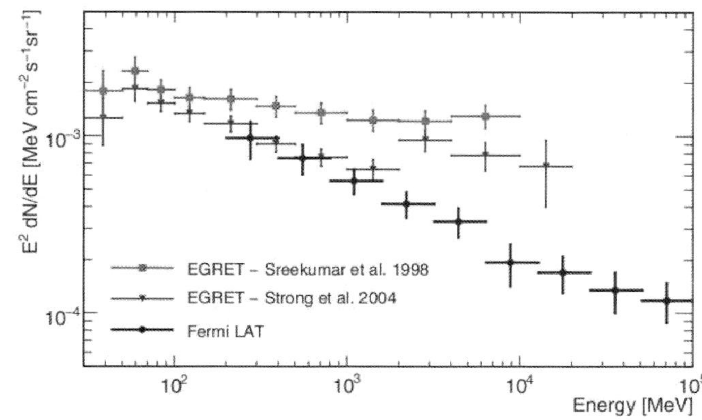


Fig. 8. — Isotropic background intensity compared with EGRET-derived intensities. The dericed spectrum is compatible with a simple power law with index $\Gamma = 2.41 \pm 0.05$ and intensity $I(> 100MeV) = (1.03 \pm 0.17) \cdot 10^{-5}\ cm^{-2}s^{-1}sr^{-1}$ where the uncertainties are systematics dominated.

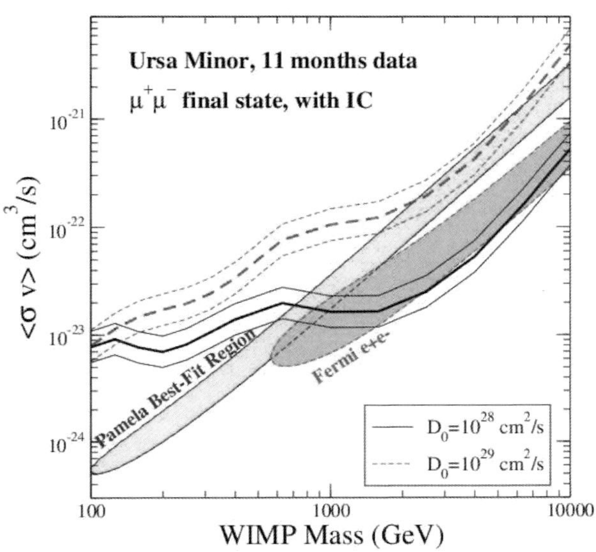


Fig. 9. — Constraints for the Ursa Minor dwarf including both gamma-ray emission from IC scattering and final state radiation. Here we consider two different diffusion coefficients, and show the effect of the uncertainties in the Ursa Minor density profile.

## 8. Conclusions

Fermi has just entered in its $3^{rd}$ year of data taking, is performing very well and is covering a very broad range of scientific topics. The mission duration goal is 10 years, with a minimum request of 5. In any case there is still much time to gain much more statistics, improve current analyses and face new challenges. Up to now very interesting results have been achieved, related to high energy acceleration mechanisms in AGNs, pulsars, SNRs and GRBs, diffuse emissions and dark matter search.

The Fermi Catalog contains 709 AGNs, with almost the same fraction of Blazars and FSRQs. Many gamma-ray pulsars have been discovered or associated. In particular Fermi discovered two new populations: millisecond and radio-quiet gamma-ray pulsars. Many sources (43%) are still unassociated.

Evidences for cosmic ray acceleration have been found in supernova remnants, consistent with the $\pi^0$ decay, and hence ruling out the leptonic models. Gamma-rays have also been detected for the first time from a nova, confirming the adronic model.

Many GRBs (250/year) have been observed, some with very high energy emission, up to 30 $GeV$. A particular GRB permitted to set very strong limits on Lorentz Invariance Violation models.

The LAT measured also the total $e^+/e^-$ spectrum, which is substantially flat, with a possible component with an injection index of -1.5 and an exponential cutoff.

The study of the galactic diffuse emission confirmed the agreement with the Galprop model, with no evidence for an extra component (like DM). The extragalactic isotropic background is flat and consistent with a power law (index$= 2.41 \pm 0.05$).

The search for DM produced no detections from any tested sites. Limits begin to constrain interesting parameters space, especially for the muon channel.


## Acknowledgements

The results reported in this work are the results of the effort of all members within the Fermi LAT and GBM collaborations. The Fermi-LAT collaboration acknowledges support from a number of agencies and institutions for both the development and the operation of the LAT as well as scientific data analysis. These institutions include NASA and DOE in the United States, CEA/Irfu and IN2P3/CNRC in France, the INFN in Italy, the Swedish Research Council and the National Space Board in Sweden. Additional support from INAF in Italy and CNES in France for science analysis during the operations phase is also gratefully acknowledged.

## References

1. Meegan C. et al., "The glast burst monitor," *AIP Conf. Proc*, 2007, 291 13.
2. The Fermi-LAT Collaboration, "The large area telescope on the fermi gamma-ray space telescope mission," *ApJ*, 2009, 697 1071.
3. The Fermi-LAT Collaboration, "Fermi large area telescope first source catalog," *ApJS*, 2010, 188 405.
4. The Fermi-LAT Collaboration, "The first catalog of active galactic nuclei detected by the fermi large area telescope," *ApJ*, 2010, 715 429.
5. The Fermi-LAT Collaboration, "The first fermi large area telescope catalog of gamma-ray pulsars," *ApJS*, 2010, 187 460.
6. The Fermi-LAT Collaboration, "A population of gamma-ray millisecond pulsars seen with the fermi large area telescope," *Science*, 2009, 325 848.
7. The Fermi-LAT Collaboration, "Detection of 16 gamma-ray pulsars through blind search frequency searchis using the fermi lat," *Science*, 2009, 325 840.
8. The Fermi-LAT Collaboration, "Fermi lat discovery of extended gamma-ray emission in the direction of supernova remnant w51c," *ApJ*, 2009, 706 L1.
9. The Fermi-LAT Collaboration, "Gamma-ray emission concurrent with the nova in the symbiotic binary v407 cygni," *Science*, 2010, 329 817.
10. The Fermi-LAT Collaboration, "A limit on the variation of the speed of light arising from quantum gravity effects," *Nature*, 2009, 462 331.
11. The Fermi-LAT Collaboration, "Fermi lat observations of cosmic-ray electrons from 7 gev to 1 tev," *arXiv:1008.3999*, 2010.
12. The Fermi-LAT Collaboration, "Searches for cosmic-ray electron anisotropies with the fermi large area telescope," *arXiv:1008.5119*, 2010.
13. The Fermi-LAT Collaboration, "Fermi large area telescope measurements of the diffuse gamma-ray emission at intermediate galactic latitudes," *PhysRevLett*, 2009, 103.251101.
14. The Fermi-LAT Collaboration, "Spectrum of the isotropic diffuse gamma-ray emission derived from first-year fermi large area telescope dat," *PhysRevLett*, 2010, 104.101101.
15. The Fermi-LAT Collaboration, "Constraints on cosmological dark matter annihilation from the fermi-lat isotropic diffuse gamma-ray measurement," *arXiv:1002.4415*, 2010.
16. The Fermi-LAT Collaboration, "Observations of milky way dwarf spheroidal galaxies with the fermi-large area telescope detector and constraints on dark matter models," *ApJ*, 2010, 712 147.
17. The Fermi-LAT Collaboration, "Constraints on dark matter annihilation in clusters of galaxies with the fermi large area telescope," *JCAP05*, 2010, 025.
18. The Fermi-LAT Collaboration, "Fermi large area telescope search for photon lines from 30 to 200 gev and dark matter implications," *PhysRevLett*, 2010, 104.091302.

# GAMMA-RAY ACTIVITY OF CYGNUS X-3 AT ENERGY RANGE OF 1 - 100 TeV DURING 15 YEAR OBSERVATIONS OF SHALON

V.G. SINITSYNA, A.Y. ALAVERDIAN, S.S. BORISOV,
S.I. NIKOLSKY, V.Y. SINITSYNA

*P.N.Lebedev Physical Institute,*
*Leninsky prospect 53, Moscow, 119991, Russia*
*E-mail: sinits@sci.lebedev.ru*

Cyg X-3 galactic binary system has been regularly observed since a 1995 by SHALON telescope. The energy spectrum of Cyg X-3 at 0.8 - 100 TeV is obtained and the average integral flux is $F(E_O > 0.8TeV) = (6.8 \pm 0.7) \times 10^{-13} cm^{-2} s^{-1}$. The binary Cyg X-3 came to period of flaring activity at radio-and X-ray energies in 2006. In May and July 2006 the significant increase of Cyg X-3 flux have detected with SHALON at TeV energy. This intensity increase was also observed by Crimea Observatory. Earlier, in 1997 and 2003 a comparable increase of the flux over the average value was also observed. In observations of SHALON telescope the formation of jets (like jet of active galactic nuclei) during the activity periods have been found. The last significant increase of very high energy gamma-quantum flux have detected in May 2009, which is correlated with flaring activity at lower energy range of X-ray and at observations of Fermi LAT.

*Keywords*: galactic binary system; Cygnus X-3; TeV gamma-rays; flaring activity.

## Introduction

The attempts of detection of TeV emission from Cygnus X-3 were first made in the mid of 1970s and continued through the mid 1980s. In 1983 the Kiel group announced that they had observed a large flux of $\gamma$ - rays with energy in excess of $10^{15}$ eV from the X-ray binary Cyg X-3. But earlier, Cocconi proposed in 1959 ICRC, Moscow an air shower array at extreme mountain altitude to detect $10^{12}$ eV $\gamma$ - rays from astrophysical sources.[1] The Cherenkov gamma-telescope SHALON[2–4] located at 3338 m a.s.l., at the Tien Shan high-mountain observatory of Lebedev Physical Institute, has been destined for $\gamma$ - astronomical observation in the energy range 1

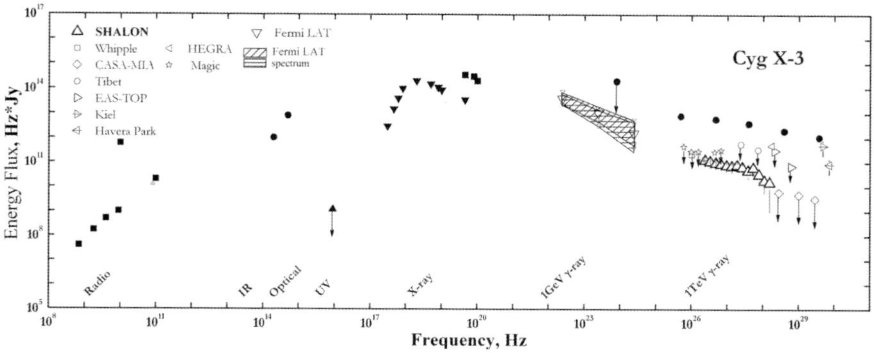


Fig. 1. The spectral energy distribution of Cyg X-3. Black points are the archival data from Cordova.[5] The high level points in radio and X-ray bands correspond to radio-frequency activity and increased x-ray activity of the source. TeV range is represented with integral spectrum by SHALON[6–18] in comparison with other experiments: Fermi,[19,20] TIBET,[21] HEGRA,[24] EAS-TOP,[25,26] Whipple,[27,28]Magic[29] CASA-MIA,[30,31]Kiel,[32] Havera Park[33]

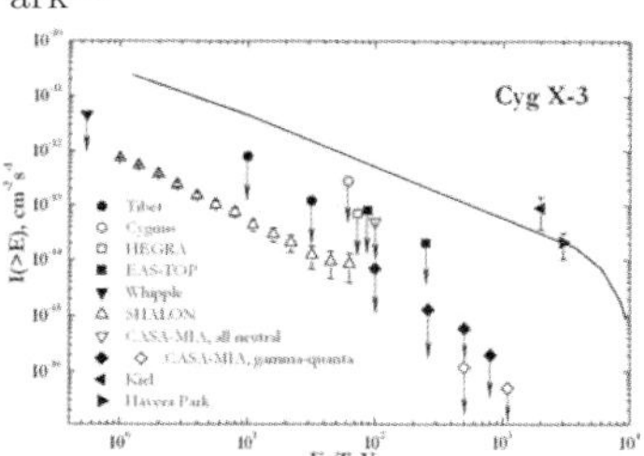


Fig. 2. The Cyg X-3 $\gamma$-quantum ($E > 0.8$ TeV) integral spectrum by SHALON in comparison with other experiments: TIBET,[21] 2 - CYGNUS,[22,23] 3 - HEGRA,[24] 4 - EAS-TOP,[25,26] 5 - Whipple,[27,28] 6 - SHALON,[6–18] diamonds - CASA-MIA,[30,31] Kiel,[32] Havera Park,[33] the solid line is the theoretical calculation (Hillas).[34,35]

– 100 TeV.[2–4,6–18] Selection of the electron-photon showers among the net cosmic rays EAS becomes possible through the analysis of a light image which, in general is emerging as an elliptic spot in light receiver matrix. The selection of $\gamma$-initiated showers from the background of proton showers is performed by applying the criteria (see[2,4]). Our analysis of the distributions of shower image parameters suggests that the background is rejected with 99.95% efficiency (see Refs.[2,4,7,15]).

## Cygnus X-3

Cyg X-3 has been observed throughout wide range of the electromagnetic spectrum (Fig. 1). It is one of the luminous X-ray sources in our Galaxy, displaying high and low states and rapid variability in X-rays. In addition

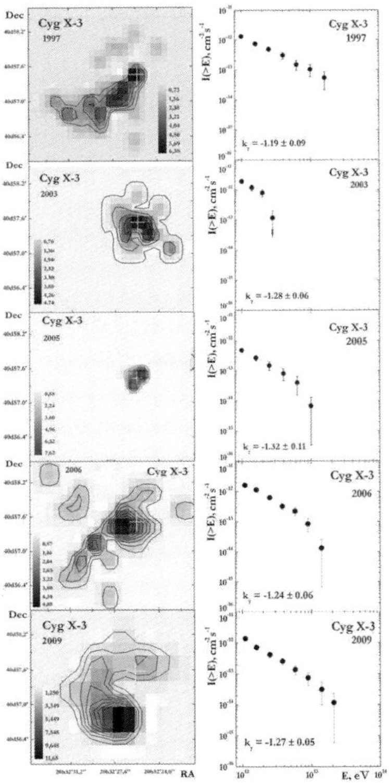


Fig. 3. The image of $\gamma$-ray emission from Cyg X-3; The Cyg X-3 in 1997, 2003, 2005, 2006 and 2009 by SHALON: The $\gamma$-ray integral spectra with the power index $k_\gamma$.

to being a powerful x-ray source, Cyg X-3 is seen in the IR and is a strong and variable radio source. It is also the strongest radio source among X-ray binaries and shows both huge radio outbursts and relativistic jets. The radio activity is closely linked with the X-ray emission and the different X-ray states. Based on the detections of ultra high energy $\gamma$-rays, Cyg X-3 has been proposed to be one of the most powerful sources of charged cosmic ray particles in the Galaxy. The searches for TeV emission had been carried out since the mid of '70s and continued through the mid '80s. Two observations were particulary important: the Kiel results[32] and contemporaneous observation at Haverah Park.[33] These results indicated a very large UHE flux from Cyg X-3. So, these results stimulated the construction of many of new detectors (see Refs.[34–36]). The upper limits of the Cyg X-3 flux are over an order of magnitude lower than the detected in the '80s levels. Figures 1 and 2 show upper limits on the steady flux from Cyg X-3 reported between

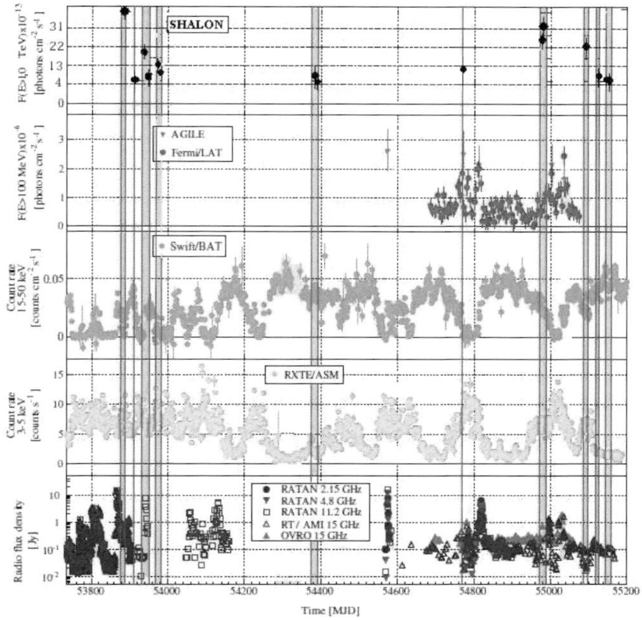


Fig. 4. The light curves of Cyg X-3 in the wide energy range.[29]

'90 and '95 compared with earlier observations. The Cyg X-3 flux obtained by SHALON is one order of magnitude lower than upper limits published before.

The observational data obtained with SHALON telescope for the Cyg X-3 point source are presented in Figures 1, 2 and 3. This galactic binary system regularly observed since a 1995 is known as a source with variable intensity (from $5 \times 10^{-12}$ to $10^{-11} cm^{-2}s^{-1}$); the average $\gamma$-quantum flux from Cyg X-3 for $E > 0.8$ TeV is estimated as $F(E_O > 0.8TeV) = (6.8 \pm 0.7) \times 10^{-13} cm^{-2}s^{-1}$. The average energy spectrum of Cyg X-3 at $0.8 - 65$ TeV can be approximated by the power law $F(> E_O) \propto E^{k_\gamma}$ , with $k_\gamma = -1.21 \pm 0.05$. This flux, measured for the first time, is several times less than the upper limits established in the earlier observations.

The binary Cyg X-3 came to the period of flaring activity at radio- and X-ray energies in 2006 (Fig. 4). In May and July 2006 the significant increase of the Cyg X-3 flux have detected with SHALON at TeV energy with flux values $(4.12 \pm 1.01) \times 10^{-12} cm^{-2}s^{-1}$ and $(1.62 \pm 0.75) \times 10^{-12} cm^{-2}s^{-1}$ respectively. The $\gamma$-ray flux in the June $(0.46 \pm 0.18) \times 10^{-12} cm^{-2}s^{-1}$. The average $\gamma$-ray flux detected by SHALON in the flaring period from May to

July 2006 was estimated as $(1.47 \pm 0.24) \times 10^{-12}\ cm^{-2}s^{-1}$ (Fig. 3).This intensity increase was also observed by Crimea Observatory and the integral flux was estimated as $F(E_O > 1TeV) \sim (3-5) \times 10^{-12} cm^{-2}s^{-1}$.[37] In the 2005 Cyg X-3 was in the quiet period in TeV energies. The average $\gamma$-quantum flux from Cyg X-3 for $E > 0.8$ TeV is estimated as $F(E_O > 0.8TeV) = (5.4 \pm 0.73) \times 10^{-13}\ cm^{-2}s^{-1}$ (Fig. 3). The images and spectra of Cyg X-3 in silent period at 2005 are shown at Fig. 3. The are no features found at flaring periods. The $\gamma$-ray flux detected by SHALON in 2003 was estimated as $(1.79 \pm 0.33) \times 10^{-12}\ cm^{-2}s^{-1}$ (Fig. 3). Earlier, in 1997, a comparable increase of the flux over the average value was also observed and estimated to be $(1.2 \pm 0.5) \pm 10^{-12}\ cm^{-2}s^{-1}$ (Fig. 3).

The last significant increase of very high energy $\gamma$-quantum flux has detected in May 2009: $(3.1 \pm 0.98) \times 10^{-12} cm^{-2}s^{-1}$ (Fig. 3). The increase is correlated with the flaring activity at lower energy range of X-ray and at observations of Fermi LAT[19,20] with average flaring flux of $F = (190 \pm 40) \times 10^{-8} phcm^{-2}s^{-1}$ above 100 MeV (Fig. 1). In the autumn of 2009 Cyg X-3 comes again in the quiet period in TeV energies. The average $\gamma$-quantum flux from Cyg X-3 for energies more than 0.8 TeV is estimated as $(0.46 \pm 0.07) \times 10^{-12} cm^{-2}s^{-1}$ with $k_\gamma = -1.10 \pm 0.12$.

During the period of observations of Cyg X-3 with SHALON 6 significant flux increases were detected at energies above 0.8 TeV. To reveal possible correlation of periods of activity in the TeV energy range with the flarings at the low energies the light curves of Swift/BAT (15 - 50 keV),[29] RXTE/ASM (3 - 5 keV),[29] the fluxes at radio-ranges from RATAN, RT/AMI, OVRO[29] and TeV fluxes from SHALON observations were analyzed (Fig. 4). The significant anticorrelation of the fluxes at TeV and hard X-rays and the correlation of very high energy flux and soft X-ray were found. It is note, that TeV flaring activities occur close (within the 4 - 5 days) to strong radio flares. Probably, it is linked with the powerful ejection from the regions are close to the centers blackhole. This ejection is accompanied with a relativistic shock where the relativistic electrons and magnetic field are generated effectively. Similar relation of TeV and soft X-ray fluxes were found in the 1997 observation period. But the flux increase of 2003 didn't obey this scheme, it was in the quite period in the soft X-rays. In general, the correlation soft X-ray and very high energy $\gamma$-ray fluxes is traced since 1996.

Confirmation of the variability of very high-energy gamma-radiation from Cyg X-3 by the future observations would be important for understanding the nature of this astrophysical object.

## References


1. J.Cocconi, in Proc VIth Int. Cosmic Ray Conf., Moscow, 2, (1960) p. 309
2. S.I. Nikolsky and V. G. Sinitsyna, VANT, Ser. TFE 1331, (1987) p. 30.
3. S. I. Nikolsky and V. G. Sinitsyna, *in Proc. Int. Workshop on VHE $\gamma$-ray Astronomy*, Crimea, ed. A.A. Stepanian, (1989) p. 11.
4. V.G. Sinitsyna, Nuovo Cim., 19C, (1996) p. 965.
5. F. A.-D. Cordova, Los Alamos Science, Spring 1986. pp. 39-54.
6. V.G. Sinitsyna, *Towards a Major Atmospheric Cherenkov Detector-V*, ed. O.C. deJager, Kruger Park, South Africa, (1997) pp. 136, 190.
7. V.G. Sinitsyna et al., Nucl. Phys. B (Proc.Suppl.) 151, (2006) p. 489.
8. V. G. Sinitsyna, S. I. Nikolsky , et al., Nucl., Phys. B. 122 (2003) 247, p. 409; 97 (2001) 215, 219; 75A (1999) p. 352.
9. V. G. Sinitsyna, Rayos Cosmicos 98, ed. Medina J. (Departamento de Fisica Universidad de Alcala) (1998) pp. 383, 367.
10. V. G. Sinitsyna, AIP 515 (1999) pp. 205, 293.
11. V.G. Sinitsyna, S.I. Nikolsky, et al., The Universe viewed in Gamma Rays, ed. R.Enomoto, M.Mori (Universal Academy Press, INC, 2003) pp. 211, 235.
12. V. G. Sinitsyna, S. I. Nikolsky et al., Chakaltaya Meeting on CR Physics, ed. O. Saavedra, et al. (Societa de Italiana di Fisica Bologna), (2000) p. 785.
13. S. I. Nikolsky, V. G. Sinitsyna, et al., Izv. RAN., ser. fiz. 66(11) (2002) 1667, 1660; 66(3) (1999) 608; 61(3) (1997) p. 603.
14. S. I. Nikolsky, V. G. Sinitsyna, Phys. of Atomic Nuclei 67(10) (2004) p. 1900.
15. V.G. Sinitsyna, S.I. Nikolsky, et al., Izv. RAN Ser. Fiz. 69(3), (2005) p. 422.
16. V. G. Sinitsyna *et al., Int. J. Mod. Phys. A* No. 29, (2005) p. 7023.
17. V. G. Sinitsyna *Rad. Phys. and Chem.*, 75, (2006) pp. 880.
18. V. G. Sinitsyna, in Proc. ICRW "Aragats-2007", ed. B. Pattison, (2007) p. 12.
19. M. Tavani et al. Nature Letteres,.462/3, (2009) p. 620.
20. A. A. Abdo et al. Science, 326, (2009) p. 1512.
21. M. Amenomori et al., in Proc. 23rd ICRC, Calgary 1, (1993) p. 342. 33.
22. D.E. Alexandreas et al.,Astrophys. J. 418, (1993) p. 832.
23. D.E. Alexandreas et al.,in Proc. 23rd ICRC, Calgary 1, (1993) p. 373.
24. A.D.Karle et al.,Astropart. Phys.4,(1995) p.1.
25. P.L. Ghia et al., Proc. 24th ICRC, Rome 2, (1995) p.421.
26. M.Aglietta et al.,Astropart.Phys. 3,(1995) p.1
27. T.C. Weekes Proc. 25th ICRC, Durban 5, (1997) p. 251.
28. M. Catanese and T. C. Weekes, Prep. Ser. No. 4811, (1999); No 4450, (1996).
29. J. Alekśic et al. ApJ, 721, (2010) p. 843
30. A.Borione et al., Phys Rev. 55, (1997) 1714.
31. A.Borione, et al., Proc.24 ICRC 2 (1995) 430
32. M. Samorscki and W. Stamm, Astrophys. J. 268 (1983) L17.
33. J. Lloyd-Evans et al., Nature 305 (1983) p. 784.
34. A.M. Hillas, Nuovo Cim. 19C, (1996) p. 701; Nature 312, (1984) p. 50.
35. J.W. Cronin, Nuovo Cim. 19C, (1996) p. 847.
36. C.M. Hofman, C. Sinnis, P. Fleury, et al., Rev. Mod. Phys. 71, (1999) p. 897.
37. Yu. T. Neshpor, et al. Bull. of RAS; Physics, l73(5), (2009) p. 694.

# Signatures of middle aged, nearby pulsars in the cosmic ray lepton spectrum?

I. Büsching

*Theoretische Physik IV, Ruhr-Universität Bochum,*
*44801 Bochum, Germany*
**E-mail: ib@tp4.rub.de*

Okker C. deJager

*Unit for Space Physics, North-west University,*
*Potchefstroom 2520, South Africa*
**E-mail:Okkie.DeJager@nwu.ac.za*

*Keywords*: Pulsars, Cosmic Rays, Leptons.

## 1. Introduction

Recent data reported by the PAMELA[1] and ATIC[2] experiments, as well as cosmic ray lepton results from FERMI[3] and H.E.S.S.[4] collaborations sparked a series of papers explaining these results either by contributions of electron positron pairs to the local interstellar cosmic ray (CR) spectrum by dark matter (DM) or pulsars. Focusing here on pulsars, we argue that at the present, our knowledge about particle acceleration at pulsars as well as of the local Galactic CR propagation is still limited, i.e. the recent results for CR electrons and positrons constrain pulsar and propagation models. We will thus not present another attempt to explain the data by contributions of pulsars to the local CR lepton flux but rather to highlight the caveats in doing so.

## 2. Pulsars as Particle Accelerators

Modelling the acceleration of particles at pulsars and subsequent injection of these particles into the interstellar medium, one may look at three scenarios: young pulsars (after the breakup of their pulsar wind nebulae),[5] mature pulsars[6] and millisecond pulsars.[7]

### 2.1. *Energy Budget*

Before discussing the acceleration of particles by pulsars, it is instructive to look for the available energy budget of a single pulsar. The energy reservoir of a pulsars is its rotational energy $1/2\Omega^2$, where $I$ is the moment of inertia. The spin down power $L_{\rm SD} = I\Omega\dot{\Omega}$ of a pulsar is given by[8]

$$L_{\rm SD}(t) = L_{\rm sd,0}\,(1 + t/\tau_0)^{-\frac{n+1}{n-1}}\,, \tag{1}$$

with the characteristic decay time $\tau_0 = P_0/((n-1)\dot{P}_0)$ and $n = 3$ for a magnetic dipole field of the pulsar.

With $P_0$, $\dot{P}_0$ being the period and its derivative at pulsar birth and assuming the magnetic field of the pulsar does not decay, the spin down power at birth can be written as

$$L_{\rm sd,0} = \left| -\frac{4\pi^2 I \dot{P} P}{P_0^4} \right| . \tag{2}$$

It is easily seen from Eq. 2 that, given a small enough birth period, one can get an arbitrary large energy reservoir. 10 ms can be seen as a reasonable lower limit for $P_0$, while initial periods in the range ≈50–150 ms are not uncommon.[9] It is estimated that up to 40% of the pulsars may be born with periods in the range 100–500 ms.[10]

### 2.2. *Source Spectra*

Several models have been put forward to describe the acceleration of CR leptons by pulsars (see e.g[11] and references therein).

The simplest approach is to assume that some fraction $f_{\rm part}$ of the spin-down power $L_{\rm SD}(t)$ is transferred to particles, which follow a power-law spectrum with index $a$ and exponential cut-off. The acceleration of leptons at $\gamma$-ray pulsars has been modeled by a number of authors. Whereas[12] assume that the positron spectrum follows the $\gamma$-ray spectrum, it is assumed by[6] that mature pulsars with ages larger than 100 kyr inject their mono-energetic wind into the interstellar medium.

From Eq. 1 it is clear that the available spin-down power (and thus the available energy that can be transferred to particles) is largest when the pulsar is still young. In fact, integrating Eq.1 over time for $n = 3$, it becomes clear that at $t = \tau_0$ the pulsar dissipated half of its available rotational energy. One can therefore expect that the majority of particle acceleration at a given pulsar takes place for $t > \tau_0$. What is indeed observed are nebula of relativistic particles around many young pulsars, i.e. pulsar wind nebulae (PWN). There is also evidence, that these highly relativistic particles are confined in the PWN.[13]

Assuming the particles from the pulsar are reaccelerated at the pulsar wind shock and then contained in the PWN until its breakup,[5] model the particle spectrum injected into the interstellar medium by the two middle aged, nearby pulsars B0656+14 and Geminga. We generalize this model by assuming a broken power law particle spectrum at the pulsar shock[14]

$$Q'(E,t) = \begin{cases} k'(t)\,(E/E_B)^{-2} & \text{for } E > E_B \\ k'(t)\,(E/E_B)^{-1} & \text{for } E < E_B \end{cases} \tag{3}$$

keeping $E_B$ a free parameter. The spectrum at the shock is normalised by the available energy deposited into particles

$$\int_{E_{\min}}^{E_{\max}} Q'(E,t)E dE = f_{\text{part}} L_{\text{SD}}(t). \tag{4}$$

The fraction of the spin down power deposited in particles

$$f_{\text{part}} = \eta/(1+\sigma) \tag{5}$$

can be expressed in terms of the magnetisation parameter

$$\sigma(t) = 0.003\,(t/1\,\text{kyr})^{3/2}\,. \tag{6}$$

With $\epsilon = r_L/r_{\text{shock}} = 0.001 \ldots 0.1$, the maximum particle energy at shock is[14]

$$E_{\max} = \epsilon e \kappa \sqrt{\sigma/(\sigma+1)}\sqrt{L_{\text{SD}}/c}. \tag{7}$$

$\kappa = 3$ is the compression ratio at the shock and $e$ the elementary charge.

The evolution of the particle spectrum in the PWN is governed by synchrotron losses

$$\frac{\partial Q(E,t)}{\partial t} - Q'(E,t) = \frac{\partial}{\partial E}\left(B_{\text{PWN}}(t)^2 E^2 Q(E,t)\right) \tag{8}$$

where the mean PWN magnetic field is assumed to decay with time

$$B_{\text{PWN}}(t) = 1200/(1+t/\text{kyr})^2 \;\; [\mu\text{G}]. \tag{9}$$

Under these assumptions, the particle spectrum in the PWN at time of breakup $T$ is:

$$Q(E,T) = \int_0^T Q'(E_0,t_0)E_0^2 E^{-2}\Theta\left(E_0 - E_{\min}\right)\Theta\left(E_{\max} - E_0\right)dt_0, \tag{10}$$

with

$$E_0 = E/\left(E\int_t^{t_0} B_{\text{PWN}}(t')^2 dt' + 1\right)$$

## 3. Propagation of Cosmic Ray leptons

The propagation of CR leptons in the Galaxy is dominated by diffusion and energy losses due to synchrotron radiation in the Galactic magnetic field. The differential density $N$ satisfies the equation

$$\frac{\partial N}{\partial t} - S = \nabla \cdot (k \nabla N) - \frac{\partial}{\partial E}\left(b_0 E^2 N\right), \tag{11}$$

where $b_0$ is determined by the Galactic magnetic field.

In the case of a pulsar, the source function is given by

$$S = Q(E,t)\delta\left(\vec{r} - \vec{r}_{\text{source}}\right), \tag{12}$$

where $Q(E,t)$ is given by a model for particle acceleration at pulsars. Above $\approx 4\,\text{GV}$, the diffusion coefficient has an energy dependence of the form

$$k = k_0 \left(E/1 \text{ GeV}\right)^{\alpha}, \tag{13}$$

where $\alpha = 0.3 \ldots 0.7$ and $k_0$ are determined by fitting observed secondary to primary and radioactive secondary data. Depending on the model, $k_0$ is in the range $0.006\,\text{kpc}^2\text{Myr}^{-1}$–$0.2\,\text{kpc}^2\text{Myr}^{-1}$ (e.g.[15]).

Although obtaining $k_0$ in this way is a well established procedure, we want to remark the following. First, fitting secondary to primary (e.g B/C) data only yields a correlation between $k_0$ and the halo height $H$.[16] This degeneracy can be broken by looking at radioactive to stable secondary ratios (e.g. $^{10}$Be/Be). In doing so, one compares life times due to radioactive decay with the times scale of escape, the latter depends on the diffusion perpendicular to the Galactic disk. Thus the fitting of the observed chemical composition of CR by a propagation model measures primarily the propagation perpendicular to the Galactic disk, which not necessarily has to be the same as that in the Galactic plane. Second, to the best of our knowledge, all studies determining $k_0$ by fitting the observed chemical composition so far use 1D or 2D models. Nevertheless, it has been shown,[17] that the primary component of locally observed CR flux may change at least by a factor of two depending on the local SN history, given these objects are indeed the main sources of Galactic CR. The flux of the secondary component on the other hand shows little or no variation. Thus, the observed secondary to primary ratios will depend on the local SN history. It may be necessary to focus on tertiary to secondary ratios when working with leaky box, 1D or 2D propagation models.

## 4. Modelling the Contribution of Pulsars to the local Cosmic Ray Lepton Spectrum

If the source function of electrons and positrons accelerated at a given pulsar and also the distance to this object is known, one can calculate the contribution of these particles to the locally observed CR spectrum. As discussed above, the magnitude of the diffusion coefficient in the Galactic plane is not well determined. Also, the spectrum of the particles injected into the interstellar medium strongly depends on the model used.

Focusing on particle acceleration at young pulsars as, according to Eq. 1, these objects promise the largest output of high energy leptons, we follow the model outlined in the previous section. According to pulsar physics, we have to allow for a large range in parameter space. The pulsar birth period can be in the range $\approx 10\,\text{ms}$ to $500\,\text{ms}$, $E_B$, is most probably in the range 10–1000 GeV, and the PWN life time $T$ smaller 100 kyr. Within this parameter space in addition to the uncertainties in $k_0$, it is possible to produce a wide range of particle spectra at earth which can fit the data. In fact, looking at Geminga or B0656+14 individually, the available CR lepton data imposes constraints to the parameter space of our model.

## 5. Experimental Discrimination between Dark Matter and Pulsar origin

If indeed the observed increase in the positron fraction is due to the pulsars B0656 and Geminga, the fact that those two pulsars are located about 7 degree apart on the sky in the outer Galaxy, we expect an anisotropy towards the galactic anticenter, as discussed in.[5] However, for DM we expect either an isotropic sky, or an anisotropy towards the galactic center region.

In the diffusion model, the anisotropy in the CR flux is given by[18] $\delta = 3\,k\,|\nabla N|/(c\,N)$. For instantaneous injection at time $t_i$ and distance $r_i$ for a diffusion coefficient as given by Eq. 13, one arrives at

$$\delta = \frac{3}{2\,c} r_i\, b_0\, (\alpha - 1)\, E^{\alpha} \left( E^{\alpha-1} - (E/\left((t - t_i)\, b_0 E + 1\right))^{\alpha-1} \right)^{-1}, \quad (14)$$

[5] find an anisotropy of a few percent above 10 GeV. We remark that one expects for $E < 10\,\text{GeV}$ a solar cycle dependent anisotropy that is due to modulation of CR electrons and positrons.

## 6. Summary

The ATIC report of an excess just below 1 TeV resulted in a wide interest in the interpretation of the signal. This signal is also supported by the

PAMELA detection of an increase in the positron to electron ratio towards the energies where ATIC saw their excess above the general cosmic ray lepton (electron and positron) spectral background. Interestingly, this effect was not seen by the detector on the FERMI satellite,[3] nevertheless the latest FERMI results indicate a flattening of the CR electron spectrum above $\approx 50\,GeV$, hinting at an additional leptonic component with a hard spectrum.[3] If we assume that the ATIC peak is real, then we are left with two competing explanations: (i) Kaluza-Klein type particle, or, (ii) a local source(s) of electrons and the best candidate for the latter is a pulsar (with its PWN) since both the Kaluza-Klein and the pulsar interpretations require both electrons and positrons, probably in equal numbers.

Whereas the Kaluza-Klein explanation required the introduction of a boost factor of 200 in the volume production rate to explain the ATIC signal,[2] the pulsar/PWN interpretation also involves a number of free parameters, such as the birth period and interstellar diffusion coefficient between the source and earth. At the moment it seems as if we are left with too many free parameters and uncertainties to make an unambiguous statement about the origin of the ATIC (or at least the PAMELA) signal. More work needs to be done on DM candidate particles, as well as the multiwavelength nature of pulsars and their wind nebulae.

## References

1. O. Adriani, et al.,Nature 458, 607 (2009)
2. J. Chang, et al. Nature 456, 362 (2008)
3. M. Ackermann et al., arXiv 1008.3999 (2010)
4. F. Aharonian et al., A&A 508, 561 (2009)
5. I. Büsching, O. C. deJager, M. S. Potgieter, C. Venter, ApJ 678, 39 (2008)
6. L. Zhang, K. S. Cheng A&A 368, 1070 (2001)
7. I. Büsching, C. Venter, O. C. deJager, AdSpR 42, 497 (2008)
8. M. J. Rees, J. E. Gunn, MmRAS 167, 1 (1974)
9. C. Faucher-Gigure, V. M. Kaspi, ApJ 643, 332 (2006)
10. N. Vranesevic, et al., ApJ 617, 139 (2004)
11. S. Profumo, arXiv:0812.4457v2 (2009)
12. A. k. Harding, R. Ramaty, Proc. 20th ICRC, 2, 92 (1987)
13. O. C. deJager, S. E. S. Ferreira, A. Djannati-Atai, AIP Conference Proceedings, 1085, 199 (2008)
14. C. Venter, O. C. de Jager, MPE Report 291, 40 (2006)
15. D. Maurin, R. Taillet, F. Donato, A&A 394, 1039 (2002)
16. D. Maurin, F. Donato, R. Taillet, P. Salati, ApJ 555, 585 (2001)
17. I. Büsching, et al., ApJ 619, 314 (2005)
18. V. L. Ginzburg, S. I. Syrovatskii, *The Origin of Cosmic Rays*, New York: Macmillan, (1964)

# Highlights from the ARGO-YBJ Experiment

P. Camarri on behalf of the ARGO-YBJ collaboration

*University of Roma "Tor Vergata" and INFN Roma Tor Vergata*
*Via della Ricerca Scientifica, 1 00133 Roma, Italy*
**E-mail: Paolo.Camarri@roma2.infn.it*

The ARGO-YBJ experiment is an air]shower detector installed at high altitude (4300 m a.s.l.) at the Yangbajing Cosmic Ray Laboratory (Tibet, China). The detector is a full-coverage single layer of Resistive Plate Chambers (RPCs) operated in streamer mode, providing a single-hit time resolution less than 2 ns. The detector structure allows unprecedented imaging of the shower front (e.g. space-time distribution and thickness). The installation of the detector was completed in October 2007 and the experiment has been running uninterruptedly since then, collecting over $2 \times 10^{11}$ events generated by cosmic rays and gamma rays with lower energy threshold of a few hundreds of GeV, over a $\sim 2$ sr solid angle. The main research topics of the ARGO-YBJ experiment are the search for astronomical gamma-ray sources and cosmic-ray physics in the energy range from a few hundreds of GeV up to above 1 PeV. The main results obtained so far are presented, emphasizing the study of the TeV emission from the Crab Nebula and the Mrk 421 AGN, the search for antimatter in the primary cosmic radiation and the search for gamma-ray bursts with a ground-based detector.

*Keywords*: Style file; LaTeX; Proceedings; World Scientific Publishing.

## 1. Introduction

The requirement of lowering the detection energy threshold of primaries (and therefore to increase the detector sensitivity) in a ground-based experiment can be fulfilled by a full-coverage detector at high altitude. Resistive Plate Chambers (RPCs)[1] were chosen for this purpose due to their high time resolution (between 1 and 2 ns, definitely suitable for a precise reconstruction of an EAS front) and to their reasonable cost which makes them a perfect choice for a large-area detector. The ARGO-YBJ experiment[2] exploits the features of RPCs for EAS detection. ARGO-YBJ was designed to investigate a large number of topics in astrophysics and cosmic-ray physics:

- $\gamma$-ray astronomy (search for point-like sources above few hundreds

of GeV);
- search for VHE tails of $\gamma$-ray bursts above $\sim 1$ GeV;
- cosmic-ray physics;
- Sun and heliosphere physics.

In this contribution the main results of ARGO-YBJ in $\gamma$-ray astronomy will be outlined.

## 2. The ARGO-YBJ detector

The ARGO-YBJ experiment is installed at Yangbajing (P.R. of China), 4300 m a.s.l., longitude 90° 31' 50'' East, latitude 30° 06' 38'' North. The detector[3] is composed of 1560 RPCs ($2.80 \times 1.25$ m$^2$ each) in the central full-coverage region, plus 276 RPCs in the "guard ring" area close to the outer wall of the experimental buiding, as shown in Figure 1.

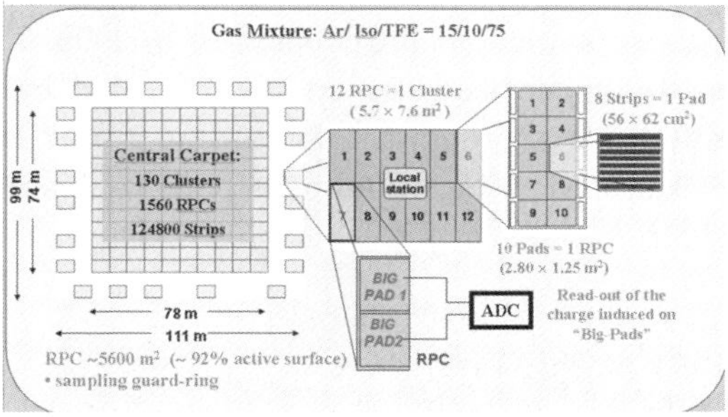


Fig. 1. Schematic drawing of the ARGO-YBJ experimental setup, with the details of the structure of one cluster and one pad.

Each of the ARGO-YBJ RPCs is composed of two 2-mm thick plates of plastic laminate with bulk resistivity of about $10^{12}$ $\Omega \cdot$ cm, enclosing a 2-mm thick gas gap. A grid of PET spacers is placed between the plates to fix the gap width over the whole chamber area. The electrical contacts for the high voltage are provided by thin graphite layers on the outer sides of the plates, protected by a PET foil. The read-out strip panel is placed on the ground face of the RPC. 80 copper strips ($6.5 \times 62$ cm$^2$ each) collect and transmit the signals generated by the passage of charged particles across the RPC gas gap. At the outer end of the strips, the front-end electronic boards[4] suitably amplify and shape the signals for subsequent processing. A "pad" is the basic acquisition unit of the ARGO-YBJ detector, and it provides the logical OR of the 8 strips. Figure 2 shows a reconstructed event, with the space pattern of the hits (left) and the space-time structure of the shower front.

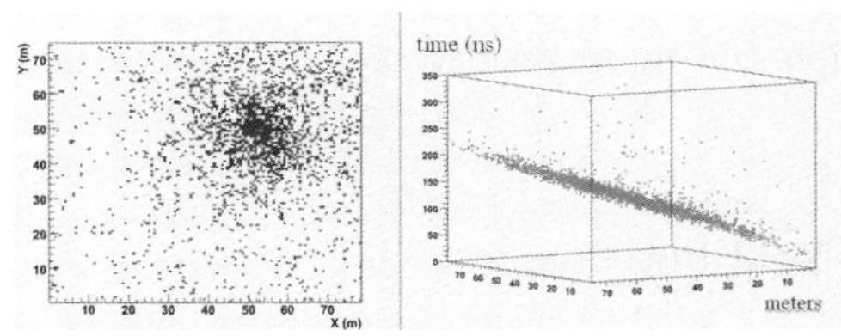


Fig. 2. ARGO-YBJ reconstructed event. Left: space map of the hit pads on the central carpet. Right: space-time structure of the shower front.

On the other side of the gas volume, two copper "big pads" are placed in order to collect the analog signal from the detector. The analog read-out of the ARGO-YBJ detector was put into operation in 2009. This information will be extremely useful for studying events generated by primaries with energy $\gtrsim$ 100 TeV, when the digital information from the read-out strip starts to become saturated and it would be hard to reconstruct the position of the air-shower core on the carpet.

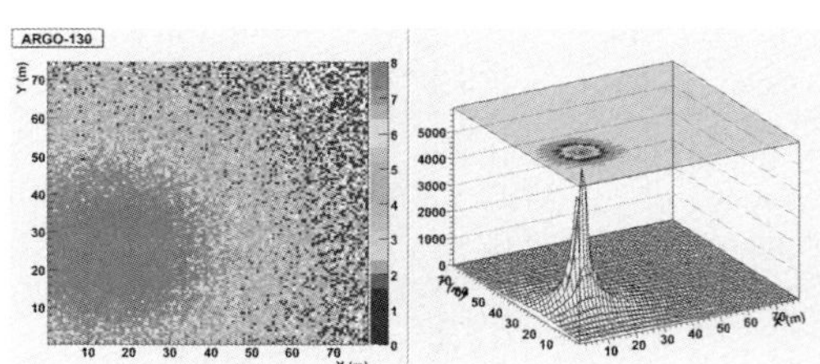


Fig. 3. Simulated ARGO-YBJ event for a proton with primary energy of 1 PeV. Left: space map of the hit pads on the detector. Right: bi-dimensional space profile for the response of the analog read-out.

Figure 3 shows a simulated event with primary energy of $10^{15}$ eV: the digital information (left) cannot easily identify the core position, while the analog information from the big pads (right) clearly does. With this technique it will be possible to extend the ARGO-YBJ energy range above 1 PeV.

ARGO-YBJ started taking data with its complete layout in October

2007. Since then it went on almost uninterruptedly with a duty cycle of 90% and trigger rate of 3.6 kHz.

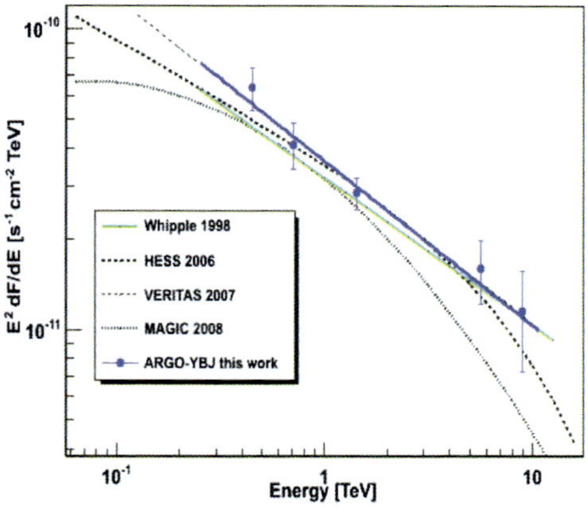


Fig. 4. Experimental energy spectrum for the $\gamma$ rays from the Crab Nebula in the range from about 100 GeV up to 10 TeV. The ARGO-YBJ experimental points are in agreement with the known results from HESS, MAGIC and Tibet AS-$\gamma$.

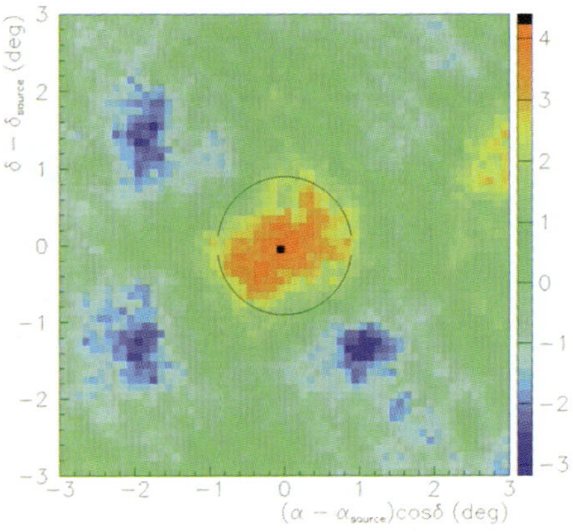


Fig. 5. ARGO-YBJ significance map of MRK421 during the flare of June 11-13, 2008. The maximum significance was 4.2 standard deviations.

## 3. $\gamma$-ray astronomy with ARGO-YBJ

The analysis of the ARGO-YBJ data in $\gamma$-ray astronomy in the first two years of running was mainly focused on the study of known sources, on the study of AGN flares, and on the limits on VHE fluence from $\gamma$-ray bursts. The background was evaluated by using two independent, widely established procedures: the "time-swapping" method and the "equi-zenith" method, which give equivalent results.

The $\gamma$-ray emission from the Crab Nebula was detected with a significance of about 14.5 standard deviations in 800 days. Figure 4 shows the experimental points measured by ARGO-YBJ on the Crab Nebula energy spectrum, in perfect agreement with the known experimental results obtained by the HESS, MAGIC, Whipple and Veritas experiments.

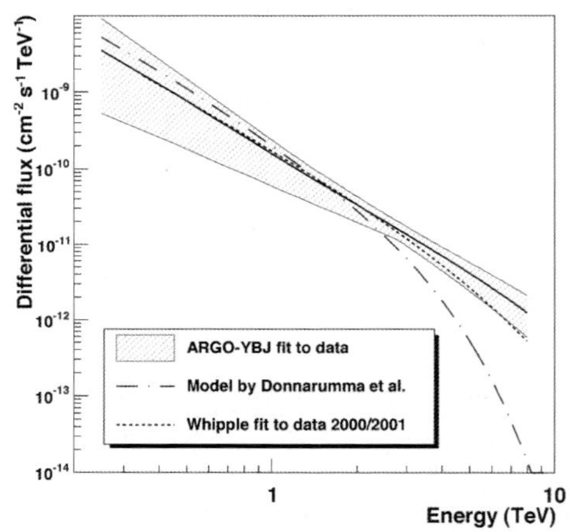


Fig. 6. Gamma-ray flux from Mrk421 measured by ARGO-YBJ on 2008, June 11-13 (solid line). The shaded band represents one standard deviation error. The dot-dashed line shows the flux according to the model by Donnarumma et al.[5] for the second flare (June 12-13). The dotted line shows the spectrum measured by Whipple[7] during a previous flare of similar intensity.

Two remarkable flares from the AGN MRK421 were observed in the X-ray range by the ASM X-ray telescope on June 4-6 and June 11-13, 2008.[5] ARGO-YBJ investigated possible gamma emission on those days from the same source.[6] In this analysis, the requirement for the selected events was a number of hit pads greater than 100. Figure 5 shows the significance map of the sky region around MRK 421 for the flare of June 11-13, 2008. The peak

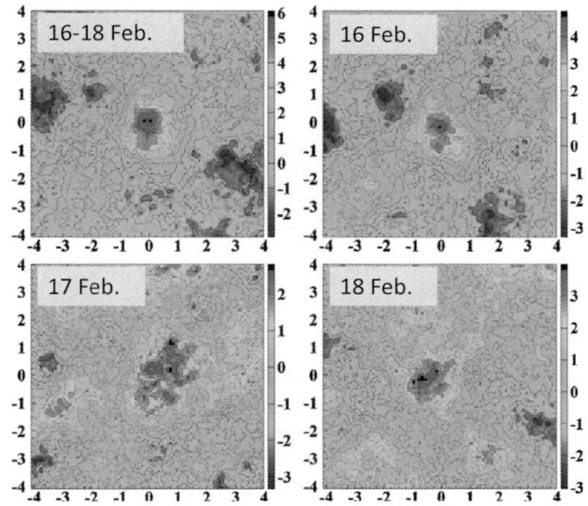


Fig. 7. ARGO-YBJ significance maps of MRK421 during the flare of February 16-18, 2010. The top-left plot is the global significance map for the whole duration of the flare. The remaining three plots refer to one single day each. This is the first measurement by a ground-based experiment with a 5 s.d. significance on a daily basis for a $\gamma$-ray flare.

significance for this flare was 4.2 standard deviations. Figure 6 shows the ARGO-YBJ experimental results for the MRK421 spectrum from the flare of June 11-13, 2008. The ARGO-YBJ data fully satisfy the relation between the spectral index and the flux resulting from the Whipple measurements of a similar flare of MRK421,[7] suggesting that this relation is an intrinsic property of the source.

A remarkable flare from MRK421 occurred on February 2010. ARGO-YBJ observed it between February 16th and February 18th at a 6 s.d. significance level. The measured flux exceeded 3 Crab units for the duration of the observation. The importance of this measurement by ARGO-YBJ lies in the fact that for the first time a ground-based experiment could detect a flare with a 5 s.d. significance level on a daily basis. Figure 7 shows the significance maps (right ascension vs. declination) related to this flare.

The $\gamma$-ray source MGRO J1908+06 was discovered by the Milagro experiment which reported a flux as high as 80% the Crab flux for it.[8] This measurement was confirmed by HESS[9] and VERITAS.[10] The Tibet AS-$\gamma$ experiment obtained a 4.4 s.d. significance level for this source.[11] The experimental energy spectra by Milagro and HESS do not agree with each other: the Milagro result is compatible with a power law corrected with an exponential cutoff, while the HESS spectrum can be described with a

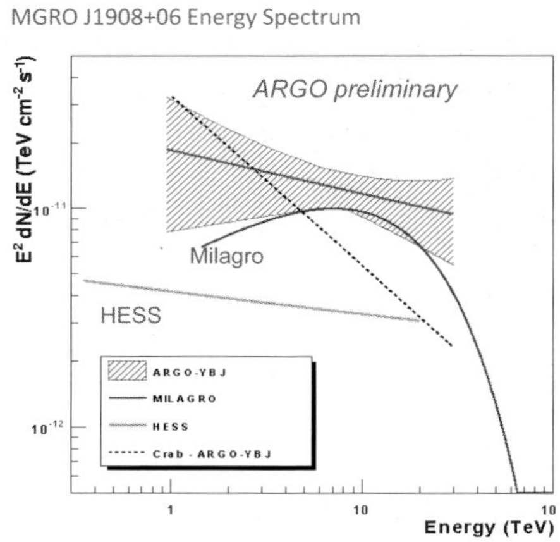


Fig. 8. Experimental energy spectrum for the $\gamma$ rays from the MGRO J1908+06 source in the range from about 300 GeV up to 50 TeV. The results of the ARGO-YBJ experiment are compared with the results obtained by Milagro and HESS for the same source. The shaded band represents the 1 s.d. uncertainty for the ARGO-YBJ spectrum.

simple power law. ARGO-YBJ detected this source in about 800 days with a 6 s.d. significance level and an experimental intrinsic angular extension $0.48^\circ \pm 0.28^\circ$. The ARGO-YBJ experimental spectrum is compatible with a power law with a spectral index 2.2 as shown in Figure 8. A direct comparison shows that the fluxes measured by ARGO-YBJ and HESS are not consistent with each other, while the ARGO-YBJ measured flux is consistent with the result obtained by Milagro at energies around 10 TeV. The ARGO-YBJ experimental spectrum of MGRO J1908+06 does not show any exponential energy cutoff.

At present ARGO-YBJ is performing a thorough full-sky search for more $\gamma$-ray sources.

## 4. Conclusions

The ARGO-YBJ experiment has been taking data with its complete layout since November 2007 with excellent performance, as confirmed by its first results in $\gamma$-ray astronomy:

- the observed energy spectrum of the Crab nebula is in great agreement with the results of other experiments;
- the recently-discovered source MGRO J1908+06 was observed with

a 6 s.d. significance level;

- a long-term monitoring of the MRK 421 AGN gave very interesting results and allowed the observation of $\gamma$-ray flares even when the Cherenkov telescopes could not operate;
- it was possible to put upper limit to the fluence from GRBs above 1 GeV, as reported by the ARGO-YBJ collaboration in two published papers.[12,13]

A detailed all-sky survey is going on, and studies to increase the sensitivity of the ARGO-YBJ analysis are in progress.

## References

1. R. Santonico, R. Cardarelli; NIM A 377 (1981) 187.
   R. Cardarelli, R. Santonico; NIM A 263 (1988) 200.
2. C. Bacci et al. (the ARGO-YBJ collaboration); Astroparticle Physics 17 (2002).
   G. Aielli et al. (the ARGO-YBJ collaboration); Astroparticle Physics 30 (2008) 85-95.
3. G. Aielli et al. (the ARGO-YBJ collaboration); NIM A 562 (2006) 92-96.
4. G. Aielli et al.; NIM A 409 (1998) 291-293.
5. I. Donnarumma et al.; ApJ 691 (2009) L13-L19.
6. G. Aielli et al. (the ARGO-YBJ collaboration); ApJ Letters 714 (2010) L208-L212.
7. F. Krennrich et al. (the WHIPPLE collaboration); ApJ 575 (2002) L9-L13.
8. A. A. Abdo et al. (the MILAGRO collaboration); ApJ 664 (2007) L91-L94.
9. F. Aharonian et al. (the HESS collaboration); Astron. Astrophys. 499 (2009) 723-728.
10. J. E. Ward on behalf of the VERITAS collaboration; arXiv:0810.0664v1 [astro-ph] 3 Oct 2008.
11. J. L. Zhang on behalf of the Tibet AS$\gamma$ collaboration; Proceedings of the 29th ICRC, Pune (India), 2005. Vol. 4, 93-96.
12. G. Aielli et al. (the ARGO-YBJ collaboration); ApJ 699 (2009) 1281-1287.
13. G. Aielli et al. (the ARGO-YBJ collaboration); Astroparticle Physics 32 (2009) 47-52.

# STATUS OF MAGIC AND RECENT RESULTS

A. DE ANGELIS

*INFN and INAF Trieste - University of Udine and LIP/IST Lisboa,*
*Udine I-33100, Italy*
**E-mail: alessandro.de.angelis@cern.ch*

V. SCALZOTTO

*INFN and University of Padova,*
*Padova I-35131, Italy*
**E-mail: villi.scalzotto@pd.infn.it*

Ground-based gamma-ray astronomy is part of a new field of fundamental research of Astroparticle Physics, that recently made spectacular discoveries. Nearly all important discoveries have been made by Imaging Air Cherenkov Telescopes (IACT) which allow one to detect Cherenkov light from extended air showers initiated by cosmic gamma-rays and charged cosmic particles. The MAGIC telescope is a new generation IACT located at La Palma, Canary Islands, Spain. Composed by two antennas with 17 m diameter each, MAGIC is equipped with the largest optical reflectors in the world, and it has the lowest threshold energy (25 GeV). MAGIC started operations in 2004 in the single-detector configuration, and in 2009 as a stereo detector. Since then, it has discovered many new sources and classes of sources, both galactic and extragalactic. Here some highlights from the most recent results are presented.

*Keywords*: Gamma-ray astrophysics.

## 1. Introduction

The Major Atmospheric Gamma Imaging Cherenkov (MAGIC) telescope is a last generation instrument for very high energy (VHE, conventionally above 30 GeV[1]) $\gamma$ -ray observations exploiting the Imaging Air Cherenkov technique. This kind of instrument images the Cherenkov light produced in the cascade initiated by a particle hitting the atmosphere.[1]

MAGIC (Fig. 1) is composed by two 17 m diameter telescopes located on the Roque de los Muchachos Observatory, in La Palma (Spain), and it is thus the largest optical reflector in the world. MAGIC incorporates a number of technological improvements in its design and achieves the lowest energy threshold (50 GeV with the standard trigger, 25 GeV with a special trigger, activated in particular conditions) among the instruments of its kind. The signal digitization utilizes 2GSample/s Flash ADCs.

The sensitivity of MAGIC is such that, at a flux peak energy of 300 GeV, a source as faint as of 0.8% of the Crab Nebula can be detected at $5\sigma$ in 50 hours of observations. The relative energy resolution above 200 GeV is below 20%; the angular resolution is about 0.08 degrees. MAGIC is also unique among Cherenkov telescopes by its capability to operate under moderate illumination (i.e., moon or twilight); this allows the increase of the duty cycle.

Operating as a single telescope since 2004, MAGIC has started stereo observations at the end of 2009, with the basic parameters summarized in the following Table (in comparison with the other large gamma-ray telescopes, H.E.S.S. in Namibia and VERITAS in Arizona).

| Instr. | Tels. # | Tel. A ($m^2$) | FoV (°) | Tot A ($m^2$) | Thresh. (TeV) | PSF (°) | Sens. (%Crab) |
|---|---|---|---|---|---|---|---|
| H.E.S.S. | 4 | 107 | 5 | 428 | 0.1 | 0.06 | 0.7 |
| MAGIC | 2 | 236 | 3.5 | 472 | 0.05(0.03) | 0.06 | 0.9 |
| VERITAS | 4 | 106 | 4 | 424 | 0.1 | 0.07 | 0.7 |

MAGIC has developed a physics program which includes topics of both fundamental physics and astrophysics. In this paper we highlight some of the MAGIC latest contributions to Galactic and extragalactic astrophysics, with emphasis on fundamental physics.

Fig. 1. The MAGIC telescopes.

## 2. Highlights of Galactic observations

In this section the latest MAGIC results on Galactic astrophysics are reviewed. MAGIC has discovered 4 new Galactic sources at VHE: MAGIC J0616+225,[2] positionally consistent with the supernova remnant IC 443; the first binary source for which periodical variability at VHE has been evidenced, LSI +61 303;[3] Cygnus X-1;[4] the pulsar from Crab.[5] They are all likely to indicate new classes of emitters. For some 20 other Galactic sources MAGIC has confirmed the discovery, performing flux and morphology measurements. Finally, MAGIC has published significant flux upper limits for interesting classes of sources, like, e.g, Wolf-Rayet binaries.[6]

### 2.1. *A shell-type Supernova Remnant with large impact on cosmic rays: MAGIC J0616+225*

MAGIC has discovered a new source of VHE $\gamma$-rays located close to the Galactic plane, which has been named MAGIC J0616+225;[2] this source is spatially coincident with the SNR IC 443. The measured energy ($E$) spectrum is well fitted by a power law with spectral index of $\Gamma = 3.1 \pm 0.3$. MAGIC J0616+225 is point-like for the MAGIC spatial resolution, and appears to be displaced from the center of the SNR shell. The observed VHE radiation may be due to $\pi^0$ decays from the interaction between cosmic rays accelerated in IC 443 and the dense molecular cloud.

Recent results from AGILE[7] and Fermi[8] support this hypothesis. In particular (Fig. 2), the centroid of emission by Fermi is significantly displaced from the centroid of emission from MAGIC (which, in turn, is consistent with a later measurement from VERITAS, and with a molecular cloud).

### 2.2. *Compact binaries: Cygnus X-1 and LS I +61 303*

Cygnus X-1 is the best established candidate for a stellar mass black-hole and one of the brightest X-ray sources in the sky. MAGIC has observed it for 40 hours along 26 different nights between June and November 2006. The MAGIC observations obtained evidence ($4.1\,\sigma$ post-trial significance) of a short-lived, intense flaring episode during 24th September 2006, in coincidence with a high flux observed in X-rays. The detected signal is point-like. The energy spectrum fits well extrapolations from Integral.

LS I +61 303 is a binary system containing a main-sequence star together with a compact object (neutron star or black hole), which displays periodic emission throughout the spectrum from radio to X-ray wavelengths. Observations with MAGIC have determined that this object produces $\gamma$-rays

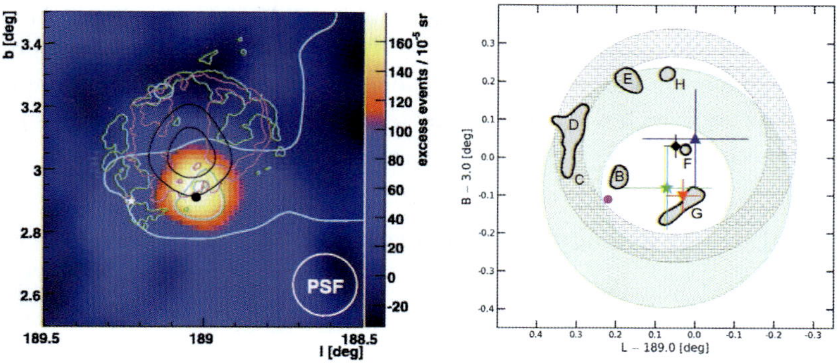


Fig. 2. Left: MAGIC skymap from IC443. Right: centroid of the emission from different gamma detectors; the position from Fermi is marked as a diamond (black), MAGIC as a downwards oriented triangle (red) and is consistent with the molecular cloud G.

up to at least $\sim$4 TeV,[3] and that the emission is periodically modulated by the orbital motion ($P_{\mathrm{TeV}} = (26.8 \pm 0.2)\,\mathrm{d}$).[9] The peak of the emission is found at orbital phases not coincident with the periastron.

In a multiwavelenght campaign involving radio (VLBA, e-EVN, MERLIN), X-ray (Chandra) and TeV (MAGIC) observations,[10] a hint of temporal correlation between the X-ray and TeV emissions and evidence for radio/TeV non-correlation were found.

### 2.3. *Crab Nebula and Pulsar*

The Crab Nebula is a SNR with a pulsar at its center. Crab is the standard candle for VHE astrophysics and as such, an important fraction of the observation time from MAGIC and from the other Cherenkov telescopes is devoted to this object. The VHE source appears point-like and its position is consistent with that of the pulsar.

Thanks to a special trigger setup, MAGIC has detected pulsed emission coming from the Crab pulsar above 25 GeV, with a statistical significance of $6.4\,\sigma$.[5] This result has revealed a relatively high energy cutoff, indicating that the emission occurs far out in the magnetosphere, hence excluding the polar-cap scenario as a plausible explanation for the high-energy origin. This is also the first time that a pulsed $\gamma$-ray emission is detected from a ground-based telescope, and opens the possibility of a detailed study of the pulsar's energy cutoff, which will help clarifying the mechanism of high energy radiation in pulsars.

The nonpulsed emission is also very relevant, and it extends up to some

100 TeV (at the limit of the sensitivity of the Cherenkov telescopes). The particular characteristics of MAGIC allow measuring the Crab energy spectrum in the continuum at low energies, overlapping with the Fermi satellite. This fact can permit cross-calibration[11] among the space-based and ground-based detectors, which will decrease the relative energy uncertainty to some 15% for the latter.

## 3. Highlights of extragalactic observations

MAGIC is the Cherenkov telescope with the lowest energy threshold (and, as a consequence, it can probe the most distant sources); it discovered 13 Active Galactic Nuclei at VHE: 1ES 1218+304[12] at $z = 0.182$, Markarian 180[13] at $z = 0.045$, BL Lacertae[14] at $z = 0.069$, 1ES 1011+496[15] at $z = 0.212$, PG 1553+113 [16] at $z > 0.09$ (co-discovered with H.E.S.S.), S5 0716+71[17] at $z = 0.31 \pm 0.08$, MAGIC J0223+430[18] at unknown redshift, possibly coincident with 3C66B, the near galaxy IC310,[19] 4C +21.35[20] at $z = 0.43$, MAGIC J2001+435[21] at unknown redshift, B3 2247+381[22] at $z = 0.12$, NGC 1275[23] at $z = 0.018$, and last but not least the farthest one, 3C 279,[24] at the unexpectedly large redshift of $z = 0.54$. IC310, 4C +21.35 and MAGIC J2001+435 have been detected in the recent months with the new stereo system. MAGIC has studied most of the AGN in its field of view, often detecting new physics features.

One of the main features of the emissions by AGN is variability. Such a variability can be very rapid, of the order of some 10 seconds or so (the typical size of the central black hole can be 100 to 1000 times larger).

### 3.1. *Blazars detected during optical outbursts*

Observations triggered by high optical states of potential VHE blazars have proven very successful with the detection of Mkn 180, 1ES 1011+496, 3C 279, and recently S5 0716+71. The observation of 1ES 1011+496 was triggered by an optical outburst in March 2007, resulting in a 6.2 $\sigma$ detection[15] of a flux $(1.58 \pm 0.32) \times 10^{-11}\mathrm{cm}^{-2}\mathrm{s}^{-1}$ above 200 GeV. An indication for a long-term optical–VHE correlation comes from the fact that in spring 2007 the VHE $\gamma$-ray flux was about 40% higher than in spring 2006, when MAGIC had observed this blazar without finding a significant signal.

In April 2008, the optical telescope supporting MAGIC, KVA, reported a bright optical state of the blazar S5 0716+71, triggering VHE observations which resulted in a 6.8 $\sigma$ detection, corresponding to a flux $\approx 10^{-11}\mathrm{cm}^{-2}\mathrm{s}^{-1}$ above 400 GeV. The source was also in a high X-ray state.

### 3.2. *Multiwavelength campaigns*

For a better understanding of the emission mechanisms of AGN, simultaneous multiwavelength (MW) observations are essential. MW observations allow the determination of spectral energy distributions (SEDs) spanning over 15 orders of magnitude in energy. MAGIC recently participated in MW campaigns carried out on several blazars, including a regular monitoring of Mkn 421 and Mkn 501. These campaigns involved the X-ray satellites INTEGRAL, Suzaku and Swift, the $\gamma$-ray telescopes H.E.S.S., VERITAS, and MAGIC, and other optical and radio telescopes.[6]

Among the main results, a 120-hour observational campaign on the giant radio galaxy M87 in spring 2008 detected two major outbursts of gamma-ray emission, while simultaneous high resolution radio observations of the activity of this radio galaxy were performed. The increase of the radio flux could be located from the innermost core of M87, associated with the immediate vicinity of the central black hole. The combined observation pins down the location of activity during the gamma-ray outbursts and thus the site of the particle accelerator.[25]

## 4. Impact of the MAGIC results on fundamental physics

MAGIC has set upper limits on possible signals from the annihilation of Dark Matter particles[26] in channels involving high-energy photons, for the Galactic Center and for the dwarf Spheroidals Draco and Willman I.

The imaging and the study of the SEDs from Galactic objects by MAGIC and by H.E.S.S. and VERITAS indicate that Galactic sources (PWN in particular) are likely to generate charged cosmic rays up to some $10^{15}$ eV, as discussed for IC443.

We summarize two additional subjects investigated thanks to the study of AGN. MAGIC is the best instrument for the discovery of sources far away, and it dominates the scene for $z > 0.2$.

### 4.1. *Does the speed of light depend on the photon energy?*

The constancy of the speed of light, $c$, is one of the postulates of the theory of relativity; it has been extensively tested in the recent years against a possible dependence on the photon energy $E$, which is predicted, for example, by theories inspired by Quantum Gravity. At first order, such a dependence is usually parametrized[27] as

$$c' = c\,[1 \pm (E/E_{s1})] \ . \tag{1}$$

The energy scale $E_{s1}$ which parametrizes the scale of violation of relativity in the above expression is usually expressed in units of the Planck mass, $M_P \simeq 1.22 \times 10^{19}$ GeV.

A favored way to search for such a dispersion relation is to compare the arrival times of photons of different energies arriving on Earth from pulses of distant astrophysical sources. The greatest sensitivities may be expected from sources with short pulses, at large redshifts $z$, of photons observed over a large range of energies: $\gamma$-ray bursts (GRBs) and AGN.

MAGIC[28] has published the results of the analysis of a giant flare of the Mrk 501 blazar, at $z = 0.034$. A correlation has been observed (with a probability of 2.5% of coming from a statistical fluctuation) between the photon energy and the arrival time. The correlation is *subluminal*, i.e., the minus sign holds in Eq.(1) (higher energy photons are delayed) with a coefficient $dt/dE = (0.030 \pm 0.012)$ s/GeV. Later, the Fermi/LAT detector has registered the most energetic photon, at $E \sim 13.2$ GeV, from the GRB080916C[29] at $z$ 4.34, 16.5 s after the primary burst. The delays from the GRB080916C are consistent within errors with the same energy scale,[30] although the sources are very different as their distance is.

However, the recent results by Fermi[31] do not fit in the exciting picture outlined above. The most conservative conclusion is that before making any statement we need to understand better the time/energy correlation of the photon emission at the source. Anyway, we the sensitivity of Cherenkov telescopes is such that we can detect effects at the Planck mass scale.

### 4.2. *Anomalies in the propagation of gamma-rays?*

The horizon of the observable Universe is expected to rapidly shrink in the VHE band as the energy increases, due to the scattering of VHE photons off background photons permeating the Universe, which produces electron-positron pairs.[32] The corresponding cross section $\sigma(\gamma\gamma \to e^+e^-)$ peaks where the VHE photon energy $E$ and the background photon energy $\epsilon$ are related by $\epsilon \simeq (500\,\mathrm{GeV}/E)\,\mathrm{eV}$. Therefore, for observations performed by Cherenkov Telescopes the resulting cosmic opacity is dominated by the interaction with ultraviolet/optical/infrared diffuse background photons, usually called Extragalactic Background Light (EBL), which is produced by galaxies during the whole history of the Universe.

The absorption is usually parametrized as $e^{-\tau(E,z)}$. The mean free path of VHE photons decreases like a power law from the Hubble radius 4.2 Gpc around 100 GeV to 1 Mpc around 100 TeV.[33] Thus, the observed flux is exponentially suppressed both at high energy and at large distances, so

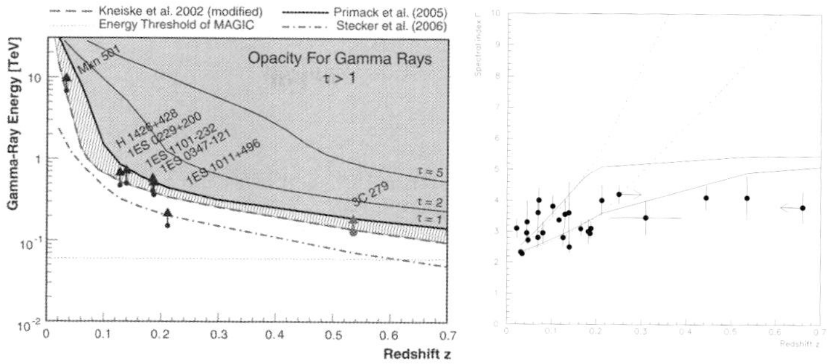


Fig. 3. Left: Gamma-ray horizon, from,[24] compared to EBL models. The dashed blue line represents an upper EBL limit based on the MAGIC 3C 279 observation. Limits obtained for other sources are shown by black arrows, most of which lie very close to the model. The narrow blue band is the region allowed between this model and a maximum possible transparency (i.e., minimum EBL level), which comes from galaxy counts. The gray area indicates an optical depth $\tau > 1$. The curves for $\tau = 2$ and $\tau = 5$ are also shown. Right: Spectral index $\Gamma_{\rm obs}$ for blazars observed in the energy band $0.2\,\text{TeV} < E_0 < 2\,\text{TeV}$ with known redshift. See ref.[34] for references. For a constant spectral index $\Gamma = 2.3$ at the emitter, the predictions from the DaRMa model are the band between the solid lines; the standard propagation predicts the band between the dashed lines.[35]

that sufficiently far away sources become hardly visible in the VHE range and their observed spectrum should be steeper than the emitted one.

Observations have not detected such a behaviour. Strong evidence comes from the observation of 3C279 (Fig. 3, left), which would be expected to be attenuated by a factor $e^{-3}$ at the maximum energies detected of about 400 GeV. Another evidence comes from the fact that one does not observe a steepening of the SED of blazars with increasing distances (Fig. 3, right).

A possible explanation[35] requires that photons oscillate into a new Axion-Like Parlicle (ALP) and vice-versa in the presence of cosmic magnetic fields (the so-called "DaRMa scenario"); once ALPs are produced close enough to the source, they travel unimpeded throughout the Universe and can convert back to photons before reaching the Earth. Since ALPs do not undergo EBL absorption, the effective photon mean free path is increased. An ALP of mass $\ll 10^{-10}$ eV could explain the observed results.

## 5. Conclusions

MAGIC has discovered many new sources and classes of sources, both galactic and extragalactic. Thanks to its low energy threshold, the lowest among Cherenkov telescopes, it can contribute to many new discoveries, also in the field of fundamental physics. The upgrade to stereoscopy is guaranteeing since end 2009 a further improvement in discovery potential.

## References


1. For a review on the Cherenkov technique and experiments see for example A. De Angelis, O. Mansutti and M. Persic, *Riv. Nuovo Cim.* **31** (2008) 187; J. Hinton and W. Hofmann, *Ann. Rev. of Astron. Astrophysics* **47** (2009) 523.
2. J. Albert et al. (MAGIC Coll.), *Astrophys. J.* **664** (2007) L87.
3. J. Albert et al. (MAGIC Coll.), *Science* **312** (2006) 1771.
4. J. Albert et al. (MAGIC Coll.), *Astrophys. J.* **665** (2007) L51.
5. E. Aliu et al. (MAGIC Coll.), *Science* **322** (2008) 1221.
6. For more details on Galactic observations and MW campaigns see J. Rico and R. Wagner, arXiv:0905.3043 and references therein.
7. AGILE Coll., arXiv:1001.5150 (January 2010), to appear in ApJ.
8. Fermi Coll., arXiv:1002.2198 (February 2010), to appear in ApJ.
9. J. Albert et al. (MAGIC Coll.) *Astrophys. J.* **693** (2009) 303.
10. J. Albert et al. (MAGIC Coll.), *Astrophys. J.* **684** (2008) 1351; H. Anderhub et al. (MAGIC Coll.), *Astrophys. J.* **706** (2009) L27.
11. D. Bastieri et al., *Astropart. Phys.* **23** (2005) 572.
12. J. Albert et al. (MAGIC Coll.), *Astrophys. J.* **642** (2006) L119.
13. J. Albert et al. (MAGIC Coll.), *Astrophys. J.* **648** (2006) L105.
14. J. Albert et al. (MAGIC Coll.), *Astrophys. J.* **666** (2007) L17.
15. J. Albert et al. (MAGIC Coll.), *Astrophys. J.* **667** (2007) L21.
16. J. Albert et al. (MAGIC Coll.), *Astrophys. J.* **654** (2007) L119.
17. M. Teshima et al. (MAGIC Coll.), ATel 1500 (2008) 1.
18. E. Aliu et al. (MAGIC Coll.), *Astrophys. J.* **692** (2009) L29.
19. MAGIC Coll., arXiv:1009.2155 (September 2010), submitted to ApJ.
20. MAGIC Coll., ATel 2684 (June 2010)
21. MAGIC Coll., ATel 2753 (July 2010)
22. MAGIC Coll., ATel 2910 (October 2010)
23. MAGIC Coll., ATel 2916 (October 2010)
24. J. Albert et al. (MAGIC Coll.), *Science* **320** (2008) 1752.
25. V.A. Acciari et al., *Science* **325** (2009) 444.
26. S. Lombardi for the MAGIC Collaboration, Contr. XXXI ICRC, Lodz 2009.
27. G. Amelino-Camelia et al., *Nature* **393** (1998) 763.
28. J. Albert et al. (MAGIC Coll.), *Phys. Lett.* **B668** (2008) 253.
29. A.A. Abdo et al. (Fermi Coll.), *Science* **323** (2009) 1688.
30. J. Ellis et al., *Phys. Lett.* **B674** (2009) 83.
31. A.A. Abdo et al. (Fermi Coll.), *Nature* **462** (2009) 331.
32. G.G. Fazio and F.W. Stecker, *Nature* **226** 135 (1970).
33. P. Coppi and F. Aharonian, *Astrophys. J.* **487** (1997) L9.
34. http://www.mppmu.mpg.de/rwagner/sources/ and references therein.
35. A. De Angelis, M. Roncadelli and O. Mansutti, *Phys. Rev.* **D76**(2007) 121301; A. De Angelis et al., *Mon. Not. R. Astron. Soc.* **394** (2009) L21 (updated in M. Roncadelli et al., Proc. SciNeGHE 2009).

# Recent H.E.S.S. results

B. Degrange* for the H.E.S.S. collaboration

*Laboratoire Leprince-Ringuet, École Polytechnique and CNRS/IN2P3, F 91128 Palaiseau Cedex, France*
**E-mail: degrange@llr.in2p3.fr*

B. Degrange

Since 2003, the H.E.S.S. array of 4 Cherenkov telescopes, located in Namibia, has detected about 60 sources of very-high-energy gamma-rays from 100 GeV to about 30 TeV, both galactic and extragalactic. Some recent H.E.S.S. results are reviewed: the detections of shell-type supernova remnants (SN 1006 and HESS J1731-347), of the starburst galaxy NGC 253 and of new Active Galactic Nuclei. In addition, deeper variability studies on the blazar PKS 2155-304 are presented, which give some insight into the emission mechanisms at play.

*Keywords*: Gamma-ray Astronomy; Imaging Atmospheric Cherenkov Telescopes; H.E.S.S.; Supernova Remnants; Starburst galaxy; Radio-galaxies; Active Galactic Nuclei; Blazars.

## 1. Introduction

H.E.S.S. (for High Energy Stereoscopic System) is an array of 4 Cherenkov imaging telescopes devoted to Gamma-Ray Astronomy above 100 GeV up to $\sim$ 100 TeV.[1] Located in the Khomas Highlands in Namibia at an altitude of 1800 m above sea level, H.E.S.S. has been continuously observing the southern gamma-ray sky since 2003. Cherenkov light from atmospheric cascades is collected by four 13 m diameter telescopes and analyzed in the focal planes by fine grain-cameras comprising 960 pixels each and covering a large field-of-view (5° diameter). Whereas the experimental setup provides a huge effective detection area ($\sim 10^5$ m$^2$), stereoscopy combined with fine-grain imaging allows for efficient selection and reconstruction of gamma-ray-induced showers, recently improved by advanced analysis methods.[2] The angular resolution is always better than 0.1°. The energy resolution is about 15% and the energy threshold varies from 120 GeV at zenith to about 700 GeV at a zenith angle of 60°. The sensitivity achieved by H.E.S.S. for a point-like source at zenith corresponds to 1% of the flux of the Crab nebula for a $5\sigma$ detection in 25 hours. The H.E.S.S. catalogue presently includes about 60 objects. In this review, a selection of recent results from H.E.S.S. is presented, first on sources in the Milky Way and in nearby galaxies, then on Active Galactic Nuclei (AGN).

## 2. Galactic sources: shell-type supernova remnants

Most of galactic sources in the H.E.S.S. catalogue have been discovered in the Galactic Plane Survey[3] which covers the interval $[-3°, +3°]$ in galactic latitude and [-85°,60°] in galactic longitude. Unlike at sub-GeV energies, the galactic diffuse gamma-ray flux is not dominant and individual gamma-ray sources stand out clearly. Supernova remnants have long been suspected to

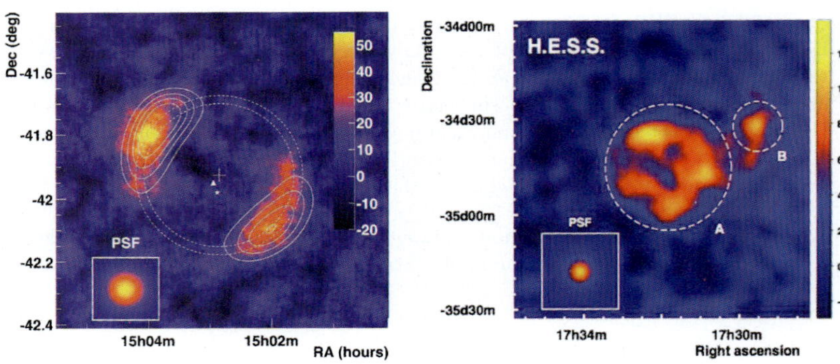


Fig. 1. Left : H.E.S.S. $\gamma$-ray image of SN 1006 (excess count map). The white contours correspond to a constant X-ray intensity as derived from the XMM-Newton flux map and smoothed to the H.E.S.S. point spread function, enclosing respectively 80%, 60%, 40% and 20% of the flux. Right : H.E.S.S. $\gamma$-ray image of the new shell HESS J 1731-347.

be the main accelerators of galactic cosmic rays by means of the diffusive shock acceleration process.[4] Some of these objects with angular radii on the order of 1° have been resolved by H.E.S.S., namely RX J1713-3946,[5] Vela Junior[6] and RCW 86.[7] Two new shells have now been detected by H.E.S.S., namely the remnant of the historical supernova SN 1006 and the formerly unidentified source HESS J1731-347. The first one was particularly interesting on theoretical grounds since the explosion took place in a relatively homogeneous low-density environment, above the Galactic Plane. X-ray observations from Chandra[8] and XMM-Newton[9] revealed a simple bipolar morphology with roughly symmetrical limbs, supporting the conclusion of a relatively uniform magnetic field oriented in the NE-SW direction. On the other hand, the gamma-ray flux from this source turned out to be on the order of 1% of that of the Crab nebula and 130 hours of observation by H.E.S.S. were necessary to yield the first morphological and spectral analysis of SN 1006 in very-high-energy (VHE) $\gamma$-rays.[10] Significant $\gamma$-ray emission is detected in two extended regions as shown in Fig. 1 (left). The most external white contour in Fig. 1 (left), obtained by smoothing the X-ray map from XMM-Newton according to the H.E.S.S. point spread function, corresponds to 80% of the X-ray flux. Within this contour, the significance of the $\gamma$-ray signal amounts to 9.3 $\sigma$ for the NE region and to 8.7 $\sigma$ for the SW one. Radial and azimuthal profiles obtained from H.E.S.S. data very well match the corresponding X-ray profiles from XMM-Newton data (2.0-4.5 keV band) smoothed to the H.E.S.S. point spread function. Therefore, particle acceleration in the narrow X-ray filaments is likely to be at the origin of TeV $\gamma$-ray emission. The differential spectra in both regions are well described by power laws $\Phi(E) \propto E^{-\Gamma}$ with comparable photon indices ($\Gamma_{NE} = 2.35 \pm 0.14_{\rm stat} \pm 0.2_{\rm syst}$ and $\Gamma_{SW} = 2.29 \pm 0.18_{\rm stat} \pm 0.2_{\rm syst}$). The broad-band spectrum from radio waves to TeV $\gamma$-rays can be accounted for by inverse Compton emission of accelerated electrons on ambient photons, assuming a magnetic field of about 30 $\mu$G, but a mixed scenario including leptonic and hadronic processes (i.e. p+p $\rightarrow \pi^0 + ...$ followed by $\pi^0 \rightarrow \gamma\gamma$) also leads to a satisfactory description, assuming a high proton-acceleration efficiency. The second new shell, HESS J1731-347, was found as an unidentified source in the H.E.S.S. Galactic Plane Survey[11] in 2008. Shortly after the publication, Tian et al.[12] found a faint non-thermal shell G353.6-0.7, both in radio and in X-rays, with a diameter of 0.4°, coinciding with HESS J1731-347. Deeper H.E.S.S. observations (60 hours of good quality data) actually reveal the shell in VHE $\gamma$-rays[13] with a significance of 22 $\sigma$ (Fig. 1, right). From the radial profile, a filled-sphere morphology can be excluded at the 3.9 $\sigma$ level. The energy spectrum of the source is found to be compatible with a power law with a photon index of $2.32 \pm 0.06$ and a $\gamma$-ray flux (1-10 TeV) of about 9% of that of the Crab nebula. A nearby secondary source (region B in Fig. 1 right) could be due to high-energy particles interacting within a nearby molecular cloud.

## 3. A nearby starburst galaxy: NGC 253

Particle acceleration can also be due to collective effects of supernova explosions in rich star-forming regions, such as the cores of starburst galaxies. NGC 253, one of the nearest spiral galaxy ($D \sim 3$ Mpc), is characterized by a very high rate of star formation in its central region and a strong production of mechanical energy through supernova explosions. In this region, the average density of the interstellar gas is estimated to be $\sim 600$ protons per $cm^3$, i.e. about 1000 times higher than in the Milky Way. The detection of NGC 253 by H.E.S.S. ($5.2\ \sigma$) required 119 hours of observation.[14] The source is found to be point-like, showing that the emission comes from the central region only. The flux corresponds to 0.3% of the Crab flux. This detection implies a high density of cosmic rays above 1.3 TeV ($4.9 \times 10^{-12}$ $cm^{-3}$), i.e. 2000 times greater than in our Galaxy.

## 4. Active Galactic Nuclei

Among the 15 Active Galactic Nuclei detected by H.E.S.S., 2 are radio-galaxies (M87 and Centaurus A), the other ones belonging to the "blazar" class. The day-scale variability of M87 ($D \sim 16.7$ Mpc) found in former H.E.S.S. observations implied that the $\gamma$-ray emission could only originate from the central core or from the innermost resolved knot HST1 in the jet. In 2008, radio (VLBA), X-ray (Chandra) and VHE $\gamma$-ray observations (VERITAS, H.E.S.S. and MAGIC) of M87 during a period of strong $\gamma$-ray flares[15] showed that HST1 was an unlikely site for the source activity. Centaurus A, the nearest radio-galaxy ($D \sim 3.7$ Mpc), has been detected by H.E.S.S. at the $5\ \sigma$ significance level after 115 hours of observation.[16] The source is found to be point-like and compatible with the radio core of the galaxy. The flux corresponds to about 0.8% of that of the Crab nebula. The differential energy spectrum is well described by a power law with a photon index of $\Gamma = 2.7 \pm 0.5_{\rm stat} \pm 0.2_{\rm syst}$. Blazars detected at very high energies generally belong to the subclass of high-frequency-peaked BL Lac objects (HBL), whose extreme properties are explained by relativistic beaming, the emission taking place within a jet directed at small angle with respect to the line of sight. Blazars emit over the whole electromagnetic spectrum from the radio band to VHE $\gamma$-rays and their broad-band spectrum can be determined by multi-wavelength campaigns such as that carried out in 2008 on PKS 2155-304[17] in the optical (ATOM), X-ray (Rossi XTE and Swift), GeV (Fermi-LAT) and VHE (H.E.S.S.) bands. All blazars from the H.E.S.S. catalogue are also detected at lower $\gamma$-ray energies by Fermi-LAT but 2, namely 1ES 0229+200[18] and the new H.E.S.S. source PKS 0548-322,[19] whose integral flux above 200 GeV is $\sim 1.3\%$ of that of the Crab nebula. In addition, 3 blazars have been recently detected with fluxes on the order of 1%-2% of the Crab flux. Two of them, 1ES 0414+009[20] and PKS 0447-439[21] belong to the HBL class. Due to their large redshifts, their emission is strongly absorbed by the extragalactic background light, resulting in steep spectra. The third new blazar, AP Libræ,[22] is a low-frequency-peaked BL Lac (LBL). Blazars are characterized by variable emissions at

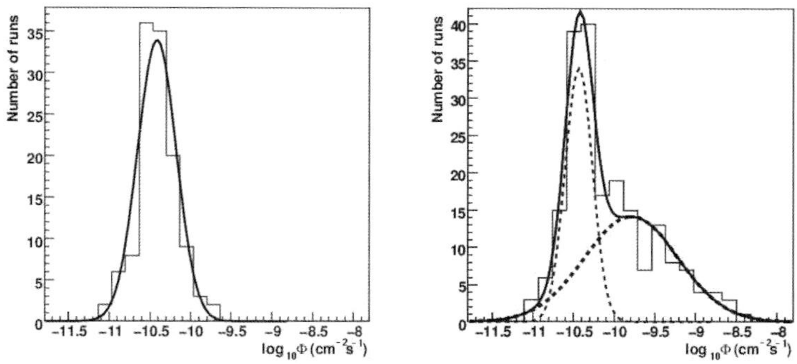


Fig. 2. Distributions of the logarithms of the integral $\gamma$-ray fluxes above 200 GeV from PKS 2155-304. Left: from 2005 to 2007 except the July 2006 period. Right: all runs from 2005 to 2007.

all wavelengths. An extreme case is that of PKS 2155-304 which, in July 2006, exhibited spectacular flares in VHE $\gamma$-rays with doubling time-scales on the order of a few minutes.[23] If this short flaring period is excluded, the distribution of the measured $\gamma$-ray flux above 200 GeV over the years 2005-2007 is very close to a log-normal one (Fig. 2 left). If the flaring period is included, the distribution of the logarithm of the flux can be accounted for by the superposition of two Gaussian distributions[24] (Fig. 2 right). Furthermore, during the flaring period, a high degree of correlation was found between the flux fluctuation (i.e. the flux r.m.s. corrected for the effect of measurement errors) and the mean flux, as measured over light-curve segments of 20 minutes sampled every minute. The preceding results suggest that the variability of PKS 2155-304 in July 2006 can be described by a random stationary lognormal process. A complete analysis of the flare data[24] shows that the corresponding Fourier power spectrum in the frequency interval $10^{-4}$ Hz to $10^{-2}$ Hz is close to that of a red noise, i.e. a power law of the frequency $P(\nu) \propto \nu^{-\alpha}$ with $\alpha = 2.06 \pm 0.21$, very comparable to those measured on PKS 2155-304 and similar blazars in the X-ray domain.[25]

## 5. Conclusion

This short review could only account for a small part of recent H.E.S.S. results. In 2012, the H.E.S.S. setup will be completed by a very large telescope (28 m diameter) in the centre of the present array. An energy threshold of 30 GeV is expected from this large telescope in a stand-alone mode. Stereoscopy including the very large telescope and at least one of the other four should allow for a threshold of 80 GeV. This new facility will provide a significant overlap in the energy ranges covered by $\gamma$-ray satellites (Fermi-LAT and AGILE) and H.E.S.S.

## References


1. F. Aharonian, et al., *A&A* **457**, 899 (2006).
2. M. de Naurois and L. Rolland, *Astropart. Phys.* **32**, 231 (2009).
3. R. C. G. Chaves, et al., in Proc. of the 4$^{th}$ International Meeting on High Energy Gamma-Ray Astronomy, Heidelberg (Germany), 2008, AIP Conference Proceedings **1085**, 219 (2008).
4. E. G. Berezhko and H. J. Voelk, *ApJ* **661**, L175 (2007).
5. F. Aharonian, et al., *A&A* **464**, 235 (2007).
6. F. Aharonian, et al., *ApJ* **661**, 236 (2007).
7. F. Aharonian, et al., *ApJ* **692**, 1500 (2009).
8. A. Bamba, et al., *ApJ* **589**, 827 (2003).
9. R. Rothenflug, et al., *A&A* **425**, 121 (2004).
10. F. Acero, et al., *A&A* **516**, A62 (2010).
11. F. Aharonian, et al., *A&A* **477**, 353 (2008).
12. W. Tian, et al., *ApJ* **679**, L85 (2008).
13. F. Acero, et al. for the H.E.S.S. collaboration, in Proc. of 38$^{th}$ COSPAR Scientific Assembly, Bremen (Germany, July 2010.
14. F. Acero, et al., *Science* **326**, 1080 (2009).
15. The VERITAS Collaboration, the VLBA 43 GHz M87 Monitoring Team, the H.E.S.S. Collaboration, the MAGIC Collaboration, *Science* **325**, 444 (2009).
16. F. Aharonian, et al., *ApJ* **695**, L40 (2009).
17. F. Aharonian, et al., *ApJ* **696**, L150 (2009).
18. F. Aharonian, et al., *A&A* **475**, L9 (2007).
19. F. Aharonian, et al., arXiv:1006.4289, accepted for publication in A&A.
20. W. Hofmann on behalf of the H.E.S.S. collaboration and S. Fegan on behalf of the Fermi-LAT collaboration, ATel #2293 on 12 Nov. 2009.
21. M. Raue, et al. for the H.E.S.S. collaboration, ATel #2350 on 16 Dec. 2009.
22. W. Hofmann for the H.E.S.S. collaboration, ATel #2743 on 17 Jul. 2010.
23. F. Aharonian, et al., *ApJ* **664**, L71 (2007).
24. A. Abramowski, et al., arXiv:1005.3702, accepted for publication in A&A.
25. J. Kataoka, et al., *ApJ* **560**, 659 (2001).

# ATMOSPHERIC EVALUATION WITH LIDAR FOR MAGIC

C. FRUCK*, J. HOSE, R. MIRZOYAN, M. TESHIMA

*MAGIC Gruppe, Max Plank Institut für Physik,*
*Föhringer Ring 6, 80805, Munich, Germany*
**E-mail: cfruck@ph.tum.de*
*www.mppmu.mpg.de*

We are operating a micro LIDAR system for lowering the influence of atmospheric conditions on the results of the 17m diameter MAGIC Imaging Atmospheric Cherenkov Telescope (IACT) for ground-based astrophysics by means of very high energy (VHE) gamma rays. The laser of our LIDAR operates at a pulse energy of $\sim 5\mu$J at ca. 1kHz repetition rate for not disturbing the operation of the other nearby Telescopes on the European North Observatory at Roque de los Muchachos on Canary island of La Palma. The low light intensity reflected from the atmosphere requires a very sensitive light detector, capable of resolving and counting single photons. A high quantum efficiency (QE) Hybrid Photon Detector (HPD) with GaAsP photo cathode in the focus of a 60cm diameter mirror does this.

*Keywords*: VHE Gamma-Ray Astronomy; Imaging Airshower Cherenkov Technique; Atmospheric Influence; LIDAR

## 1. Introduction

The Imaging Air-shower Cherenkov Technique (IACT) became a powerful observation method for doing VHE astrophysics by means of $\gamma$-rays. The two 17m MAGIC (Major Atmospheric Gamma-ray Imaging Cherenkov) telescopes, located in 2200m on the Roque de los Muchachos on the Canary island La Palma, are the biggest instruments in the world, that apply this technique.[1] Using the earth's atmosphere as a calorimeter for cosmic rays has the advantage of having a very big effective detection area. This way it is possible to collect good statistics in the 100GeV to 100TeV regime, where the fluxes of most sources are getting too low for space-borne experiments. However, this ground-based method, even if used in geographic locations with good meteorologic conditions, suffers from uncertainties due to varying transmission of the Cherenkov light through the atmosphere. In contrast to the optical astronomy, here not only the total transmission is important, but also the altitude of possible haze or aerosol layers plays a role. This is due to the fact, that the Cherenkov light is produced all the way down in an Extended Air Shower (EAS) and thus, different fractions of the light pass through different paths in the atmosphere. The total amount of Cherenkov light produced in such an air shower is proportional to the energy of the primary particle. If the attenuation by the atmosphere is varying and this effect is not taken into account, it can distort the measured energy spectral distribution and reduce its resolution. To counteract this, the MAGIC experiment uses a sensitive radiometer, that can detect even a thin cloud or a haze layer. If the weather conditions are adverse, the collected air shower data cannot be used in the analysis. To improve the evaluation of the atmospheric conditions on one hand and to be able to correct the effects of atmospheric attenuation in the data on the other hand, a single wavelength 'micro'-LIDAR (LIght Detection And Ranging) system has recently been installed at the MAGIC site.

Fig. 1. The LIDAR telescope inside the open dome in front of the MAGIC II telescope and the LIDAR control room below the dome.

## 2. Characteristics of the LIDAR system

The laser we use is a frequency doubled, passively Q-switched Nd:YAG operating at 532nm with a pulse energy of 5.1$\mu$J and a pulse width of 0.5ns. It can be triggered with an external pulse to a frequency of up to 2kHz. A beam expander lens is used for reducing the beam divergence down to 0.1°. The light detecting module is located In the focal plane of a 60cm diameter and 150cm focal distance telescope. The laser is aligned parallel to the axis of the 60cm optical telescope, but it is mounted 45cm off from the mirror center. The whole system is mounted on a NTM-500 commercially available robotic mount, originally designed for ESO's site-search for the EELT. The communication with the mount is via ASCOM interface. The custom made detector module uses a HAMAMATSU R9792U-40 Hybrid Photo Detector (HPD). It consists of a GaAsP photo cathode and an avalanche diode as anode. The latter is operated under the applied voltage of 400V while the photo cathode is operated at ~ -8kV. Thus the HPD reaches a charge amplification of over 100000 with very good multiple photo-electron (ph.e.) resolution. The measured Quantum Efficiency (QE) is of ~50% for the laser wavelength of 532nm (fig. 2). The HPD is placed behind a system of aperture and lenses with a 3nm bandwidth interference filter in between. This is done for centering on the bandwidth of the interference filter, for reducing the field of view and thus the the background from the Light of the Night Sky (LoNS). The signal is then amplified by a charge-to-current amplifier (~5000 V/A) and digitized by using a 200 MSamples/s FADC card installed in a computer. The latter is used for counting the ph.e.s. The raw data of 50000 laser shots uses 2GB of memory and is analyzed directly after the measurement. The photon counting method works best for the very low rate from large distances (high altitudes). It is counting single ph.e. events exceeding a pre-defined threshold closely above the system noise level. After that one subtracts a background rate (LoNS), determined from the end of each sample. However, this method has problems with some pulse pile-up, mostly occurring on the first 1000m from the laser, where the signal is still strong.

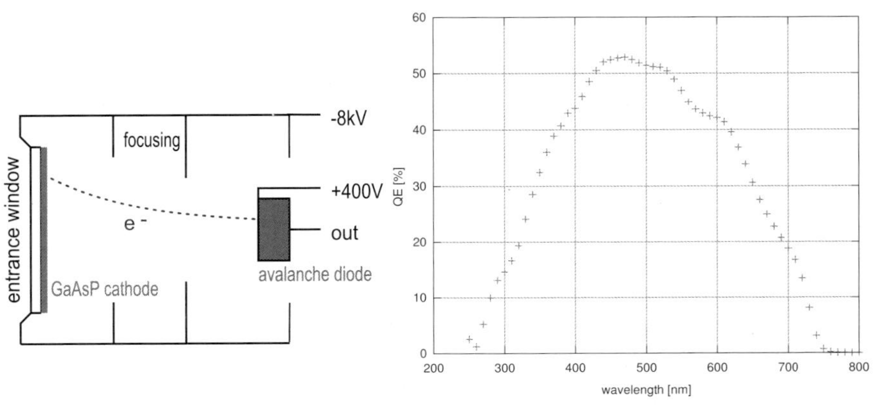


Fig. 2. Working principle and measured QE of the HPD, used in the LIDAR

## 3. First data and performance

For the first time the new LIDAR system was operated on the MAGIC site in July and August 2010. On August 2nd, the return signal did not show any signs of clouds or aerosol layers. So we have assumed, that mostly molecular Rayleigh scattering contributed to the back-scattering.[2]

$$\frac{\mathrm{d}\sigma_R(\theta)}{\mathrm{d}\Omega} = \frac{\pi^2(n^2-1)^2}{N^2\lambda^4}\,\frac{\cos^2(\theta)+1}{2} \tag{1}$$

In such a case it is possible to apply the back-integration method, suggested by J. D. Klett in 1981[3] for determining the range dependent total volume scattering coefficient

$$\sigma(r) = \frac{\exp\left(\frac{S(r)-S_m}{k}\right)}{\frac{1}{\sigma_m} - \frac{2}{k}\int_r^{r_m} \exp\left(\frac{S(r)-S_m}{k}\right)\mathrm{d}r'} \tag{2}$$

where $S(r) = \ln\left(P(r)\, r^2\right)$. $S_m$ and $\sigma_m$ are the logarithmic range corrected signal and the assumed input total volume-scattering coefficient at the reference range $r_m$. This method works well for pure Rayleigh scattering, because the assumption of a defined and stable LIDAR-ratio, i.e. $k = 1$, is true in this case. Because for a pure Rayleigh atmosphere, $\sigma(r)$ should only depend on the number density of air molecules $n(r)$, the determined differential optical thickness should roughly follow the barometric formula.

$$\sigma(h) = \sigma_0\ \exp\left(\frac{h}{\frac{R\ T}{Mg}}\right) \approx \sigma_0\ \exp\left(h\cdot(-1.2\cdot 10^{-4}\mathrm{m}^{-1})\right) \tag{3}$$

As it can be seen in figure 3, this works quite well beyond 2km range, meaning that we can trust our data in this region. However, for data with an aerosol fraction in the back-scattered signal, the ratio of volume back-scattering to attenuation is not anymore known and can vary over a large range. In most cases, for aerosols, the forward scattering dominates. Thus the attenuation is bigger compared to the pure Rayleigh effect, meaning, that the attenuation would be underestimated, using the Klett method.

## 4. Possible application to MAGIC data

A quite common weather phenomenon on the Canary islands, that can disturb the observations of the telescopes of the Observatoiro del Roque de los Muchachos (ORM), is the so called Calima, a warm wind from the African continent, carrying Sahara dust and also enough water vapor for condensation. This layer usually ranges from ~1km to 5km above sea level

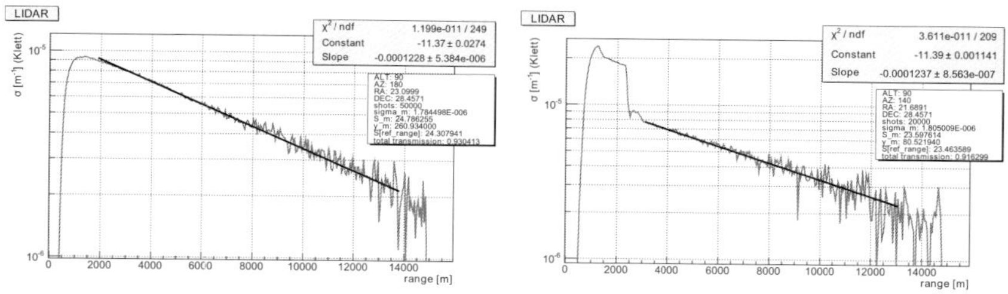


Fig. 3. Exponential fit to the total volume-scattering coefficient derived with the Klett back-integration method from LIDAR data of August 2nd in zenith direction. The exponential coefficient of $-1.24 \cdot 10^{-4}\mathrm{m}^{-1}$ fits quite well the assumption of a barometric atmosphere (left). This also works well for the upper part of the signal, if there is an aerosol layer in the lower part (right).

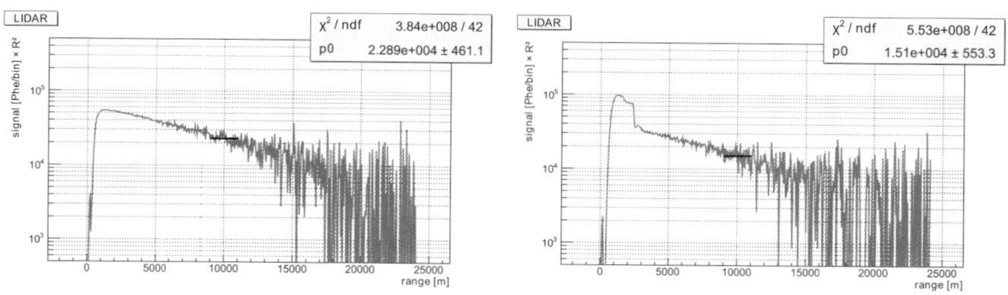


Fig. 4. Range corrected LIDAR returns (zenith, 10000 laser shots each) of the 02.08.2010 (left) and 03.08.2010 (right). The mean of the range corrected LIDAR return between 9km and 11km range was calculated. The light transmission $T \approx 80\%$ through the Calima layer, that was present on August 3rd, was determined by taking the square root of their quotient.

(a.s.l.)[4] That agrees with our LIDAR measurements, in which the layer ends at $\sim$ 4.7km a.s.l. (see fig. 4). Under the assumption, that the overall efficiency of the LIDAR system stays constant over time, one could use reference measurements during clear air conditions for determining the total attenuation in the aerosol layer for single transition. Having the photon counts from a range beyond the aerosol layer $N_{r_{\mathrm{ref}}}$, for the actual and a reference measurement, light transmission is the square root of their quotient $T = \sqrt{N_{r_{\mathrm{ref}},\mathrm{aerosol}}/N_{r_{\mathrm{ref}},\mathrm{clear}}}$ (fig. 4). Now, the excess from pure Rayleigh scattering, caused by such a layer is known. Having this information, two methods for correcting the energy spectra seem possible. One is preparing a large set of Monte Carlo data, considering different atmospheric situations. The other would make use of the range information contained in the shower images and apply differential corrections to the collected amount of

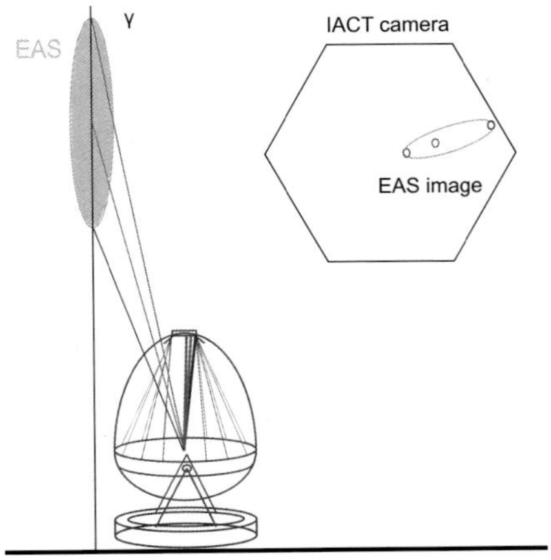


Fig. 5. The images of air-showers from $\gamma$-particles contain the information about the altitude, where the Cherenkov light was emitted. This makes it possible to apply differential, range dependent corrections to the collected amount of Cherenkov light.

Cherenkov light (fig. 5).

## 5. Conclusion

We have started operating a 'micro'-LIDAR system installed on the MAGIC site, between the two telescopes. By using the LIDAR data during routine observations, we intend to lower the influence of atmospheric conditions on the final results of our observations and enhance the precision of differential energy measurements. We are hoping that the LIDAR data will provide a better insight about the taken data quality, especially when taken under adverse atmospheric conditions. Moreover, a good parametrization of the momentary atmospheric conditions shall allow us to optimize our data selection strategy.

## References

1. J. C. et al. (2009), `http://arxiv.org/abs/0907.1211`.
2. R. M. Measures, *Laser Remote Sensing* (Wiley, 1984).
3. J. D. Klett, *Applied Optics* **20**, 211 (1981), `http://www.opticsinfobase.org/abstract.cfm?URI=ao-20-2-211`.
4. G. L. et al., *Astronomy & Astrophysics* **483**, 651(March 2008), `http://arxiv.org/abs/0802.3947`.

# THE AMS-02 SILICON TRACKER

S. Haino*

*INFN Sezione di Perugia, I-06100 Perugia, Italy*
**E-mail: Sadakazu.Haino@pg.infn.it*

*On behalf of the AMS-02 Tracker Collaboration*

The Alpha Magnetic Spectrometer (AMS) is a large acceptance cosmic-ray detector which will be installed as an independent module on the International Space Station (ISS). The instrument will provide a precise measurement of the cosmic-ray energy spectra and extensive antimatter search up to several TeV for particle charges up to $Z = 26$. The spectrometer will be delivered to the ISS by STS-134 flight in February 2011. In August 2010 the calibration and performance evaluation of the spectrometer were performed with test beam at CERN.

*Keywords*: Charged particle spectrometers; Tracking and position sensitive detectors; Cosmic-ray detectors.

## 1. Introduction

The Alpha Magnetic Spectrometer is a particle physics experiment in space. The purpose is to perform accurate, high-statistics, long-duration measurements of the spectra of charged cosmic rays from GeV to multi-TeV energy. A preliminary version of the detector AMS-01, with a permanent magnet and six-layer silicon tracker, was flown on NASA space shuttle *Discovery* in June 1998 [1]. The second generation detector AMS-02 contains a superconducting magnet and an eight-layer silicon tracker.

Recently the question of dark matter in the Universe has been put into focus, as data from a new generation of cosmic ray detectors have been tentatively interpreted in terms of indirect signature of dark matter annihilation (or decay). In particular, the positron fraction measured by PAMELA [2] and the energy spectra of sum of electrons and positrons by ATIC [3] and Fermi-LAT [4] show an excess compared to the standard prediction of background spectrum which consists of secondary products of high energy interactions between primary cosmic rays and interstellar medium [5]. Up to now, AMS-02 will be the only experiment which can extend the positron fraction measurement by PAMELA up to nearly 1 TeV. AMS data should give us many clues to understand the source of these excesses. AMS-02 will also extend the antiproton energy spectrum measurement by PAMELA [6] up to several hundreds of GeV. The antiproton spectrum by PAMELA shows a good agreement with the secondary production model indicating no dark matter signature.

Another important physics of AMS-02 is the extensive search of antimatter. The strong evidence that supports the Big Bang theory of the origin of the Universe requires matter and antimatter to be equally abundant at the very hot beginning [7]. The absence of sharp annihilation gamma-ray peaks excludes the presence of large quantities of antimatter within our local super cluster of galaxies [8]. However only a single detection of antinucleus in cosmic ray is a strong evidence on the existence of the antimatter domain in the Universe. Up to now, the most stringent upper limit on the antihelium to helium ratio has been obtained by the combined results of BESS and BESS-Polar experiments as $2.7 \times 10^{-7}$ [9]. AMS-02 will have the sensitivity of $10^{-9}$ level on the antihelium to helium ratio, more than two orders of magnitude higher than the current searches, and in much wider energy range up to nearly 1 TeV.

It is now getting more important to have good prediction of the cosmic-ray propagation to carefully study the indirect signature of dark matter. AMS-02 will collect $10^9$ nuclei up to $Z = 26$ (Fe) and their isotopes. Among the interesting issues, the accurate determination of the secondary to primary cosmic-ray ratio such as boron to carbon ratio over a wide energy range provides crucial information on the material thickness which the cosmic rays pass through during the propagation in Galaxy.

## 2. The Spectrometer

To perform a high accuracy measurement of the energy spectra of charged particles in space, the AMS detector is based on experience and observation from experiments to study rare signals among intense backgrounds.

It was the strict adherence to these techniques that ensured that a background rejection of $10^{10}$ was indeed possible and made these experiments successful. AMS is designed following the same principles. Figure. 1 shows the AMS-02 detector configuration for the ISS. AMS-02 will contain the following components:

(1) a 20-layer Transition Radiation Detector (TRD) which identifies elec-

trons and positrons with a rejection factor against hadrons of $10^3$–$10^2$ from 1.5 to 300 GeV,

(2) four layers of time-of-flight (TOF) hodoscopes that provide precision TOF measurements (~120 ps) and energy loss measurements,

(3) the permanent magnet with the central field of 0.14 T,

(4) nine layers (6.45 $m^2$) of silicon Tracker, which provides a proton rigidity ($R \equiv pc/Ze$, momentum divided by electric charge) resolution of 25 % at 0.5 TV and charge discrimination of nuclei up to $Z = 26$ (Fe),

(5) veto, or anti coincidence, counters (ACC) which ensure that only particles passing through the magnet aperture will be accepted,

(6) a Ring Imaging Cherenkov counter (RICH), which measures the velocity with 0.1 % resolution and provides charge discrimination of particles or nuclei, and

(7) a 3D sampling calorimeter (ECAL) made out of 16.7 $X_0$ (radiation lengths) of lead and scintillating fibers, which measures the energy of gamma rays, electrons and positrons and distinguishes electrons and positrons from hadrons with a rejection of $10^4$ in the range between 1.5 GeV and 1 TeV.

## 3. The Silicon Tracker

The AMS-01 Silicon Tracker [10] was the first application in space of the high-precision silicon technology developed for position measurements in accelerator experiments. The high modularity, low voltage levels (< 100 V), and gas-free operation of the device is well suited to operation in space. The AMS-01 test flight demonstrated both the successful adaptation of the technology to the space environment and the feasibility of large area detectors. Silicon micro-strip sensors were originally developed for vertex detectors in colliding beam experiments in order to provide a few high-precision position measurements near the interaction point. The AMS application differs considerably. The tracking information is provided uniquely by the silicon sensors, which implies a large surface area and higher inter-strip capacitances. The major challenges were to maintain the required mechanical precision and low-noise performance in the large-scale application in space.

The tracker is composed of 2264 double-sided silicon microstrip sensors positioned with high precision in 192 electrically and mechanically independent modules called ladders with a total active area 6.75 $m^2$. The 192 ladders are arranged in eight layers on five support planes to fill the circular plane geometry. The two outermost silicon layers are separated by a

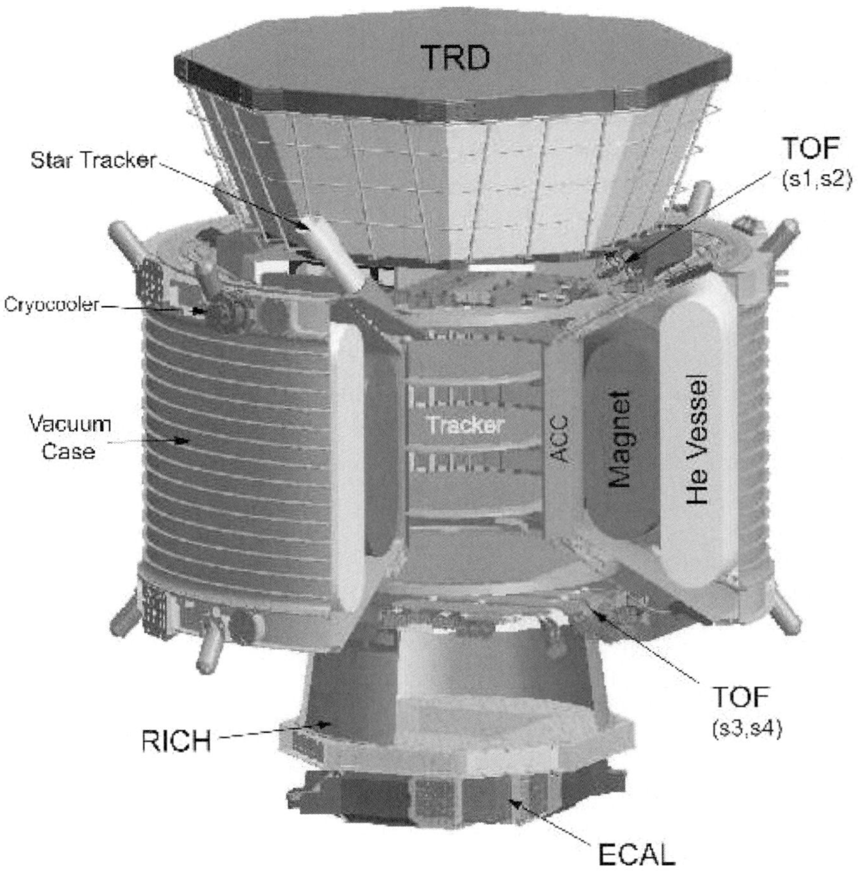


Fig. 1. The AMS-02 spectrometer.

distance of 1.06 m and located on the first and last support planes at both ends of the spectrometer, just outside the magnet. Double silicon layers are disposed on both sides of the three support planes located inside the magnetic volume as shown in Fig. 2 All the support planes are made of an aluminum honeycomb structure enclosed within thin carbon fiber skins. The total thickness of one internal plane including two silicon layers and one support plane is about 1 % $X/X_0$.

On opposite sides of the silicon sensor, $p^+$ (junction) and $n^+$ (ohmic) strips are implanted along orthogonal directions with implantation (readout) pitches of 27.5 (110) $\mu$m and 104 (208) $\mu$m, respectively. The p-side strips provide the measurement of the bending coordinate in the spectrometer: running along the ladder length, the strips from different sensors are daisy chained up to the ladder end where they are connected to an hybrid, which provide bias voltages and front-end readout chips. The n-side strips, providing the measurement of the non-bending coordinate, are also routed to the ladder end by means of a kapton cable which distributes signals of alternating sensors to the same readout channel.

A limited number (4 to 11) of ladders have been exposed to several particle beams to evaluate their performance. A detailed description of the ladders, as well as results for the position resolution are found in Ref. [11]. We have also evaluated the tracker performance with cosmic-ray muons after the pre-integration of the whole AMS spectrometer. Figure. 3 shows the estimated Tracker position resolution as a function of projected inpact angle. The detail of the analysis is found in Ref. [12].

## 4. The upgrade with a permanent magnet

The AMS spectrometer was originally designed with superconducting magnet. It has an advantage of strong magnetic field of 0.8 T, while the cryogen system has a limited life time from two to three years. In February 2010, the US goverment had determined to extend the operation of the ISS up to 2020–2029. At the same time, according to the thermo vacuum test at ESTEC the life time of the cryogen was estimated to be 20 months. In order to maximize the available time of the ISS we had made a decision to replace the superconducting magnet by a permanent magnet. The permanent magnet was used in AMS-01 flignt and it has a central field of 0.14 T. We have also reconfigured the Tracker layers to keep high momentum resolution even with the lower magnetic field as follows:

(1) Move one Tracker layer at the bottom on top of TRD
(2) Tek four ladders from the layer on TRD and eight ladders from the layer on top of the magnet to build a new layer by adding four spare ladders.
(3) The new layer is placed between RICH and ECAL

The track path length with all the nine layers is about 3 m, which is much longer than the original 1 m. The performance of a magnetic spectrometer can be characterized by the maximum detectable rigidity (MDR) corresponding to a rigidity measurement error of 100 %. Based on the simulation, MDR with the nine-layer tracking with the new Tracker configuration

Fig. 2. The three internal planes in the tracker assembly jig. The conical sections, which support the two external ladder planes at the entrance and exit of the magnet, are visible at the top and bottom of the cylindrical volume.

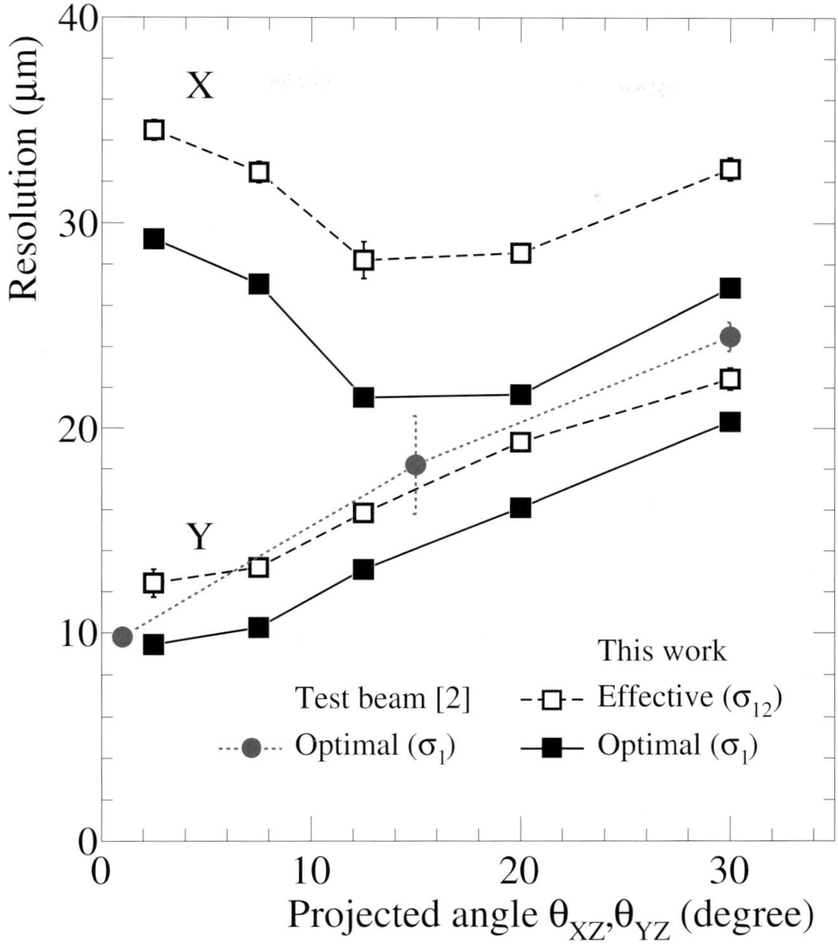


Fig. 3. The position resolution in the $x$- (upper points) and $y$- (lower points) coordinates estimated as a function of the projected angle, $\theta_{xz}$ and $\theta_{yz}$. Results of two different Gaussian width, $\sigma_{12}$ and $\sigma_1$ are shown, where $\sigma_{12}$ and $\sigma_1$ indicate the effective and optimal performance, respectively. Proton test beam results [11] are also shown.

is estimated as 2.1 TV, which is almost same as 2.2 TV with the original design with the superconducting magnet.

## 5. Conclusions

The AMS-02 spectrometer has successfully passed the "pre-integration" phase, the expected performance levels have been attained. The rest of 2009 will be devoted to the final integration with the superconducting magnet, and a vibration and thermal-vacuum qualification of the spectrometer at ESTEC, Noordwijk (The Netherlands). A test beam at CERN is also programmed prior to the shipment of the detector to the Kennedy Space Center early next year. The shuttle flight STS-134 has been confirmed by NASA, with the possible launch dates of July or September 2010.

## References


1. M. Aguilar et al., Phys. Rep. 366 (2002) 331.
2. O. Adriani et al., Nature 458 (2009) 607.
3. J. Chang et al., Nature 456 (2008) 362.
4. A.A. Abdo et al., Phys. Rev. Lett. 102 (2009) 181101.
5. I.V. Moskalenko and A.W. Strong, Astrophys. J. 493 (1998) 694.
6. O. Adriani et al., Phys. Rev. Lett. 102 (2009) 051101.
7. A.D. Dolgov, Phys. Rep. 222 (1992) 309.
8. A.G. Cohen et al., Astrophys. J. 495 (1998) 539.
9. M. Sasaki et al., Adv. Space Res. 42 (2008) 450.
10. R. Battiston, Nucl. Phys. Proc. Suppl. 44 (1995) 274; W.J. Burger, Nucl. Phys. Proc. Suppl. 113 (2002) 139.
11. J. Alcaraz, et al., Nucl. Instr. and Meth. A 593 (2008) 376.
12. B. Alpat et al., Nucl. Instr. and Meth. A 613 (2010) 207.

# From the Knee to the Ankle: From Galactic to Extragalactic Origin of Cosmic Rays?

Andreas Haungs*

*Karlsruhe Institute of Technology, KIT-Campus North, Institut für Kernphysik, Karlsruhe, Germany*
**E-mail: haungs@kit.edu*
*www.kit.edu*

The detection of high-energy cosmic rays above a few hundred TeV is realized by the observation of extensive air-showers. By applying different detection techniques with ground-based observatories the spectrum and the elemental composition of high-energy cosmic rays is determined. The most distinct feature of the spectrum, the 'knee', is thought to be the beginning of the end of the galactic origin of cosmic rays. As the highest energies (above the 'ankle') are most probably of extragalactic origin, between 10 PeV to 1 EeV one expects the transition of galactic to extragalactic origin. Whereas in the last decade many experiments have shown results on the knee and above the ankle, dedicated experiments to explore the transition region only recently went into operation or are going to be in operation. Recent experimental results, in particular from the KASCADE-Grande experiment, are shown and their astrophysical implications discussed. In addition, perspectives of future experimental efforts for this energy range are shortly discussed.

*Keywords*: Cosmic Rays; Extensive Air Showers; Energy Spectrum; Astrophysical Models

## 1. Introduction

The all-particle energy spectrum of cosmic rays exhibits a distinctive discontinuity at few PeV, known as the knee, where the spectral index changes from $-2.7$ to approximately $-3.1$ (Fig. 1). This feature has been discovered more than half a century ago by Kulikov and Khristiansen of the Moscow State University[1] within studies of the intensity spectrum of the content of charged particles of extensive air showers, which roughly reflects the primary energy. Another feature in the striking power-law behavior of the spectrum is visible around $4 - 10 \cdot 10^{18}$ eV, known as the ankle, where the spectral index of the particle flux changes back to approximately $-2.7$.[2]

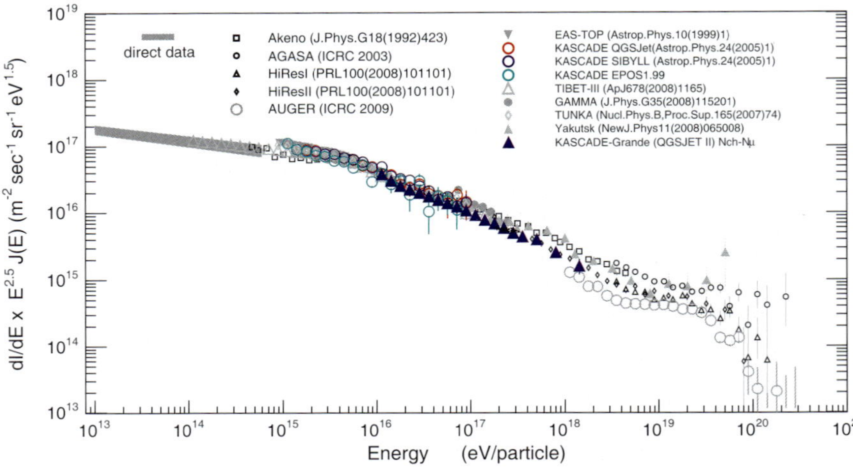


Fig. 1. The all-particle energy spectrum as measured by various experiments. References to the data points are given within the plot; the results of KASCADE-Grande refer to reference 3.

At energies above $10^{1}$ eV direct measurements are hardly possible due to the low flux. Thus indirect measurements observing extensive air showers (EAS) attempt to reveal the structure of the spectrum. Figure 1 sketches the various possibilities in measuring extensive air showers. Sophisticated experiments are characterized by applying more than one technique and observing several EAS techniques simultaneously in order to obtain as much information on the individual EAS as possible.

The key questions of the origin of this knee are still not convincingly solved. Astrophysical scenarios like the change of the acceleration mechanisms at the cosmic ray sources (supernova remnants, pulsars, etc.) or effects of the transport mechanisms inside the Galaxy (diffusion with escape probabilities) are conceivable for the origin of the knee as well as particle physics reasons like a new kind of hadronic interaction inside the atmosphere or during the transport through the interstellar medium. An overview on the zoo of these theoretical models were given in reference 4. It is obvious that only detailed measurements covering the full energy range of the knee from $10^{1}$ eV to $10^{18}$ eV and analyses of the primary energy spectra for the different incoming particle types can validate or disprove some of these models.

Despite of numerous EAS studies with various different experimental set-ups in the last decades, this demand could be never accomplished,

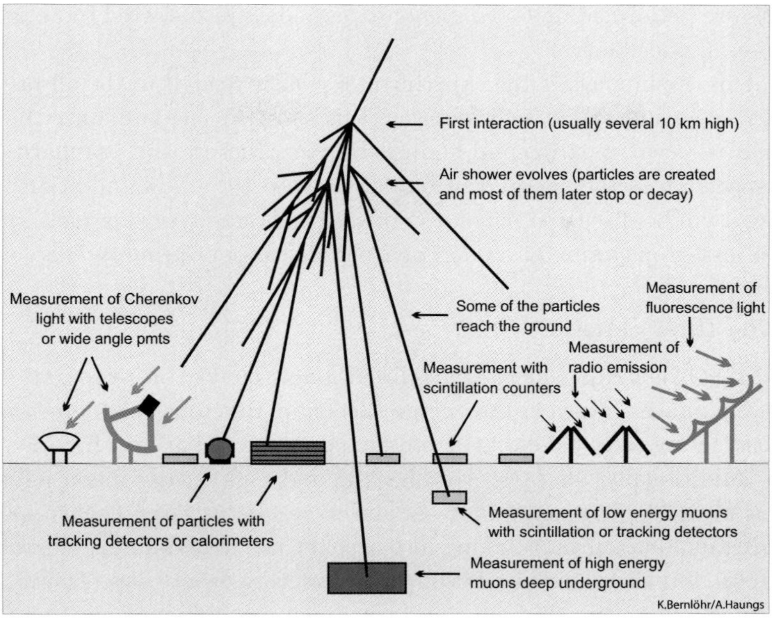


Fig. 2. Sketch of various ways to study experimentally extensive air showers.

mainly due to the weak mass resolution of the measured shower observables.[5]

The highest energies above the so called ankle at a few EeV are believed to be exclusively of extragalactic origin. The extragalactic component and the possible cut-off (GZK) at c. 100 EeV is subject of the Pierre Auger Observatory, see[6] Thus, in the experimentally scarcely explored region between the (first) knee and the ankle there are two more peculiarities of the cosmic ray spectrum expected: (i) A knee of the heavy component which is either expected (depending on the model) at the position of the first knee scaled with Z (the charge) or alternatively with A (the mass) of iron. (ii) A transition region from galactic to extragalactic origin of cosmic rays, where there is no theoretical reason for a smooth crossover in slope and flux. Dependent on the considered astrophysical model the second knee is allocated to case (i) or (ii), respectively. In order to investigate different scenarios precise EAS measurements of high statistical accuracy are needed for the energy range of $10^{16}$ and $10^{18}$ eV, which appears a little bit ignored in the last decades. In recent years some interest arose from results of sophisticated experiments like KASCADE-Grande. In addition, new experimental efforts are performed and a couple of experiments just started to be or will be soon in operation.

Figure 1 summarizes the experimental results related to the all-particle energy spectrum. What will follow in this short review is mainly the discussion of recent experimental findings on the structure and peculiarities of this spectrum in the energy range of $10^{15}$ eV to $10^{18}$ eV. In addition these findings will be discussed in context of various astrophysical models for the origin and propagation of cosmic rays of this range of primary energies.

## 2. The (first) knee

The KASCADE experiment[7] introduced a new quality of investigations of this well known feature in the energy spectrum, due to its high precision in reconstructing the total electron and muon number separately in individual EAS. Additionally, for each event hadron and high-energy muon information is available. The data analyses aim to reconstruct the energy spectra of individual mass groups taking into account not only different shower observables, but also their correlation on an event-by-event basis. The content of each cell of the two-dimensional spectrum of reconstructed electron number vs. muon number $g(y)$ is the sum of contributions from the individual primary elements. Hence the inverse problem $g(y) = \int K(y,x)p(x)dx$ with $y = (N_e, N_\mu^{\mathrm{tr}})$ and $x = (E, A)$ has to be solved, and with known $p(x)$ the spectra of individual masses can be built. The kernel function is obtained by Monte Carlo simulations on basis of (different) hadronic interaction models as options embedded in CORSIKA. CORSIKA is a sophisticated air shower simulation tool package,[8] which was developed in the frame of the KASCADE experiment and which meanwhile is used worldwide.

By applying these procedures (with the assumption of five primary mass groups) to the experimental data energy spectra for 5 individual mass groups are obtained and the sum of these spectra results in the all particle spectrum (fig. 1).[9] Concerning the individual spectra a knee like feature was clearly detected in the spectra of primary proton and helium. This demonstrated that the knee feature originates from a decreasing flux of the light primary particles. The described analysis was repeated by using different hadronic interaction models where these findings were confirmed.[10,11]

Comparing the unfolding results based on the different hadronic interaction models, similar structures of the individual spectra as well as the all-particle spectrum agree nicely, but the relative abundances of the primary mass groups (composition) differ significantly. Thus presently it is not yet possible to pin down definitely the mass composition. Modeling

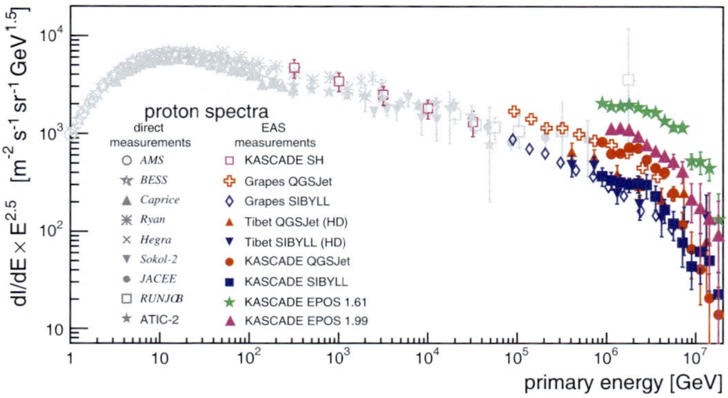


Fig. 3. Compilation of direct and indirect measurements of the cosmic ray energy spectrum of primary protons. In case of KASCADE the results of different models are included.[11]

the hadronic interactions underlies assumptions from particle physics theory and extrapolations resulting in large uncertainties, which are reflected by discrepancies in the resulting composition. The differences in the total energy appear to be comparatively smaller and are less model dependent.

Crucial parameters of the modeling of hadronic interaction models which can be responsible for these inconsistencies are the total nucleus-air cross-section and the parts of the inelastic and diffractive cross sections leading to shifts of the position of the shower maximum in the atmosphere, and therefore to a change of the muon and electron numbers as well as to their correlation on single air shower basis. The multiplicity of the pion generation at all energies at the hadronic interactions during the air shower development is also a 'semi-free' parameter in the air-shower modeling as accelerator data have still large uncertainties. Arbitrary changes of free parameters in the interaction models will change the correlation of all shower parameters. Tests using KASCADE observables,[1 ,1] which are measured independently of those used in the unfolding procedure, give further constraints, in particular by investigating correlations of the hadronic shower component with electron or muon numbers. The aim is to provide hints for the model builder groups how the parameters (and the theory) should be modified in order to describe all the data consistently. Combining all the correlations, the final observables of the showers have not to be different by more than 15% to be consistent with the data. This requires rather a fine-tuning of the free parameters in the simulation of the hadronic interactions than a need on new physics. That this is true can be seen in figure 3 where the resulting proton spectra are compared with direct measurements, where many data are available. Here, most of the results are in the range of the statistical uncertainty of the extrapolation of direct measurements.

In summary, the KASCADE results provide clear hints for a rigidity dependence of the knee feature, which automatically leads to an increasing heavier mean mass of the composition passing the knee.[14] But what is the dominant primary mass group at the knee? This is still an open question, and the KASCADE and TIBET[15] collaborations interpret their data in a different way despite their proton fluxes (see fig. 3) are in good agreement. Direct measurements of high-energy cosmic rays, which are able to determine the composition in a model independent way, may be possible in future and could give an answer to this question. In addition, the GRAPES[16] experiment in India, whose detector installation is similar to KASCADE, but located at higher altitude, will reach lower primary energies, providing an overlap with direct measurements.

Another open question is the 'sharpness' of the knee. Some experiments (TIBET,[15] TUNKA,[17] e.g.) claim explicitly that they see a sharp knee which can be explained by a considerable contribution to the overall bulk of cosmic rays stemming from a near-by single source.[18] The knee of the underlying 'standard' spectrum should be smeared out as it is originating from many sources (SNR's) and also by propagation effects.

## 3. Approaching the ankle: The transition region

For the still less explored energy range between knee and ankle there are now new results available from the KASCADE-Grande[19] experiment, which is an extension of the former KASCADE installation, still allowing the independent measurement of the muon content of the EAS. In particular this possibility to reconstruct the total muon number for Grande measured showers is the salient feature of KASCADE-Grande compared to other experiments in this energy range.

Figure 1 shows the reconstructed all-particle spectrum which was communicated this year as a first result of KASCADE-Grande.[3] For displaying the very interesting details of the shape of the spectrum a residual plot was constructed in order to see the deviation from a flux proportional to $E^{-3.015}$ (see fig. 4) which is the index obtained by fitting the middle part of the spectrum. Two interesting features show up, one is a concavity above $10^{16}$ eV and another one is a small break at $\approx 10^{17}$ eV.

For a discussion of the astrophysical implications of these findings in some more details, figure 5 (adapted from Michael Hillas) depicts a pos-

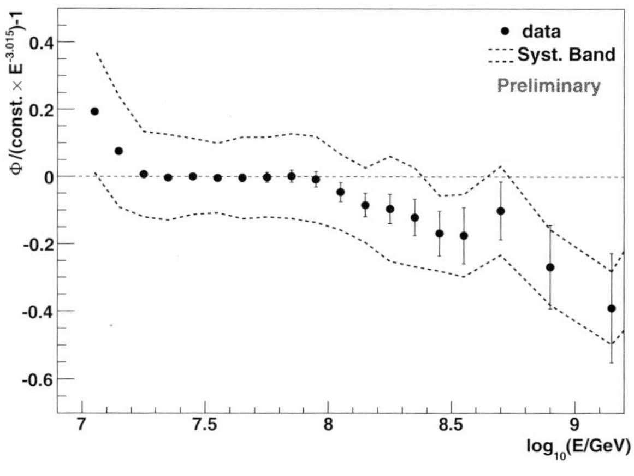


Fig. 4. Residual plot for the reconstructed all-particle energy spectrum obtained from KASCADE-Grande. The systematic error band is also shown (dotted lines). The plot is taken from reference.[3]

sible scenario. In order to reproduce the measured spectrum, before the KASCADE-Grande results appeared, Hillas proposed a 'component B' of the spectrum additionally to the rigidity dependent main galactic component.[20] It fills the gap between the knee and the ankle before the extragalactic component gets efficient. This model is in contradiction to the 'dip model',[21] where the rigidity dependent main galactic component is directly overtaken from a proton dominant extragalactic component already at $10^{17}$ eV. Here, the ankle is generated by an extragalactic feature.

The concavity or hardening of the spectrum seen by KASCADE-Grande can be explained in terms of the rigidity dependent main bulk of cosmic rays responsible for the knee, see figure 5. According to the relative abundance of the heavy component the significance of this hardening can increase or decrease. By that, the KASCADE-Grande finding also can serve as a hint to the composition around $10^{16}$ eV. The weak knee-like feature close to $10^{17}$ eV seen by KASCADE-Grande is also in agreement with the rigidity model as the position of the structure is at an energy where one expect the knee of the iron component. But, both findings are still in agreement with both contradicting astrophysical models. Crucial for clarifying the situation are detailed studies of the composition in this energy range where KASCADE-Grande has a rich potential.

For completion of the present experimental situation, results of the GAMMA experiment[22] will be mentioned. They reported about a peak

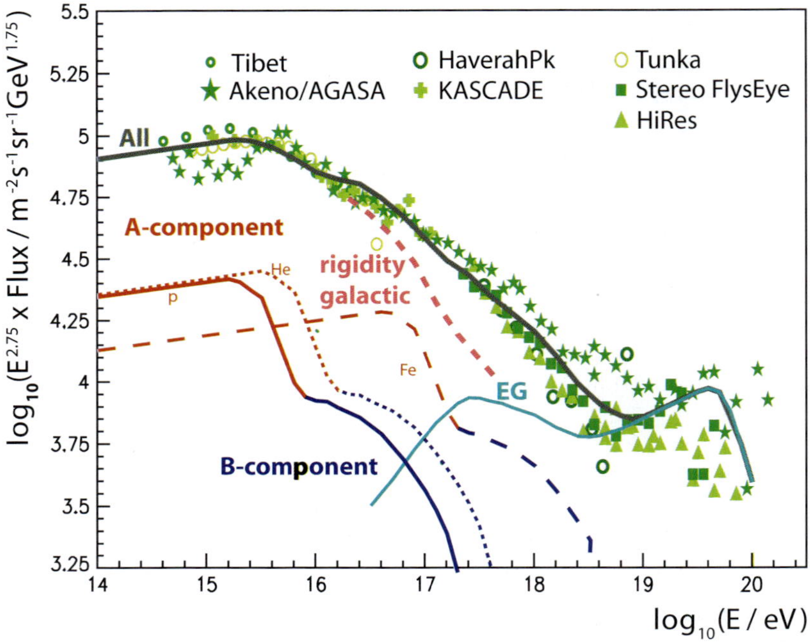


Fig. 5. Possible scenario for explaining structures of the cosmic ray all-particle energy spectrum. To the standard galactic component 'A' a component B is added closing the gap to the extragalactic component (EG) at highest energies.[20]

in the spectral shape close to $10^{17}$ eV (see fig. 1) which can be explained by the iron component of the mentioned near-by single source of cosmic ray origin. Such a strong peak is not confirmed by the KASCADE-Grande result and its existence, at least in such a significant appearance, is still under discussion.

In near future, new experiments will be in operation (or are already in operation) which will give new insights in the energy range of the transition region. We can expect a deeper insight in the physics of cosmic rays for example by

- KASCADE-Grande: Sophisticated analysis methods of the available data enables detailed composition studies.[3]
- TUNKA: An extension of the Cherenkov-Array TUNKA is in operation and allows access to energies up to a few hundred PeV.[17]
- ICETOP/ICECUBE: A combination of the on-ice particle detector array with the muon-sensitive in-ice neutrino detector will allow detailed

composition studies for this energy range.[23]

- Pierre Auger Observatory Enhancements: Specialized fluorescence telescopes in combination with a denser surface array, underground muon detectors and possibly a radio antenna array will serve as a super-hybrid detector for the energy range below the ankle.[6]

## Acknowledgments

The author wants to express that the present article reflects a personal view on the subject, which is far from completeness. The author would like to thank all the colleagues in the field, in particular from the KASCADE-Grande Collaboration, for stimulating discussions.

## References

1. G.V. Kulikov and G.B. Khristiansen *Soviet Physics JETP* **35(8)** No.3 (1958).
2. M. Nagano and A.A. Watson, Rev. Mod. Phys. **72**, 689 (2000).
3. J.C. Arteaga et al. (KASCADE-Grande Coll.), ISVHECRI (2010), arXiv:1009.4716
4. J. Hörandel *Astropart. Phys.* **21** 241 (2004).
5. A. Haungs, H. Rebel, M. Roth **Rep. Prog. Phys.** 66, 1145 (2003).
6. J. Blümer et al. (Pierre Auger Coll.), this volume
7. T. Antoni *et al.*, KASCADE Coll., NIM A **513**, 490 (2003).
8. D. Heck et al. *Report FZKA 6019, Forschungszentrum Karlsruhe* (1998).
9. W.-D. Apel et al. (KASCADE Coll.), Astrop. Phys. **24** (2005) 1.
10. W.-D. Apel et al. (KASCADE Coll.), Astrop. Phys. **31** (2010) 86.
11. A. Haungs et al. (KASCADE-Grande Coll.), Proc. ARENA 2010, to be published in NIM A proceedings Supplementary (2010).
12. T. Antoni et al.-KASCADE Coll., J.Phys.G:Nucl.Part.Phys. **25** 2161 (1999).
13. W.-D. Apel et al.-KASCADE Coll., J.Phys.G:Nucl.Part.Phys. **36** (2009) 035201.
14. C. De Donato, G.A. Medina-Tanco, Astrop. Phys. **32** (2009) 253.
15. M. Amenomori et al., ApJ **678** (2008) 1165
16. H. Tanaka et al., Nucl.Phys.B, Proc.Sup. **280** (2008) 175.
17. V. Prosin et al., Nucl.Phys.B, Proc.Sup. **165** (2007) 74.
18. A.D. Erlykin and A.W. Wolfendale, J.Phys.G:Nucl.Part.Phys. **31** (2005) 1475.
19. W.-D. Apel et al. (KASCADE-Grande Collaboration), NIM A **620** (2010) 202.
20. A.M. Hillas, J. Phys. G: Nucl. Part. Phys. 31 (2005) R95.
21. V. Berezinsky, A. Gazizov, S. Grigorieva, Phys. Rev **D74** (4) (2006) 043005.
22. A.P. Garyaka et al., J.Phys.G:Nucl.Part.Phys. **35** (2008) 115201.
23. T. Waldenmaier et al. (ICETOP Coll.), this volume

# HIGH ENERGY COSMIC-RAY PROTONS AND HELIUM*

STANISLAV BORISOV, SERGEY VORONOV, ARKADY GALPER AND ALEXANDER KARELIN†
FOR PAMELA COLLABORATION

*Experimental Nuclei Physics and Cosmic Physics, NRNU MEPhI, Kashirskoe shosse 31 Moscow, 115409, Russia*
*karelin@hotbox.ru*

The work is done within the framework of the PAMELA project [1]. The PAMELA is a space experiment for study of charged particles in the cosmic radiation that has been performed since the launch June 15 2006. This work is dedicated to the high energy cosmic ray proton and helium flux determination (an energy region from 50 GeV/nucleon) based on using a sampling electromagnetic calorimeter. The energy estimation was built on a measurement of a total deposited energy in the calorimeter and an applying of certain event selection criteria connected with an electromagnetic component of a hadronic shower. The helium-proton separation was performed by scintillation detectors of the PAMELA Time of Flight system. This method is indispensable to widen a measurable energy range and to receive additional information in the PAMELA experiment. The obtained energy spectra of protons and helium are presented.

## 1. Primary Cosmic-ray Helium and Proton Spectra

Helium nuclei and Protons are the most abundant elements in cosmic rays. The primary cosmic-ray proton and helium spectrum knowledge is very important to understand the fundamental phenomena of cosmic ray physics. They include cosmic rays acceleration mechanisms, particle propagation, cosmic ray origin and the matter distribution in interstellar space. In addition the proton and helium fluxes contain an essential part of information to be known for a neutrinos absolute flux calculation for example in the movement a neutrino oscillations analysis.

The energy range between 1TeV up to 10 TeV remains the most scantily explored one for cosmic ray protons and helium till now. There have been very few measurements for the high energy protons and helium spectra in previous time in balloon borne or satellite experiments, beginning from the PROTON satellites [2] till CREAM [3], but some questions have kept so yet.

---

* This work is supported by RFFI grant 10-02-11301a.
† The work was carried out in frame of FTP "Scientific and educational professionals of innovation Russia" for 2009-2013.

## 2. The PAMELA experiment

PAMELA – the instrument abroad the Resurs-DK1 satellite was launched into Earth's orbit in June 2006. The PAMELA is a space experiment for study charged particles and antiparticles in the cosmic radiation. The main scientific goal of the experiment is a precise measurement of the energy spectra of antiprotons, protons, positrons, electrons, light nuclei and their isotopes of cosmic rays in the wide energy range. The PAMELA instrument consists of several specialized detectors as shown in Fig.1: a magnetic spectrometer with a silicon tracking system, a time of flight (ToF) system with three double planes, an anticoincidence system, a neutron detector, a bottom shower scintillator detector and a tungsten/silicon sampling electromagnetic calorimeter.

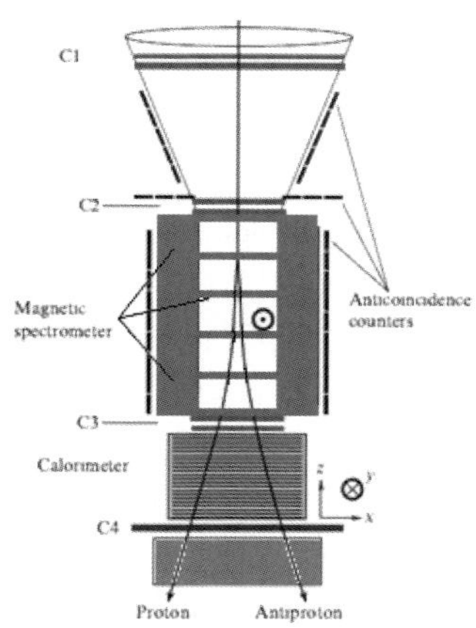


Figure 1. Overview of PAMELA

The ToF system comprises six layers of fast plastic scintillators arranged in three planes (S1, S2 and S3). Each detector layer is segmented in strips. Strips placed in alternate layers orthogonal to each other. The distance between S1 and S3 is 77.3 cm.

The total calorimeter thickness is 16.3 radiation lengths and 0.6 nuclear interaction length. The calorimeter composed of 44 silicon layers interleaved by 22 tungsten plates 0.26 cm thick. Each silicon plane is segmented in 96 strips. 22 planes are used for the X view and 22 for the Y view in order to provide a topological and energy information of the shower development inside the calorimeter [4].

The incident proton and helium energy measurement by a determination of a declination angle value in a magnetic field inside the magnetic spectrometer is possible just up to the energy about 1 TeV, while for the energy range more than 1 TeV/nucleon such ability is available for the calorimeter only.

## 3. The method

The method is based on calorimetric measurements. Each thin for nuclear interaction calorimeter (e. g. thickness less than 2 nuclear lengths) as the PAMELA electromagnetic one has an ability to measure an energy of hadrons and nuclei. The absorption of hadronic particles, which undergo strong interaction, develops as a cascade process. In thin calorimeters, the energy dependent signals are near linear as mainly the first hadron interaction is contained inside calorimeter matter. The measured energy of the hadronic shower is an electromagnetic fraction almost all. The existence of energy-independent fluctuations in an undetectable part of the energy leads up to an energy resolution value which is worse than 50%. The Monte Carlo simulation code GPAMELA *(http://www.ba.infn.it/ambriola/gpamela/)* and experimental data was used to assess the performances of the PAMELA calorimeter for high-energy protons and high-energy electrons.

### 3.1. *The charge discrimination. The electrons rejection in case of the Proton analysis*

Protons, helium nuclei and electrons are dominating charged components in cosmic rays. One of the main tasks of any cosmic ray experiment is to distinguish between one particle kind and others. For the helium and proton separation the ionization loss values in different planes of ToF scintillators have been used. Energy deposits in ToF planes provide information about a particle charge value. In general the knowledge of signal amplitudes from each PMT in an active paddle is enough to discriminate between helium and protons without a charge value calculation. After such procedure a contamination of another charge component is small and is included in systematic errors.

Though the electron portion in the common cosmic ray flux is only 1% it can increase among selected events. Moreover electrons of the specified energy will be reconstructed as protons with a higher energy.

More than 97% electrons generate the electromagnetic showers in the first three layers of the calorimeter. For this reason the energy release along a reconstructed shower axis (at a distance within 5 strips from the reconstructed axis) was used for the first four calorimeter planes to suppress electrons. A deposited energy threshold was installed as 4 mip [5].

### 3.2. *The events selection*

Without any selection the total energy deposit distribution in the calorimeter for hadrons is similar to one in Fig 2a (for 1 TeV protons with in an aperture incidence on the top calorimeter surface) obtained from simulation data. However, after applying described mentioned below criteria the peak on the left-side of the histogram, produced by non-interacting hadrons disappears (see Fig.2b).

The following selection criteria were implemented:

1. A 4000 mip threshold for total deposited energy in the calorimeter was installed to eliminate events with a low energy release inside the calorimeter.
2. Particles with incident angles within an aperture were selected by a shower axis reconstruction inside the calorimeter [6].
3. Events with a starting point of a shower within 12-22 calorimeter layers were cut, as they have a big value of an energy leakage.

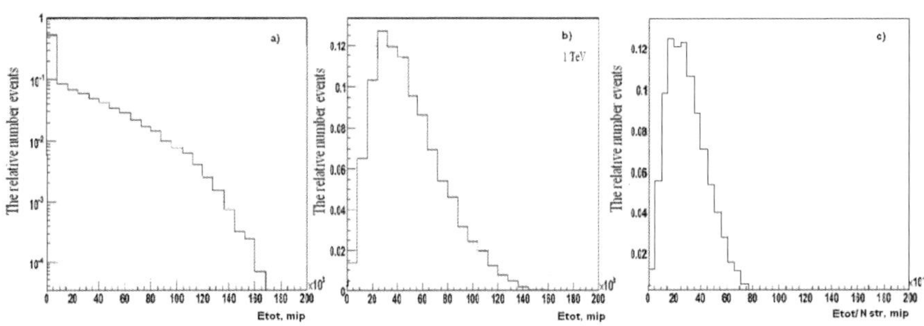


Figure 2. a) Etot distribution before the selection b) Etot distribution after the selection c) Etot/Nstr distribution after the selection (1TeV simulated protons were used in all cases)

### 3.3. *The energy determination*

The total energy release distributions in the calorimeter for incident protons/helium different energies at the same time are the distributions of a probability to obtain the certain response from the calorimeter by a protons/ helium with the certain energy. These distributions can be approximated by the universal curve reflecting the pion production multiplicity in hadron interactions with tungsten planes. To obtain more symmetric distributions an energy released density was used instead of the total deposited energy. This value was calculated as a ratio between the total deposited energy *Etot* and a total number of hit strips

*Nstr*. The relative asymmetry index for this variable *Etot/Nstr* one order less than for *Etot* only (see Fig.2c) [7].

### 3.4. *The flux reconstruction*

To reconstruct the primary energy proton and helium spectra the approximation was used known as a multidimensional unfolding method based on the Bayes theorem [8]. The Bayesian unfolding method has been applied to the experimental data for an inference problem of a poor value of the energy resolution. To use it the simulated proton spectrum with the slope index 2 was applied additionally. For helium nuclei in absence of a simulation the recalibrated simulated proton data supply with the flight data were involved in a process of the flux reconstruction. The selection criteria efficiency was calculated on the base of the simulated data for proton events, and then recalculated for helium ones taking into account the strong interaction cross section values for He nuclei.

## 4. Results

The cosmic ray proton and helium spectra obtained by this method are shown in Fig 3. The results were compared with ones obtained from the tracker data in the energy range between 50 and 1200 GeV and other experiments: CAPRICE98 [9], ATIC [10], CREAM [3], Ryan [11].

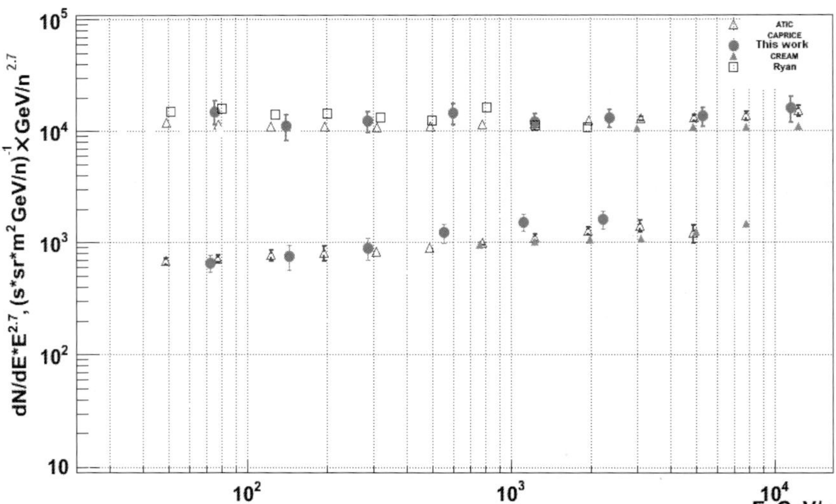


Figure 3. The helium and proton fluxes.

The figure displays fluxes multiplied by $E^{2.7}$. The obtained differential proton spectrum was fitted by power law with the spectral index 2.75±0.05. Helium spectrum was approximated by power law as well with index 2.65±0.9.

## 5. Conclusion

The reconstruction method of the cosmic ray proton and helium energy with the electromagnetic calorimeter of the PAMELA instrument was developed. This method uses the value of the total deposit energy in the calorimeter and the total number of hit strips. The mean spectral index is 2.75±0.05 in the energy range from 50 GeV/nucleon up to 15 TeV/nucleon for protons and 2.65±0.09 for helium from 50 GeV/nucleon up to 4 TeV/nucleon. The obtained spectra contain the important information for cosmic ray physics from point of view the propagation process into interstellar medium and the cosmic ray source nature and can be used to improve the neutrino atmospheric calculation as well.

## References

1. M. Casolino, P. Picozza, F. Altamu, *Adv. In Space Res.*, **42**, 455 (2008).
2. N. L. Grigorov, V.E. Nesterov, I.D. Rapoport et al., *The Space research*, **5, 3**, 395 (1967)
3. H. S. Ahn, P. Allison, M. G. Bagliesi, et al., *The Astrophysical Journal Letters,* **L89**, 714 (2009).
4. P. Papini, O. Adriani, M. Ambriola, et al., *Nucl. Instr. And Meth. In Phys. Res.*, **A 588**, 259 (2008).
5. M. Boezio, M. Pearce, M. Albi, *et al., Astroparticle Physics, **26**,111* (2006)
6. S. V. Borisov, S. A. Voronov, A. V. Karelin, et al., *Physics of Particles and Nuclei Letters,* **Vol. 7, No. 1,** 39 (2010).
7. A.V. Karelin The measurement of cosmic ray proton energy with electromagnetic calorimeter of PAMELA instrument, *31th ICRC, Lodz, Poland,* **HE.0493** (2009).
8. G.D. Agostini, *Nucl. Ins. And Meth. In Ph. Res. A*, **V. 362**, 487 (1995).
9. M. Boezio, P. Carlson, T. Francke, et al., *The Astrophysica Journal*, **518**, 457 (1999).
10. A. D. Panov , J. H. Adams Jr. , H. S. Ahn, Bulletin of the Russian Academy of Sciences: Physics, **Vol. 73, No. 5**, 564 (2009).
11. M.J Ryan., J.F Ormes., and V.K Balasubrahmanyan, *Phys. Rev. Lett.*, **V 28**. 985(1972).

# STATUS OF UHE CR ORBITAL FLUORESCENCE DETECTOR TUS

P. KLIMOV*, G. GARIPOV, B. KHRENOV, N. KALMYKOV, V. MOROZENKO, M. PANASYUK, S. SHARAKIN, A. SHIROKOV, I. YASHIN

*Skobeltsyn Institute of Nuclear Physics,*
*Moscow, 119991, Russia*
**E-mail: pavel.klimov@gmail.com*
*www.sinp.msu.ru*

S. BIKTEMEROVA, A. GRINYUK, D. NAUMOV, L. TKACHEV, A. TKACHENKO

*Joint Institute for Nuclear Research, Dubna, Moscow Region, Russia*

O. SAPRYKIN

*Consortium "Space Regatta", Korolev, Moscow Region, Russia*

I. PARK, J. LEE, G. NA

*Physics Department, EWHA Woman University, Seoul, Korea*

O. MARTINEZ, H. SALAZAR

*University of Puebla, Puebla, Puebla, Mexico*

The pioneer space fluorescence ultra high energy cosmic rays (UHE CR) detector TUS, is preparing in SINP MSU, Russia. The main goal of this project is study of cosmic rays beyond GZK cut-off (50 EeV). It consists of segmented Fresnel–type mirror concentrator (area 1.86 $m^2$ , focal distance 1.5 m) and photo receiver (256 pixels – PMT R1463). Total FOV of detector is $9 \times 9$ degrees. Developed electronics allow events measurements in various time scales, and provide PMT gain control to operate with any atmosphere UV background. Night atmosphere radiation depends mainly on moon phase and atmosphere air-glow and was studied in "Tatiana-1" and "Tatiana-2" experiments. The electronics of TUS were tested in space conditions during these experiments. Recently, detector TUS was included into "Mikhailo Lomonosov" satellite scientific payload. This satellite should be launched at the end of 2011 in honour of Russian scientist and founder of Moscow State University M.V. Lomonosov. TUS will study various extreme phenomena which produce UV light in the atmosphere (UHECR, transient luminous events) and will be a "pathfinder" for further projects (JEM-EUSO, MEMS Space Telescope).

*Keywords*: ultra high energy cosmic rays; fluorescence detector.

## 1. Introduction

One of the most important astrophysical problem of last decades is Cosmic Rays (CR) with energies more than 50 EeV (so called UHE CR) and all questions connected with them (energy spectrum, chemical composition, astrophysical sources etc.). Recently, in ground-based fluorescence detector measurement, the GZK–type energy spectrum (suppression of CR at energy above 50 EeV) is confirmed,[1] but still the available statistics is not enough, especially at energies $E \geq 100$ EeV. There is obvious difficulties in chemical composition understanding and latest results seems to be contradictory,[2,3] as well as in UHE CR asymmetry search, where there is no satisfactory interpretation of experimental results.[4,5] Further success could be done by orbital fluorescence detector, which can provide long-time observation of large area at the atmosphere with high transparency of its upper layers, uniform measurements all over the celestial sphere with one detector on board the satellite.

Nowadays several projects are developed, among which "Tracking SetUp" (TUS) and JEM–EUSO should be launched in near future.

## 2. TUS detector

The TUS detector design on board future MSU satellite "Mikhailo Lomonosov" is presented on fig 1. This satellite will have scientific payload for study various extreme phenomena in our Universe (gamma ray bursts, transient luminous events and UHE CR).

TUS detector consists of Fresnel–type mirror concentrator, photo receiver and system of photo receiver moving.

The mirror concentrator composed from the central parabolic segment and 6 identical surrounding segments with rings of parabolic shape. The technological model is produced and passes test in SINP MSU nowadays. The model is presented on fig 2. The first probe of surface quality and mirror focusing capability were made with parallel beams of light from lasers. The distribution of signal on the focal plane from beams parallel to mirror optical axis is shown on fig 3. It is seen that focusing properties are not perfect, but even in this technological model most part of reflected light falls in one TUS pixel. Further improvements of mirror quality are going on.

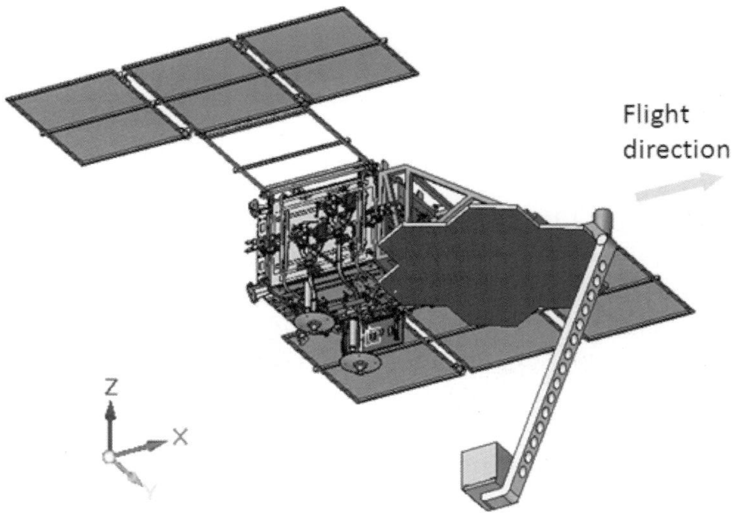


Fig. 1. TUS detector on board “Mikhailo Lomonosov” satellite.

Fig. 2. Fresnel–type mirror concentrator in SINP MSU.

Photo receiver comprises 256 pixels combined in 16 pixel clusters. One cluster with PMT and electronics board shown on fig 4. The UV sensor

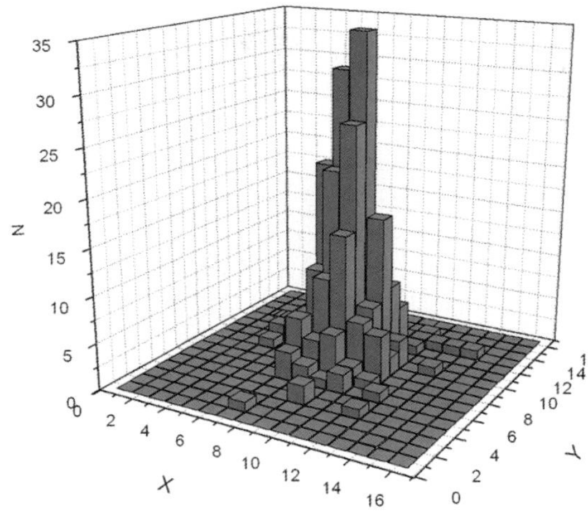


Fig. 3. Intensity distribution (in arbitrary units) on the focal plane from light beam parallel to mirror optical axis. Bin size is 2.5 × 2.5 mm.

(pixel) of the detector is the photomultiplier tube (PMT) of Hamamatsu type R1463 (13 mm tube diameter, multi-alkali cathode, UV glass window). At the PMT's cathode, a filter is placed to cut radiation with wavelength larger than 400 nm. Light guides with square windows are used for having higher fill-in-factor in pixel area. The pixel sensitivity is restricted to the near UV wavelength range of $\lambda = 300$–$400$ nm. Signals from every PM tube anode are coming to the multiplexer and then to 10-bits ADC. The main feature of the electronics is the use of FPGA for digital analysis of the signals after ADC. The fast signals from an UHE CR event are collected every 0.8 $\mu$s. Digital integration is used for selection and measurements of slower event types (micro meteors, sub-relativistic dust grains, transient luminous events). The important feature of TUS electronics as well as UV detectors on "Tatianas" satellites is using high voltage controls system. Electronics always follow the UV background variations and adopt PMT gain. It saves power consumption and allows one to operate in the whole range of UV atmosphere radiation intensity. For details of the electronics operation see.[6]

Fig. 4. TUS PMT cluster

## 3. UV night atmosphere radiation data obtained by UV detectors on board "Universitetsky–Tatiana-1" and "Universitetsky–Tatiana-2" satellites.

On board of two MSU satellites "Tatiana-1" and "Tatiana-2" were installed detectors of UV and IR radiation. These detectors could measure slow variations of UV and IR background and transient events. UV flashes were selected and their temporal profiles were measured by digital oscilloscopes. Two oscilloscopes were used for "Tatiana-1": 1) with the trace length 4 ms and time sample 16$\mu$s, and 2) with the trace length 64 ms and time sample 256$\mu$s and one oscilloscope for "Tatiana-2": 128 ms trace length and 1 ms time sample. It is important to mention that these detectors represent TUS photo sensor units, which has the same electronics and the same photomultiplier tube Hamamatsu R1463. Two years of "Tatiana-1" operation proved stable in sensitivity of PMT (in 10% error) and reliable work of electronics. More details about detectors see in.[7,8]

### 3.1. *UV night atmosphere radiation intensity monitoring*

The main factor which determines UV light amount is the scattered Moon light. The UV intensity depends on Moon phase and varies from $3\times10^7$ photon cm$^{-2}$sr$^{-1}$s$^{-2}$ during moonless nights up to $3\times10^9$ photon cm$^{-2}$sr$^{-1}$s$^{-2}$ when the Moon phase is near 100 %.[9] This fact determines monthly variation of UHE CR detector threshold. For TUS project the range of threshold variation is from 70 EeV to 300 EeV. "Tatiana-1" satellite orbit allows one to measure night light from aurora oval. The intensity of UV radiation in this region is compatible with full moon scattered light (up to $2\times10^9$ photon $\text{cm}^{-2}\text{sr}^{-1}\text{s}^{-2}$), and long-continued UHE CR measurements are undesirable there because of high energy threshold.

The map of UV intensity obtained by "Tatiana-2" satellite is shown on fig 5. It is seen that Sahara desert and some regions above Pacific and Atlantic oceans are darkest places on this map. Also it is obvious that some industrial regions produce huge amount of UV radiation, about $2 \times 10^8$ photon $\text{cm}^{-2}\text{sr}^{-1}\text{s}^{-2}$ (Europe, North America, Japan). It will influence TUS threshold as well as Moon phase but in smaller time scale (within one satellite circulation). During moonless night the UV noise variation is one order of magnitude and TUS threshold will changes from 70 to 200 EeV. The careful and continuous background monitoring should be provided in UHE CR orbital experiment for correct detector exposition estimations for permanently changeable energy threshold.

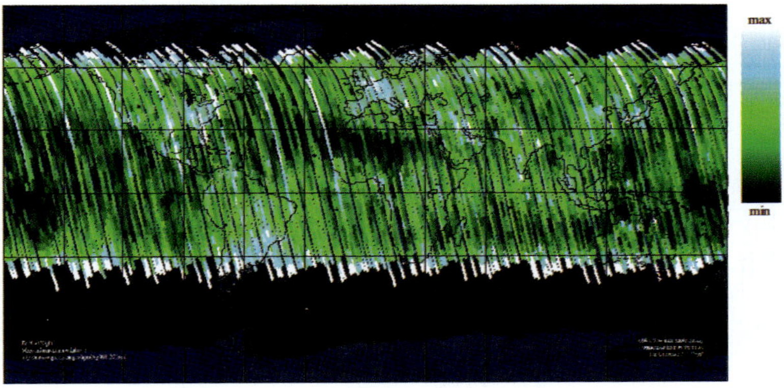


Fig. 5. UV intensity map, obtained in "Tatiaba-2" experiment

### 3.2. *Transient luminous events measurements*

During 2 years of "Tatiana-1" operation it was measured near 300 of transient UV flashes. Such low statistics are explained with low transmitting facilities of satellite. Detector had to record one the most powerful event per orbit circulation. In "Tatiana-2" experiment the data transmittion possibility was higher and it recorded one event per minute with trigger criteria that maximal UV pulse in 1 msec time interval in observation period of 1 minute. This allowed one to enlarge statistics and during only three months of proper work it gathered more than 1000 flashes.

UV detector on board “Tatiana-2” satellite measured only temporal profiles of flashes, but it didn’t show spatial structure and geometry of event in the field of view. Structures of measured events are various. We made the classification of events based on temporal features into 4 types (fig 6).The first two classes are events with clear structure of short (1–3 msec) pulses. The 1-st class are events with more than 50% of UV brightness was concentrated in the trigger pulse . The 3-d class are events with longer pulse structure and the 4-th class are events with saturated signals in part of time samples. Percentage of various UV flash classes is presented in Table 1.

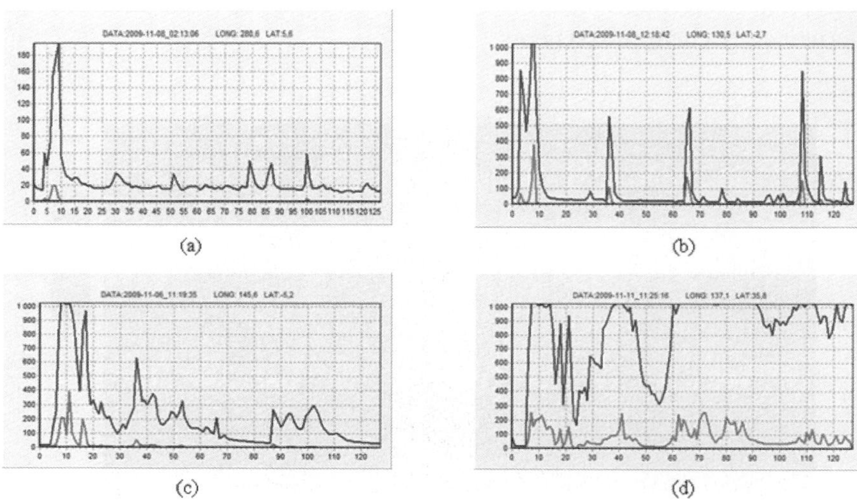


Fig. 6. UV flashes measured by “Tatiana-2” satellite classification ((a) – 1-st class, (b) – 2-d class, (c) – 3-d slass, (d) – 4-th class)

Table 1. UV flashes class distribution.

| Class | 1 | 2 | 3 | 4 |
|---|---|---|---|---|
| Number of flashes, % | 26 | 31 | 33 | 10 |

The important information is energy spectrum of measured events and their rate, because even small UV flashes (tens J) will trigger UHE CR detector. Unfortunately it is not easy to obtain energy spectrum, because measured distribution reflects not only flashes physics, but also several factors: triggering system influence, which decrease low energy events number and geometry of transient in detectors field of view. The energy distribution of measured events is shown on fig 7. It is seen the large amount of low energy events. It is worth attention the geographical distribution of UV flashes. The most powerful (more than 10 kJ) are attracted to well-known thunderstorm regions, but low energy events (less than 5 kJ) have more even distribution (fig 8). The region of flashes observation is limited by satellite night side orbit part.

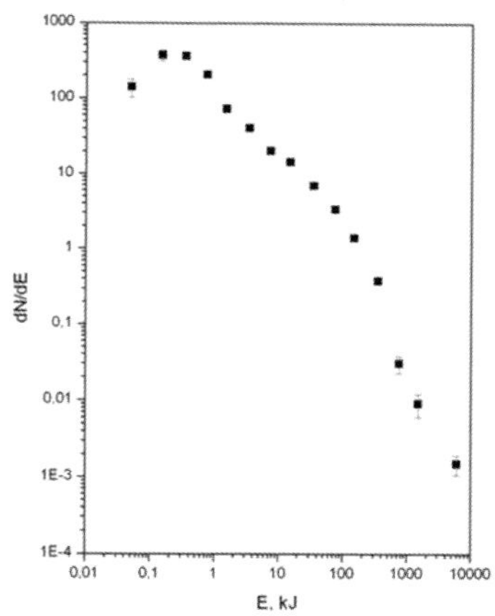


Fig. 7. UV flashes energy distribution

## 4. Expected TUS results

In consideration of TUS orbit close to orbit of ISS we estimated effective operation time for detector with different thresholds. It was found that one year operation time at moonless local nights (with moon-phase less 40%) is 15 % (~ 1400 hours) of all operation time, while all nights time is 3000 hours. Operation of the TUS pixel in the "Universitetsky-Tatiana" mission showed that introduction of feedback control of the PMT's voltage allows one to operate the detector at moonlit nights with reduced gain on PMT. This method applied to the TUS operation will provide the measurement of UHE CR events with operation time of 3000 hours for

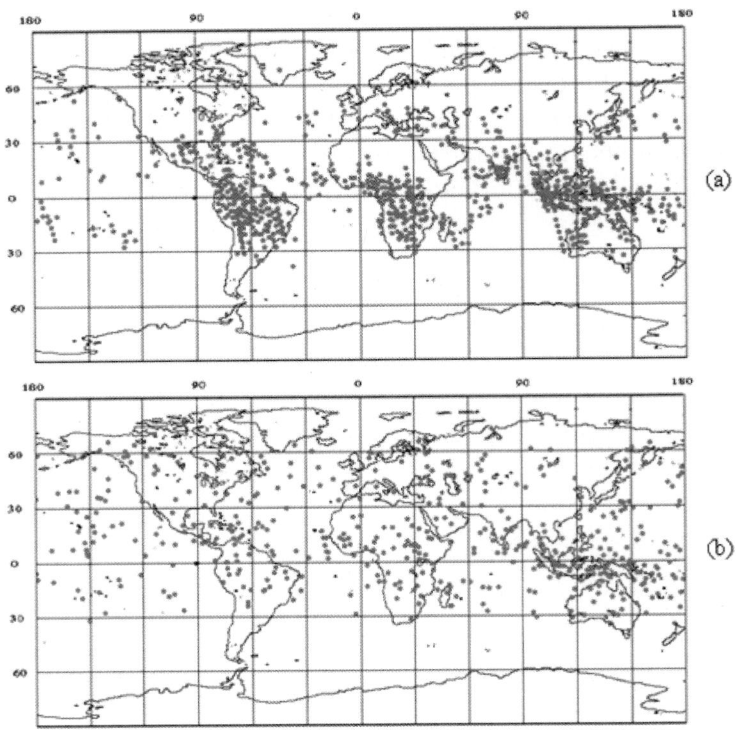


Fig. 8. UV flashes measured by "Tatiana-2" geographical distribution for various energy ranges. (a) – E > 10 kJ, (b) – E < 5 kJ.

energy threshold $E_{\rm thr} = 300$ EeV for "horizontal" primary particle directions. In 3 years of operation in space, the TUS exposure for those events will be 12000 km$^2$ year sr – comparable with the exposure of the largest ground-based UHE CR detectors. If we consider spectrum, obtained on Pierre Auger Observatory, we can expect about 20 events with threshold 70 EeV, or 10 events with threshold 100 EeV during three years of TUS operation.

On the other hand, the GZK cut-off of usual cosmic ray particles (protons, nuclei, photons) gives way for detecting exotic relativistic dust grains, which at lower energies are over numbered by cosmic ray particles. Relativistic dust grains (if they exist) will produce signals, different from usual EECR signals: extensive air showers generated by dust grains are developed much higher in the atmosphere and tracks of them in TUS are longer in time and path.

Very important result of TUS mission will be test of UHE CR space measurements principles, in various conditions, UV background, noise from atmospheric phenomena. TUS detector will contain pin-hole camera for TLE research. In pin-hole camera will be tested JEM-EUSO photo sensor (multi-anode PMT).

## 5. Conclusions

During several years, after 2011, of operation TUS detector will answer several important questions. How really do operate the fluorescence detector in space environment with variable night UV background, and extreme events in the atmosphere such as TLE. It will enlarge worldwide statistics of UHE CR with the exposition comparable to PAO. TUS will provide first uniform monitoring of the whole sky in highest energy range. It will measure first stages of TLE development in the atmosphere with high temporal resolution and signal sensitivity and further discharge development and spatial structure with the pin-hole camera and produce PMT test for JEM–EUSO project.

## Acknowledgments

This work is supported by Russian Fund of Basic Research grant 09-02-12162.

## References

1. J. Abraham et al, Observation of the suppression of the flux of cosmic rays above 4 x $10^{19}$ eV, Phys. Rev. Lett. 101, 061101, 2008.
2. J. Abraham et al, Correlation of the highest-energy cosmic rays with the positions of nearby active galactic nuclei, Astropart. Phys. 29, 188-204, 2008.
3. R. U. Abbasi et al, Analysis of large-scale anisotropy of ultra-high energy cosmic rays in HiRes data,The Astrophysical Journal Letters, 713:L64L68, 2010.
4. J. Abraham, et al., Measurement of the Depth of Maximum of Extensive Air Showers above $10^{18}$ eV, Phys.Rev.Lett, 104, 091101, 2010.
5. R. U. Abbasi et al, Indications of Proton-Dominated Cosmic Ray Composition above 1.6 EeVPhys.Rev.Lett., 104, 161101, 2010.
6. V.I. Abrashkin et al, Updated TUS space fluorescence detector for study of UHECR. Advances in Space Research 41, pp. 2079–2088, 2008.
7. G. K. Garipov et al., Ultraviolet Radiation Detector of the MSU Research Educational Microsatellite Universitetsky-Tatiana. Instruments and Experimental Techniques, Vol. 49, No. 1, pp. 126-131, 2006.

8. G.K. Garipov et al., Program of transient UV event research at Tatiana-2 satellite, J. Geophys. Res. V. 115, 2010, doi:10.1029/2009JA014765.
9. G.K. Garipov et al., UV radiation from the atmosphere: results of the MSU "Tatiana" satellite measurements, Astroparticle Physics, V. 24, pp. 400-408, 2005.

# The observation of the light component spectrum in the 5–250 TeV region by the ARGO-YBJ experiment

S.M. Mari and P. Montini* on behalf of the ARGO-YBJ Collaboration

*Dipartimento di Fisica, Università Roma Tre and INFN, Sezione di Roma Tre, Via della Vasca Navale 84, I-00146 Rome, Italy*

**email: montini@fis.uniroma3.it*

The ARGO-YBJ experiment detects extensive air showers by means of a full coverage detector located at the YBJ International Cosmic Ray Observatory in Yangbajing (4300 m a.s.l, Tibet, P.R. China). In this work the light component spectrum in the energy region between 5 - 250 TeV is presented. The results are compared with data provided by the CREAM experiment in the same energy region. The ARGO-YBJ data are fully consistent with the CREAM data. This analysis is based on a Bayesian procedure.

*Keywords*: Cosmic Rays, EASs, light component spectrum

## 1. Introduction

Cosmic rays in the energy range $10^{14} - 10^{20}$ eV are investigated via ground based experiments that detect showers of secondary particles produced in the interaction of cosmic rays with atmospheric nuclei. Ground based experiments can provide a large collecting area which is necessary to observe the low intensity flux at higher energies. The measurements based on the detection of EASs with surface arrays, however, rely on Monte Carlo simulations in order to unfold the cosmic rays energy spectrum. Direct measurements with satellite or balloon-borne detectors can determine the composition and the energy of cosmic rays. Due to small detector size and reduced exposure time they are not suitable to detect cosmic rays with energies greater than $10^{15}$ eV. A better understanding of the comparison between direct and indirect data can be achieved by extending the indirect measurements in the low energy region. Cosmic ray physics in the 1–$10^4$ TeV energy range is one of the main goals of the ARGO-YBJ experiment.[1–3] The detector is operating in the YBJ International Cosmic Ray Observatory located at Yangbajing, in the Tibet region (P. R. China) at an altitude of about 4300

m above sea level, which corresponds to a vertical depth of 606 g cm$^2$. The apparatus is a single layer full coverage detector logically divided into 130 clusters ($5.72 \times 7.64$ m$^2$) each made by 12 Resistive Plate Chambers (RPCs),[3] 24 additional clusters form a guard ring surrounding the central carpet in order to improve the core reconstruction. Each RPC is read out by using 10 pads ($55.6 \times 61.8$ cm$^2$) which are further divided into 8 strips ($6.75 \times 61.80$ cm$^2$). The system provides a single hit time resolution of about 1.8 ns,[4] allowing a complete and detailed reconstruction of the shower front. A system for the analog charge readout of RPCs is now being implemented. This will allow one to extend the detector operating range up to energies of about 1 PeV.
In this work the light component spectrum in the energy region between 5–250 TeV is presented. This analysis is based on a Bayesian unfolding procedure.[5,6]

## 2. Data analysis and results

The Monte Carlo data sample used in this analysis was generated by using the CORSIKA (v. 6.7.10) code including QGSJET-II and GEISHA hadronic interaction models. The data have been generated in the energy range 1–10$^4$ TeV with an energy distribution given by a power law. The showers were sampled at an altitude of 4300 m, which corresponds to the ARGO-YBJ site. A full detector simulation based on GEANT3 package was then applied, taking into account backgrounds, trigger and RPC's efficiency. Only protons and helium nuclei have been generated, because the heavier components give a minor contribution in this energy interval[7–9] and the contribution of the CNO component is suppressed by the selection criteria used in this analysis and it turns out to be less then 2 %.

The data used in this analysis were collected by the central 130 clusters and by using an inclusive trigger requiring at least 20 fired pads in a time window of 420 ns. A first selection based on the reconstruction quality was applied then additional cuts have been applied in order to estimate with good accuracy the probabilities used in the unfolding procedure and to make negligible the contamination of external events. The following selection criteria were adopted.

The data and Monte Carlo events were selected requiring that the zenith reconstructed angle is in the range $0° \div 30°$. The Monte Carlo events used in this analysis were generated with the zenith angle in the range $0° \div 45°$ (See Fig. 1). The events were selected requiring that the strip multiplicity M, measured for each event, is in the range $500 \leq M \leq 50000$. This cut

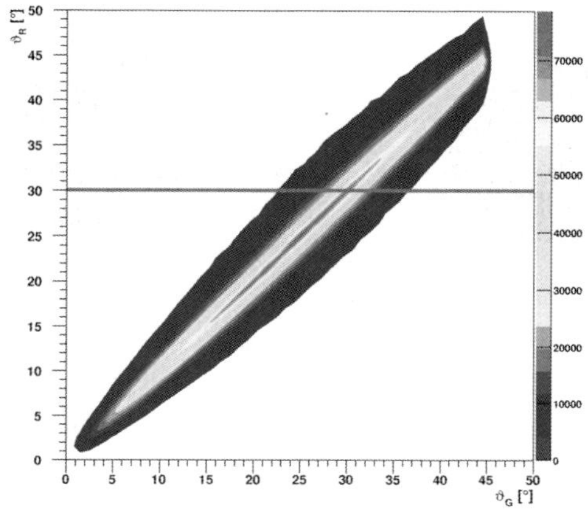


Fig. 1: The distribution of the reconstructed zenith angle $\theta_R$ versus the generated zenith angle $\theta_G$. The line shows the cut used in this analysis.

selects events in an energy range internal to the range ($1 \div 10^4$ TeV) used for the events simulation. The rejection of events falling far from the detector is achieved by comparing the average particle density measured by the innermost clusters ($\rho_{in}$) to the one measured by the outermost ones ($\rho_{out}$) and requiring that $\rho_{in} > \rho_{out}$.

In Fig. 2 the distribution of the core position ($x_{core}, y_{core}$) is shown for Monte Carlo events and data. The figure shows clearly that all the events surviving the cuts used in this analysis are located in the innermost area of about $50 \times 50\ m^2$.

In order to measure the light-component spectrum the Monte Carlo events have been sorted in seven energy bins and twelve multiplicity bins. The same multiplicity bins have been used to analyze a sample of data. The Bayesian unfolding was performed by using the probabilities computed by means of the Monte Carlo events.

The results are shown in Fig. 3. The statistical error on each point turns out to be less than 1 %. The sources of the systematic effects considered in this work are: effects due to the selection cuts, effects due to the reliability of the simulation of the detector response, effects related to the fraction of helium component used to evaluate the Bayesian probabilities. The uncertainty on the measured spectrum, due to the selection cuts, is about $\pm 5\%$. An estimate of the uncertainty due to the effect of different run conditions

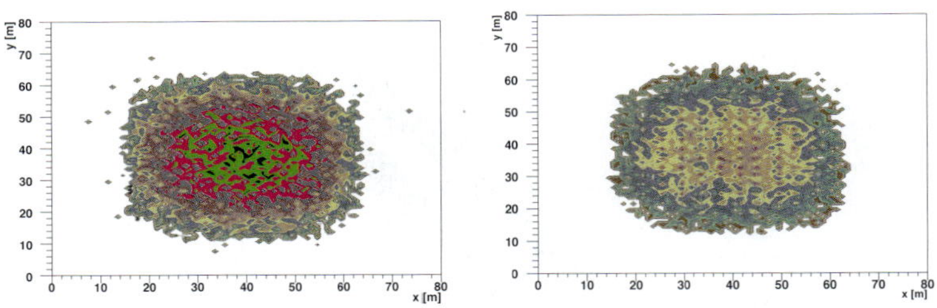


(a) Monte Carlo (b) Data

Fig. 2: Distribution of the true core position $(x_{core}, y_{core})$ for Monte Carlo events (a) and of the reconstructed core position for data events (b) selected by the criteria used in this analysis.

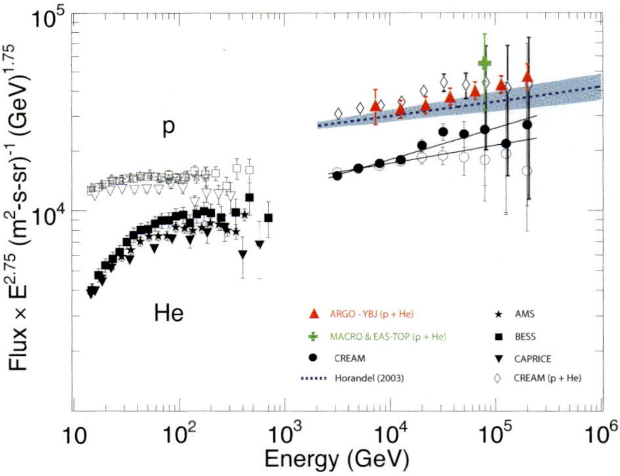


Fig. 3: The differential energy spectrum of the light-component (proton and helium) measured by ARGO-YBJ (filled triangles) compared with the results provided by other experiments. The dashed line represents the best fit to proton and helium data performed by Hörandel.[10] The open diamonds represent our sum of the proton (open circles) and helium (filled circles) data measured by CREAM experiment.[11] The spectra measured by AMS (stars), BESS (squares), CAPRICE (inverted triangles) are shown.

has been obtained by analyzing three runs separately and it turns out to be about ±4%. The effect of the variation of the helium component in the range ($7 \div 11\%$) has been evaluated as negligible. We estimate that the obtained result is affected by a total uncertainty not exceeding $\pm 10\%$. In order to have a very preliminary estimation of the systematic uncertanties due to different hadronic interaction model the relevant quantities used in this analysis has been evaluated using QGSJET-II and SYBILL high energy interaction models, showing not significant differences.

The cosmic ray light-component energy spectrum measured in this work is found in agreement within the systematics with the one quoted by Hörandel,[10] however it shows a harder spectral index. On the contrary, the ARGO-YBJ data are in good agreement with the values obtained by adding up the the proton and helium fluxes measured by CREAM experiment.[11]

## 3. Conclusions

The peculiar characteristics of the ARGO-YBJ experiment allow one to detect showers produced by cosmic rays in the TeV region. The present analysis does not allow the determination of the individual proton and helium contribution to the measured flux. However it suggests an hardening of the proton spectrum at energies $> 5$ TeV respect to the one obtained by low energy data. The Bayesian unfolding procedure used to analyze the experimental data recorded by ARGO-YBJ can be successfully applied to infer the cosmic ray spectra in air shower experiments.

## Acknowledgments

This work is supported in China by NSFC (Contract No. 101201307940), the Chinese Ministry of Science and Technology, the Chinese Academy of Science, the Key Laboratory of Particle Astrophysics, CAS, and in Italy by the Istituto Nazionale di Fisica Nucleare (INFN), and Ministero dell'Istruzione, dell'Università e della Ricerca (MIUR).

## References

1. C. Bacci et al., *Nucl. Phys. Proc. Suppl.* **78**, 38 (1999).
2. C. Bacci et al., *Astropart. Phys.* **17**, 151 (2002).
3. G. Aielli et al., *Nucl. Instrum. Meth.* **A562**, 92 (2006).
4. G. Aielli et al., *Nucl. Instrum. Meth.* **A608**, 246 (2009).
5. G. D'Agostini, *Nucl. Instrum. Meth.* **A362**, 487 (1995).
6. S.Bussino, E. De Marinis and S.M.Mari, *Astropart. Phys.* **22**, 81 (2004).
7. M. Amenomori et al., *Astrophys. J.* **678**, 1165 (2008).

8. M. Ave et al., *Astrophys. J.* **678**, 262 (2008).
9. A. V. Apanasenko et al., *Astropart. Phys.* **16**, 13 (2001).
10. J.R. Hörandel, *Astropart. Phys.* **19**, 193 (2003).
11. H. S. Ahn et al., *The Astrophysical Journal Letter* **714**, L89 (2010).

# Status and plans of the LUCIFER Experiment

F. Orio

*Sapienza Università di Roma and Sezione INFN di Roma,*
*Roma, 00185, Italy*
*E-mail: filippo.orio@roma1.infn.it*

Neutrinoless Double Beta Decay searches are the only way to give an answer to the neutrino mass nature, Dirac or Majorana. These experiments are extremely delicate and the greatest obstacle to improve their sensitivity is the background level that can be achieved. LUCIFER is a project, financed by a ERC-AdG, that would like to build a demonstrator of a technique based on the double read-out (scintillation light + heat) of ZnSe crystals used as bolometers. The goal is to reach a background lower than 0.001 counts/kg/keV/year. Along the way we would like to learn how practical is the enrichment of Se into $^{82}$Se, how efficient is the process of crystallization and how radiopure the crystals can be grown. In this talk we will discuss the properties of ZnSe crystals and sketch the layout of the project.

*Keywords*: Neutrino mass; Double Beta Decay; Scintillating Bolometers; LUCIFER

## 1. Introduction

In the last decade much progress has been made in neutrino physics. Oscillation experiments created a clear picture of this elusive particle and we are now entering the era of precision measurements. There are however questions that cannot be addressed by oscillation experiments. The absolute mass scale of the neutrino is one of these, and it is considered a key quantity in many theories beyond the Standard Model of particle physics. Being only sensitive to square mass differences, oscillation experiments are not able to measure this parameter. Moreover, the mechanism that is responsible for the generation of neutrino masses is still unknown. Neutrinos are electrically neutral particles, and the only carried charge is that of weak interactions. In the Standard Model there is no symmetry constraining the conservation of lepton number, even though a non-conservation has never been observed. If lepton number is not conserved, neutrinos could be equal to their antiparticles, thus being Majorana particles.

The Double Beta decay without emission of neutrinos ($0\nu$DBD) violates the lepton number by two units, and has never been observed. The observation of this nuclear decay would imply that neutrinos are Majorana particles, a breakthrough in the picture of nature we have. Moreover it could provide information on the absolute mass scale of neutrinos because a virtual neutrino is exchanged and the propagator is proportional to its mass $m_{\beta\beta}$.

Neutrinoless Double Beta decay is, at present, the most sensitive method to study neutrino properties and bolometers[1] are - together with germanium diodes - the detectors which have provided the best results so far. Bolometers allow the application of the so-called "source=detector" approach, where the detector is composed of the same material candidate to decay, and in the mean time allow the study of many isotopes by means of high resolution devices (FWHM around 0.2-0.5 % above 2500 keV), this latter feature being necessary to resolve the searched peak from background.

The purpose of the $0\nu$DBD new generation experiments is to reach a sensitivity on $m_{\beta\beta}$ of the order of 50 meV.[2–5] The high sensitivity required implies excellent energy resolution, a low number of spurious counts within the region of interest (low background) and a high quantity of the isotope on which the study focuses (large mass).

Further improvements cannot rely simply on the mass increase or on a better energy resolution but will require the implementation of innovative techniques for background discrimination.

In the case of a scintillating bolometer the double independent readout (heat and scintillation) will allow, thanks to the different scintillation yield between $\alpha$ and $\beta$ - $\gamma$ particles, the suppression of $\alpha$ background events. These have been identified as the most important background source in bolometers dedicated to $0\nu$DBD searches.[2,6] Furthermore, the contribution of environmental gammas can be strongly reduced using a scintillating bolometer containing a DBD emitter whose transition energy exceeds the highest gamma line due to natural radioactivity (the 2615 keV line of $^{208}$Tl). Examples are $^{116}$Cd ($Q_{\beta\beta}$ = 2805 keV), $^{82}$Se ($Q_{\beta\beta}$ = 3000 keV) and $^{100}$Mo ($Q_{\beta\beta}$ = 3030 keV). Finally, this technique is also extremely helpful for rejecting other unavoidable sources of background such as direct interactions of neutrons.

## 2. Bolometric detectors

Bolometers are calorimeters operating at low temperature, in which the energy deposited by a particle is converted into phonons and is detected as a temperature variation. In this approach, the energy deposited by a single particle into an energy absorber (weakly connected to a heat sink) determines an increase of its temperature $T$. This variation corresponds simply to the ratio between the energy released by the particle and the heat capacity of the absorber. The only requirements are therefore to operate the device at low temperatures (less than 0.015 K), in order to make the heat capacity low enough, and to have a sensitive thermometer coupled to the energy absorber. The thermometer is usually a thermistor, a resistive device with a steep dependence of the resistance on the temperature.

The energy absorbing part of the detector is made usually of diamagnetic, dielectric crystal in order to avoid electron contributions to the specific heat in addition to the Debye term, proportional to $T^3$ at low temperatures.

Semiconductor thermistors consist usually of Ge or Si small crystals with a net dopant concentration slightly below the metal-insulator transition. This implies a strong dependence of the sensor resistivity on the temperature at low temperatures, where the Variable Range Hopping conduction mechanism dominates.

## 3. Scintillating bolometers

When the bolometer crystal is an efficient scintillator at low temperatures, a small but significant fraction of the deposited energy (up to a few %) is converted into scintillation photons while the remaining dominant part is detected in the form of heat. The emitted light can be measured by a light detector facing the scintillating bolometer[7] (see Fig 1).

The idea to use a bolometer as light detector was first developed by Bobin et al.[8] and then optimized[9,10] for Dark Matter (DM) searches. The first simultaneous detection of phonons and photons was achieved more than a decade ago by the Milano group with a $CaF_2$ bolometer and a conventional Si photodiode as light detector, developed as a pilot device for the search of 0$\nu$DBD in $^{48}$Ca.

Nowadays it is clear that the effective method to detect scintillation photons in such a very low temperature environment is to develop a dedicated bolometer, in form of a thin slab, opaque to the emitted light and provided with its own phonon sensor. This auxiliar bolometer, normally a

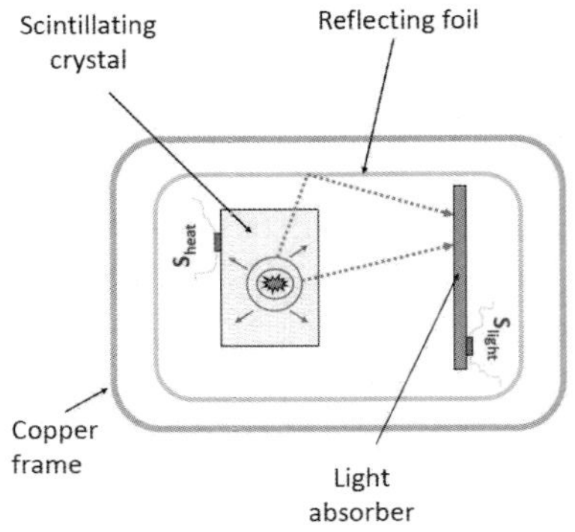


Fig. 1. Scheme of a scintillating bolometer.

Si or Ge slab, is positioned very close to a flat, optically polished, side of the main scintillating bolometer. In order to maximize the light collection, it is convenient to surround the whole set-up with a reflecting foil.

## 4. Environmental Background

In a bolometric experiment, the signature of the 0$\nu$DBD is a peak at the $Q_{\beta\beta}$ value of the transition, while in ordinary Double Beta decay (2$\nu$DBD) is a continuous beta-like spectrum. However, thanks to the excellent energy resolutions of bolometers, 2$\nu$DBD is seldom a significant source of background.

There are various sources that give rise to spurious counts in the region of interest such as external $\gamma$ background, neutrons, surface contaminations, $^{238}$U and $^{232}$Th internal contaminations and cosmogenic activity. As already discussed, a way to eliminate the problem of $\gamma$ lines is to study isotopes with transition energy above the 2615 keV line. This is the $^{208}$Tl line that is the highest energy $\gamma$-ray line from natural radioactivity. Above this energy there are only extremely rare $\gamma$'s from $^{214}$Bi: the total Branching Ratio in the energy window from 2615 up to 3270 keV is 0.15 % in the $^{238}$U decay chain. All the 0$\nu$DBD isotopes with $Q_{\beta\beta}$ >2615 keV are listed in Table 1.

The use of scintillating crystals as absorbers can further overcome the external background limitations, providing a powerful tool to improve the experimental 0$\nu$DBD sensitivities. In fact, thanks to the different scintillation yield of different particles (namely $\beta$ - $\gamma$, $\alpha$ and neutrons) they can be very efficiently discriminated.

In particular, if the scintillating crystal is made of a 0$\nu$DBD candidate,

Table 1. Double Beta decay isotopes with endpoint energies above the $^{208}$Tl line.[11]

| Isotope | $Q_{\beta\beta}$ [MeV] | Natural abundance |
|---|---|---|
| $^{116}$Cd | 2.80 | 7.5 % |
| $^{82}$Se | 3.00 | 9.2 % |
| $^{100}$Mo | 3.03 | 9.6 % |
| $^{96}$Zr | 3.35 | 2.8 % |
| $^{150}$Nd | 3.37 | 5.6 % |
| $^{48}$Ca | 4.27 | 0.19 % |

the $0\nu$DBD signal (i.e. the energy deposition produced by the two electrons emitted in the decay) can be distinguished from an $\alpha$ signal. This makes feasible the rejection of $\alpha$ induced background, opening a new window in the future of $0\nu$DBD with bolometers.[12] In fact, it is true that, in the recently concluded bolometric experiment CUORICINO, the major source of background is identified in $\alpha$ particles due to surface contaminations.[13]

## 5. The LUCIFER project

LUCIFER is a Double Beta decay pilot project based on scintillating bolometers, that has been recently financed by a ERC-AdG. The experimental basis for this project is the R&D activity performed by S. Pirro at Laboratori Nazionali del Gran Sasso (LNGS) in the framework of the programs BOLUX and ILIAS-IDEA.

### 5.1. *Isotope choice*

Not all the $0\nu$DBD isotopes can be used to fabricate a good scintillating crystal. In the last years different crystals were tested in dedicated runs to study their characteristics such as their thermal response, light yield and radio purity. Practical scintillators can be built out Mo, Cd and Se.
In Table 2 the three LUCIFER candidates are compared with respect to the physics sensitivity achievable with 48 125 cm$^3$ crystals. Since sensitivity is the figure of merit of the experiment, LUCIFER baseline choice is $^{82}$Se. In Table 3 the same three candidates are compared with respect to the measured light yields. As can bee observed in the table, $\alpha$ particles in ZnSe have higher light yield than $\beta$ particles. Although not welcome, this unexpected characteristic, according to results of preliminary investigations, does not degrade the discrimination power of this material (see Fig. 2). Preliminary studies, in particular, reveal the capability of an $\alpha$ rejection in

excess of 99%. Since the characteristic $\alpha$ background of such experiments is about 0.1 counts/kg/keV/year, LUCIFER aims to reach a value lower than 0.001 counts/kg/keV/year. In order to confirm these statements, a careful study on the prototype crystals for the experiment is foreseen.

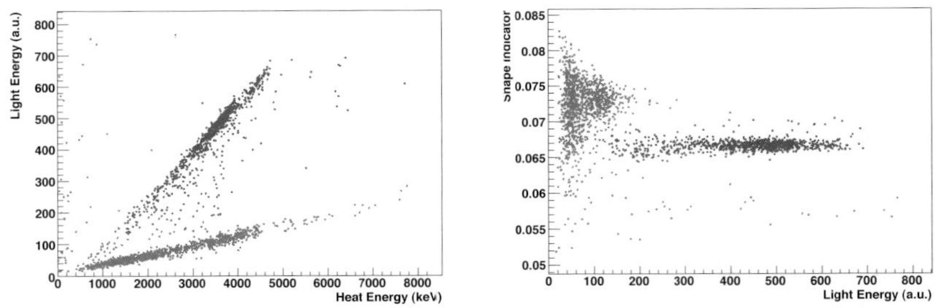


Fig. 2. Comparison between $\alpha$ events from $^{234}$U and $^{238}$U sources (in blue) and $\gamma$ events generated by neutrons from AmBe source (in red). Combining the informations of the two plots, more than 99% of $\alpha$ particles can be rejected, maintaining a $\gamma$ identification efficiency in excess of 98%. Left: Scatter plot of light signal amplitudes vs. heat signal amplitudes (calibrated in energy). Right: Scatter plot of shape sensitive parameter values vs light signal amplitudes.

Table 2. Sensitivity of the three experimental options for LUCIFER.

| Crystal | Isotope weight[a] | Useful material[b] | $Q_{\beta\beta}$ value [keV] | Sensitivity[c] to $m_{ee}$ [meV] |
|---|---|---|---|---|
| $CdWO_4$ | $^{116}$Cd 15.1 kg | 32% | 2809 | 65-80 |
| $ZnMoO_4$ | $^{100}$Mo 11.3 kg | 44% | 3034 | 67-73 |
| ZnSe | $^{82}$Se 17.6 kg | 56% | 2995 | 52-65 |

*Note*: [a] It has been considered a detector made of 48 125 cm$^3$ crystals.
[b] The percentages take into account an enrichment level of the 0$\nu$DBD candidate isotope as high as 97%. The feasibility and the cost of the enrichment procedure are under control thanks to the experience gathered by the NEMO3 collaboration and the investigation performed in the framework of ILIAS.
[c] The 1 $\sigma$ sensitivity is calculated with the Feldman Cousin approach for 5 y running, considering a background of $10^{-3}$ counts/keV/kg/y, the matrix element coming from the two most recent QRPA calculations[14–17] and a resolution of 5 keV.

Table 3. $\beta$ light yields (LY) and quenching factors (QF) for interesting scintillating bolometers.

| Crystal | $LY_\beta$ [keV/MeV] | QF ($LY_\alpha/LY_\beta$) |
|---|---|---|
| $CdWO_4$ | 34 | 0.19 |
| $ZnMoO_4$ | 1.4 | 0.16 |
| ZnSe | 7.4 | 4.2 |

### 5.2. *Single module structure*

LUCIFER will consist of an array of tens of individual bolometers arranged in a close-packed configuration and housed by the same cryogenic infrastructure. This approach allows one to achieve large masses keeping the single-crystal mass reasonable and providing a rough granularity, very helpful for background identification and rejection.

The foreseen elementary module structure is shown in Fig. 3. Four ZnSe cubic crystals are arranged in each module. The cubes will have 5 cm side, leading to a single crystal mass of 660 g ($\rho_{ZnSe} = 5.27\,\mathrm{g/cm^3}$).

Two copper frames, connected by four copper columns, bound the module above and below. These frames, which act as a 10 mK heat sink for the bolometers, hold 14 PTFE elements, serving as mechanical support and thermal link for the crystals.

One Neutron Transmutation Doped (NTD) Ge thermistor is glued on each crystal. Two Au wires are ball-bonded from the thermistor contacts to pads located on the frames for the signal read-out. A resistive heater, to be use for the offline correction of the gain variations, is glued on the crystal as well and read out as the thermistors.

The module upper frame is designed so as to house an intrinsic-Si disk-shaped slab 0.5 mm thick and 9.5 cm diameter, acting as a photon absorber for the light collection. The slab is placed above the ZnSe crystals and the PTFE elements supporting them, and will be kept by four PTFE clamps fixed at the frame. A NTD Ge thermistor will be glued on the slab. This configuration allows the light detector to see the four crystals underneath and to offer its upper face to the four crystals of the adjacent module.

The module can be inscribed in a cylinder with 18 cm diameter and a total height of 7 cm. A reflecting foil will be placed laterally, connecting the two frames, to improve the light collection.

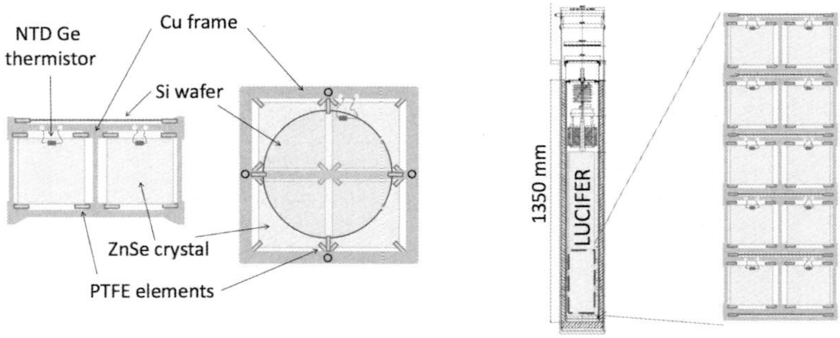


Fig. 3. Left: Elementary module of LUCIFER, with four ZnSe scintillating crystals read out by a single light detector. Right: The LUCIFER tower inside the cryostat. A detail of the tower, comprising 5 elementary modules, is shown on the right.

### 5.3. *Full detector structure*

The described modules will be stacked so as to form a tower of 12 floors as in Fig. 3. The sequence will start from below with an isolated frame similar to the upper frame of a single module, containing a light detector. The 12 modules will be then assembled one above the other, giving a total tower height of 85 cm.

The detector array will be placed inside a dilution cryostat, yet to be defined. One of the most feasible options is the one previously used for the CUORICINO experiment at LNGS. The tower will be then surrounded by a copper thermal shield and suspended using a stainless steel spring connected to the 50 mK plate of the fridge.

The front end electronics, the post-amplifiers and the DAQ system will be the same that were used for CUORICINO.[18]

## 6. Present status and future plans

The project is at the phase of procuring the enriched material and establishing a vigorous effort for the production of high quality, reproducible ZnSe crystals. As for the enrichment, the project is at the negotiation phase with potential companies in western Europe and Russia. The crystals will be produced at ISMA (Ukraine) where a detailed R&D program has been agreed.

Great effort is dedicated to the study of the ZnSe stoichiometry, looking for the solution that maximizes the light yield. Concerning the light detectors, ultrapure Ge thin slabs are the baseline choice but different options, like Si slabs, are currently in phase of study.

**Acknowledgments**

The project LUCIFER has received funding from the European Research Council under the European Unions Seventh Framework Programme (FP7/2007-2013) / ERC grant agreement n. 247115.

**References**

1. E. Fiorini and T. O. Niinikoski, Nucl. Instr. Meth A **224** (2004) 83
2. C. Arnaboldi *et al.*, Nucl. Instr. Meth. A **518** (2004) 775
3. S. Scheonert *et al.*, Proc. Neutrino 2006 (Santa Fe, 2006)
4. C. E. Aalseth *et al.*, Yad. Fiz. **67** (2004) 11
5. A. Piepke *et al.*, Proc. Neutrino (Santa Fe, 2006)
6. C. Arnaboldi *et al.*, Phys. Rev. Lett. **95** (2005) 142501
7. A. Alessandrello *et al.*, Physics Letters B **420** (1998) 109
8. C. Bobin *et al.*, Nucl. Instr. Meth. A **386** (1994) 453
9. G. Angloher *et al.*, Astropart. Phys. **23** (2005) 325
10. S. Cebrian *et al.*, Phys. Lett. B **563** (2003) 48
11. S. Y. F. Chu *et al.*, nucleardata.nuclear.lu.se/nucleardata/toi (2005).
12. S. Pirro et al., Phys. Atom. Nucl. **69** (2006) 2109
13. C. Arnaboldi *et al.*, Phys. Rev. C **78**(2008) 035502
14. J. Menendez *et al.*, NPA **818** (2009) 139
15. O. Civitarese *et al.*, JoP:Conference series **173** (2009) 012012
16. F. Simkovic *et al.*, PRC **77** (2008) 045503
17. J. Barea and F. Iachello, PRC **79** (2009) 044301
18. C. Arnaboldi *et al.*, IEEE Trans. Nucl. Sci. **49**, (2002) 2440

# In-flight measurement of the absolute energy scale of the Fermi Large Area Telescope

M. Pesce-Rollins on behalf of the Fermi LAT collaboration

*Istituto Nazionale di Fisica Nucleare*
*Sezione di Pisa*
*Largo B. Pontecorvo, 3*
*Pisa, Italy 56127*
*E-mail: melissa.pesce.rollins@pi.infn.it*

The Large Area Telescope (LAT) on-board the Fermi Gamma-ray Space Telescope is a pair conversion telescope designed to survey the gamma-ray sky from 20 MeV to several hundreds of GeV. The geomagnetic cutoff in the Cosmic-Ray Electron (CRE) spectrum in low-Earth orbit is a spectral feature that can be potentially exploited to perform an in-flight calibration of the absolute energy scale of the instrument. The absolute position and shape of this cutoff can be predicted with the aid of a numerical code tracing charged particles in the Earth's magnetic field. By comparing these values with the those measured by the LAT in different geomagnetic positions it is possible to obtain several calibration points between ~7 and ~15 GeV/c.

*Keywords*: Trajectory tracing; Fermi LAT; Geomagnetic cutoff rigidity

## 1. Introduction

Validating the performance of an instrument is always a delicate and important task for any experiment. This task is greatly complicated when the instrument is in orbit, as is the case for the Fermi Large Area Telescope (LAT). To perform an in-flight validation of the absolute energy scale of the LAT it is necessary to find an astronomical source with a spectral feature whose absolute position and shape are well known. Given the orbital position of the Fermi observatory (inclination of 25.6° and altitude of 565 km) the geomagnetic rigidity cutoff in the Cosmic Ray electron plus positron (CRE) spectrum can serve as such a source in the range between $\sim 7$ and $\sim 15$ GeV/c. In articles such as[1,2] the extraordinary capabilities of the LAT to measure the CRE spectrum (including the cutoff rigidity feature) have been demonstrated. The value for the cutoff rigidity can be predicted by numerically tracing particle trajectories in the Earth's magnetic field. The comparison of these values with those measured in flight provide the opportunity to perform this validation. In this paper I present the analysis performed and the preliminary results of the in-flight validation of the absolute energy scale of the Fermi LAT.

## 2. Particle tracing

Geomagnetic cutoff rigidities can be obtained by tracing the orbits of cosmic ray trajectories in a model of the Earth's magnetic field. The standard mathematical description of the Earth's magnetic field is given by the International Geomagnetic Reference Field models (IGRF)[3a] and a popular code to perform the particle tracing (hereafter tracer) has been written by D. F. Smart and M.A. Shea.[4]

Given that it is difficult to trace the trajectory of an incoming particle through the magnetic field and expect it to intersect the precise location which is desired, it is commonly accepted to calculate the trajectory in the reverse direction. Therefore the starting point of the trajectory is given by the geographical coordinates, direction and altitude of the location in question and the trajectories are calculated outward towards interplanetary space. The output of the code (amongst other things) tells you whether a particle with these initial conditions escaped from the Earth's B field or re-enters the atmosphere and becomes trapped. Where the escaped condition implies that the particle is at a distance larger than the pre-defined escaped radius (taken to be $20R_{\oplus}$ for this analysis) and a re-entering particle is at a distance closer than 20 km from the Earth surface.

The energy distribution of the electrons and positrons has been extracted following a pure power law with the same spectral index as measured by the Fermi LAT.[1] The angular distributions ($\theta$,$\phi$), in instrument coordinates, of the electrons and positrons are extracted from Monte Carlo simulations used to evaluate the instrument response functions. However, as stated above, to correctly calculate the trajectories it is necessary to provide the initial coordinates with respect to the Earth therefore, I perform a coordinate transformation with the aid of the spacecraft pointing history. In figure 1 are the distributions of the reconstructed angle with respect to local zenith (left panel) and the reconstructed heading with respect to the Earth (right panel) for the flight data (black) and for the tracer output (red). The overall agreement is very good demonstrating that the angu-

[a]IGRF-11 is used in this analysis

lar distribution have been well described in the tracer code. In addition to the measurements made by Fermi, I also included the positron fraction measured by the PAMELA experiment.[5]

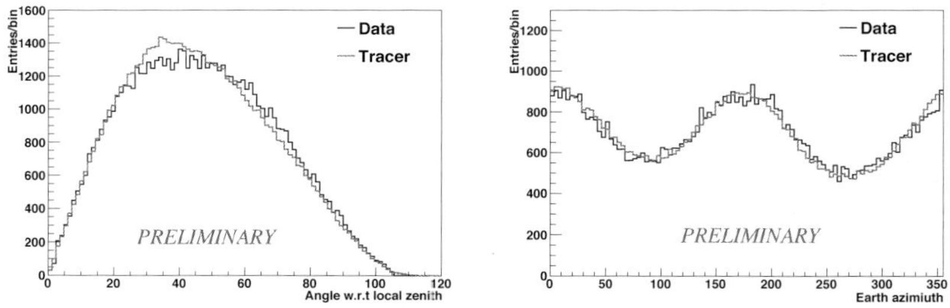


Fig. 1. The reconstructed angle with respect to local zenith (left panel) and the reconstructed heading with respect to the Earth (right panel) for data (black) and tracer (red). Both are orbital averaged and for energies greater than 20 GeV in order compare the distribution of primary CREs. There is a good agreement between data and tracer.

## 3. Estimating the fraction of primary Cosmic Ray electrons

Given that the trajectories of the trapped particles is essentially chaotic, the output from the tracer code only provides information on escaped (or primary) particles. It is therefore necessary to estimate the fraction of trapped particles, as a function of energy, in the flight data. To this aim it is possible to exploit the fact that primary and trapped particles have different measured azimuthal distributions in Earth coordinates and perform a template fitting. By assuming that the population at low energy ($E<<E_c$) represents that of pure trapped particles, I can use this distribution as the template of the trapped particles. While the template for the primaries is taken from the tracer code.

In figure 2 is an example of the template fitting performed to estimate the fraction of primaries and trapped particles for the interval 1.20 < McIlwain L < 1.28. From these figures, the gradual transition from a pure sample of trapped particles (top left panel, for energies between 2 and 4 GeV) to that of only primaries (bottom right panel, for energies greater than 14 GeV) is evident. For reference, the rigidity cutoff in this McIlwain L[4] interval is $\sim$ 9 GeV/c. Only four energy intervals are shown for simplicity, however the analysis has been performed in 16 overlapping energy intervals.

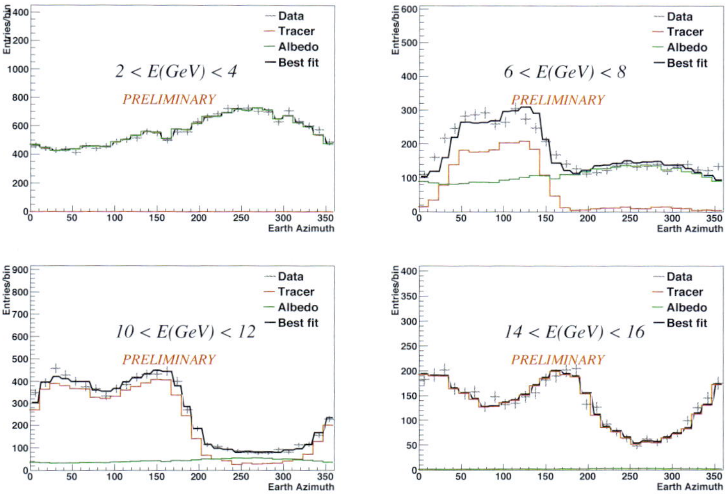


Fig. 2. Distribution of Earth Azimuth for flight data (dark gray), tracer (red) and trapped (green) in the McIlwain L interval 1.20 < L < 1.28. The black distribution is the linear combination of tracer and trapped which best fits the data. Each panel depicts the distributions in a given energy interval (labeled in the panel).The rigidity cutoff in this McIlwain L range is $\sim$ 9 GeV/c.

### 3.1. *Measuring the cutoff rigidity*

The primary $e^- + e^+$ spectrum can be parameterized by:[1]

$$cE^{-\Gamma}/(1 + (E/E_c)^{-6}) \tag{1}$$

Where the $\Gamma$ is the spectral index and $E_c$ is the value for the rigidity cutoff. By measuring the spectrum in different McIlwain L intervals one can obtain several rigidity cutoff values to compare with the reference values from the tracer code. To properly compare the rigidity cutoffs, the tracer spectra have been binned in measured energy (using the energy resolution of the Fermi LAT to electrons and positrons).

To reconstruct the primary spectrum from flight data I remove the trapped particles from the counts spectrum using the fraction of primaries estimated as described in section 3, divide by the geometry factor and the width of the energy interval. In the left column of figure 3, the resulting spectra for both data (blue) and tracer (red) are shown while the fraction of primaries (blue) and trapped particles (red) are shown in the right column of the same figure. Only three McIlwain L intervals are shown, however the analysis has been performed in five intervals in the range [1.0,1.56]. The

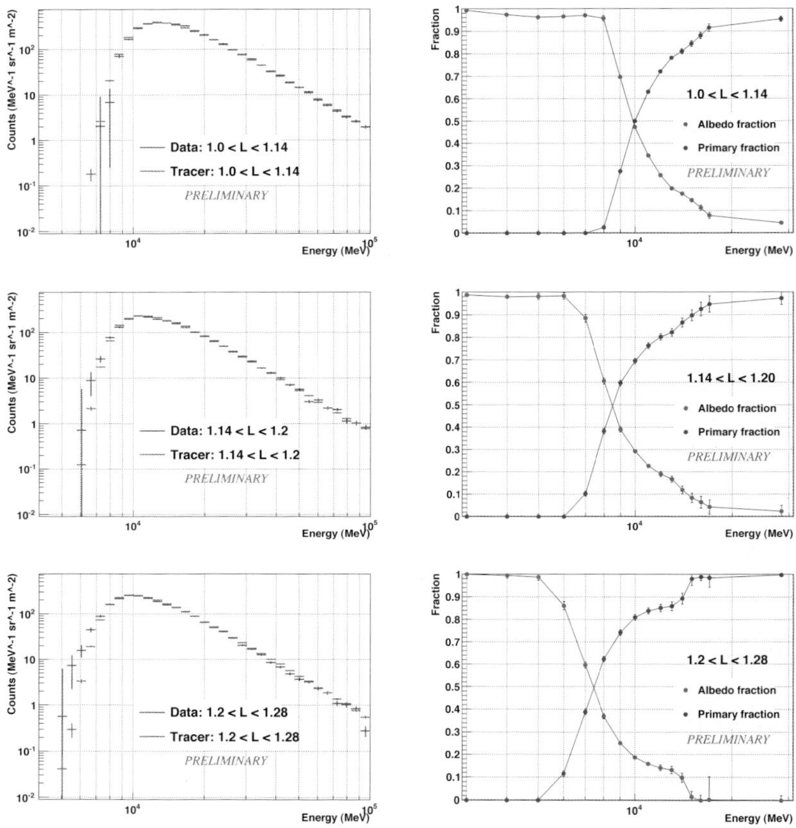


Fig. 3. Left panel: Spectrum of primary $e^- + e^+$ from flight data (blue) and tracer (red). Right panel: Fraction of secondaries (red) and primaries (blue) as a function of energy.

agreement between the data and tracer spectra is good and the only region in which the two spectra show a slight deviation is for energies well below the cutoff rigidity. These regions are however excluded when fitting the spectrum.

The choice of the energy interval in which to fit the spectrum can influence the final value of the cutoff rigidity and therefore be cause of systematic uncertainty. To estimate its value I performed a fit scan over several energy intervals around the optimal one. The mean of the resulting distribution gives the value for the uncertainty on the absolute energy scale, while the rms gives the estimate for the systematic uncertainty. To quantify the total

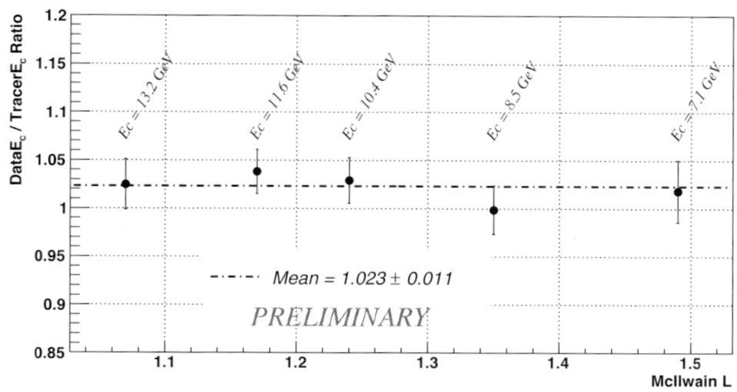


Fig. 4. Distribution of cutoff rigidity ratios as a function of McIlwain L. The ratio values are constant over the entire range. Quoted value for the ratio is the weighted mean. As a reference, the value for the cutoff rigidity in each McIlwain L interval is also shown.

error, the systematic and fit errors are summed in quadrature. The systematic uncertainties associated to the tracer code (e.g. from magnetic field intensity and stability) still need to quantified.

## 4. Conclusions

From this analysis, the uncertainty on the absolute energy scale for the LAT is found to be $-2.3 \pm 1.1\%$. As can be seen in figure 4, the resulting value for the ratios of $E_c$ from tracer and data is constant over the entire McIlwain L range ($\sim 7$ - $\sim 15$ GeV/c) considered. The fit and the systematic errors have been summed in quadrature and the quoted value for the ratio is the weighted mean. A more in-depth study of the systematic uncertainties, particularly those associated to the tracer code, still needs to be performed. As a reference, this value is considerably smaller than the [−10,+5%] found during the 2006 beam test campaign performed on the Calibration Unit at CERN.[1] These results are preliminary.

## References

1. Fermi LAT collaboration, *ArXiv e-prints* (August 2010).
2. A. A. Abdo *et al.*, *Phys. Rev. Lett.* **102**, p. 181101(May 2009).
3. International Geomagnetic Reference Field http://www.ngdc.noaa.gov/IAGA/vmod/igrf.html.
4. D. Smart and M. Shea, *Advances in Space Research* **36**, 2012 (2005), Solar Wind-Magnetosphere-Ionosphere Dynamics and Radiation Models.
5. O. Adriani *et al.*, *Nature* **458**, 607(April 2009).

# THE SYNERGY BETWEEN ASTROPARTICLE AND COLLIDER PHYSICS IN THE SEARCH FOR DARK MATTER

JAMES L PINFOLD

*Centre for Particle Physics, University of Alberta, Edmonton, Alberta T6G 2N6, Canada.*

Research into the fundamental nature of matter at the high-energy frontier takes place in three main areas: accelerator based particle physics; high-energy astrophysics; and, astroparticle physics. As a consequence the study of astroparticle physics and astrophysics can have significant implications for collider physics at the LHC. Likewise, the LHC project provides the laboratory for measurements of importance to astroparticle physics, astrophysics and cosmology. This paper reviews some of the synergistic links between astroparticle physics, collider physics, astrophysics & cosmology.

## 1. Introduction

The first collisions of protons were observed by the four main LHC experiments on the 23rd of November 2009. By the 30th of November the LHC [1] machine became the world's highest energy particle collider achieving a centre-of-mass energy ($E_{cm}$) of 2.36 GeV, beating the record of 1.96 GeV held by the TEVATRON for eight years. In the early hours of the 30th of March the LHC restarted after a shutdown of a few months at the unprecedented $E_{cm}$ of 7 TeV. This date can be regarded as the start of the LHC research program.

To date the LHC has seven approved experiments. Detectors located underground in large caverns excavated at the LHC's intersection points. Two of them, ATLAS [2] and CMS [3], are large general purpose collider detectors designed to have as large acceptance as possible for new particle physics in all its imagined forms, including: the origins of mass, new symmetries of nature, extra dimensions, the nature of dark matter, etc. for clues to the nature of dark matter. ALICE [4] is dedicated to the study of heavy-ion collisions and will search for and study a new a "liquid" form of matter called the quark–gluon plasma that would have existed at. LHCb [5] is dedicated to the study of the physics of the bottom quarks, particularly the nature of CP violation on the b-sector. This could give clues as to the origin of the matter anti-matter asymmetry in the very early universe. Last, but not least, we have smaller, new experiments: TOTEM [6] dedicated to forward physics particularly a precision measurement of the p-p total cross-section; LHCf [7] that will make measurements in the very forward direction that are of use in understanding extended air shower development arising from high energy cosmic ray collisions with the upper atmosphere. MoEDAL [8] is the latest LHC experiment to be approved. It is designed to search for highly ionizing particle production at the LHC, including: magnetic monopoles, dyons, Q-balls and massive stable particles.

Some of the main cutting edge areas of investigation of the microcosm engendered by the interplay between particle physics, astrophysics and cosmology are: the elucidation of the nature of dark matter (DM); new symmetries; extra dimensions; searches for new particles; QCD, the flavour problem; cosmic ray physics, including the origin and nature of the highest energy cosmic rays. In this very brief review we will concentrate on a sketch of how a partnership between astrophysics astroparticle physics and collider physics would be necessary in order to reveal the nature of dark matter.

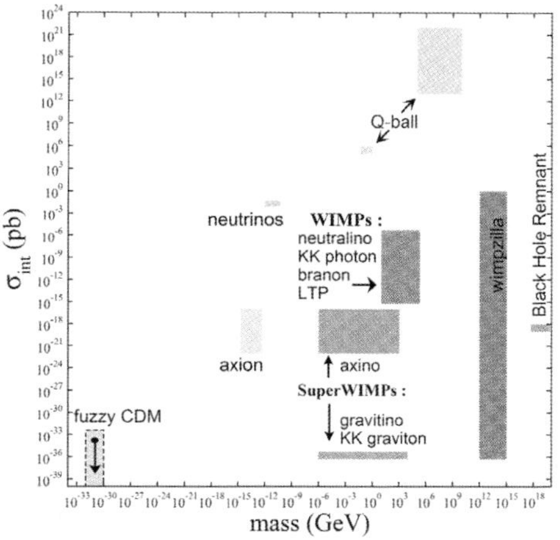

Figure 1 The mass versus interaction cross-section for a number of particle dark matter candidates.

## 2. The Search for Dark Matter

Given the few constraints, it is not surprising that there are many candidates for the particles that constitute Dark Matter (DM): axions, axinos, thermal gravitinos, neutralinos, Kaluza-Klein particles, wimpzillas, self-interacting particles, self-annihilating particles, mirror matter DM, fuzzy DM, superWIMPs, etc. The situation is summarized in Figure 1 where it can be seen that Masses and interaction strengths span many, many orders of magnitude.

If a stable, weakly interacting, electroweak-scale particle exists, then it is likely to be present in the universe today with an abundance similar to the measured dark matter density. Thus the focus of our dark matter search strategy is centred on this particularly well motivated scenario in which the dark matter particle has electroweak interactions and a mass near the electroweak scale.

## 2.1. *Recent Direct Dark Matter Search Results*

WIMP scattering off of ordinary matter proceeds through the weak interaction. The direct detection strategy consists in recording the recoil of target nuclei scattered by a WIMP. Considering the recoil energy of the nucleus in the laboratory frame (given by non-relativistic kinematics):

$$E_R = \frac{|q|^2}{2m_N} = \frac{\mu^2 v^2 (1-\cos\theta)}{m_N}$$

where, $v$ is the velocity of the WMPI relative to the target, N is the number of the target nuclei, $m_N$ is the mass of the target nucleus, $\theta$ is the scattering angle in the centre-of-mass system and $\mu$ is the reduced mass ($\mu = m_N m_{DM}/(m_N + m_{DM})$). The scattering rate is approximately:

$$R \sim \frac{N\sigma_s v \rho_{DM}}{m_{DM}}$$

where, $\sigma_s$ is the scattering cross-section and $\rho_{DM}$ is the local DM density. Obviously, on top of experimental uncertainties, the extraction of constraints on the WIMP parameter space is sensitive to the assumptions related to WIMP phenomenology, such as the mean velocity, and the local DM density.

The nuclear recoil produced by the WIMP scattering can be measured by detecting the induced scintillation light, the charge induced by ionization or phonons resulting from lattice heating. Current direct detection experiments detect one or a combination of two of these signals. The most recent results from direct detection experiments are summarized below.

*CDMS:* The CDMS [9] (Cryogenic Dark Matter Search) experiment measures phonons and ionization excitation of with Ge-Si detectors. They have recently detected 2 events with an expected background of 0.8 events. This statistical significance (P=23%, 1.5σ) of the "excess" is too low to allow any conclusion. However, the interpretation in terms of WIMP scattering with $m_{DM} \sim 10$ GeV is consistent with DAMA results, at the 2σ level as can be seen in Figure 2 [10].

*COGENT:* [11] detected a number of cosmogenic peaks, plus an exponential-like signal not immediately identifiable with background. It is a Ge-detector with lower threshold (0.4 keV) than CDMS, but larger background contamination. Noting that detecting only the ionization signal does not allow the experiment to discriminate between nuclear and electron recoils. The region of interpretation in terms of WIMP scattering has some overlapping with DAMA and CDMS regions as shown in Figure 2 [10]

*CRESST:* In the meantime, unofficial reports preliminary data from the CRESST experiment using Calcium Tungstate crystals, discuss a few events, not immediately ascribable to background. This has led to speculation – again no official analysis has been reported by the CRESST team - indicating that the parameter space for the interpretation of these data in terms of WIMPs can overlap with DAMA, CDMS, and COGENT as shown in Figure 2 [2010].

*DAMA:* Over the last few years: the DAMA experiment, and its successor DAMA/LIBRA, experiment [12] has investigated a WIMP signal with scintillating sodium iodide (NaI) revealed by the annual modulation of WIMP signal arising from the Earth's motion around the Sun. The DAM group detected a signal with proper modulation features (i.e., cosine like with period ~1 year and phase ~ June 2nd), no modulation above 6 keV and in the 2-6 keV multiple-hits residual rate (NB. WIMPs do not induce multiple-hits events), and high statistical significance (~9$\sigma$.). This makes hard to reconcile it with other sources of background and the DM interpretation is plausible. On the other hand, the "standard interpretation" of spin-independent elastic scattering of WIMPs, is in disagreement with other experiments, unless the WIMP is very light ($m_{DM}$ ~ 10 GeV). The DAMA excess can also be explained within the context of more exotic scenarios even for larger masses (see references in [12]).

XENON100: [13] The XENON collaboration is aiming to reach ultra-low background levels using a liquid Xenon (~100 kg) target. They record both the scintillation and ionization resulting from the recoil events. To date no events they have reported no signal events. Thus, they have excluded signals from spin-independent, elastic, light mass WIMP scattering. However, this result has been has been criticized [14] since the relative scintillation efficiency at low recoil energy - crucial in the derivation of constraints for low mass WIMPs - was derived from an extrapolation. This was necessary due to the lack of experimental data for liquid Xenon below 4 keV. The XENON collaboration stands by their analysis [15]. But, the original criticism was reiterated [16] with the claim that a more "realistic" efficiency would show that the constraints from current XENON100 analysis becomes less restrictive at low masses.

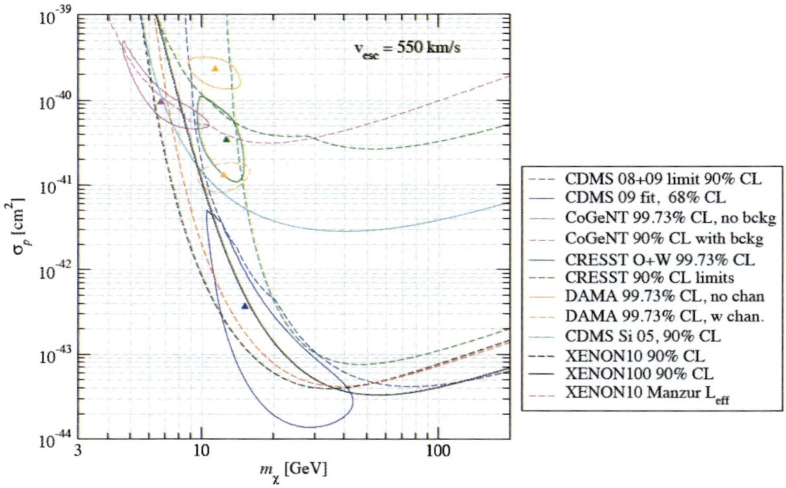


Figure 2 WIMP parameter space selected by direct detection experiments in the case of spin-independent elastic scattering, from [10].

## 2.2. *Indirect Dark Matter Search Results*

### 2.2.1. *Antimatter*

Positrons, anti-protons, and anti-deuterium, induced by WIMP annihilations/ decays in the galactic halo, can be observed as an anomalous contribution in the local spectra of cosmic-rays (CRs).

Measurements of the positron fraction up to 100 GeV by the PAMELA satellite borne experiment [17] has triggered a lot of speculation that there may be a dominant contribution to the positron flux induced by WIMPs. The data shown in Figure 3 show a sharp raise at high-energy that is puzzling if the source is given by interactions of cosmic rays (CRs) with the interstellar medium (under "standard" propagation assumptions as expressed by the GALPROP [18] program indicated by the dotted line), which could indicate the presence of a nearby source of positrons within a few kpc.

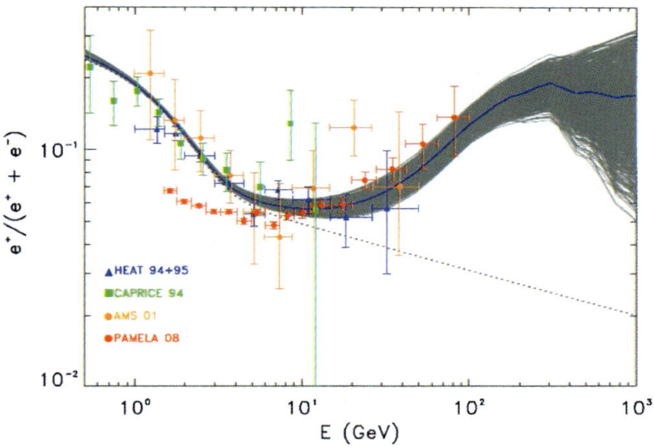


Figure 3 Positron fraction as measured by PAMELA in comparison with the GALPROP prediction (GALPROP). The gray region shows the range of the predicted positron positron fraction assuming it is due to local pulsers.

The FERMI-LAT electron spectrum is shown in Figure 4. FERMI data are in good agreement with the measurement by HESS at higher energies [19], but do not exhibit prominent excesses as recently reported by ATIC [20]. If compared with a conventional model of CR propagation based on the GALPROP code [18], LAT electrons show a spectrum harder than expected. This spectrum, combined with the measurement by PAMELA of an increase in the $e^+$ fraction up to 100 GeV [17], is hard to reconcile with such conventional models.

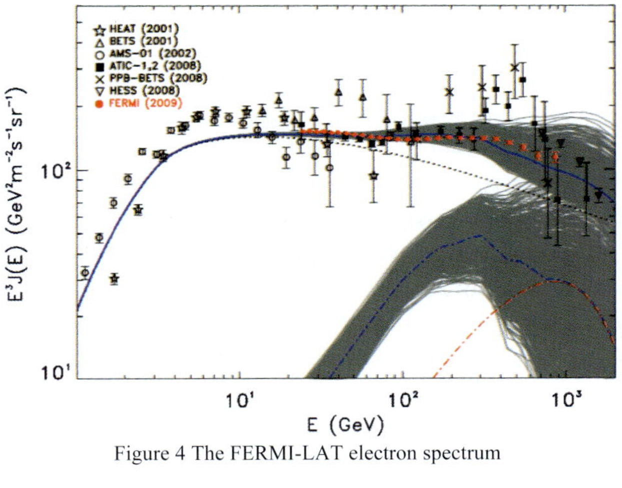


Figure 4 The FERMI-LAT electron spectrum

In order to ascribe the excess to WIMPs, the DM particle has to be heavy - in the range 100 GeV/$c^2$ to 1 TeV/$c^2$. The DM also has to have a dominant annihilation branching ratio into leptons (leptophilic), and with a greatly enhanced annihilation rate compared to that expected from standard WIMP annihilations, in order to explain the recent PAMELA FERMI $e^+$ - $e^-$ data.

Anisotropy in the $e^+$ - $e^-$ spectrum and bumpiness of the e+ spectrum can be expected to be sizable in case of dominant contribution from pulsars [21], or DM substructures [22]. We note that the change in the positron fraction data below ~10 GeV is expected to be due to the solar modulation. At energies between 100 GeV and 1 TeV the electron flux reaching the Earth may well be the sum of an almost homogeneous and isotropic component produced by Galactic supernova remnants and the local contribution of a few pulsars with the latter expected to contribute with increasing significance as the energy increases. Indeed, assuming the escape rate and/or the number of old and nearby pulsars to be free parameters such pulsars can nicely explain the PAMELA and FERMI data, as shown in Figures 3 and 4 [23].

If only a few nearby sources contribute, one could try to search for an anisotropy in the arrival direction of the positrons. However, even with only a few sources the diffusive scattering washes out anisotropies efficiently. How would one distinguish between the contributions of pulsars and dark matter annihilations? Most likely, a confirmation of the dark matter signal will require a consistency between different experiments and new high statistics measurements of the reported excesses. The observed excess in the positron fraction should be consistent with corresponding signals in absolute positron and electron fluxes in the PAMELA data and all lepton data collected by Fermi.

Fermi has a large effective area and a 5 years nominal (with a goal of 10 years mission) lifetime, which makes it an excellent detector of cosmic-ray electrons up to ~1 TeV. Future Fermi measurements of the total lepton flux with large statistics will be able to clearly distinguish a gradual change in slope with a sharp cutoff [24]. This sharp cutoff, is evidence in favor of the dark matter hypothesis. A strong leptonic signal should be accompanied by a boost in the $\gamma$-ray yield providing a distinct spectral signature detectable by Fermi.

### 2.2.2. *Photons*

A phenomenon of great interest for particle physics, astrophysics and cosmology is the diffuse gamma-ray emission. The Galactic diffuse emission (GDE) is produced by the interactions of cosmic rays (CRs) with the interstellar gas (via neutral pion decay and Bremsstrahlung) and with the interstellar radiation field (via Inverse Compton scattering). Prompt emissions of gamma-ray photons in the DM halo include continuum emission due to the production and decay of neutral pions, final state radiation and mono-energetic spectral signatures related to direct production at loop level. Electrons and positrons can be directly or indirectly produced by WIMP annihilations. They act as sources for radiative processes generating a multi-wavelength spectrum, mainly involving synchrotron emission at radio/infrared frequencies and inverse Compton (IC) scattering on CMB and starlight in the X- and gamma-ray bands. Thus, Multi-wavelength searches can be a very powerful technique in WIMP searches [25].

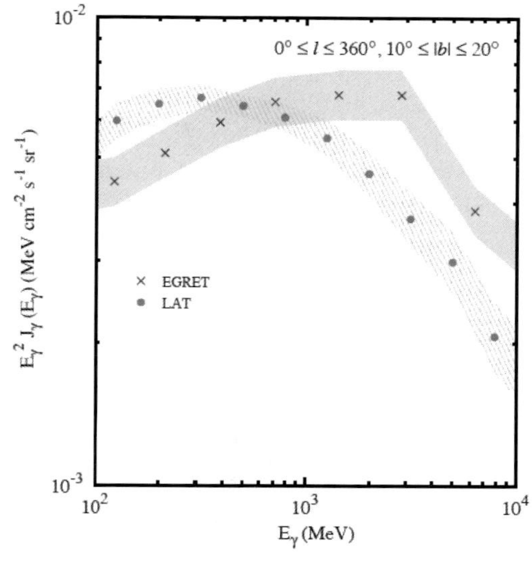


Figure 5 Spectrum of the gamma-ray emission measured by the LAT for energies between 100MeV and 10 GeV, averaged over all longitudes in the region with Galactic latitude 10_ _|*b*|_2. The measurement by the LAT (red) is compared with the measurement by EGRET (blue).

The search for prompt and radiative emissions of gamma-ray photons from WIMP annihilations and decays constitutes one of the prime goals of the FERMI collaboration, e.g., [26]. The objects of investigation include: Galactic center, Milky Way halo, extra-galactic regions, satellites (dwarf spheroidal and DM substructures), and clusters of galaxies, with focuses on both spectral line and continuum signatures. The current (after 1.5 year of mission) main conclusion from the team is that there has been no discovery so far, but, however, constraints on the nature of DM have been significantly improved. For the near future, prospects are still very promising since the ongoing analysis for a better understanding of the detector response and of astrophysical backgrounds will significantly improve the potential for DM discovery.

The spectrum of photons measured by EGRET showed an excess of a factor ~ 2 in all directions on the sky at energies greater than around 1 GeV with respect to conventional models based on the locally measured CR spectra, interpreted by some authors as the long-awaited signature of dark matter. The first target of study for the LAT was the emission at intermediate Galactic latitudes, where most of the emission comes from interactions of CRs with the local gas. In Figure 5 the spectrum measured by EGRET is compared with the spectrum measured by the LAT that is significantly softer.

The Wilkinson Microwave Anisotropy Probe (WMAP) has been used to provide the best measurements to date of the expected foregrounds due to standard interstellar medium emission mechanisms, including thermal dust, spinning dust, ionized gas and synchrotron. In addition to these expected foregrounds, the observations have revealed an excess of microwave emission in the inner $20^{\circ}$ around the center of the Milky Way that has roughly spherical morphology and radius ~4 kpc. This excess is known as the "WMAP Haze" [27]. Conventional explanations such as thermal bremsstrahlung from hot gas were ruled out allowing speculations [28] that the signal could be due to annihilating dark matter particles produce relativistic electrons and positrons that travel under the influence of the Galactic magnetic field. As they do, they will emit synchrotron photons, which naturally fall within the frequency range measured by WMAP. The WMAP Haze requires only standard scenario WIMP annihilations. The observed angular distribution, intensity and spectrum are consistent with standard WIMP scenario particles produced thermally in the early universe and are distributed with a cusped halo profile in our Galaxy.

The Fermi Gamma-Ray Space Telescope has reported a diffuse inverse Compton (IC) signal in the inner Galaxy with a similar spherical spatial morphology to the microwave haze observed by WMAP [29]. Taking out "backgrounds" from $\pi^0$ gammas, as well as IC and bremsstrahlung components associated with known soft-synchrotron counterpart Fermi found a significant gamma-ray excess towards the Galactic center with a spectrum that is significantly harder than other sky components and is most consistent with IC from a hard population of electrons. The morphology and spectrum are consistent with it being the IC counterpart to the electrons that generate the microwave haze seen at WMAP frequencies. In addition, the implied electron spectrum is hard; electrons accelerated in supernova shocks in the disk that then diffuse a few kpc to the haze region would have a softer spectrum. Although Fermi is making no claims this haze could be due to the decay or annihilation of dark matter, or new astrophysics.

### 2.3. *Colliders and Dark Matter*

A program to reconstruct sparticle masses, couplings and quantum numbers will undoubtedly follow the discovery of Supersymmetry (SUSY). Assuming that DM particles arise from SUSY type scenarios that are addressable at the LHC then the discovery and elucidation process will take place in three stages. First, inclusive Studies (comparison of significance in inclusive channels etc.) will give rise to the first rough predictions of the relic density ($\Omega_\chi h^2$) within a specific model framework. The second phase utilizes exclusive studies (calculation of model dependent masses, etc.) where the data is interpreted within a specific model dependent the framework. At this stage Model-dependent calculation of LSP mass for comparison with e.g. direct searches; detailed model-dependent calculations of DM quantities ($\Omega_\chi h^2$, $\sigma_{\chi p}$, indirect observations etc.). The last stage would involve an approach to model-independent measurement of, for example, $\Omega_\chi h^2$. through measurement of all relevant masses etc.

What will we be able to say about dark matter in light of these measurements? As an illustration, the precision with which the neutralino relic density is constrained by collider measurements for the LCC2 point, which is in the HB/FP region of the mSUGRA model, is shown show in Figure 6 [30]. Measurements at the LHC cannot fix the neutralino composition, and so are unable to resolve the degeneracy between a wino-LSP solution (which gives a tiny relic density) and the true solution with mixed Higgsino-bino DM. But, the determination of the chargino production cross sections at the ILC would easily resolve the degeneracy. However, it is striking that up to this degeneracy ambiguity, experiments at the LHC can still pin down the relic density to within ~ 50%. This improves to 10-20% if we can combine LHC and ILC measurements.

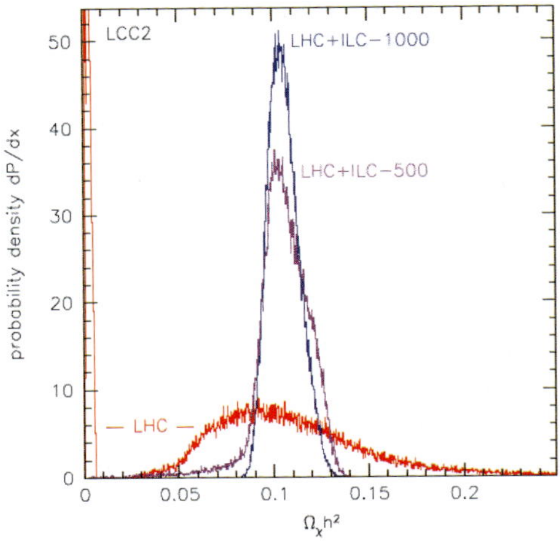


Figure 6 Determination of neutralino relic abundance via measurements at the LHC and ILC, taken from Baltz et al. (Baltz et al.).

A collider determination of the relic density is of crucial importance. If it agrees with the cosmological measurement it would indicate that cold DM (CDM) is formed from dominantly thermal neutralinos from the Big Bang. If the neutralino relic density from colliders falls significantly below the measured CDM density, it would provide direct evidence for multi-component DM– perhaps neutralinos plus axions or other exotica. On the other hand, if the collider determination gives a much larger value of the relic density then, it could point to a long-lived but unstable neutralino and/or non-thermal DM.

The collider determination of model parameters would also pin down the direct neutralino nucleon scattering cross section. Thus, if a WIMP signal is observed in direct detection experiments, one might be able to determine the local DM density of neutralinos - and aspects of their velocity distribution - based on the observed signal rate. This density should agree with that obtained from astrophysics if the DM in our Galaxy is comprised only of neutralinos.

Last but not least, the collider determination of the neutralino annihilation cross section times relative velocity, in the limit that $v \to 0$, would eliminate uncertainty on then particle physics side of projections for any indirect signal from annihilation of neutralinos in the galactic halo. Thus, the observation of a gamma ray and/or anti-matter signal from neutralino halo annihilations would facilitate the determination of the galactic dark matter density profile.

## 3. Conclusion

The quest for a microscopic understanding of the dark universe is a fundamentally important challenge. The WMAP data opened up an era of precision cosmology that has reinforced the case for cold dark matter in the universe. If the predictions for 'standard' supersymmetric scenarios are found to agree with observations at the LHC and are consistent with the interaction strengths and relic density as determined by astroparticle physics and cosmology, this concurrence will provide strong evidence that the dark matter is supersymmetric. However, if this agreement is not seen a whole host of possibilities arise. It is only through the combination of approaches in particle physics, astrophysics and cosmology that the identity of dark matter will be uncovered. The task requires many diverse experiments, and will likely take decades to complete. Nevertheless, if we can establish nature's connection between the weak scale and cosmology, an exciting vista of discovery physics will open up before us.

## References


1. LHC Study Group 1995. The large hadron collider conceptual design, CERN/AC/95-052.H. Müller and B. D. Serot, *Phys. Rev.* **C52**, 2072 (1995).
2. ATLAS Collaboration 1999. *Detector and Physics Performance Technical Design Report* CERN/LHCC/99-15.
3. CMS Collaboration.The compact muon solenoid *Technical Proposal* CERN/LHCC/94-38 (2004).
4. ALICE Collaboration. A large ion collider experiment *Technical Proposal* CERN/LHCC/95-71 (2005).
5. LHCb Collaboration. A large hadron collider beauty experiment for precision measurements of CP-violation and rare decays *Technical Proposal* CERN/LHCC/98-4 (1998).
6. TOTEM Collaboration 1999 Total cross-section elastic scattering and diffractive dissociation at the LHC, *Technical Proposal* CERN-LHCC 99-7 (1999).
7. Technical Proposal for the CERN LHCf Experiment, CERN-LHCC-2005-032, LHCC-P-007 (2005.)

8. MoEDAL TDR, CERN-LHC-2009-006 (2009).
9. Z. Ahmed *et al.*, CDMS-II Collaboration, arXiv:0912.3592 (2009)
10. T. Schwetz talk at GGI conference, http://www.ggi._.infn.it/, (2010)
11. C. E. Aalseth *et al.*, CoGeNT collaboration, arXiv:1002.4703 (2010).
12. R. Bernabei *et al.*, DAMA Collaboration, arXiv:1007.0595, (2010).
13. E. Aprile *et al.*,XENON100 Collaboration, arXiv:1005.0380 (2010).
14. J. I. Collar and D. N. McKinsey, arXiv:1005.0838 (2010).
15. Xenon Collaboration, arXiv:1005.2615 (2010)
16. J. I. Collar and D. N. McKinsey, arXiv:1005.3723 (2010).
17. O. Adriani *et al.*, PAMELA Collaboration, Nature 458 (2009) 607.
18. A.W. Strong, and I.V.Moskalenko, ApJ, 509 (1998) 212; arXiv:astro-ph/9807150 and http://galprop.stanford.edu/A.W. Strong, I.V. Moskalenko and O. Reimer, ApJ, 53; 7 (2000) 763 [Erratum-ibid. 541, 1109]; arXiv:astroph/9811296 (1998).
19. F. Aharonian, *et al.*, Phys. Rev. Lett. 101 261104 (2008).
20. J. Chang, Nature 456 (362)
21. D. Hooper, P. Blasi and P. D. Serpico, JCAP 0901, 025 (2009).
22. M. Regis and P. Ullio, Phys. Rev. D 80, 043525 (2009).
23. D. Grasso, *et al.*, Astropart. Phys. 32, 140–151, 0905.0636 (2009)
24. E. Baltz *et al.*, JCAP07 (2008) 013, arXiv:0806.2911 (2008).
25. M. Regis and P. Ullio, Phys. Rev. D 78 043505 (2008).
26. S. Murgia talk at GGI conference, http://www.ggi.fi.infn.it/
27. D. P. Finkbeiner, Astrophys. J. 614, 186 (2004).
28. D. P. Finkbeiner, arXiv:astro-ph/0409027 (2004).
29. G.Dobler *et al.* Astrophys.J.717:825-842, (2010)
30. E. Baltz *et al.*, Phys. Rev. D 74 103521 (2006).

# PICASSO: SEARCH FOR DARK MATTER IN THE SPIN-DEPENDENT SECTOR

M.-C. PIRO

*For the PICASSO Collaboration,*

*Department of Physics, University of Montréal*

*E-mail: piro@lps.umontreal.ca*

The PICASSO project is using superheated droplets of $C_4F_{10}$ for the direct detection of Dark Matter candidates in the spin-dependent (SD) sector. The total setup includes 32 detectors installed in the SNOLAB underground laboratory in Sudbury (Ontario, Canada). The present level of sensitivity is at 0.16 pb on protons at 90% C.L. ($M_W$= 24GeV/$c^2$) following an analysis of two detectors only. A concentrated effort in detector purification and a new fabrication procedure allowed an additional background reduction of about a factor of ten. In order to increase the sensitivity of the detectors, new discrimination tools were developed to distinguish between WIMP induced nuclear recoils and alpha decay background. We report preliminary results where an alpha background rejection of 80% could be achieved in the region where WIMP induced nuclear recoils are expected.

## 1. Introduction

Astronomical and cosmological observations strongly suggest the presence of Dark Matter and show that only 1% of the matter of the Universe is luminous and 23% of all the matter should be of a new exotic kind [1]: Cold Dark Matter (CDM). The preferred candidate for particle physicists is the neutralino, the Lightest Supersymmetric Particle (LSP). It is a very Weakly Interacting Massive Particle (WIMP) and is a natural candidate in the Minimal SuperSymmetric Model (MSSM) [2]. PICASSO (Project In CAnada to Search for Supersymmetric Objects) is one of the many worldwide efforts to hunt for dark matter through the direct detection of neutralinos via their spin dependent (SD) interactions with $^{19}F$ nuclei [3, 4, 5]. The goal of the PICASSO project at SNOLAB is to exploit the favourable properties of $^{19}F$ by using the superheated droplet detection technique, which is based on the operation principle of the classic bubble chamber [6, 7].

## 2. Detector Principle and Response

The active detector material in PICASSO consists of droplets of $C_4F_{10}$, a liquid which is at ambient temperature and pressure in a metastable, superheated state. The total mass of $C_4F_{10}$ amounts to 85g per detector module. A present generation PICASSO detector is a 4.5L cylindrical acrylic module, closed on top by a stainless steel lid sealed with polyurethane O-rings. Thirty-two of these modules are presently installed at SNOLAB.

When an ionizing particle deposits enough energy in a droplet, it creates a heat spike that induces a phase transition, changing the droplet into a vapour bubble. This phase transition is accompanied by an acoustic signal which is recorded by nine piezoelectric transducers mounted on the outside of the detector walls. Calibrations with mono energetic neutron beams allow to determine precisely the energy thresholds of the detector as a function of pressure and temperature [8].

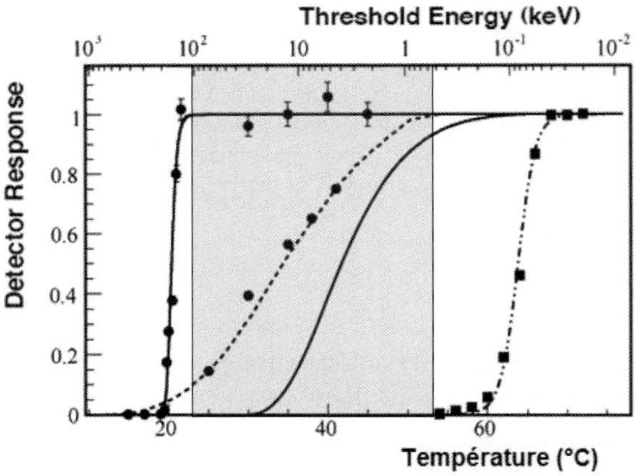


Figure 1: Response to different kinds of radiations as a function of temperature for a detector based on superheated $C_4F_{10}$ [9]. From left to right: response to alpha particles of 5.6 MeV in a detector spiked with $^{226}Ra$ (data points fit with continuous line); response to nuclear recoils from fast neutrons of an AmBe source compared to simulations (dotted line); expected response to nuclear recoils following scattering of a 50 GeV/$c^2$ WIMP (continuous line); response to 1.275 MeV gamma rays of a $^{22}Na$ source (dashed line). All the responses are normalized to one at full detection efficiency and temperatures are converted into threshold energies. The grey area is the nuclear recoils region.

These calibration measurements were performed at the Tandem accelerator facility at University of Montreal. The use of the light target nucleus $^{19}F$, together with a low detection threshold of 2 keV for recoil nuclei, makes PICASSO particularly sensitive to low-mass WIMPS. This is important as few experiments are sensitive below 15 GeV/$c^2$ and it is precisely in this mass range where recent interpretations of the DAMA/LIBRA annual modulations in terms of SD interactions remain partially unchecked [10, 11].

The detector response to various types of particles as a function of temperature is shown in Figure 1. The different detector responses as a function of temperature allow us to discriminate a possible neutralino signal from backgrounds when we expose the detectors in the mine for counting. The operating temperature in each sub-unit can be varied independently from 20°C to 55°C and the temperature is regulated with a precision of ± 0.1°C. In this range of temperatures, the detectors are mainly sensitive to alpha-particles and neutrons.

## 3. Detector fabrication and purification

The sensitive constituent of a PICASSO detector is an emulsion of superheated droplets dispersed in water based polymerized gel. The active liquid $C_4F_{10}$ is dispersed in a gel of poly-acrylamide which forms a smooth interface around the superheated droplets. The emulsion of the droplets is created by a magnetic stirrer with a rotation speed adjusted to create a droplet size distribution centred at 200 microns in diameter.

In order to reduce to the possible minimum all contamination sources, the fabrication and purification of the detectors are performed in a class 10 000 clean room environment at University of Montreal. A reverse osmosis system is used to provide an Ultra Pure Water (UPW) with a resistivity of 18 Mohm.cm at 25°C. The use of UPW also ensures the cleanliness of equipment employed during the fabrication of the detectors. The purification itself is performed using Hydrous Zirconium Oxide (HZrO), a precipitate that attracts actinides like U and Th. Since $C_4F_{10}$ has a boiling temperature of -1.7°C, the fabrication process of the suspension is done at -20°C at ambient pressure. Two generations of detectors were developed to create a homogenous and uniform distribution of active droplets in the detector. First, a series of 4.5 L of detectors was produced that contains a gel of poly-acrylamide loaded with high density Cesium Chloride (CsCl) solution to equalize the high density of the $C_4F_{10}$ droplets, ($\rho$=1.6g/cm$^3$), and that of the gel solution. But it turned out that this heavy salt is a major background source because of its intrinsic contamination with alpha-emitters from the U/Th decay chains.

Therefore, a new recipe of gel was developed: 4.5 L detectors based on a gel of poly-ethylene glycol loaded with viscous glycerol, to suspend the $C_4F_{10}$ without density matching. In addition, fabrication and purification were performed under a constant flow of $N_2$ cover gas to avoid the radon ingress. All these new improvements allowed a background reduction of about a factor 10! (Figure 2).

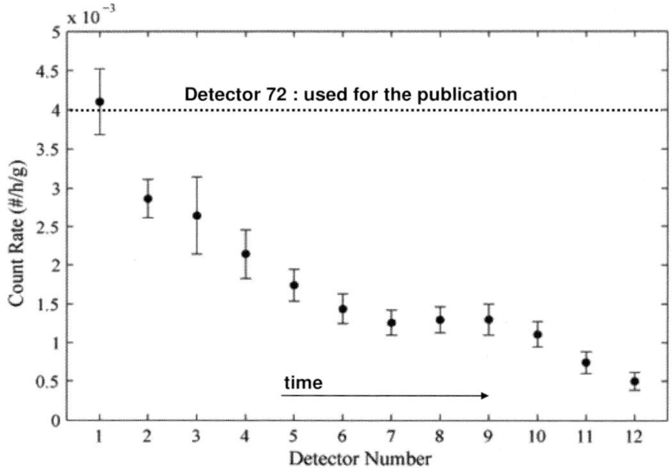


Figure 2: Evolution of background count rate over time (40°C) following improvements in production and detector purification. The new generation is compared with the best detector of the publication 2009 in *Phy. Lett. B* [9].

## 4. Results

The experimental set-up is installed 2070 meters underground at SNOLAB in the Creighton mine, Sudbury. At this depth, the neutron flux coming from the ambient cosmic muon flux is largely reduced, with an expected count rate at the level of $10^{-5}$ counts/h/g which is two orders of magnitude below the count rate of the detectors which are the subject of this analysis. The set-up is now completed with all 32 detectors installed yielding a total active mass of 2.6 kg of $C_4F_{10}$. The first results published in 2009 has been done with only two detectors (71 and 72) of the CsCl loaded generation with a total target mass of $^{19}F$ target mass of

65.06 ± 3.2 g and 69.0± 3.5 g, respectively. For WIMP masses around 24 GeV/$c^2$□and a total exposure of 13.75 ± 0.48 kgd, limits on the SD cross section were obtained on $^{19}$F of $\sigma_F$ = 13.9 pb (90% C.L.) which can be converted into cross section limits on protons and neutrons of □$_p$ = 0.16 pb and □$_n$ = 2.60 pb respectively at 90% C.L. (Figure 3). The analysis is presently being extended to the remaining 30 detectors and we expect that a very competitive limit will be obtained when the detectors have been fully characterized and the analysis has been completed.

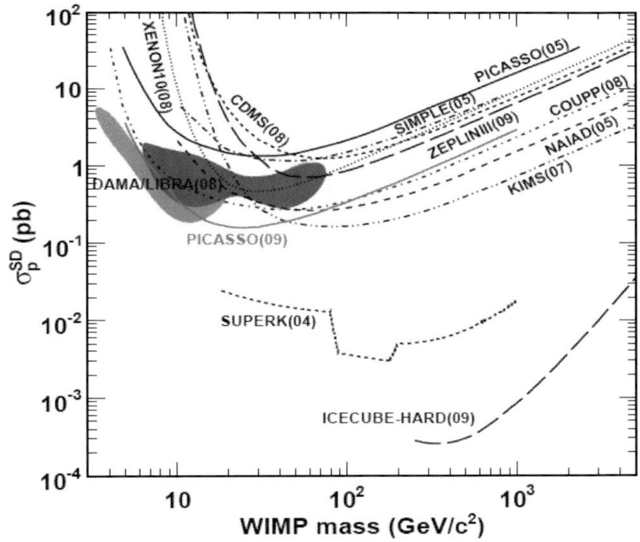


Figure 3: Exclusion curve of PICASSO are shown as full lines for only two detectors [9].

## 5. New analysis and discrimination tools

A major advance in the field of dark matter detection with superheated liquids was made when PICASSO discovered a significant difference between the acoustic signals of neutron and alpha-particle induced events. In fact, a search for signatures in the recorded transducer signals showed that the intensity and frequency content of the acoustic signals contains information about the nature of the primary event. A detailed discussion of this phenomenon can be found in [12].

From this information, variables can be constructed, which allow first of all an efficient event by event discrimination between particle and non-particle induced backgrounds. The variable used is the acoustic energy, which is constructed from a measure of the integrated sound intensity of the fast component of an event signal. Calibration data with fast neutrons of AcBe, AmBe and Cf sources showed that the amplitude distributions of the high frequency content of the signals (>18 KHz) concentrate in a well defined peak.

In addition it was found that alpha particle induced signals are significantly more intense than those created by neutron induced recoils (and therefore also of WIMP induced events). In both cases the energy released during vaporisation increases with temperature, but remains well defined for a given temperature (Figure 4).

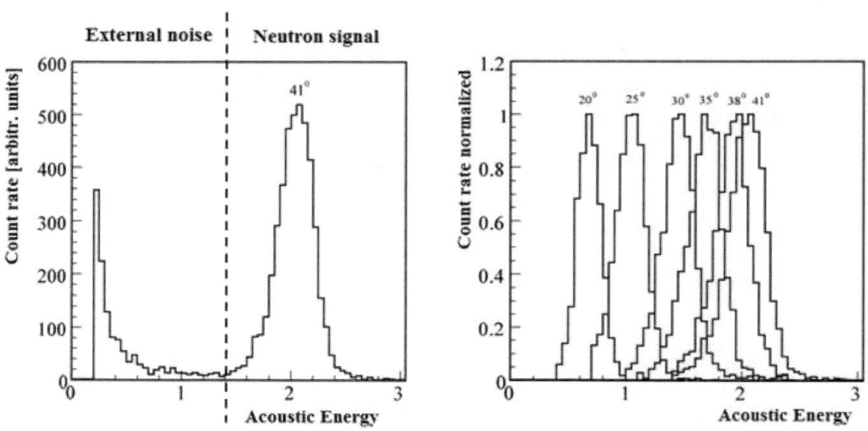


Figure 4: Acoustic energy distribution recorded in calibrations with polyenergetic neutrons from an AmBe source. Neutron induced recoils accumulate in a peak (the same peak where WIMP induced recoils are expected) (left); this peak is well separated from acoustic and electronic noise and varies as a function of temperature (right).

By adapting the range of the applied frequency filters to the operating temperature, the discrimination between alpha-particles and neutrons can be improved substantially: e.g. by applying at 45°C a band pass filter between 130kHz and 160kHz, alpha particle induced events can be rejected with an efficiency of 80%, keeping full acceptance for nuclear recoils; in contrast at the same temperature an 18 kHz high pass filter shows no discrimination at all (Figure 5). This work of optimization is in progress and it is expected that a further improvement in discrimination can be achieved by an increase of the DAQ sampling rate from the presently 400 kHz to 1MHz.

Since an improved discrimination between WIMP induced recoils and alpha particles can significantly increase the sensibility of PICASSO detectors, a lot of effort has been spent in understanding the details and the origin of this effect.

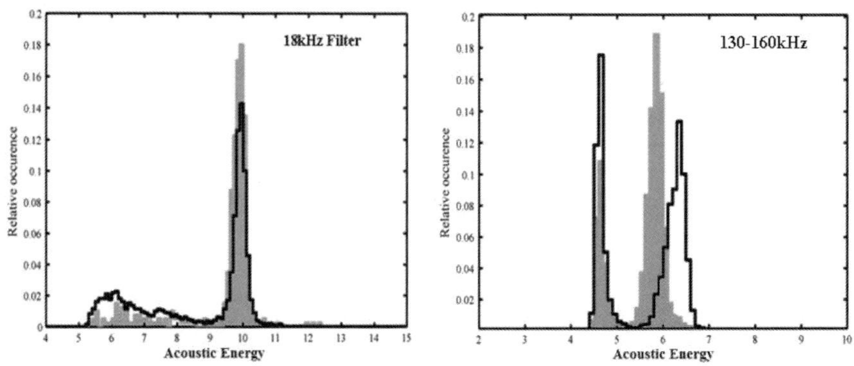


Figure 5: Comparison between the acoustic energy distribution of neutrons (shaded histogram) and alpha-particles (unshaded histogram) for one detector at 45°C with an applied high pass filter at 18kHz (left) and band pass filter between 130kHz and 160kHz (right).

In particular two detectors where spiked with alpha emitters of known types and energies ($^{241}$Am and $^{226}$Ra) and the emitters where located deliberately outside and inside the droplets. The normalized count rates are shown in Figure 6. The count rate of the Am spiked detectors shows a threshold at 22° C, corresponding to 71 keV. Since the alpha emitters are predominantly located outside the droplets, the only alpha particles which can trigger a phase transition at threshold are those with the highest specific energy loss at the Bragg peak.

However, if the detector is spiked with $^{226}$Ra a lower threshold is recorded which starts at 19°C: in this case $^{226}$Ra decays into $^{222}$Rn and its short-lived daughter nuclei. The gaseous Rn diffuses all over the detector and also into the droplets. Therefore the observed activity is now caused by $^{210}$Pb nuclei recoiling inside the droplets with energy of 146 keV, since this is the nucleus with the highest recoil energy in the chain. By further increasing the temperature above threshold, the $^{226}$Ra spiked detector becomes also sensitive to alpha particles themselves, but since the detector is already fully sensitive to alpha emitters, the count rate remains unaffected.

However, above 25°C, alpha particles contribute to the amplitude of the acoustic signal. This can be seen in Figure 7, where the acoustic energy distribution is shown for an alpha spiked detector for three temperatures. Between threshold at 21 and below 25, only recoil nuclei create a peak which coincides with the location of the neutron induced recoils during exposure to an AcBe source. With further increase in temperature, when the detector becomes sensitive also to alpha particles a second peak arises at higher acoustic energy and the first peak gradually diminishes. During this redistribution between the peaks the sum of the count rates remains constant. This effect is described in details in [8].

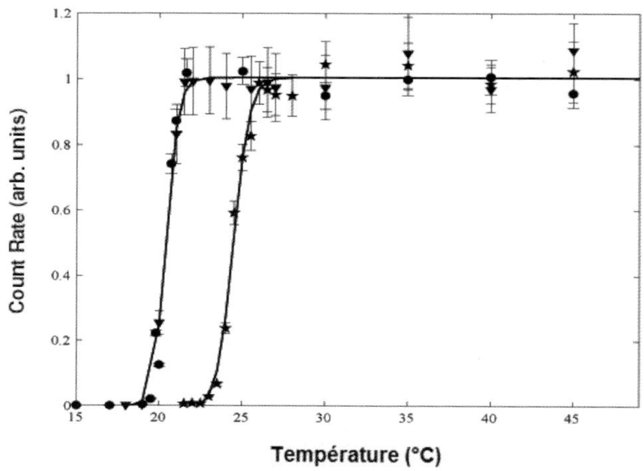


Figure 6: Normalized response of three spiked detectors. From left to right: response to alpha particles of 5.6 MeV in two detectors spiked with $^{226}$Ra for comparison (data points (dots and inverted triangle) fit with continuous line); response to alpha particles of 5.64 MeV in a detector spiked with $^{241}$Am (data points (stars) fit with continuous line).

## 6. Conclusion

PICASSO developed a new generation of detectors with a substantially reduced background by at least a factor 10 and which will allow obtaining competitive limits in the sector of spin–dependent WIMP interactions. Presently, half of the set-up has been replaced by these new detectors and the full installation will be completed before the end of this year. The analysis of the new generation of detectors is in progress and studies of an improved acoustic discrimination between WIMP induced nuclear recoils and alpha-particles promises a further substantial reduction in background.

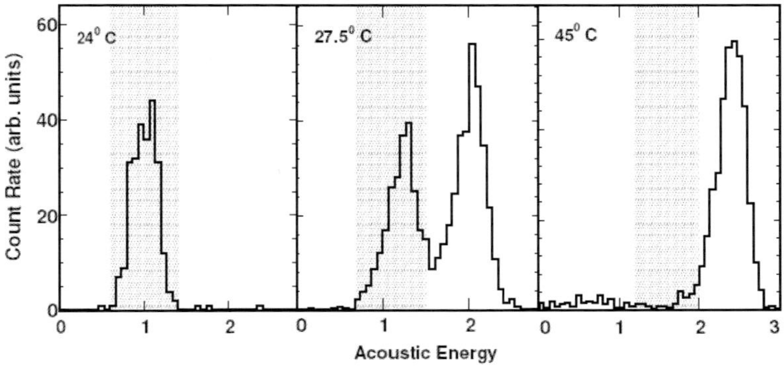


Figure 7: Acoustic energy distribution recorded of the alpha background measurements in the mine for a high rate detector at different temperatures: 24°C, 27.5°C and 45°C. The three grey shaded areas define the nuclear recoils region in the alpha response [8].

## References


1. C.L. Bennett et al., *Astrophys. J. Suppl.* **148** (2003) 97.
2. S. P. Martin et al., arXiv:hep-ph/9709356v4 (2006).
3. J. Ellis, R. Flores, *Phys. Lett.* **B263** (1991) 259.
4. M.T. Divari et al., *Phys. Rev.* **C61** (2000) 054612.
5. V. A. Bednyakov et al., arXiv:hep-ph/0412067 (2004).
6. D.A. Glaser, *Phys. Rev.* **87** (1952) 665.
7. V. Zacek, *Il Nuovo Cimento* **A107** (1994) 1247.
8. S. Archambault et al., PICASSO collaboration, New Insights into Particle Detection with Superheated Liquids, (2010) , arXiv: 1011.4553v1 (2010).
9. S. Archambault et al., PICASSO collaboration, *Phys. Lett.* **B621** (2009) 85.
10. R. Bernabei et al., *Eur. Phys. J.* **C56** (2008) 333, arXiv: 0804.2741 (2008).
11. C. Savage et al., arXiv: 0808.3607 (2009).
12. F. Aubin et al., PICASSO Coll., *New J. Phys.* **10** (2008) 103017.

# Recent Results from VERITAS

John Quinn* for the VERITAS Collaboration

*School of Physics, University College Dublin,*
*Belfield, Dublin 4, Ireland*
**E-mail: john.quinn@ucd.ie*
*http://veritas.sao.arizona.edu*

VERITAS is an array of four imaging atmospheric Cherenkov telescopes, located in southern Arizona, for $\gamma$-ray astronomy above 100 GeV. VERITAS has been fully operational since September 2007 but was reconfigured in 2009 through the relocation of a telescope to provide enhanced sensitivity. To date, VERITAS has detected 33 sources of TeV $\gamma$ rays including active galactic nuclei, shell-type supernova remnants, pulsar wind nebulae, a binary system, a starburst galaxy, and several unidentified sources. This paper provides a status update on VERITAS and its science results as of September 2010, with emphasis on recent discoveries.

*Keywords*: gamma ray; atmospheric Cherenkov telescopes; active galactic nuclei; starburst galaxies; supernova remnants; dark matter.

## 1. Introduction

Very high energy (VHE, $\gtrsim$ 100 GeV) $\gamma$ rays are signatures of non-thermal relativistic processes in the Universe and may be used to probe sites of particle acceleration and environments associated with the production of cosmic rays. VERITAS (the Very Energetic Radiation Imaging Telescope Array System, fig. 1) is an array of four 12 m diameter imaging atmospheric Cherenkov technique[1] (IACT) telescopes for VHE $\gamma$-ray astronomy, located at the Fred Lawrence Whipple Observatory in Arizona, that has been fully operational since Autumn 2007. The VERITAS collaboration consists of approximately 90 scientists from 22 institutions in four countries (Canada, Ireland, U.K., U.S.A.), plus associate and non-affiliated members from Argentina, Finland, Germany and the U.S.A. The collaboration operates a strong scientific observing program that encompasses both Galactic and extragalactic science, with emphasis on new source discovery and detailed studies of established sources. The success of this program has been demonstrated by the detection of 33 VHE sources with VERITAS, in addition to sensitive limits being placed on the emission levels of many other objects of interest.

## 2. VERITAS Technical Details, Performance and Upgrades

VERITAS consists of four 12 m diameter IACT telescopes, each equipped with a 499-pixel photomultiplier-tube (PMT) camera. The array employs a sophisticated three-level trigger system, with the PMT pulses digitised by a 500 MS/s Flash-ADC system. Image cleaning and parameterisation techniques are applied to characterise the images, and stereoscopic 'cuts' based on Monte Carlo simulations of extensive air showers are applied to select candidate $\gamma$-ray events and reject background cosmic ray events.

Fig. 1. The VERITAS Array at the base camp of the Fred Lawrence Whipple Observatory.

The telescopes are currently arranged with a typical baseline of $\sim$100 m. The array was changed to this configuration in summer 2009 to provide an optimum baseline as previously telescopes 1 and 4 had been close together with a separation of $\sim$30 m. Simultaneous with the move of telescope 1, a new mirror facet alignment system[2] was implemented to give better image definition. The result of these changes is that VERITAS has improved its flux sensitivity and has decreased the time it takes to detect a 1% Crab Nebula source from $\lesssim$ 50 hrs to $\lesssim$ 30 hrs (fig. 2). VERITAS has an angular resolution (68% containment) of $< 0.1^\circ$, a pointing accuracy of $< 50$ arc-secs, an energy range of 100 GeV-50 TeV, and an energy resolution (above 200 GeV) of 15-20%. Future planned upgrades of VERITAS include replacement of the camera PMTs with high quantum efficiency PMTs, an improved array-level trigger, improved atmospheric monitoring with a LI-

DAR system, and a possible drive replacement to improve slew speed.

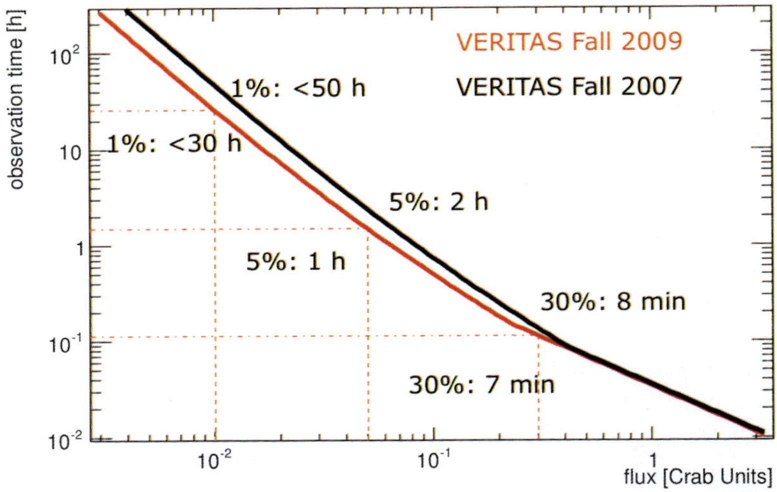


Fig. 2. VERITAS sensitivity before and after the move of telescope 1 and implementation of an improved mirror-alignment tool..

## 3. VERITAS Science Program and Results

The VERITAS collaboration has key science projects that receive guaranteed observing time as well as a competitive time allocation program where collaboration members, and scientists from outside the collaboration who collaborate with VERITAS scientists, can submit proposals for observing time on objects of interest. In addition, the collaboration has a $\gamma$-ray burst program where follow-up observations are made as soon as possible after notification of a suitable burst from the GCN network. VERITAS typically obtains $\sim$800 hours of good-quality dark-time observations per year, plus $\sim$200 hours under moonlight conditions. VERITAS science results to date are discussed below with emphasis on the most recent.

### 3.1. *Starburst Galaxies*

VERITAS detected >700 GeV $\gamma$-ray emission from the starburst galaxy M82 through 137 hours of observations made in 2008 & 2009.[3] The significance of the detection was $4.8\sigma$ (post-trials) with a $\gamma$-ray rate of 0.7 photons per hour, making it one of the weakest VHE sources so-far detected. The VHE $\gamma$-ray emission is believed to be produced by the interaction of cosmic rays with interstellar dust and gas, and the flux is in agreement with theoretical predictions. The VERITAS detection implies a cosmic ray density of 250 eV $cm^{-3}$ in M82, almost 500 times that of our own galaxy.

### 3.2. *Radio Galaxies*

M87 is one of only three radio galaxies detected in VHE $\gamma$-rays, the others being Centaurus A[4] and IC 310 (ATEL#2510). M87 exhibits flaring episodes across the electromagnetic spectrum and detailed analysis of the VERITAS observations, including a 4-day flare observed in February 2008 with peak flux $\sim$8% of the Crab Nebula flux have recently been published.[5] A joint multi-wavelength monitoring campaign in 2008 revealed that the TeV emission is associated with the core and not features in the jet.[6] VERITAS observations of other radio galaxies, including NGC 1275[7] and 3C 111 have yielded upper limits.

### 3.3. *Blazars*

Blazars are the subset of radio-loud AGN that happen by chance to have their jets oriented towards the Earth and are the most prominent extragalactic objects in the $\gamma$-ray sky. The study of blazars is one of VERITAS's key science projects, and has four components: a VHE blazar discover program, a target-of-opportunity program for flaring blazars, multi-wavelength studies of VHE blazars, studies of distant blazars to constrain the extragalactic background light.

To date, VERITAS has detected 17 VHE blazars, of which 10 have been discoveries, with two additional unidentified objects that are probably blazars (table 1). All discoveries are announced by ATEL and have accompanying multi-wavelength data. Of the 17, four are classified as intermediate BL Lac (IBL) objects while the remainder are classified as high-energy-peaked BL Lac objects (HBL).The successful VHE blazar discoveries of PKS 1424+240,[8] 1ES 0502+675, VER J0521+211, RX J0648.8+1516 and 1ES 1440+122 were motivated by results from Fermi/LAT, but RGB J0710[9] was the first Fermi/LAT source found after first being identified by a VHE instrument.

Multi-wavelength studies of the VERITAS VHE blazars have revealed that the synchrotron self-Compton (SSC) models fit the HBL broadband spectra well, including flare states (e.g. RGB J0710+591,[9] 1ES2344+514[10]).

Table 1. VHE blazars detected by VERITAS.

| Name | Class | Redshift | ATEL # |
|---|---|---|---|
| Markarian 421 | HBL | 0.030 | |
| Markarian 501 | HBL | 0.034 | |
| 1ES2344+514 | HBL | 0.044 | |
| 1ES1959+650 | HBL | 0.047 | |
| W Comae | IBL | 0.102 | 1422 |
| RGB J0710+591 | HBL | 0.125 | 1941 |
| H 1426+428 | HBL | 0.129 | |
| 1ES 0806+524 | HBL | 0.138 | 1415 |
| 1ES 0229+200 | HBL | 0.139 | |
| 1ES 1440+122 | IBL | 0.162 | 2786 |
| 1ES 1218+304 | HBL | 0.182 | |
| RBS 0413 | HBL | 0.190 | 2272 |
| 1ES 0414+113 | HBL | 0.287 | |
| PG 1553+113 | HBL | 0.34 - 0.47 | |
| 1ES 0502+675 | HBL | ? | 2301 |
| 3C 66A | IBL | ? | 1753 |
| PKS 1424+240 | IBL | ? | 2084 |
| VER J0521+211 | ? | ? | 2260, 2309 |
| VER J0648.7+152 | ? | ? | 2486 |

Monitoring of Markarian 421 with the Whipple 10 m telescope and VERITAS revealed large flares in 2008 and 2010 (ATEL #2443) which initiated multi-wavelength campaigns. Variability on 5-10 min. time-scales was observed during a massive flare on 17th February 2010 and detailed multi-wavelength spectra were obtained for a variety of different flux levels. Modeling of the broadband spectra again reveals that the SSC model works well.[11,12] However, the SSC model does not appear to work quite as well for IBLs, and suggests that an inverse-Compton model where additional photons that originate external to the jet might be needed (e.g. W Comae,[13,14] 3C 66A[15,16]).

### 3.4. *Indirect Dark Matter Search Results*

Indirect dark matter searches with ground-based $\gamma$-ray observatories provide an alternative for identifying the particle nature of dark matter that is complementary to that of direct searches or accelerator production experiments. As part of the VERITAS Dark Matter key science project, four dwarf spheroidal galaxies (Draco, Ursa Minor, Boötes 1 and Willman 1) were selected as targets based on proximity to Earth and favorable estimates for the dark matter density profile based on stellar kinematic data. No significant $\gamma$-ray excess was observed from any of the four targets and the derived $\gamma$-ray flux upper limits were used to constrain $< \sigma v >$ (the thermally averaged product of the total self-annihilation cross section and the velocity of the WIMP) for neutralino pair annihilation as a function of neutralino mass to be $\lesssim 10^{-23}$ cm$^3$ s$^{-1}$ for neutralino masses $M_\chi \gtrsim$300 GeV.[17] The derived limits are well above generic predictions for minimal supersymmetric Standard Model (MSSM) models.

### 3.5. *Survey of the Cygnus Region*

Between April 2007 and December 2008 VERITAS conducted a survey for sources of VHE $\gamma$-ray emission in the Cygnus region of the Galactic plane between 67° and 82° in Galactic longitude and -1° and 5° in Galactic latitude. 112 hours of quality observations were made in the base survey with an additional 56 hours in followup observations. The survey analysis had four pre-defined sets of cuts to provide optimum sensitivity to hard and soft point-like and moderately extended sources. A sensitivity of ~3% of the Crab Nebula flux for point-like sources was achieved. The survey led to the rediscovery of the unidentified source TeV J2032+4130 and the discovery of a new source, VER J2019+407. Further observations of VER J2019+407 have confirmed at the 7.5$\sigma$ level a source with flux level ~3% of the Crab Nebula and an extension of $\sim$ 0.2°. VER J2019 is located in the northwest side of the $\gamma$-Cygni supernova remnant and could be associated with the nearby Fermi pulsar PSR J2021+4026, or be possible shock/molecular cloud interaction.

### 3.6. *Supernova Remnants and Pulsar Wind Nebulae*

In addition to the Crab Nebula, VERITAS has detected five SNR/PWN systems as VHE $\gamma$-ray sources: Tycho, G54.1+0.3,[18] Cassiopeia A,[19] IC443[20] and Boomerang (G106.3+2.7).[23] The three most recent results are discussed below.

#### 3.6.1. *Tycho*

Tycho's SNR is the historical shell-type remnant of the Type Ia supernova observed in 1572. VERITAS observations taken in 2008 to 2010 provided 67 hours of data which result in a detection of Tycho with a pre-trials significance of 5.8$\sigma$. The emission is weak, at $\sim$ 1% of the Crab Nebula flux above 1 TeV, and peaks somewhat to the north of the remnant, overlapping a region with enhanced CO emission (fig. 3). Detailed studies of multiwavelength morphology and $\gamma$-ray emission mechanisms are currently under way.

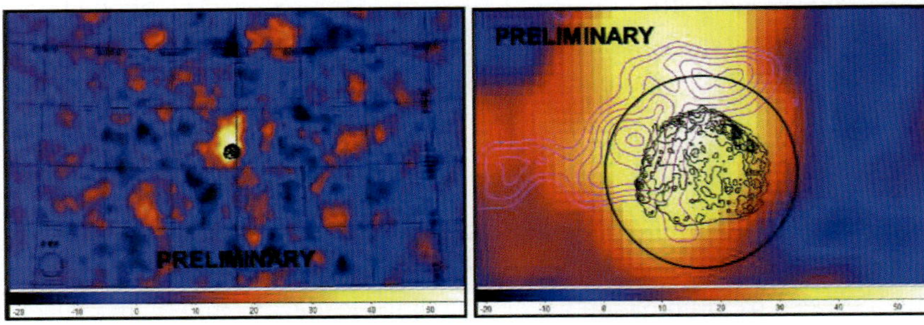


Fig. 3. Preliminary VERITAS excess maps of Tycho, full map (left) and zoomed map (right). Black contours are the Chandra ACIS[21] and magenta contours are $^{12}CO$ data from the FCRAO Survey.[22]

### 3.6.2. *G54.1+0.3*

The SNR G54.1+0.3 is a young ($\sim$3 kyr) X-ray PWN which, by virtue of its similarities to that object, has been called a "cousin to the Crab".[24] VERITAS observed G54.1+0.3 for $\sim$36 hours in 2008/2009, after partial-moonlight observations in 2007 revealed a hint of signal. These observations resulted in a $7\sigma$ detection of a point-like object, centered on the PWN location. The integral flux of the source above 1 TeV is 2.5% of the Crab Nebula flux, and the energy spectrum follows a power-law with index $2.39 \pm 0.23_{stat} \pm 0.3_{sys}$.

### 3.6.3. *Cassiopeia A*

Cassiopeia A (Cas A) is a compact young shell-type remnant of a massive star, located 3.4 kpc away, and is the brightest radio source in the sky with a synchrotron spectrum which extends all the way up to hard X-rays. Cas A had been previously detected in VHE $\gamma$-rays by HEGRA[25] and MAGIC.[26] VERITAS observed Cas A for 22 hours, resulting in an $8.3\sigma$ detection. The source, which has a flux of $\sim$3% that of the Crab Nebula above 200 GeV, shows no evidence for extension beyond the VERITAS point spread function, and has an energy spectrum which is well fit by a power-law of index $2.61 \pm 0.24_{stat} \pm 0.2_{sys}$. The Fermi/LAT team has modeled the high-energy emission from Cas A and finds that a hadronic emission model with a hard proton population (spectral index = 2.1) and a cutoff energy of 10 TeV fits the combined GeV/TeV excess well.[27]

### 3.7. *Binary Systems*

Binary systems are the Galactic VHE $\gamma$-ray sources that exhibit the greatest degree of variability and there are now four binary systems detected in VHE $\gamma$ rays: Cygnus X-1,[28] LS 5039,[29] LS I +61 303,[30] and PSR B1259-63.[31] Of these, VERITAS has detected LS I +61 303.[32] In addition, it has been suggested that variable point source HESS J0632+057,[33] which has recently been detected by VERITAS after previous observations failed to detect it,[34] is possibly a binary system.[35]

#### 3.7.1. *LS I +61 303*

LS I +61 303 is a high-mass X-ray binary system consisting of a compact object, either black hole or neutron star, orbiting a B0Ve companion with a mass of $\sim$12.5 $M_{\odot}$ and a circumstellar disk. VERITAS has made continued observations of LS I +61 303 in 2008, 2009 and 2010, having previously detected it with a statistical significance of 8.4$\sigma$ with clear orbital modulation of the flux and peak emission near apastron.[36] LS I +61 303 has not been detected in the recent VERITAS observations but it has been detected by Fermi/LAT which showed emission peaks near periastron and a cut-off in the spectrum at 6 GeV. However, since a 40% increase in the Fermi/LAT-measured flux in March 2010, there is no evidence for orbital modulation of the flux. Detailed simultaneous multi-wavelength observations of this object are needed to gain an insight into the $\gamma$-ray emission mechanism.

## 4. Summary and Conclusions

VERITAS, especially after the recent move of Telescope 1, has excellent sensitivity and is performing very well. The scientific program is a mixture of discovery and deep observations of established sources, and has proved highly successful with the detection of 33 VHE sources. With the planned upgrades of VERITAS in the coming years we expect the scientific output to continue to flourish.

## Acknowledgements

This research is supported by grants from the US Department of Energy, the US National Science Foundation, and the Smithsonian Institution, by NSERC in Canada, by Science Foundation Ireland (SFI

10/RFP/AST2748), and by STFC in the UK. We acknowledge the excellent work of the technical support staff at the FLWO and the collaborating institutions in the construction and operation of the instrument.

## References

1. T.C. Weekes., *AIP Conf. Proc.*, **1085**, 3 (2008)
2. McCann *et al.*, *Astroparticle Physics*, **22**, 6, 325 (2010)
3. The VERITAS Collaboration, *Nature*, **462**, 770 (2009)
4. Aharonian *et al.*, *ApJ*, **695**, L40 (2009)
5. Acciari *et al.*, *ApJ*, **716**, 819 (2010)
6. Acciari *et al.*, *Science*, **325**, 444 (2009)
7. Acciari *et al.*, *ApJ*, **706**, L275 (2009)
8. Acciari *et al.*, *ApJ*, **708**, 100 (2010)
9. Acciari *et al.*, *ApJ*, **715**, 49L (2010)
10. Acciari *et al.*, *ApJ*, Submitted
11. Acciari *et al.*, *ApJ*, **703**, 169 (2009)
12. Acciari *et al.*, *ApJ*, Submitted
13. Acciari *et al.*, *ApJ*, **684**, L73 (2008)
14. Acciari *et al.*, *ApJ*, **707**, 612 (2009)
15. Acciari *et al.*, *ApJ*, **693**, 104 (2009)
16. Acciari *et al.*, *ApJ* submitted
17. Acciari *et al.*, *ApJ*, **720**, 1174 (2010)
18. Acciari *et al.*, *ApJ*, **719**, L69 (2010)
19. Acciari *et al.*, *ApJ*, **714**, 163 (2010)
20. Acciari *et al.*, *ApJ*, **698**, 133 (2009)
21. Warren *et al.*, *ApJ*, **634**, 376 (2005)
22. Heyer *et al.*, *ApJS*, **115**, 241 (1998)
23. Acciari *et al.*, *ApJ*, **703**, L6 (2009)
24. Lu *et al.*, *ApJ*, **568**, L49 (2002)
25. Aharonian *et al.*, *A&A*, **370**, 112 (2001)
26. Albert *et al.*, *A&A*, **474**, 937 (2007)
27. Abdo *et al.*, *ApJ*, in press
28. Albert *et al.*, *ApJ*, **665**, L51 (2010)
29. Aharonian *et al.*, *Science*, **309**, 746 (2005)
30. Albert *et al.*, *Science*, **312**, 1771 (2006)
31. Aharonian *et al.*, *A&A*, **442**, 1 (2004)
32. Acciari *et al.*, *ApJ*, **679**, 1427 (2008)
33. Aharonian *et al.*, *A&A*, **469**, L1 (2007)
34. Acciari *et al.*, *ApJ*, **698**, L94 (2009)
35. Hinton *et al.*, *ApJ*, **690**, L101 (2009)
36. Acciari *et al.*, *ApJ*, **700**, 1034 (2009)

# Recent results from the PAMELA experiment

S. B. Ricciarini* on behalf of the PAMELA collaboration

*INFN, Structure of Florence,*
*via G. Sansone 1, Sesto Fiorentino, 50019, Italy*
**E-mail: ricciarini@fi.infn.it*

Since July 2006 the PAMELA experiment on satellite is continuously collecting data on several cosmic ray species, in the energy range from tens of MeV to hundreds of GeV. Here results will be presented of the most recent and refined data analyses, in particular those involving observations of interest for indirect dark-matter searches, and with a specific aim at illustrating how the various sources of systematic errors have been taken into account in the measurements.

*Keywords*: PAMELA; Space Experiment; Cosmic Rays; Antiparticles.

## 1. Introduction

PAMELA[1] is a satellite-borne experiment designed to study charged particles in the cosmic rays, specifically aimed at identifying the antiparticle component. It has been designed to obtain an unprecedented precision in the measurement of the energy spectra of antiprotons between 80 MeV and 180 GeV, positrons in the range 50 MeV – 270 GeV, and additionally to obtain high-precision spectra of protons, light nuclei ($Z < 8$) and electrons. The main purposes of these measurements are to test cosmic-ray propagation models and to search for evidence of dark-matter particles annihilations or other unknown primary sources.

PAMELA is also able to search for direct evidence of anti-nuclei, with a design sensitivity in the $\overline{\mathrm{He}}/\mathrm{He}$ ratio better than $\sim 10^{-7}$. Moreover, the quasi-polar orbit and low geomagnetic cut-off experienced by the PAMELA apparatus, combined with its intrinsic ability to measure low momenta, allows phenomena connected with Sun activity and Earth magnetosphere to be investigated.

This paper constitutes a review of the recently obtained results, published or near to publication. Only a short summary of the various items involved can be given here; for further details, the reader is referred to the papers cited in the bibliography and to future publications. Sec. 2 describes the main characteristics of the PAMELA apparatus and of the data analysis criteria employed. The recently published results on positrons and antiprotons are presented and discussed in Sec. 3 and Sec. 4, respectively. A general discussion of the analysis methods developed for the measurement of the absolute fluxes of H and He is given in Sec. 5. Finally, the last section contains some concluding remarks and a general outlook on the items currently under analysis, which will form the argument of future publications by the PAMELA collaboration.

## 2. PAMELA apparatus and performances

PAMELA (see Fig. 1) is housed inside a pressurized Al vessel filled with $N_2$ at 1 atm, with thickness of 2 mm in correspondence of the PAMELA acceptance window, and installed on-board of the Resurs-DK1 satellite, which follows an orbit with altitude varying between 350 and 600 km and inclination of 70°. PAMELA is in continuous data acquisition mode since July 2006; the mission is foreseen to last until the end of 2011.

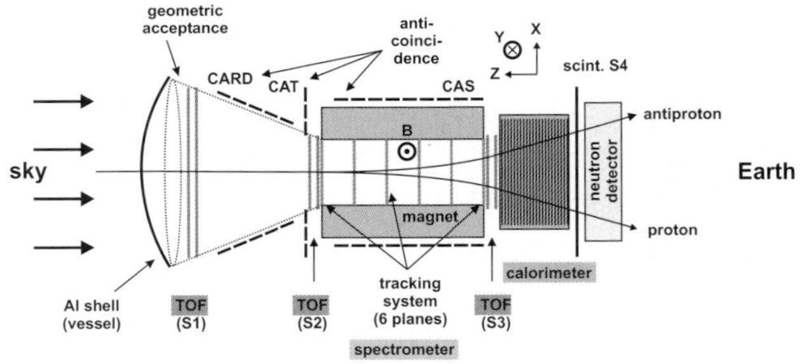


Fig. 1. Schematic view of the PAMELA apparatus, with the indication of the main direction of the magnetic field inside the spectrometer cavity. PAMELA is 1.3 m long and less than 1 m wide.

The core of the PAMELA apparatus is formed by a high-precision magnetic spectrometer: a 140 kg permanent magnet produces an approximately uniform and dipolar magnetic field of $\approx 0.4$ T in a $16 \times 13 \times 44$ cm$^3$ rectangular cavity, equipped with a tracking system composed of six equidistant planes of double-sided microstrip Si detectors. Each Si sensor is 300 $\mu$m thick, $5.3 \times 7.0$ cm$^2$ wide, with a read-out pitch of 51.0 and 66.5 $\mu$m on the junction and ohmic side, respectively, thus providing two independent impact coordinates. The cavity dimensions and magnetic field define a geometrical factor of 21.6 $cm^2$ sr.

The magnetic spectrometer determines the sign of the electric charge Ze of the incident particle and its magnetic deflection $\eta = Ze/Pc$ (where Pc is the particle momentum in GeV) and rigidity $R = 1/|\eta|$. The deflection resolution depends on the spatial resolution of the Si detectors and on the precision in the knowledge of the relative positions of the Si sensors. The former has been directly measured at beam tests and amounts to 3 (12) $\mu$m for the junction (ohmic) side, for normally incident trajectories; the Si sensor positions after launch have been determined with an *in-flight* alignment procedure using the collected relativistic p, $e^-$ and $e^+$, validated at beam tests and characterized by a precision better than 1 $\mu$m.

For relativistic particles, the average deflection uncertainty $\delta\eta$ is constant with particle energy (corresponding to a $\delta R/R \propto R$) and it is usually expressed in terms of the maximum detectable rigidity MDR, defined as the rigidity for which the relative uncertainty is 100% ($MDR = 1/\delta\eta$). The MDR is evaluated through the track fitting algorithm and varies on an event basis: it depends on the number and distribution of fitted points and on the magnetic field intensity along the trajectory; it also varies with the track inclination in the Si sensors and the noise of the hit microstrips, both affecting the spatial resolution of the single position measurements. From measurements at beam tests, the MDR distribution exceeds 1 TV.

The time-of-flight (ToF) system comprises six layers of fast plastic scintillators, segmented into paddles with independent PMT read-out, and arranged in three double planes (S1, S2 and S3). Coincidental energy deposits in combinations of layers provide the main trigger for the experiment. Time-of-flight information is combined with spectrometer track-length information to determine particle velocities. The measured time-of-flight resolution, better than 300 ps, allows $e^-/e^+$ to be separated from $p/\overline{p}$ up to an energy of $\sim 1$ GeV. Upward-going (i.e. *albedo*) particles are rejected with a significance of 60 standard deviations.

Ionization measurements in the scintillator layers and Si sensors allow the absolute value of particle charge to be determined at least up to $Z = 7$ and 5, respectively.

The sampling electromagnetic calorimeter is composed of 22 modules, each one formed by a central plate of W absorber surrounded by 2 single-sided Si strip planes, for a total depth of 16.3 radiation lengths and 0.6 nuclear interaction lengths. Each Si plane is composed of nine 380 $\mu$m thick, $8 \times 8$ $cm^2$ wide sensors with 2.4 mm strip read-out pitch. The main task of the calorimeter is to discriminate $e^+$ ($e^-$) from p ($\overline{\mathrm{p}}$); the longitudinal and transverse segmentation of the calorimeter, combined with the measurement of the energy deposited on each Si strip, allows a high identification (or rejection) power for electromagnetic showers against interacting and non-interacting hadrons (see also Sec. 3).

The anti-coincidence (AC) system consists of four plastic scintillators (CAS) surrounding the sides of the magnet, one (CAT) covering the top and four (CARD) delimiting the volume between the first two ToF planes. The main purpose of the AC system is to identify events characterized by secondary particles produced in the apparatus.

The following list summarizes the "basic" event selection criteria which have been generally applied to identify clean events with primary particle of galactic origin:

- single reconstructed trajectory in the spectrometer, entirely contained in a chosen fiducial acceptance, with good quality (number of hits, track fit $\chi^2$) for a proper rigidity measurement;
- cut on the event MDR, depending on the particle rigidity R, R<MDR/n (or equivalently $\delta\eta < |\eta|/\mathrm{n}$), to select events with enough precisely measured deflection;
- no spurious signals from multi-particle events in the ToF and AC scintillators above the tracking system;
- ionization releases consistent with particle energy, for the selected Z;
- down-going particle, with rigidity exceeding the vertical Stoermer geomagnetic cut-off (estimated on the basis of the satellite position) by a safety factor (1.3), thus taking into account the uncertainties associated to the model of the Earth magnetosphere.

Additional cuts are required for rare particles, as discussed in Sec. 3 and 4. All the employed analysis techniques have been validated through extended cross-checks between *in-flight* measurements and custom Monte Carlo simulations, based on GEANT3[2] code (and also FLUKA[3] for what concerns hadronic interactions) and tuned by means of data from several beam tests.

## 3. Positron fraction

The most important aspect to be considered for the measurement of the positron fraction (i.e. the $e^+$ flux divided by the sum of $e^+$ and $e^-$ fluxes) is to keep under control the proton contamination in the positron sample; this contamination is related to $\pi_0 \rightarrow \gamma\gamma$ decays in hadronic showers, which can mimic electromagnetic showers; moreover, the $p/e^+$ ratio increases with the particle rigidity, being $\sim 10^3$ at 1 GV and $\sim 10^4$ at 100 GV.

In our conservative approach, the residual proton contamination is quantified by using only *in-flight* data, without any dependence on simulations or beam-test data. Precisely, a nearly pure proton sample is obtained by requiring non-interacting particles in the upper 2 (out of 22) calorimeter modules; the selection cuts for $e^+$ are then applied to calorimeter variables evaluated in the "restricted" calorimeter formed by the lower 20 modules. On the other side, $e^+$ (with residual p contamination) and $e^-$ samples, with charge sign given by the spectrometer, are selected by using the same calorimeter variables and selection cuts, applied to the completely equivalent "restricted" calorimeter formed by the upper 20 modules.

For what concerns the chosen selection criteria, studies with particle beams and simulations showed[4] that PAMELA achieves a proton rejection power better than $10^5$ up to an energy of $\sim$ 200 GeV, with a positron selection efficiency of 80%, when employing suitably strong cuts, based on the topology of the shower inside the calorimeter, and when requiring the match between the total energy release detected in the calorimeter and the momentum given by the spectrometer.

A parametric bootstrap analysis with maximum likelihood fitting is performed on the three distributions obtained with the previously outlined method (see Fig. 2), for a number of rigidity intervals, and the amounts of detected electrons, positrons and contaminating protons are thus determined. As a cross-check, a non-parametric statistical analysis gives consistent results, well within one standard deviation.

A first positron fraction measurement from PAMELA was presented in 2009, with data collected up to February 2008 (see Ref. 5); recently, a new paper[6] has been published, containing data up to December 2008 (with an increase of 30% in statistics) and moreover a better understanding of systematic uncertainties related to the subtraction of p background: the new results are fully consistent with the previous ones, as shown in Fig. 3.

The positron fraction, as measured by PAMELA, is characterized by one order of magnitude improvement in statistics over previous measurements; besides, it covers the most extended energy range ever achieved by a single apparatus. The positron fraction above 10 GeV is in agreement with previous measurements but for the first time it clearly shows a significant increase with energy; this measurement had and is still having an unprecedented impact on the cosmic-ray community, since it can not be described in the framework of the standard theoretical predictions of sec-

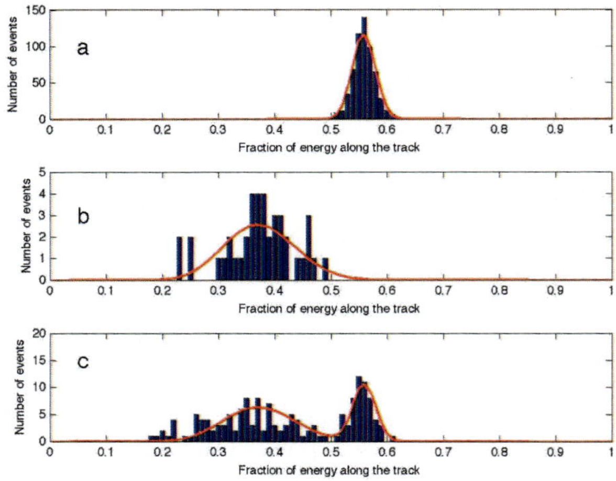


Fig. 2. Distribution of the fraction of energy measured in the calorimeter along the trajectory extrapolated from the spectrometer, for particles selected respectively as (a) electrons, (b) protons, (c) positrons, with the method outlined in the text, in the rigidity range 28–42 GV. In (c) the residual proton contamination (left) is clearly separated from the $e^+$ peak (right).

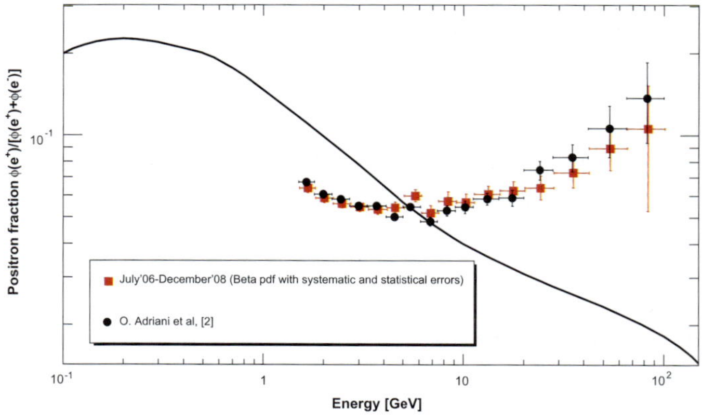


Fig. 3. Black points: the positron fraction measured by PAMELA as published in 2009.[5] Red points: the recently published new measurement,[6] where the vertical bars include both statistical and systematic errors, summed in quadrature. Line: a classical pre-PAMELA theoretical calculation[7] for pure secondary production of positrons during the propagation of cosmic rays in the galaxy. Figure from Ref. 6.

ondary production of positrons in the galaxy, even taking into account the large uncertainties in the models involved. Therefore either a significant modification in the theories of cosmic-rays acceleration and propagation is needed, or primary sources like dark-matter annihilation or astrophysical sources (e.g. pulsars), have to be taken into account.

At low energies the PAMELA positron fraction is lower than most of the other measurements, collected during the previous solar cycle; this is interpreted as the observation of a charge-dependent solar modulation effect, hypothesis supported by the fact that the PAMELA measurement is in agreement with the results of a contemporary balloon-borne experiment[8] which flew in June 2006 and also observed a similar positron fraction at low energies, but with much larger statistical uncertainties.

See e.g. Ref. 9 for a review of the possible interpretations of PAMELA positron data.

## 4. Antiproton analysis

The main issue to be faced in the antiproton analysis is the background of *spillover* protons, i.e. protons for which the charge sign is wrongly determined because of the uncertainty in the deflection measurement performed by the spectrometer. The *spillover* contamination is reduced to the desired level by introducing stronger selection criteria on the quality of the reconstructed trajectory ($\chi^2$ and MDR), and by rejecting tracks with low-resolution position measurements related to noisy read-out channels or emission of $\delta$-rays.

The first PAMELA measurement[10] of the $\overline{\mathrm{p}}/\mathrm{p}$ flux ratio, in the kinetic energy range 1.5–100 GeV, has been followed by a new publication[11] including more recent data and more refined analysis methods, extending the measurement down to 60 MeV and up to 180 GeV kinetic energies (see Fig. 4).

The higher energy limit of the new analysis is obtained by means of a looser (but still strong) MDR cut on the precision of the measured rigidity, i.e. requiring $\mathrm{R} < \mathrm{MDR}/6$ instead that $\mathrm{R} < \mathrm{MDR}/10$ (see Fig. 5); this is coupled to a refined study of the residual contamination from *spillover* protons: a fine-tuned simulation of the apparatus, including also variations with time of the detector performances, has been validated by comparing the distributions of several representative variables (e.g. coordinate residuals, $\chi^2$ and covariance matrix for parameters of the track fit) with those obtained for *in-flight* data. The *spillover* background thus estimated is subtracted from the data and the discrepancy between the simulated and real behaviour of the deflection distribution in the high-energy region is used to derive the residual systematic error.

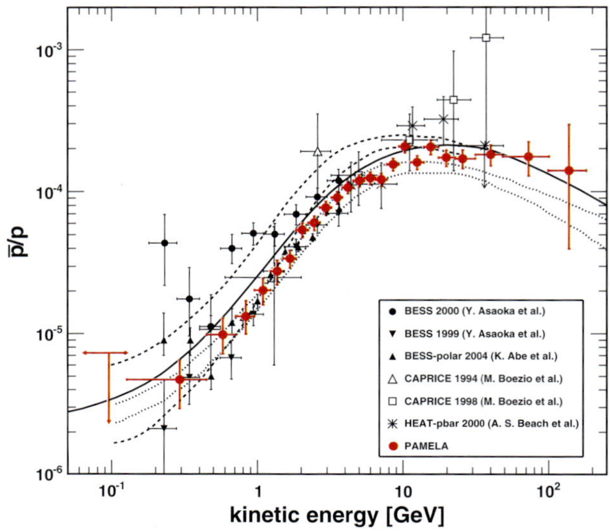


Fig. 4. Measurements of the antiproton/proton flux ratio from PAMELA and contemporary experiments; the lines represent different theoretical calculations for a pure secondary production of antiprotons during the propagation of cosmic rays in the galaxy. Figure from Ref. 11.

Other important issues are also taken into account in the antiproton analysis, e.g. the different calorimeter efficiency for p and $\overline{\mathrm{p}}$, the loss of particles in the instrument, the contamination from secondary pions produced in the materials above the spectrometer. For what concerns pions, a dedicated simulation has been validated by comparison with *in-flight* data below 1 GV, where pions are identified by relating the velocity and rigidity measurements from ToF and spectrometer, respectively; above 1 GV, the residual pion background is estimated to be negligible, apart from the range 1–3 GV where it is $\sim$ 10% and hence is subtracted from the data.

The measured $\overline{\mathrm{p}}/\mathrm{p}$ ratio increases smoothly with kinetic energy up to about 10 GeV and then levels off. The data follow the trend expected from models of secondary production in the galaxy, but at the same time these results are enough precise to place strong constraints on theoretical calculations and on possible contributions from exotic sources, e.g. dark-matter particle annihilations. The new publication[11] also contains the first

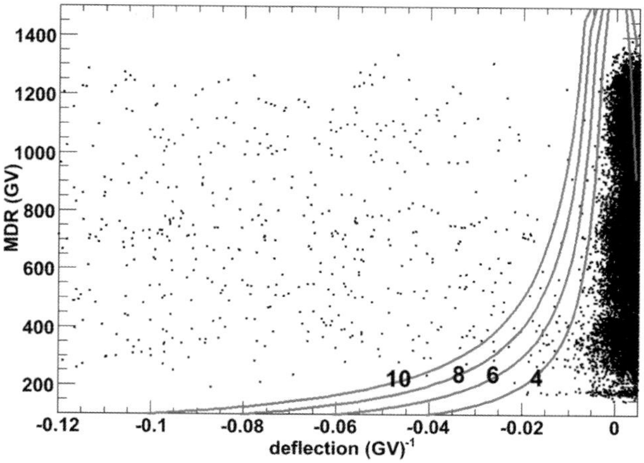


Fig. 5. Distribution in the $\eta$–MDR plane of a sample of selected antiproton and proton events before applying the event-dependent MDR cut. Events with negative $\eta$ are mostly $\overline{\mathrm{p}}$; on the right side (positive $\eta$) the proton distribution is visible, with its *spillover* tail extending to negative deflections. The lines define different possible MDR cuts R<MDR/n (or equivalently MDR>n/$|\eta|$), with n = 10,8,6,4.

PAMELA measurement of the absolute antiproton flux. The additional issues concerning absolute flux measurements are illustrated in Sec. 5.

## 5. Absolute fluxes of Hydrogen and Helium

Cosmic-ray Hydrogen (mostly protons, with a $\sim 1\%$ deuterium component) and Helium ($\sim 10\%$ of H) are overwhelming with respect to other particle species; therefore pure samples of events can be obtained, which are not significantly affected by contamination backgrounds. In fact, this measurement of absolute flux is mostly affected by the instrumental systematics in the determination of the selection efficiencies and, above few hundreds GeV, in the spectrometer measurement of the particle rigidity; specifically, possible energy-dependent systematic biases, which can alter the shape of the measured spectra, have to be kept under control.

Given the performances of the PAMELA detectors, the purity of the *in-flight* collected samples of H/He and their high statistics (more than $\sim 10^8$ proton events), it has been possible to measure the absolute fluxes of H and He over more than three decades of rigidity, up to $\approx 1$ TV, with an unprecedented total (statistical and systematic) precision of few %. This approach exploits the abundant *in-flight* acquired data to adequately characterize and minimize each single mechanism contributing to the instrumental systematics, with precision ranging from $\sim 0.1\%$ to few % over the energy range of interest.

A dedicated paper with the results of this work is currently in preparation; here only the main aspects of the analysis procedure are summarized and discussed. The PAMELA measurement of H and He fluxes will contribute to put further strong constraints on theoretical cosmic-ray models, and in particular to reduce the uncertainties on the calculations of the expected secondary antiparticles spectra. Moreover, most of the knowledge thus obtained on the PAMELA instrumental performances has been suitably exploited for the analysis of absolute fluxes of the other, much rarer, species ($\overline{\mathrm{p}}$, $\mathrm{e}^-$, $\mathrm{e}^+$, heavier nuclei).

### 5.1. *Selection efficiencies*

With the aim of obtaining a precise ($\sim$ %) measurement of the selection efficiencies, selection criteria have been grouped in subsets for which the efficiency can be directly measured with *in-flight* data; the small (not more than few %) instrumental biases introduced with this procedure are then corrected by means of the Monte Carlo simulation of the apparatus, validated through extensive cross-checks with *in-flight* collected data.

Concerning direct measurement of selection efficiencies with *in-flight* data, biases can arise from several mechanisms.

- **Inhomogeneities**. The dependence of a given selection-cut efficiency on time, particle energy and spatial/angular parameters of the trajectory in the geometrical acceptance of the apparatus, can affect the result, due to possible differences between time, energy or geometrical-acceptance composition of the data sample prepared for the efficiency measurement (indicated briefly as "efficiency data sample" in what follows) and the sample over which the selection cut is applied to obtain the flux.
- **Correlations**. Different selection cuts can be mutually correlated, i.e. each of them can have an efficiency, which is less than 100%, in the selection of the *same* specific category of H/He events. Therefore, to properly express the combined efficiency of all the above cuts as a product of single efficiency measurements, it is necessary that each affected category of events is included only in *one* of the corresponding "efficiency data samples".
- **Contaminations**. An "efficiency data sample", if not properly prepared, can contain spurious events, i.e. events which are by construction rejected by the selection cut whose efficiency is to be measured, and therefore

should not be part of the sample. Because of these contaminations, *in-flight* measured efficiencies can be underestimated.

Concerning Monte Carlo simulation of the apparatus, biases at the % level arise mostly from the practical limits to the possibility of building an accurate model including all the "fine" geometrical and physical characteristics of detectors and read-out electronics, and especially their possible variations with time, e.g. caused by changes of thresholds or gains or by failures of single sections.

The H/He selection cuts and the choice of fiducial geometrical acceptance have been fine-tuned to find an optimal compromise to minimize the net effect of all the different sources of instrumental systematics. Finally, several kinds of self-consistency checks have been performed on the whole procedure, with the specific purpose of estimating the residual systematic uncertainty in the efficiency measurement (shown in Fig. 6 together with other contributions):

- by varying the selections of "efficiency data samples" (the weights of the different inhomogeneity/correlation/contamination effects change, usually with conflicting trends);
- by using different time and energy bins;
- by cross-checking the stability of the high-rigidity flux (R $>$ 50 GV), for which solar modulation is negligible, obtained for different time intervals (months, years);
- by varying the fiducial acceptance; in particular, by cross-checking the measured flux with the one obtained with a restricted 10% acceptance corresponding to the inner part of the apparatus with respect to its cross-sectional plane.

### 5.2. *Rigidity measurement*

The uncertainty in the rigidity measurement performed by the spectrometer can cause a given event to be assigned to the wrong rigidity bin; because of the power-law behaviour of the spectra, this adds a spurious contribution to the flux measured for high-rigidity bins, which amounts to $\sim$ 1% at 100 GV and $\sim$ 10% at 1 TV, and is therefore not negligible for the H and He analysis.

This effect, varying from event to event with the MDR as discussed in Sec. 2, is accounted for in the measured spectra by means of a statistical unfolding approach based on the detailed simulation of the spectrometer:

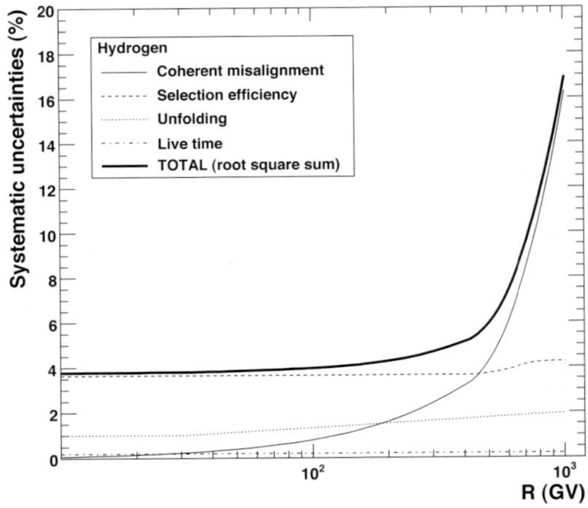


Fig. 6. The significant relative systematic uncertainties $\Delta\phi/\phi$ estimated in the PAMELA measurement of H flux $\phi$, as a function of particle rigidity.

afterwards, the residual systematic error is estimated, which is mainly associated to the adopted unfolding technique.

Moreover, the uncertainty ($\sim 1\ \mu$m) of the spectrometer alignment procedure, implies that it can not account for a small "coherent" shift between the real and estimated positions of all the Si sensor planes; this spurious "bending" of the whole spectrometer system, in turn introduces a constant, event-independent shift $\Delta\eta$ in the measured deflections, or equivalently a $\Delta R \approx \pm\Delta\eta \cdot R^2$ in the measured rigidities for particles of negative/positive charge, respectively.

This "coherent misalignment" contribution is determined and corrected for by means of *in-flight* acquired samples of relativistic $e^-$ and $e^+$, by comparing the distribution of energies resulting from the spectrometer and calorimeter measurement (taking into account possible emission of *brehmsstrahlung* photons in the spectrometer): while the calorimeter uncertainty does not depend on charge sign, the spectrometer energy measurement for $e^-$ ($e^+$) is systematically lower (higher) because of the deflection shift superposed to the other random fluctuations ($\Delta\eta$ turns out to be negative). The limited $e^+$ statistics implies a residual uncertainty $\delta\eta \approx 1 \cdot 10^{-4}$ $(\mathrm{GV})^{-1}$ after the correction is applied; this is propagated on the measured flux and considered as an additional systematic uncertainty.

The "coherent misalignment" systematics estimated for the H flux is shown in Fig. 6 together with the other main contributions. It is worth to note that the resulting overall systematic uncertainty in the H flux amounts to less than 5% up to 400 GV. Similar, even if slightly worse, results are obtained for He.

## 6. Conclusions and outlook

In this paper several aspects of the recent analysis work, done by the PAMELA collaboration, have been presented. The results of this work, i.e. the measurements of antiparticle ($e^+$ and $\overline{p}$) flux ratios and of H, He and $\overline{p}$ absolute fluxes, performed with unprecedented precision, constitute a further important step toward a better understanding of cosmic-ray physics.

Moreover, additional items, not discussed here, are currently under analysis and will be addressed in forthcoming publications. These include: the absolute fluxes of electrons, positrons and light nuclei; the study of solar modulation, affecting the low-energy end of the particle spectra, over a period of more than three years of data taking; the search for $\overline{\mathrm{He}}$ events or other exotic signals (i.e. strangelets).

## References

1. P. Picozza *et al.*, *Astropart. Phys.* **27**, 296 (2007).
2. R. Brun *et al.*, *Computer code GEANT*, version 3.21, CERN program library (1994).
3. A. Fassò *et al.*, *FLUKA: Status and Prospective for Hadronic Applications*, in *Proceedings of the MonteCarlo 2000 Conference*, Springer-Verlag, Berlin, 955 (2001).
4. M. Boezio *et al.*, *Astropart. Phys.* **26**, 111 (2006).
5. O. Adriani *et al.*, *Nature* **458**, 607 (2009).
6. O. Adriani *et al.*, *Astropart. Phys.* **34**, 1 (2010).
7. I.V. Moskalenko and A.W. Strong, *Astrophys. J.* **493**, 694 (1998).
8. J. Clem and P. Evenson, in *Proc. $30^{th}$ ICRC* (Merida, Mexico, 2007).
9. M. Boezio *et al.*, *New J. Phys.* **11**, 105023 (2009).
10. O. Adriani *et al.*, *Phys. Rev. Lett.* **102**, 051101 (2009).
11. O. Adriani *et al.*, *Phys. Rev. Lett.* **105**, 121101 (2010).

# FIRST RESULTS OF LHCF; VERY FORWARD PARTICLES AT LHC COLLISION

T. SAKO* for the LHCf Collaboration

*Solar-Terrestrial Environment Laboratory, Nagoya University,*
*Nagoya, 464-8601, Japan*
**E-mail: sako@stelab.nagoya-u.ac.jp*

The LHC forward experiment (LHCf) is specifically designed for measurements of the very forward ($\eta$>8.4) production cross sections of neutral pions and neutrons at Large Hadron Collider (LHC) at CERN. LHCf has successfully finished data taking at LHC proton-proton collisions of $\sqrt{s}$=900 GeV and $\sqrt{s}$=7 TeV in 2009–2010. During this operation, LHCf has collected 113k events of high energy showers (corresponding to ~7M inelastic collisions) at 900 GeV and ~400M showers (~350 $nb^{-1}$ of delivered integrated luminosity) at 7 TeV. Analysis results with the limited sample of data demonstrate that LHCf will provide crucial data to improve the interaction models to understand ultra-high-energy cosmic-ray air showers.

*Keywords*: ultra-high-energy cosmic-ray; hadron interaction; LHC

## 1. Introduction

There are remarkable improvements in the observation of ultra-high-energy cosmic-rays (UHECR; >$10^{18}$ eV) in the past years.[1–4] However we have not understood their nature. One of the reasons that puzzle compilation of observed data is a lack of knowledge on the hadron interaction at such high energy. To understand the development of extensive air-showers, it is necessary to know cross sections especially in the forward region. Although many accelerator data have been taken into account to construct hadron interaction models used in the air-shower simulations,[5] only limited number of very forward cross sections are available.[6,7]

## 2. The LHCf Experiment

LHCf is a dedicated experiment to measure the cross sections of neutral pions and neutrons emitted very forward direction ($\eta$>8.4) at LHC collisions. Because the proton-proton collision at the designed collision energy of

LHC (14 TeV) corresponds to $10^{17}$ eV, LHCf is expected to provide crucial calibration data for the improvement of the hadron interaction models.

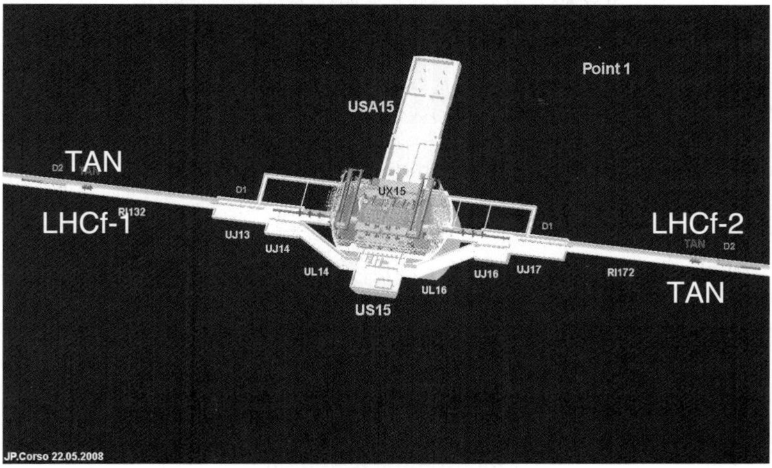


Fig. 1. Location of the LHCf detectors. Figure shows the long straight section of LHC at IP1. The ATLAS detector is at the center. Two LHCf detectors were installed ±140 m from from IP.

Two detectors, called Arm1 and Arm2, have been installed in the slot of the TANs (Target Neutral Absorber) located ±140 m from the ATLAS interaction point (IP1) as shown in Figure 1. Inside TAN the beam vacuum chamber makes a Y shaped transition from a single common beam tube facing the IP to two separate beam tubes joining to the arcs of LHC. Charged particles from the IP are swept aside by the inner beam separation dipole D1 before reaching the TAN. This unique location covers the pseudo-rapidity range from 8.4 to infinity (zero degree).

Each detector has two sampling imaging calorimeters, hereafter called 'towers.' The longitudinal structure of the towers is shown in Figure 2. Each tower is composed of 44 radiation lengths (1.7 hadron interaction lengths) of tungsten and 16 sampling layers of plastic scintillators. The transverse sizes of the towers are 20 mm×20 mm and 40 mm×40 mm in Arm1, and 25 mm×25 mm and 32 mm×32 mm in Arm2. Four X-Y layers of position sensitive detectors, scintillating fiber (SciFi) belts in Arm1 and microstrip silicon sensors in Arm2, are inserted in order to provide incident positions of the showers.

More detail of scientific goal, technical detail and performance of the

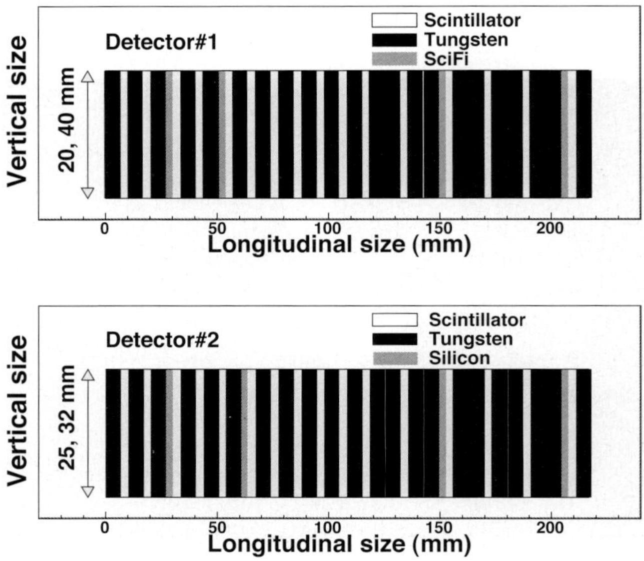


Fig. 2. Longitudinal structure of the LHCf calorimeters (towers).

detectors are found elsewhere.[8–14]

## 3. LHCf Operation at LHC

When the LHC provided first stable beams (colliding beams for physics) in December 2009 at $\sqrt{s}$=900 GeV, LHCf has successfully started data taking. After short winter shutdown, LHC has increased the collision energy up to 7 TeV in the end of March 2010. LHCf has continued data taking during 7 TeV collisions and finished operation on 19 July when delivered integral luminosity at IP1 reached at 350 $nb^{-1}$. Meanwhile LHCf has increased statistics of 900 GeV data when LHC operated at this energy in May. On 20 July during technical stop, LHCf has removed the detectors from the LHC tunnel.

The trigger of the LHCf was generated at two stages. The first level trigger was generated based on the beam pickup signals (BPTX). The second level trigger was generated when signals from any successive 3 layers in any tower exceed predefined threshold level. The threshold is chosen to achieve almost 100% efficiency for >100 GeV gamma-rays.

During the 900 GeV and 7 TeV collisions, LHCf has accumulated 113k and 400M high energy shower events, respectively, in two arms. At 7 TeV LHCf has also recorded 2M $\pi^0$'s in two arms. Figure 3 shows an example of $\pi^0$ candidate event observed in the Arm2 detector. Assuming the $\pi^0$'s decay into two gamma-rays at IP, invariant mass of gamma-ray pairs can be reconstructed using the measurements of energy and position. The invariant mass spectrum observed in the Arm2 detector is shown in Figure 4. Peak at 135 MeV due to $\pi^0$'s is clearly seen. By selecting events with invariant mass around 135 MeV, $\pi^0$ events can be extracted.

The results shown in this paper are mainly from the data in May because the data taking was stabilized and luminosity was low enough ($\sim 10^{28}\,\mathrm{cm}^{-2}\ \mathrm{s}^{-1}$) to avoid pile-up (more than two collisions in a single bunch crossing) events.

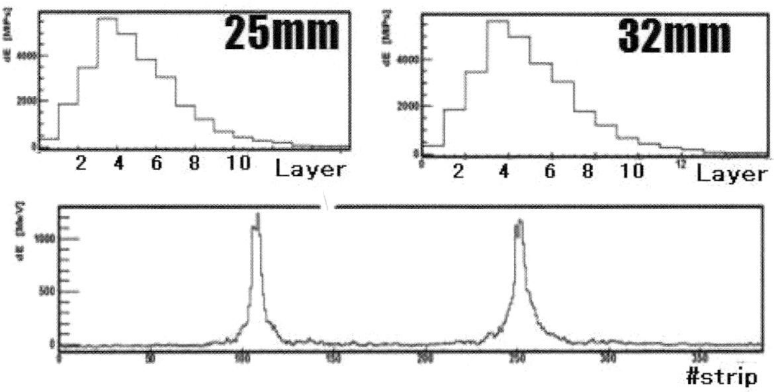


Fig. 3. A candidate event of $\pi^0$ observed in the Arm2 detector. Top: transition curves of electromagnetic-shower observed in the small tower (left) and the large tower (right). Bottom: lateral distribution of the two showers measured by the silicon strip detector. From the energy and position of two showers, invariant mass of the pair is calculated.

## 4. Data Analysis

Impact point of the particle is determined using the information of the position sensitive detectors. Based on the position, non-uniformity of the light collection and shower leakage are corrected. Because of the small transverse size of the LHCf calorimeters, the shower leakage from the towers that is only a function of position is not negligible. Events fell within 2 mm from the edge are removed from the analysis due to the large uncertainty in the

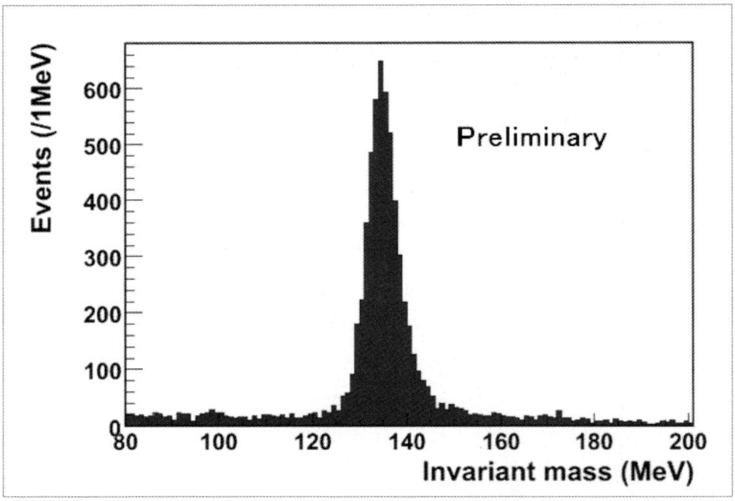


Fig. 4. Invariant mass spectrum of gamma-ray pairs observed in the Arm2 detector. Clear peak at 135 MeV is due to the events produced by $\pi^0$'s decayed at IP.

energy determination. Sum of the energy deposit in the $2^{nd}$–$13^{th}$ layers are converted to the primary energy based on the function determined by Monte Carlo (MC) simulation using the EPICS detector simulation package.[15] The validity of MC simulation is confirmed by beam test at <200 GeV. Here we note that this function is studied only for electro-magnetic showers. The energy assigned to the hadron-like showers below is 'gamma-ray equivalent energy' at this moment.

To identify electromagnetic and hadronic showers, a simple parameter, $L_{90\%}$ is defined. $L_{90\%}$ is the longitudinal position in radiation length measured from the entrance of the tower where 90% of total energy deposit is contained.

## 5. Results

Figure 5 shows distribution of $L_{90\%}$ for the 900 GeV data and MC simulation using QGSJET2 as a generator of collision products. From this figure, we can find as follows. 1) Two distinct peaks correspond to gamma-ray and hadron events. We can set a criterion at around $L_{90\%} = 20$ to separete them. To keep high purity in the gamma-like spectrum, here energy dependent criterion between 16 and 20 is used. 2) MC simulation reproduces the data fairly well.

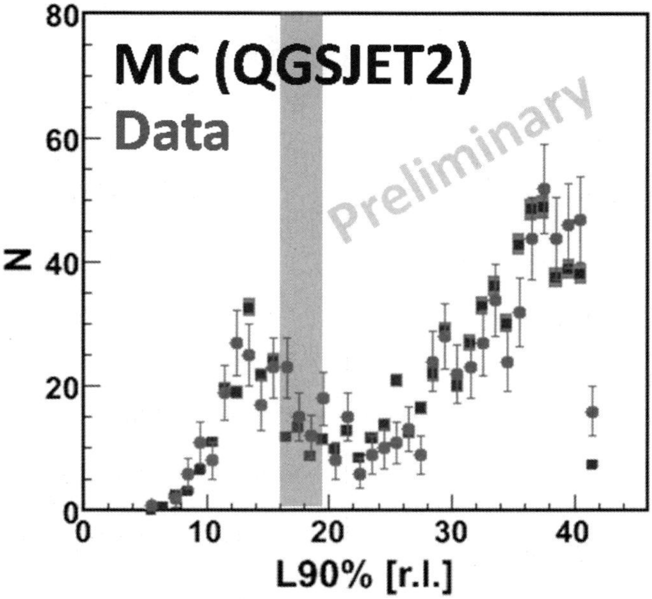


Fig. 5. Distribution of $L_{90\%}$ for the 900 GeV data (circle) and MC (square). Grey band at 16-20 radiation lengths shows energy dependent criterion to separate gamma-like and hadron-like events.

Energy spectra of gamma-like and hadron-like events observed in the both detectors are shown in Figure 6 and Figure 7 for 900 GeV data and 7 TeV data, respectively. Because of the different geometry and live time in data acquisition, direct comparison of two detectors is not relevant. By selecting events with common rapidity range in both arms, we found agreement of two detectors within the systematic error described below. Careful study to combine two data is in progress. In the 900 GeV result we used all data and also compared with the prediction of different interaction models widely used in the cosmic-ray study. The MC results are normalized to the experimental data to adjust total number of event. We can find a wide variation in the prediction and the data is well reproduced by QGSJET models. In the plots only statistical errors are assigned and we estimate a conservative systematic error in the energy scale to be 10%. To make statistical arguments possible, careful study of the energy scale is now on going. In the 7 TeV result, we used only 2% of all data we collected. High statistics data allows us to produce spectra in different $P_T$ ranges.

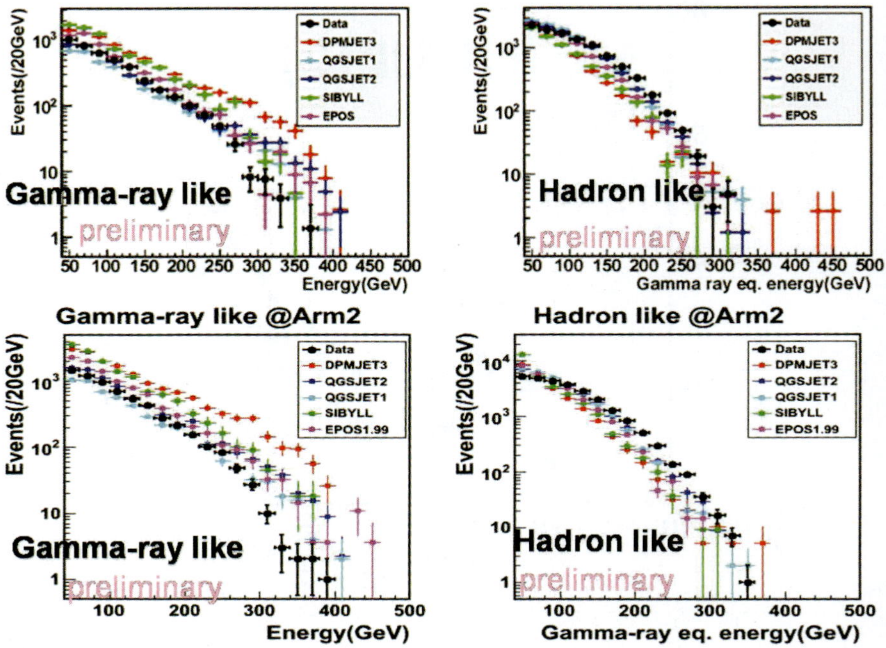


Fig. 6. Energy spectra for 900 GeV collision data. Gamma-like events (left) and hadron-like events (right) observed in the Arm1 (top) and Arm2 (bottom) detectors are shown. Model predictions are normalized to the data with the total number of event in each Arm. Because of the different geometry of two detectors, direct comparison between two arms is not relevant.

## 6. Future Plan

In the analysis, we are working to understand systematic error, PID bias and to determine the cross section based on the measured luminosity. Statistical comparison of spectra between data and model predictions and differential cross section will be available soon.

LHC will increase collision energy presumably to 14 TeV in 2013 and LHCf is assured to take data there. Because the plastic scintillators used in the current detectors are not hard enough to radiation, GSO crystal scintillators are ready to be replaced. Beam test of basic property is planned in 2010 and construction of new detectors and overall performance test are planned in 2011 to 2012. The 14 TeV data will be important not only due to the highest energy, but also because we can study energy dependence of the forward cross sections. It is crucial to extrapolate to the ultra-high-energy cosmic-ray range.

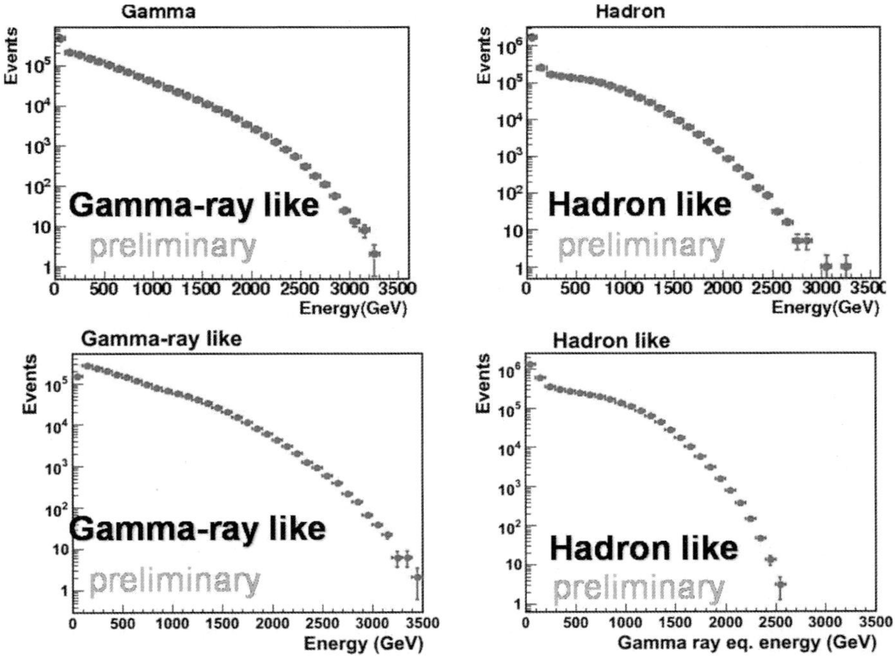


Fig. 7. Energy spectra for 7 TeV collision data. Gamma-like events (left) and hadron-like events (right) observed in the Arm1 (top) and Arm2 (bottom) detectors are shown. Because of the different geometry of two detectors, direct comparison between two arms is not relevant.

## 7. Summary

LHCf has successfully finished its data taking at 900 GeV and 7 TeV proton-proton collisions at LHC. LHCf has accumulated sufficient data at ideal low luminosity condition and analysis is now ongoing. After finalizing the analysis, LHCf will provide crucial data to calibrate hadron interaction models those are important to understand the nature of ultra-high-energy comic-rays. LHCf is also assured to take data when LHC increases collision energy up to ~14 TeV in 2013. The data at different collision energies will be very useful to extrapolate into the UHECR energy range.

## 8. Acknowledgments

We thank to all the staff in CERN, especially staff to operate the accelerator complex for providing excellent beams. We also thank to the ATLAS collaboration to provide us various infrastructure at IP1. The experiment is supported by Grant-in-Aid (Kakenhi) by the Ministry of Education, Culture, Sports, and Technology (MEXT) of Japan and by Istituto Nazionale di Fisica Nucleare (INFN) in Italy.

## References

1. J. Abraham *et al.*, Phys. Rev. Lett., **104**, 091101 (2010)
2. The Pierre Auger Collaboration, Science, **318**, 938-943 (2007)
3. R.U. Abbasi, *et al.*, Phys. Rev. Lett., **104**, 161101 (2010)
4. G. Thomson, presented in this conference
5. R. Engel, Nucl. Phys. B (Proc. Suppl.), **151**, 437-461 (2006)
6. E. Paré *et al.*, Phys. Lett. B, **242**, 531-535 (1990)
7. W. Flauger and F. Mönnig, Nucl. Phys. B, **109**, 347-356 (1976)
8. LHCf Technical Design Report, CERN-LHCC-2006-004 (2006)
9. O. Adriani *et al.*, JINST, **3**, S08006 (2008)
10. R. D'Alessandro *et al.*, Acta. Physica, Polonica B, **38**, 829 (2007)
11. H. Menjo *et al.*, Astropart. Phys., submitted (2010)
12. O. Adriani *et al.*, JINST, **5**, P01012 (2010)
13. T. Sako, *et al.*, NIM, **A578**, 146 (2007)
14. O. Adriani, *et al.*, in preparation
15. K. Kasahara, EPICS web page, http://cosmos.n.kanagawa-u.ac.jp/

# STATUS AND RECENT RESULTS FROM THE CREAM EXPERIMENT

E. S. SEO[*†], H. S. AHN, P. BHOYAR, J. EATON, O. GANEL, J. H. HAN, A. HAQUE, K. C. KIM, M. H. KIM, M. H. LEE, S. E. LEE, L. LUTZ, A. MALININ, O. OFOHA, S. S. RYU , B. P. SMITH, A. VARTANYAN, P. WALPOLE, J. WU, J. H. YOO, Y. S. YOON[†]

*Institute for Physical Science and Technology, University of Maryland, College Park, MD 20742, USA*

T. ANDERSON, N. B. CONKLIN, S. COUTU, M. GESKE, S. I. MOGNET

*Department of Physics, Penn State University, University Park, PA 16802, USA*

L. BARBIER, J. T. LINK,[‡] J. W. MITCHELL

*Astrophysics Space Division, NASA Goddard Space Flight Center, Greenbelt, MD 20771, USA*

A. BARRAU, M. BUÉNERD, B. COSTE, L. DEROME, M. MANGIN-BRINET, A. PUTZE, Y. SALLAZ-DAMAZ

*USA Laboratorire de Physique Subatomique et de Cosmologia, Grenoble, France*

R. BAZER-BACHI

*Center d'Etude Spatiale des Rayonnements, UFR PCA-CNRS-UPR 8002, Toulouse, France*

J. J. BEATTY, T. J. BRANDT

*Department of Physics, Ohio State University, Columbus, OH 43210, USA*

G. BIGONGIARI, P. MAESTRO, R. ZEI

*Department of Physics, University of Siena & INFN, Via Roma 56, 53100 Siena, Italy*

---

[*] e-mail: seo@umd.edu

[†] also Department of Physics, University of Maryland, College Park, MD 20742, USA

[‡] CRESST/USRA, Columbia, MD 21044, USA

J. A. JEON, G. NA, J. NAM, S. NAM, I. H. PARK, N. H. PARK, J. YANG
*Department of Physics, Ewha Womans University, Seoul 120-750, Republic of Korea*

A. MENCHACA-ROCHA
*Instituto de Fisica, Universidad Nacional Autonoma de Mexico, Mexico*

S. NUTTER
*Department of Physics, Northern Kentucky University, Highland Height, KY 41099, USA*

The Cosmic Ray Energetics And Mass (CREAM) balloon-borne experiment has accumulated ~156 days of exposure during five successful flights over Antarctica. Energy measurements are made with a transition radiation detector and an ionization calorimeter. Charge measurements are made with timing scintillators, pixelated Si, and Cherenkov detectors to minimize the effect of backscattered particles. High energy cosmic-ray data were collected over a wide energy range from ~ $10^{10}$ to ~ $10^{15}$ eV at an average altitude of ~38.5 km, with ~3.9 g/cm$^2$ atmospheric overburden. All cosmic-ray elements from protons (Z = 1) to iron nuclei (Z = 26) are separated with excellent charge resolution. The instrument performance, results from the ongoing data analysis, and their implications on cosmic-ray origin, acceleration and propagation are discussed.

## 1. Introduction

It is generally believed that the bulk of cosmic rays are accelerated in outer blast-wave shocks of supernova remnants (SNR). Supernovae can provide the power for sustaining the galactic cosmic-ray intensity, but details of the acceleration mechanism and propagation of cosmic rays remain to be solved. Observations of non-thermal X-rays and GeV to TeV gamma rays provide convincing evidence that the shock acceleration of cosmic rays in SNRs is nonlinear, which could amplify the ambient magnetic field. Understanding the origin and interactions of cosmic rays is essential for understanding cosmic sources, since they can alter significantly the source nature and dynamics [1].

Cosmic rays entering the atmosphere interact with atmospheric nuclei to produce secondary particles. The secondaries from low-energy particles are absorbed in the atmosphere, but the cascade of secondary particles can reach the ground if the incident energy is high enough. Cosmic rays in the energy range $10^{14}$ – $10^{20}$ eV have been detected via ground-based observations of the particle showers they initiate in the atmosphere. Such indirect measurements have the advantage of large collecting power to observe the rapidly decreasing cosmic-ray flux with increasing energy, but they cannot unambiguously identify the primary particle that initiated the shower. On the other hand, cosmic rays can be measured directly by flying particle detectors on spacecraft or high-altitude balloons. In this case, the primary particle identity and its energy can be determined unambiguously, but the energy reach is limited by the detector size and exposure time.

Primary cosmic-ray energy spectra have been measured with rather good precision at energies up to ~ $10^{11}$ eV. The pioneering measurements above this energy [2, 3] have rather large uncertainties. Consequently, the exact shape of the elemental spectra, e.g., whether or not the spectral index is the same for all elements including protons, has remained a tantalizing question. Precise measurements of the energy dependence of elemental spectra from ~$10^{12}$ to ~ $10^{15}$ eV, where the rigidity-dependent supernova acceleration limit could be reflected in a composition change, provide a key to understanding cosmic-ray acceleration and propagation.

The Cosmic Ray Energetics And Mass (CREAM) [4] investigation was conceived to measure the detailed energy dependence of elemental spectra to the highest energy possible with a balloon-borne instrument. The goal was to understand the acceleration and galactic propagation of the bulk of cosmic rays. This included whether and how the "knee" structure in the all-particle spectrum observed by air shower experiments is related to the mechanisms of acceleration, propagation, and confinement.

The CREAM project has had five successful flights, respectively called CREAM-I, -II, -III, –IV, and –V. The launch and termination dates and the flight duration for each flight are summarized in Table 1. A 40 million cubic foot (MCF), 1.1 million cubic meter, balloon carried each payload to its float altitude between ~38 and ~40 km, with an average atmospheric overburden of ~3.9 g/cm$^2$. The balloons were launched from McMurdo, Antarctica, and each flight subsequently circumnavigated the South Pole two or three times. CREAM has accumulated ~156 days of flight data, the longest known exposure for a single balloon project. These direct measurements in near-space overlap the abundant indirect measurements made from the ground.

Table 1. Previous Flight Summary.

| | CREAM-I | CREAM-II | CREAM-III | CREAM-IV | CREAM-V |
|---|---|---|---|---|---|
| Launch | 2004.12.16 | 2005.12.15 | 2007.12.19 | 2008.12.18 | 2009.12.1 |
| Termination | 2005.1.27 | 2006.1.13 | 2008.1.17 | 2009.1.7 | 2010.1.8 |
| Duration | 41 days<br>22 hrs | 28 days<br>10 hrs | 28 days<br>22 hrs | 19 days<br>13 hrs | 37 days<br>10 hrs |

## 2. Instrument

The instrument is configured with redundant and complementary charge identification and energy measurement systems. The CREAM-I and II instruments and their performance are discussed elsewhere [5]. The CREAM-III, -IV and V instrument consists of a Timing Charge Detector (TCD), a Cherenkov Detector (CD), an imaging Cherenkov Camera (CherCam), a double layer Silicon Charge Detector (SCD), carbon targets, and an ionization calorimeter comprised of a stack of tungsten plates with interleaved scintillating fiber layers. See Fig. 1. The TCD defines the trigger geometry of 2.2 $m^2$sr and measures the incident particle charge using fast electronics before the backscattered particles hit the detector. The CD with a 1 cm thick wavelength shifted Cherenkov radiator vetoes low-energy background particles at balloon altitude due to the low geomagnetic cutoff over Antarctica. The CherCam consists of a silica aerogel Cherenkov radiator plane and a photon detector plane with an array of 1,600 photomultiplier tubes (PMT's). The planes are separated by a 10 cm ring expansion gap to ensure that most Cherenkov photons are collected in 8 tubes surrounding the tube hit by the incident particle. Since upward moving particles are absorbed in the radiator, the CherCam provides efficient discrimination against backscattered particles. The SCD is segmented into small pixels (2.12 $cm^2$) to minimize hits of accompanying backscattered particles in the same segment as the incident particle. A single layer of 2 x 2 $mm^2$ square scintillating fibers, S3, measures the time at which backscattered particles start their way back to the TCD. The carbon target induces hadronic interactions for the showers to develop in the calorimeter. The particle energy was determined from the energy deposited in the calorimeter, which also provides tracking information to determine which segment(s) of the charge detectors to use for the charge measurement. In CREAM-I the energy determination was augmented by a second measurement derived from signals in a transition radiation detector (TRD).

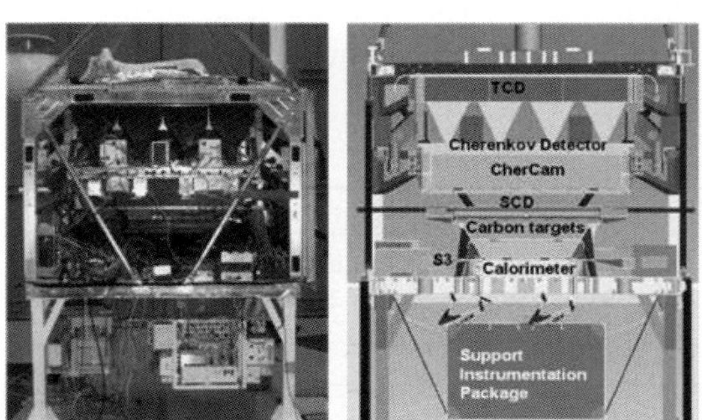


Figure 1. A photo of CREAM-V instrument (left) and its cross sectional view with subsystem names (right).

## 3. Current Results

The exceptional performance of both the science instrument and the flight support systems can be attributed to the fact that they were developed for 100-day ultra long duration balloon (ULDB) flights. The instrument was improved for each flight. A major improvement for CREAM-III and subsequent flights over CREAM-I and -II was the addition of a redundant Science Flight Computer, which was a potential single point failure during the previous flights. Two computers were accommodated successfully with a USB interface [6]. Another significant upgrade for CREAM-III and subsequent flights was the addition of the CherCam optimized for charge measurements [7]. The finely segmented SCD flown successfully in CREAM-II was refurbished for CREAM-III and subsequent flights. The detector performance did not change for these flights. Individual nuclei were measured with excellent charge resolution (e.g. $\sigma \sim 0.2e$ for p and He).

The calorimeter trigger required 6 consecutive layers (3x and 3y coordinates) to be active for a high-energy shower trigger. A layer is considered active if at least one (out of 50) ribbon's low-energy range registered a signal exceeding a commandable threshold, typically set at about 60 MeV for CREAM-I and –II, and about 15 MeV for CREAM-III -IV and V. For the high energy triggered events, i.e., for the calorimeter trigger, the shower axis was reconstructed by a linear fit of the scintillating fiber strip with the maximum energy deposit in each layer. This reconstructed trajectory was required to traverse the SCD active area and the bottom of the calorimeter active area. The particle energy was determined from energy deposit in the calorimeter assuming the same fraction of energy was deposited for all energies. All particle counts are shown as a function of incident energy in Fig. 2. This plot is not intended for spectral index or absolute flux measurements, but for a quick consistency check. It does not include the energy dependent shower leakage corrections for the energy scale and no corrections have been made for charge dependent efficiencies. The CREAM-III and later flight data indicate an energy threshold significantly lower than in the previous two flights, due to improved readout electronics [8]. The slight difference in the threshold energy between CREAM-I and -II was due to a slightly higher high voltage setting for CREAM-II. All five flight data sets follow a consistent power law above the calorimeter threshold (low energy roll off). The CREAM calorimeter data cover 3 decades in energy.

The CREAM flight duration exceeds the cumulative flight time of all prior balloon experiments, including the multi-flight emulsion chamber experiments JACEE [2] and RUNJOB [3]. The number of particles > 6 TeV measured by

CREAM also exceeds the total of all the prior experiments. The CREAM payload is relatively light as a Long Duration Balloon (LDB) payload, and it maintained high altitude. The corresponding atmospheric overburden was ~3.9 g/cm$^2$ throughout all four flights. That implies about 6.8 g/cm$^2$ for the maximum angle of acceptance, which is smallest among the comparable experiments. For example, the average vertical atmospheric overburden for RUNJOB was more than twice that of CREAM, due to its lower flight altitude. Considering the RUNJOB acceptance of particles at large zenith angles, its effective atmospheric depth is as large as 50 g/cm$^2$. For that depth, large corrections are required to account for the fact that 41% of protons and 84% of Fe nuclei would have interacted before reaching the detector.

The effective geometry of the calorimeter, after taking into account the interaction fraction, is about 0.3 m$^2$sr for protons. It is increasingly higher for heavier nuclei, due to their shorter interaction mean free path. The collecting power of CREAM is about a factor of two larger than that of ATIC [9] for protons and He and, considering the much larger geometry factor of the TRD in CREAM-I, about a factor of 10 larger for heavier nuclei. TRACER [10] has a larger geometry factor than CREAM, but a smaller dynamic charge range (Z = 8 – 26) was reported for its 10-day Antarctic flight. Its dynamic charge range was improved to Z = 3 – 26 for its ~ 4-day flight from Sweden to Canada in 2006. We estimate that the latter should have approximately the same amount and quality of data as CREAM-I for precise high-energy B/C measurements. The CREAM-I secondary to primary ratio, e.g., B/C, has been reported elsewhere [11]. With excellent particle charge and energy resolutions, and relatively large collection factor, each CREAM flight will extend the reach of precise composition measurements to energies not previously possible.

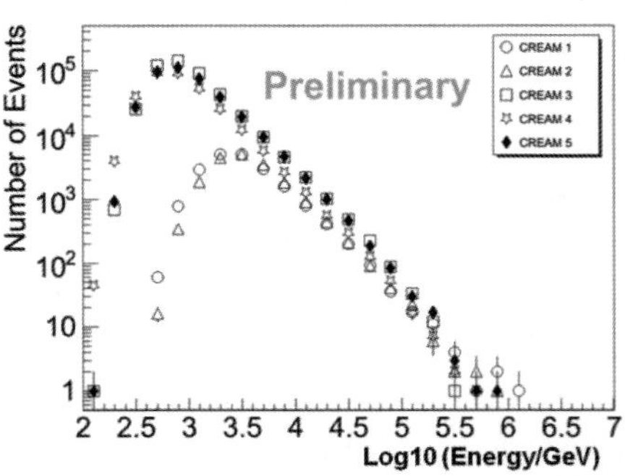


Figure 2. Preliminary all particle counts as a function of incident energy for CREAM-I (circles) CREAM-II (triangles), CREAM-III (squares), CREAM-IV (stars), and CREAM-V (filled diamonds).

Elemental spectra from CREAM-I and -II are compared with previous measurements in Fig. 3. The absolute fluxes were obtained by correcting the measured spectra for attenuation due to interactions in the atmosphere above the balloon altitude (3.9 g/cm$^2$), trigger and selection efficiencies (~ 80%), background from misidentified charges (e.g., 5% for protons and 9% for helium) due to interactions above the charge detector or by particles backscattered from the calorimeter, geometry factor, and live time. Further details of the detectors, their performance, and analysis are discussed elsewhere [12, 13]. Overall, the CREAM data are consistent with previous measurements, namely JACEE [2] and ATIC-2 [14] for protons and helium and, where they overlap, HEAO-3 [15], CRN [16] and TRACER [10] for heavy nuclei. Note that JACEE and RUNJOB did not have sufficient resolution to measure individual elements above helium.

Figure 3 shows the elemental fluxes multiplied by $E^{2.75}$ as a function of energy per nucleon. All elements appear to have the same spectral shape with a peak at the same energy per nucleon (rigidity). A single power law fit to our O – Fe data agrees with the TRACER power law fit [12], but the fluxes above 200 GeV/nucleon tend to be higher than the single power-law fit. A broken power law gives a better fit to the CREAM data. The TRACER instrument utilized three distinctly different detectors to cover the widest energy range, but its O – Fe data have a gap in the middle energy range where the spectral shape changes. Figure 14 of Ref. [10] shows that it has only one data point from a dE/dx

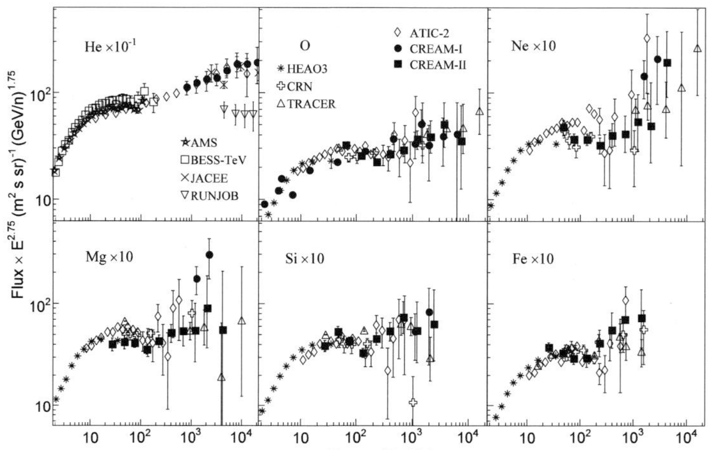


Figure 3. The elemental spectra of He and heavier nuclei measured by CREAM-I and II are compared with previous measurements.

relativistic rise measurement between ~ 10 GeV/ nucleon and ~ 400 GeV/nucleon, where we observe the spectral change.

The overall trend of the compiled data indicates harder spectra above ~200 GeV/n. As discussed in Ref. [12], it may represent different source components [17, 18] or concavity of the spectra predicted from the diffusive shock acceleration mentioned above [19]. Whatever the explanation, these results are intriguing, and they point to the need to obtain high statistics data at the highest energies, which is a key to understanding the SNR acceleration process for cosmic rays in detail. The hardening of elemental spectra must be accounted for in an explanation of the high energy electron enhancement [20,21]. As reported in a separate paper [13], preliminary results from subsequent flights are consistent with the CREAM-I and -II data.

## 4. Project Status and Plan

The CREAM-V instrument survived landing without serious damage. It was fully recovered with Twin Otter and Basler airplane flights, and the recovered instrument came back from Antarctica in March 2010. The sixth flight will incorporate an instrument essentially identical to CREAM-V, including a calorimeter, graphite targets, SCD, and CherCam. A major change for CREAM-V and -VI is a new support system. All four previous flights were supported with the Command and Data Handling Module (CDM) developed by the NASA Wallops Flight Facility for ULDB flights. There is some concern that the CDM may be nearing the end of its useful lifetime without a spare. To mitigate the risk of not having CDM available, the CREAM data acquisition system (DAQ) has been modified to accommodate the Support Instrumentation Package (SIP) normally used by the Columbia Scientific Balloon Facility (CSBF) to support LDB payloads. The main difference between the CDM and SIP from the view point of the interface with the instrument is that the SIP is serial-based while the CDM is ethernet-based. The modified CREAM DAQ with the serial interface was successfully tested during the CREAM-V flight. Following a beam test at CERN the instrument was delivered to CSBF in July for final integration with a SIP. After a successful hang test and burn-in test, the instrument and the ground support equipment were shipped to Antarctica to be ready for launch in December 2010.

In parallel with the CREAM-VI flight preparation, refurbishment of the recovered CREAM-V items was begun for use in CREAM-VII. The CREAM-VII instrument will be exposed to particle beams for calibration at CERN, Geneva, Switzerland, in October 2010. CREAM has achieved the longest exposure for a single balloon project with four successful LDB flights, but the exposure from 3-4 LDB flights is equivalent to only one nominal ULDB flight. It should be noted that a 7 MCF super-pressure balloon (i.e., ULDB) was flown successfully for 54 days. As ULDB becomes available, long duration exposures can be achieved faster and more efficiently without multiple refurbishment and launch efforts. The data from each flight will reduce the statistical uncertainties and extend the reach of the measurements to energies higher than previously possible.

## Acknowledgments

The authors thank the NASA Wallops Flight Facility Balloon Program Office, Columbia Scientific Balloon Facility, National Science Foundation Office of Polar Programs, and Raytheon Polar Services Company for the successful balloon launches, flight operations, and payload recoveries. This work was supported in the U.S. by NASA grants NNX08AC11G, NNX08AC15G, NNX08AC16G and their predecessor grants, in Italy by INFN, and in Korea by the Creative Research Initiatives of MEST/NRF.

## References

1. M. Hillas, *J. Phys. G: Nucl. Part. Phys.* **31**, R95 (2005).
2. K. T. Asakimori, et al., *Astrophys. J.* **502**, 278 (1998).
3. A. V. Apanasenko, et al., *Astropart. Phys.* **16**, 13 (2001).
4. E. S. Seo et al., *Adv. in Space Res.* **42**, 1656 (2008).
5. H. S. Ahn et al., *Nucl. Instrum. Meth. A* **579**, 1034 (2007).
6. J. H. Yoo et al., *Proc. 30th Int. Cosmic Ray Conf., Merida*, **2**, 401 (2007).
7. L. Derome et al., *Proc. 30th Int. Cosmic Ray Conf., Merida*, **2**, 453 (2007).
8. M. H. Lee et al., *Proc. 30th Int. Cosmic Ray Conf., Merida*, **2**, 409 (2007).
9. H. S. Ahn et al., *Adv. Sp. Res.*, **37**, 1950 (2006).
10. M. Ave et al., *Astrophys. J.*, **678**, 262 (2008).
11. H. S. Ahn et al., *Astropart. Phys.*, **30/3**, 133 (2008).
12. H. S. Ahn et al., *Astrophys. J.,* **714**, L89 (2010).
13. Y. S. Yoon et al. *Proc. 31st Int. Cosmic Ray Conf. Toronto*, in press (2009).
14. A. D. Panov et al. *Bull. Russ. Acad. of Sci., Phys.* **71/4**, 494 (2007).
15. J.J. Engelmann et al., *Astron. & Astrophys.* **233**, 96 (1990).
16. D. M¨uller, et al., *Astrophys. J.* **374**, 356 (1991).
17. P. L., Biermann, *Astron. & Astrophys.,* **271**, 649 (1993).
18. Y. M. Butt, A. M. Bykov, *Astrophys. J.*, **677**, L21 (2008).
19. D. C. Ellison, E. G. Berezhko, M. G. Baring, *Astrophys. J.* **540**, 292 (2000).
20. J. Chang et al., *Nature*, **456**, 362 (2008).
21. A. A. Abdo et al., *Phys. Rev. Lett.*, **102**, 181101 (2009).

# ON THE POSSIBILITY OF REGISTERING UHE EAS CHERENKOV LIGHT BY THE TUS DETECTOR

O. P. SHUSTOVA

*Space Chair, Physics Department, Lomonosov Moscow State University, Moscow, 119991, Russia*
*E-mail: olga.shustova@gmail.com*

N. N. KALMYKOV and B. A. KHRENOV

*Lomonosov Moscow State University Skobeltsyn Institute of Nuclear Physics, Moscow, 119991, Russia*

We have estimated the components of Cherenkov light arising when an extensive air shower with energy $\sim 10^{20}$ eV propagates over ocean. Photons reflected from ocean surface and scattered in air make a main contribution to the Cherenkov light. Our model calculations have shown that the TUS detector can register a flash of the Cherenkov photons reflected if zenith angles of the shower are less than $\approx 20^\circ$.

*Keywords*: Ultrahigh energy cosmic rays; Extensive air showers; Cherenkov light; the TUS detector.

## 1. Introduction

A particle of ultrahigh energy cosmic rays (UHE CR) generates an extensive air shower (EAS) of secondary particles, accompanied by huge fluxes of Cherenkov and fluorescence radiation. The latter is isotropic so it allows one to register shower tracks over a wide area of the atmosphere by arrays like Fly's Eye[1] and its improved variant HiRes[2].

At present, the new space experiments TUS[3] and JEM-EUSO[4] are in active preparation to study the UHE CR by recording the fluorescence from the EAS. An important advantage of the space investigation in comparison with the ground one is the possibility of detecting the UHE CR on the entire celestial sphere by the same detector.

There is another way to analyze the UHE CR, which consists in registering Cherenkov light reflected from a surface with high albedo. With a space detector spending roughly 2/3 of its productive time over the ocean, an estimate of the Cherenkov signal from the shower propagating over the ocean surface is of interest. In this case there arise several components of the Cherenkov light, directed upwards. These are 1) photons generated in the atmosphere (below "atmospheric") and reflected from the ocean surface, 2) photons produced in water and appearing on the ocean surface through scattering, 3) atmospheric photons entering in the ocean and appearing on its surface again. Besides, a portion of the atmospheric Cherenkov light rises to upper semisphere owing to scattering in air.

In Sec. 2 we describe the procedure of the calculations performed and present the estimates of the Cherenkov light components, mentioned above. In Sec. 3 the possibility of registering the Cherenkov photons reflected is discussed.

## 2. Calculations

Monte-Carlo simulations are applied to generate Cherenkov photons by the electron of the shower and spread them in medium. We use the laboratory system in which the origin is located at the point where the shower axis enters in the ocean, the plane $Oxy$ coincides with the flat ocean surface, and the axis $Oz$ is directed upward.

Generating the Cherenkov photon implies fixing coordinates of its "place of birth" and direction of its motion. The position of the photon is determined by the depth of the parental electron, modeled from the Gaisser-Hillas formula with the parameters appropriate to the primary energy of the shower $E_0 = 10^{20}$ eV. If the energy of the electron and its angle relative to the shower axis are known, the direction of the photon in the laboratory system can be found. For these energy and angle of the parental electron to be modeled, we use the distribution functions given in Ref. 5.

After the coordinates and direction of the photon generated are found, its propagation in medium is simulated. The path length of the photon to the next scattering event, direction, and scattering angle relative to its previous direction in the laboratory system are determined in every scattering point. As to the photons spreading in air, only Rayleigh scattering is taken into account. The atmosphere is assumed to be isothermal. In case of water we use scattering and absorption mean free path data given in Ref. 6. The parameter value of the Henyey-Greenstein distribution function according to which the scattering angle of the photon in seawater is modeled, is presented ibid. Calculations of reflection and refraction angles of the photon going from one medium to another are based on geometric optics.

It is pertinent to note that we associate each photon generated with a definite weight depending on the energy of the parental electron. It decreases in further modeling the propagation of the photon, as a result of absorption. The calculations are executed till the value of the weight is less than a predetermined number.

The direction and weight of the Cherenkov photon reflected are determined on the assumption of wave ocean surface. We apply a simple model in which the wave is described by parametric equation of truncated cycloid. The wave front is assumed to be parallel to the axis $Oy$. To set values of the parameters characterizing length $\lambda_w$ and amplitude $h_w$ of the wave we use the following information. From the satellite observations[7] it appears that wind speeds from 2 m/s to 8 m/s dominate over an ocean surface, and they correspond, according to the Beaufort scale, to wave heights from 0.3 m to 2 m. So using the joint distribution for half-wave length and amplitude, presented in Ref. 8, we choose the following parameter values: $\lambda_w = 40$ m, $h_w = 0.7$ m.

Table 1. Contributions of the Cherenkov light components.

| $\theta_{\rm sh}$ (deg) | $\lambda$ (nm) | $N_\lambda^{\rm a}/10^{13}$ (nm$^{-1}$) | $W_{\rm refl}$ (%) | $W_{\rm scat}$ (%) | $W_{\rm a-w}$ (%) | $N_\lambda^{\rm w}/10^{12}$ (nm$^{-1}$) | $W_{\rm wat}$ (%) |
|---|---|---|---|---|---|---|---|
| | 300 | 5.44 | 1.5 | 1.4 | 0.02 | 26.1 | 0.2 |
| 0 | 340 | 4.24 | 1.5 | 1.3 | 0.03 | 20.4 | 0.3 |
| | 390 | 3.22 | 1.6 | 1.0 | 0.05 | 15.5 | 0.4 |
| | 300 | 6.43 | 1.5 | 1.8 | 0.02 | 12.6 | 0.3 |
| 30 | 340 | 5.00 | 1.6 | 1.6 | 0.04 | 9.83 | 0.4 |
| | 390 | 3.81 | 1.6 | 1.3 | 0.06 | 7.47 | 0.4 |
| | 300 | 7.11 | 1.8 | 2.5 | 0.02 | 2.81 | 0.5 |
| 45 | 340 | 5.54 | 2.0 | 2.4 | 0.04 | 2.20 | 0.6 |
| | 390 | 4.21 | 2.1 | 1.9 | 0.06 | 1.67 | 0.7 |
| | 300 | 7.20 | 2.3 | 4.0 | 0.02 | 0.02 | 0.9 |
| 90 | 340 | 5.60 | 3.1 | 4.1 | 0.03 | 0.01 | 1.0 |
| | 390 | 4.26 | 3.8 | 3.4 | 0.06 | 0.01 | 1.2 |

Contributions of the Cherenkov light components to the total number of the Cherenkov photons with wavelength $\lambda$, generated in air ($N_\lambda^a$) or water ($N_\lambda^w$) by the shower with energy $10^{20}$ eV and zenith angle $\theta_{\rm sh}$ over the ocean, are given in Table 1. We use the following notations here: $W_{\rm refl}$ for the photons reflected; $W_{\rm scat}$ for the photons rising to the height of the satellite orbit owing to Rayleigh scattering; $W_{\rm a-w}$ for the photons entering in the ocean and appearing on its surface through scattering; $W_{\rm wat}$ for the photons produced in the ocean and appearing on its surface.

## 3. Signal from Cherenkov photons reflected

As seen from Table 1, "reflected" and "scattered" components make a main contribution to the total Cherenkov light produced by the shower propagating over the ocean. In this paper we consider the possibility of recording the reflected component by the TUS detector. The registration of its rather short signal at the end of the fluorescence signal stretched out in time would allow us to define the parameters of the shower more exactly.

The following parameters of the TUS detector are used in calculating. Its field of view and mirror area are equal to $\approx 9.5°$ and $1\,{\rm m}^2$, respectively. The height of the satellite orbit averages 400 km. The photoreceiver consists of $16\times16$ photomultiplier cells, each one covering an area of $4\times4$ km$^2$.

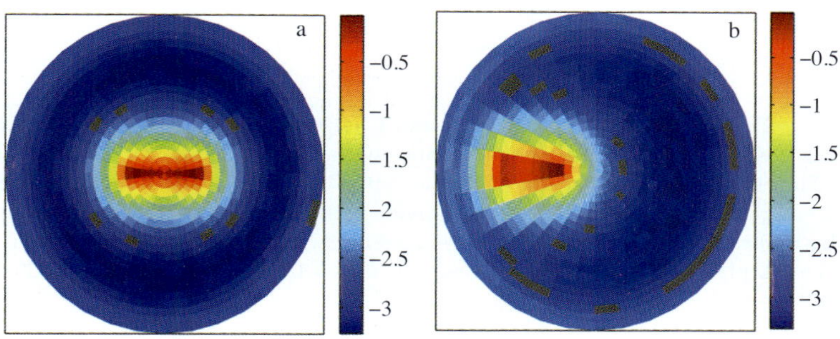


Fig. 1. The angular distribution of the reflected Cherenkov photons near to the wave ocean surface in case of the shower angles $\varphi_{\rm sh} = 0°$, $\theta_{\rm sh} = 0°$ (a) and $\theta_{\rm sh} = 20°$ (b). The surface density of the photons on hemisphere is projected on the circle of unit radius. The direction of the photon is defined by azimuth $\varphi$ and zenith $\theta$ angles. The value of $\sin\theta$ is counted along the circle radius, the value of $\varphi$ anticlockwise. The colorbars present the common logarithm of the relative photon density.

As shown in Fig. 1, the range of zenith angles $\theta$ of the photons reflected is extended when their azimuth angles $\varphi$ are close to $0°$ and $180°$. The TUS detector will be able to register the photons hitting its mirror under the condition $\theta \leq 4.75°$ that corresponds to the central circlet with radius $\approx$0.083. Our calculations demonstrate that roughly 96% of the reflected component are concentrated at the area of $4\times4\,{\rm km}^2$ with the center at the origin of the laboratory system (see Sec. 2), so nearly all its signal will focus at one cell of the photoreceiver. The value of the signal changes in accordance with the position of the TUS detector over the spot of the reflected photons on the ocean surface.

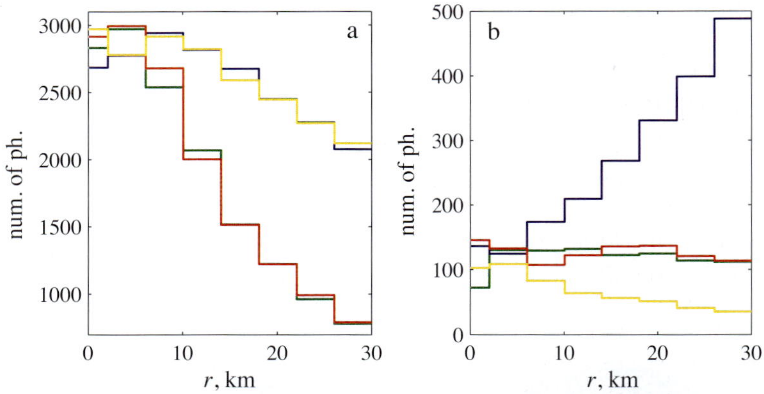


Fig. 2. The mean number of the reflected Cherenkov photons that can be registered by the TUS detector in case of the shower angles $\varphi_{\rm sh} = 0°$, $\theta_{\rm sh} = 0°$ (a) and $\theta_{\rm sh} = 20°$ (b). The circular area at a height of 400 km with the radius $r$ and center projecting itself to the origin of the laboratory system, is divided into four sectors. The number of photons with azimuth angle $\varphi$ in each sector, averaged over the annular sector with the thickness of 4 km, is presented in different colors: $\varphi = 45° \div 135°$ (green), $\varphi = 135° \div 225°$ (blue), $\varphi = 225° \div 315°$ (red), $\varphi = 315° \div 45°$ (yellow).

The number of the reflected Cherenkov photons that reach the mirror of the TUS detector almost at the same time is illustrated in Fig. 2. As we can see, their signal decreases with increasing the zenith angle of the shower. Note that in our wave model zero azimuth angle of the shower corresponds to the case when the shower axis and the wave direction lie in the same plane. After investigating all the possible orientations of the wave front we conclude that the TUS detector can reveal a flash of reflected Cherenkov photons at the end of the fluorescence signal from the shower with zenith angles $\leq 20°$.

It's worth mentioning the scattered component of Cherenkov light in a few words. According to calculations, the corresponding signal overlaps the fluorescence signal and partly blurs its maximum. So one has to study this effect in detail and take it into account in processing future data.

## 4. Conclusions

Our calculations indicate that photons reflected from the ocean surface and photons scattered in the atmosphere make a main contribution to the Cherenkov light from the shower of ultrahigh energy. But the TUS detector with field of view $\approx 9.5°$ will be able to reveal a flash of the reflected photons if zenith angles of the shower are less than $\approx 20°$. The JEM-EUSO experiment is assumed to extend this angle range.

## Acknowledgments

This work was supported by the Russian Foundation for Basic Research under grants 09-02-12162-ofi_m and 10-02-09729-mob_z.

## References

1. R. M. Baltrusaitis, R. Cady, G. L. Cassiday et al., *Nucl. Instrum. Methods Phys. Res., A.* **240** (2), 410 (1985).
2. R. U. Abbasi et al. (High Resolution Fly's Eye Collaboration), *Phys. Rev. Lett.* **92** (15), 151101 (2004).
3. B. A. Khrenov, V. V. Alexandrov, D. I. Bugrov et al., *Phys. Atom. Nucl.* **67** (11), 2058 (2004).
4. Y. Takahashi and the JEM-EUSO Collaboration, *New Jour. Phys.* **11**, 065009 (2009).
5. S. Lafebre, R. Engel, H. Falcke et al., *Astropart. Phys.* **31** (3), 243 (2009).
6. C. D. Mobley, *Light and Water: Radiative Transfer in Natural Waters* (Academic Press, London, 1994).
7. *Winds. Measuring Ocean Winds from Space.* `http://winds.jpl.nasa.gov`.
8. J. Ryden, S. van Iseghem, M. Olagnon, I. Rychlik, *Appl. Ocean Res.* **24**, 189 (2002).

# TeV GAMMA-RAYS FROM NGC 1275 DETECTED IN 15 YEAR OBSERVATION OF SHALON TELESCOPE

V.G. SINITSYNA, S.I. NIKOLSKY, V.Y. SINITSYNA

*P.N.Lebedev Physical Institute,*
*Leninsky prospect 53, Moscow, 119991, Russia*
*E-mail: sinits@sci.lebedev.ru*

In 1996 year a new metagalactic source was detected by SHALON at TeV energies. This object was identified with Seyfert galaxy NGC1275; its image is presented. The maxima of the TeV gamma-ray, X-ray and radio emission coincide with the active nucleus of NGC 1275. But, the X-ray and TeV emission disappears almost completely in the vicinity of the radio lobes. The correlation TeV with X-ray emitting regions was found. The integral gamma-ray flux of NGC1275 is found to be $(0.78 \pm 0.13) \times 10^{-12} cm^{-2} s^{-1}$ at energies of $> 0.8$ TeV. Its energy spectrum from 0.8 to 40 TeV can be approximated by the power law with index $k = -2.25 \pm 0.10$. The NGC1275 has been also observed with the Tibet Array (5TeV) and then with Veritas telescope at energies about 300 GeV at 2009. The recent detection by the Fermi LAT of gamma-rays from the NGC1275 makes the observation of the energy $E > 100$ GeV part of its broadband spectrum particularly interesting.

*Keywords*: Active Galactic Nuclei; BL Lacs; Mkn 421; Mkn 501; Extragalactic Supernova explosion.

## Introduction

The $\gamma$-astronomical researches are carrying out with SHALON[1–5] mirror Cherenkov telescope at the Tien-Shan high-mountain observatory. During the period 1992 - 2010, SHALON has been used for observations of galactic and metagalactic sources; among them are the known blazars Mkn421 (z=0.031), Mkn501 (z=0.034) and Seyfert galaxy NGC1275 (z=0.0179). The observation results of two type of metagalactic sources: Seyfert galaxy NGC1275[1–4] and SN 2006gy are presented. The explosion of SN2006gy extragalactic supernova (z = 0.019) was detected at TeV energies with SHALON telescope during the observations of NGC1275.

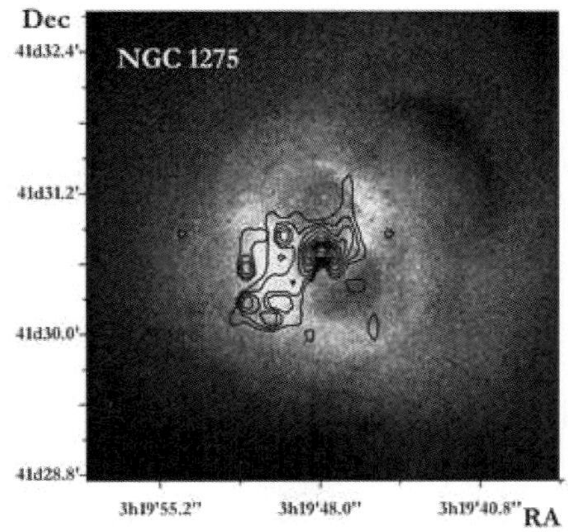


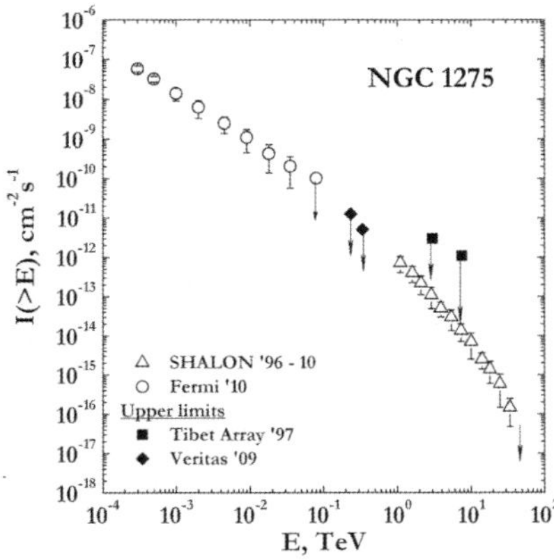


Fig. 1. Chandra X-ray image of NGC 1275 together with SHALON data. The contour lines show the TeV - structure by SHALON observations. The TeV energy spectrum of NGC 1275 from SHALON, 15 year observations in comparison with other experiments: Fermi LAT,[6] Veritas,[7] Tibet Array.[8]

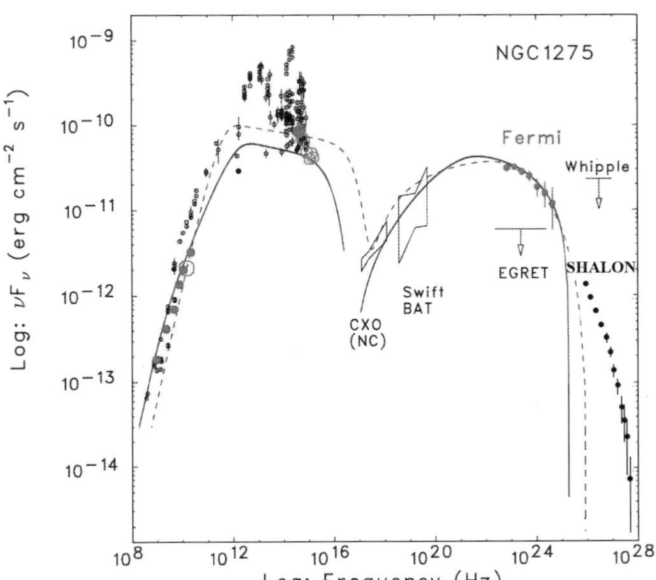


Fig. 2. Overall spectral energy distribution (SED) of NGC 1275. The low energy data from.[6] The SED is fitted with a one-zone synchrotron/SSC model (blue dashed curve) and a decelerating flow model (blue solid curves).[9]

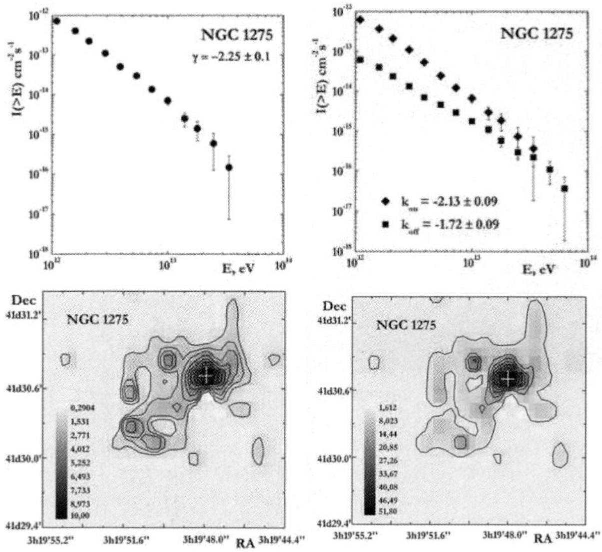


Fig. 3. The NGC 1275 observational data by SHALON: **Top** left: the $\gamma$-quantum integral spectrum; The NGC 1275 image at energy range of $> 0.8$ TeV; right: The event spectrum from the source with background with index of $k_{ON}$ and spectrum of background events observed simultaneously with the object with index $k_{OFF}$; **Bottom** left: The NGC 1275 image at TeV energy range; right: The energy image (in TeV units) of NGC 1275.

## NGC 1275

Galaxy clusters have been consider as sources of TeV $\gamma$-rays emitted by high-energy protons and electrons accelerated by large scale structure formation shocks, galactic winds, or active galactic nuclei. The Perseus cluster of galaxies is one of the best studied clusters due to its proximity and its brightness. Galaxy NGC 1275 is the central dominant galaxy of the Perseus Cluster of Galaxies and is of Seyfert galaxy class. NGC 1275 is known as powerful X-ray and radio source. Many studies explored correlations of X-ray, radio,(fig. 1) optical and ultraviolet emission (see e.g.[10]).

In 1996 year a new metagalactic source are detected by SHALON at TeV energies (figs. 1, 2, 3). This object was identified with Seyfert galaxy NGC 1275[1–4] (with redshift z=0.0179); its image is shown in fig. 1, 3. The maxima of the TeV gamma-ray, X-ray[11] and radio emission coincide with the active nucleus of NGC 1275. In contrast, the X-ray and TeV emission disappears almost completely in the vicinity of the radio lobes. The correlation TeV with X-ray emitting regions was found. The integral $\gamma$-ray flux for this source is found to be $(0.78 \pm 0.13) \times 10^{-12} cm^{-2} s^{-1}$ at energies of $> 0.8$ TeV. The energy spectrum of NGC 1275 at 0.8 to 30 TeV can be approximated by the power law $F(> E_O) \propto E^{k_\gamma}$, with $k_\gamma = -2.25 \pm 0.10$. The variations of the gamma-ray flux on the year scale are no more than 20% of average value.

The flux increase was detected from the region NGC 1275 in autumn 2006. The detailed analysis of gamma-shower direction turned out the detection of metagalactic object. This object was identified with the supernova SN 2006gy (see fig. 4)[12] that is about 10 minutes away from NGC 1275. Observations had been done in cloudless nights of moonless periods of 2006

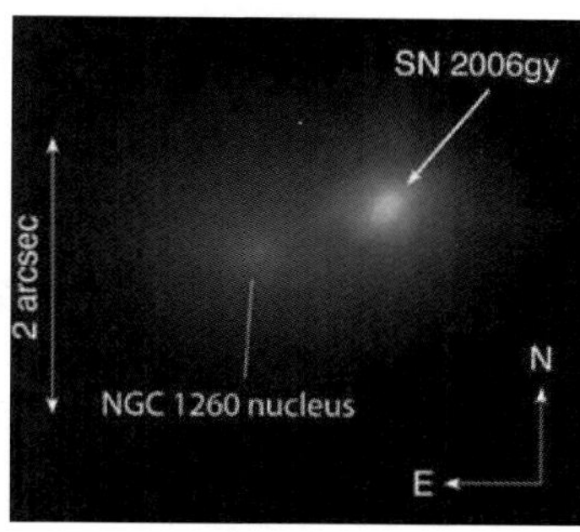


Fig. 4. The image of SN 2006gy (Chandra) and the nucleus of NGC 1260 at three wavebands: J band(1.25 m), H band (1.65 m), and Ks band (2.2 m).

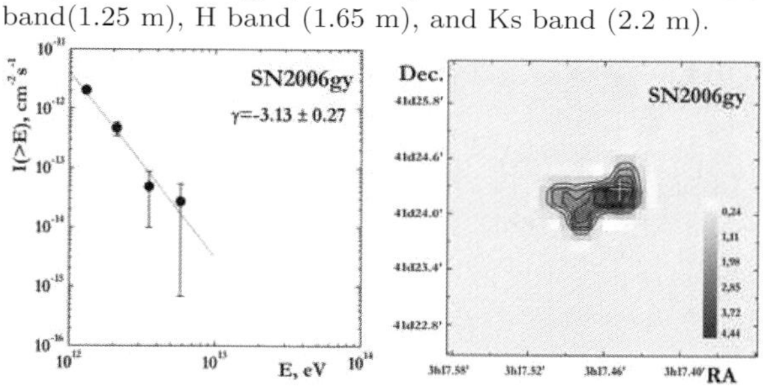


Fig. 5. The SN2006gy by SHALON: **left:** the $\gamma$-quantum integral spectrum; **right:** The SN2006gy image at TeV energy range.

Sep., Oct., Nov. Dec. and then during the winter of 2007. No flux increase was found in September observations. In the flare, observed on Oct. 22, the flux increased 6 times from the NGC 1275 and stayed on this level all Oct. moonless period. The integral gamma-ray flux for SN 2006gy is found to be $(3.71 \pm 0.65) \times 10^{-12} cm^{-2} s^{-1}$ at energies of $> 0.8$ TeV. The energy spectrum of SN2006 gy at 0.8 to 7 TeV can be approximated by the power law $F(> E_O) \propto E^{k_\gamma}$, with $k_\gamma = -3.13 \pm 0.27$ (fig. 5). An image of gamma-ray emission from SN2006 gy by SHALON telescope is shown in Fig. 5. Follow-up observations on end of Nov. showed that the flux of SN2006 gy had dropped to a flux level of about $(0.69 \pm 0.17) \times 10^{-12} cm^{-2} s^{-1}$ and was constant during the Nov. Dec. period. The results of observation analysis of 2007 have no revealed TeV gamma-ray emission from region of SN 2006gy. So, the explosion of extragalactic supernova was observed at TeV energies for the first time with SHALON Cherenkov telescope.

No other significant variations of the gamma-quantum flux during the observations of NGC1275 were found during all the period of observations.

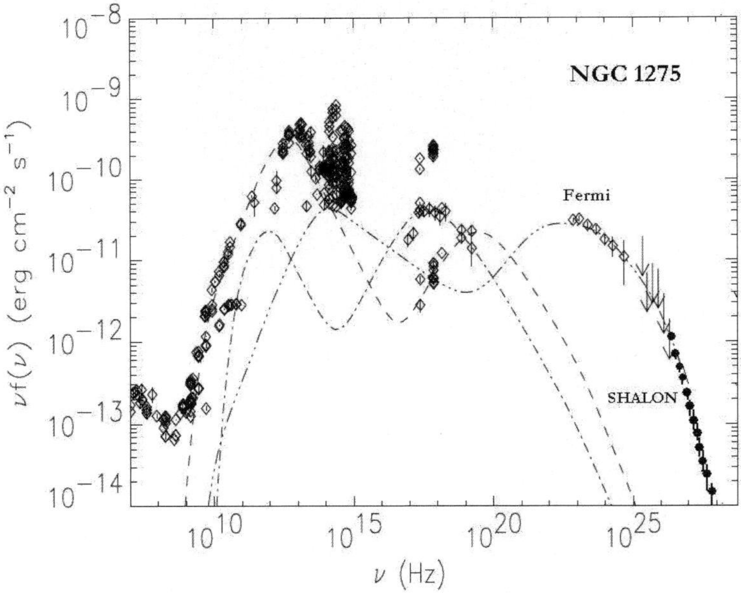


Fig. 6. Overall spectral energy distribution (SED) of NGC 1275. The TeV energy spectrum of NGC 1275 from SHALON, 15 year observations in comparison with other experiments: Fermi LAT,[6] and upper limits: Veritas,[7] MAGIC.[13]

Experimental limits on the very high energy gamma-ray emission from the Perseus cluster and its central galaxy NGC1275 have been obtained by the other experiments.[6–8,13,14] The Seyfert galaxy NGC 1275 has been also observed with the Tibet Array[8] (about 5 TeV) and then with Veritas[7] telescope at energies about 300 GeV at 2009. The radio-galaxy NGC 1275 has been recently detected by Fermi[6] as a source of high-energy gamma rays with an average flux and power-law photon index of $F(> 100MeV) = (2.31 \pm 0.13) \times 10^{-7} cm^{-2} s^{-1}$ and $\gamma = 2.13 \pm 0.02$, respectively. The recent detection by the Fermi LAT of high-energy gamma-rays from the radio galaxy NGC 1275 makes the observation of the very high energy ($E > 100$ GeV) part of its broadband spectrum particularly interesting. The overall spectral energy distribution of NGC 1275 from the low energies to the TeV energies is presented (fig. 2, 6). The spectrum of NGC 1275 from SHALON 15 year observations is also shown.

The Perseus cluster as a many other galaxy clusters is expected to be a source of gamma-ray emission due to various emission mechanisms [15] In the model discussing[15] the photon emission at the wide energy range is produced by three components with a specific jet-blob structure that is able to reproduce the observed multi-frequency spectral energy distribution (fig. 6). The available high and very high energy data for NGC 1275 are well fitted in this model with three components, in which the second, most energetic and smaller blob produces a SSC emission observed at gamma-ray energies[15] (fig. 6 ).

## References

1. V. G. Sinitsyna, Proc. of Towards a Major Cherenkov Detector-V ed. O.C. de Jager, South Africa, 1997, p.136.
2. V. G. Sinitsyna, et al.,16th European Cosmic Ray Symposium, ed. J.Medina, Spain, 1998, p. 367.
3. V. G. Sinitsyna, et al., Nucl. Phys. (Proc. Suppl.), 75A (1999) 352-355
4. V. G. Sinitsyna , AIP 515, (2000) 293 - 297.
5. V. G. Sinitsyna *Rad. Phys. and Chem.*, 75, (2006) pp. 880.
6. A. A. Abdo et al., The Astrophysical Journal, 699 (2009) 31-39
7. V. A. Acciari et al., The Astrophysical Journal, 706, (2009) L275-L280
8. M. Amenomori in Proc. 26th ICRC, Salt Lake City, 3, (1999) p. 418.
9. M. Georganopoulos and D. Kazanas, ApJ, 594, (2003) p. L27
10. B. R. McNamara, R. W. O'Connell, C. L. Sarazin, *AJ* 112 (1996) 91.
11. A. C. Fabian, J. S. Sanders, et.al. Mon. Not. R. Astron. Soc. 318, (2000) L65–L68.
12. E. O. Ofek, P. B. Cameron, *et. al., arXiv:astro-ph/0612408v1*
13. J. Alekśic, C. Pfrommer, A. Pinzke, et al., (2009), *arXiv:arXiv:0909.3267*
14. D. J. Thompson *et al., Astrophys. J. Suppl.*, 10,(1995) p.209, 259.
15. S. Colafrancesco et al., (2010) *arXiv:1006.2333v1*

# CONSTRAINS ON EXTRAGALACTIC BACKGROUND LIGHT FORM DISTANT QUASARS 3C454.3 (z = 0.859) AND 1739+522(z = 1.375) DETECTED BY SHALON

V.G. SINITSYNA, S.I. NIKOLSKY, V.Y. SINITSYNA

*P.N.Lebedev Physical Institute,*
*Leninsky prospect 53, Moscow, 119991, Russia*
*E-mail: sinits@sci.lebedev.ru*

Extragalactic diffuse background radiation blocks the propagation of TeV gamma-ray over large distances ($z > 0.1$) by producing $e^+e^-$ pairs. As a result, primary spectrum of gamma-source is changed, depending on spectrum of background light. So, a hard spectra of Active Galactic Nuclei with high red shifts of 0.03 - 1.8 allow the determination an absorption by Extragalactic Background Light and thus spectrum of EBL. The redshifts of SHALON very high energy gamma-ray sources range from z=0.0183 to z=1.375. Among them bright enough AGNs of BLLac type: Mkn421 (z=0.031), Mkn501 (z=0.034) and FSRQ type: 3c454.3 (z=0.895), 1739+522 (z=1.375) those spectra are resolved in the TeV energy band from 1 to $\sim 20 - 30$ TeV. Spectral energy distributions and images of distant AGNi are presented. Spectral energy distribution of EBL constrained from observations of Mkn421, Mkn501, 3c454.3 and 1739+522 together with models and measurements are presented.

*Keywords*: Active Galactic Nuclei; BL Lacs; Mkn 421; Mkn 501; Flat Spectrum Radio Quasars; Extragalactic Background Light.

## 1. Introduction

The cosmological processes, connecting the physics of matter in active galactic nuclei will be observed in the energy spectrum of electromagnetic radiation. The understanding of mechanisms in active galactic nuclei requires the detection of a large sample of very high energy gamma-ray objects at varying redshifts. The redshifts of very high energy gamma-ray sources observed by SHALON range from z=0.0179 to z=1.375.

The gamma - astronomical researches are carrying out with SHALON[1] mirror telescope at the Tien-Shan high mountainous station since 1992. During the period 1992 - 2010 SHALON has been used for observations of metagalactic sources: Mkn 421, Mkn 501, Mkn 180, NGC 1275, SN2006 gy,

Table 1. The metagalactic gamma-quantum sources catalogue, observed by SHALON; at the column Relative intensity of source the Crab Nebula intensity is taken as a unit – **R**

| Sources | Observable flux ($\times 10^{-12} cm^{-2}s^{-1}$) | Distance, (Mpc) | **R** |
|---|---|---|---|
| Mkn 421 | $(0.63 \pm 0.14)$ | 124 | $3.8 \times 10^9$ |
| Mkn 501 | $(0.86 \pm 0.13)$ | 135 | $4.46 \times 10^9$ |
| Mkn 180 | $(0.65 \pm 0.23)$ | 182 | $6.2 \times 10^9$ |
| NGC 1275 | $(0.78 \pm 0.13)$ | 71 | $1.2 \times 10^9$ |
| SN2006 gy | $(3.71 \pm 0.65)$ | 83 | $4.2 \times 10^9$ |
| 3c4543 | $(0.43 \pm 0.13)$ | 4685 | $5.3 \times 10^{12}$ |
| 1739+522 | $(0.53 \pm 0.10)$ | 7500 | $1.4 \times 10^{13}$ |

3c454.3, 1739+522 and galactic sources: Crab Nebula, Cyg X-3, Tycho's SNR, Geminga, 2129+47XR.[2–8] Our method of the data processing is described in.[1–4] Some representative results are shown in[2–8] and figures in these proceedings. Our data for Crab, Mkn 421 and Mkn 501 are compared with those from other experiments,[9–14] within a wide energy range $10^8 - 10^{14}$ eV.

## 3c454.3

In 1998 year a new metagalactic source 3c454.3 (z=0.859) has been detected by SHALON at TeV energies. The integral gamma-ray flux above 0.8 TeV was estimated as $(0.43 \pm 0.13) \times 10^{-12}$ $cm^{-2}s^{-1}$ (Table 1, Figs. 1, 2). It is consistent with the upper limit $0.84 \times 10^{-11}$ $cm^{-2}s^{-1}$ obtained by Whipple telescope at energy more than 0.5 TeV.[11,12] Taking into account that the spectrum from 3c454.3 measured by EGRET in the energy range $\sim$ 30 MeV to 50 GeV can be approximated as $E^{-1.2}$.[13]

Recently, 3c454.3 has observed with Fermi LAT at energies 200 MeV - 300GeV.[14] The spectrum by Fermi is fitted with a a broken power-law with photon indices $\gamma_{low} = 2.27 \pm 0.3$, $\gamma_{high} = 3.5 \pm 0.05$ with an average flux of $\sim 3 \times 10^{-6} photons\ cm^{-2}s^{-1}$, for energies $> 100 MeV$.

Figure 2 presents spectral energy distributions of 3c454.3. Black circles is SHALON data. The dashed lines represent the best fits to the Synchrotron and Inverse Compton part of the quasi-simultaneous SED.[15]

## 2. 1739+522

One more remote metagalactic gamma - source was detected by SHALON in 1999 and is being intensively studied since then. This object was identified with the active galactic nucleus 1739+522; its image is shown in fig. 3. This the most distant object (with redshift z=1.375) is also the most powerful:

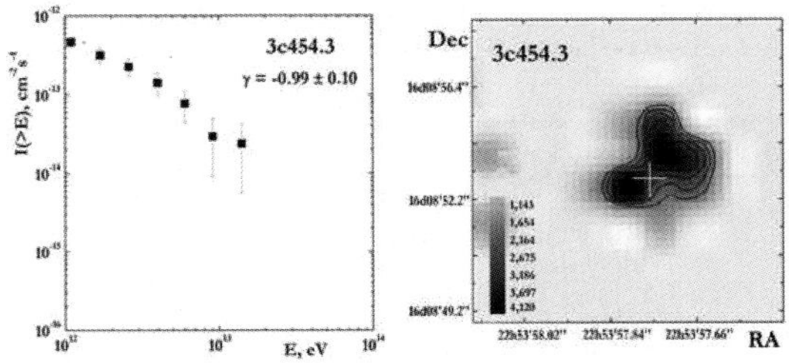


Fig. 1. **left:** The 3c454.3 gamma-quantum integral spectrum with power index $k_\gamma = -0.99 \pm 0.10$; **right:** The 3c454.3 image at energy range of more then 0.8 TeV by SHALON.

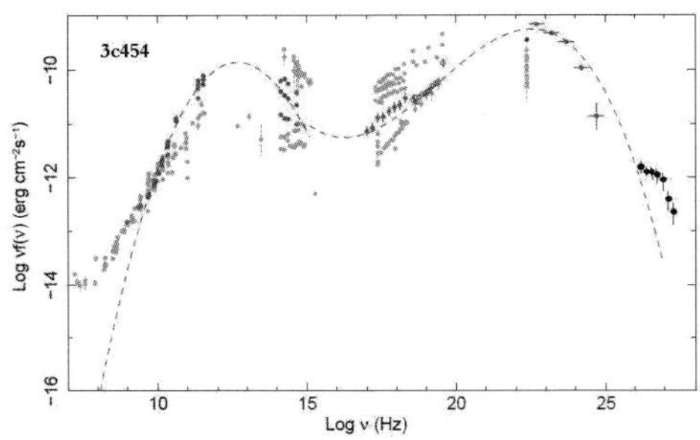


Fig. 2. Spectral energy distributions of 3c454.3. Black circles (at TeV energies) are SHALON data. The dashed lines represent the best fits to the Synchrotron and Inverse Compton part of the SED[15]

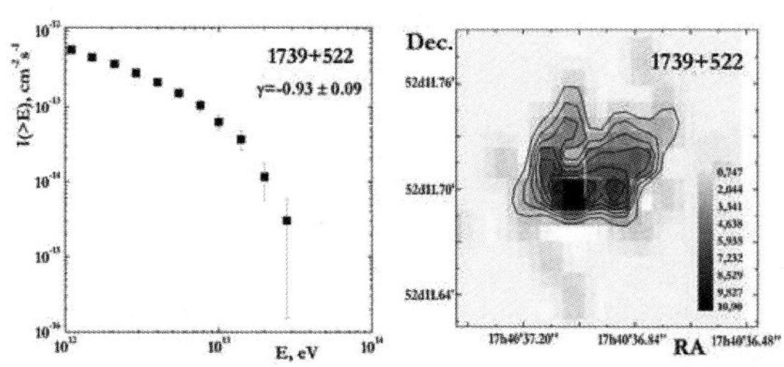


Fig. 3. **left:** The 1739+522 gamma-quantum integral spectrum with power index of $k_\gamma = -0.93 \pm 0.09$; **right:** The image of gamma-ray emission from 1739+522 by SHALON.

its integral gamma-ray flux is found to be $(0.53 \pm 0.10) \times 10^{-12}$ at energies of $> 0.8$ TeV. Within the range 0.8 - 7 TeV, the integral energy spectrum is well described by the single power law $I(> E_\gamma) \propto E_\gamma^{-0.93 \pm 0.09}$ (fig. 3). The average gamma-flux measured by EGRET telescope of Compton Observatory (CGRO)in the range $\sim$ 30 MeV to 50 GeV is about $2 \times 10^{-8}\ cm^{-2}s^{-1}$

Table 2. The integral spectrum indices of SHALON spectra of AGNi

| Sources | z | $k_\gamma$ | $k_{ON}$ | $k_{OFF}$ |
|---|---|---|---|---|
| NGC 1275 | 0.0179 | $-2.25 \pm 0.10$ | $-2.13 \pm 0.09$ | $-1.72 \pm 0.09$ |
| SN2006 gy | 0.019 | $-3.13 \pm 0.27$ | $-2.54 \pm 0.16$ | $-1.73 \pm 0.11$ |
| Mkn 421 | 0.031 | $-1.87 \pm 0.11$ | $-1.85 \pm 0.10$ | $-1.76 \pm 0.09$ |
| Mkn 501 | 0.034 | $-1.85 \pm 0.11$ | $-1.83 \pm 0.06$ | $-1.72 \pm 0.06$ |
| 3c4543 | 0.859 | $-0.99 \pm 0.10$ | $-1.13 \pm 0.08$ | $-1.71 \pm 0.08$ |
| 1739+522 | 1.375 | $-0.93 \pm 0.09$ | $-1.10 \pm 0.08$ | $-1.71 \pm 0.08$ |

with integral spectrum index about −1.2.[13]

According to our analysis, the energy spectra of distant quasars 3c454.3 and 1739+522 differ from those of the known blazars Mkn 421 (z=0.031) and Mkn 501 (z=0.034): $F_{Mkn\ 421}(> E_\gamma) \propto E_\gamma^{-1.87\pm0.11}$ and $F_{Mkn\ 501}(> E_\gamma) \propto E_\gamma^{-1.85\pm0.11}$. Hence, the average energy spectrum of these two metagalactic sources differs from spectra of remote objects 1739+522 and 3c454.3 within the energy range $10^{12} - 10^{13}$ eV. This observation does not contradict to unified energy spectrum $F(> E_\gamma) \propto E_\gamma^{-1.2\pm0.1}$.

## Extragalactic Background Light

As the TeV gamma rays can be absorbed due to interaction of low-energy photons of Extragalactic Background Light (EBL), the observations of active galactic nuclei can also be used for the study background light from UV to far infrared and even cosmic microwave background. The EBL spectrum contains information about star and galaxy formation on early stages of Universe evolution. TeV gamma-rays, radiated by distant sources, interact with photons of background via $\gamma + \gamma \rightarrow e^+e^-$ resonant process, then relativistic electrons can radiate gamma-ray with energies less than of primary gamma-quantum. As a result, primary spectrum of gamma-source is changed, depending on spectrum of background light. So, a hard spectra of Active Galactic Nuclei with high red shifts of 1 - 1.8 allow the determination an absorption by Extragalactic Background Light and thus spectrum of EBL. The redshifts of SHALON very high energy gamma-ray sources range from z=0.0179 to z=1.375. Among them bright enough AGNs of BLLac type (Mkn421, Mkn 501) and FSRQ type (3c454.3, 1739+522) those spectra are resolved in the TeV energy band from 1 to ~20-30 TeV. The fit of a simple power law function to the observational data presented in table 2 and discussed above. As it is seen from the figs. 1, 3 and from[16] the measured spectrum can be fitted by a power law with an exponential cutoff: $F(> E) \propto E^{-\gamma} \times exp(-E/E_{cutoff})$ with hard power

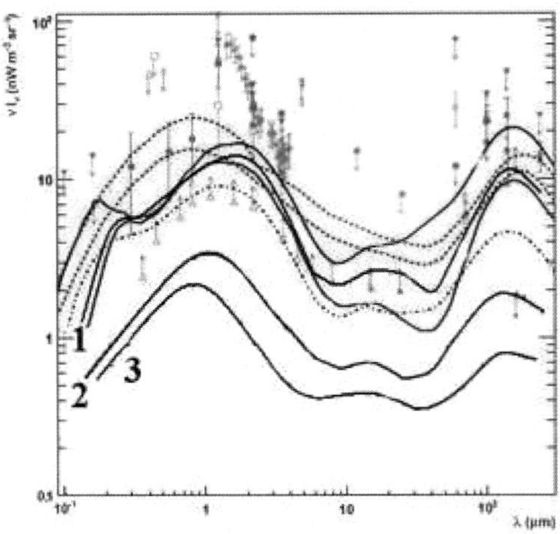


Fig. 4. Spectral energy distribution of Extragalactic Background Light: measurements[17] and models;[18,19] 1 - averaged EBL shape from best-fit model and Low-SFR model,[19] 2 - EBL shape from constrained from observations of 3c454.3 (z=0.859); 3 - 2 - EBL shape from constrained from observations of 1739+522 (z=1.375)

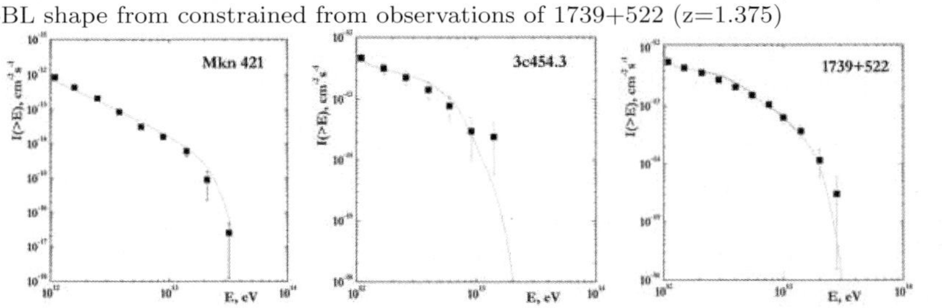


Fig. 5. The measured spectra for Mkn 421, 3c454.4 and 1739+522 (black squares) together with spectra attenuated by EBL (see text)

indices of about $\gamma \sim 1.55$ for Mkn 421 and Mkn 501 and $\gamma \sim 0.6$ for 3c454.3 and 1739+522. The value of $E_{cutoff}$ ranges from $11 \pm 2$ TeV for Mkn421, Mkn 501 and to $7 \pm 2$ TeV for distant sources. It has mentioned that the observed spectra are modified by gamma-ray attenuation, i.e. $F_{observed}(E) = F_{intrinsic}(E) \times exp(\tau(E, z))$ where $\tau(E, z)$ is optical depth for pair creation for a source at redshift $z$, and at an observed energy $E$. According to the definition of the optical opacity the medium influences on the primary source spectrum at $\tau \geq 1$, but for $\tau < 1$ the medium is transparent, so the measuring of source spectrum in the both range of $\tau$ can give the intrinsic spectrum of the source to to constrain the EBL density. The optical depth for sources at redshifts from 0.031 to 1.375 was calculated with assumption of EBL shapes shown in fig 4. We used the averaged EBL shape from best-fit model and Low-SFR model[19] (see fig. 4 upper black curve) to calculate the attenuated spectrum of Mkn 421 in assumption of simple power low intrinsic spectrum of the source with spectrum index of $\gamma = 1.5$, taken from the range of $\tau < 1$. The result is shown at fig. 5 with thin line; the black squares are observational data for Mkn 421. The shapes of EBL density constrained from the spectra of the high redshift sources 3c454.3 (z=0.859) and 1739+522 (1.375) are shown in fig 4 with curves 2 for and 3, accordingly. For these FSRQ sources the slope of intrinsic spectrum is taken $\gamma = 0.4$. The attenuated spectra for 3c454.3 and 1739+522 are also presented at fig. 5 (thin lines) together with observational data. Observations of distant metagalactic sources have shown that the Universe is more transparent to very high-energy gamma-rays than previously believed.

## References


1. S. I. Nikolsky and V. G. Sinitsyna, *in Proc. Int. Workshop on VHE $\gamma$-ray Astronomy*, Crimea, ed. A. A. Stepanian *et al.*, p. 11, 1989.
2. V. G. Sinitsyna *et al., Int. J. Mod. Phys. A* no. 29, p. 7023, 7026, 7029, 2005.
3. V. G. Sinitsyna *et al., Nucl. Phys. B (Proc. Suppl.)*, vol. 151, p, 108, 2006; *ibid.* vol. 122, pp. 247, 409, 2003; *ibid.* vol. 97, p. 215 and 219, 2001; *ibid.* vol. 75A, p. 352, 1999.
4. S. I. Nikolsky and V. G. Sinitsyna, *Phys. Atom. Nucl.* vol. 67, p. 1900, 2004.
5. V. G. Sinitsyna, *AIP Conf. Proc.* p. 515, 205 and 293, 1999.
6. V. G. Sinitsyna, *in Proc. Toward a Major Atmospheric Cherenkov Detector-IV*, ed. M. Cresti (Papergraf PD), p. 133, 1995; *in Detector-V*, ed. O. C. De Jager (Wesprint, Potcherfstroom), p. 136 and 190, 1997; *in Detector-VII*, ed. B. Degrange, G. Fontain, p. 57, 105, 11, 2005 .
7. V. G. Sinitsyna,*et al., Izv. Ross. Akad. Nauk Ser. Fiz.* vol. 71, no. 7, p. 94, 2007; *ibid.* vol. 69, no. 3, p. 422, 2005; *ibid.*, vol. 66, no. 11, p. 1667 and 1660, 2002; *ibid.*,vol. 63, no. 3, p. 608, 1999; *ibid.* vol. 61, no. 3, p. 603, 1997.
8. V. G. Sinitsyna, *et al., in Proc. The Universe Viewed in Gamma-Rays*, ed.R. Enomoto, M. Mori (Universal Academy Press, Inc.), pp., 235, 383, 503, 2003.
9. F. Krennrich, T.C. Weekes *et al., Univ. Acad. Press, Inc.*, p. 157, 2003.
10. M. Catanese and T. C. Weekes, *Preprint Series No4811*, 1999.
11. A. D. Kerrick *et al., ApJ*, L59, p. 438, 1995.
12. J. H. Buckley, *Astropart. Phys.*, vol. 11, p. 119, 1999.
13. R. Mukherjee *et al., Astrophys. J.*, vol. 490, p. 116, 1997.
14. A. A. Abdo, et al., ApJ, 699, p. 817, 2009.
15. A. A. Abdo, et al. 2009, *arXiv:0912.2040v1*
16. V. G. Sinitsyna *et al.,* Proc. 31th ICRC, Lodz, Poland, 2009, icrc1479
17. T. M. Kneiske, K. Mannheim, and D. H. Hartmann, *arXiv:0202104v1*
18. F.W. Stecker, M.A. Malkan & S.T. Scully ApJ, v. 648, p.774, 2006;
19. T. M. Kneiske, K. Mannheim, and D. H. Hartmann, *A&A* v. 1, p. 386, 2002.

# STATUS OF THE HIGH ALTITUDE WATER CHERENKOV (HAWC) GAMMA RAY OBSERVATORY

WAYNE SPRINGER[1]

*Department of Physics and Astronomy, University of Utah, 115 S 1400 E Salt Lake City, UT 84112, USA*

FOR THE HAWC COLLABORATION

The High Altitude Water Cherenkov (HAWC) gamma ray observatory is a high duty cycle, wide field of view gamma ray observatory sensitive to gamma rays in the energy range from below approximately 100 GeV to above 100 TeV. HAWC will observe the extensive air showers initiated by gamma rays using a water Cherenkov technology developed by the Milagro experiment. The detector has been optimized for sensitivity to gamma rays with good rejection of the cosmic ray background. Construction of the HAWC observatory located at an elevation of approximately 4100 m on Sierra Negra mountain in Mexico began in Fall 2009. A discussion of the detector design, science capabilities as well as the current status of the experiment will be presented.

## 1.1. *Gamma Ray Astronomy with HAWC*

Gamma ray observatories utilize either a satellite based direct detection method [1][2] or an indirect method that observes the gamma ray induced extensive air shower. The satellite-based detectors are similar to small high-energy physics experiments, complete with tracking systems and calorimeters. The Imaging Atmospheric Cherenkov Technique (IACT) detectors [3][4][5] use large mirrors to collect and focus onto an array of light sensitive detectors the Cherenkov light produced by the charged particles of an air shower. Ground based surface detectors [6][7] use a detector to sample the charged particles of extensive air showers as they strike the ground. These three techniques are complementary in that they cover somewhat exclusive energy ranges with differing sensitivities and field of view. Table 1 shows a qualitative comparison for the three techniques for several parameters. Satellite based detectors have the advantages of a wide field of view, excellent background rejection, high duty cycle, excellent detector resolution for energy and arrival direction estimation, sensitivity to low energy gamma rays (~10MeV--~100GeV). Satellite based detectors have the disadvantages of a small collection area and low sensitivity to

[1] springer@physics.utah.edu

| | Satellite | IACT | Surface Detector |
|---|---|---|---|
| Collection Area | small | large | intermediate |
| Field of View | large | small | large |
| Background rejection | excellent | Good-excellent | Good-excellent |
| Duty Cycle | Nearly 100% | ~10% | Nearly 100% |
| Energy Range | 10MeV-100GeV | ~10GeV-100TeV | ~10GeV-100TeV |
| Energy Resolution | Excellent | Good | Good |
| Pointing Resolution | Excellent | Good-Excellent | Good |

Table 1 Chart comparing features of gamma ray observation techniques

high energy (>100's of GeV) gamma rays. IACT gamma ray observatories have the advantages of large collection area and sensitivity to intermediate energy gamma rays (~10GeV--~100 TeV).IACT gamma ray observatories have the disadvantages of a low duty cycle and relatively narrow field of view. Surface detector based gamma ray observatories have the advantages of wide field of view and sensitivity to high energy gamma rays (~10GeV--~100 TeV). Surface detector based gamma ray observatories have the disadvantages of a smaller collection area than IACT and slightly poorer resolution. Figure 1 shows the point source sensitivity vs energy. For the IACT observatories this sensitivity corresponds to a 10 event (5 σ ) flux limit for 50 hours on source for an IACT. The sensitivity for HAWC[8] is for a point source occurring anywhere in the field of view, which is nearly half of the sky, for a 1 year/5year exposure.

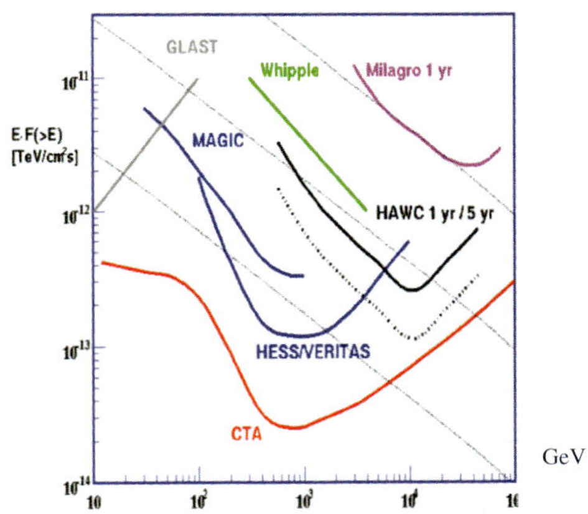


Figure 1: HAWC sensitivity for 1 year (dotted black line) and 5 year (solid black line) observations of a point source with a Crab-like spectrum. The sensitivities of current generation IACT instruments (MAGIC, HESS, VERITAS) are plotted in blue.

## 1.2. *Scientific Motivation*

The HAWC observatory will be used to study various astrophysical phenomena such as active galactic nuclei (AGN), radio galaxies, X-ray binaries, shell type supernova remnants, pulsar wind nebulae, starburst galaxies, pulsar emission, galactic diffuse emission, and star forming regions. Many questions associated with the detailed mechanisms occurring in these astrophysical phenomena remain unanswered. Observations of gamma rays at the highest possible energies are critical to answering these questions. HAWC nicely complements existing gamma ray observatories such as FERMI and VERITAS in its ability to continuously monitor a large fraction of the sky with excellent sensitivity. The following sections briefly describe some of the major scientific goals and tasks of HAWC.

### 1.2.1. *All Sky Survey*

HAWC will be sensitive to gamma rays coming from a large fraction of the visible sky at any given moment. Its continuous operation, while the earth rotates, will enable it to view a large fraction of the entire sky enabling it to perform a complete, unbiased survey for TeV gamma-ray sources over ½ of the sky with a source detection threshold of 50 mCrab in a single year. Imaging Cherenkov gamma ray telescopes (IACTs) are not as well suited for performing such all-sky surveys due to their limited field of view. Figure 2 shows the sensitivity of HAWC for 1 year of operation as a function of declination. HAWC's sensitivity is more than an order of magnitude greater than MILAGRO and comparable to HESS for point sources with hard spectra.

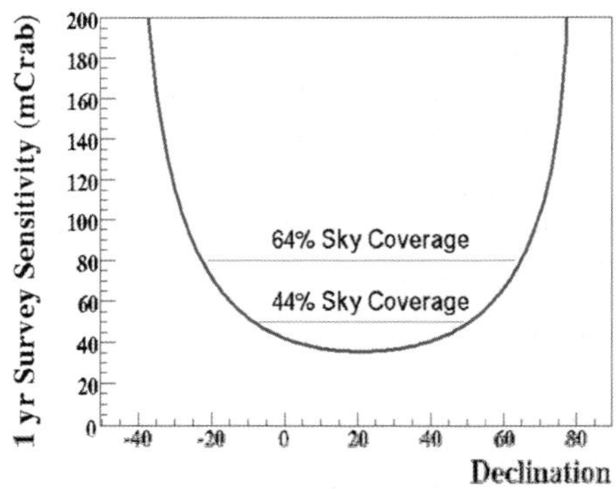


Figure 2: HAWC 1 year survey sensitivity vs source declination. 44% of $4\pi$ sr will be surveyed with a $5\sigma$ sensitivity of 50 mCrab in one year.

### 1.2.2. *Extended TeV Sources*

Many sources of TeV gamma rays are extended over large regions of space involved with the acceleration of charged particles. Due to longer exposure times on each source, HAWC has higher sensitivity than typical IACT detectors for extended objects. HAWC has an expected angular resolution of 0.25°. For extended objects smaller than this angular size, there is no increase in sensitivity. With the increased sensitivity to extended objects, HAWC should detect many such extended sources of both galactic and extragalactic origin. Additional follow-up observations using IACTs of the extended sources found by HAWC will provide an efficient manner for mapping the entire sky over all angular scales. The deep exposures made possible by HAWC will allow measurements of the energy spectra of gamma rays from extended sources that are potential cosmic ray accelerator sites over an energy range from ~1 TeV to ~100 TeV.

### 1.2.3. *Galactic Plane*

HAWC will examine the properties of gamma ray emission in the galactic plane in detail. Much of the gamma ray emission is of a diffuse nature resulting from interactions of cosmic rays with matter and radiation fields in the galactic plane. Measurements by HAWC of the energy spectra of gamma rays from various regions in the galactic plane will provide valuable information to further understanding of the production and propagation of cosmic rays in our galaxy. These measurements may also aid in the measurement of the relative abundances of CO to molecular hydrogen in various regions of the galaxy. Measurements of the diffuse emission of gamma rays could also probe other, more esoteric, processes such as self-annihilating super-symmetric dark matter.

### 1.2.4. *Transient Sources*

HAWC will be able to monitor a large fraction of the sky for gamma ray sources that are transient in nature. Active Galactic Nuclei are known to flare and are one of the three known types of transient TeV gamma ray sources. Gamma-ray bursts (GRBs), and x-ray binaries/micro-quasars are the other two types of known transient TeV sources of gamma ray emission. An all-sky continuously operating observatory such as HAWC will detect and monitor these sources over long time periods.

## 1.3. *HAWC Observatory design*

HAWC will be built at Sierra Negra, Mexico (18° 59' 41"N,97° 18' 28" W, 4100 m a.s.l.). The HAWC detector is based upon the proven technology of the MILAGRO detector but at a much higher altitude leading to a factor of 15 increase in sensitivity over MILAGRO. The HAWC WCD array will be built at Sierra Negra, Mexico (18° 59' 41"N,97° 18' 28" W, 4100 m a.s.l.). As a cost saving, the existing PMTs of the MILAGRO detector will be re-deployed in HAWC among an array of 300 Water Cherenkov Detectors (WCD). The water tanks are constructed using off-the-shelf water storage products. A prefabricated steel structure provides the support for a water and light-tight bladder that will hold the water and the PMTs. A single WCD consists of a water tank holding approximately 45,000 gallons of water instrumented with three 8" Hamamatsu R5912 PMTs from MILAGRO. A single RG-59 cable provides high voltage to each PMT and carries the signal back to the front-end electronics. The Milagro front-end electronics will also be reused. These boards shape the PMT pulses and generate analog edges at two discriminator levels: ~1/4 photoelectron (PE) and ~5 PEs. The Time-Over-Threshold (TOT) method is used to measure both pulse arrival time and amplitude with a single multi-hit TDC channel. In HAWC, the single deep layer of PMTs is used for triggering, angular reconstruction, and calorimetry. The HAWC single layer forms a high efficiency detector with an area of 20,000 $m^2$. The optical isolation afforded by the 300 individual water tank reduces the noise rate in HAWC PMTs to ~20 kHz, despite the higher elevation. The complete HAWC data set will be 150-300 TB/year (with compression) when fully operational.

The continuous operation, and wide field of view afforded by the MILAGRO Water Cherenkov Detector technology results in a survey instrument of excellent sensitivity. The HAWC detector will see a 5$\sigma$ signal from a Crab-like source in one 4-hr transit maintaining this sensitivity over 2$\pi$ sr HAWC will be capable of expanding the limited catalog of TeV gamma ray sources and provide nearly all sky monitoring for transient activities such as GRBs or AGN flares. The monitoring capability for HAWC will provide a trigger to initiate multi-wavelength observing campaigns involving other instruments including neutrino detectors.

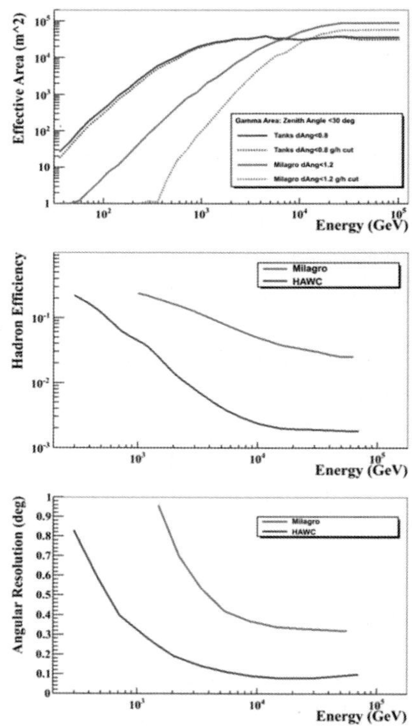


Figure 3: Simulated Energy dependence of Effective detection area (upper), hadronic background rejection (middle) and angular resolution (bottom) for Milagro (red) and HAWC (blue).

The response to gamma and hadron induced extensive air showers of the HAWC detector is simulated using the standard software packages CORSIKA and Geant4. The identification of gamma ray induced air showers among the significantly more numerous hadronic (cosmic ray) induced air showers is achieved by exploiting the “compactness” and smoothness of the electromagnetic shower produced by gamma rays. The compactness parameter, C, is defined as the total number of PMTs hit divided by the largest pulse amplitude that is more than 40m from the reconstructed core position. Gamma ray induced showers have only small hits far from the core and therefore have large values of C. Hadron induced showers with muons and hadrons and multiple clumps of EM energy and have low values for C. This background rejection capability of HAWC improves with increasing energy. At similar energies, HAWC can reject hadronic backgrounds ~10x better than Milagro. Figure 3 shows the expected energy dependence of the effective area, background rejection and angular resolution of HAWC and Milagro. HAWC achieves a sensitivity improvement of 15x compared to Milagro for a Crab-like point source. The sensitivity gain is a direct result of the optimized detector design and the higher observatory elevation. For low energies ($E < 1$TeV), the effective area of HAWC is approximately 100x that of Milagro. This improvement will provide a substantial aperture for detection of possible VHE gamma-ray emission from transient sources such as gamma-ray bursts. At the highest energies ($E>30$ TeV), the effective area of HAWC is similar to/slightly smaller than that of Milagro. However, the large improvements in angular resolution and background rejection of HAWC combine to create a substantial increase in gamma-ray sensitivity compared to Milagro.

### 1.4. *Status of the HAWC Detector in Fall 2010*

Funding for construction of the HAWC gamma ray observatory has been secured from both the U.S. and Mexican funding agencies. Construction of site infrastructure began in Fall 2009. Deployment of the first WCD tank occurred in June 2010. It is foreseen that the full array of 300 WCD tanks will be deployed by Fall 2004.

## References

1. http://fermi.gsfc.nasa.gov/
2. http://swift.gsfc.nasa.gov/docs/swift/swiftsc.html
3. http://wwwmagic.mppmu.mpg.de/
4. http://veritas.sao.arizona.edu/
5. http://argo.na.infn.it/
6. http://www.lanl.gov/milagro/
7. http://hawc.umd.edu/
8. http://hawc.umd.edu/science.php

# Light Nuclei and Isotope Abundances in Cosmic Rays. Results from AMS-01

N. Tomassetti*, for the AMS-01 Collaboration

*Perugia University and INFN Perugia,*
*Perugia, I-06122, Italy*
**E-mail: nicola.tomassetti@pg.infn.it*
*www.pg.infn.it*

Observations of the chemical and isotopic composition of light cosmic-ray nuclei can be used to constrain the propagation models. Nearly 200,000 light nuclei ($Z > 2$) have been observed by AMS-01 during the 10-day flight STS-91 in June 1998. Using these data, we have measured Li, Be, B and C in the kinetic energy range 0.35 - 45 GeV/nucleon. In this proceeding, our charge and isotopic composition results are presented and discussed.

*Keywords*: cosmic rays — charge composition — isotopic composition

## 1. Introduction

Cosmic rays detected on Earth with kinetic energies from 100 MeV to 100 GeV per nucleon are believed to be produced by particle acceleration mechanisms occurring in galactic sites such as supernova remnants. Most of our knowledge on the insterstellar propagation of the galactic cosmic rays comes from the study of secondary species, i.e. spallation products of C-N-O and Fe nuclei that are almost absent in the cosmic ray (CR) sources.

The relation between secondary CRs and their primary progenitors allows us to determine the propagation parameters, such as the diffusion coefficient and its energy dependence [1]. Among the ratios B/C and Sub-Fe/Fe, it is of great interest to determine the propagation history of the lighter nuclei Li, Be, and their isotopes, which are mostly of spallogenic origin. Their abundances depend not only on interactions of the primary species C, N and O, but also on tertiary contributions like Be$\rightarrow$Li or B$\rightarrow$Li. Therefore the Li/C and Be/C ratios may provide further restrictions on propagation models [2].

AMS-01 observed cosmic rays at $\sim$380 km of altitude during a period close to the solar minimum, providing data free from atmospheric-induced background and little influenced by the solar modulation. Using these data, we present a measurement of the ratio Li/C, Be/C and B/C.

## 2. Instrument

The AMS-01 precursion mission of the Alpha Magnetic Spectrometer (AMS) project [3] operated successfully during the 10-day flight STS-91 onboard *Discovery*.

The AMS-01 spectrometer was composed of a cylindrical permanent magnet (analyzing power $BL^2 = 0.14\,\mathrm{Tm}^2$), a silicon microstrip tracker (six layers of double sided silicon sensors with resolution of $10\,\mu$m in the bending coordinate), four time of flight (TOF) scintillator planes (time resolution of $\sim$90 psec for $Z > 1$ ions) an aerogel Cerenkov counter (threshold velocity $\beta$=0.985) and anti-coincidence counters [4].

The detector response was simulated with `GEANT3` [5]. The effects of energy loss, multiple scattering, nuclear interactions and decays were included in the code, as well as efficiencies, resolutions and reconstruction algorithms. Further simulations have been performed using `GEANT4` [6] and `FLUKA` [7] within the framework of the `Virtual MC` tool-kit [8].

## 3. Data Analysis

The identification of a cosmic-ray particle with the AMS-01 spectrometer was perfomed through the combination of independent measurements provided by the various sub-detectors. The particle rigidity $R$ (momentum per unit charge $p/Z$) was provided by the deflection of the reconstructed particle trajectory. The velocity $\beta = v/c$ was measured from the particle transit time between the four TOF planes along the track length. The particle charge magnitude $|Z|$ was obtained by the analysis of the multiple measurement of energy deposition in the four TOF scintillators (up to $Z = 2$ [4]) and the six silicon layers (up to $Z = 8$ [9]). Figure 1 shows the charge histogram of CR particles with $Z > 2$ obtained with the silicon tracker. The particle mass was then determined from the resultant charge $Z$, velocity $\beta$ and rigidity $R$:

The differential energy spectrum of the $Z$-charged particles measured by AMS-01 in the energy bin $E$ of width $\Delta E$ is related to the measured counts $\Delta N^Z$ by:

$$\Phi^Z(E) = \frac{\Delta N^Z(E)}{A^Z(E) \cdot \Delta T^Z(E) \cdot \Delta E} \tag{1}$$

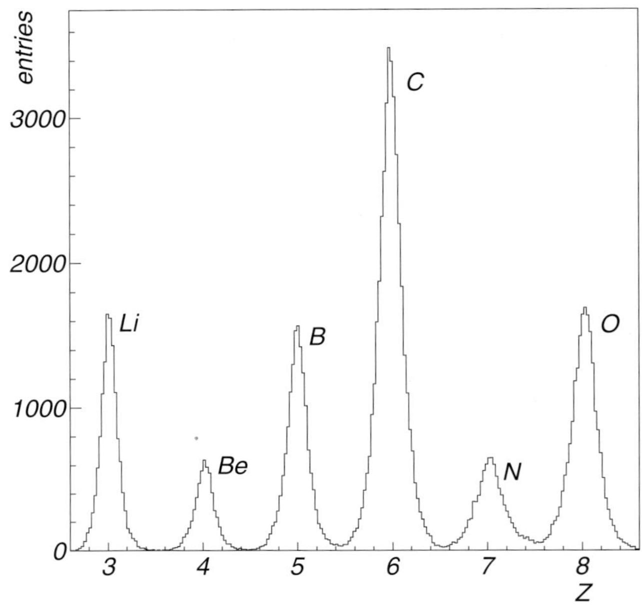


Fig. 1. Charge spectrum of the selected $Z > 2$ data. The velocity-dependent signal amplitudes of are equalized to $\beta \equiv 1$ and shown in units of charge. Different nuclear species fall in distinct charge peaks of widths 0.1-0.16 charge units.

where $\Delta T^Z(E)$ is the effective exposure time and $A^Z(E)$ is the detector acceptance. In order to compute bin-to-bin ratios, all the elemental fluxes were determined in a common grid of kinetic energy per nucleon $E$, obtained by the rigidity measurements performed by the tracker. The relation between the measured energy of detected particles and their true energy were studied using unfolding techniques [10]. Most of systematic uncertainties arising from many steps of the analysis are suppressed in the ratios. Differences in the trigger efficiencies of the two species are present, as expected, from the charge dependence of delta ray production and fragmentation effects in the detector material. This study was performed through extensive simulations employing the transport codes `GEANT3`, `GEANT4` and `FLUKA`, Uncertainties of 2-10 % were estimated for the trigger efficiency. The spill over from adjacent charges also produces net effects on the measurements, leading to errors of up to 5 %. The MC determination of the acceptance gave a statistical uncertainty of $\sim$1$-$3 %, increasing with energy.

## 4. Results and Discussions

Figure 2 shows the isotopic ratio $^7$Li/$^6$Li as a function of the rigidity. On-going measurements of other isotopic fractions of H, He, Li, Be and B will be discussed during the conference. Results for the Li/C, Be/C and B/C

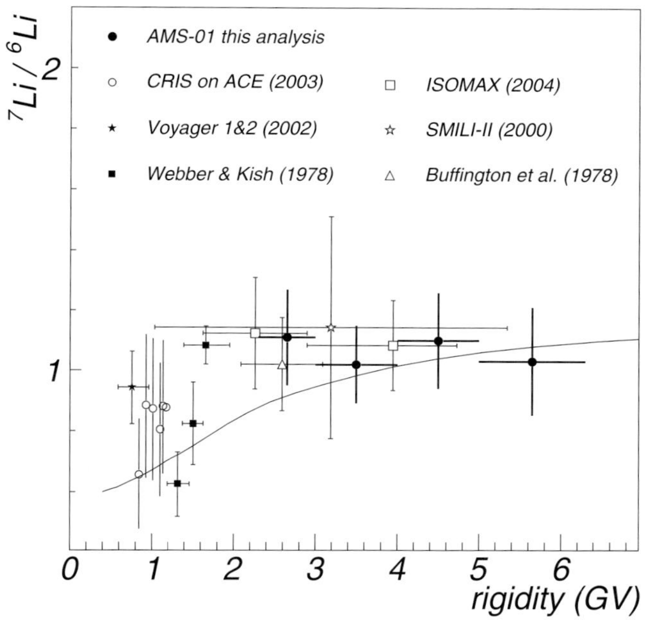


Fig. 2. The rigidity dependence of the ratio $^7$Li/$^6$Li. The other experimental values are converted from kinetic energy to rigidity [16–21]. The solid line is a diffusion model prediction obtained with GALPROP (§4).

ratios are presented in Figure 3 with the existing experimental data [11–16]. The error bars in the figure represent the sum in quadrature of statistical errors with the systematic uncertainties.

Our B/C ratio measurement agrees well with the results from the first flight of CREAM in 2004 [15] and with the data collected by HEAO-3-C2 [14] from October 1979 and June 1980. The Be/C ratio is consistent, within errors, with the HEAO data, but not with balloon data [12]. Our Li/C data have unprecedented accuracy in a poorly explored energy region. In comparison with balloon data from Ref. [12], our data indicate a quite different trend in the high energy part of the Li/C ratio. In these ratios,

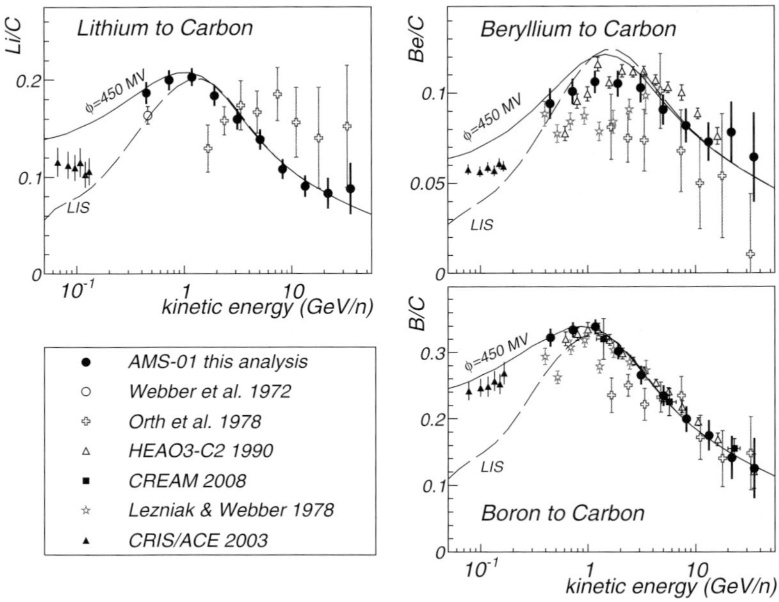


Fig. 3. Results for the ratios Li/C, Be/C and B/C from this work (solid circles) and previous measurements [11–16]. Model calculations, obtained with GALPROP 22, are reported for the interstellar (LIS) and solar modulated ($\phi$=450 MV) cosmic ray fluxes.

the main progenitors of boron nuclei are primary cosmic rays (CNO). On the contrary, the abundances of Li and Be depend also on secondary progenitors Be and B through tertiary contributions like B$\rightarrow$Be, Be$\rightarrow$Li and B$\rightarrow$Li. However, the observed shapes of the measured ratios Li/C and Be/C suggest their suitability in constraining the propagation parameters: their decreasing behaviour with increasing energy is a direct consequence of the magnetic diffusion experienced by their progenitors, while the characteristic peak around $\sim$1 GeV/n is a strong indicator of low energy phenomena like stochastic reacceleration or convective transport with the galactic wind.

To describe our results, we make use `GALPROP`[a], a numerical code that incorporates all the astrophysical inputs of the CR galactic transport. In thus proceeding, we use the latext version `GALPROP-v54` through the web-based interface `WebRun` [22]. We describe the propagation in the framework of the diffusive-reacceleration model, that has been very successful in the description of the cosmic ray nuclei fluxes.

[a]http://galprop.stanford.edu

Our results for the B/C ratio, the Li/C ratio and the $^7$Li/$^6$Li isotopic ratio of Figiure 2 are described quite well within the uncertainties. It is difficult, however, to accomodate the Li and B description with the Be/C ratio by only means of astrophysical parameters. Beryllium appears to be overproduced in the model by a factor $\sim$10$-$15 %. This discrepancy is also apparent in the previous measurements. It should be noted that spallation processes of light nuclei over interstellar medium are not well understood quantitatively. The lack of cross section measurements limits the model predictions to uncertainties of $\sim$10–20 % in the Li/C and Be/C ratios [2].

Cosmic rays are expected to be measured with high precision in the near future. Understanding fragmentation is a key factor in establishing final conclusions concerning cosmic ray propagation.

## References

1. Strong, A. W., et alii, 2007, Annual Rev. of Nucl. and Part. Sci., 57, 285–327.
2. de Nolfo, G. A., et alii (CRIS), 2006, Adv. Space Res., 38, 1558–1564.
3. Battiston, R., 2010, this proceedings.
4. Aguilar, M., et alii (AMS-01), 2002, Phys. Rep., 366, 331–405.
5. Brun, R., et alii (GEANT3), 1987, DD/EE/84-1, CERN.
6. Agostinelli, S., et alii (GEANT4), 2003, Nucl. Inst. Meth. A, 506, 250–303.
7. Fassò, A., et alii (FLUKA), 2005, CERN-2005-10.
8. Hrivnacova, I., et alii (VMC), 2003, Proc. CHEP La Jolla (CA).
9. Aguilar, M., et alii (AMS-01), 2010, to appear in ApJ.
10. D'Agostini, G., 1995, Nucl. Inst. Meth. A, 362, 487.
11. Webber, W. R, et alii, 1972, Astrophys. and Sp. Sci., 15, 245–271.
12. Orth, C. D., et alii, 1978, ApJ, 226, 1147–1161.
13. Lezniak, J. A., & Webber, W. R., 1978, ApJ, 223, 676–696.
14. Engelmann, J. J., et alii (HEAO3-C2), 1990, A&A, 233, 96–111.
15. Ahn, H. S., et alii (CREAM), 2008, Astropart. Phys., 30–3, 133–141.
16. de Nolfo, G. A., et alii (CRIS), 2003, Proc. 28th ICRC (Tsukuba), 2, 1667.
17. Webber, W. R., et alii (VOYAGER), 2002, ApJ, 568, 210–215.
18. Webber, W. R., & Kish, J., 1979, Proc. of 16th ICRC (Kyoto), 1, 389.
19. Hams, T., et alii (ISOMAX), 2004, Astrophys. J., 611, 892.
20. Ahlen, S. P., et alii (SMILI), 2000, Astrophys. J., 534, 757–769.
21. Buffington, A., et alii, 1978, Astrophys. J., 226, 335–371.
22. Vladimirov, E., et alii(GALPROP), 2010, arXiv:astro-ph/1008.3642.

# Cosmic Rays Propagation and Environment

| | |
|---|---|
| P. Bobik | The AMS-02 proton spectra and the geomagnetic field |
| R. Bucik | Stereo observations of the energetic heavy ions during the minimum of solar cycle 23 |
| S. Della Torre | Electron and Positron solar modulation and prediction for AMS02 |
| S. Gabici | How to use molecular clouds to study the propagation of cosmic rays in the galaxy |
| M. Gervasi | Proton modulation in the heliosphere for different solar conditions and prediction for AMS-02 |
| D. Grandi | Proton and antiproton modulation in the heliosphere for different solar conditions and AMS-02 measurements prediction |
| D. Grasso | A consistent interpretation of recent CR nuclei and electron spectra |
| F. Jansen | Cosmic rays for heliospheric space weather storm prediction |
| K. Kudela | Energetic particles in the magnetosphere of earth: selected results and problems |
| R.A. Lineros | Cosmic rays of leptons from Pulsars and Supernova Remnants |
| B. Mailyan | High energy phenomena in the low atmosphere; particle fluxes from thunderstorm clouds |
| P. Martin | The cosmic-ray populations of nearby galaxies |
| D. Maurin | USINE: a new public cosmic ray propagation code (Basic phenomenology, sample results, and a bit of USINE) |
| M. Pato | Propagation of galactic cosmic rays and the AMS-02 experimen |
| M. Potgieter | Galactic cosmic rays in the dynamic heliosphere |
| A. Putze | A Markov Chain Monte Carlo technique to sample transport and source parameters of Galactic cosmic rays |
| J.P. Roberts | PAMELA through a Magnetic Lens |

| | |
|---|---|
| O. Strelnikova | Analysis of possibility of cosmic rays proton anisotropy phase and amplitude and electron spectra description at TeV-region within the bounds of the same set of sources |
| A.W. Strong | Interstellar gamma rays and cosmic rays: new insights from FERMI-LAT and integral |
| M. Tacconi | Energy loss for electrons in the heliosphere and local interstellar spectrum for solar modulation |

# THE AMS-02 PROTON SPECTRA AND THE GEOMAGNETIC FIELD[1]

P. BOBIK[1], M.J. BOSCHINI[2,4], C. CONSOLANDI[2], S. DELLA TORRE[2,5], M. GERVASI[2,3], D. GRANDI[2], K. KUDELA[1], S. PENSOTTI[2,3] AND P.G. RANCOITA[2]

[1] *Institute of Experimental Physics, Kosice ( Slovak Republic)*
[2] *Instituto Nazionale di Fisica Nucleare, INFN Milano-Bicocca, Milano (Italy)*[2]
[3] *Department of Physics, University of Milano Bicocca, Milano (Italy)*
[4] *CILEA, Segrate, (MI) (Italy)*
[5] *Department of Physics and Maths, University of Insubria, Como (Italy)*
*E-mail: bobik@saske.sk*

We evaluate a the AMS-02 spectra for the year 2012 using the HelMod a software that reproduces the effect of the Heliospheric Modulation for different periods and polarities. Once a the Local Interstellar Spectrum (LIS) is given, we can obtain a modulated spectrum at Earth (1 AU) using the measured solar conditions (tilt angle apha, solar wind velocity V) and the extrapolated diffusion coefficients $K_0$ (from Sunspot numbers). This calculations provide a flux of the primary (PCR) cosmic rays, protons, outside the geomagnetic field, that must be compared to the measured one at the low Earth orbit (about 400 km altitude). A back-tracking procedure of simulated protons entering the AMS-02 spectrometer has provided the fraction of allowed (and hence, forbidden) trajectories of the PCRs. Consequently, it has allowed the determination of the so-called Transmission Function (TF) which was able to describe the properties of the PCR transport from the Earth's magnetopause (i.e. the modulated primary spectrum at 1 AU) to the atmosphere and finally the fluxes of the PCRs in 10 geomagnetic regions for the AMS-02 observations.

## 1. Introduction

The AMS-02 experiment will fly on the ISS from/starting next year and will very precisely measure a spectrum of Galactic cosmic rays (GCR hereafter). The spectrum of the GCR is constant outside of Heliosphere. On the way from the border of heliosphere to a low orbit where the AMS-02 will be located is a cosmic ray spectrum modulated first by the heliosphere magnetic field and in the Earth vicinity by a geomagnetic field. The heliospheric modulation is time dependent through the diffusion coefficient $K_0$, tilt angle and solar wind velocity. The HelMod is a time dependent model [1][2] using a given LIS spectra and empirically assigned time dependent parameters to evaluate a

[1] * This work is supported by the Slovak VEGA grant agency project 2/0081/10.

spectra inside the Heliosphere. We use the HelMod simulation to get spectrum at 1AU. This result is then used as an input spectrum of cosmic rays entering the magnetosphere. We evaluate the effect of the magnetosphere shielding on cosmic rays by a backtracing method. This method is described in the second part (Method) of the article.

## 2. Method

The backtracing method is used for evaluation of particle trajectories between a selected location inside the magnetosphere and magnetopause. The method [3] is based on the inversion of charge sign (Zq) and velocity vector (v) in the equation of motion for a particle with relativistic mass m in the magnetic field B:

$$m\frac{dv}{dt} = Ze[v \times B] \tag{1}$$

In all simulations we used a combined model of the geomagnetic field B. The total magnetic field B consisted from the external geomagnetic field described by the model Tsyganenko 2004 [4] and an internal geomagnetic field described by the IGRF model [5], $\vec{B} = \vec{B}_I + \vec{B}_E$. The IGRF model of an internal geomagnetic field gave prediction for January 2012. For the external geomagnetic field Tsyganenko 2004 we used values from october 1987 as input parameters. For January 2012 we expect the same condition as in October 1987, for details about this prediction see [1][2]. Inversion in equations (1) means that an antiparticle with rigidity R shoted from a selected location and direction has in magnetic field the same trajectory as a particle with the same rigidity and direction coming to this location. The evaluation of trajectories for all possible incoming directions and all energies allow us to construct the so called transmission function (TF hereafter) for a selected location. The transmission function is a probable function of particle energy (rigidity) and position inside the magnetosphere. The TF shows the probability of particles from magnetopause with rigidity in the range R±dR to reach a selected point (or region) inside the magnetosphere. The locations (3600) of the particles to be back-tracked are distributed uniformly over a complete sphere surrounding the Earth at an altitude of 400km. 78.9% of back-tracked locations are within the geographic latitudes of the ISS orbits, i.e $|\theta_{lat}| < 51.6°$. South-Atlantic anomaly region i.e. the region with latitude between -55° and 0° and with longitude between -80° and 20° is excluded from results. The particle directions (90) are isotropically distributed within the outward hemisphere and inside the 45° acceptance cone (around the local geocentric Zenith) of the AMS-02 spectrometer. In addition, more particle directions (at all 288) covering up to the full outward hemisphere have been back-tracked to investigate the TF dependence on the acceptance cone. The TF has been computed for the same 31 rigidity intervals of the AMS-01 data (see Table 4.5 of [6]), i.e. the lowest rigidity value is about 0.37 GV and the largest about 200GV. In the present article the TF has been determined for the 10 geomagnetic regions (see Table 1 in [7]), regions M, for which the AMS-01 data are available and was indicated by $TF_M(R_b)$.

$$TF_M(R_b)=\frac{\sum_{i_M} TF_M(R_b, i_M)}{\sum i_M} \tag{2}$$

where $R_b$ is particle rigidity in the b-th rigidity interval, $TF_M(R_b,i_M)$ is a transmission function for the position $i_M$ inside the geomagnetic region M and $\Sigma i_M$ is the total number of locations for the same region.

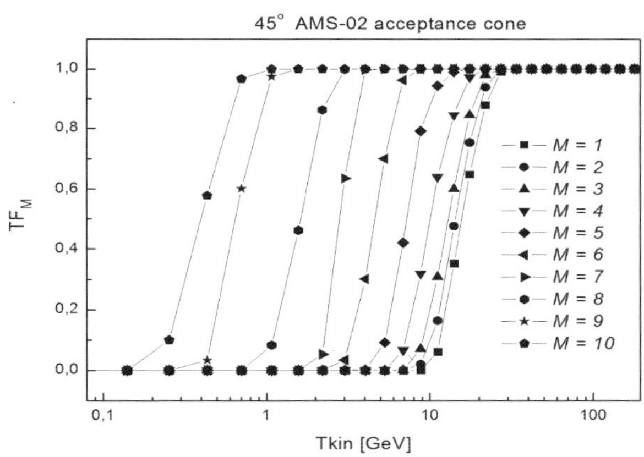


Figure 1. $TF_M$ for all ten AMS geomagnetic regions as a function of the proton kinetic energy in GeV for detector acceptance cones of 45° around the local geocentric Zenith.

For the AMS-02 observations, the predicted primary cosmic ray fluxes per unit of solid angle ($\Phi_M(R_b)$) are obtained by convolving the transmission function of each geomagnetic region (M) with the estimated proton flux at 1AU (i.e. outside the magnetosphere) from Hellmod simulations ($\Phi_{1AU}(R_b)$), as function of the proton rigidity.

$$\Phi_M(R_b)=\Phi_{1AU}(R_b)\, TF_M(R_b) \tag{3}$$

For more detail about the TF and flux evaluation method see [7].

## 3. Results

The $TF_M$ evaluated for 45° acceptance cone of detector for all ten AMS geomagnetic region for January 2012 (expected same condition as in October 1987 [1][2]) is presented at figure 1. At figure 2 we show the $TF_M$ for the all ten AMS geomagnetic regions evaluated for the full outward hemisphere (i.e. acceptance cone of detector 90°).

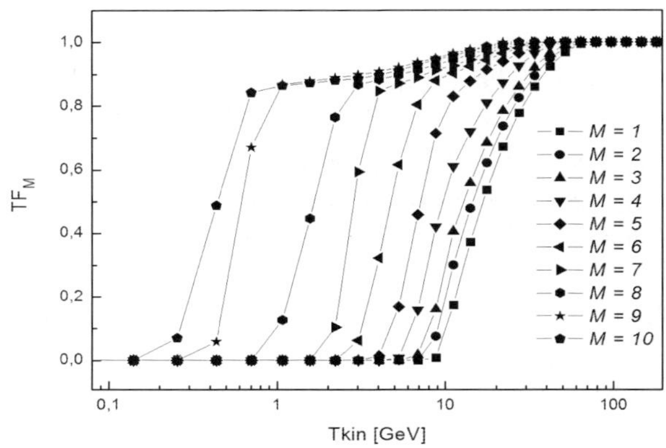


Figure 2. $TF_M$ for all 10 AMS geomagnetic regions as a function of the proton kinetic energy in GeV for the full outward hemisphere (i.e. acceptance cones of 90° around the local geocentric Zenith).

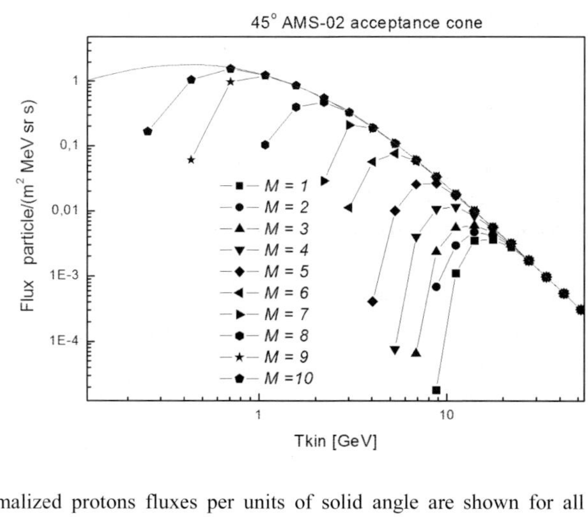


Figure 3. Normalized protons fluxes per units of solid angle are shown for all the geomagnetic regions as functions of the proton kinetic energies.

Dependence of the $TF_M$ on acceptance cone illustrated by comparison of figure 1. and 2. is result of cut-off rigidity dependency on local zenith and azimuth angles of incoming particles. With smaller acceptance cone we obtain the steeper $TF_M$ and smaller penumbra region of transmission functions.

Primary cosmic rays fluxes for the AMS-02 flight for January 2012 are presented at figure 3. The blue line on figure 3. is proton spectrum outside the magnetosphere, the result of the HelMod software simulation.

## 4. Conclusions

We evaluate the AMS-02 fluxes of the PCRs for January 2012 at a low Earth orbit using a back-tracking procedure and the HelMod software for the estimation of a cosmic ray spectra at the 1AU outside the magnetosphere. The results of back-tracing simulations allowed the determination of the so-called Transmission Function (TF) which was able to describe the properties of the PCR transport from the Earth's magnetopause to the low orbit and finally the fluxes of the PCRs in the 10 geomagnetic regions of AMS-02 observations.

## Acknowledgments

This work was supported by the Slovak VEGA grant agency project 2/0081/10.

## References

1. S. della Torre et al., *Proc. of the ICATPP,* **xxx** (2010).
2. D. Grandi et. al., *Proc. of the ICATPP,* **xxx** (2010).
3. D. F. Smart, M. A. Shea, E. O. Flückiger, *Space Science Reviews*, **93**, 305 (2000).
4. N. A. Tsyganenko and M. I. Sitnov, *J. Geophys.Res*., **110**, A3, A03208 (2005).
5. IGRF-11, web: *http://www.ngdc.noaa.gov/IAGA/vmod/igrf.html*
6. M. Aguilar et al., *Phys. Reports*, **366**, 331 (2002).
7. P. Bobik et. al., . *Geophys.Res*.. **111**, A5, A05205 (2006).

# STEREO OBSERVATIONS OF THE ENERGETIC HEAVY IONS DURING THE MINIMUM OF SOLAR CYCLE 23

R. BUČÍK*, U. MALL, A. KORTH

*Max-Planck-Institut für Sonnensystemforschung, Katlenburg-Lindau, Germany*

**E-mail: bucik@mps.mpg.de*

G. M. MASON

*Applied Physics Laboratory, Johns Hopkins University, Laurel, Maryland, USA*

In this paper we present observations of the two STEREO s/c for the first 2.5 years of the mission. During this period corotating interaction regions (CIRs) were the main source of the energetic heavy ions at 1 AU. Thanks to the unique s/c orbit the two STEREO can sample CIRs twice at different heliolatitudes during periods which are shorter than one solar rotation. We find that the occurrence of the strong CIR events was the most frequent at the beginning of the period. The tilt angle of the heliospheric current sheet (HCS) was quite high in the first stage of the mission and gradually flattened with time followed by decrease in the CIR activity. By examining the differences between measurements on the two STEREO s/c we discuss how the changes in the position of the s/c relative to the CIRs affect the energetic particle observations.

## 1. Introduction

As the Sun rotates, fast solar wind, emerging from coronal holes, overtakes the flow emanating from the belt of the slow solar wind, producing compressive interaction regions [1]. These regions corotate with the Sun and are commonly known as corotating interaction regions (CIRs). The CIRs can accelerate particles up to a few MeV/n in energy [2].

A three-dimensional CIR model has been developed for a tilted-dipole geometry at the Sun [3]. The model predicts that the strength of the compression regions changes with the latitude and the tilt angle. In addition, the model shows that during minimum phase of the solar cycle, CIRs are organized around the HCS. Both latitude and tilt angle effects were seen in the solar wind plasma and energetic particle observations [4, 5, 6, 7].

In this paper we investigate in more detail energetic ions in the tilted CIRs with the two STEREO s/c during the minimum of Solar Cycle 23.

## 2. Experiment

Both STEREO s/c are placed on a heliocentric orbit, STEREO-A preceding the Earth and STEREO-B trailing behind. The separation of the two s/c increases in longitude gradually in time. The latitude separation between the s/c periodically changes with the amplitude increasing in time. Because of unique configuration of s/c orbits, the two STEREO can sample the same CIR twice at different heliolatitudes during periods which are shorter than one solar rotation.

The measurements reported here were made with the Suprathermal Ion Telescope (SIT) [8] instruments on the STEREO-A and -B s/c. The SIT instrument is a time-of-flight mass spectrometer which measures ions from H to Fe in the energy range from 20 keV/n to several MeV/n. In our study we also make use of solar wind measurements made by the STEREO/PLASTIC instrument [9], together with magnetic field measurements obtained by the magnetometer on STEREO [10]. In addition, we use ground-based observations of the solar magnetic field made by the Global Oscillation Network Group (GONG). We use data measured over the extended solar minimum phase of cycle 23 from January 2007 to September 2009.

## 3. Observations

Figure 1 provides an overview of the data during our investigated period. Panels (A) and (D) show 1-hr He ion intensities (particles/cm$^2$ s sr MeV/n) for 0.16-0.23 MeV/n measured by SIT-A and -B, respectively. The ion enhancements coincide with the high-speed streams shown in panels (B) and (E) indicating an association with CIRs. Vertical dashed lines in panel (A) and (D) mark the start times of the Interplanetary Coronal Mass Ejections (ICMEs) obtained from a list compiled by the STEREO magnetometer team. Black curves in panels (C) and (F) show the heliographic latitude of the HCS seen from STEREO-A and -B, respectively, derived from the Sun's neutral line position provided by the GONG (http://gong. nso.edu/data/magmap/). The heliographic latitude of the s/c is marked by a red line. Over-plotted by blue dots in the panels (C) and (F) is the magnetic field azimuth in Radial-Tangential-Normal (RTN) coordinates.

The neutral line was delayed by five days, to represent the current sheet at 1 AU. This delay corresponds to a propagation speed of ~350 km/s. Earlier investigations reported that the simple constant-speed mapping technique is a good approximation of the current sheet at 1 AU [4]. Careful inspection of the

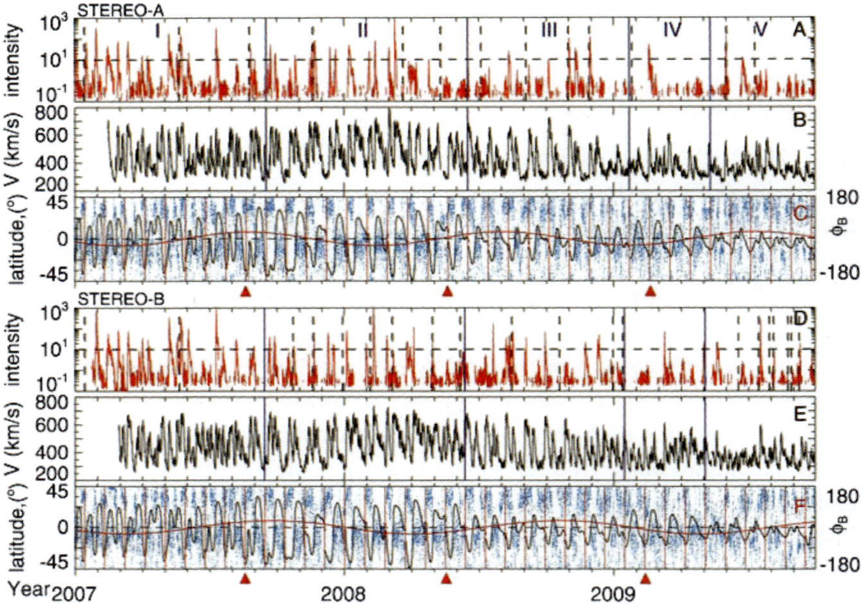


Figure 1. Panel (A) SIT-A 1-hr He intensities (particles/cm$^2$ s sr MeV/n) in the energy range 0.16-0.23 MeV/n. (B) Solar wind speed from STEREO-A. (C) Heliographic latitude of the current sheet for STEREO-A (black trace), heliographic latitude of the STEREO-A (red trace), and magnetic field azimuth in RTN (blue dots). (D, E, F) same as (A, B, C) but for STEREO-B. Roman numbers in panel (A) refer to the different periods (separated by solid blue vertical lines) identified in the text. Dashed vertical black lines in panels (A) and (D) indicate start times of the ICMEs. Red vertical lines in panels (C) and (F) mark start times of the Carrington rotations (CRs). Red triangles mark from left to right the CR 2060, 2070 and 2080.

panels (C) and (F) in Fig. 1 reveals that the crossings of the inferred current sheet (intersections of the red and black curves) perfectly match the reversals in magnetic field polarity observed by STEREO.

Further exploration of Fig. 1 shows that the two STEREO s/c regularly cross or approach the current sheet at different latitudes. We divide this period into the five different intervals, based on the HCS shape. In the interval (I) the HCS was warped. A warp comparable to the main current sheet latitude extent was visible at the beginning of the period with the amplitude slowly decreasing in time. We observe that the occurrences of the ion enhancements, as well as the ion intensities were high. In the period (II) the HCS was almost flat with a somewhat higher heliolatitude compared to the previous period. The frequency of the occurrence of the CIR events and interestingly peak intensities as well subsided a bit. In the interval (III) the distance of the HCS from the heliographic equator notably decreased and small distortion in the current sheet developed again. The peak intensities as measured by STEREO-A are somewhat decreased, but no change is seen on STEREO-B. In the period (IV) the current sheet heliolatitude temporary increased and the HCS was again flat. This period was quiet in the CIR event activity, notably on STEREO-A. In the period (V), the current sheet was almost aligned with the heliographic equator, but no marked difference in particle intensities compared to previous period was observed. The periodicity of the measured increases of particle intensities is very similar on the two s/c in 2007, when STEREO-A and -B trace almost the same latitudes relative to the HCS. In 2008 the phase shift between s/c latitudes is starting to increase. It is accompanied by the differences in particle intensities between STEREO-A and -B.

## 4. Discussion

In the following we discuss in more detail the differences between energetic particle increases on STEREO-A and -B. In our study we use the list of the CIR events observed by STEREO in the interval January 2007- September 2008 which has been compiled by Mason *et al.* [11]. For the reasonable statistical accuracy the authors required that the 1-hr He ion intensity in the 0.16-0.23 MeV/n energy range exceed 10 particles/(cm$^2$ s sr MeV/n) on at least one of the two STEREO s/c. We use the same selection criterion to identify CIR enhancements through October 2008 and September 2009.

### 4.1. *Rate of the Occurrence of CIR Enhancements*

Panels (A) and (C) of Fig. 2 show SIT instrument counts in 0.16-0.23 MeV/n energy and He mass (3-8 amu) range along the Carrington longitude between 2053 and 2087 CRs. Panel (A) shows data for STEREO-A and panel (C) data for STEREO-B. The counts are collected in Carrington longitude bins with the width of 1°. Our data show three recurrent source regions responsible for the CIR events in period (I), two in period (II) and one stable recurrent region in period (III). In periods (IV) and (V) we observe quite irregular pattern in the distribution of the number of counts. Furthermore, only little difference is seen in the structure between data on the two s/c in period (I) and in the most part of period (II). From the end of period (II) the difference in the occurrence of the CIR enhancement start to be more striking.

Panels (B) and (D) in Fig. 2 show histograms of the number of CIR enhancements per CR observed by STEREO-A and -B, respectively. Only He

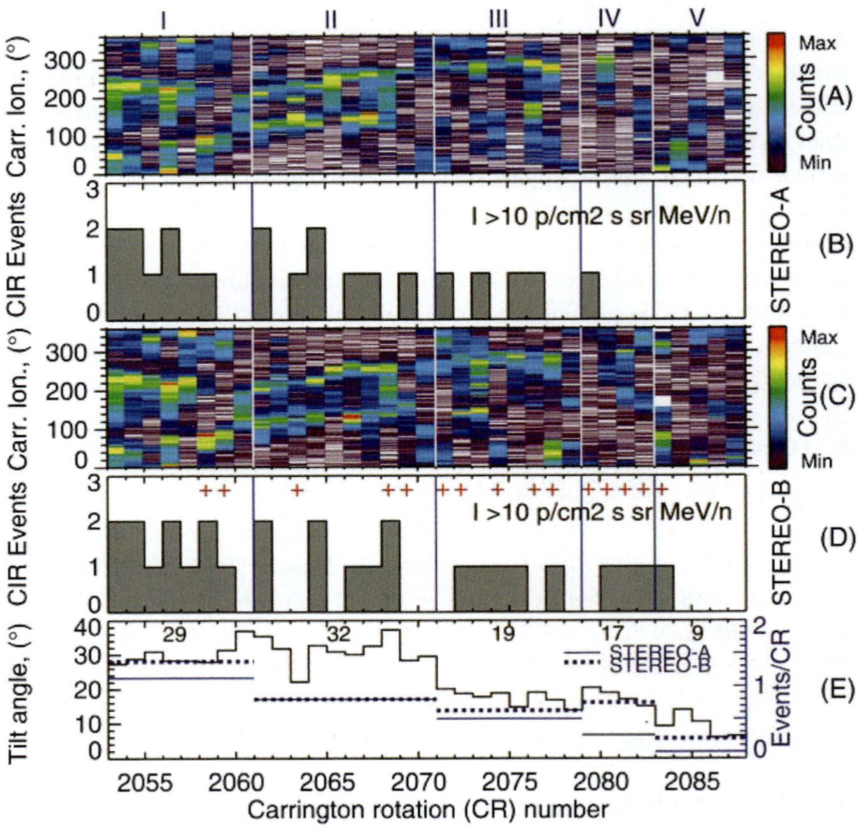


Figure 2. Panel (A) SIT-A He counts in Carrington longitude *vs.* Carrington rotation number diagram. (B) Histogram of the number of He ion enhancements per Carrington rotation on STEREO-A. Panels (C, D) same as (A, B) but for STEREO-B. Red crosses match the CRs with different number of CIR events on STEREO-A and -B. (E) Tilt angle. Numbers indicate medians of the tilt in intervals (I)-(V). Average number of events per CR in intervals from (I) to (V) for STEREO-A (solid blue lines) and -B (dashed blue lines).

ion enhancements with an intensity > 10 p/(cm$^2$ s sr MeV/n) in the range of 0.16-0.23 MeV/n on each s/c are taken into account. To avoid any possible interference between CIR and the transient disturbances driven by ICMEs, which might intensify the strength of the compression regions [12]; the pairs of particle events with mixed ICME-CIR at least on one s/c are not considered in the following analysis. Panel (E) shows the HCS tilt angle (black line). The tilt angle is the same as the maximum heliolatitude attained by the HSC. By definition (http://wso.stanford.edu/) we calculate the tilt angle as an average of the maximum northern and southern extent of the HCS during the CR.

Panels (B) and (D) in Fig. 2 show that the Carrington rotations with two CIR events only occurred during the period of enhanced tilt angles. In intervals from (IV) to (V) there exist periods lasting for 8 (panel B) and 4 (panel D) solar rotations without any CIR activity with ion intensities above 10 p/cm$^2$ s sr MeV/n. The average number of events during one CR, shown by solid (STEREO-A) and dotted (STEREO-B) blue lines in panel (E), decreases from interval (I) to interval (V). Through the period (II) to (V) this decrease follows the tilt angle. Enhanced rate of the occurrence of the CIR intensity increases could also be related to large warps in the current sheet, which may give rise to another interaction region. We see in Fig. 1 that some warp crossings in the interval (I) are associated with strong He peaks. This is consistent with the results reported in Sanderson *et al.* [4]. High occurrence of the CIR events during periods of high tilt angles may be interpreted as a result of the dependence of the strength of the compression and consequently the particle intensity on the tilt angle. Thus, where the tilt angle is high the events are strong enough to exceed the He ion intensity threshold selected for this analysis.

A comparison of the histograms in panels (B) and (D) in Fig. 2 shows there are periods with duration of 5 and 4 CRs in intervals (I) and (II) with no differences between the number of events on STEREO-A and -B. In contrast to the period (I) and (II), the period (IV) shows differences for every CR. Red crosses in panel (D) mark the CRs with different number of corotating increases on the two STEREO s/c. These observations indicate the large differences between the number of events on STEREO-A and -B in a period of low tilt angles and almost no differences when the tilt angle increases to higher values. An exception is interval (V) where due to weak increases the differences in number of events above the threshold intensity were not determined.

### 4.2. *Intensity Differences on STEREO-A and -B*

In this section we examine the differences between the ion intensities on the two s/c with respect to the changes in relative position between the two STEREO s/c. Figure 3 shows ratio between 1-hr CIR event peak intensities on STEREO-A and -B for He ions in the energy range of 0.16-0.23 MeV/n plotted versus the s/c latitude (panel A) and longitude (panel B) differences. The ratios are computed in way that the higher intensity is divided by the lower one.

Panels (A) and (B) show a positive correlation for s/c latitude and longitude differences, which are highly significant for a sample of 34 events. The increase

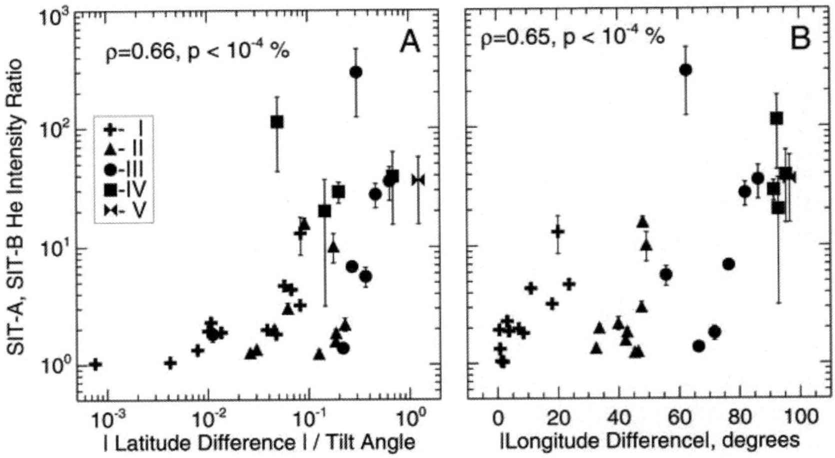


Figure 3. Ratio between 1-hr CIR event peak intensities on STEREO-A and STEREO-B for He ions in 0.16-0.23 MeV/n energy range *vs.* (A) heliographic latitude difference between the s/c relative to the tilt angle and (B) heliographic longitude difference between the s/c.

in the CIR ion intensity differences with increase in s/c latitude separation would be expected from the latitude variations in CIRs. The increase in the CIR ion intensity differences with increasing s/c longitude separation might be either due to temporal changes over the growing corotation time between the two s/c or just a reflection of the simultaneous increase of the latitude differences as the longitudinal separation between s/c becomes larger. To clarify this we compare the temporal profile of the intensity differences with the temporal profiles of the s/c longitude and latitude differences in CIR events.

Figure 4A shows time profile of the ratios between 1-hr CIR event peak intensities on STEREO-A and -B for He ions in the energy range of 0.16-0.23 MeV/n. The red crosses mark corresponding absolute s/c latitude difference relative to the tilt angle. Figure 4B displays the s/c longitude separations for the intensity ratios shown in panel (A). All quantities in Fig. 4 are plotted at times of peak intensities on STEREO-A.

One interesting feature seen in Fig. 4A is that the intensity ratio time profile appears to be well ordered with the time profile of the latitude difference in the period (I), (II) and (III). Through these three intervals the ratio periodically increases and decreases following the changes in the time profile of the latitude differences. In periods (IV) and (V) the time profile of the intensity ratios is unclear due to high statistical errors. In contrast the to latitude differences, the

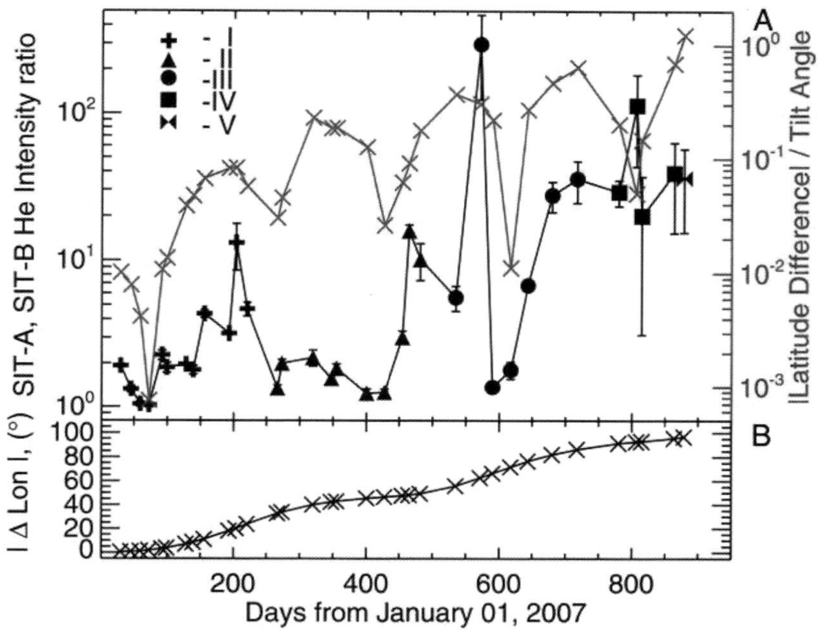


Figure 4. Panel (A) Ratio between 1-hr CIR event peak intensities on STEREO-A and -B for He ions in 0.16-0.23 MeV/n energy range. Over-plotted is heliographic latitude difference between s/c relative to the tilt angle (red curve). (B) Heliographic longitude differences between STEREO-B and STEREO-A.

differences between s/c longitudes show only increase over the investigated period of the mission. This suggests that at least in period from (I) to (III) the variations in differences between ion intensities observed on the two s/c might not be related to the increasing s/c longitude separation. In the intervals (IV) and (V) where the ratio time profile is not clear and the longitudinal separation between two STEREO s/c is large (at that time > 90°) the temporal differences in the strength of the CIRs on STEREO-A and -B can not be ruled out. The color diagrams in Fig. 2 demonstrate that two strips of enhanced counts spreading over the intervals (I)-(III) persisted for a period of > 20 solar rotations. Lacking of any stable pattern in the intervals (IV) and (V) suggests the temporal changes in the CIRs may play a role in the later stage of the investigated period.

## 5. Conclusions

From examining He ions in the 0.16-0.23 MeV/n energy range from STEREO during the minimum of Solar Cycle 23 we found the following results:

1. The high rate of the occurrence of the CIR ion enhancements is observed during periods of high tilt angles and large warps in the current sheet. The rate decreased with the current sheet tilt angle. The differences between the number of CIR events on STEREO-A and -B were very small when the tilt angle was high. Conversely, during periods of lower tilt angles the differences increased.
2. The CIR intensity differences between STEREO-A and -B are well ordered with latitudinal differences between the two s/c relative to the tilt angle. The temporal changes in CIRs over the corotation time between the two STEREO s/c might have contributed in the end of the investigated period which is characterized by the absence of the stable source regions for the CIRs.

**Acknowledgements**

We acknowledge the STEREO PLASTIC team (NASA Contract NAS5-00132) for the use of the solar wind plasma data. This work utilizes data obtained by the Global Oscillation Network Group (GONG) program. This work was supported by the Max-Planck-Gesellschaft zur Förderung der Wissenschaften and the Bundesministerium für Bildung und Forschung (BMBF) under grant 50 OC 0501. The work at the Johns Hopkins University/Applied Physics Laboratory was supported by NASA under contract SA4889-26309 from the University of California, Berkeley.

**References**

1. J. T. Gosling *et al.*, *Geophys. Res. Lett.* **20**, 2789 (1993).
2. C. W. Barnes, and J. A. Simpson, *Astrophys. J.* **210**, 9 (1976).
3. V. J. Pizzo, *J. Geophys. Res.* **99**, 4173 (1994).
4. T. R. Sanderson *et al.*, *J. Geophys. Res.* **103**, 17235 (1998).
5. T. R. Sanderson *et al.*, *Geophys. Res. Lett.* **26**, 1785 (1999).
6. J. T. Gosling *et al.*, *Geophys. Res. Lett.* **24**, 309 (1997).
7. M. I. Desai *et al.*, *J. Geophys. Res.* **104**, 6705 (1999).
8. G. M. Mason *et al.*, *Space Sci. Rev.* **136**, 257 (2008).
9. A. B. Galvin *et al.*, *Space Sci. Rev.* **136**, 437 (2008).
10. M. H. Acuña *et al.*, *Space Sci. Rev.* **136**, 203 (2008).
11. G. M. Mason *et al.*, *Solar Phys.* **256**, 393 (2009).
12. E. Keppler *et al.*, *Science* **268**, 1013 (1995).

## Electron and Positron solar modulation and prediction for AMS02.

P. Bobik[1], M.J. Boschini[2,4], C. Consolandi[2], S. Della Torre[2,5]*,
M. Gervasi[2,3], D. Grandi[2], K. Kudela[1], S. Pensotti[2,3], and P.G. Rancoita[2]

[1] *Institute of Experimental Physics, Kosice (Slovak Republic)*
[2] *Istituto Nazionale di Fisica Nucleare, INFN Milano-Bicocca, Milano (Italy)*
[3] *Department of Physics, University of Milano Bicocca, Milano (Italy)*
[4] *CILEA, Segrate (MI) (Italy)*
[5] *Department of Physics and Maths, University of Insubria, Como (Italy)*
**E-mail: stefano.dellatorre@mib.infn.it*

The solar modulation, a combination of diffusion, convection, magnetic drift and energy loss inside the heliosphere is usually seen as a depletion in the Galactic cosmic ray (CR) flux at low energy (less than 10 GeV/nuc). Antiparticles such as antiprotons or positrons undergo the same processes of respective particles but with a different magnitude depending on the Solar magnetic field polarity. For electrons and positrons, due to the small mass, energy loss mechanisms as inverse compton, synchrotron, bremsstrahlung and ionization have to be taken into account, together with the typical adiabatic losses considered in the heliosphere. We developed a Monte Carlo stochastic simulation with the aim to compare the solar modulation of particles and antiparticles in the same observation period. We are able to estimate the different behaviours associated to the charge sign dependent processes of the heliospheric modulation. We compared the simulated positron fraction with measurements performed by AMS-01 and PAMELA. We also present the prediction for the AMS-02 experiment.

*Keywords*: cosmic rays; heliosphere; positron ratio; solar modulation;

### 1. Introduction

Solar modulation of Cosmic Rays (CRs) is actually described by model integrating the Parker's equation[1]. The CRs propagation is depending on the charge sign of CRs and to the solar magnetic polarity[2,3]. In this work we introduce a description of our present modulation model, then we summarize the different behaviours with particles of opposite charge sign in same observation periods. Finally we compare results with experimental data providing also a prediction for AMS-02.

## 2. Model Description

Present model[4] is based on the 2D (radius and colatitude) approximation of Parker equation[1]: a Fokker-Planck equation including diffusion, convection, adiabatic energy loss and magnetic drift. A rigorous proof by Ito[5] demonstrate the equivalence with a set of ordinary stochastic differential equations (SDEs). The main advantage is that SDEs could be easly integrated with Montecarlo techniques. The model depends on some parameters related to the solar activity and is fine-tuned by comparing results with experimental data[6]. We used the Heliosferic Magnetic Field (HMF) introduced by Parker and modified according to Jokipii and Kóta[7]: the magnitude of the HMF in the polar regions is increased without modifying the topology of the field. The major effect is a reduced CR penetration along the polar field lines in the inner part of the heliosphere, due to a lower magnetic drift velocity. According to Ulysses data[8], for periods of low solar activity, we used a latitudinal dependence of the solar wind speed: the value increases by a factor of two moving from the ecliptic plane to the polar region. For the drift description we use the model by Potgieter & Moraal[9,10] that describes periods of solar maximum as well solar minimum. The model requires as input a Local Interstellar Spectrum (LIS). We use the model proposed by Zhang & Cheng[11], which has been corrected comparing it with the experimental data[12].

## 3. Results

### 3.1. *Drift effect on modulation*

The charge dependence of propagation is due to the drift term of the Parker equation. The fundamental parameter that fix the direction of particle magnetic drift is $qA$, where $q$ is the particle charge and $A$ is the normalization factor of the magnetic field ($B$) that gives its polarity. We made a simulation of a typical solar minimum in both polarities for both electrons and positrons. We found, in agreement with theory[13], that the flux is lower when $qA < 0$ than for $qA > 0$ conditions. Results are shown in fig. 1 (left panel). In the right panel of fig. 1 we show the modulated positron fraction above 1 GeV: for a period with $A > 0$ the modulation does not affect $e^-/(e^+ + e^-)$ ratio significantly; for a period with $A < 0$ the effect of solar modulation is relevant up to $\sim$ 10 GeV, reducing the modulated fraction by a factor of two or more.

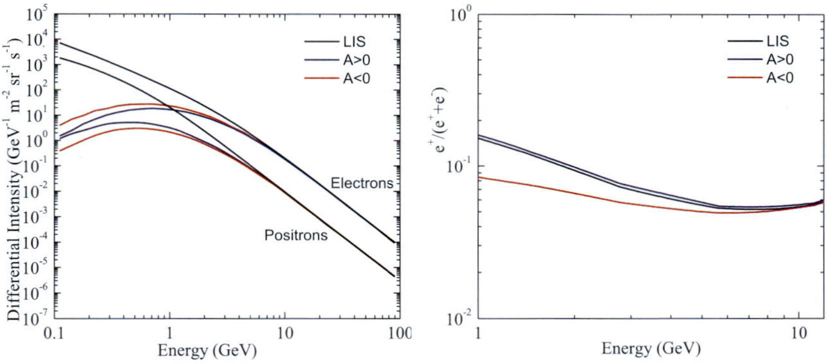


Fig. 1. Electron and positron modulation (left) and positron ratio for a typical solar minimum period (we use conditions related to June 1998) with both magnetic field polarities.

### 3.2. *Measurements by AMS-01 and PAMELA*

We selected two data set in similar solar conditions but opposite magnetic field polarity. AMS-01[14] operated on board the Space Shuttle in June 1998, at the end of a solar minimum occured during a period with $A > 0$. PAMELA[15] is a space born experiment working since July 2006. Published data have been taken between 2006 and December 2008, therefore during the last long solar minimum with $A < 0$.

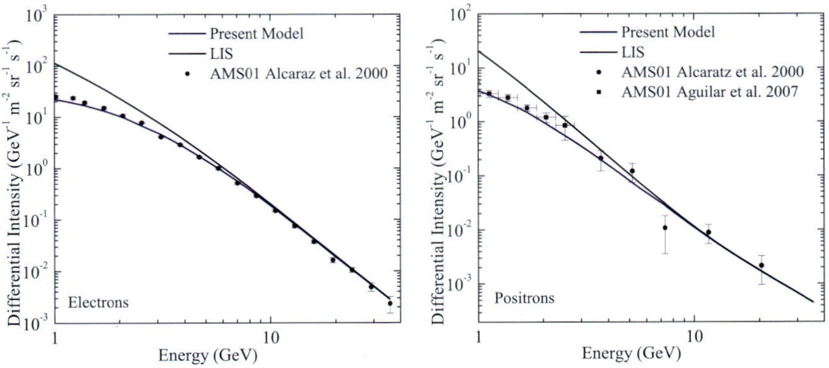


Fig. 2. Simulated spectra for AMS-01 mission.

We reproduce, within the errors bars, the AMS-01 data for electrons and positrons (see figure 2). We find a good agreement between simulation and experimental data also for $e^+$ fraction. Despite of the large solar modulation of electron and positron flux the modulated ratio is very close to the interstellar ratio (figure 3) as obtained in previous results. We used the same model of propagation to reproduce data published by PAMELA collaboration. We simulated several heliospheric conditions covering the periods of the data taking. The results have been averaged and shown in figure 3. We reproduced PAMELA data and find that modulated ratio is lower than the interstellar one, as expected.

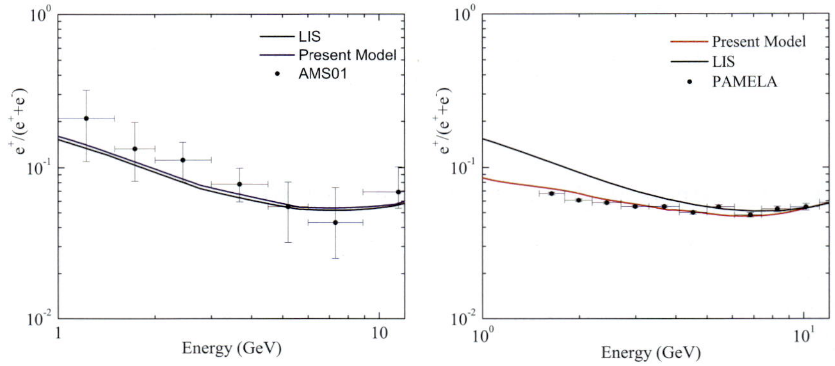


Fig. 3. Simulations of positron ration for AMS-10 and PAMELA mission.

### 3.3. *Prediction for AMS-02*

The simulation code was used to predict CR fluxes for future measurements. The periodic behavior of the heliosphere allows us to predict, with a certain level of precision, the solar modulation parameters needed for the simulation. The assumption is that diffusion coefficient, tilt angle and solar wind speed show a near-regular and almost periodic trend. The periodicity is two consecutive 11-years solar cycles. In order to get these data we considered the prediction of Smoothed Sunspot Numbers from IPS (Ionospheric Prediction Service) of the Australian Bureau of Meteorology. Using SIDAC data (Solar Influences Data Analysis Center) we selected periods with a similar solar activity conditions and same solar field polarity of the simulation time: therefore approximately 22 years before. Under the previous assumpion, we used the values measured in that periods as an estimation of the conditions of the heliosphere in the near future. We concentrate the simulations on the AMS-02 mission that will be installed on the ISS in February 2011, and, in particular, on one period, during its data taking, approaching the solar maximum: January 2012. results are shown in figure 4. This period is still with $A < 0$ and the modulated positron fraction is still below the interstellar one.

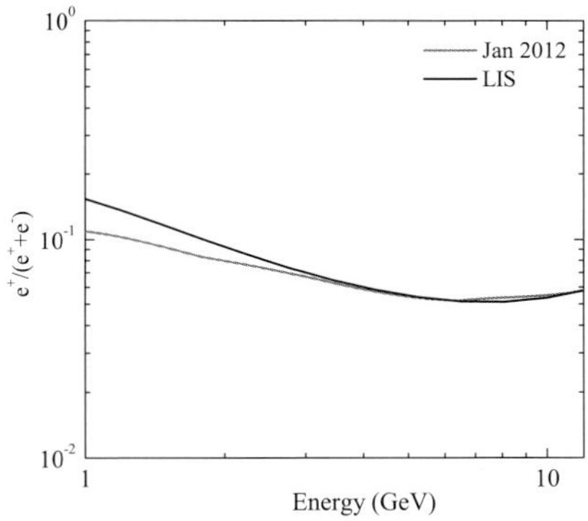


Fig. 4. Prediction of positron ratio for January 2012.

## 4. Conclusions

We built a 2D stochastic Monte Carlo code for particles propagation across the heliosphere. Present model takes into account drift effects and shows quantitatively a good agreement with measured values, both for positive and negative periods and for different particle's charge. We found that for periods with $A > 0$ the modulated positron ratio is similar to the interstellar one. During periods with opposite field polarity we found instead a significative reduction of the modulated positron ratio respect to the interstellar ratio. We compared simulations with positron ratio measured by AMS-01 and PAMELA. We found that the discrepancy of PAMELA positron ratio respect to the previous experiments can be well explained with a polarity-dependent effect of modulation, due to magnetic drift of particle. The prediction for AMS-02 shows that, in the first period of data taking, it will probably observe a modulation effect similar to the one measured by PAMELA. It will be interesting to compare the measurements of AMS-02 in the following years to confirm the behavior shown in fig. 1.

## References


1. E. N. Parker, *Plan. & Space Sci.* **13**, p. 9 (1965).
2. G. Boella, M. Gervasi, S. Mariani, P. Rancoita and I. Usoskin, *J. Geophys. Res. A21* **106**, p. 29355 (2001).
3. e. a. J.M. Clem, *Astrophys. J.* **464**, p. 507 (1996).
4. Bobik, P, Boschini, M.J., Della Torre, S., Gervasi, M., Grandi, D., Kudela, K. and Rancoita, P. G., *Proton and antiproton modulation in the heliosphere for different solar conditions and AMS-02 measurements prediction - Proc. of the ICATPP Conference on Cosmic Rays for Particle and Astroparticle Physics, Como 7–8/10/2010, World Scientific, Singapor, ISBN 978-981-4329-02-6; arXiv:1012.3086v3 [astro-ph.EP], available at the web site: http://arxiv.org/abs/1012.3086* (2011).
5. C. Gardiner, *Handbook of stochastic methods: for physics, chemistry and natural sciences* (Springer Edition, 1985).
6. P. Bobik, G. Boella, M. J. Boschini, S. Della Torre, M. Gervasi, D. Grandi, K. Kudela, S. Pensotti and P. Rancoita, *Proc. of the 31st ICRC, Lodz* (2009).
7. J. R. Jokipii and J. Kota, *Geophys. Res. Lett.* **16**, 1 (1989).
8. Ulysses online database http://cohoweb.gsfc.nasa.gov/form/ulysses.html.
9. M. S. Potgieter and H. Moraal, *Astrophis. J.* **294**, 425 (1985).
10. Bobik, P, Boella, G., Boschini, M.J., Della Torre, S., Gervasi, M., Grandi, D., Kudela, K., Pensotti, S. and Rancoita, P. G., *Drift models and polar field for cosmic rays propagatin in the heliosphere, Proc. of the 11th International Conference on Particle Physics and Advanced Technology, Como 5–9/10/2009, World Scientific, Singapore* , 760(october 2009).
11. Zhang, L. and Cheng, K. S., *Astron. & Astrophys.* **368**, 1063 (2001).
12. Tacconi, M, Bobik, P, Boschini, M.J., Della Torre, S., Gervasi, M., Grandi, D., Kudela, K., Pensotti, S. and Rancoita, P. G., *Energy Loss for electrons in the heliosphere and local interstellar spectrum for solar modulation, Proc. of the ICATPP Conference on Cosmic Rays for Particle and Astroparticle Physics, Como 7–8/10/2010, World Scientific, Singapore* (2010).
13. Alanki-Huotari K., Usoskin I.G., Mursula K. and Kovaltsov G.A., *J. Geophys. Res.* **112** (08/2007).
14. Aguilar *et al.*, *Physics Lett. B* **646**, 145 (03/2007).
15. Adriani *et al.*, *Nature* **458**, 607 (04/2009).

# HOW TO USE MOLECULAR CLOUDS TO STUDY THE PROPAGATION OF COSMIC RAYS IN THE GALAXY

S. GABICI*

*Astroparticule et Cosmologie (APC), CNRS, UMR7164, Paris, France*
**E-mail: stefano.gabici@apc.univ-paris7.fr*

Observations of molecular clouds in the gamma ray domain provide us with a tool to study the distribution of cosmic rays in the Galaxy. This is because cosmic rays can penetrate molecular clouds, undergo hadronic interactions in the dense gas, and produce neutral pions that in turn decay into gamma rays. The detection of this radiation allows us to estimate the spectrum and intensity of cosmic rays at the cloud's position. Remarkably, this fact can be used to constrain the cosmic ray diffusion coefficient at specific locations in the Galaxy.

*Keywords*: cosmic rays: origin and propagation; gamma rays; molecular clouds

Galactic Cosmic Rays (CRs) are believed to be accelerated by SuperNova Remnants (SNRs).[1] The main (but not conclusive) arguments supporting this idea are the fact that SNRs can provide the total power required to maintain the galactic CR population, and the belief that an efficient acceleration mechanism, diffusive shock acceleration,[2] operates in these objects. The detection of some SNRs in TeV gamma rays,[3] expected if they indeed are the sources of CRs,[4] also supports this scenario, though the emission might have a leptonic origin and not be related to the acceleration of CRs.[5]

Besides the total CR power of the Galaxy, a theory of CR origin must also explain the spectrum, isotropy, and chemical composition of CRs.[6] This requires a thorough knowledge not only of the nature of CR sources, but also of the way in which CRs propagate in the Galaxy. This is because the observed properties of CRs results from the equilibrium between the injection rate of CRs from sources and their diffusive escape from the Galaxy.

Measurements of the amount of spallation suffered by CRs allow us to infer the average residence time of a CR of energy $E_p$ in the Galaxy as $t_{res} \propto E_p^{-\delta}$, with $\delta \sim 0.3-0.7$. If $h$ is the length a CR has to move away from its source before escaping the Galaxy (i.e. the Galaxy's thickness), then the diffusion coefficient reads: $D_{gal} \approx h^2/t_{res} \approx 10^{28}(E_p/10\ \text{GeV})^{\delta}\ \text{cm}^2/\text{s}$. However, this has to be intended as the average diffusion coefficient in the Galaxy, and local variations (both in time and space) might exist.

In particular, the diffusion coefficient might be suppressed close to CR sources. This is because CRs can excite magnetic turbulence while streaming away from their acceleration site. This would enhance the scattering rate of CR themselves and thus reduce the diffusion coefficient.[7] The problem of estimating, on theoretical grounds, the diffusion coefficient around CR sources is far from being solved, mainly because of its intrinsic non-linearity and because various mechanisms might damp the CR–generated waves and thus affect the way in which CRs diffuse.[8,9]

In principle, gamma ray observations can provide us with constraints on the diffusion coefficient close to CR sources. Once escaped from their sources, CRs undergo hadronic interactions with the surrounding gas and produce gamma rays. The characteristics of such radiation (in particular its spectrum and intensity as a function of the time elapsed since CRs escaped the source) depend on the value of the diffusion coefficient that can thus be constrained, if a reliable model for CR acceleration at the source is available. The presence of massive Molecular Clouds (MCs) close to the source would enhance the gamma ray emission, making its detection more probable. Studying such radiation is of great importance not only in order to reach a better understanding of how CRs diffuse, but also because its detection can provide an indirect way to identify the sources of galactic CRs.[10–14]

## Molecular clouds as cosmic ray barometers

Consider a MC with mass $M_{cl}$ at a distance $d$, located in a region in the Galaxy where the CR (protons) intensity is $J_{CR}$. For illustrative purpose, we set here $J_{CR} = K_{CR} E_p^{-\alpha}$. We further assume that the high energy CRs (the ones with energy above the threshold for $\pi^0$-production) can freely penetrate the cloud.[15] Under these assumptions the expected gamma ray flux from the MC due to proton-proton interactions is given by:

$$F_{\gamma}(E_{\gamma}) \sim Y_{\gamma} \frac{\sigma_{pp}}{m_p} J_{CR}(E_{\gamma}) \left( \frac{M_{cl}}{d^2} \right) \propto E_{\gamma}^{-\alpha} \tag{1}$$

where $\sigma_{pp} \approx 34$ mb is the interaction cross section, $m_p$ is the proton mass, and $Y_{\gamma}$ depends on $\alpha$ and is tabulated in Ref. 6. Equation 1 is valid at high energies only ($E_{\gamma} \gtrsim 10$ GeV), while at lower energies the spectrum (in log-log scale) is symmetric with respect to the energy $E_{\gamma} = m_{\pi^0}/2 \sim 70$ MeV.

Assume now that the CR intensity in the region under exam differs by a factor $\delta(E_p) = J_{CR}(E_p)/J_{bg}(E_p)$ from the one measured at the Earth (which is:[16] $(4\pi/c)E_p^2 J_{bg}(E_p) \sim 6 \times 10^{-3}(E_p/\text{TeV})^{-0.7}\text{eV cm}^{-3}$), so that Equation 1 can be rewritten as:

$$E_\gamma^2 F_\gamma(E_\gamma) \sim 2.5 \times 10^{-13} \left(\frac{Y_\gamma f_{0.1}^{2.7-\alpha}}{0.0275}\right) \left(\frac{M_5}{d_{kpc}^2}\right) \delta(E_p) E_\gamma^{-0.7} \text{TeV/cm}^2/\text{s} \quad (2)$$

where $M_5$ is the mass of the MC in units of $10^5 M_\odot$, $d_{kpc}$ is the distance in kpc, and $f_{0.1} = (f/0.1) \approx 1$ takes into account the fact that on average CRs with energy $E_p$ produce gamma rays with energy $E_\gamma \approx f \times E_p$. Moreover, a multiplicative factor of 1.5 has been applied to account for the contribution to the emission from nuclei heavier than H both in CRs and in the MC gas.

If gamma rays are detected from a MC, and its mass and distance are known from its CO emission,[17] then Equation 2 allows one to measure both the spectrum and intensity of CRs at the MC's location, and thus the quantity $\delta(E_p)$. It follows that MCs can be effectively used as probes of the energy density of CRs at different locations in the Galaxy, and for this reason have been sometimes referred to as *CR barometers*.[18–20] Since observations in the GeV range of the Galaxy suggest that, on large spatial scales, CR variations are not very large,[21–23] $\delta(E_p)$ is normally interpreted as the excess above the galactic CR background. Some examples of how to use this fact to estimate the CR diffusion coefficient are given in the following sections.

### The galactic centre ridge

The H.E.S.S. collaboration reported on the detection of diffuse gamma ray emission from the Galactic Centre (GC) ridge.[24] Remarkably, the emission correlates spatially with a complex of giant MCs, suggesting that the emission itself is likely to be originated by CR hadronic interactions in the dense ambient gas. The TeV gamma ray spectrum extracted from the region of galactic coordinates $|l| < 0.8°$, $|b| < 0.3°$ can be fitted by a power law with index $\alpha \sim 2.29$ and normalization $\sim 1.73 \times 10^{-8}\text{TeV}^{-1}\text{cm}^{-2}\text{s}^{-1}\text{sr}^{-1}$. At the distance of the GC ($\sim$ 8.5 kpc) the emitting region has a size of $\approx 240 \times 90$ pc, and encloses a total gas mass of $1.7 - 4.4 \times 10^7 M_\odot$.[24]

By assuming that all the observed gamma ray emission has an hadronic origin, we can use Equation 2 to show that, close to the GC, the CR intensity at 10 TeV is enhanced with respect to the local one by a factor of $\delta \approx 3 - 9$, and grows at higher energies due to the hardness of the measured emission. Moreover, the total energy in form of CRs needed to explain the emission is $\approx 10^{50}$erg, if the observed spectrum is extrapolated from 1 GeV to 1 PeV.

Thus, a single SNR with explosion energy $E_{SN} \approx 10^{51}$erg and acceleration efficiency $\eta \approx 10\%$ might explain the whole CR excess in the region.[24]

One possible CR accelerator is the SNR Sgr A East, located close to the GC and with an estimated (but uncertain) age of $t_{age} \approx 10$ kyr. If the CRs responsible for the observed emission are assumed to be released during the earliest phase of the SNR evolution, then it is possible to estimate the diffusion coefficient by requiring them to fill the emitting region of size $l_\gamma \approx 120$ pc in a time $t_{age}$. This gives $D \approx l_\gamma^2/4\ t_{age} \approx 10^{29}\text{cm}^2/\text{s}$, which, taking into account the large uncertainties in the assumptions made, is in rough agreement with (maybe somewhat smaller than[24]) the value of the diffusion coefficient of $\sim$ TeV CRs in the galactic disk. The presence of a massive MC at $l = 1.3°$ ($\sim 200$ pc) which does not show prominent gamma ray emission suggests that CRs accelerated at the GC have not diffused yet up to that distance, and this justifies our assumption for $l_\gamma$.[24]

This estimate of the diffusion coefficient has to be regarded as a very rough one, due to the presence of many uncertainties in its derivation. For example, another source (e.g. the black hole Sgr A*[25]), or many sources (e.g. an ensemble of SNRs[26]) might be the accelerators of the CRs responsible for the TeV emission. Moreover, proton energy losses (ignored here) might affect the estimate of the diffusion coefficient by a factor of $\approx 2$.[27]

Better constraints on the diffusion coefficient and on the nature of the source(s) of the accelerated particles could be obtained if the spatial variation of the spectrum of the gamma ray emission was known. Measuring such variation, though challenging for currently operating instruments, is within the capabilities of future ones like the Cherenkov Telescope Array. It seems thus appropriate to discuss the future perspectives in this direction.

According to the scenario summarized above, the SNR Sgr A East, located close to the GC, released about $10^{50}$ erg in form of CRs about 10 kyr ago. Since high energy CRs are believed to be accelerated by young SNRs (aged up to few thousands years), we are allowed to consider this as an impulsive injection event. If CRs are injected with a power law spectrum $\propto E^{-2}$, their time dependent spatial distribution around the source is:[12]

$$N_{CR}(R, E_p, t) = \frac{\eta\ E_{SN}}{\pi^{3/2} \ln(E_p^{max}/E_p^{min})} \frac{E_p^{-2}}{R_d^3}\ e^{-\left(\frac{R}{R_d}\right)^2} \tag{3}$$

where $E_{p,max}(E_{p,min})$ is the maximum (minimum) CR energy, $R_d = \sqrt{4\,D(E_p)\,t}$ is a diffusion length and $D$ the diffusion coefficient. The CR intensity is the sum of Equation 3 and the galactic CR background.

The expected excess over the background for CRs with energy 3 TeV is

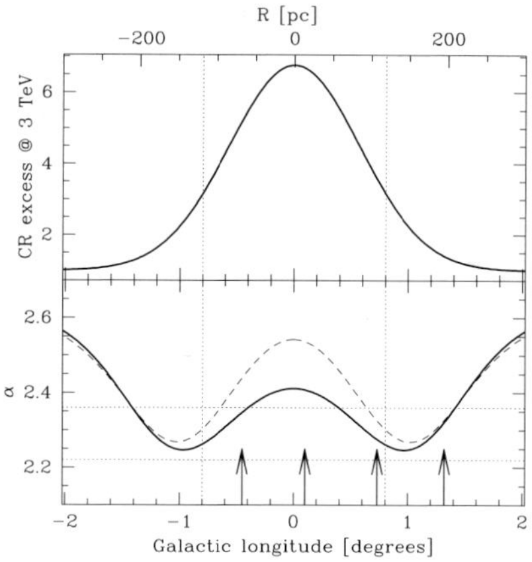


Fig. 1. Upper panel: excess of 3 TeV CRs over the CR background. Lower panel: slope of the gamma ray emissivity at 300 GeV for $\delta = 0.3$ (solid) and 0.4 (dashed). Horizontal lines represent the observed slope (measured value ± error). Dotted vertical lines delimit the emitting region. Arrows refer to the positions of the peaks in the gas distribution.

shown in Figure 1 (top panel) as a function of galactic longitude. These are the CRs that produce most of the emission detected above $\sim 300\,\mathrm{GeV}$ in the region $|l| < 0.8°$ (marked by dotted vertical lines in Figure 1). The slope of the gamma ray emissivity at 300 GeV and for a diffusion coefficient $D \propto E^{\delta}$ with $\delta = 0.3$ is plotted as a solid line in the bottom panel, where it is compared with the measured one, represented by the horizontal dotted lines (measured slope ± statistical error). The arrows represent the (projected) position of the four prominent peaks in the gas density distribution. If the total gas mass is concentrated at these locations and if projection effects in the gas distribution do not play an important role, the spectral slope of the emissivity should roughly represent the observed slope also.

The results in Figure 1 can be interpreted as follows. Within roughly one diffusion length, the gamma ray spectrum hardens with the distance from the source. This is because higher energy CRs occupy a larger volume of space around the source, and thus the low energy emission is progressively suppressed at larger and larger distances. On the other hand, at distances much larger than one diffusion length, the CR excess vanishes and the steep spectrum characteristic of the CR background is recovered. Though these calculations are far too crude, they might still serve to estimate the order of magnitude of the expected spectral variations within the emitting region. From Figure 1 one sees that variations of the order of $\Delta\alpha \approx 0.1 - 0.2$ are expected for a diffusion coefficient with slope $\delta = 0.3$. A stronger dependency on energy would produce more pronounced variations, as illustrated by the dashed line which refers to $\delta = 0.4$. Values of $\delta$ significantly larger than 0.3 would require a harder injection spectrum of CRs in order to preserve the observed slope in gamma rays. Such variations in the slope of the emission, if detected, would support the scenario of a localized injection of CRs close to the GC and provide better constraints on the diffusion coefficient. Complementary studies of the three dimensional distribution of the gas in the region are also needed in order to control projection effects.[28]

### Molecular clouds in the W28 region

W28 is a SNR in its radiative phase of evolution, located at a distance of $\sim 2$ kpc, in a region rich of dense molecular gas. Gamma ray emission has been detected from the surroundings of W28 both at TeV[29] and GeV energies,[30,31] by HESS, FERMI, and AGILE, respectively. The TeV emission correlates quite well with the position of three massive ($\approx 10^5 M_\odot$) MCs, one of which is interacting with the north-eastern part of the shell (and corresponds to the TeV source HESS J1801-233), and the other two being located to the south of the SNR (TeV sources HESS J1800-240 A and B) .

Let us assume that this gamma ray emission is the result of hadronic interactions of CRs that have been accelerated at the SNR and then escaped in the surrounding medium.[32] In this scenario, the distribution of runaway CRs around the SNR is still given by Equation 3. A diffusion coefficient $D = \chi D_{gal} \propto E^{0.5}$ is assumed, where $\chi$ represents possible deviations with respect to the average diffusion coefficient in the Galaxy. Equation 3 tells us that, up to a distance equal to the diffusion radius $R_d$, the spatial distribution of CRs around the source is roughly constant, and given by $N_{CR} \propto \eta E_{SN}/R_d^3$. On the other hand, the observed gamma ray flux from each one of the MCs is: $F_\gamma \propto N_{CR} M_{cl}/d^2$. By using the definitions of $N_{CR}$ and $R_d$ we can finally write the approximate equation, valid within a distance $R_d$ from the SNR:

$$F_\gamma \propto \frac{\eta \; E_{SN}}{(\chi \; D_{gal} \; t_{age})^{3/2}} \left( \frac{M_{cl}}{d^2} \right) . \tag{4}$$

Estimates can be obtained from observations for all the physical quantities in Equation 4 except for the CR acceleration efficiency $\eta$ and the diffusion coefficient $\chi D_{gal}$. By fitting the TeV data we can thus attempt to constrain a combination of these two parameters (namely $\eta/\chi^{3/2}$). Given all the uncertainties above, our results have to be interpreted as a proof of concept

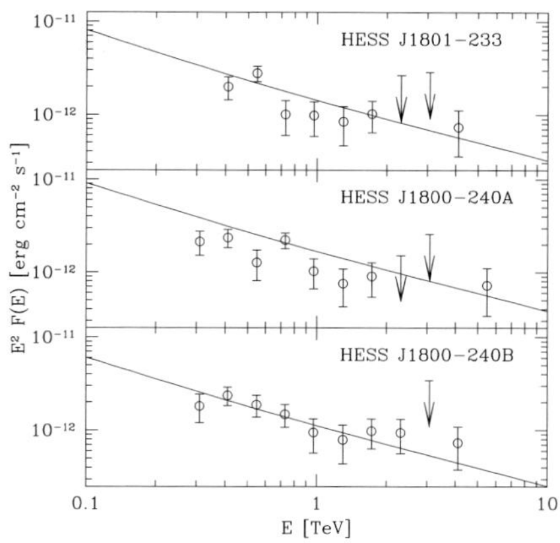


Fig. 2. Simultaneous fit to the three TeV sources detected by HESS in the W28 region. Figure from Ref. 32. See text for more details.

of the fact that gamma ray observations can serve as tools to estimate the diffusion coefficient. More detections of SNR/MC associations are needed in order to check whether the scenario described here applies to a whole class of objects and not only to a test-case as W28. Future observations from the Cherenkov Telescope Array will most likely solve this issue.

Figure 2 shows a simultaneous fit to the HESS data for the three massive MCs in the W28 region, obtained by setting $\eta/\chi^{3/2} \approx 20$. This implies that the normalized diffusion coefficient $\chi$ has to be much smaller than 1 for any reasonable value of $\eta < 1$. For example, an acceleration efficiency $\eta = 30\%$ corresponds to $\chi = 0.06$, which in turn gives a diffusion distance for TeV particles of $R_d \approx 60$ pc. This means that the results in Figure 2 are valid if the physical (not projected) distances between the MCs and the SNRs do not significantly exceed $R_d$. Small values of the diffusion coefficient have been also proposed in Ref. 31,33,34. Note that, since we are considering gamma rays in a quite narrow energy band around $\approx 1$ TeV, we are actually constraining the diffusion coefficient of CRs with energy $\approx 10$ TeV. With some caveats,[32,35] observations in the GeV range might be used to constrain the diffusion coefficient down to GeV particle energies.

## IC 443 and other examples

The SNR IC443 has an estimated age of $t_{age} \approx 3 \times 10^4$ yr, explosion energy of $E_{SN} \gtrsim 4 \times 10^{50}$erg, and distance $\approx$ 1.5 kpc.[36] It has been detected at both GeV[37] and TeV[38] energies. A power law fit to the TeV emission is:

$F_\gamma \sim 10^{-11}(E_\gamma/0.4\,\mathrm{TeV})^{-3.1}\mathrm{cm}^{-2}\mathrm{s}^{-1}\mathrm{TeV}^{-1}$. Observations of the CO line indicate that about $10^4 M_\odot$ of molecular gas are present in the region.[36] Though the GeV emission might be explained as the emission from shocked clouds,[39] this scenario cannot account for the very high energy CRs that originate the TeV photons. If these photons are produced by runaway CRs we can attempt to estimate the diffusion coefficient, as done for W28. If the TeV radiation is produced in a MC of mass $\sim 10^4 M_\odot$, then from Equation 2 we can infer an excess in the intensty of multi-TeV CRs of the order of $\sim 100$. If these particles had time to diffuse up to a distance $R_d$, their total (integrated over volume) energy is $\pi^{3/2} E_p^2 N_{CR}(0, E_p, t_{age}) R_d^3 \approx 3 \times 10^{47}(R_d/10\ \mathrm{pc})^3$ erg (see Equation 3). This number has to be multiplied by $\approx \ln(E_p^{max}/E_p^{min})$ to account for the extension of the spectrum of CRs injected by the SNR and then equated to $\eta\ E_{SN}$. This gives (very roughly) a diffusion coefficient for multi-TeV CRs of $D \approx 10^{27}(\eta/0.1)^{2/3}\ \mathrm{cm}^2/\mathrm{s}$. This is, again, much smaller than the average diffusion coefficient in the Galaxy,[36] and corresponds to $R_d \approx 20$ pc.

Other SNR/MC associations detected in TeV gamma rays that would deserve attention include CTB 37A, HESS J1745-303, and W51.[40] The approach described here might now be extended to the GeV range, where several SNR/MC associations are being detected by FERMI.[41] The Cygnus region, rich of dense gas and detected at multi-TeV energies, also promises to be an ideal region of the sky for this kind of studies.[42]

## Conclusions

MCs, when detected in gamma rays, can serve as probes of the CR intensity throughout the Galaxy. If they are located close to CR sources, the intensity of the gamma ray emission might impose constraints on the local diffusion coefficient. In at least one case (SNR W28), preliminary evidence has been reported for a significant suppression of the diffusion coefficient with respect to the average galactic one. This suppression might be the result of an enhancement in the magnetic turbulence due to the streaming of CRs away from the source. Observations with future facilities such as the Cherenkov Telescope Array will conclusively test the feasibility and reliability of the approach to constrain CR diffusion that has been presented here.
*Support from the EU is acknowledged [FP7 - grant agr. n° 256464].*

## References

1. A.M. Hillas, *J. Phys. G: Nucl. Phys.* **31**, 95 (2005)

2. A.R. Bell, *Mon. Not. R. Astron. Soc.* **182**, 147 (1978)
3. F. Aharonian, et al., *Rep. Prog. Phys.* **71**, 096901 (2008)
4. L.O'C. Drury, et al., *Astron. Astrophys.* **287**, 959 (1994)
5. D. C. Ellison, et al., *Astrophys. J.* **712**, 287 (2010)
6. V.S. Berezinskii, et al. 1990, *Astrophysics of Cosmic Rays* (Amsterdam: North-Holland)
7. R. Kulsrud, W.P. Pearce, *Astrophys. J.* **156**, 445 (1969)
8. A.J. Farmer, P. Goldreich, *Astrophys. J.* **604**, 671 (2004)
9. V.S. Ptuskin, et al., *Adv. Space Res.* **42**, 486 (2008)
10. T. Montmerle, *Astrophys. J.* **231**, 95 (1979)
11. M. Cassé, J.A. Paul, *Astrophys. J.* **237**, 236 (1980)
12. F.A. Aharonian, A. Atoyan, *Astron. Astrophys.* **309**, 917 (1996)
13. S. Gabici, F.A. Aharonian, *Astrophys. J.* **665**, L131 (2007)
14. S. Gabici, et al., *Mon. Not. R. Astron. Soc.* **396**, 1629 (2009)
15. For a discussion on CR exclusion from MCs see: J. Skilling, A.W. Strong, *Astron. Astrophys.* **53**, 253 (1976); S. Gabici, et al., *Astrophys. Space Sci.* **309**, 365 (2007); R.J. Protheroe, et al., *Mon. Not. R. Astron. Soc.* **390**, 683 (2008)
16. K. Nakamura et al. (Particle Data Group), *J. Phys. G* **37**, 075021 (2010)
17. T.M. Dame, et al., *Astrophys. J.* **547**, 792 (2001)
18. M.R. Issa, A.W. Wolfendale, *Nature* **292**, 430 (1981)
19. F.A. Aharonian, *Astrphys. Space Sci.* **180**, 305 (1991)
20. S. Casanova, et al., *Pub. Astron. Soc. Japan* **62**, 769 (2010)
21. A.W. Strong, et al., *Astron. Astrophys.* **207**, 1 (1988)
22. S.D. Hunter, et al., *Astrophys. J.* **481**, 205 (1997)
23. A.A. Abdo, et al., *Astrophys. J.* **710**, 133 (2010)
24. F.A. Aharonian, et al., *Nature* **439**, 695 (2006)
25. D.R. Ballantyne, et al., *Astrophys. J.* **657**, L13 (2007)
26. A.D. Erlykin, A. Wolfendale, *J. Phys. G: Nucl. Part. Phys.* **34**, 1813 (2007)
27. S. Dimitrakoudis, et al., *Astropart. Phys.* **31**, 13 (2009)
28. M. Tsuboi, et al., *Astrophys. J. Suppl.* **120**, 1 (1999); M.J. Reid, et al., *Astrophys. J.* **705**, 1548 (2009); G. Ponti, et al., *Astrophys. J.* **714**, 732 (2010)
29. F.A. Aharonian, et al., *Astron. Astrophys.* **481**, 401 (2008)
30. A.A. Abdo, et al., *Astrophys. J.* **718**, 348 (2010)
31. A. Giuliani, et al., *Astron. Astrophys.* **516**, L11 (2010)
32. S. Gabici, et al., to appear in the Proc. of SF2A-2010 – arXiv:1009.5291
33. Y. Fujita, et al., *Astrophys. J.* **707**, L179 (2009)
34. H. Li, Y. Chen, *Mon.Not. R. Astron. Soc.*, in press - arXiv:1009.0894
35. Y. Ohira, et al., *Mon. Not. R. Astron. Soc.*, in press - arXiv:1007.4869
36. D.F. Torres, et al., *Mon. Not. R. Astron. Soc.* **387**, L59 (2008)
37. A.A. Abdo, et al., *Astrophys. J.* **712**, 459 (2010)
38. J. Albert, et al., *Astrophys. J.* **664**, L87 (2007)
39. Y. Uchiyama, et al., arXiv:1008.1840
40. F.A. Aharonian, et al., *Astron. Astrophys.* **483**, 509 (2008); *ibid.*, **490**, 685 (2008); A. Fiasson, et al. 2009, Proc. XLIV Rencontres de Moriond, p. 81
41. D. Castro, P. Slane, *Astrophys. J.* **717**, 372 (2010)
42. A.A. Abdo, et al., *Astrophys. J.* **688**, 1078 (2008)

# PROTON MODULATION IN THE HELIOSPHERE FOR DIFFERENT SOLAR CONDITIONS AND PREDICTION FOR AMS-02

P. Bobik[1], G. Boella[2,3], M.J. Boschini[2,4], C. Consolandi[2], S. Della Torre[2,5], M. Gervasi[2,3], D. Grandi[2], K. Kudela[1], E. Memola[2], S. Pensotti[2,3], P.G. Rancoita[2] and M. Tacconi[2]

[1] *Institute of Experimental Physics, Kosice (Slovak Republic)*
[2] *Istituto Nazionale di Fisica Nucleare, INFN Milano-Bicocca, Milano (Italy)*
[3] *Department of Physics, University of Milano Bicocca, Milano (Italy)*
[4] *CILEA, Segrate (MI) (Italy)*
[5] *Department of Physics and Maths, University of Insubria, Como (Italy)*
*E-mail: Massimo.Gervasi@mib.infn.it*

Spectra of Galactic Cosmic Rays (GCRs) measured at the Earth are the combination of several processes: sources production and acceleration, propagation in the interstellar medium and propagation in the heliosphere. Inside the solar cavity the intensity of GCRs is reduced due to the solar modulation, the interaction which they have with the interplanetary medium. We realized a 2D stochastic simulation of solar modulation to reproduce CR spectra at the Earth, and evaluated the importance in our results of the Local Interstellar Spectrum (LIS) model and its agreement with data at high energy. We show a good agreement between our model and the data taken by AMS-01 and BESS experiments during periods with different solar activity conditions. Furthermore we made a prediction for the differential intensity which will be measured by AMS-02 experiment.

*Keywords*: Heliosphere, Cosmic Rays, Solar Magnetic Field

## 1. Introduction

Models of the Heliosphere need to be very accurate in order to reproduce the complex structure of the solar cavity and its effect on the Cosmic Rays propagation. In particular simulations of CR intensity have to be compared with experimental values, which have been measured in the last years by several space experiments (e.g. AMS-01, BESS) and will be measured even more accurately in the near future (e.g. AMS-02 on the ISS). We have developed a two dimensional (radius and helio-colatitude) model of GCR propagation in the Heliosphere[1,2] that is a function of measured values of the Solar Wind velocity in the ecliptic plane ($V_0$) and of the neutral sheet tilt angle ($\alpha$), as well as of stimated values of the diffusion parameter ($K_0$). This model is including curvature, gradient and current sheet drifts, which are depending on the charge sign of particles and magnetic field polarity[3]. Particles modulation strictly depends on the Local Interstellar Spectrum (LIS), up to now not measured and supposed to be constant with time outside of the heliosphere. In the model we include also the effects of the evolving solar activity conditions experienced by a CR particle inside the heliosphere, due to the time spent by the solar wind to reach the outer border of the solar cavity. Here we deal with the differential intensity observed at 1 AU: we do not take into account the effects of the Earth magnetosphere[4].

## 2. Stochastic 2D Monte Carlo

GCR transport in the Heliosphere is described by the Parker equation[5,6]:

$$\frac{\partial U}{\partial t} = \frac{\partial}{\partial x_i}\left(K^S_{ij}\frac{\partial U}{\partial x_j}\right) - \frac{\partial}{\partial x_i}(V_{sw_i}U) + \frac{1}{3}\frac{\partial V_{sw_i}}{\partial x_i}\frac{\partial}{\partial T}(\varrho T U) - \frac{\partial}{\partial x_i}(\langle v_{d_i}\rangle U) \quad (1)$$

where $U$ is the CR number density per unit interval of particle kinetic energy with time ($t$), $T$ is the kinetic energy (per nucleon), $V_{sw_i}$ the solar wind velocity along the axis $x_i$, $\langle v_{d_i}\rangle$ is the drift velocity[7], $K^S_{ij}$ is the symmetric part of the diffusion tensor and $\varrho=(T+2T_0)/(T+T_0)$, where $T_0$ is particle's rest energy.

As demonstrated by Ito[8], eq. 1 is equivalent to a set of ordinary stochastic differential equations that could be integrated with Monte Carlo techniques. The integration time step ($\Delta t$) is related to the accuracy of the integrating process and is taken to be proportional to $r^2$ ($r$ is the distance from the sun) to save CPU time[9]. We considered[10] the 2D (radius and colatitude) approximation of eq. 1. The drift velocity is split[7] in regular drift ($v_{Di}$) and neutral sheet drift ($v_{ns}$). The model is depending on some parameters related to the solar activity which are fine-tuned comparing simulations with experimental data[11].

The perpendicular diffusion coefficient has two components, radial $K_{\perp r}$, and polar $K_{\perp\theta}$. The parallel diffusion coefficient[12] is evaluated by the relation $K_{||}=K_0\beta K_P(P)(B_\oplus/3B)$. Moreover $(K_\perp)_0$ represents the ratio between perpendicular and parallel diffusion coefficients: $K_{\perp r}=(K_\perp)_0 K_{||}$. Here $K_0 = 1-6\times10^{22}$ cm$^2$s$^{-1}$GV$^{-1}$, $\beta$ is the particle velocity, $P$ is the CR particle's rigidity, the $K_P(P)$ term takes into account the dependence on rigidity (in GV), $B_\oplus \sim 5$ nT is the value of heliospheric magnetic field at the

Earth orbit, and $B$ is the magnitude of the Parker field[13]. $K_0$ is estimated following the procedure discussed in Section 2.2 of[14]. The perpendicular diffusion coefficient in the polar direction $K_{\perp\theta}$ has been enhanced with respect to $K_{\perp r}$ to reproduce the correct magnitude and rigidity dependence of the latitudinal cosmic ray proton and electron gradients[15].

We use as Heliospheric Magnetic Field (HMF) the Parker field model[16]. The Parker field has been modified[17] introducing a small latitudinal component. This modification increases the magnitude of the HMF in the polar regions without modifying the topology of the field. The main effect is a lower CR penetration along polar field lines in the inner part of the heliosphere, caused by a lower magnetic drift velocity in this region, as expected from measured data.

According to Ulysses data, for periods of low solar activity, we use a latitudinal dependence[18] of the solar wind velocity on the solar colatitude. Solar wind speed values range from $V_{min} \simeq 400$ km/s in the ecliptic plane, to $V_{max} \simeq 750$ km/s in the polar regions. Drift effects are included through analytical effective drift velocities: in the Parker spiral field we evaluated the three components of drift (gradient, curvature and neutral sheet[19]) that modify the integration path inside the heliosphere. We adopted the approach of Potgieter & Moraal[7], because, in comparison with other wavy neutral sheet models, it is able to reproduce the effects in both quite and active solar periods.

## 3. Parameters and experimental constraint

### 3.1. *Proton LIS*

One of the inputs of our code is the proton LIS. In order to estimate the reliability of LIS available in literature we have considered the proton spectra measured in several periods by several experiments for energy greater than 20 GeV. In this energy region we can assume negligible the effects of solar modulation. Therefore the spectrum, which can be represented by a power law ($N(E)_{E\geq 20GeV} = \Phi_0 E^\gamma$) with a spectral index $\gamma \simeq -2.7$, can be also considered the LIS. We focused our analysis on 4 data sets: CAPRICE-1994[20], BESS-1998[21], AMS-1998[22] and BESS-2002[23]. We estimated the spectral index ($\gamma$) and normalization constant ($\Phi_0$) for each experiment, then we evaluated an error-weighted average on these results. We obtained: $\gamma = -2.77 \pm 0.01$ and $\Phi_0 = 15.8 \pm 0.6$. In figure 1 we compare our results with the model by Burger & Potgieter[13], where it seems to overestimate a bit the experimental data. We repeated the analysis fitting all together the data sets obtaining: $\gamma = -2.77 \pm 0.04$ and $\Phi_0 = 17 \pm 3$. Therefore in the following calculation we adopt the model by Burger & Potgieter, corrected by a scale factor in order to fit experimental data.

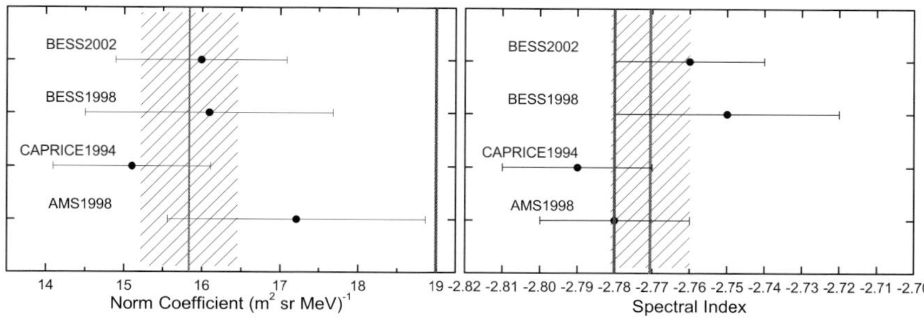


Fig. 1. Fit (red line) of high energy proton spectrum measured by several experiments (black dots) and comparison with Burger & Potgieter[13] LIS (blue line). The spectral index is consistent with data, but the normalization constant is too large.

### 3.2. *Latitudinal Gradient*

Observations made by the Ulysses spacecraft in the inner heliosphere have shown that the latitudinal dependence of CR protons is significantly less than predicted by classical drift models[25]. Our model, due to the modification of the $K_{\perp\theta}$[14] parameter in the polar regions can reproduce the gradient observed by Ulysses (see Fig. 3 in Heber et al.[25]) between the poles and the ecliptic plane. We find, for the period of AMS-01 data taking (June 1998), a difference of $\sim$ 16 % between the ecliptic plane and a colatitude of $\sim 30°$ from the poles. In Figure 2 a simulation of the proton differential intensity on the ecliptic plane is compared with the one at 30 deg from the poles (left panel). The latitudinal gradient of the integral intensity is shown in the right panel of Figure 2.

## 4. Results

### 4.1. *Proton differential intensity*

We performed the simulations using dynamic values of $K_0$, $\alpha$ and $V_{sw}$: we consider present time parameters in the inner shell of the heliosphere, but we use the values assumed by the parameters up to 14 months back in the past at the heliopause. Results are shown in Figure 3. Simulated differential intensity with dynamic values shows a very good agreement with measured

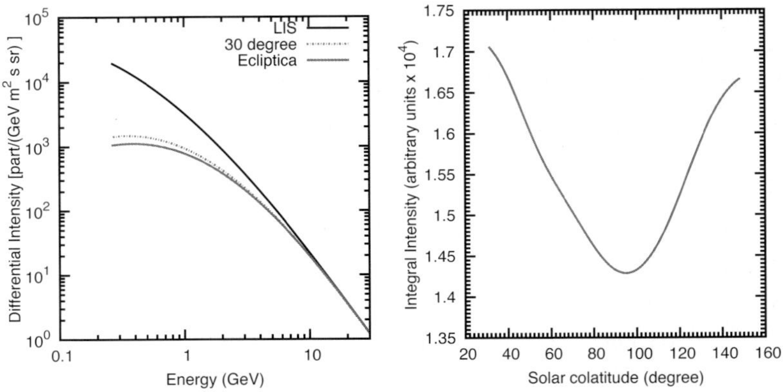


Fig. 2. Left panel: simulation of the proton differential intensity on the ecliptic plane is compared with the same quantity at 30 deg from the poles. Right panel: the latitudinal gradient, i.e. the integral intensity for $E \geq 200$ MeV vs the solar colatitude. Simulations have been done using solar parameters occurred in June 1998.

data, within the quoted error bars. This happens in periods with low solar activity and A>0, in comparison with AMS-01[22], BESS-97[21] and BESS-99[21]; in periods with high solar activity, in comparison with BESS-00[21]; and in periods with a lower solar activity and A<0, in comparison with BESS-02[23] and BESS-04[24]. This means that our description of the Heliosphere improves the understanding of the complex processes occurring inside the solar cavity.

### 4.2. *AMS-02 Predictions*

Our simulation code has been used to predict CR differential intensity for future measurements. The assumption is that diffusion parameter, tilt angle and solar wind speed show a near-regular and almost periodic trend. The periodicity occurs after two consecutive 11-years solar cycles. Using SIDAC (Solar Influences Data Analysis Center) data we selected periods with a nearly similar solar activity conditions and same solar field polarity of the simulation time: therefore approximately 22 years in advance. We concentrate our simulations on the AMS-02 mission, that will be installed on the ISS in February 2011, for a period approaching the solar maximum: January 2012. Results are shown in Fig. 4.

The Sun is currently in an unpredicted long duration solar minimum, that forced scientists to review all their estimations for the next solar cycle, therefore this prediction could be object of revision in the next future. AMS-02 is expected to collect, in a few years of operation, more than $10^{10}$

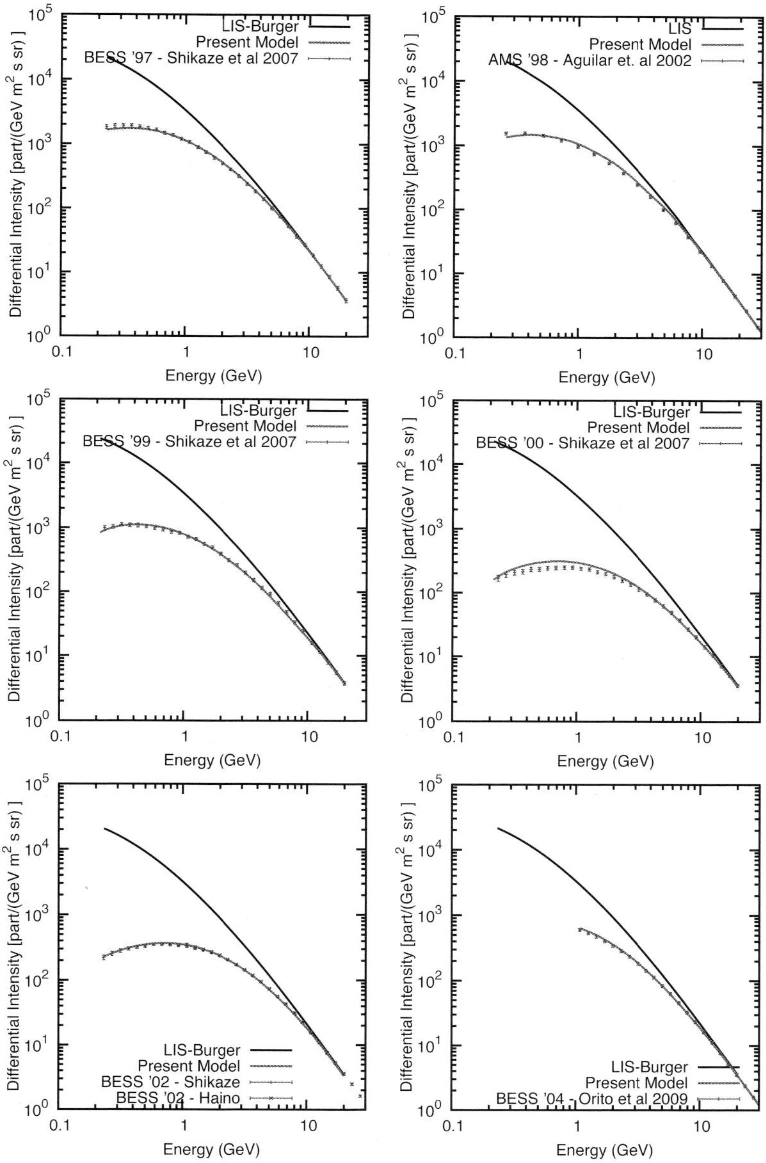


Fig. 3. Comparison of simulated galactic proton differential intensity at 1 AU and experimental data: AMS-01[22] (1998) and BESS[21,24] (1997, 1999, 2000, 2002, 2004).

protons with energy $\geq 1$ GeV, and $\sim 10^6$ with energy $\geq 1$ TeV.

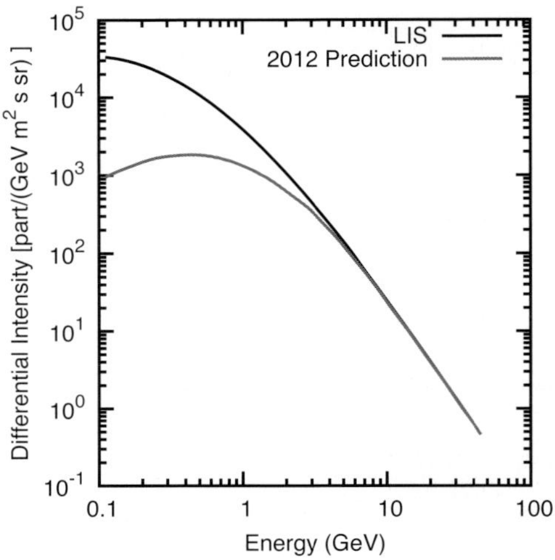


Fig. 4. Prediction of modulated proton differential intensity as will be measured by AMS-02 experiment in January 2012.

## 5. Conclusions

We built a 2D stochastic Monte Carlo code for particles propagation inside the heliosphere. Our model takes into account drift effects and shows a good agreement with measured values, in periods with positive as well as with negative polarity. Proton spectra, as predicted by the model, are decreasing with increasing tilt angle and solar wind velocity. We use as input LIS the model published by Burger & Potgieter[13], corrected in order to fit the high energy measured intensity. Recent measurements have pointed out the needs to reach a high level of accuracy in the modulation of the differential intensity, in relation to the charge sign of the particles and the solar field polarity[26]. This aspect will be even more crucial in the next generation of experiments.

## References


1. P. Bobik, M. Gervasi, D. Grandi, P.G. Rancoita, and I.G. Usoskin, *Proceedings of ICSC*, 2003, ESA SP-533 (2003).
2. P. Bobik, et al., Proc. of the 11th ICATPP, Como 5–9/10/2009; *Astroparticle, Particle, Space Physics, Detectors and Medical Physics Applications*, **5**, 210 (2010); *World Scientific*, Singapore, ISBN-13 978-981-4307-51-2.
3. G. Boella, M. Gervasi, S. Mariani, P.G. Rancoita, and I.G. Usoskin, *J. Geophys. Res.*, **106**, A12, 29355 (2001).
4. P. Bobik, G. Boella, M.J. Boschini, M. Gervasi, D. Grandi, K. Kudela, S.

Pensotti, and P.G. Rancoita *J. Geophys. Res.*, **111**, A05205 (2006).
5. M.S. Potgieter, et al., *Astroph. J.*, **403**, 760 (1993).
6. L.A. Fisk, *J. Geophys. Res.*, **81**, 4646 (1976).
7. P. Bobik, et al., Proc. of the 11th ICATPP, Como 5–9/10/2009; *Astroparticle, Particle, Space Physics, Detectors and Medical Physics Applications*, **5**, 760 (2010); *World Scientific*, Singapore, ISBN-13 978-981-4307-51-2.
8. C.W. Gardiner, *Handbook of Stochastic Methods*, Springer Verlag, Berlin (1989).
9. K. Alanko-Huotari, I.G. Usoskin, K. Mursula, and G.A. Kovaltsov, *J. Geophys. Res.*, **112**, A8, CiteID A08101, (2007).
10. P. Bobik, et al., "2D Stochastic Montecarlo to evaluate the modulation of GCR for positive and negative periods", *Proc. 21th ECRS 2008 (Košice, Slovakia)* (2009).
11. P. Bobik, M. Boschini, M. Gervasi, D. Grandi, K. Kudela and P.G. Rancoita "Heliosphere modulation of Primary Comsic Rays for the AMS-02 mission", *Proceedings of the 31$^{st}$ ICRC*, Lodz, Polonia, 2009.
12. M.S. Potgieter, and J.A. Le Roux, *Astroph. J.*, **423**, 817 (1994).
13. R.A. Burger, and M.S. Potgieter, *Astroph. J.*, **339**, 501 (1989).
14. P. Bobik, et al., 2011, Proton and antiproton modulation in the heliosphere for different solar conditions and AMS-02 measurements prediction, *Proc. of the ICATPP Conference on Cosmic Rays for Particle and Astroparticle Physics*, October 7–8 2010, Villa Olmo, Como, Italy, Giani, S., Leroy, C. and Rancoita, P.G., Editors, World Scientific, Singapore, ISBN 978-981-4329-02-6; arXiv:1012.3086v3 [astro-ph.EP], available at the web site: *http://arxiv.org/abs/1012.3086*
15. M.S. Potgieter, et al., *Adv. Space Res.*, **19**, No. 6, 917 (1997).
16. R.A. Burger, and M. Hitge, *American Geophysical Union* SH71A-04, Fall Meeting 2002.
17. J.R. Jokipii, and J. Kóta, *Geophys. Res. Lett.*, **16**, 1 (1989).
18. J.P.L. Reinecke, C.D. Steenberg, H. Moraal, F.B. McDonald, *Adv. Space Res.*, **19**, No. 6, 901 (1997).
19. M. Hatting, and R.A. Burger, *Adv. Space Res.* **16**, No. 9, 213 (1995).
20. G.M. Boezio, et al., *Astrophys. J.*, **518**, 457 (1999).
21. Y. Shikaze, et al., *Astropart. Phys.*, **28**, 154 (2007).
22. J. Alcaraz, et al., *Physics Lett. B*, **490**, 27 (2000).
23. S. Haino, et al., *Physics Lett. B*, **594**, 35 (2004).
24. K.C. Kim, et al., *Proc. 30th ICRC 2007 (Mexico City, Mexico)*, **2**, 71 (2009).
25. B. Heber, et al., *J. Geophys. Res.*, **103**, A3, 4809 (1998).
26. G. Boella, M. Gervasi, M.C.C. Potenza, P.G. Rancoita and I.G. Usoskin, *Astropart. Phys.*, **9**, 261 (1998).

# PROTON AND ANTIPROTON MODULATION IN THE HELIOSPHERE FOR DIFFERENT SOLAR CONDITIONS AND AMS-02 MEASUREMENTS PREDICTION.

P. Bobik[1], M.J. Boschini[2,4], C. Consolandi[2], S. Della Torre[2,5], M. Gervasi[2,3], D. Grandi[2], K. Kudela[1], S. Pensotti[2,3] and P.G. Rancoita[2]

[1] *Institute of Experimental Physics, Kosice (Slovak Republic)*
[2] *Istituto Nazionale di Fisica Nucleare, INFN Milano-Bicocca, Milano (Italy)*
[3] *Department of Physics, University of Milano Bicocca, Milano (Italy)*
[4] *CILEA, Segrate (MI) (Italy)*
[5] *Department of Physics and Maths, University of Insubria, Como (Italy)*
*E-mail: Davide.Grandi@mib.infn.it*

Galactic Cosmic Rays (GCRs) are mainly protons confined in the galactic magnetic field to form an isotropic flux inside the galaxy. Before reaching the Earth orbit they enter the Heliosphere and undergo diffusion, convection, magnetic drift and adiabatic energy loss. The result is a reduction of particles flux at low energy (below 10 GeV), called solar modulation. We realized a quasi time-dependent 2D Stochastic Simulation of Solar Modulation that is able to reproduce CR spectra once known the Local Interstellar Spectrum (LIS). We were able to estimate the different behaviors associated to the polarity dependence of the Heliospheric modulation for particles as well as for antiparticles. We show a good agreement with the antiproton/proton ratio measured by AMS-01, Pamela, BESS, Heat and Caprice and we performed a prediction for the AMS-02 Experiment.

*Keywords*: Heliosphere, Cosmic Rays, Solar Magnetic Field

## 1. Introduction

The effect of the heliospheric structure on GCRs propagation can be reproduced by a two dimensional (radius and helio-colatitude) Stochastic model solving numerically the Parkers's equation[1]. If we do not take into account the effects of the Earth magnetosphere[2] modulated fluxes depends not only on the level of solar activity but also on particles charge sign and solar magnetic field polarity[3]. The study of the modulation of $\bar{p}/p$ ratios is particularly important, because it includes explicitly the combination of charge sign and polarity dependence. The Local Interstellar Spectra (LIS) used as input of the code, both for protons and antiprotons, are taken by the Galprop model[4].

## 2. Stochastic Model, Parameters and Data Sets

### 2.1. *Main parameters of the Model*

The present code simulates the interactions of a GCR entering the heliosphere which extends up to a fixed distance of about 100 AU from the Sun.

One of the main parameters of the model is the diffusion tensor. The parallel diffusion coefficient is: $K_{||}=k_1\beta K_P(P)(B_\oplus/3B)$, where $P$ is the particle rigidity (usually expressed in GV), $k_1$ is the diffusion parameter discussed in the next section and $K_P \approx P$. The perpendicular diffusion coefficient has two components, radial $K_{\perp r}$, and polar $K_{\perp\theta}$. We used the relation: $K_{\perp r} = (K_\perp)_0 K_{||}$, where $(K_\perp)_0 = 0.05$. We also considered $K_{\perp\theta} = K_{\perp r}$ in the equatorial region, while we enhanced its value in the polar regions of the heliosphere[5]: $K_{\perp\theta} = 10K_{\perp r}$.

We used the tilt angle $\alpha$ of the heliospheric current sheet (HCS) as a parameter for the level of the solar activity[6]: the higher the value of $\alpha$ the lower the expected GCR flux, for both solar field polarities. Values of the tilt angle are computed using two different models: the usual model uses a line-of-sight boundary condition at the photosphere and includes a significant polar field correction; an alternative model uses a radial boundary condition at the photosphere, and requires no polar field correction. As suggested by Ferreira and Potgieter[7], the classical model is used for periods of increasing solar activity (for example 2007–2012, AMS-01 data, AMS-02 data), while the new model fits better for periods of decreasing solar activity (for example 2000–2007, BESS data).

The three drift components do not depend on external parameters, except the solar polarity[8] (A>0 for positive periods and A<0 for negative periods[6]). The general drift expression is locally unlimited for a quasi-isotropic distribution[9,10], therefore we limit all the drift components below $(\pi/4)v$, which is the spatially averaged maximum value.

### 2.2. *Data Sets*

We selected GCR proton and antiproton data from 5 different experiments in order to compare and tune model results: AMS-01[11], Caprice[12], BESS[13–15], HEAT and Pamela[16]. The first two experiments took data in a period of positive solar polarity, BESS in both solar polarities, and the last two in A<0 period. The corresponding periods of measurements are: June 1998 (AMS); May 1998 (Caprice); from July 1997 to December 2004 (BESS); June 2000 (HEAT); and from 2007 to 2008 (Pamela). Solar wind values for these periods have been obtained from omniweb[17] by 27 daily averages, while tilt angle values from the Wilcox Solar Laboratory[18]. We can estimate the diffusion coefficient from a long term study[19,20] of neutron monitor measurements and the *Force Field model*[21] (FFM) approach to the treatment of Heliosphere transport of GCRs, using $\sim$ 40 years of data ending in 2004.

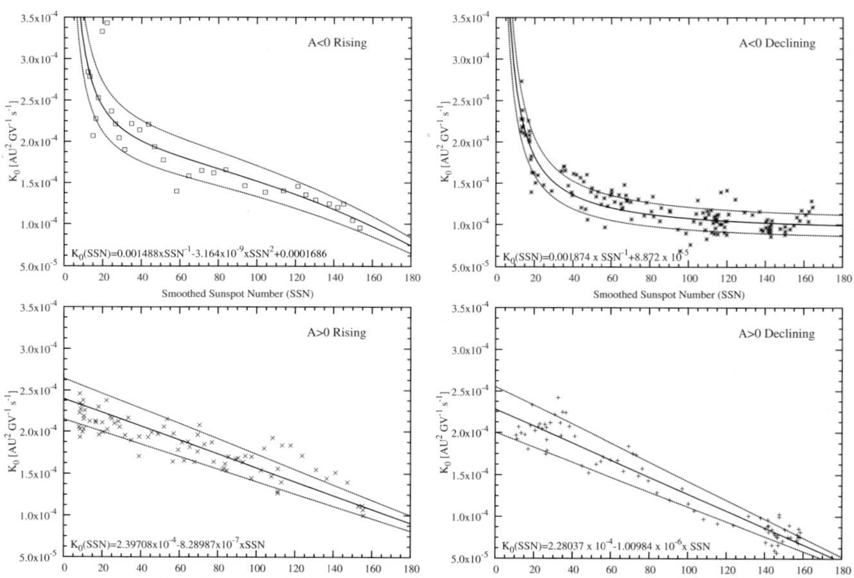


Fig. 1. Diffusion parameter $K_0$ - rising and declining phases for both negative and positive solar magnetic-field polarities - as a function of the SSN value; the continuous lines are obtained from a fit of $K_0$ with respect to SSN values.

In FFM[21] [e.g., see also Section 4.1.2.3 of Leroy and Rancoita (2009)], Gleeson and Axford (1968) assumed that i) modulation effects can be expressed with a spherically symmetric modulated number density $U$ of GCRs - the so-called differential density - with kinetic energy between $T$ and $T + dT$, ii) the diffusion coefficient at the time $t$ is given by a separable function of $r$ (the radial distance from the Sun) and $P$ (the particle rigidity in GV):

$$\mathcal{K}(r,t) = \beta k_1(r,t) K_P(P,t),$$

with $\beta = v/c$ , $v$ the particle velocity, $c$ the speed of light, $K_P(P,t) \approx P$ for particle rigidities above $\approx 1\,\mathrm{GV}$ and iii) the modulation occurs in a steady-state condition, i.e., the relaxation time of the distribution is short with respect to the solar cycle duration so that

$$\frac{\partial U}{\partial t} = 0.$$

They derived that the differential intensity at a radial distance $r$ is given by the expression

$$J(r, E_{\mathrm{t}}, t) = J(r_{\mathrm{tm}}, E_{\mathrm{t}} + \Phi_{\mathrm{p}}) \left[ \frac{E_{\mathrm{t}}^2 - m_{\mathrm{r}}^2 c^4}{(E_{\mathrm{t}} + \Phi_{\mathrm{p}})^2 - m_{\mathrm{r}}^2 c^4} \right], \tag{1}$$

where $J(r_{\mathrm{tm}}, E_{\mathrm{t}} + \Phi_{\mathrm{p}})$ is the undisturbed intensity beyond the solar wind termination located at a radial distance $r_{\mathrm{tm}}$ from the Sun; $E_{\mathrm{t}}$ is the total energy of the particle with rest mass $m_{\mathrm{r}}$ and, finally, $\Phi_{\mathrm{p}}$ is the so-called force-field energy loss[21,23]. When modulation is small[21,23] - i.e., $\Phi_{\mathrm{p}} \ll m_{\mathrm{r}}c^2, T$ -, they determined that

$$\Phi_{\mathrm{p}} = \frac{ZeP}{K_P(P,t)}\,\phi_{\mathrm{s}}(r,t) \approx Ze\,\phi_{\mathrm{s}}(r,t),$$

where $Ze$ is the particle charge and $\phi_{\mathrm{s}}(r,t)$ is the so-called modulation strength (or modulation parameter). Assuming that $v_{\mathrm{w}}$ (the solar wind speed) and $k_1$ are almost constant, $\phi_{\mathrm{s}}(r,t)$ - usually expressed in units of GV (or MV) - reduces to

$$\phi_{\mathrm{s}}(r,t) \approx \frac{v_{\mathrm{w}}(t)\,(r_{\mathrm{tm}} - r)}{3k_1(t)}, \tag{2}$$

from which one gets

$$k_1(t) \approx \frac{v_{\mathrm{w}}(t)\,(r_{\mathrm{tm}} - r)}{3\phi_{\mathrm{s}}(r,t)}, \tag{3}$$

i.e., $k_1$ is linearly dependent on $(r_{\mathrm{tm}} - r)$. In the FFM, the diffusion coefficient $\mathcal{K}(r,t)$ is scalar quantity and, as a consequence, does not account for effects related to the charge sign of the transported particles. $\phi_{\mathrm{s}}(r,t)$ is independent of the species of GCR particles [e.g., see discussion at page 1014 of Gleeson and Axford (1968) or Equation (1) of Usoskin and collaborators (2005)]. The values of the modulation strengths $[\phi_{\mathrm{s}}(r_{\mathrm{Earth}})]$ were monthly determined for the time period[20] from 1951 up to 2004 using measurements of neutron monitors (i.e., located at $r_{\mathrm{Earth}} = 1\,\mathrm{AU}$); while those for the solar wind speeds are available on the web[24]. It has to be remarked that $k_1$ depends on the value of the solar wind termination located at a radial distance $r_{\rm tm}$ related, in turn, also to the solar wind speed [e.g., see Chapter 7 of Meyer-Vernet(2007) and Section 4.1.2.2 of Leroy and Rancoita (2009)]. However, because the present simulation code assumes a fixed solar wind termination at 100 AU to calculate the modulated differential intensities at $r_{\rm Earth}$, one has to derive from the diffusion parameter $k_1$ that ($K_0$) for an effective heliosphere with a radial extension of 100 AU (see Sect. 2.1). Thus, using Eq. (3) one can obtain

$$K_0 \approx k_1 \frac{99\,\text{AU}}{(r_{\rm tm} - r_{\rm Earth})} = 99\,\text{AU} \left[ \frac{v_{\rm w}}{3\,\phi_{\rm s}(r_{\rm Earth})} \right], \tag{4}$$

where 99 AU is the distance of the Earth from the border of the effective heliosphere as defined in the current simulation code. In Fig. 1, the diffusion parameter $K_0$ - obtained from Eq. (4) - is shown as a function of the corresponding Smooted Sunspot Number (SSN) value.

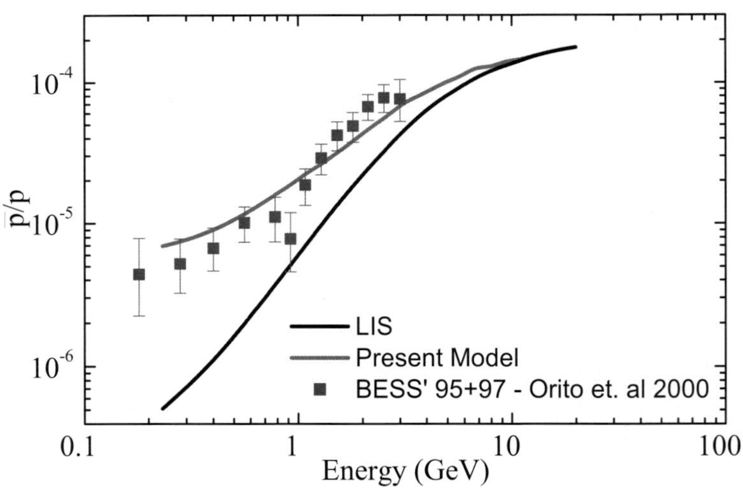


Fig. 2. Simulated $\bar{p}/p$ ratio as a function of the kinetic energy in GeV at 1 AU in comparison with experimental data: BESS (1995-1997).

The $K_0$ data had to be subdivided in four sets, i.e., rising and declining phases for both negative and positive solar magnetic-field polarities. For each set, the data i) could be fitted with a relationship - indicated in in Fi. 1 - between $K_0$ and Smoothed Sunspot Numbers SSN[26] values and ii)

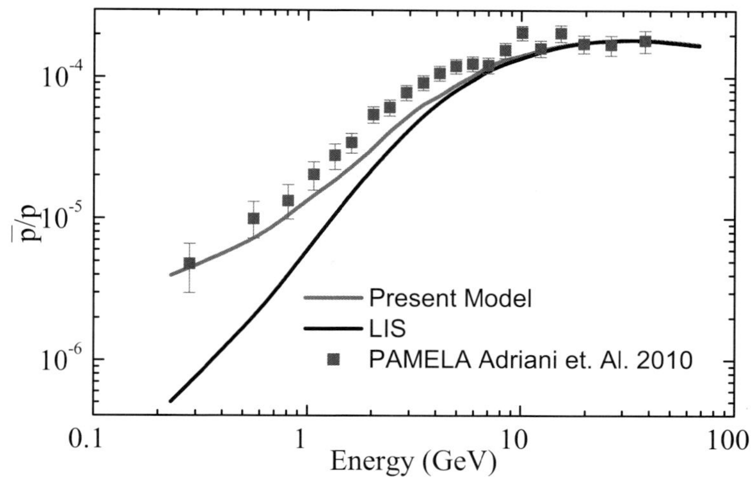


Fig. 3. Simulated $\bar{p}/p$ ratio as a function of the kinetic energy in GeV at 1 AU in comparison with experimental data: Pamela (2007-2008).

exhibited a Gaussian distribution of difference of $K_0$ values from the corresponding fitted values. The RMSs of the Gaussian distributions were found to be $\approx 0.1339, 0.1254, 0.1040, 0.1213$ for the phases rising with $A < 0$, declining with $A < 0$, rising with $A > 0$, declining with $A > 0$, respectively. In this way - i.e., from the relationships found -, we can use the estimated SSN values to obtain the diffusion parameter $K_0$ at times beyond 2004. In practice, this procedure allows one to extend the $\approx 40$ years period by exploiting the linear relationship between the fitted $K_0$ values and the SSN values (one of the main parameters related to the solar activity). Thus, we introduced in our code a Gaussian random variation of $K_0$ with RMSs corresponding to those found for each subset of data. Results of the simulation with and without the Gaussian variation are consistent inside the uncertainties of the code[1].

## 2.3. *Dynamic Parameters*

Our code simulates the interactions of a GCR entering the heliosphere from its outer limit, the helio-pause, located - as already mentioned - approximately at 100 AU[27], and moving inwards to the Earth located at 1 AU. We evaluated the time $t_{sw}$ needed to SW to expand from the outer corona up to the helio-pause. Considering an average speed of 400 km/s it takes nearly 14 months. While the time interval $\tau_{ev}$ of the stochastic evolution of a quasi particle inside the heliosphere from 100 AU down to 1 AU is $\sim$ 1 month at 200 MeV and few days at 10 GeV. This scenario, where $\tau_{ev} < t_{sw}$ and $t_{sw} >> 1$ month, indicates that we can not use fixed parameters (monthly averages) to describe the conditions of heliosphere in the modulation process. In fact at 100 AU, where particles are injected, the conditions of the solar activity are similar to the conditions present at the Earth roughly 14 months before. Therefore we consider $\tau_{ev}$ negligible with respect to $t_{sw}$ and divide the heliosphere in 14 regions, as a function of the radius. For each region we evaluated[28] $K_0$, $\alpha$ and $V_{sw}$, in relation to the expansion velocity, in a dynamic way. In the future the time spent by a GCR particle inside the heliosphere, as a function of the stochastic path and of the particle energy, will be also taken into account.

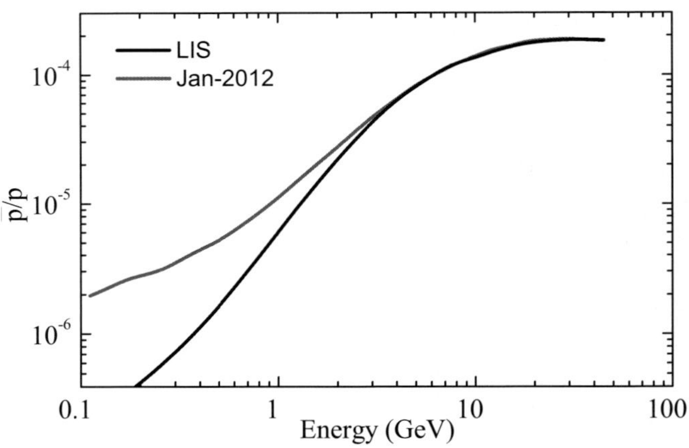


Fig. 4. Prediction of modulated $\bar{p}/p$ ratio as a function of the kinetic energy in GeV at 1 AU for AMS-02 experiment (January 2011).

### 2.4. *Antiproton/Proton: Comparison with Data and Prediction for AMS-02*

We performed the simulations using dynamic values of $K_0$, $\alpha$ and $V_{sw}$. Results are shown in Fig. 2 and 3. Simulated fluxes with dynamic values show a very good agreement with measured data, within the quoted error bars. This happens both in periods with A>0, in comparison with BESS, and in periods with A<0, in comparison with Pamela. This means that our dynamic description of the Heliosphere improves the understanding of the complex processes occurring inside the Solar Cavity.

The periodic behavior of the heliosphere allows us to predict, with a certain level of precision, the parameters needed for a simulation related to a time in the near future. In order to get these data we considered the prediction of SSN from IPS (Ionospheric Prediction Service) of the Australian Bureau of Meteorology[29].

We concentrate our simulations on the AMS-02 mission[30,31] that will be installed on the ISS in February 2011, and, in particular, at a time approaching the solar maximum: January 2012. We show in Fig. 4 the predictions of GCR modulation for the antiproton/proton ratio.

## 3. Conclusions

We built a 2D stochastic Monte Carlo code for particles propagation across the heliosphere. Present model takes into account drift effects and shows quantitatively a good agreement with measured values, both for positive and negative periods and for different particles and charge sign. This is relevant because particles with opposite charge sign undergo a different solar modulation[3]. We compared our simulations with antiproton/proton ratios measured by BESS and PAMELA. We used *dynamic* parameters values ($K_0$, $\alpha$ and $V_{sw}$) for the related periods, in order to reproduce the propagation of incoming GCR through magnetic disturbances carried by the outgoing solar wind. The *dynamic* description of the heliosphere and the forward approach seem to reproduce better the real physical propagation of GCR in the solar cavity. In order to have a more sophysticated model we need to introduce a dependence on the particle time spent in the heliosphere and a larger statistics of measured data during negative solar field periods, as AMS-02 will provide in the next years. Recent measurements[16] have pointed out the needs to reach a high level of accuracy in the modulation of the fluxes, in relation to the charge sign of the particles and the solar field polarity[32]. This aspect will be even more crucial in the next generation of experiments[30,31].

## References

1. P. Bobik, G. Boella, M.J. Boschini, M. Gervasi, D. Grandi, K. Kudela, S. Pensotti, and P.G. Rancoita, "Proton Modulation In The Heliosphere For Different Solar Conditions And Prediction For AMS-02", Proc. of the ICATPP Conference on Cosmic Rays for Particle and Astroparticle Physics, October 7–8 2010, Villa Olmo, Como, Italy, Giani, S., Leroy, C. and Rancoita, P.G., Editors, World Scientific, Singapore, ISBN 978-981-4329-02-6; arXiv:1012.0938 [astro-ph.EP], available at the web site: *http://arxiv.org/abs/1012.0938.*

2. P. Bobik, G. Boella, M.J. Boschini, M. Gervasi, D. Grandi, K. Kudela, S. Pensotti, and P.G. Rancoita, *J. Geophys. Res.*, **111**, A05205 (2006).
3. G. Boella, M. Gervasi, S. Mariani, P.G. Rancoita, and I.G. Usoskin, *J. Geophys. Res.*, **106**, A12, 29355 (2001).
4. http://galprop.stanford.edu/webrun/
5. M.S. Potgieter, *J. Geophys. Res.*, **105**, A8, 18295 (2000).
6. P. Bobik, M. Gervasi, D. Grandi, P.G. Rancoita, and I.G. Usoskin, *Proceedings of ICSC*, 2003, ESA SP-533 (2003).
7. S.E.S. Ferreira, and M.S. Potgieter, *Astrophys. J.*, **603**, 744 (2004).
8. P. Bobik, G. Boella, M.J. Boschini, S. Della Torre, M. Gervasi, D. Grandi, K. Kudela, S. Pensotti, and P.G. Rancoita, "Drift models and polar field for cosmic rays propagatin in the heliosphere", Proc. of the 11th ICATPP, Como 5-9/10/2009, World Scientific, Singapore , 760 (2009).
9. R.A. Burger, et al., *Proc. 20th ICRC* (Moscow), **3**, 283 (1987).
10. R.A. Burger, and M.S. Potgieter, *Astroph. J.*, **339**, 501 (1989).
11. M. Aguilar, et al., *Phys. Rep.*, **366**, 331 (2002).
12. M. Boezio, et al., *Astrop. J.*, **487**, 415 (1997).
13. T. Maeno, et al., *Astropart. Phys.*, **16**, 121 (2001).
14. Y. Asaoka, et al., *Phys. Rev. Lett.*, **88**, 051101-1 (2002).
15. K. Abe, et al., *Phys. Lett. B*, **670**, 103 (2008).
16. O. Adriani, et al., *Phys. Rev. Lett.*, 105, 121101 (2010).
17. http://omniweb.gsfc.nasa.gov/form/dx1.html
18. http://quake.stanford.edu/∼wso/wso.html
19. I.V. Moskalenko, et al., *Astrophys. J.*, **565**, 280 (2002).
20. I.G. Usoskin, et al., *J. Geophys. Res.*, **110**, A12108 (2005).
21. L.J. Gleeson, and W.J. Axford, *Astrophys. J.*, **154**, 1011–1026 (1968).
22. C. Leroy and P.G. Rancoita, Priciples of Radiation Interaction in Matter and Detection - 2nd Edition, *World Scientific*, Singapore (2009), ISBN-978-981-281-827-0.
23. L.J. Gleeson and I.H. Urch, *Astrophys. Space Scie.*, **11**, 288 (1971); **25**, 387 (1973).
24. http://omniweb.gsfc.nasa.gov/form/dx1.html
25. N. Meyer-Vernet, Basics of the Solar Wind, Cambridge University Press, Cambridge (2007).
26. http://www.sidc.oma.be/sunspot-data/
27. R.B. Decker, et al., *AGU - Fall Meeting 2007* (2007).
28. P. Bobik, et al., "2D Stochastic Montecarlo to evaluate the modulation of GCR for positive and negative periods", *Proc. 21th ECRS 2008 (Košice, Slovakia)* (2009).
29. http://www.ips.gov.au/Solar/1/6
30. P. Zuccon and the AMS-02 Collaboration, *The AMS-02 experiment on the ISS: status and perspectives*, *Proc. 31$^{st}$ ICRC*, Lodz (2009).
31. J. Casaus, *J. Phys. Conf. Ser.*, **171**, 012045 (2009).
32. G. Boella, M. Gervasi, M.C.C. Potenza, P.G. Rancoita and I.G. Usoskin, *Astropart. Phys.*, **9**, 261 (1998).

# A consistent interpretation of recent CR nuclei and electron spectra

Giuseppe Di Bernardo
*Department of Physics, University of Gothenburg, SE-412 96, Gothenburg, Sweden*

Carmelo Evoli
*SISSA, via Bonomea 265, 34136 Trieste, Italy*

Daniele Gaggero, Dario Grasso*
*INFN, Sezione di Pisa, Largo Bruno Pontecorvo 3, 56127 Pisa, Italy*

Luca Maccione
*Deutsches Elektronen-Synchrotron, Notkestraße 85, 22607 Hamburg, Germany*

Mario Nicola Mazziotta
*Istituto Nazionale di Fisica Nucleare, Sezione di Bari, 70126 Bari, Italy*

We try to interpret the recently updated measurement of the cosmic ray electron (CRE) spectrum observed by Fermi-LAT, together with PAMELA data on positron fraction, in a single-component scenario adopting different propagation setups; we find that the model is not adequate to reproduce the two datasets, so the evidence of an extra primary component of electrons and positrons is strengthened. Instead, a double component scenario computed in a Kraichnan-like diffusion setup (which is suggested by B/C and $\bar{p}$ data) gives a satisfactory fit of all exisiting measurements. We confirm that nearby pulsars are good source candidates for the required $e^{\pm}$ extra-component and we show that the predicted CRE anisotropy in our scenario is compatible with Fermi-LAT recently published constraints.

## 1. Introduction

Last year the Fermi-LAT Collaboration published the $e^+ + e^-$ spectrum in the energy range between 20 GeV and 1 TeV, measured during the first

*dario.grasso@pi.infn.it

six months of the Fermi mission.[1] That result came in the middle of an interesting debate which arose after that ATIC and PAMELA collaborations reported some anomalous results: ATIC observed a bump in the $e^+ + e^-$ spectrum at around 600 GeV, while PAMELA found the positron fraction $e^+/(e^+ + e^-)$ to increase with energy above 10 GeV: these features are hardly compatible with the standard scenario in which CR electrons are accelerated in supernova remnants (SNRs) and positrons are predominantly of secondary origin, and were interpreted by many authors as a possible signature of decay or annihilation of Dark Matter (DM) particles, even if a more conventional interpretation in terms of astrophysical sources (namely pulsars) was also considered. Fermi-LAT spectrum does not display the feature seen by ATIC, being compatible with a single power law; the absence of that feature was later confirmed by the H.E.S.S. Cherenkov telescope whose spectrum below 1 TeV is in agreement with Fermi-LAT's. Soon after that measurement was published, some of us, with other members of Fermi collaboration, showed that an interpretation of the $e^+ + e^-$ spectrum is possible within a conventional model in which SNRs are the only primary sources of CREs.[2] Below 20 GeV, however, we found the predictions of that model to be in tension with pre-Fermi data; furthermore, the positron fraction measured by PAMELA was not reproduced within that framework. We therefore proposed, in the same paper, a scenario which invoked the presence of an extra component of $e^- + e^+$: it was argued that such extra-component can naturally be produced by near middle-aged pulsars. An alternative interpretation based on the annihilation of DM particles was also discussed but considered disfavoured. Other possible origins of the required $e^+ + e^-$ extra-component were proposed by other Authors: e.g. enhanced secondary production in standard SNRs[3] or an inhomogeneous distribution of sources in the Galaxy.[4] Recently, the Fermi-LAT Collaboration released a new measurement of the CRE spectrum based on one year data. The observed spectrum extends down to 7 GeV[5] and is confirmed to be compatible with a single power-law with spectral index $3.08 \pm 0.05$. Hints of a deviation from a pure power-law behavior, already found in the six month data, are still present in the updated spectrum, whose low energy part is compatible with pre-Fermi measurement and not reproduced by the models we proposed in Ref. 2: this consideration calls for some revisions of our scenario. Moreover, after the release of PAMELA antiproton data some of us used (see Ref. 6) the recently developed `DRAGON` [b] package[7] to

[b]Code available at http://www.desy.de/~maccione/DRAGON/

perform a combined maximum likelihood analysis on B/C and antiproton data and found that the "conventional model", based on the assumption of the existence a Kolmogorov-like turbulence in the ISM, is not the most adequate to describe the measurements: we therefore proposed a new model based on a Kraichnan-like turbulence. This result gives another reason to revise the models described in Ref. 2.

## 2. The propagation setups

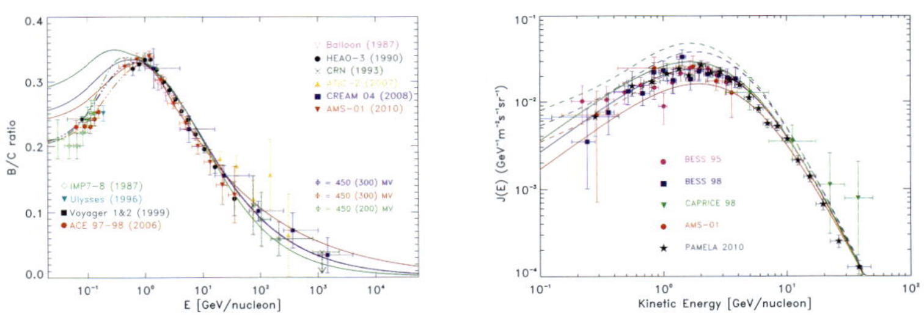


(a) (b)

Fig. 1. Panel a) Predicted B/C compared with data. Red lines: Kolmogorov model (KOL); blue: Kraichnan model (KRA); green: plain diffusion (PD). Solid lines: $\Phi =$ 450 MV; triple-dotted lines: $\Phi = 300$ MV. Panel b) Theoretical $\bar{p}$ spectra are compared with data; solid lines: $\Phi = 550$ MV; dashed lines: LIS.

The CR propagation in the Galaxy is described by a well known diffusion-loss equation which can be solved analitically, under simplifying assumptions, or numerically, making use of `GALPROP` or `DRAGON` packages. This equation includes several free parameters which need to be tuned by comparison with data: $D_0$ and $\delta$, i.e. the normalization and energy dependence of the diffusion coefficient, the Alfvén velocity $v_A$ which parametrizes the level of reacceleration, the height of the Galactic diffusion region $z_h$, and the injection index of the CR species $\gamma_p$. Moreover, when considering data below a few GeV/n also the modulation due to solar activity plays a significant role and must be taken into account. The first step of our work, before starting to investigate the $e^+ + e^-$ spectrum, consists in fixing the propagation parameters, i.e. $D_0$ and $\delta$, and the level of reacceleration. The datasets which constrain these data are secondary-to-primary ratios, in particular B/C and $\bar{p}/p$. In the following we will consider three propagation models, which are defined as follows. 1) The PD model is a plain diffusion one, in which we tried to reproduce CR data with no reacceleration, hence setting $v_A = 0$. In this model $\delta = 0.6$. 2) The KOL model is the so called "conventional model", built assuming a Kolmogorov-like turbulence, which fixes $\delta = 1/3$; in this model the reacceleration is quite high: $v_A = 30\,\mathrm{km/s}$. 3) The KRA model assumes a Kraichnan spectrum for the galactic turbulent magnetic field, hence setting $\delta = 1/2$. Reacceleration is lower: $v_A = 15\,\mathrm{km/s}$. At low energy a modified behaviour of the diffusion coefficient is adopted to reproduce the peak at 1 GeV in B/C data. This model was presented in Ref. 6 and obtained as the result of a combined maximum likelihood analysis based on B/C and $\bar{p}/p$ measurements. The comparison between data and model predictions for B/C and $\bar{p}/p$ can be seen in Fig. 1(a) and Fig. 1(b). We refer to Ref. 6 for the complete list of parameters we chose for the 3 models.

## 3. Modeling the CRE spectrum with a single Galactic component

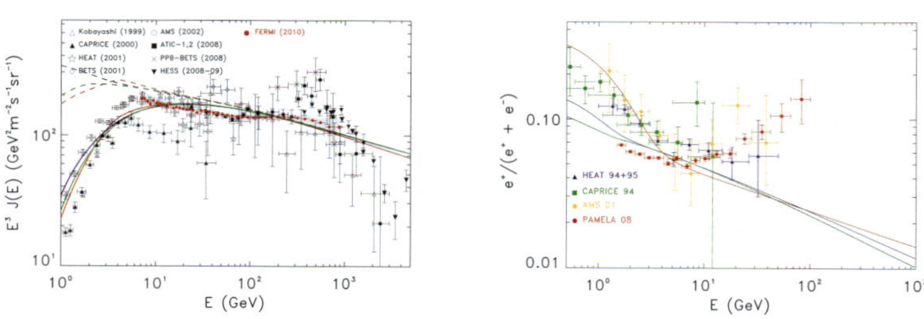


(a) (b)

Fig. 2. Panel a): the $e^- + e^+$ spectrum for the KOL (red), KRA (blue) and PD (green) diffusion setups. The electron source spectral indexes are 1.60/2.50 below/above 4 GeV for the KOL model and 2.00/2.43 and 2.0/2.40 below/above 2 GeV for the KRA and PD models respectively. Panel b): $e^+/(e^- + e^+)$. $\Phi = 550$ MV.

We start our analysis by trying to interpret CRE spectrum with single-component models. We evaluated these model with `DRAGON` numerical package[6] and verified that our results are reproduced by `GALPROP`[9,10] under the same physical conditions. We consider the three diffusion setups (KOL, KRA and PD) discussed in the previous section and for each of them we tune the free parameters involved in the calculation against Fermi-LAT data. We find that a reasonable fit of Fermi-LAT spectrum can be obtained

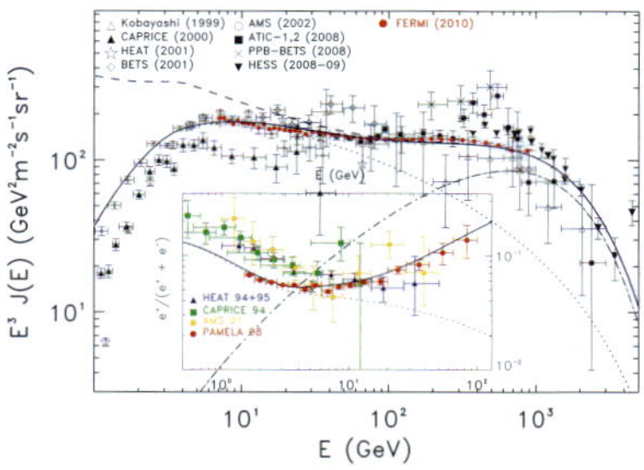

Fig. 3. The $e^- + e^+$ total spectrum and positron fraction (in the box) for our two-component model. Dotted line: propagated standard component with injection slope $\gamma_{e^-} = 2.00/2.65$ above/below 4 GeV and $E_{\rm cut}^{e^-} = 3$ TeV; dot-dashed line: $e^\pm$ component with $\gamma_{e^\pm} = 1.5$ and $E_{\rm cut}^{e^\pm} = 1.4$ TeV. Blue solid line: modulated total spectrum ($\Phi = 550$ MV). Blue dashed: LIS total.

normalizing the models to data at $\simeq$ 10 GeV and adopting the following injection indexes: 2.50 in the KRA model, 2.43 for the KOL, 2.40 for the PD (see Fig. 2(a)). It is worth noticing that KOL model requires a sharp spectral break (i.e. an injection index of 1.60 below 4 GeV) to avoid an anomalous behavior which would otherwise arise in the propagated spectrum: the combined effect of reacceleration and energy losses creates a pronounced bump in the unmodulated spectrum at low energy, which is only partially smeared out by solar modulation. This effect is smaller in KRA model and absent in PD setup, so in these cases a much softer break is required: 2.0 instead of 1.60 is the index below 2 GeV. However, the main drawback of a single component approach, concerning Fermi-LAT data, is that – no matter what propagation model is adopted – it is impossible to reproduce all the features revealed in the CRE spectrum, in particular the flattening observed at around 20 GeV (which was also recently found by PAMELA[11]) and the softening at $\sim$ 500 GeV. Concerning the positron fraction, the results can be seen in Fig. 2(b). Clearly, below 10 GeV the $e^+/(e^- + e^+)$ measured by PAMELA can be reproduced in the KRA and PD models while the fit is unsatisfactory for the KOL: again, low reacceleration models seem to provide a better description of low energy data. None of the single component realizations, however, can reproduce the rising behaviour observed by PAMELA above 10 GeV.

## 4. Two components models

In the following we try to reproduce all data in a two-component scenario. The first component (standard) consists of electrons accelerated in SNRs and a secondary contribution of $e^-$ and $e^+$ originated by the interaction of the hadronic part of CRs with interstellar gas. The second one (extra-component) is made of $e^-$ and $e^+$ injected in the ISM with a common spectrum of this kind: $Q_{e^\pm}(E) = Q_0 \left(\frac{E}{E_0}\right)^{-\gamma_{e^\pm}} e^{-E/E_{\rm cut}^{e^\pm}}$, where the injection index is harder: $\gamma_{e^\pm} < 2$. We also assume that both source classes have the same continuous spatial distribution. The normalization of the two components is performed in two steps. 1) We tune the standard component to reproduce both the $e^- + e^+$ spectrum measured by Fermi-LAT and the $e^+/(e^- + e^+)$ measured by PAMELA below 20 GeV, where the effect of the extra component is supposed to be negligible. Remarkably, this is possible only if we use propagation setups with low reacceleration, namely either the KRA or the PD, because the low-energy part of PAMELA positron ratio can't be reproduced in KOL setup. Since the KRA[6] also provides the best combined fit of B/C and antiproton data (see Sec. 2), we will stick to this model from now on. The required source spectral slopes for the electron standard component is $\gamma_{e^-} = 2.00/2.65$ below/above 4 GeV for this propagation model. From Fig. 3 the reader can see as such a set-up allows a remarkably good fit of Fermi-LAT as well as other experimental data. Such index is quite steep if compared to theoretical predictions regarding Fermi acceleration mechanism, but we remind the reader that we modeled the standard component in the approximation of a cylindrically symmetric source distribution, which may be less realistic for high energy electrons where the local distribution is relevant. Accounting for the spiral arm distribution of SNRs may result in a different requirement for the injection index. Indeed, being the Sun in the so-called "local spur" situated in a interarm region, the average distance from SNRs is larger than in the smooth case: as a consequence, a harder injection spectrum may be required to compensate for the larger energy losses and reproduce the observed spectrum. Clearly, in the absence of the extra $e^\pm$component, high energy CRE and positron fraction data would completely be missed (see dotted line in Fig. 3). 2) We tune the extra-component to reproduce Fermi-LAT and H.E.S.S. high energy CRE data. We find here that this is possible by taking $\gamma_{e^\pm} = 1.5$ and $E_{\rm cut} = 1.0 \div 1.5$ TeV (see Fig. 3). This is similar to what done in Ref. 2 but in that paper a KOL diffusion was used (so low energy positron data were not reproduced in that case). It is interesting that preliminary $e^-$ spectrum

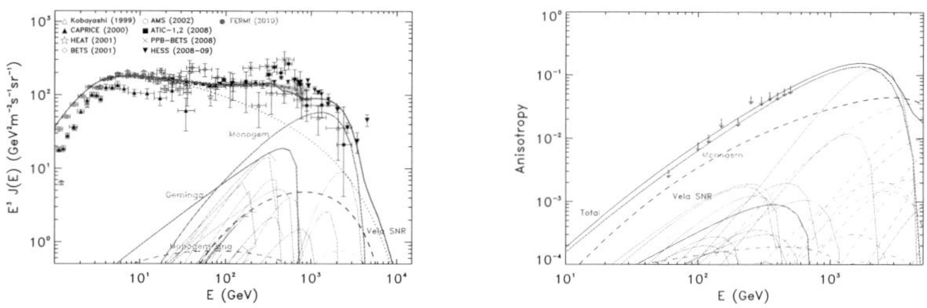


(a) (b)

Fig. 4. Standard component + nearby pulsars and SNRs. Panel a): $e^+ + e^-$ spectrum. Energy release by each SNR: $2 \times 10^{47}$ erg . Pulsar efficiency: $\simeq 30\%$. $\Phi = 500MV$. Panel b): the integrated expected anisotropy, as a function of minimum energy, is compared to Fermi-LAT the upper limits.[12]

measured by PAMELA[11] is also nicely reproduced by our extra-component models: above 100 GeV this spectrum is softer than the $e^- + e^+$ measured by Fermi-LAT by the exact amount which is required to leave room the $e^+$ extra-component.

## 5. The role of astrophysical nearby sources

The nature of the extra-component of primary electrons and positrons that we invoked in the previous section is an intriguing matter of debate, and the possible scenarios include both an exotic explanation (involving annihilation or decay of Particle Dark Matter) or a purely astrophysical interpretation. Here we concentrate on the second possibility. Differently from the previous section, we treat the extra component as originating from a discrete collection of sources; we then treat electron and positron propagation from those sources to the Solar System by solving analytically the diffusion-loss equation similarly to what done in Ref. 2. The large scale (standard) component is instead modeled with `DRAGON` as done in the previous section; for consistency, we treat analytical and numerical propagation under the same physical conditions. The propagation setup adopted is the KRA one. There are basically two classes of objects we consider: SNRs (as sources of *electrons only*) and pulsars (as sources of *electron+positron pairs*). SNRs are the well known natural candidates as CR accelerators, and also the conventional component is expected to originate from objects of this kind; considering individual SNRs in the nearby ISM is important due to the fact that at high energy (above some hundred GeV) diffuse models don't give a proper description of reality since the loss length due to synchrotron and IC becomes comparable to the average SNR mutual distance so that only few sources within few hundred parsecs are expected to dominate. Pulsars, instead, are very interesting candidates as primary sources of both electrons and positrons. They are extreme astrophysical environments that release very large amounts of energy ($\sim 10^{52} \div 10^{54}$ erg) during their lifetime. The $e^- + e^+$ pairs are expected to be produced in the magnetosphere by the interaction of curvature photons with the pulsar magnetic field; those particles are then expected to be accelerated at the termination shock of the pulsar wind nebula (PWN). The possibility that electrons and positrons from nearby pulsars can dominate the high energy tail of the CRE spectrum and explain the rising behavior of the positron fraction was already proposed in Ref. 13 and studied in more recent papers (e.g. Refs 14 and 15). Similarly to the approach taken in Ref. 2, we model the emission from SNRs and pulsars as a point-like burst; for pulsars we also introduce a time delay with respect to the birth of the object: such a delay is motivated by the fact that electrons and positrons are expected to be trapped in the PWN until it merges with the ISM. We assume that for both SNRs and pulsars particles are injected with a power law spectrum up to an exponential cutoff; the injection index is however set in a different way for the two classes of objects. We considered all observed SNRs within 2 kpc as taken from the Green catalogue[16] and the pulsars within 2 kpc from Earth taken from the ATNF catalogue.[17] We verified that more distant objects give a negligible contribution. In Fig. 4(a) we represent the CRE spectrum obtained for a reasonable combination of parameters, namely: for SNRs: spectral index $\gamma_{e^-\,\mathrm{SNR}} = 2.2$, cutoff energy $E_{\mathrm{cut}}^{\mathrm{SNR}} = 2$ TeV, electron energy release per SN $E^{\mathrm{SNR}} = 2 \times 10^{47}$ erg; for pulsars: $\gamma_{e^\pm} = 1.5$, cutoff energy $E_{\mathrm{cut}} = 1$ TeV, efficiency $\eta_{e^\pm} \simeq 30\%$. We see from Fig. 4(a) that the main contribution comes from Monogem pulsar; this is due to the proximity of this source and to the introduction of the delay between the pulsar birth and the actual injection of the pairs in the ISM; without considering such delay, Monogem contribution would decrease significantly and other nearby young pulsar, such as Vela, would be the most important sources in the high energy region; instead, in our framework electrons and positrons from Vela are expected to be still trapped in the surrounding nebula and therefore not yet observed in the CRE spectrum. It is important to check if these results are compatible with recently published upper limits on the anisotropy in $e^- + e^+$ flux[12] . Our prediction is plotted in Fig. 4(b). The important result is that our scenario is not excluded by anisotropy measurements;

the reader may notice that Monogem pulsar (red solid line) and Vela SNR (black dashed line) contribute most to the total anisotropy (the black solid line). However, the total expected anisotropy is very close to the measured upper limit, so a future detection at level $\sim 1\%$ at $\sim 1$ TeV towards the portion of the sky where Vela and Monogem are located is to be expected.

## 6. Conclusions

The spectacular data on CR electrons and positrons, together with measurements on light nuclei and antiprotons, suggest that a double-component approach to the leptonic part of CRs computed in the framework of a Kraichnan-like turbulence provides a good self-consistent scenario which satisfactorily reproduces all existing observations. In this picture, in addition to the conventional component accelerated in SNRs, a contribution from pulsars, as emitters of $e^- + e^+$, permits to correctly fit both the features revealed by Fermi-LAT in the CRE spectrum and the $e^-/(e^+ + e^-)$ measured by PAMELA. The expected anisotropy in the direction of the most prominent CRE candidate source, Monogem pulsar, is compatible with the present upper limit just released by Fermi-LAT collaboration and may be detectable in a few years.

## References


1. A. A. Abdo *et al.* [The Fermi LAT Collaboration], Phys. Rev. Lett. 102 (2009) 181101
2. D. Grasso *et al.* [FERMI-LAT Collaboration], Astropart. Phys. 32 (2009) 140
3. P. Blasi, Phys. Rev. Lett. 103 (2009) 051104
4. N. J. Shaviv, E. Nakar and T. Piran, Phys. Rev. Lett. 103 (2009) 111302
5. M. Ackermann *et al.* [Fermi LAT Collaboration], arXiv:1008.3999 [astro-ph.HE], accepted for publication in Phys. Rev. D
6. G. Di Bernardo, C. Evoli, D. Gaggero, D. Grasso and L. Maccione, arXiv:0909.4548 [astro-ph.HE]; Astropart. Phys., in press doi:10.1016/j.astropartphys.2010.08.006 .
7. C. Evoli, D. Gaggero, D. Grasso and L. Maccione, *JCAP* 0810 (2008) 018
8. G. Di Bernardo, C. Evoli, D. Gaggero, D. Grasso, L. Maccione and M.N. Mazziotta arXiv:1010.0174 [astro-ph.HE]; submitted to Astropart. Phys.
9. A. W. Strong, I. V. Moskalenko, T. A. Porter, G. Johannesson, E. Orlando and S. W. Digel, arXiv:0907.0559 [astro-ph.HE].
10. A. E. Vladimirov *et al.*, arXiv:1008.3642 [astro-ph.HE].
11. O. Adriani, talk at ICHEP conference 2010, Paris (http://www.ichep2010.fr/)
12. M. Ackermann *et al.* [Fermi-LAT collaboration] arXiv:1008.5119 [astro-ph.HE]; accepted for publication in Phys. Rev. D.
13. F. A. Aharonian, A. M. Atoyan and H. J. Volk, Astron. Astrophys. 294 (1995) L41.
14. D. Hooper, P. Blasi and P. D. Serpico, JCAP 0901 (2009) 025.
15. S. Profumo, arXiv:0812.4457 [astro-ph].
16. D. A. Green, Bull. Astron. Soc. Ind., 37 (2009) 45; arXiv:0905.3699 [astro-ph.HE].
17. R. N. Manchester, G. B. Hobbs, A. Teoh, & M. Hobbs, ApJ 129 (2005) 1993

# COSMIC RAYS FOR HELIOSPHERIC SPACE WEATHER STORM PREDICTION

FRANK JANSEN, JÖRG BEHRENS

*German Aerospace Center (DLR), Institute of Space Systems, Robert-Hooke-Str.7 28359 Bremen, Germany*

The classical imaging of CME's by solar UV light is successfully applied on many satellites like for instance on SOHO, STEREO A and B or SDO. CME detection via UV light from ground based telescopes can't be carried out, because of the absorption of UV light in the Earth atmosphere. However CME detection is done on ground by means of cosmic ray muon telescopes in the Global Muon Detection Network (GMDN). Therefore the astrophysics and solar physics for GMDN are summarized in chapter 1. GMDN data and results are described in chapter 2. A space situational awareness (SSA) satellite, which has also the purpose of CME imaging by means of a telescope for cosmic rays are sketched in chapter 3.

## 1. Astrophysics and Solar Physics for Cosmic Ray related Space Weather Storm Prediction

Cosmic rays with charges Z=1 (protons) to Z=92 (uranium nuclei) from astrophysical sources like supernovae, Wolf-Rayet stars and other star types in the Galaxy are detected at Earth for decades. For example (see [1-4] among others) the plastic scintillator detector stacks like CR-39 of the Dublin-ESTEC Ultra-Heavy Cosmic Ray Experiment (UHCRE), retrieved in 1990 on NASA's Long Duration Exposure Facility (LDEF) after almost 6 years of space exposure, have collected cosmic ray nuclei with charge Z>65 including actinides and energy E>1.5 GeV/n. The main objective of the UHCRE analysis of LDEF was to search for signatures of nucleosynthesis processes like the s- and r-process. For instance, comparison of the abundances of Os, Ir and Pt, which represent the heaviest stable r-process nuclei, with that of Rb (considered to be at least 50% of s-process origin) provides information on the relative importance of these two basic nucleosynthesis processes for ultra heavy cosmic ray nuclei. Summarised: between about one GeV up to $10^5$ GeV the cosmic rays with Z = 1…92 are of galactic origin. According to the halo-diffusion model (e.g. [5-6] with references therein) the cosmic ray protons, antiprotons and ultra-heavy nuclei are primarily originated respectively produced in the inner Galaxy and propagated via the galactic halo to the heliosphere. Due to long time propagation (between about $10^6$ year in the leaky-box model to about $10^8$ years in cosmic ray diffusion models) of galactic cosmic rays (GCRs) in the galactic magnetic field, the cosmic ray particles from interstellar space arriving nearly isotropic distributed (in the GeV energy range) at the heliospheric boundary. This isotropic distribution of GCRs is nearly preserved during their propagation from heliospheric boundary towards Earth and therefore the GCRs are also isotropic distributed in the interplanetary space between the Sun and Earth. However, only in the time period, if a solar plasma cloud (CME - Coronal Mass Ejection) propagates from the Sun towards Earth and interact with GCRs and the density of comic rays become anisotropic distributed. This process for the cosmic ray anisotropy is caused by GCR interaction near the CME shock and was studied in detail [7] in 1999. After the interaction the GCR leave the CME shock region, which travels only up to about 2000 km/s. In contrast, the velocity of light propagating GCR immediately arrives the upper Earth atmosphere, in which the GCRs produce secondary cosmic ray muons. Because of the muons do not change the propagation direction of the GCRs, a global network of ground based cosmic ray muon telescopes observe in real time the CME propagation towards Earth. Already in 2000 this method was used as precursors of geomagnetic storms prediction [8].

Section 2 contains the principles, some details and results of the Global Muon Detector Network (GMDN). A space based cosmic ray telescope for CME imaging is sketched in chapter 3.

## 2. Principles and Results of the GMDN for Heliospheric Space Weather Storm Prediction

The principles, details and results of the GMDN can be summarized as follows [9]:

1) <u>CME distant determination.</u> The most distant detection of CME signature is given by

$$r \sim 0.1\, \lambda_P \cos \beta \qquad (1)$$

($\lambda_P$ scattering mean free path of GCRs, $\beta$ angel between Sun – Earth line and the mean interplanetary magnetic field at Earth). The distance r can be up to an order of 50 Million km. Therefore the GMDN measures CMEs already from the Venus orbit or about 24 hours before CMEs arrive at Earth.

2) <u>Principles of muon telescopes.</u> The cosmic ray muon telescopes on ground consist of two detector layers. The muon signals in the layers are measured in coincidence to receive the arrival direction of the cosmic ray muon (left side of figure 1).

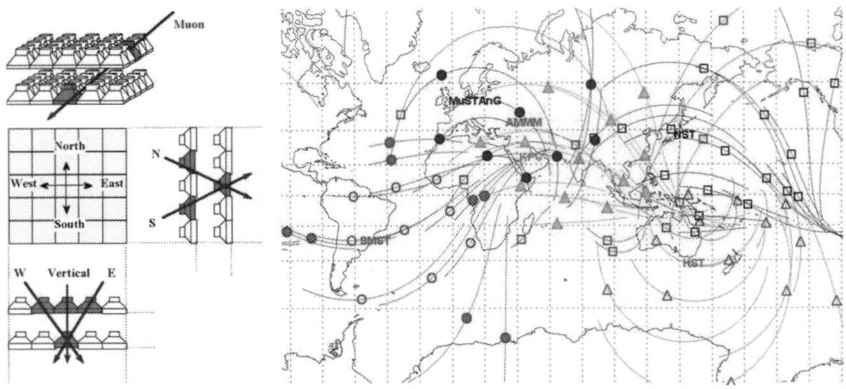


Fig. 1. Left: the telescopes dispose of a combination of viewing directions due to the analysis of muon data from different detectors in both layers. Right: the geographic location of muon telescopes (NST – Nagoya Scintillator Telscope, HST – Hobart Scintillator Telescope, SMST – Sao Martinho Scintillator Telescope, KPC – Kuwait Proportional Counter, GST - German MuSTAnG Scintillator telescope and AMMM (Aragats Multichannel Muon Monitor)) with the median primary rigidity, after correction for geomagnetic bending of GCR orbits. The track through each symbol represents the spread of viewing directions corresponding to the central 80% of each telescope's energy response [10].

NST in Japan has 17, HST in Australia has 13, SMST in Brazil 13 and KPC in Kuwait 13 observational directions [11]. The ESA / DLR funded MuSTAnG telescope (see figure 2 or in [12],[13]) in Germany has 17 and AMMM in Armenia 11 viewing directions.

| Local viewing direction | Count rate ($hr^{-1}$) | Median rigidity (GV) | Asym. Lat. (°) | Asym. Lon. (°) |
|---|---|---|---|---|
| V | 2577718 | 57.33 | 32.79 | 51.98 |
| N | 1182664 | 61.49 | 50.20 | 79.92 |
| S | 1182690 | 61.53 | 8.28 | 38.68 |
| E | 1176049 | 61.30 | 13.34 | 78.20 |
| W | 1185480 | 61.69 | 45.17 | 15.05 |
| NE | 577119 | 65.42 | 26.71 | 95.00 |
| NW | 580052 | 65.78 | 69.48 | 37.04 |
| SE | 576419 | 65.48 | -2.39 | 64.21 |
| SW | 579831 | 65.74 | 18.04 | 11.86 |
| N2 | 591681 | 79.99 | 58.23 | 106.89 |
| S2 | 591146 | 79.99 | -8.90 | 32.92 |
| E2 | 589315 | 80.04 | 3.41 | 90.82 |
| W2 | 592255 | 80.07 | 43.65 | -16.21 |
| N3 | 147937 | 103.91 | 58.00 | 133.70 |
| S3 | 147910 | 103.91 | -23.29 | 30.36 |
| E3 | 147697 | 103.95 | -4.34 | 101.13 |
| W3 | 147978 | 103.88 | 36.58 | -37.20 |

Tab. 1. MuSTAnG telescope asymptotic viewing direction, muon count rate, median rigidity, asymptotic latitude and longitude [12].

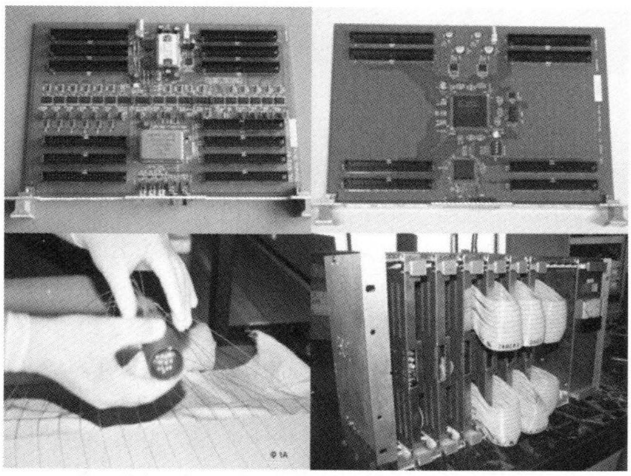

Fig. 2. The electronic block of GST contains a recording system as a coincidence unit based on advanced logical circuit using several boards with Field Programmable Gate Array (FPGA). GST and four telescopes (NST, HST, SMST and KPC) are equipped with this recording system developed by Shinshu University. The wavelength shifting fibers in the plastic scintillators (lower left image) of GST is an application of the L3+C detector technology at CERN.

3) GMDN data processing and calibration. The five muon telescopes NST, HST, SMST, KPC and GST belong to the GMDN, because of the usage of the same data recording system, identical software for data processing and the calibration of cosmic ray muon data between the five telescopes. The following five results are feasible by the GMDN for a real time space weather service

a) real time GCR density in nearby Earth $I_{i,j}{}^{cal}$,
b) real time interplanetary CME detection by GCR anisotropy $\zeta^{GEO}$ ,
c) real time status, while Earth is within a CME by measurement of bidirectional GCR streaming within the CME $\zeta^{GEO}$,
d) real time GCR wind direction in nearby Earth in and out of the ecliptic plane by measurement of GCR first-order anisotropy $\zeta^{GSE}$ and
e) real time GCR wind strength in nearby Earth in and out of the ecliptic plane by measurement of GCR first-order anisotropy $\zeta^{GSE}$.

These results are obtained by fitting in real time the function to the observed hourly muon counting of all five GMDN telescopes by means of the numerical solution of the equation

$$I_{i,j}^{cal}(t) = c_{0i,j}^{0} I^{0}(t) + \xi_{x}^{GEO}(t)\left(c_{li,j}^{l}\cos\omega t_i - s_{li,j}^{l}\sin\omega t_i\right)$$
$$+ \xi_{y}^{GEO}(t)\left(s_{li,j}^{l}\cos\omega t_i + c_{li,j}^{l}\sin\omega t_i\right)$$
$$+ \xi_{z}^{GEO}(t) c_{li,j}^{0} \qquad (2)$$

with $\omega = 2\pi/24$ and $I^0(t) = I'_0(t) + I''_0(t) \bullet P_{mi,j} / P_{m1,1}$ (best-fit parameters $I'_0(t)$ and $I''_0(t)$, $P_{mi,j}$ is the median primary rigidities of i-th muon telescope with it's j-th observational channel (for GST see in table 1), $P_{m1,1}$ for the NST vertical channel (i=1,j=1)). The coupling coefficients c, s, which connect the measured muon count rates of the i-th telescope (i = 1 to 5) with it's j-th viewing channel of the telescope. The best-fit anisotropies $\zeta^{GEO}{}_{x}$, $\zeta^{GEO}{}_{y}$ and $\zeta^{GEO}{}_{z}$ are defined in the local geographical coordinate system (GEO) and transformed into geocentric solar ecliptic coordinate system (GSE) $\zeta^{GSE}{}_{x}$, $\zeta^{GSE}{}_{y}$ and $\zeta^{GSE}{}_{z}$. The local time of the telescope is $t_i$ and t the universal time. Instead of usage the theoretical value of 45° for the interplanetary magnetic field angle at Earth orbit, we use the real time measured value by the ACE satellite (ACE-Level2 data) to realize real time propagation conditions between ACE and Earth.
The real time processing and display of graphic results (source codes are written in Fortran, C++ and IgorPro) of equation (2) is done for five telescopes at the DLR Institute of Space Systems in Bremen and for four telescopes (i = 4) at the University of Delaware. Details of equation (2) and the solution are explained in [14],[11] and [9].

4) Display of GMDN results. The five displays a) to e) mentioned in sub-chapter 3) are updated every hour and are available for all five GMDN telescopes since 1 December 2006 – that means in the solar minimum period. Since this date only moderate (6 – 16 December 2006, 3 CMEs, see figure 3) and weak CMEs (12 – 15 February 2010, 8 – 11 April 2010, 1 – 3 August 2010 (last day: arrival at Earth according to SOHO/CELIAS/MTOF measurements) propagated towards Earth. Insofar the GMDN data displayed in this paper are tests for the sensibility of the network as a preparation for the nearby solar maximum. Therefore for purposes of space weather service the real time, absolute value of the interplanetary magnetic field strength |B|, the absolute value and direction of $B_z$ and the geomagnetic index $K_P$ (Costello respectively Wing) are displayed. All plots are available as 3 days, 27 days or for selected periods. The display of the data is focused on space weather user needs and therefore with plotted with colour codes in green, yellow and red.

a) GMDN data and moderate CMEs. Form 6 December to 14 December 2006 three moderate CMEs arrived at Earth from Sun.

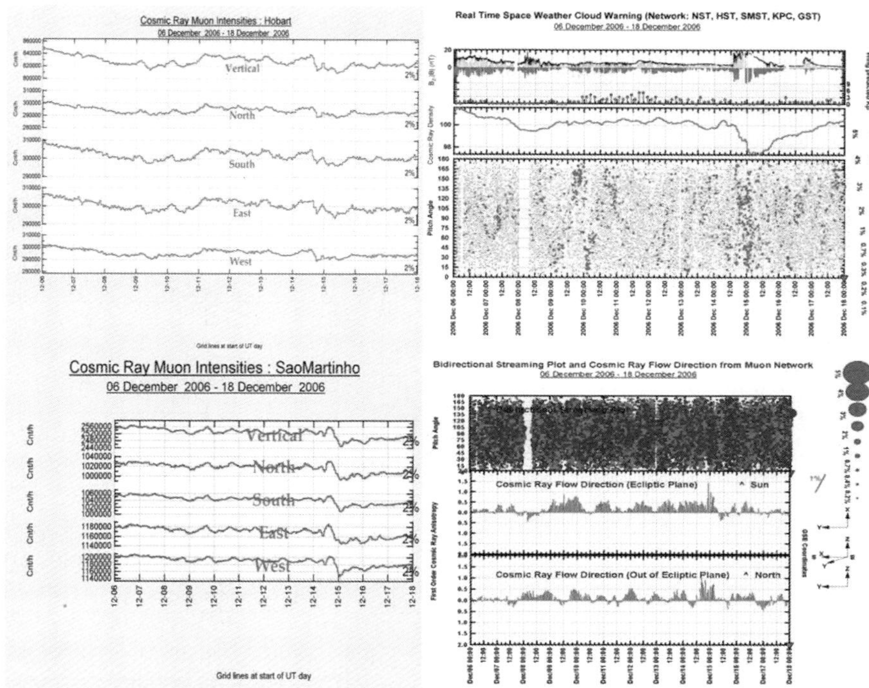


Fig. 3 CME detection by GCR measurement from 6 – 18 December 2010 (see also in [15]). Two left figures: hourly muon counts from HST and SMST, with a Forbush decrease measured in both telescopes and in all directions on 14 - 15 December 2010. Upper right image (from top to bottom): ACE magnetic field and Kp data, GCR density of all five telescopes with Forbush decrease on 14 December 12:00 UT to 15 December 03:00 UT. All three CMEs observed by increased GCR anisotropy around 8 – 10 December and 14 – 16 December as red circles distributed over nearly 180 degree. Lower right image: stronger changes in GCR wind strength and direction from Sun and north direction (0.5% to 1.5%) towards Earth occurred when the CMEs arrived at Earth and the Earth was within the CMEs.

b) GMDN data, weak CMEs and CME free interplanetary space. On 1 August 2010 2 pm a weak full halo CME started towards Earth. SOHO / CELIAS / MTOF measured on 3 August 4:52 pm a weak forward shock (30 km/s solar wind speed changes and an order of magnitude proton density increase). Hints for this weak event were also observed in GMDN data. On late 1 until 3 August 2010 the weak interplanetary CME was also detected as GCR anisotropy with smaller than 1%. HST observed a weak Forbush decrease in the vertical channel during the time of CME arrival. The GCR wind strength and direction changes were low as well (less than 0.5%).

As an example the GMDN data for a CME free interplanetary space are displayed in figure 4.

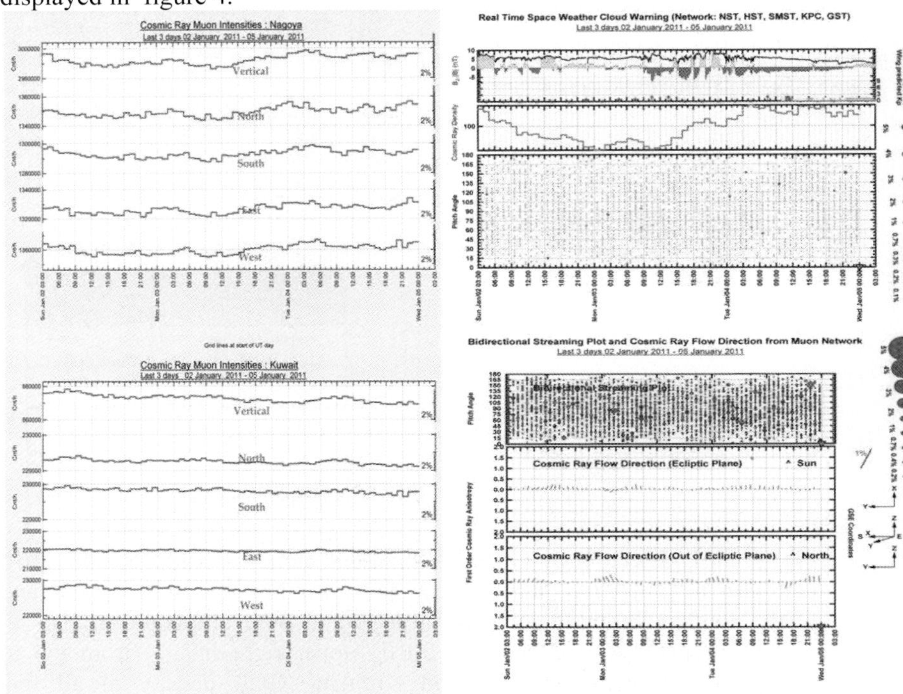


Fig.4 In case of a CME free interplanetary space, like between 2 and 5 January 2011 no Forbush decrease, no GCR anisotropy and negligible very low GCR wind strength (smaller than 0.5%) and changes are detected. The bidirectional streaming plot shows no hints for the Earth to be within a CME.

## 4. Space Situational Awareness related Space Weather Satellite

In the SSA satellite proposed to ESA in 2010 (figure 5) the cosmic rays undertake the role of UV light as a mean for CME imaging. Insofar the cosmic ray rested CME detection is complementary to the UV light rested CME detection on satellites. In addition the space based cosmic ray telescope will not only produce images of CMEs, but will also measure the GCR density, anisotropy, wind strength and direction like the ground based GMDN. In contrast to the ground based GMDN telescopes with plastic scintillators or proportional counter tubes, it is foreseen to install a Medipix-2 based cosmic ray imaging telescope to visualize the CMEs ([16], [17]) by means of GeV GCRs.

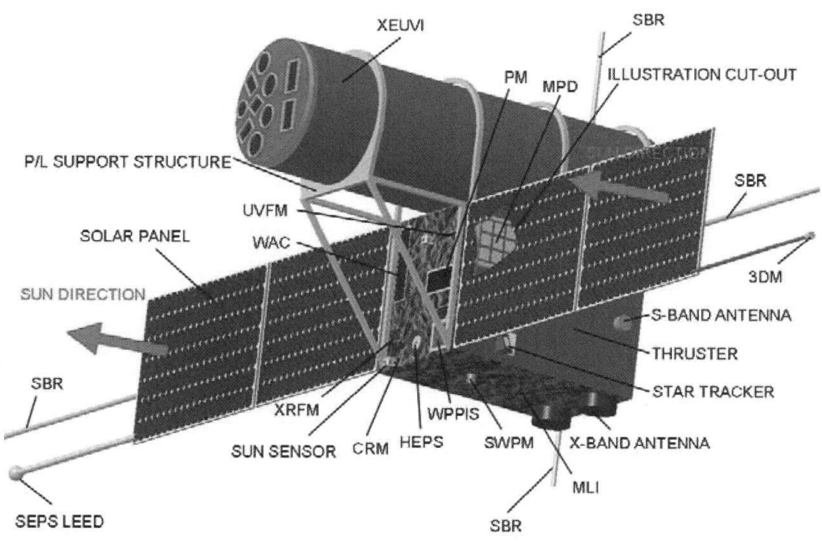


Fig.5 Proposed SSA satellite (based on DLR Bremen CEF (Concurrent Engineering Facility)) with Medipix based Multi Particle Detector (MPD) for CME imaging by GCRs. A solar $\gamma$- and X-ray telescope (XEUVI) / flux monitor (XRFM), particle monitor (PM), magnetometer (3DM), solar radio burst antennae (SBR) and others including spacecraft bus, power, telemetry, navigation and propulsion are sketched.

**Acknowledgments**

The UHCRE / LDEF results were received in close cooperation and strong support by Dr. K.-P. Wenzel (ESA/ESTEC, The Netherlands), Prof. D. O'Sullivan (DIAS, Ireland). The GMDN data became usable by long-term fruitful cooperation with Profs. K. Munakata (Shinshu University, Japan), Prof. J. Bieber (University of Delaware U.S.A), I. Sabbah (Kuwait), Drs. M. Duldig (Australian Antartic Division Hobart), N.J. Schuch (CRS/INPE Santa Maria, Brazil), T. Kuwabara (University of Delaware, U.S.A) and G. Bartling (Berlin, Germany). The acknowledgement has to be extended to W. Bauer (DLR Bremen, Germany) and Dr. S. Bedrich (Kayser-Threde Muinch, Germany) related to the close and flexible cooperation for the SSA satellite in an ESA proposal (2010). Last but not least Profs. S. Pospisil (CTU Prague, Czech Republic) and K. Kudela (IEPSAS Kosice, Slovakia) for discussions of several Medipix and cosmic ray aspects.

**References**

1. A. Thompon, D. O'Sullivan, J. Bosch, R.Keegan, K.-P. Wenzel, F. Jansen, C. Domingo, Progress Report on the UHCRE (AO178), *2nd LDEF Post-Retrieval Symposium,* NASA CP-3194, (1992).

2. D. O'Sullivan, A. Thompson, J.Bosch,R.Keegan, K.-P.Wenzel, F. Jansen, C. Domingo, *3rd LDEF Post Retrieval Symposium*, NASA CP-3275, (1993).

3. A. Thompson, D. O'Sullivan, K.-P. Wenzel, J. Bosch, R. Keegan, C. Domingo, F. Jansen, *Proc 23rd ICRC (Calgary)* Vol 1, 603, (1993).

4. D. O'Sullivan, A. Thompson, J. Bosch, K.-P. Wenzel, F. Jansen, *Advances in Space Research I5*, 6, 25, (1995).

5. F. Jansen, K.-P.Wenzel, D.O'Sullivan, A. Thompson, *Advances in Space Research*, 15, (6) 49 – (6) 58, (1995).

6. I. Halm, F. Jansen, D. deNiem, *Astron. Astrophys.*, 269, 601-607, (1993).

7. D. Ruffolo, *Astrophys. J.*, Vol. 515, 787 - 800, (1999).

8. K. Munakata, J.W. Bieber, Y. Shin-ichi, C. Kato, M. Koyama, S. Akahane, K. Fujimoto, Z. Fujii, J.E. Humble, M.L. Duldig, *J. Geophys. Res.*, Vol. 105, 27457 - 27468, (2000).

9. F. Jansen, J. Behrens, *Proc. 21st European Cosmic Ray Symp.*, ISBN 978-80-968060-5-8,626-631, (2008).

10. T. Kuwabara, J. W. Bieber, J. Clem, P. Evenson, R. Pyle, K. Munakata, S. Yasue, C. Kato, S. Akahane, M. Koyama, Z. Fujii, M. L. Duldig, J. E. Humble, M. R. Silva, N. B. Trivedi, W. D. Gonzalez, N. J. Schuch, *Space Weather*, Vol. 4, S08001, doi:10.1029/2005SW000204, (2006).

11. Y. Okazaki, A. Fushishita, T. Narumi, C. Kato, S. Yasue, T. Kuwabara, J. W. Bieber, P. Evenson, M. R. Da Silva, A. Dal Lago, N. J. Schuch, Z. Fujii, M. L. Duldig, J. E. Humble, I. Sabbah, J. Kota, K. Munakata, *Astrophys. J.*, 681, 693–707, (2008).

12. F. Jansen, K. Munakata, M.L. Duldig, R. Hippler, *Proc. Space Weather Workshop: Looking Towards a Future European Space Weather Programme*, ESTEC, (2001).

13. F. Jansen, *Proc. 2nd Int. Symp. Solar Extreme Events*, IBN 99941-0-165-X, 252-255, (2006).

14. Fujimoto et. al., *Cosmic Ray Res. Lab. Rep.9*, Nagoya, (1984).

15. T. Kuwabara, J. W. Bieber, P. Evenson, K. Munakata, S. Yasue, C. Kato, A. Fushishita, M. Tokumaru, M. L. Duldig, J. E. Humble, M. R. Silva, A. Dal Lago, N. J. Schuch, *Journal Geophysical Research*, VOL. 114, A05109, doi:10.1029/2008JA013717, (2009).

16. F. Jansen, J. Behrens, *Astroparticle, Particle and Space Physics, Detectors and Medical Physics Applications, Proc. of the 11th Conference on ICATPP-11*, Eds.: C. Leroy, P.-G. Rancoita, M. Barone, A. Gaddi, L. Price, R. Ruchti, 770 – 779, (2010).

17. F. Jansen, J. Behrens, S. Pospisil, K. Kudela, Nucl. Instr. and Meth. A, doi:10.1016/j.nima.2010.06.175, (2010).

# ENERGETIC PARTICLES IN THE MAGNETOSPHERE OF EARTH: SELECTED RESULTS AND PROBLEMS

KAREL KUDELA[†]

*Institute of Experimental physics, Slovak Academy of Sciences, Watsonova 47, 04001 Kosice, Slovakia, kkudela@kosice.upjs.sk*

LEONID L. LAZUTIN, YURI I. LOGACHEV

*Skobeltsyn Institute of Nuclear Physics, Moscow, Russia*

Magnetosphere of the Earth is well known reservoire of the energetic particles and the cause of cosmic ray trajectory changes. Radiation belt models succsesfully describe spatial, pitch-angle and energy particle distribution in undisturbed conditions. During the magnetospheric disturbances, especially during strong magnetic storms there are still many unsolved problems, concerning energetic auroral particles acceleration during substorms, relativistic electron acceleration and losses, solar proton trapping and acceleration in the inner magnetosphere. Recently energetic particle acceleration in the atmosphere was registered as well. We illustrate some events when the strong redistribution of energetic particles in the magnetosphere is observed. Selected results of such type, especially the experimental ones, as well as the problems remaining to be solved, are discussed in our short review.

## 1. Introduction. Classification of energetic particles in Earth's magnetosphere.

Cosmic rays as well as energetic particles energized in space without the response on Earth's surface, are important part of space weather research (see e.g. reviews [1-4] among others). State of the magnetosphere is one of the key point in Space Weather effects understanding and forecasting. Monitoring of energetic particles at various orbits in space and during different periods is important for producing the probabilistic models of radiation hazards not only near Earth but also for interplanatery missions. On the other hand the energetic particles monitored may serve as a tool for the description of magnetosphere state especially during strong geomagnetic disturbances initialized from the Sun and from interplanetary space.

[†] Work supported by project ITMS Grant No. 26220120009.

There is a large range of kinetic energies of the charged particles observed within the Earth's magnetosphere. On one side of the spectra there are ionospheric plasma particles with energy below 1 eV. On the other side there are cosmic rays with very high energies for which the magnetosphere is just a small episode on the trajectory. Classification of particles is possible in comparison of the characteristic dimensions of their trajectories with respect to dimension of magnetosphere. Stormer length, $l = (\mu_o \, . \, M \, . \, |q| \, / \, ( 4 \, p \, m \, v))^{1/2}$ where $M$, $q$, $p$, $m$, $v$ are magnetic moment of the dipole, elementary charge, momentum, mass and velocity of particle, serves as a key parameter for trajectory classification in comparison with the characteristic dimension of the magnetosphere, $r$. Trajectories of particles with $l >> r$ can be described with using of adiabatic invariants and usually (if the phase of motion is not important) it can be described by almost independent superposition of three cyclicities with different periods, namely gyromotion, bounce along the field line and azimuthal drift. On the other hand, particles with $l \leq r$ have the trajectory which must be computed numerically in the model geomagnetic field. This is the case of cosmic rays (CR). The periodicities of cyclic motion of protons with energy above ~ 100 MeV in the outer magnetosphere are comparable and thus the adiabatic approach fails.

We briefly overview some of the recent results on the question of magnetospheric field model checking by cosmic rays; on particle penetration into the magnetosphere; on the contribution of particles accelerated behind the magnetosphere to the magnetospheric population; and on dynamics of energetic electrons being important element for the damage for satellites.

## 2. Can CR serve as a tool for geomagnetic field model checking in strongly disturbed intervals?

Nowadays there exist several models of geomagnetic field, including both internal and external current sources. A review on the selected models in the context of cosmic ray research can be found e.g. in [5].

Most significant changes of the geomagnetic field configuration are observed during the strong geomagnetic disturbances. Cosmic rays is a specific tool for checking validity of the models. At energies of primaries to which neutron monitors are sensitive, the characteristic trajectories dimensions are comparable to the size of magnetosphere. Thus, before arriving to a specific point above the measurement device at the top of atmosphere, the primary particle "feels" large scale changes of the geomagnetic field.

The changes of external current systems are parametrized in the newer models (e.g. [6] with references therein). For rather strong, almost isolated in time geomagnetic storm in November 20-23, 2003, the different predictions for cosmic ray penetration into magnetosphere corresponding to different field models, is studied in [7]. The improvement of magnetospheric transmissivity is clearly seen at neutron monitors with high cut-off rigidities as the peak around the time of maximum Dst depression. For that period it was possible to compute parameters of different external current systems including the "prehistory" of the storm based on interplanetary characteristics. Thus in addition to Tsyganenko'89 model [8] and the model described in paper [9] also the Tsyganenko and Sitnov TS4 model [10] was possible to include into trajectory computations. The computations were done similarly to the procedures used long time ago and summarized in [11] with specific control of trajectory smoothness and other improvements described by [12]. It was shown that the expectations of the peak in neutron monitor data is better corresponding to the model [10] than to [8,9]. The transmissivity function, being the probability that the particle in given finite rigidity interval is "allowed" for acceptance at given station [13], is only slightly shifted towards lower rigidities for [8]. The shift using computations in the model [9] is unusually large. Model [10] gives intermediate shift of transmissivity function to lower rigidities.

The variations of vertical cut-off rigidities for the same geomagnetic storm was studied in [14]. The authors determined the cut-off rigidity changes by two methods, namely (a) by trajectory calculations and (b) by the spectrographic global survey method described e.g. in [15]. The cut-off rigidity changes obtained by both methods are correlated with Dst, IMF and solar wind plasma parameters with most significant importance of Dst and the model [16] is found to be well representing the neutron monitor data during the storm.

However, cosmic ray profiles on neutron monitors during another storm, namely November 7-11, 2004, do not seem to be better described by the model [10] in comparison with [8,9]. It seems that not only single global parameter like Dst is the decisive for the transmissivity of the cosmic rays during the storms. For strong storms even penumbra structure at high nominal cut-off stations is different if one uses different geomagnetic field models.

In addition to cut-off changes, asymptotic direction reconstruction and transmissivity function variation during geomagnetic storms, which are changing only due to the geomagnetic field variability, there must be assumed also the interplanetary anisotropy of cosmic rays which evolves several hours before and after the time when minimum Dst is reached. For using the middle cut-off rigidity neutron monitors as tools for checking validity of geomagnetic field models during the storms, the estimates of interplanetary anisotropy are important. One such possibility is the observation of anisotropy at low energies (at high latitudes with geomagnetic cut-offs below the atmospheric ones and with the asymptotic directions only slightly changing) – Spaceship Earth [17]. At high energies where muon detectors provide the information (the asymptotic directions are not strongly varying), the anisotropy could be obtained from the network of telescopes used for geomagnetic activity alerts [18].

## 3. Penetration of Solar Cosmic Rays (SCR) to the magnetosphere during strong disturbances.

The simplest approach to energetic particle penetration into geomagnetic field from outside is the Stormer cut-off discussed e.g. by [19] (all particles with rigidity below that are forbidden). For vertically incident particles the cut-offs behave as a function of McIlwain's $L$ parameter ~ $L^{-n}$, with n approximately 2 [20].

Low altitude polar orbiting satellites with devices having large geometrical factor for energetic particles may serve as a spectrometer using the geomagnetic filter during the events with solar and/or interplanetary accelerated ions penetrating to magnetosphere. This was done in recent years e.g. with use of experiments on CORONAS-F [21] or PAMELA [22].

More difficult becomes the situation during the geomagnetic disturbances. Although many crossings of the low altitude satellites through the penetration boundary for various geomagnetic situations lead to very detailed cut-off rigidity maps for different levels of geomagnetic activity and local time [23], the large variability of the position of the boundary for the same Dst is observed [24]. Dispersion of the invariant latitudes Λ with the proportionality L ~ $1/\cos^2(\Lambda)$ presented as a dependence on Kp and Dst in [24] confirms large uncertainty of the boundary position for 1-5 and 50-90 MeV protons, if only one parameter (geomagetic activity) is used. Figure 1 indicates the change of boundary position for proton and electron penetration from interplanetary space to magnetosphere for different periods of geomagnetic activity according to data from CORONAS-F satellite.

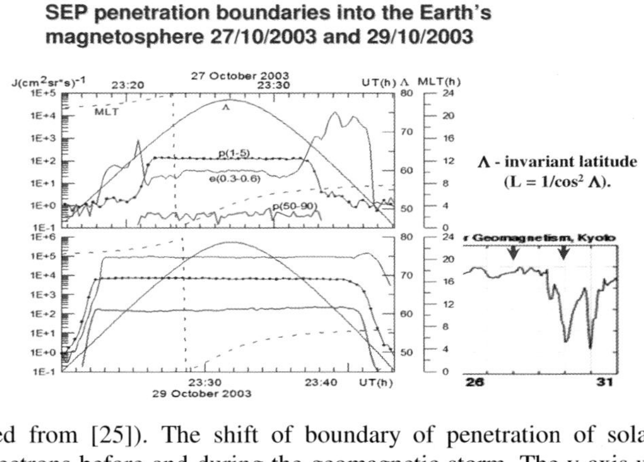


Fig. 1 (adapted from [25]). The shift of boundary of penetration of solar energetic protons and electrons before and during the geomagnetic storm. The y-axis units for the right panel is 100 nT.

During some events there are found structures more complicated than just a single boundary of particle penetration. Solar protons with energy 1–5 MeV might be trapped, creating temporary solar CR belts on L = 2–3 or providing additional flux to the previously existed population. This trapping action is observed as a double boundary effect in the CORONAS-F measurements [26]. It should be mentioned that the value of $L$ at given position during the storms depends on the geomagnetic field model used. 1-100 MeV SCR penetrate into the magnetosphere to lower latitudes as deep as it is not allowed by any magnetic field models. During some strong magnetic storms the penetration boundaries coincide for wide energy range in comparison with normal (expected) penetration structure. It is not clear: is it created by some special "transmissivity" of the magnetospheric boundary during strong magnetic storms or it is a consequence of some changes of the magnetic field structure of the inner magnetosphere.

## 4. Contribution of SCR to the Formation of the Earth's Proton Radiation Belt

Trapped population of particles in the magnetosphere is very important element of radiation in the near Earth's environment with consequencies of space weather effects especially for satellites as well for atmosphere (for events of precipitation of protons and electrons). There are static models working well for the radiation hazard estimates (e.g. [27]). A review of processes affecting the dynamics of trapped radiation, along with the cosmic rays and solar energetic particles can be found e.g. in [28].

There is some mutual relation between the trapped populations of proton dynamics and the penetration of solar particles into the magnetosphere. Paper [29] indicates the significance of the contribution of solar protons to fluxes of trapped radiation in the Earth's outer radiation belt ($L > 2$). During high solar activity periods proton fluxes (1–5 MeV) at $L = 2–3$ for the bulk of time have solar cosmic rays as a source. During solar activity minimum, trapped proton fluxes are controlled by conventional diffusive mechanism under the action of sudden IMF impulses. The variability of protons 27-45 MeV measured on HEO-3 on long time scale is reported at L = 2 -3 in the detailed study [30]. The inward extent and energy independence of the observed loss events are inconsistent with the scattering in a static magnetic field. The geomagnetic cut-off suppresion observed during storms may be the cause of significant losses. Both solar proton injections and radial inward diffusion of preexisting trapped protons is related to the drift resonance with pulses of electric field caused by rapid magnetospheric compression.

The specific orbit of the CRESS satellite allowed to observe the energetic particle injection into inner magnetosphere [31]. Measurements by two satellites, namely CORONAS-F and SERVICE-1 with different altitude have shown that the belt with a maximum at L = 2 formed due to capture of 1 MeV solar protons during the recovery phase of a strong geomagnetic storm October 29 – 31, 2003 remained to be apparent for rather long time [32]. At higher energies the newly formed belt is decaying faster.

Paper [33] presents the two satellite studies of the penetration, trapping, and acceleration of solar protons in the magnetosphere during magnetic storms in November 2004 and January 2005. Acceleration of protons by up to 2 orders was observed after trapping of solar protons 1-15 MeV during the recovery phase of the magnetic storm of November 7-8, 2004. The acceleration was accompanied by an earthward shift of the particle flux maximum for several days, during which the series of magnetic storms continued. After that the intensification was terminated by the precipitation, and a new proton belt split with the formation of two maxima at L similar to 2 and 3 was observed. Reconfiguration of the proton belts during strong magnetic storms changes the intensity and structure of the belts for a long time. The consequences of changes observed first during July 2004 storm did not disappear until November disturbances.

## 5. Energetic electron dynamics during geomagnetic storms

It is well known that the relativistic electrons due to their penetration ability through the satellite systems are very important element of radiation hazard. The failures of the satellites in coincidence with increase of energetic electron fluxes while the high energy proton fluxes were not strongly increasing was indicated e.g. in [34]. Pitch angle diffusion due to interactions with waves and radial diffusion are important mechanisms for dynamics of trapped electrons. Recent paper [35] indicates that the pitch-angle scattering rates of electrons as well as loss timescales depend on the magnetic field model used.

The penetration boundary of 0.3 – 0.6 MeV electrons before and during a strong storm is seen from figure 1. Russian satellite Tatyana had the chance to observe significant reconstructions of the trapped electron fluxes during December 2006 when strong geomagnetic activity occurred. Paper [36] is discussing these variations. Figure 2 is illustrating the variability of high energy electron flux.

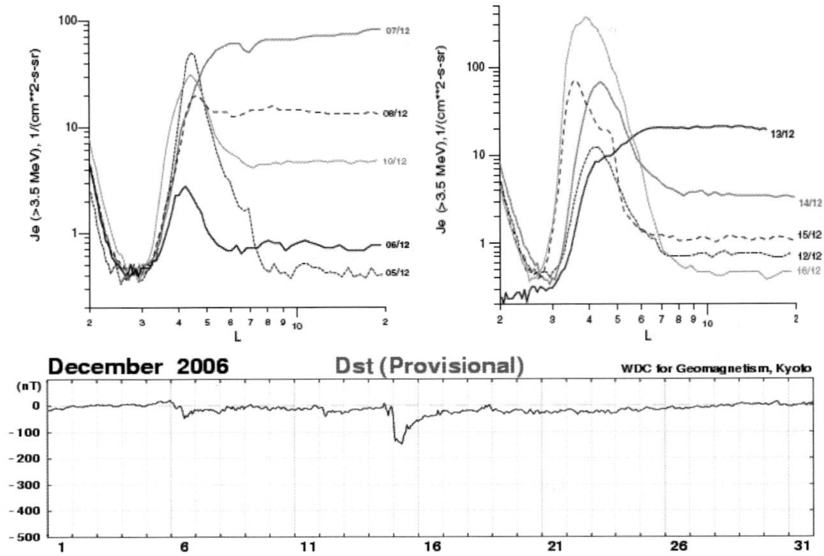


Fig.2. (adopted from [36] and Kyoto WDC). Radiation belt electron (>3.5 MeV) fluxes observed during different days of interval December 5 – 10, 2006 (left) and December 12 – 16 (right panel). The Dst for December 2006 is in the lower panel.

## 6. Concluding remarks.

1. For checking validity of geomagnetic field model by cosmic ray profiles at many neutron monitors during the storms one has to assume of (a) trajectory tracing in the disturbed field and (b) the estimates of the evolving in time of cosmic ray interplanetary anisotropy.
2. Low altitude measurements, namely those by CORONAS-F and SERVICE-1 have shown:
   a. The boundary position of penetrating protons have some features (double structure, single position in wide energy range, lower invariant latitude observed in comparison with the existing models) which needs further study.
   b. The contribution of protons accelerated outside the magnetosphere to the trapped population via their penentration inside is clear and in selected cases the newly established trapped population feature decays slowly and remains at low energies for long time period.
   c. The relativistic electron fluxes important for radiation hazard are highly variable in the magnetosphere at low altitudes during and after strong geomagnetic storms.

All items mentioned above require further experimental and theoretical studies including various models of geomagnetic field now available. Point (2) is relevant for improvements of trapped population models updates.

**Acknowledgments**

This article was created by the realisation of the project ITMS No. 6220120009, based on the supporting operational Research and development program financed from the European Regional Development Fund.

**References**

1. K. Kudela, *Proc. ICRC 2009, 6, 195* (2009).
2. F. Jansen, J. Behrens, *Proc. ECRS 2008*, Kosice, Slovakia, 626 (2008).
3. M. Storini, ECRS 2010, Turku, Finland, http://ecrs2010.utu.fi/done/presentations/EDU1/2A_PA1_Wednesday/2_Storini.pdf
4. A. K. Singh, D. Siingh, R. P. Singh, *Surveys in Geophysics*, in press (2010).
5. L. Desorgher, K. Kudela, E.O. Flückiger et al. *Acta Geophysica, 57 (n°1),* 75 (2009).
6. N. A. Tsyganenko, http://modelweb.gsfc.nasa.gov/magnetos/tsygan.html
7. K. Kudela, R. Bučík, P. Bobík. *Adv. Space Res.*, 42, 1300 (2008).
8. N.A. Tsyganenko, *Planet. Space Sci.* **37**, 5 (1989).
9. P.R. Boberg, A.J. Tylka, J.H. Adams Jr. et al. *Geophys. Res. Lett.*, 22, 9, 1133 (1995)
10. N.A. Tsyganenko and M.I. Sitnov. *J. Geophys.Res.*, 110, A3, 10.1029/2004JA010798 (2005).
11. D.F. Smart and M.A. Shea, *Adv. Space Res.*, 44, 10, 1107 (2009).
12. P. Bobík, PhD thesis, P.J. Safarik University, Kosice (2001).
13. K. Kudela and I. G. Usoskin, *Czech. J. Phys.*, 54, 2, 239 (2004).
14. M.I. Tyasto, O.A. Danilova, L.I. Dorman et al. *Adv. Space Res.*, 42, 1556 (2008).
15. L.I. Dorman, Cosmic Ray Variations and Space Exploration. North-Holland Publ. Co., Amsterdam (1974).
16. N.A. Tsyganenko, H.J. Singer and J.C. Kasper. *J. Geophys. Res.* 108 (A5), 1209, doi:10.1029/2002JA009808 (2003).
17. J.W. Bieber and P. Evenson, *Proc. 24th ICRC*, Rome, 4, 1316 (2005).
18. K. Munakata, J.W. Bieber, S.-I. Yasue et al. *J Geophys Res.*, 105 (27), 457 (2000).
19. D.J. Cooke, J.E. Humble, M.A. Shea et al. *Nuovo Cimento* 14C, 213 (1991).
20. M.A. Shea, D.F. Smart and L.C. Gentile. *Phys. Earth and Planet. Interiors*, 48, 200 (1987).

21. S.N. Kuznetsov, B.Y. Yushkov and K. Kudela. *Cosmic Res* 45(4) 373 (2007).
22. N. De Simone, O. Adriani, G.C. Barbarino et al. Proc. 31st ICRC, Lodz, icrc0794 (2009).
23. D.F. Smart, M.A. Shea, A.J. Tylka et al. Adv Space Res 37, 1206 (2007).
24. I.N. Myagkova, A.V. Bogomolov, B.Y. Yushkov et al. *Bull Russian Acad Sci: Phys* 73(3), 322 (2009).
25. S.N. Kuznetsov. In Model of cosmos, KDU, Moscow, ed. by M.I. Panasyuk and L.S. Novikov, V 1, 2007 (in Russian).
26. L.L. Lazutin, S.N. Kuznetsov and A.N. Podorolsky. *Geomagn. Aeron.*, 47, 2, 187-197 (2007).
27. http://modelweb.gsfc.nasa.gov/models/trap.html
28. R. Vainio, RL. Desorgher, D. Heynderickx et al. *Space Sci. Rev.*, 147, 3-4, 187 (2009).
29. L.L. Lazutin and Yu.I. Logachev. *Cosmic Res.*, 47, 5, 24 (2010).
30. R.S. Selesnick, M.K. Hudson and B.T. Kress. *J. Geophys. Res.*, 115, A08211, doi:10.1029/2010JA015247 (2010).
31. J.B. Blake, W.A. Kolasinski, R.W. Fillius et al. *Geophys. Res. Lett.*, 19, 821 (1991).
32. L.L. Lazutin. *Cosmic Res.,* 48, 1, 108 (2010).
33. L. L. Lazutin, Y.V. Gotselyuk, E.A. Murav'eva et al. *Geomagn. Aeron.*, 50, 2, 168-180 (2010).
34. D. N. Baker, J. H. Allen, S. G. Kanekal, G. D. Reeves, *Eos* 79, 477 (1998).
35. K.G. Orlova and Y.Y. Shprits. *Geophys. Res. Lett.*, 37, L05105, Mar 10 (2010).
36. I.N. Myagkova, M.I. Panasyuk, L.L. Lazutin et al. *Adv. Space Res.*, 43, 4, 489 (2009).

## Cosmic rays of leptons from Pulsars and Supernova Remnants

Roberto A. Lineros*

*INFN, sezione di Torino, I-10122 Torino, Italy and Dipartimento di fisica teorica, Università di Torino, I-10122 Torino, Italy.*

The latest results from PAMELA and FERMI experiments confirm the necessity to improve theoretical models of production and propagation of galactic electrons and positrons. There are many possible explanations for the positron excess observed at energies larger than 10 GeV and for some features around 1 TeV in the total flux of electrons and positrons. Supernovae are astrophysical objects with the potential to explain these observations. In this work, we present an updated study of the astrophysical sources of lepton cosmic rays and the possible and the possible explanation of the anomalies in terms of astrophysical sources.

*Keywords*: cosmic rays, astrophysical sources, supernovae, pulsars

### 1. Introduction

During the last decades, cosmic ray experiments like HEAT[1] and AMS01[2] have found very interesting features in the cosmic rays measuremnts, specially in the component of *electrons*[a] . The theoretical prediction of the positron fraction (number of positron divided by the number of *electrons* per unit of energy) based on standard sources mismatched the experimental results. This feature corresponds to an increment in the fraction for energies larger than 10 GeV. Recently, newer and refined experiments like PAMELA[3] and FERMI[4,5] have presented more accurate measurements of the positron fraction and the *electrons* flux, respectively, confirming the positron excess and other anomalies present in the observations. Let us point that these anomalies have not been completely explained by standard theoretical predictions.

*email:lineros@to.infn.it

[a]in this work, we refer to electrons and positrons as *electrons*

In this work, we analyze the potential of astrophysical sources (supernova remnants, pulsars, and secondary *electrons*) to explain these anomalies. Moreover, we consider the effect of uncertainties in the propagation model and in the sources to verify the robustness of a possible discovery.

## 2. Propagation of electrons and positrons

Since *electrons* are injected in the interstellar medium, these cosmic rays are affected by the inhomogeneous galactic magnetic field, radiation fields from stellar activity and from the microwave background, and the interstellar gas. The continuous interaction with the environment makes them to loose (or gain) energy and to propagate following a diffusion pattern.

The model of propagation of *electrons* is based on a continuity equation for the number density of cosmic rays, known as transport equation, which contains most of the physical processes that those have to face:

$$\frac{\partial \psi}{\partial t} - \nabla\big(D(\varepsilon)\nabla\psi\big) - \frac{\partial}{\partial \varepsilon}\big(b(\varepsilon)\psi\big) = q(\vec{x}, \varepsilon)\,, \tag{1}$$

where $\psi$ is the number density of electrons or positrons per unit of energy, $D(\varepsilon)$ is the diffusion term, $b(\varepsilon)$ the energy loss term, and $q(\vec{x}, \varepsilon)$ is the source term.

The diffusion term is usually considered homogeneous and isotropic in space with a power law dependence in energy:

$$D(\varepsilon) = K_0 \left(\frac{\varepsilon}{\varepsilon_0}\right)^{\delta}, \tag{2}$$

where $K_0$ and $\delta$ are phenomenological parameters inspired by magnetohydrodynamics models of the interstellar medium. The normalization energy scale $\varepsilon_0$ here is fixed at the value of 1 GeV.

The energy loss term depends on the interaction between the galactic environment and the type of cosmic rays. The process that dominates the energy evolution of *electrons* at the GeV–TeV scale is the inverse Compton scattering with the Interstellar Radiation Field and the synchrotron emission related to the Galactic magnetic field.

For simplicity, the inverse Compton scattering can be described by the Thomson regime which is valid for *electrons* with energy lower than a certain threshold (details in Refs. 6 and 7). In this regime, the energy loss term corresponds to:

$$b(\varepsilon) = \frac{\varepsilon_0}{\tau_E} \left( \frac{\varepsilon}{\varepsilon_0} \right)^2 , \tag{3}$$

where $\tau_E$ is the energy loss scale time which is calculated from the total photon energy density.
At higher energies, the Thomson regime is not longer valid to describe the interaction between *electrons* and the photon bath. The Klein-Nishina regime[8] becomes more accurate than the Thomson regime for the interaction between GeV–TeV *electrons* with the ultraviolet, infrared and microwave component of the interstellar radiation field[6,7].

For generic functions of the diffusion and energy loss terms, solutions of the transport equation (Equation 1) are fully analytical. We highlight the solutions for continuous and burst injection cases (details in Refs. 9 and 6) because those are fully of physical meaning. We remark that the analytical approach allows us to study the importance of different sources and to scan the propagation space of parameters in rather short time compared to numerical methods.

### 2.1. *Uncertainties*

Most of the processes included in the propagation models present theoretical and observational uncertainties. The principal propagation parameters are the diffusion parameters ($K_0$, $\delta$), and the half-thickness ($L_z$) of the propagation zone[10]. This zone corresponds to a cylinder of radius equal to the galactic one (20 kpc) and half-thickness $L_z$. The model also includes the escape of cosmic rays from the propagation zone by imposing that cosmic rays density vanishes at the boundaries.

One method to constrain the propagation space of parameter is based on observation of the Boron/Carbon ratio (B/C)[11]. In principle, this method is less sensitive to injection properties of the source, allowing a better determination of propagation parameters. For the scope of our analysis, we assume that propagation space of parameters is common for all species of cosmic rays.

## 3. Astrophyscial sources of *electrons*

In the GeV–TeV range of energy, *electrons* can be produced in many different ways. We highlight the astrophysical sources: spallation with the interstellar medium (secondaries), and supernovae.

### 3.1. *Secondaries*

This component is the result of the spallation of nuclear cosmic rays (protons and alpha particles) off the interstellar gas, which is mainly composed by hydrogen and helium.
The environmental factors that determine the flux of secondaries are the amount of interstellar gas, the cosmic ray density, and the production rate of *electrons* related to the nuclear scattering. Moreover, secondary electrons and positrons are produced in similar amounts. The secondary contribution, in the case of electrons, is not dominant compared to the observed flux. On the other hand, secondary positrons encompass the low energy positron flux.[12] We highlight that the compatibility among observation and propagation space of parameter reaffirm the idea that propagation models are common for every specie of cosmic rays[12]. However, the propagation uncertainties produce bigger uncertainties in the positron flux than in the case of nuclear cosmic rays calculations.

### 3.2. *Supernova remnants and pulsars*

Supernovae are the astrophysical objects responsible of most of the galactic cosmic rays. Moreover, those may give birth to two astrophysical objects with potential to produce *electrons*: Supernova remnants and pulsars.
Supernova remnants can expel a big fraction of the electron cosmic rays which is contained inside the former star, but very few positrons. Different mechanisms have been proposed to enhance the positron production[13], however, it seems to be not enough to explain the anomalies. On the other hand, pulsars are able to produce in same amount positrons and electrons due to the interaction of pulsar's magnetic field with ambient photons.

In both cases, the injection spectra is expected to follow a power-law like function:

$$q_{\rm SN} \propto Q_0 \varepsilon^{-\gamma} \exp\left(-\frac{\varepsilon}{\varepsilon_c}\right), \tag{4}$$

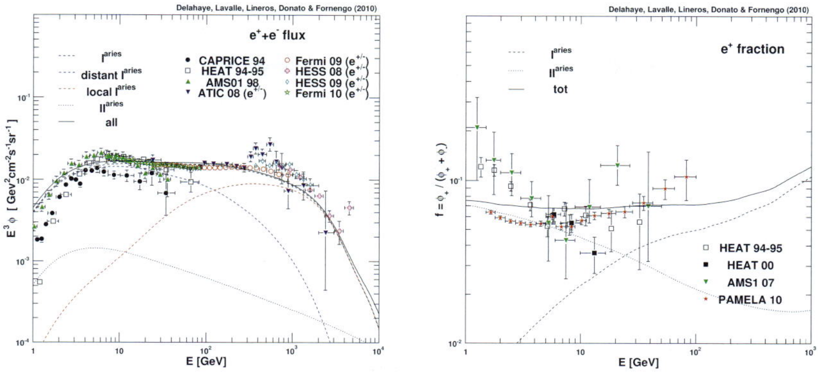


Figure 1. *Electrons* flux (left) and positron fraction (right) versus energy. *Electrons* are calculated using information about supernova remnants, pulsars, and also contribution from secondaries. Both observables are encompassed by the conservative assumption about astrophysical sources. Further details in Ref. 6.

where $Q_0$ is the injection intensity of electrons/positrons, which is calculated from the averaged energy released by the supernova (or pulsar) and the efficiency to convert this energy into electron kinetic energy (details in Ref. 6), $\gamma$ is the power index, and $\varepsilon_c$ is a cut-off energy ($\sim$1–10 TeV) suggested by FERMI and HESS[14].

Let us stress that supernovae are distributed non-smoothly in space and time. The diffusive behavior of propagated *electrons* smooths any evidence of discrete source distribution at low energy, because this range is dominated by the older and farther supernovae. On the other hand, it is expected to be observed some features at high energies due to the contribution of younger and closer objects.

In figure 1, we present some of the results of a detailed study of astrophysical sources of *electrons* at the GeV–TeV scale[6]. We remark that under conservative assumption of injection profiles, the observations by FERMI and PAMELA can be encompassed with just the astrophysical component i.e. supernova remnants, pulsars and secondaries. Also, the presence of inhomogeneities in the local source distribution naturally explains the features observed by FERMI at the TeV range.

## 4. Conclusions

The cosmic rays anomalies observed by PAMELA and FERMI, have triggered a revolution in our understanding about the galactic environment. Dark matter is far the most exiting solution to the puzzle, but this supposition considers that the astrophysical background is absolutely known. Uncertainties in the propagation model and in the sources make harder the task to discriminate a possible non-astrophysical source. After some detailed study regarding galactic supernova population, it seems natural that supernova remnants and pulsars may be the solution to these anomalies. It is indispensable to refine the propagation model by using other complementary observables like diffuse gamma emission and radio observations.

## Acknowledgements

R.L. is grateful to T. Delahaye, J. Lavalle, N. Fornengo, P. Salati, and F. Donato for the rich collaboration and discussion at different stages of the publications referred in this proceeding. Also, R.L. acknowledges financial support given by Ministero dell'Istruzione, dell'Università e della Ricerca (MIUR), by the University of Torino (UniTO), by the Istituto Nazionale di Fisica Nucleare (INFN) within the Astroparticle Physics Project, and by the Italian Space Agency (ASI) under contract Nro: I/088/06/0.

## Bibliography

1. S. W. Barwick and The HEAT Collaboration, *ApJL* **482**, L191+(June 1997).
2. J. Alcaraz *et al.*, *Physics Letters B* **484**, 10(June 2000).
3. O. Adriani *et al.*, *Nature* **458**, 607 (2009).
4. Fermi LAT collaboration, *ArXiv e-prints* (August 2010).
5. The Fermi LAT Collaboration, *ArXiv e-prints* (August 2010).
6. T. Delahaye *et al.*, *ArXiv e-prints* (February 2010).
7. G. R. Blumenthal *et al.*, *Reviews of Modern Physics* **42**, 237 (1970).
8. O. Klein and T. Nishina, *Zeitschrift für Physik A Hadrons and Nuclei* **52**, 853 (1929), 10.1007/BF01366453.
9. R. A. Lineros, *ArXiv e-prints* (December 2008).
10. V. L. Ginzburg, *et al.*, *Astrophysics and Space Science* **68**, 295(April 1980).
11. D. Maurin *et al.*, *ApJ* **555**, 585(July 2001).
12. T. Delahaye *et al.*, *A&A* **501**, 821(July 2009).
13. P. Blasi, *Physical Review Letters* **103**, 051104(July 2009).
14. F. Aharonian *et al.*, *Physical Review Letters* **101**, 261104(December 2008).

# HIGH ENERGY PHENOMENA IN THE LOW ATMOSPHERE; PARTICLE FLUXES FROM THUNDERSTORM CLOUDS

ASHOT CHILINGARIAN, BAGRAT MAILYAN*

*Cosmic Ray Division, Alikhanyan National Laboratory, Alikhanyan brothers 2*
*Yerevan, 0036, Armenia, *E-mail: mbagrat@gmail.com*

The huge flux of thunderstorm correlated particles was detected by the experimental facilities of Aragats Space environmental Center (ASEC) in 2009. Using ASEC detectors with various energy thresholds, spectra of thunderstorm correlated electrons were obtained. Using GEANT4 based detector response calculation we reconstruct gamma ray spectra from the energy release spectra measured by the Aragats Solar Neutron Telescope (ASNT). We have compared very large September 19, 2009 event with smaller events on May 21 and July 9, 2009. The obtained energy spectra prove the validity of the relativistic runaway electron avalanche (RREA) process in the thunderstorm atmospheres.

## 1. Introduction

A relativistic runaway electron avalanche in the atmosphere is believed to be the major driving force of the detected high-energy phenomena connected with thunderstorm activity [1]. The RREA electron energy spectrum believed to have a universal exponential shape independent on electrical field embedded in the thunderstorm atmosphere [2]. Dwyer et al., [3] and references therein claim that the electron energy spectrum is approximately exponential with an e-folding energy of 7.3 MeV.

Terrestrial γ-ray flashes (TGFs), observed by gamma-ray observatories aboard satellites, have been firmly established during the last decades [4]. Recently, it was reported that substantial fraction of TGF events observed by the CGRO/BATSE instrument are, in fact, not gamma-rays at all, but instead are high energy electrons [5], [6]. The source of TGFs was located in the space just above or even within thunderclouds (12 - 20 km, see [7]). Surprisingly, amount of the surface detection of the electron and gamma-ray fluxes correlated with thunderstorms is not large. Only Baksan group was investigated surface particle enhancements correlated with thunderstorms in a systematic way [8]. Unfortunately the location of surface array in Baksan valley did not allow registration of large fluxes. The array is located in deep narrow valley and thunderclouds are rather high.

However, if electron and gamma-ray fluxes are unambiguously detected by orbiting gamma-ray observatories near 500 km from the source, we can expect very intensive fluxes on highland altitudes from thunderclouds located few hundred meters above.

Aragats high mountain research station [9] of the Yerevan Physics Institute is located on the highland, 3250 m above sea level, 500 m below from the southern peak of Aragats (3750 m.), near large lake; other 3 peaks of Aragats are located from 10 to 15 km far. The thunderstorm activity on Aragats is extremely strong in May-June. Sometimes, lightning hits the ground in the vicinity of the station continuously during an hour and more. Thunderstorm clouds are usually below southern peak (i.e. not higher than 500 m above) and some times 100-200 m above the station. In 2009, we have detected several intense fluxes of particles at Aragats detected by various particle detectors [10].

## 2. RREA Electron and Gamma Energy Spectra on 19 September 2009

The electron spectrum was obtained using count rates measured by the same type 5 cm thick scintillators with various energy thresholds. Using enhancements (peaks) detected in these five cm thick scintillators, we select groups corresponding to 4 diverse energy thresholds. The energy thresholds were determined by comparing the mean count rates of scintillators with simulations of the secondary cosmic ray fluxes. Proceeding from the gamma-ray energy spectrum, unfolded from the measurements of the ASNT "calorimetric" thick scintillator energy releases spectrum [11], we correct the electron energy spectrum for the gamma-ray contamination.

Thereafter, the obtained spectra were fitted by an exponential function. The electron integral spectrum of the event on 19 September 2009 is shown in Figure 1 along with gamma spectrum. The spectral index is ~ -0.18. Electron spectrum abruptly ended at 25 - 30 MeV, as there is no evidence of additional electrons detected by "11" coincidence of ASNT (the energy threshold 25 - 30 MeV). The electron spectrum is fitted by exponential function - $A \cdot e^{-b \cdot E}$ and gamma spectrum is fitted by power function - $A \cdot E^{-b}$ in the range 7-30 MeV. The third is the spectrum of electrons at height 3400 m, just below the thundercloud, before reaching the detectors. The estimation of this spectrum is given below.

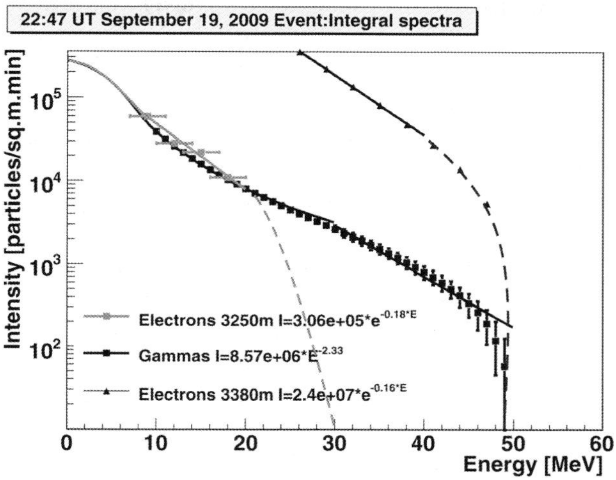


Figure 1. Unfolded electron spectrum at 3250 m and 3380 m and gamma ray spectrum fitted by exponential and power functions respectively, see error bars of the electron and gamma spectra in Table 1.

Table 1. Parameters of the fit for September 19 event

| | a | b | Mean Energy | $\chi^2$/ndf |
|---|---|---|---|---|
| Electron $e^{a-b*E}$ 3250 m | 3.06e+05 ± 0.3e+05 | -0.18 ± 0.001 | 5.5 ± 0.04 | 0.34 |
| Electron $e^{-b*E}$ 3380 m | 1.86e+07 ± 8.2e+04 | -0.152 ± 0.005 | 6.6 ± 0.2 | |
| Gamma $a*E^{-b}$ 3250 m | 1.86e+07 ± 81,853 | -2.33 ± 0.02 | | 2.6 |

From the measured electron energy spectrum we can conclude that at least 20 MeV electrons have reached ground level and were detected by the 5 cm scintillators. We can assume that maximal energy of the RREA electrons can reach 40-50 MeV[a]. Therefore, if we assume average ionization losses ~200 KeV per meter, the most probable height of the cloud will be 100-150 m, well coinciding with the observations of the Armenian meteorological service. Taking for granted the height of 130 m, we extrapolate the RREA electrons estimated spectrum up to the thundercloud by selecting the trial electron

[a] at least AGILE gamma observatory detects one gamma ray with energy 40 MeV, Marisaldi et al., [12], see also [6].

spectrum at 3380 m and simulating with GEANT4 the electron-gamma avalanche till 3250 m.

By trying different test spectra, evaluating the avalanche from 3380 till 3250 m and comparing each of the obtained spectra with experimental one, we have found that the best spectrum above 130 m has a functional form of exp(-E/6.6).

Ionization loses of electrons traversing 130 m of atmosphere at ~3250 m are about 25MeV. So electrons <25MeV can't reach the detectors and we reconstruct the RREA electron spectra only for energies greater than 25MeV.

The number of electrons with energies greater than 25MeV at 3380 m is 350,000/sq.m.min, only 90,000 /sq.m.min electrons with energies greater than 7MeV survive near the ground at 3250 m. The mean electron energy is ~6.6 ± 0.2 MeV at lower edge of the thundercloud (we assume that electrical field is present only in the cloud) and 5.5 ± 0.04 MeV at the ground (130m lower) where ASEC detectors are located. This value is in a good agreement with the value calculated from simulations - 7.3MeV [13].

From the measured RREA electron spectrum and spectrum of secondary cosmic ray electrons at 3400m[b], we have calculated the multiplication rate (avalanche growth factor in the thundercloud electric field). The obtained RREA electron spectrum was extrapolated until 7MeV and the number of RREA electrons was divided by the secondary cosmic ray electron number. The calculations were done for the electrons with energies greater than 7MeV. Obtained factor equals to 330, corresponding to ~6 e-folding lengths. Assuming an area of electric field with radius 500m (maximal distance between ASEC detectors measuring the event), we will get a total number of electrons above 7 MeV ~$3.8*10^{12}$.

The maximal energy gain of electrons per one avalanche length is approximately independent of the atmosphere density and electrical field value and equals ~ 7 MeV [13]. Thus, based on the calculations of the multiplication rate we obtain maximal energy of electrons ~40 MeV in a good agreement with our assumptions and measurements.

## 3. July 9 and May 21, 2009 events

September 19, 2009 thunderstorm event was the largest event observed by ASEC detectors so far. Another strong thunderstorm event was detected on July

---

[b] *we accept here that RREA seed particles are electrons from the Extensive Air Showers –EAS, initiated by primary proton and stripped nuclei; the seed electrons were simulated by the WEB generator of secondary cosmic rays **(T.Sato [14],** http://phits.jaea.go.jp/expacs/).*

9, 2009. We have solved the “inverse” problem for July 9, 2009 event and reconstructed the incident gamma-ray spectrum from energy releases spectrum. In Figure 2 the integral gamma spectrum of July 9, 2009 event is shown.

After reconstruction it was found that there are fewer gammas with energies greater than 7MeV on July 9 than on September 19, 2009. The simulated and measured energy releases spectra are in a good agreement if the gamma spectrum, incident on the roof of laboratory building, has the form of ~ $\mathbf{E^{-4.2}}$. Gamma spectrum of July 9 event is steeper than for September 19 event. The gamma ray intensity is obtained to be ~130,000 /min.m.sq. for particles with energies >7 MeV. Proceeding from obtained gamma-ray spectrum and from calculated gamma-ray registration efficiency in 5cm detector, we estimate number of gammas to be detected in the ASNT 5 cm scintillators (electron energy threshold > 15 MeV) to be ~3600, very close to experimentally detected number 3506. Thus, no additional electrons with energy greater than 15 MeV were observed on July 9, 2009.

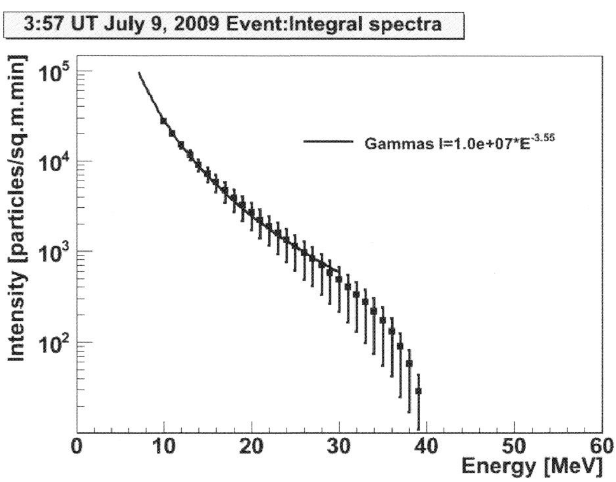


Figure 2. Unfolded gamma ray spectrum of July 9, 2009 event fitted by power function.

On May 21, 2009 another thunderstorm correlated enhancement was observed. Gamma contamination of electron spectrum was calculated and no electrons were seen with energies greater than 15 MeV. In Figure 3 electron and gamma spectra of May 21 event are presented. The gamma intensity of the event is ~30,000 /min.m.sq. for gamma-rays >7 MeV.

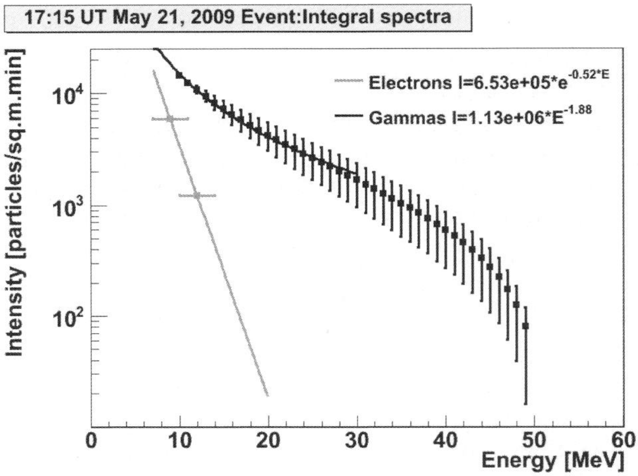


Figure 3. Unfolded gamma ray and electron spectra of May 21, 2009 event.

## 4. Discussion

Measured gamma ray spectra are believed to be the result of the bremsstrahlung produced by runaway electrons. Due to ionization losses and rather large distance to be covered by electrons till detection, there is only one report on the direct detection of RREA electrons. Tsuchiya et al. [15], have simultaneously observed electrons and gamma rays on the top of the mountain Norikura, 2700 m above sea level. The RREA electron source is estimated to be ~90m above the detector and spectral index of the gamma spectrum is $-1.15 \pm 0.09$ and extend till 10 MeV; no energy spectrum of electrons was presented.

We measure for the first time the energy spectra of RREA electrons and gamma rays with ASEC facilities. Detailed calculations of the detector response allow the unfolding of the both spectra from energy release spectrum measured by 60 cm thick scintillator and from count rates of plastic scintillators operated under different thresholds.

Measured electron spectra follow the exponential law and gamma ray spectra - the power law. On the other hand, we expect that if "parent" electron spectrum is exponential the bremstrahlung gamma-ray spectrum should be also exponential. This looks like contradiction; however, gamma rays are emitted already from the beginning of electron-photon avalanche, where electron spectrum is not formed yet. According to simulations of [16], the RREA electrons starting in electrical field of ~200 kV/m strength and using as seed power-law EAS electrons, follow power law spectrum during several e-folding lengths. And gamma rays born during these first stages of the RREA process will follow the parent electron spectrum, i.e., the power law spectrum. But higher energy electrons gaining energy in the electric field give more bremsstrahlung radiation than lower energies. Large energy electrons loose more energy than low energy electrons and spectrum becomes steeper (power law turns to exponential).

Proceeding from the maximal energy of electrons detected and from the simulations of electron-gamma avalanche below the thunderstorm cloud, we recover the electron spectrum just below thunderstorm cloud in which the plausible values of electrical field strength and structure were assumed. Calculations were used to estimate the electron growth factor and total number of electrons originated from RREA process. If thunderclouds are low enough, RREA electrons may reach the detectors. It's just the case of September 19, 2009 event; we locate the cloud height on 100-150 m above the Aragats research station. The mean RREA electron energy equals 6.6 MeV just below the thundercloud and it diminished to 5.5 MeV at ground after energy losses in the atmosphere. These values are in a very good agreement with the simulation results [3].

In the Figure 4 the differential gamma rays energy spectra of the three biggest events in 2009 are presented. Spectral indices vary in a large range (2.5-4.7); maximal gamma ray energy is 40-50 MeV.

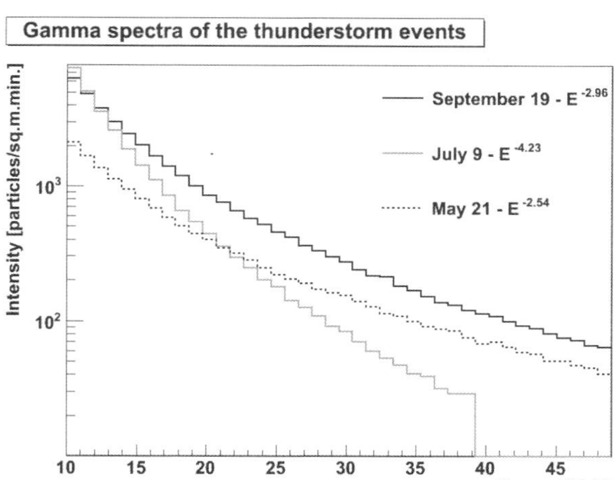


Figure 4. Differential spectra of gammas for three various events.

## 5. Conclusions

For the first time we present the energy spectra of RREA electrons and gamma rays, using the variety of ASEC detectors including the enhanced possibilities of ASNT to measure energy releases in the thick scintillator. Gamma ray spectra are described by power law and electron spectra by exponential law. The difference in spectral shape is confirmed by simulations of RREA development in atmospheric electric fields. Seed electrons spectrum obeys the power ànd becomes exponential only after avalanche leave behind several e-folding lengths in thunderstorm cloud. The energy spectra of electrons and gamma rays are analyzed for 3 energetic events, which differ by intensity, power indices and maximal energy. Every event is unique due to different cloud height above the detector, controlling the number of electrons reaching the detector, and due to the electric field structure and intensity. The obtained spectra are in a good agreement with theory of the RREA process in the thunderstorm atmosphere.

**Acknowledgments**

This work was funded by CRDF-NFSAT-SCS MES RA grant ECSP09-69/A-27.

**References**

1. A. V. Gurevich, G. M. Milikh, and R. A. Roussel-Dupre, *Phys. Lett.* **A165**, 463 (1992).
2. L. P. Babich, E. N. Donskoy, R. I. Il'kaev et al., *Plasma Physics Reports* **30(7)**, 616 (1995).
3. J. R. Dwyer, D. M. Smith, M. A. Uman et al., *J. Geophys. Res.***115**, D09206, doi:10.1029/2009JD01203910 (2010).
4. G. J. Fishman, P. N. Bhat, and R. Mallozzi et al., *Science* **264,** 1313 (1994).
5. J. R. Dwyer, B. W. Grefenstette, and D. M. Smith, *Geophys. Res. Letters* **35**, L02815, doi:10.1029/2007GL032430 (2008).
6. G. J. Fishman and M. Briggs, *Report to TEPA-2010 conference*, Nor Amberd, Armenia (2010).
7. J. R. Dwyer and D. M. Smith, *Geophys. Res. Letters,* **32**, L22804, doi:10.1029/2005GL023848 (2005).
8. N. S. Khaerdinov, A. S. Lidvanskiy and V. B. Petkov, *Atmospheric Res.* **76**, 346 (2005).

9. A. A. Chilinagarian, R. Mirzoyan and M. Zazyan, *Advances in Space Res.***44**, 1183 (2009).
10. A. A. Chilingarian, A. Daryan, K. Arakelyan et al., *Phys. Rev. D* **82**, 043009 (2010).
11. A. A. Chilingarian, L. Melkumyan, G. Hovsepyan and A. Reymers, *NIM* **A574**, 255 (2007).
12. M. Marisaldi, F. Fuschino, C. Labanti et al., Detection of Terrestrial Gamma-Ray Flashes up to 40 MeV by the AGILE satellite, *Journal of Geophysical Res.* **115**, CiteID A00E13 (2010).
13. J. R. Dwyer, *Geophysical Res Lett.* **31**, L12102, doi:10.1029/2004GL019795 (2004).
14. T.Sato, EXPACS: Excel-based Program for calculating Atmospheric Cosmic-ray Spectrum. User's Manual, (http://phits.jaea.go.jp/expacs/), (2009).
15. H. Tsuchiya, T. Enoto, T. Torii et al., *Phys. Rev. Lett.* **102**, 255003 (2009).
16. L. Vanyan, *Report to TEPA-2010 conference,* Nor, Amberd, Armenia, (2010).

# The cosmic-ray populations of nearby galaxies

P. MARTIN* on behalf of the *Fermi*/LAT collaboration

*Max-Planck Institut für extraterrestrische Physik,*
*85748 Garching, Germany*
**E-mail: martinp@mpe.mpg.de*
*www.mpe-garching.mpg.de*

Every star-forming galaxy is thought to host a large-scale population of cosmic rays accelerated in the various astrophysical shocks that accompany the evolution of massive stars from the main sequence to the compact remnant. As they propagate in the interstellar medium, these non-thermal particles radiate in the radio and gamma-ray bands through interactions with matter and radiation fields. The resulting diffuse glow bears the marks of the cosmic-ray acceleration and transport processes, and the comparison of the emissions from different galactic systems thus can provide insights into both aspects of the cosmic-ray phenomenon. Such a population study was however not possible before the LAT gamma-ray telescope onboard the *Fermi* satellite came into operation about two years ago. In this paper, we review the detections of external star-forming systems achieved by the *Fermi*/LAT so far, and we emphasise how these observations hold potential for improving our understanding of galactic cosmic rays.

*Keywords*: Cosmic rays – Galaxies: M31, M33, Milky Way, LMC, SMC, M82, NGC253, NGC4945 – Gamma rays: galaxies

## 1. Introduction

Cosmic rays (CRs) with energies $< 10^{15}$ eV are very likely energised by the powerful outflows that accompany massive star evolution, from main-sequence stellar winds to supernova explosions and even beyond with the fast rotation/accretion-driven plasma ejections from compact objects. Yet, CRs are not simply a side-product confined to these phenomena. They actually diffuse out from their sources to eventually form an essential component of the entire Milky Way (MW) and as such, they have observational consequences on quite large scales. The sky in gamma rays with energies >100 MeV is dominated at the 80-90% level by diffuse emission from CRs illuminating the interstellar medium (ISM) of our Galaxy through the processes of Bremsstrahlung and inverse-Compton scattering for CR leptons, and neutral pion production and decay for CR nuclei. In the same time, at the other end of the electromagnetic spectrum, the <1 GHz radio flux is dominated by synchrotron emission from CR leptons spiralling in the Galactic magnetic field.

Extended radio emission of that nature has been observed in countless galaxies and the relation to the massive-star-forming activity seems to be evidenced by the so-called FIR-radio correlation, at least to first order.[1–3] Conversely, the gamma-ray sky has remained nearly devoid of external galaxies whose emission in the >100 MeV range result primarily from stellar activity and the associated CR-ISM interactions. At the end of the 1990s, the LMC was the only such source of high-energy gamma-rays detected by the EGRET telescope onboard the CGRO satellite.[4]

This was regrettable since the characteristics of the diffuse gamma-ray emission from a given galaxy can inform us about its CR population and particularly about its hadronic component, which cannot be directly inferred from the radio observations, and so comparing the emission from galactic systems with different geometries, contents, and physical conditions provide the possibility to make further progress in understanding the origin and propagation of CRs. In addition, studying externally-viewed systems allows to get rid of the complications inherent to line-of-sight accumulation in the Milky Way and from the possible bias due to our particular position in the Galactic disk and in the local bubble especially.[5,6]

The higher sensitivity of the LAT telescope onboard the *Fermi* satellite launched in 2008 was predicted to extend the sample of star-forming gamma-ray-emitting galaxies to several nearby objects, thus allowing for the first time a population study. In this paper, we review the detections achieved by the *Fermi*/LAT over its first two years of operation and we emphasise how these observations hold potential for challenging and/or improving our understanding of galactic cosmic rays.

## 2. The *Fermi*/LAT sample of star-forming galaxies

External galaxies actually are a major source class in the >100 MeV sky, as they account for about half of the entries in the 1FGL catalog, the rest being unassociated objects.[7] Yet, the vast majority of these galaxies emit in gamma-rays through the activity of an accreting central black hole or so-called galactic nucleus. In this work, we are interested only in galaxies that emit in gamma-rays through CR-ISM interactions resulting from stellar activity and in the following we will therefore refer to these systems as star-forming galaxies.

After about two years of operation, the *Fermi*/LAT data have allowed the detection of 6 external star-forming galaxies. Among these are the Small and Large Magellanic Clouds (hereafter SMC and LMC), and the starburst galaxies M82 and NGC253.[8–10] More recently the Andromeda galaxy M31 was observed just above the sensitivity limit of *Fermi*/LAT for about two years of data, while its neighbour the Triangulum galaxy M33 remained undetected.[11] Gamma-ray emission from the composite starburst/Seyfert 2 galaxies NGC4945 and NGC1068 was reported by,[12] but the authors concluded that the emission from NGC1068 may be dominated by the active nucleus activity. In the following, we will discuss only the LMC, SMC, M31, M82 and NGC253. For the population study, we also considered the Milky Way (MW) for comparison and we used as its total gamma-ray luminosity the value computed from the GALPROP model, which is based on cosmic-ray, *Fermi*/LAT and other experimental data.[13]

## 3. Morphology

The proximity of the Magellanic Clouds (hereafter MCs) and their relatively large angular sizes (about 6-8° and 8-10° in optical for the SMC and LMC respectively) allowed to unambiguously resolve their gamma-ray emission using the *Fermi*/LAT. From only 11 months of data, the LMC was already spatially resolved at a level enabling to correlate the gamma-ray emission with known structures in the galaxy, while 17 months of data did not provide enough statistics for such correlations to be conclusively established in the SMC. The emission from the other galaxies in our sample is consistent with point-source emission (although there is marginal evidence for source extension in the case of M31), so the morphology information comes from the MCs only.

In the LMC, the distribution of the gamma-rays reveals a strong emission from the 30 Doradus direction. An estimated ~25% of the total >100 MeV flux indeed come from a rather compact region ~ 200 pc in size coincident with this very active star-forming region, one of the most active of the Local Group of galaxies. On a larger scale, the gamma-ray emission is poorly correlated with the gas distribution but instead strongly correlates with tracers of massive stars like ionised gas or WR stars. These observations therefore establish an unambiguous connection between CR acceleration sites and massive-star-forming regions. They also suggest a short diffusion length for 1-10 GeV protons, which apparently remain confined close to their production sites. Related to that, it is also interesting to note that the gas ridge south of 30 Doradus, which hosts ~20% of the total gas mass of the galaxy, is definitely not a bright source of high-energy emission, thereby suggesting that 1-10 GeV protons do not so easily diffuse through the gas.

In the SMC, the situation is less clear-cut because of the more limited statistics. The emission is not clearly correlated with the gas distribution or with the active star-forming regions, but a certain correlation with supergiant shells is observed, as in the case of the LMC. More *Fermi*/LAT data are needed here to firmly establish the morphology of the gamma-ray emission from the SMC. For both the LMC and SMC, however, the situation seems to contrast with the MW, where gas distributions (like HI, CO or E(B-V) maps) were found to be pretty good tracers of the gamma-ray diffuse emission.[14]

## 4. Spectrometry

The spectra derived for our sample of star-forming galaxies look quite similar and are actually rather flat, with a spectral index ~2. For most objects the spectra are not really constraining above ~10 GeV because of limited statistics and so the slope of the spectrum up to TeV energies remains rather ill-defined from *Fermi* measurements.

For the Magellanic Clouds, a dedicated spectral modelling was done from a one-zone model, taking into account the specificities of the MCs in terms of gas mass, size...etc, but assuming a electron-to-proton ratio and CR spectrum similar to the local MW. The modelled spectra fit the data quite well (with an underdensity of CRs compared to locally, see further down) and the distribution of the different spectral component is similar to what is obtained for the MW. In both cases, the assumptions of electron-to-proton ratio and CR spectrum similar to the local MW do not conflict with the radio synchrotron measurements, but the uncertainties are large. In the case of M31, the measured spectrum has a shape consistent with the one obtained for the MW from the GALPROP model.

The cases of M82 and NGC253 deserve more attention as both galaxies were detected at TeV energies by VERITAS and HESS respectively.[15,16] The measured intensities at TeV energies suggest for both objects a spectral slope of $\sim 2.2 - 2.3$ over the GeV-TeV range.[8,17] If the gamma-ray emission is dominated by hadronic processes (pion decay), then the slope of the gamma-ray spectrum above ~10 GeV reflects the slope of the underlying CR nuclei population. That would mean that the population of CR nuclei in both starburst galaxies has a spectral distribution close to the injection one (slope ~2.2-2.4), much harder than the slope ~ 2.7 inferred for the CR nuclei population of our Galaxy. Such a difference may be indicative of a different transport scenario in M82 and NGC253. The transport of CR nuclei in the MW is dominated by the energy-dependent diffusive escape that steepens the CR injection spectrum by ~ 0.5. In contrast, the transport of CR nuclei in starburst galaxies may be dominated by convection by strong galactic winds and/or energy losses by pion production in their very dense cores, both energy-independent processes that can preserve the injection spectrum. Alternatively, or in addition, diffusion in the very active starburst cores might be somehow reduced in the hot medium where most CRs are injected, as was observed to be the case in 30 Doradus.

## 5. Photometry

From the observed > 100 MeV photon fluxes, it is possible to estimate the average CR density in the galaxies of our sample. Under the assumption that the gas content of a galaxy is uniformly irradiated by CRs, the > 100 MeV emissivity can be derived. Then, a comparison to the locally measured value gives an upper-limit on the average CR nuclei density in the galaxy, under the implicit assumption that the CR nuclei spectrum is similar (which appears to be the case from the spectral measurements for the Local Group galaxies at least, see above).

The inferred average CR densities for the SMC, LMC and M31 are smaller than the value measured locally in the MW by factors of 6-7, 2-4 and 3 respectively. In the case of the SMC, such a lower CR density may be due to a smaller CR confinement volume and the same conclusion can probably be reached for the LMC. For M31, the smaller CR density may be due to a lower CR injection rate (per unit volume). In contrast, M82 and NGC253 have average CR densities that are larger than the local one by factors of ~100-1000 but here M82 and NGC253 cannot be directly compared to the MW since their CR population seems to have a harder spectrum (see above).

Yet, the resolved emission from the LMC seems to show a more complicated picture than these average estimates. In the Fig. 10 of Ref. 9, which shows the emissivity (or equivalently CR density) distribution across the LMC, it can be seen that, although the emissivity in the LMC is on average 2-4 times smaller than the local value, it can reach values twice as high over a significant fraction of the galaxy. This naturally comes from the poor correlation of the gamma-ray emission with the gas distribution, which means that in places the emissivity should be pushed to high values to compensate for the low gas column density.

## 6. Population study

The variation of the high-energy gamma-ray luminosity from one galaxy to another may inform us about how the CR population is affected by the global galactic properties, and this can bring constraints on the CR production and transport processes. In Ref. 11, we computed the >100 MeV gamma-ray luminosities $L_\gamma$ for the galaxies of our sample from the measured gamma-ray fluxes and the distance to each object, and then searched for a correlation with global galactic parameters like the total gas mass ($M_{gas}$) of the star formation rate ($SFR$). The distances, gas masses and star formation rates were extracted from the literature and the ranges of values considered in our population study reflect the current scatter in the estimation of these parameters from various methods.

We found no correlation of $L_\gamma$ with $M_{gas}$ but instead obtained a strong correlation of $L_\gamma$ with $SFR$, with $L_\gamma$ increasing as the power 1.4 $\pm 0.3$ of $SFR$. Such a link is not totally unexpected since $SFR$ can be considered as a proxy for the CR injection rate, which will obviously influence the gamma-ray luminosity, but such a tight correlation over such a variety of galaxies is surprising. If determined from the Local Group galaxies only (SMC, LMC, M31, MW and the upper limit on M33), the correlation still holds for the starburst galaxies M82 and NGC253. In addition, statistical tests based for instance on the Kendall tau coefficient confirm that the $L_\gamma - SFR$ correlation is significant and not an effect of the distance-squared multiplication that affect both variables.

The $L_\gamma - SFR$ correlation is reminiscent of the FIR-radio correlation that is thought to arise, at least to first order, from CR electron calorimetry. Yet, CR proton calorimetry can definitely not be invoked to explain the $L_\gamma - SFR$ correlation since galaxies like the MW have proton calorimetric efficiencies of the order of $\sim 1\%$.[13] The observed trend could be explained by an increasing CR proton calorimetric efficiency or CR grammage as we move from small galaxies like the SMC, that may have relatively small CR confinement volume, to larger galaxies like NGC253, that have very dense cores where pionic losses are very likely enhanced. A significant contribution of CR electrons to the gamma-ray luminosity would tend to favour a correlation of $L_\gamma$ with $SFR$ because CR electron calorimetry is very likely achieved with high efficiencies in most galactic systems (as suggested by the FIR-radio correlation). In some GALPROP models of the MW, CR electrons can account for up to $\sim$40% of the gamma-ray luminosity through the inverse-Compton process. Such a high contribution would obviously drive some correlation of $L_\gamma$ with $SFR$ if it were a widespread characteristic of the gamma-ray emission of star-forming galaxies.

Whatever the origin of the $L_\gamma - SFR$ correlation, assuming that it holds for M33 allows estimating the gamma-ray luminosity of the galaxy and its associated flux at Earth. From the latter, we computed that M33 may be detected by *Fermi*/LAT within the next few years. The ongoing survey of the sky by *Fermi*/LAT will probably lead to the detection of new objects and very likely allow defining constraints on others, which will enable us to confirm the existence of such a global scaling relation for the gamma-ray luminosity of star-forming galaxies.

## 7. Conclusion

After two years of operation, the LAT instrument onboard the *Fermi* satellite has allowed the detection of 6 star-forming galaxies emitting in high-energy, GeV gamma-rays as a result of their stellar activity and the associated CR-ISM interactions. The current sample of such objects includes the SMC, LMC, M31 and the starburst M82, NGC253 and NGC4945, with the Magellanic Clouds being spatially resolved at GeV energies and M82 and NGC253 being detected also at TeV energies by VERITAS and HESS, respectively. These observations allow for the first time to compare the emission from CR-ISM interactions in galactic systems with different geometries, contents, and physical conditions, hence providing the possibility to make further progress in our understanding of the origin and propagation of CRs.

The LMC displays a gamma-ray emission that is poorly correlated with the gas distribution but instead strongly correlates with massive stars, and in particular with the very active 30 Doradus region. This confirms the role of massive stars in CR acceleration and suggest a short diffusion length of ~1-10 GeV CR protons around their injection sites. The emission from the SMC does not seem to be correlated with neither the gas or the massive star forming regions, but the current photon statistics are too limited to conclusively relate the gamma-ray emission with particular features in the galaxy. The spectra of the galaxies in our sample are quite similar in the GeV range and consistent with that inferred for our Galaxy. The starbursts M82 and NGC253 seem to have a relatively hard spectrum over the GeV-TeV range, possibly indicative of a harder CR proton distribution and thus of a different transport scenario compared to the Milky Way. The average CR densities in the Local Group objects are lower than the local Galactic value, but the resolved emission from the LMC shows that such average estimates may not be representative. Finally, we found a surprisingly tight correlation of the gamma-ray luminosity and the total star formation rate for the objects in our sample. Such a scaling relation, if confirmed by future *Fermi*/LAT observations of star-forming galaxies, holds potential for defining constraints on CR production and transport processes.

## References


1. G. Helou, B. T. Soifer and M. Rowan-Robinson, *ApJl* **298**, L7 (1985).
2. M. S. Yun, N. A. Reddy and J. J. Condon, *ApJ* **554**, 803 (2001).
3. E. J. Murphy, G. Helou, R. Braun, J. D. P. Kenney, L. Armus, D. Calzetti, B. T. Draine, R. C. Kennicutt, *et al.*, *ApJl* **651**, L111 (2006).
4. P. Sreekumar, D. L. Bertsch, B. L. Dingus, C. E. Fichtel, R. C. Hartman, S. D. Hunter, G. Kanbach, D. A. Kniffen, *et al.*, *ApJl* **400**, L67 (1992).
5. I. A. Grenier, *ArXiv Astrophysics e-prints* (2004).
6. A. Putze, L. Derome and D. Maurin, *A&A* **516**, A66 (2010).
7. A. A. Abdo, *et al.*, *submitted to ApJs* (2010).
8. A. A. Abdo, *et al.*, *submitted to ApJl* (2009).
9. A. A. Abdo, *et al.*, *A&A* **512**, A7 (2010).
10. A. A. Abdo, *et al.*, *A&A* **523**, A46 (2010).
11. A. A. Abdo, *et al.*, *A&A* **523**, L2 (2010).
12. J. Lenain, C. Ricci, M. Türler, D. Dorner and R. Walter, *ArXiv e-prints* (2010).
13. A. W. Strong, T. A. Porter, S. W. Digel, G. Jóhannesson, P. Martin, I. V. Moskalenko, E. J. Murphy and E. Orlando, *ApJl* **722**, L58 (2010).
14. A. A. Abdo, *et al.*, *ApJ* **710**, 133 (2010).
15. F. Acero, *et al.*, *Science* **326**, 1080 (2009).
16. V. A. Acciari, *et al.*, *Nature* **462**, 770 (2009).
17. B. C. Lacki, T. A. Thompson, E. Quataert, A. Loeb and E. Waxman, *ArXiv e-prints* (2010).

# USINE: a new public cosmic ray propagation code

## Basic phenomenology, sample results, and a bit of USINE

D. MAURIN

*Laboratoire de Physique Subatomique et de Cosmologie (LPSC), Université Joseph Fourier Grenoble 1, CNRS/IN2P3, Institut Polytechnique de Grenoble, 53 avenue des Martyrs, Grenoble, 38026, France*
*E-mail: dmaurin@lpsc.in2p3.fr*

Sample results obtained with the USINE code are presented. We focus on the nuclear component of cosmic-rays. We pay special attention to the determination of the secondary-to-primary ratio, the radioactive nuclei, the primary species, and finally antinuclei. We provide throughout this contribution simple toy-model descriptions of the phenomenology of these quantities. We conclude by emphasising on the structure and some features of the soon-to-be-release USINE propagation code.

*Keywords*: Cosmic Rays — Propagation — Code

Dead stars pervade the Galaxy and pinpoint the locations where cosmic-ray spectra built up in the past. Numerous young sources, that are still in the process of accelerating charged particles, have be recently detected with the H.E.S.S. telescope, through hadronic (pionic decay from $pp$ collisions with the surrounding gas) and electronic (inverse Compton of the local high radiation field on high energy electrons) induced TeV energy $\gamma-$ray emissions. These charged particles then diffuse away from the sources and pervade the Galaxy, scattered off by the magnetic inhomogeneities. During this wandering, some of them escape from the Milky Way while others interact with the radiation fields and/or the interstellar gas. In the latter case, the charged particles either loose a large fraction of their energies or simply inelastically react (e.g. $pp \to \bar{p} + X$), leading to a secondary contribution to the GCR fluxes (as opposed to the primary contribution from sources). A detector seated on top of the atmosphere (TOA) will have access to the charged particles cruising around the solar neighbourhood.

The CR fluxes are largely dominated in number by protons ($\sim 90\%$), helium ($\sim 10\%$) and heavier species ($\lesssim 1\%$). The subdominant electronic component ($\sim 1\%$) plays nonetheless an important role regarding the diffuse $\gamma$−ray background. Secondary fluxes of $\bar{p}$, $\bar{d}$, positrons and $\gamma$ provide the most sensitive probes for indirect detection: the annihilation of GeV to TeV mass DM particles might distort the background induced by standard secondary processes. These fluxes are calculated as a convolution of the primary $p$, He (and $e^-$) source spectra, the cross sections for the processes involved, using the gas (and the radiation field) distribution in the Galaxy and a description of the diffusion process. The charged fluxes are isotropised by the diffusion. As a consequence, i) these fluxes vary smoothly with the position in the Galaxy, ii) contrarily to $\gamma$ and $\nu$, hadronic astronomy is not possible, except at the highest energies (i.e. $\gtrsim 10^{19}$ eV). These fluxes can be directly measured at Earth, or indirectly evaluated at different locations through the so-called diffuse $\gamma$ and $\nu$ emission they produce by interacting with the interstellar medium. The $\gamma$−ray emission is currently measured by the FERMI instrument, providing new insights and challenges for cosmic-ray propagation models (see Andy Strong's contribution). The $\nu$ diffuse emission is very interesting, since it has a purely hadronic origin. However, the sensitivity of existing (ANTARES, AMANDA) and future experiments (km$^3$, ICECUBE) might barely be sufficient to detect it. The GCR journey, from sources (with $\gamma$−ray astronomy) to their detection (fluxes of charged particles, diffuse emission...), is summarised in the schematic Fig. 1.

This contribution gives a brief overview of the results obtained with the propagation code USINE in the last few years. It focuses on the nuclear and standard component of the cosmic-ray fluxes (see Pierre Salati's contribution for the exotic component, and many other contributions for the leptons and $\gamma$-rays). We discuss i) secondary-to-primary ratios that are used to determine the value of the transport coefficients; ii) $\beta$-radioactive species that allow one to break the degeneracy between the diffusion coefficient and the halo size of the Galaxy; iii) primary species that are being measured by many recent experiments (BESS, CREAM, PAMELA, TRACER) and are deeply connected to the source spectrum; and iv) standard (and a few words on exotic) antiprotons and antideuterons. We finish with some features of the code USINE that should soon become public[a].

[a]Note that we do not have time in this contribution to talk about the many other very interesting astrophysics that can be made using all the nuclei found in the CRs (heavy nuclei, EC-decay nuclei, anomalous isotopic abundances, etc.).

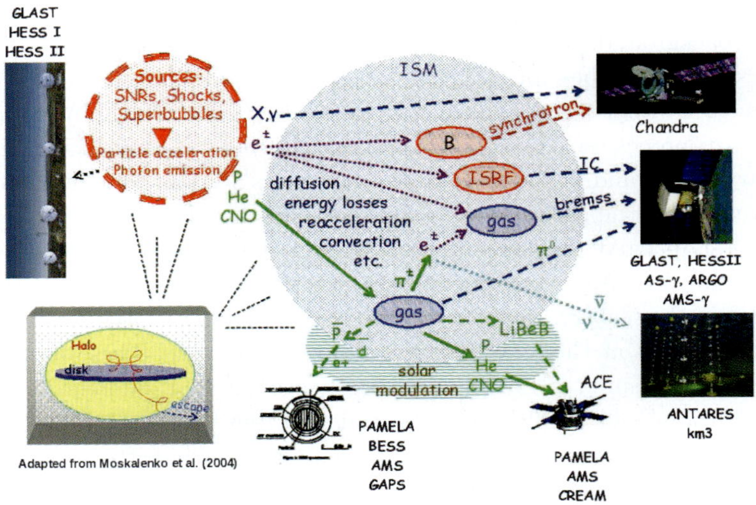

Fig. 1. Schematic view and history of GCRs: i) acceleration in sources, ii) diffusion in the Galaxy along with secondary production of antiparticles and diffuse emission, and iii) solar modulation for charged particles before detection in balloon-borne or satellite experiments. Adapted from Moskalenko et al. (2004).

## 1. Secondary to primary ratio

The transport equation is described in standard textbook. It generally contains a source term (standard or exotic source/production), a diffusion term, and possibly convection, energy gain and losses, and catastrophic losses (fragmentation and decay). A full numerical treatment is generally required to solve the transport equation. However, analytical (or semi-analytical) solutions may be derived assuming a simplified description of the spatial dependence of the ingredients (e.g. the gas distribution) and some transport parameters.

### 1.1. *Toy model*

We consider an infinite plane whose density and source distributions do not depend on $r$ (see Fig. 2). The thin-disc approximation is used ($h \ll L$). The diffusion/convection transport equation (without energy redistributions) for a source term $q(z, E)$ reads

$$-\frac{d}{dz}\left\{K(z)\frac{dN}{dz}\right\}+\frac{d}{dz}[V_{\rm gal}(z)N]+nv\sigma 2h\delta(z)N=q(z,E). \tag{1}$$

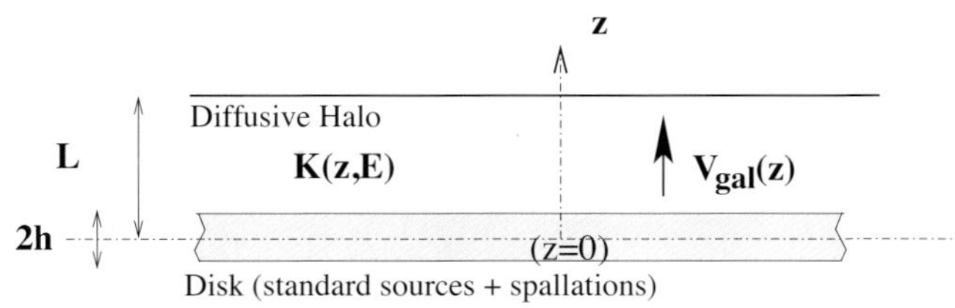


Fig. 2. Upper half-plane of the 1D infinite disc along $r$ (half-thickness $h$). Cosmic rays diffuse in the disc and in the halo, with a diffusion coefficient $K(z, E)$.

The diffusion coefficient is generally assumed not to depend on the position: $K(z, E) = K(E)$. In this thin disc model, if we further drop the convective term, the transport equation reads ($N$ is the differential number density):

$$-KN'' + nv\sigma 2h\delta(z) \times N = 2h\delta(z)Q(E).$$

For $z = 0$, the solution is

$$N(0) = \frac{2hQ(E)}{2K/L + nv\sigma} \, .$$

If we push further this simple model and neglect spallation, we have for a species present in the source term (so-called *primary* contribution) $N^p \propto Q(E)/K(E)$. For a species only produced by interaction of a primary with the interstellar medium (so-called *secondary* species), we have $N^s \propto N^s/K(E)$, such as

$$\frac{N^s}{N^p} \propto K(E).$$

The exact shape of the source term is still debated (see P. Blasi and L. Drury's contributions), but it is generally assumed to be a power-law of index $\sim 2$. The diffusion coefficient is generally taken to be a power-law $K(E) = \beta K_0(z)\mathcal{R}^\delta$ ($\mathcal{R}$ is the rigidity, $\beta = v/c$).

From the above equation, we note that by measuring a secondary-to-primary ratio, such as B/C, we are able in principle to determine the slope and the normalisation of the diffusion coefficient.

### 1.2. *Degeneracy $K_0/L$*

The Leaky-box equation has been used for more than forty years to successfully describe most of the nuclear components, and also the B/C ratio. The Leaky-box equation states that the CR flux is a balance between the source term and an energy-dependent probability to escape from the Galaxy. This leads to an equation, the solution of which has the same structure to that of the diffusion one above. Hence, the diffusion model parameters are related to the escape probability by

$$\tau_{\rm esc} \propto \frac{2K}{L}.$$

If a given value of the escape probability reproduces the B/C ratio, then it fixes the value of the ratio $K_0/L$ in diffusion models. But it does not specify $L$ or $K_0$ (see next section). This is the so-called degeneracy of the diffusion coefficient normalisation and the halo size of the Galaxy. This degeneracy has been known for a long time and is illustrated in Fig. 3, obtained from a study using a scan of the parameter space with the full diffusion equation.

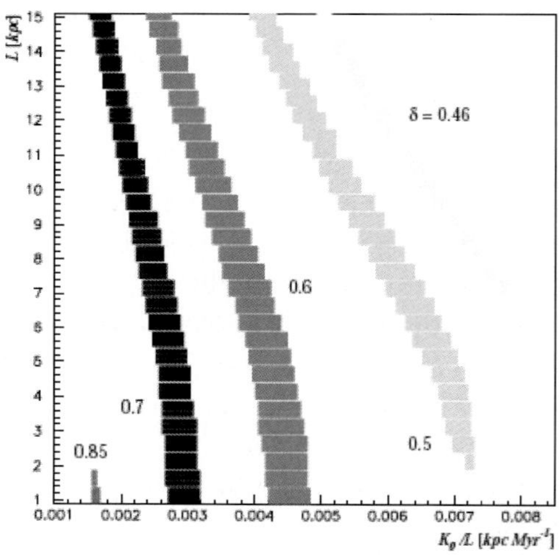


Fig. 3. For fixed values of $\delta$, the large band show all the models consistent with B/C data. Using only stable nuclei, the degeneracy $K_0/L$ is illustrated on this figure based on the diffusion/convection/reacceleration configuration.[1]

## 1.3. *Status of transport parameter determination*

Given the solution of the transport equation, it should be, in principle, simple enough to fit the B/C ratio to get a range of preferred values for the transport parameters. However, as shown in a series of studies we carried out, first based on a parameter scan analysis,[1,2] then on a more robust

Markov Chain Monte Carlo analysis,[3,4] the value of the spectral index $\delta$ of the diffusion coefficient is hardly constrained to be in the range $[0.3 - 0.8]$. Similar studies confirmed this result.[5–7] Moreover, even using radioactive species (see next section) we cannot conclude yet on the value of the halo size $L$.[8,9]

To understand the difficulty of pinpointing the value of $\delta$, we remind that the most precise data fall below 100 GeV/n. In this range, the asymptotic pure diffusive regime where the toy-model equations apply is not reached: energy gains and losses as well as spallations affect this simple picture. The B/C fit is also very sensitive to the assumptions made for the input parameters (cross-sections, gas density, etc.). Indeed, the origin of the large range of values for $\delta$ can mostly be explained as follows:[9]

- if $K(E) = \beta K_0 \mathcal{R}^\delta$, models with reacceleration prefer $\delta \approx 0.3$ (as found in Ref. 10, but as soon as convection is added, $\delta \approx 0.7$;[1]
- if $K(E) = K_0 \mathcal{R}^\delta$ (leaky-box inspired), $\delta \approx 0.5$, as found in Ref. 11;
- using the parameterisation $K(E) = \beta^{\eta_T} K_0 \mathcal{R}^\delta$, negative values of $\eta_T$ are possible,[12] and depending on the exact value of $\eta_T$, almost any value of $\delta$ is allowed.[9]

To summarise, extracting tight constraints on the transport parameters is already challenging using simple propagation models. It could be, as too-many-studies-to-be-quoted propose, that these simple models are far from reality: the transport coefficients may very well depend on the position, and/or diffusion is anisotropic, and/or time-dependant effects have to be taken into account, and/or secondary production during the acceleration stage is relevant at a high enough energy... Hence, despite the fact that various models in the literature can satisfactorily fit secondary-to-primary ratios such as B/C, sub-Fe/Fe, etc., we are still far from a complete and convincing picture of the details of CR propagation. Data at higher energy as well as employing other ratios (Li/C, and d/p and $^3$He/$^4$He) may help us in this task.

## 2. Radioactive species

The typical distance travelled in a diffusive process in a finite time is given by $l_{rad} \sim \sqrt{Dt}$. For a $\beta$-decay unstable secondary species, it means that at low energy (plugging the typical value of the diffusion coefficient), these nuclei cannot travel farther than a few hundreds of parsecs: they do not feel the halo size $L$ and are only sensitive to the diffusion coefficient $K(E)$. In principle, this lifts the degeneracy between $K_0$ and $L$ discussed in the previous section. However, things are not as simple when we have a closer look at the hundred-of-parsec scale. It happens that there is no target to produce these species in the solar neighbourhood: we live in a local bubble.[13,14] This lack of targets affects their flux.[4,8,15]

### 2.1. *Toy model*

We consider a radioactive nucleus that diffuses in an unbound volume and decays with a rate $1/(\gamma\tau_0)$. In spherical coordinates, appropriate to describe this situation, the diffusion equation reads

$$-K\triangle_{\vec{r}}G + \frac{G}{\gamma\tau_0} = \delta(\vec{r}) \ . \tag{2}$$

The solution for the propagator $G$ (the flux is measured at $\vec{r} = \vec{0}$ for simplicity) is

$$G(\vec{r'}) \propto \frac{e^{-r'/\sqrt{K\gamma\tau_0}}}{r'} \ . \tag{3}$$

Secondary radioactive species, such as $^{10}$Be, originate from the spallations of the CR protons (and He) with the ISM. We model the source term to be a thin gaseous disc, except in a circular region of radius $r_h$ at the origin. In cylindrical coordinates (see Fig. 4)

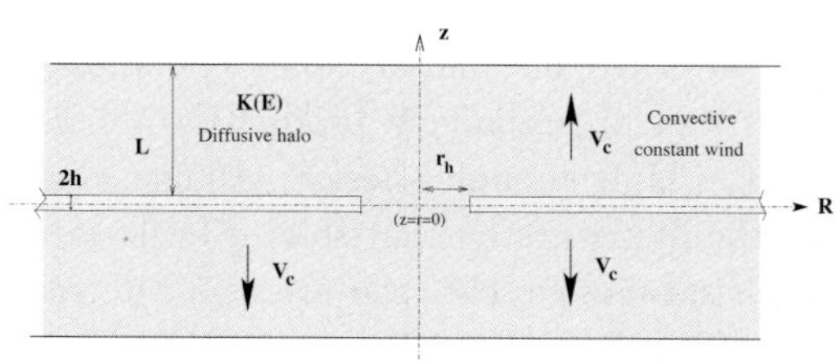


Fig. 4. Hole toy-model for the radioactive nuclei ($V_c$ is taken to be 0 here).

$$Q(r, z) \propto \Theta(r - r_h)\delta(z) \ , \tag{4}$$

where $\Theta$ is the Heaviside function. The flux of a radioactive species is thus given by (we rewrite the propagator in cylindrical coordinates)

$$N(r\!=\!z\!=\!0) \propto \int_0^\infty \int_{-\infty}^{+\infty} G(\sqrt{r'^2\!+\!z'^2})\, Q(r', z')\, r'dr'dz'. \tag{5}$$

### 2.2. *Attenuation of the flux*

The ratio of the flux calculated for a cavity/hole $r_h$ to that of the flux without hole ($r_h = 0$) is

$$\kappa \equiv \frac{N_{r_h}}{N_{r_h=0}} = \exp\left(\frac{-r_h}{\sqrt{K\gamma\tau_0}}\right) = \exp\left(\frac{-r_h}{l_{\rm rad}}\right) . \quad (6)$$

Table 1 gathers, at a few hundreds MeV/n energy, the attenuation $\kappa$ for a typical value of the diffusion coefficient (i.e. $\sim 10^{28}$ cm$^2$ s$^{-1}$) and a hole size $r_h \sim 100$ pc. This attenuation is energy dependent because it decreases

Table 1. Attenuation factor due to the local bubble for various radioactive species.

| Species | $\tau_0$ (Myr) | $l_{\rm rad}$ (pc) | $\kappa$ |
|---|---|---|---|
| $^{10}$Be | 2.17 | 351 | 0.57 |
| $^{26}$Al | 1.31 | 273 | 0.48 |
| $^{36}$Cl | 0.443 | 159 | 0.28 |
| $^{54}$Mn | 2.9 | 406 | 0.61 |

with the energy when the time-of-flight of a radioactive nucleus is boosted by both its Lorentz factor and the increase in the diffusion coefficient value. It is also species dependent because nuclei half-lives for the standard $Z < 30$ cosmic-ray clocks range from 0.307 Myr for $^{36}$Cl to 1.51 Myr for $^{10}$Be.

### 2.3. *Consequences and results*

In a grid exploration of the parameter space, we found that when included in the fit, radioactive isotopic data favour a value of $r_h \sim 80$ pc,[8] consistent with the direct measurement in the local ISM.[13] However, because of this extra parameter, the halo size $L$ remains loosely constrained. This is discussed further in A. Putze's contribution who presents a more thorough analysis by means of an MCMC analysis.[4] Note that in Ref. 4, for the first time, the probability density function of $L$ was given.

To summarise, the radioactive isotopic ratios are for the moment well fitted by any simple model (w/wo reacceleration, w/wo convection). However, because of the possible effect of the local underdensity, the degeneracy $K_0/L$ is not completely lifted. Furthermore, the fit to the radioactive elemental ratios, such as Be/B is not satisfactory, regardless of the model used.[4] This is shown in Fig. 5, and the discrepancy may lie in the production cross-section of Be, or in the data, or in the propagation models themselves. The results from PAMELA should be available soon to further inspect this issue.

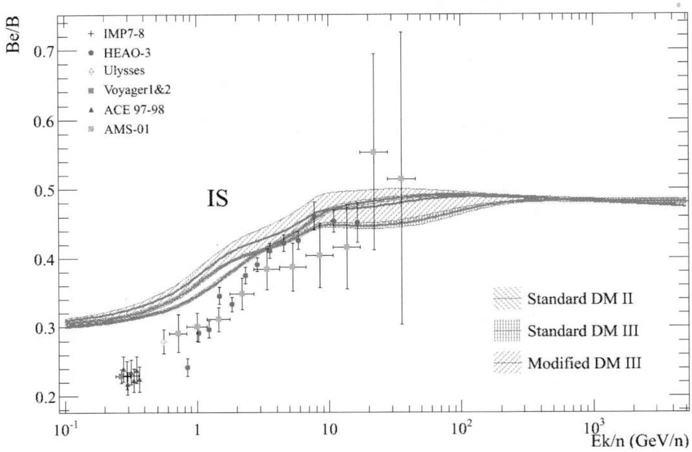


Fig. 5. Envelopes (confidence level) of Be/B for models consistent with B/C data and the radioactive isotopic ratio $^{10}$Be/Be. Adapted from Ref. 4, with the recently published AMS-01 data[16] added (courtesy A. Putze).

## 3. Primary species

A natural question to ask is whether primary species can tell us something about the transport coefficients, or if they can only tell us something about the source parameters.

To quickly understand the phenomenology of primary species, let us start with the simplest expression for the primary flux, and then add the other ingredients:

- For diffusion only,

$$\Phi(E) \propto \frac{Q(E)}{K(E)} \propto \frac{q}{K_0} E^{-(\alpha+\delta)}.$$

  There is a degeneracy in $q/K_0$ ($q$ is the source term normalisation) and $\alpha + \delta$ (where $\alpha$ is the source spectral index): only the total spectral index is relevant in the fit, not the separate indices.
- When spallations are taken into account

$$\Phi(E) = \frac{v}{4\pi} \frac{Q(E)}{\frac{K(E)}{hL} + n\sigma v}.$$

In this Equation, $\alpha + \delta$ is now also coupled to $K_0$, to balance the effect of nuclear destruction at low energy: a steeper source spectrum requires more spallations to still fit the low-energy data, hence a smaller value of the diffusion coefficient.

- When energy losses and gains are considered, there is also a subtle change of the concavity and the shape of the primary spectrum, related to the strength of these effects. It can balance the need for a steeper or flatter source spectrum for instance.

From this simple analysis, we can conclude that primary species cannot alone constraint the transport parameters. A more thourough analysis shows that most of the time, it is possible to find some source parameters that allow a good match of the data, regardless of the values of the transport parameters.[17]

Figure 6 shows the typical spectrum of $p$ and He data. At high energy ($>$ 100 GeV/n), current experiments seem to indicate a deviation from a single power-law. Waiting for more accurate data, we can still assume a

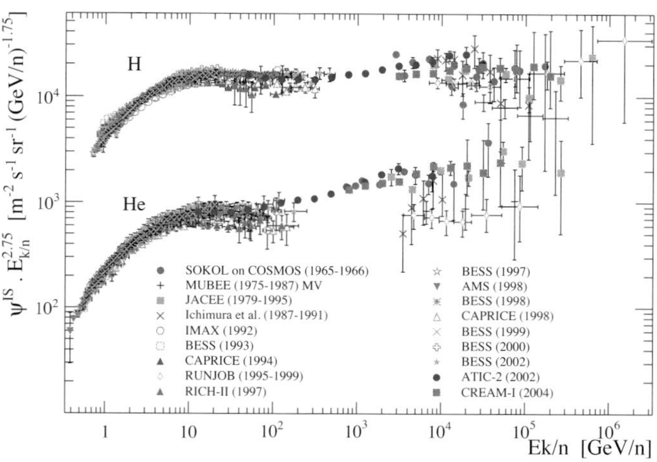


Fig. 6. Compilation of $p$ and He demodulated data using the Force-Field (from Ref. 17).

simple power-law (no break in the slope) and, fixing the transport coefficients, try to constrain the source parameters $q$ and $\alpha$, and inspect how sensitive they are to these transport parameters. This can be achieved by means of an MCMC analysis, where the transport parameters are chosen so as to reproduce the B/C ratio. The median value and confidence levels for $\alpha$ for $p$ and He are shown in Fig. 7. For a given set of data and a given configuration of the transport model (w/wo reacceleration, w/wo convection), we find that $p$ and He have the same source spectral index within the uncertainties.[17] However, there is a large spread in the overall values. Still, whereas these different models correspond to a range of diffusion slope $\delta \approx [0.3 - 0.8]$, the source slope $\alpha$ only varies in the range $\sim [2.25 - 2.5]$.

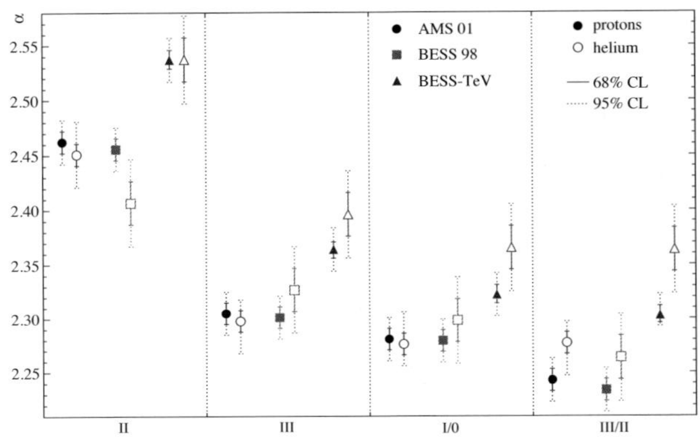


Fig. 7. Median value and confidence level on the spectral index $\gamma$ fitted on $p$ and He below-1-TeV data (from Ref. 17).

This MCMC approach is powerful and can also be applied to heavier primary species (C, O, Si, Fe...), though the uncertainties are larger than for $p$ and He (lower statistics and larger systematics on the data). This is discussed in A. Putze's contribution. These kinds of studies have the potential to constrain the source slope and ascribe any difference in the various species, as well as inspect their source abundances. This could provide useful information for the acceleration models. Conversely, if the breaks in the data are confirmed, i.e. if source spectra are not pure power-laws, some further thinking is in order to decide how to tackle this new feature in the CR spectra (when fitting the forthcoming high-precision data).

## 4. Antiprotons and antideuterons

### 4.1. *Standard production*

Even if it happens that all the models discussed here are only *effective* models, the standard antiprotons and antideuterons fluxes can be confidently evaluated, as the propagation history of these species closely resembles that of the B production from the C primary. Indeed, $p$ an C are both primary species located in the thin disc of the Galaxy, and they produce the secondary species by interacting in the ISM.

Such an approach was successfully followed in Ref. 18 to show that the antiprotons calculated without further fine-tuning (of the model) matched very well the data of the time. This flux was updated in Ref. 19 (but the final results is mostly unchanged) fro comparison with the recent PAMELA data, as reproduced in Fig. 8.

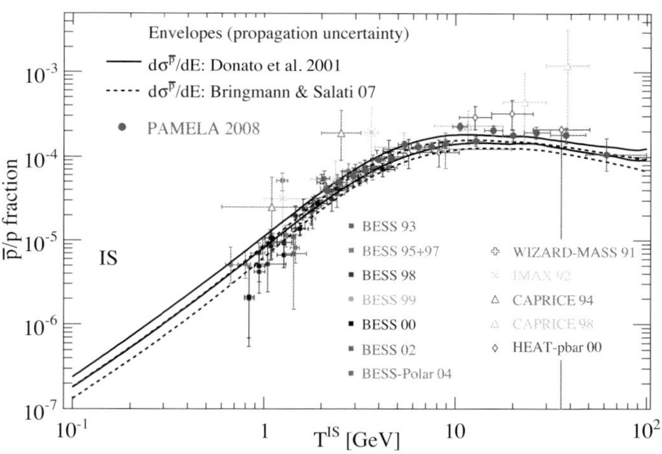


Fig. 8. Antiproton-to-proton ratio (from Ref. 19).

The same kind of calculation can be repeated for the antideuterons[20–22] for which only an upper limit exists (set by the BESS experiment). For both cases, the main source of uncertainties is from the nuclear production cross-sections ($\sim 20\%$ for antiprotons, but as large as a factor of 10 for low energy antideuterons).

### 4.2. *Exotic production*

In contrast, exotic sources are located in the whole diffusive halo. As a result, the propagation history is not the same as for the B/C production or for a standard primary species.

Adopting a crude model where the dark matter source term is constant in the diffusive halo of the Galaxy[23] the transport equation (no spallations) now reads

$$-KN'' = q_{\rm Dark},$$

whose solution is given by

$$N^{pure-DM}_{(\bar{p},\,\bar{d})}(z=0) = \frac{q_{Dark}L^2}{2K} = \frac{q_{Dark}\lambda_{\rm esc}L}{\mu v}\,. \qquad (7)$$

For a given value of $\lambda_{\rm esc}$, i.e. a given B/C ratio, the primary flux depends linearly on $L$. The bigger the box containing the dark matter, the stronger the signal, which seems very reasonable. This linear dependence was observed in more realistic propagation models for decaying or annihilating dark matter.[24–27] We will not discuss this topic further and refer the reader to P. Salati's contribution for more details.

## 5. USINE features and public release

The package USINE deals with the propagation of Galactic Cosmic Ray nuclei (all existing nuclei) and antinuclei (antiprotons and antideuterons) in various models (LB, diffusion models...). The energy range validity (unmodulated) is between a few hundreds of MeV/nuc to PeV energies. USINE is a C++ object-oriented code that makes use of the ROOT package[b] to provide many display facilities that ease the understanding, allows quick checks, and provides graphical results (along with the data) of the propagated fluxes. The code is documented with the ROOT THtml class.

A *propagation model* is defined as i) the propagation processes themselves (diffusion, convection...); ii) the description of the medium in which CR nuclei propagate (Galaxy geometry, gas distribution in the Galaxy, spatial dependence of propagation processes); and iii) the set of equations (solutions) for the system (formed by the two previous items), which can be solved analytically or numerically. Most of the ingredients required to achieve the calculation are shared by all models.

The USINE code has been developed to calculate background and signals, for cosmic ray physics and dark matter indirect searches. However, it has also been conceived in a way to be very flexible and user-friendly for

- any experimentalist who wants to see how models can be constrained by his data;
- a nuclear physicist who wants to check how his particular set of cross sections changes the predictions of models;

[b]http://root.cern.ch/

- for a CR phenomenologist who derived a new set of equations for propagation and would like to apply it but who is usually discouraged by the huge amount of work required to compile all necessary ingredients.

The blend of semi-analytical and numerical approaches is also an advantage for validating the code.

USINE should be made public soon. Along with the code, we will also provide a database (mySQL web-interfaced database), a web-run interface, and a Graphical User Interface to ease the use of the code. We are also working on including electron, positrons and $\gamma$-rays flux for the second release.

## 6. Conclusion

We have presented an overview of some results obtained in the framework of diffusion/convection/reacceleration models. Such calculations were performed by means of the USINE package. A first public release of this code will appear soon.

## Acknowledgments

I would like to thank my colleagues F. Donato, P. Salati, and R. Taillet (early stage of our propagation studies), L. Derome and A. Putze (implementation the MCMC analysis), and many others F. Barao, C. Combet... for past, present and future developments of the USINE package.

## References

1. D. Maurin, F. Donato, R. Taillet and P. Salati, *ApJ* **555**, 585(July 2001).
2. D. Maurin, R. Taillet and F. Donato, *Astronomy and Astrophys.* **394**, 1039(November 2002).
3. A. Putze, L. Derome, D. Maurin, L. Perotto and R. Taillet, *A&A* **497**, p. 991 (Paper I)(April 2009).
4. A. Putze, L. Derome and D. Maurin, *A&A* **516**, A66+(June 2010).
5. F. C. Jones, A. Lukasiak, V. Ptuskin and W. Webber, *ApJ* **547**, 264(January 2001).
6. A. M. Lionetto, A. Morselli and V. Zdravkovic, *Journal of Cosmology and Astro-Particle Physics* **9**, p. 10(September 2005).
7. G. Di Bernardo, C. Evoli, D. Gaggero, D. Grasso and L. Maccione, *ArXiv e-prints* (September 2009).
8. F. Donato, D. Maurin and R. Taillet, *Astronomy and Astrophys.* **381**, 539(January 2002).

9. D. Maurin, A. Putze and L. Derome, *A&A* **516**, A67+(June 2010).
10. A. W. Strong and I. V. Moskalenko, *ApJ* **509**, 212(December 1998).
11. W. R. Webber, A. Soutoul, J. C. Kish and J. M. Rockstroh, *ApJS* **144**, 153(January 2003).
12. V. S. Ptuskin, I. V. Moskalenko, F. C. Jones, A. W. Strong and V. N. Zirakashvili, *ApJ* **642**, 902(May 2006).
13. R. Lallement, B. Y. Welsh, J. L. Vergely, F. Crifo and D. Sfeir, *A&A* **411**, 447(December 2003).
14. B. Y. Welsh, R. Lallement, J. Vergely and S. Raimond, *A&A* **510**, p. A54(February 2010).
15. V. S. Ptuskin and A. Soutoul, *A&A* **337**, 859(September 1998).
16. The AMS-01 Collaboration, *ArXiv e-prints* (August 2010).
17. A. Putze, F. Donato and D. Maurin, *to be submitted* (2010).
18. F. Donato, D. Maurin, P. Salati, A. Barrau, G. Boudoul and R. Taillet, *ApJ* **563**, 172(December 2001).
19. F. Donato, D. Maurin, P. Brun, T. Delahaye and P. Salati, *Physical Review Letters* **102**, p. 071301(February 2009).
20. F. Donato, N. Fornengo and P. Salati, *Phys. Rev. D* **62**, p. 043003(August 2000).
21. R. Duperray, B. Baret, D. Maurin, G. Boudoul, A. Barrau, L. Derome, K. Protasov and M. Buénerd, *Phys. Rev. D* **71**, p. 083013(April 2005).
22. F. Donato, N. Fornengo and D. Maurin, *Phys. Rev. D* **78**, p. 043506(August 2008).
23. D. Maurin, R. Taillet and C. Combet, *ArXiv Astrophysics e-prints astro-ph/0612714* (December 2006).
24. A. Barrau, G. Boudoul, F. Donato, D. Maurin, P. Salati and R. Taillet, *Astronomy and Astrophys.* **388**, 676(June 2002).
25. A. Barrau, G. Boudoul, F. Donato, D. Maurin, P. Salati, I. Stéfanon and R. Taillet, *A&A* **398**, 403(February 2003).
26. F. Donato, N. Fornengo, D. Maurin, P. Salati and R. Taillet, *Phys. Rev. D* **69**, p. 063501(March 2004).
27. A. Barrau, P. Salati, G. Servant, F. Donato, J. Grain, D. Maurin and R. Taillet, *Phys. Rev. D* **72**, p. 063507(September 2005).

# Propagation of galactic cosmic rays and the AMS-02 experiment

Miguel Pato*

*Institut d'Astrophysique de Paris,*
*98bis bd Arago, 75014, Paris, France*
**E-mail: pato@iap.fr*

*Dipartimento di Fisica, Università degli Studi di Padova,*
*via Marzolo 8, I-35131, Padova, Italy*

Dan Hooper

*Center for Particle Astrophysics,*
*Fermi National Accelerator Laboratory, Batavia, IL 60510*

*Department of Astronomy and Astrophysics,*
*University of Chicago, Chicago, IL 60637*

Melanie Simet

*Department of Astronomy and Astrophysics,*
*University of Chicago, Chicago, IL 60637*

A precise determination of cosmic ray spectra up to TeV energies and light isotope separation are crucial steps in understanding the origin and propagation of galactic cosmic rays. Using the expected capabilities of the upcoming Alpha Magnetic Spectrometer (AMS-02), we anticipate the measurements of B/C, $^{10}$Be/$^{9}$Be and the proton flux. This projected data set, which represents a great improvement upon the current status of GeV-TeV cosmic ray data, is then used to constrain models of injection and propagation of cosmic rays in the Milky Way. Minimal or next-to-minimal setups turn out to be tightly constrained. Nevertheless, the precise AMS-02 anticipated measurements will not be sufficient to distinguish between models with different assumptions regarding the rigidity dependence of the diffusion coefficient, source distribution and abundances or stochasticity.

*Keywords*: galactic cosmic rays, cosmic ray propagation

## 1. Introduction

The origin and propagation of cosmic rays are puzzling issues of major importance for astrophysics. Over the years many composition studies and the complementarity between different measurements have taught us a lot about the environment where cosmic rays are accelerated as well as how they travel through the interstellar medium until the Earth. Nevertheless, a fully-consistent picture has not yet been assembled. At low energies – meaning energies below the knee at $\sim 10^{15}$ eV – the bulk of the cosmic ray flux is believed to be of galactic origin. Once injected in the galactic medium, charged cosmic rays – unlike photons or neutrinos – undergo several processes capable of steering, degrading energy or determining their extinction along the path towards us. Besides diffusion through magnetic irregularities and spallation on the interstellar medium, cosmic rays may also be affected by energy losses, radioactive decays, reacceleration and convection winds. All these processes and their relative importance may be studied using the measured local spectra of individual cosmic ray species. In particular, stable secondary-to-primary ratios (e.g. B/C) and unstable ratios (e.g. $^{10}$Be/$^{9}$Be) are useful tools in constraining injection and propagation parameters. While stable secondary-to-primary ratios provide information about the effective column density cosmic rays pass through before reaching the Solar System, unstable ratios roughly fix the time interval since spallation. These two observables have been used before to constrain the properties of different propagation setups. There are some degeneracies between the parameters so that the present data do not point to a single propagation scheme. But the situation is expected to improve greatly with data from the Alpha Magnetic Spectrometer AMS-02. Besides largely improved statistics, AMS-02 will also provide cosmic ray flux measurements up to $\sim$ TeV/n and hopefully separate Be isotopes up to $\sim$ 10 GeV/n.

In the present work we focus on the expected performances of AMS-02, mainly regarding B/C, $^{10}$Be/$^{9}$Be and the proton flux, and investigate the impact that its data may have on our knowledge about cosmic ray physics. Using GALPROP numerical code[1,2] to model injection and propagation of galactic cosmic rays, we address the question of whether AMS-02 will give us enough information to single out an unique propagation scheme and whether we will be able to distinguish between setups with different assumptions.

## 2. Prospects For AMS-02

The second and final version of the Alpha Magnetic Spectrometer (AMS-02)[3,4] is a large acceptance cosmic ray detector scheduled to be placed on-board the International Space Station soon. Over its mission duration, it will measure with unprecedented statistics and precision the spectrum of cosmic rays over an energy range of approximately 100 MeV to 1 TeV. The AMS-02 instrument will be able to detect and identify nuclei as heavy as iron ($Z \lesssim 26$) with rigidity up to a few TV, and separate isotopes of light elements (namely, H, He and Be) over a kinetic energy range of 0.5–10 GeV/n. High precision measurements of the ratios B/C and sub-Fe/Fe (D/p, $^3$He/$^4$He and $^{10}$Be/$^9$Be) up to energies of ~1 TeV/n (~10 GeV/n) are anticipated. We try to estimate in a reasonable manner the capabilities of the AMS-02 instrument to measure B/C, $^{10}$Be/$^9$Be and the proton flux – details of the adopted procedure may be found in Ref. 5. To reduce the dependence on solar modulation, we apply an energy cut $T > 5$ GeV/n throughout our analysis.

## 3. Results

We start by fixing the Alfvén speed to $v_A = 36$ km/s, and neglecting the effects of convection ($V_{c,0} = dV_c/dz = 0$). Proceeding in a fashion similar to Ref. 6, we run GALPROP 245 times, in a 7x7x5 grid of the parameters $(D_{0xx}, L, \alpha)$ [a] over the following ranges: $D_{0xx} = 4.54 - 8.03 \cdot 10^{28}$ cm$^2$/s, $L = 3.5 - 6.5$ kpc and $\alpha = 0.39 - 0.43$. Figure 1 shows the 1, 2 and $3\sigma$ contours from the combination of B/C and $^{10}$Be/$^9$Be projected AMS-02 data. As this figure demonstrates, the upcoming measurements on B/C and $^{10}$Be/$^9$Be are sufficient (within the context of the simple models presently being considered) to determine the underlying propagation parameters with an accuracy of $\Delta D_{0xx} \sim 1.4 \cdot 10^{28}$ cm$^2$/s, $\Delta L \sim 1.0$ kpc, and $\Delta\alpha \sim 0.02$ (at $1\sigma$). This precision is much greater than that obtained with present (pre-AMS-02) data; see Refs. 6 and 7. In particular the degeneracy between $D_{0xx}$ and $L$ is broken as can be seen in the upper left frame of Figure 1 where we have overplotted in dashed the $3\sigma$ contour from Ref. 6.

Thus far, we have not considered any effects of convection and/or variations in the Alfvén velocity from our default value of $v_A = 36$ km/s. As the quantity and quality of cosmic ray data improves, however, it will become increasingly possible to test these assumptions, and determine the related

[a] $D_0xx$ is the normalisation of the diffusion coefficient at a rigidity of 4 GV, $\alpha$ is its slope, and $L$ is the half-height of the cylindrical diffusive region.

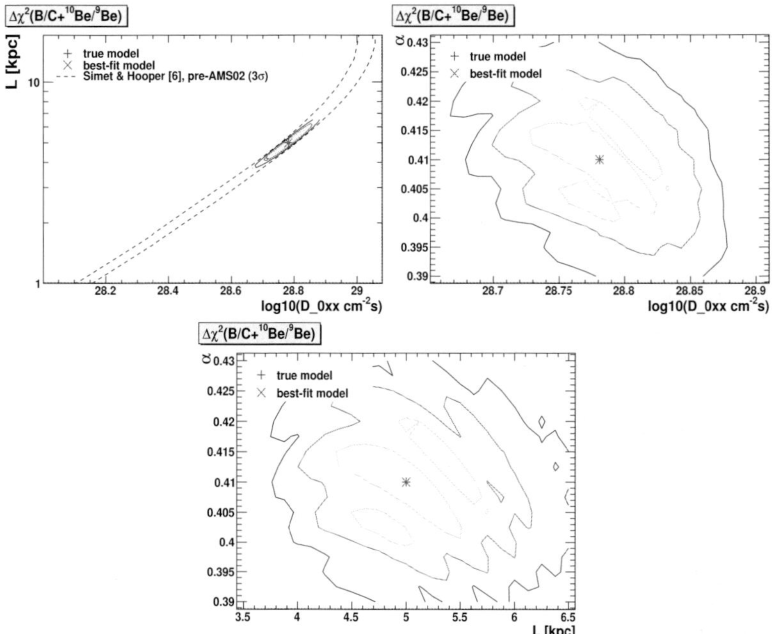


Fig. 1. Regions consistent (within 1, 2 and $3\sigma$) with projected B/C and $^{10}$Be/$^{9}$Be AMS-02 data in the $L$ vs. $D_{0xx}$, $\alpha$ vs. $D_{0xx}$, and $\alpha$ vs. $L$ planes. We have assumed $v_A = 36$ km/s, $V_{c,0} = dV_c/dz = 0$. In the top left frame we show in dashed the $3\sigma$ contour from Ref. 6.

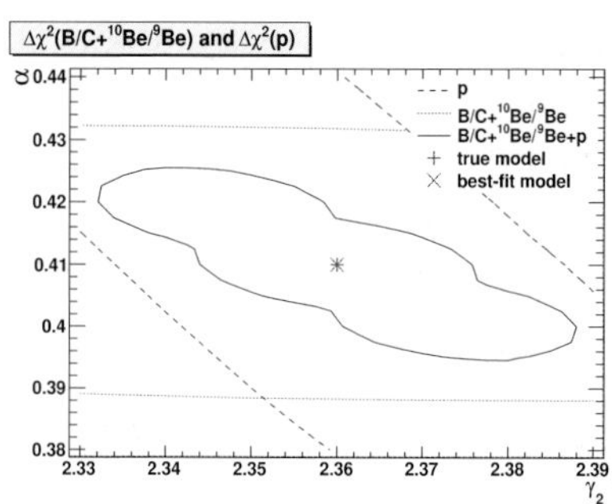


Fig. 2. Inferred parameter regions (within $3\sigma$) in the plane $\alpha$ vs. $\gamma_2$. The dashed (dotted) [solid] lines refer to the case where the AMS-02 projected data set consisted of the proton flux (B/C and $^{10}$Be/$^{9}$Be) [proton flux, B/C and $^{10}$Be/$^{9}$Be]. We have assumed $v_A = 36$ km/s, $V_{c,0} = dV_c/dz = 0$.

parameters. In order to study such possibility, we have repeated the procedure using the following values for the Alfvén velocity and the convection velocity: $(v_A[\text{km/s}], dV_c/dz[\text{km/s/kpc}]) = \{(36,0),(15,0),(0,0),(0,10)\}$, while leaving $V_{c,0} = 0$. As greater departures from our default assumptions are considered, the fits to the projected data become considerably worse. In particular, even modest ($\sim$10 km/s/kpc) amounts of convection lead to very poor fits to the projected data. Large variations in $v_A$ also lead to observable effects, thus enabling AMS-02 to be sensitive to the details of diffusive reacceleration.

In order to make the questions addressed in this study tractable, we have relied on a number of simplifying assumptions. In particular, we have assumed homogeneity of the diffusion coefficient $D_{xx}$ over the volume of the diffusion zone, considered cylindrical symmetry for the system, and adopted a smooth distribution of cosmic ray sources with universal injected chemical composition and spectra. While such assumptions are reasonable and have been useful up to this point in time, they will eventually have to be discarded or revised as they break down under the increasing precision of future cosmic ray data. We have loosen a few of the usual assumptions and found that AMS-02 precise data will not be sensitive to probe stochasticity, source abundances and distribution, and the detailed energy-dependence of the diffusion coefficient (unless an abrupt break is present well inside AMS-02 energy range).

Finally, we turn our attention to the source spectral index. Up to now the injection spectrum at the sources was assumed to be a double power law in rigidity with indices $\gamma_1 = 1.82$ and $\gamma_2 = 2.36$ respectively below and above $\tilde{R}_0 = 9$ GV. We now release this assumption by taking several values for $\gamma_2$ – the low energy rigidity index $\gamma_1$ and the break $\tilde{R}_0$ are kept fixed since we are focussing on high energy data only. To proceed further and quantify how sensitive AMS-02 will be to the cosmic ray injection spectrum, we scan the parameter space $(D_{0xx}, \gamma_2, \alpha)$ and fix $v_A = 36$ km/s, $V_{c,0} = dV_c/dz = 0$, and $L = 5$ kpc (notice that $D_{0xx}$ and $L$ are approximately degenerate). Figure 2 shows the $3\sigma$ regions in the $\alpha$ vs. $\gamma_2$ plane from the projected AMS-02 measurements of the *(i)* proton flux (dashed lines), *(ii)* B/C and $^{10}$Be/$^{9}$Be (dotted lines), and *(iii)* B/C, $^{10}$Be/$^{9}$Be and proton flux (solid lines). The ratios B/C and $^{10}$Be/$^{9}$Be are efficient probes of the diffusion index but insensitive to $\gamma_2$, while the proton flux constrains essentially the quantity $\alpha+\gamma_2$. Hence, within minimal propagation models, AMS-02 has the potential to pinpoint both the diffusion index $\alpha$ and the high energy source spectral index $\gamma_2$ with good accuracy. Note that this result stems exactly from the combination of the B/C and $^{10}$Be/$^{9}$Be ratios and the proton flux. While a reasonably precise high energy proton flux has been measured by past and present instruments (e.g. PAMELA) – allowing correspondingly precise estimates of $\alpha+\gamma_2$ –, only AMS-02 (or future detectors) will be able to break down the degeneracy between $\alpha$ and $\gamma_2$ with high energy quality cosmic ray ratios.

## 4. Conclusions

In this work, we have considered the ability of the upcoming AMS-02 experiment to measure stable secondary-to-primary and unstable ratios (such as boron-to-carbon and beryllium-10-to-beryllium-9), and studied to what extent this information could be used to constrain the model describing cosmic ray propagation in the Milky Way. We find that in propagation models with 3 free parameters – namely, the normalisation and slope of the diffusion coefficient and the half-thickness of the cylindrical diffusive region – AMS-02 data will effectively break existing degeneracies and provide measurements of the parameters with much better accuracy than currently possible. The same holds for models where the normalisation and slope of the diffusion coefficient and the source spectral index are varied freely. Furthermore, a significant degree of sensitivity to convection (and, to a lesser extent, reacceleration) is also predicted. However, when certain assumptions are loosen – including the rigidity dependence of the diffusion coefficient, source distribution and abundances and stochasticity –, AMS-02 upcoming measurements will likely be insufficient to pinpoint propagation parameters correctly. Inaccurate assumptions may, in particular, lead to a very well-fit, but incorrect, cosmic ray propagation model from AMS-02 data.

*Acknowledgements:* MP acknowledges a scholarship from Fundação para a Ciência e Tecnologia (Ministério da Ciência, Tecnologia e Ensino Superior).

## References

1. A. W. Strong and I. V. Moskalenko, Astrophys. J. **509** (1998) 212 [arXiv:astro-ph/9807150].
2. http://galprop.stanford.edu/web_galprop/galprop_home.html
3. http://ams.cern.ch/
4. P. Zuccon [AMS-02 Collaboration], *Prepared for 31st International Cosmic Ray Conferences (ICRC 2009), Łódź, Poland, 7-15 Jul 2009.*
5. M. Pato, D. Hooper and M. Simet, JCAP **1006** (2010) 022 [arXiv:1002.3341 [Unknown]].
6. M. Simet and D. Hooper, JCAP **0908**, 003 (2009) [arXiv:0904.2398 [astro-ph.HE]].
7. G. Di Bernardo, C. Evoli, D. Gaggero, D. Grasso and L. Maccione, arXiv:0909.4548 [astro-ph.HE].

# GALACTIC COSMIC RAYS IN THE DYNAMIC HELIOSPHERE

MARIUS POTGIETER[†]
*Centre for Space Research, North-West University,
2520 Potchefstroom, South Africa*

STEFAN FERREIRA
*Centre for Space Research, North-West University,
2520 Potchefstroom, South Africa*

DU TOIT STRAUSS
*Centre for Space Research, North-West University,
2520 Potchefstroom, South Africa*

The relation between the dynamics of the heliosphere, its shape and geometry, and solar activity has become increasingly important. The global features of the heliosphere influence what happens inside its boundaries on a variety of time-scales. Galactic cosmic rays serve as excellent probes for this purpose, conveying vital information on global heliospheric changes in the way that they respond. By observing galactic cosmic rays, also neutral and other charged particles, over a wide range of energies on various spacecraft and at Earth, a better understanding is gained about heliospheric features, the basics of particle transport and modulation, and various heliospheric phenomena, including space weather and space climate. Significant progress is made in this field, stimulated by observations in the outer heliosphere by the two Voyager spacecraft and in the inner heliosphere by ULYSSES, PAMELA and other space missions.

## 1. The Dynamic Heliosphere

Interstellar space between the Sun and nearby stars is filled with plasmas, magnetic fields, neutral and charged particles. The Sun moves through the local interstellar medium (ISM) with a velocity of ~26 km.s$^{-1}$ so that an interface with the ISM is formed. The solar wind prevents this medium from flowing into the large volume dominated by the Sun, called the heliosphere. Because it is moving through the ISM, the heliosphere is asymmetrical with respect to the Sun, with the 'tail' region much larger than the 'nose' region, the direction in which it is moving. It extends over at least 500 AU in its equatorial regions and at least 250

[†] Email address: marius.potgieter@nwu.ac.za.

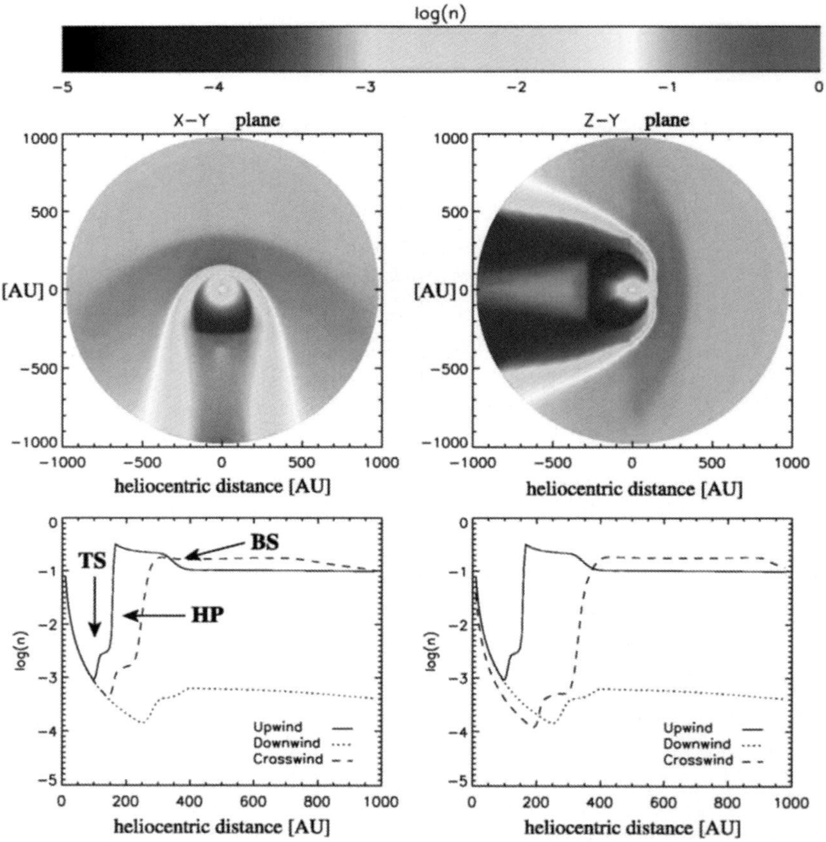


Figure 1. Illustration of the heliospheric geometry, structure and boundaries using contour plots of the proton number density in the X-Y and Y-Z planes (with the Sun in the middle), obtained with a 3-dimensional hydrodynamic model with the associated number density profiles in the heliospheric upwind, downwind, crosswind and polar directions shown in the lower panels [3].

AU in the polar plane. The main constituents of this interface with the local ISM are shown in Figure 1. They are the solar wind termination shock (TS), the heliopause (HP) and perhaps also a bow shock (BS) in the 'nose' direction, with the region between the TS and the HP defined as the inner heliosheath, with the outer heliosheath located between the HP and the BS. The latter is expected to be rather weak. For an overview see [1,2,3].

The HP separates the solar wind and local ISM so that it may be considered the outer boundary of the heliosphere. A prediction of MHD-HD modeling is that the solar wind forms a TS where it goes from supersonic to subsonic speeds. It is predicted that the heliospheric structure is dynamically asymmetric,

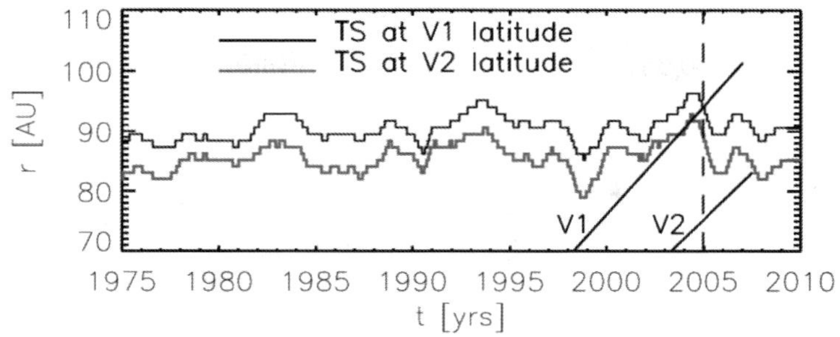


Figure 2. The computed heliocentric radial position (in AU) of the TS at 34.1° heliolatitude (solid black line) and at −27.0° (solid blue line) for the period 1975 to 2010. Also shown are the radial positions of the Voyager 1 and Voyager 2 spacecraft at the time. The vertical dashed line indicates the time when Voyager 1 crossed the TS in December 2004 [6].

yielding a ratio for the upwind-to-downwind TS distance of ~1:2. The heliosphere is elongated in its polar directions because of the latitudinal variation of the solar wind momentum flux. This asymmetry may become more pronounced during solar minimum conditions when the solar wind can go from 400 km.s$^{-1}$ in the equatorial regions to 800 km.s$^{-1}$ in the polar regions. Recently, Richardson et al. [4] reported that the radial flow speeds in the inner heliosheath vary between 80 and 200 km.s$^{-1}$, with an average speed of 138 km.s$^{-1}$, as measured by Voyager 2. These flows suggest that the heliosphere is wider than it is high.

The Voyager 1 spacecraft encountered the TS [5] in December 2004 at a distance of ~94 AU from the Sun, and at a polar angle of ~60°. Its twin spacecraft Voyager 2 crossed the TS in August 2007 at 83.7 AU in the southern hemisphere at ~120° and ~10 AU closer to the Sun, in general accord with the predictions shown in Figure 2. The TS position oscillates significantly with solar activity. In this respect, Snyman [6] modeled the distance to the TS as it varies over an 11 year cycle in comparison with the positions of Voyager 1 and Voyager 2. This is shown in Fig. 2. It follows that the TS could move inwards as close as ~75 AU and as far out as 95 AU from the Sun over a solar cycle. Because the recent solar minimum activity period had been very quiet and prolonged, the TS is probably now at its closest position to the Sun.

Asymmetries in the HP could indicate an asymmetric pressure from the interstellar magnetic field, from transient-induced shock motion, or from the solar wind dynamic pressure. Recently, Opher et al. [7] reported on the strength and orientation of the magnetic field in the interstellar medium near the heliosphere and found that the field strength in the local interstellar medium is 3.7–5.5 μG (previous estimates were 1.8–2.5 μG). This field is also tilted 20–

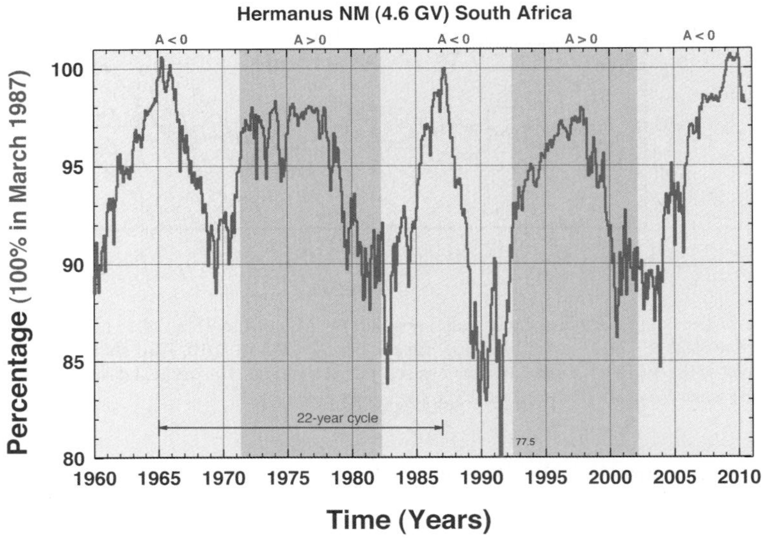


Figure 3. The long-term GCR record observed by the Hermanus Cosmic Ray Monitor in South Africa with the intensity in March 1987 taken as 100%, corrected for atmospheric pressure to get rid of daily and seasonal variations. Notice the 11-year and 22-year cycles and the large step-like decreases and recoveries. Peaks are formed during the A < 0 solar magnetic field polarity cycles in contrast to A > 0 cycles when the modulation profiles are flattish. GCRs have reached the highest level over five solar cycles in 2009. The recent solar minimum period was almost two years longer than the previous ones.

30° from the interstellar medium flow direction and is at an angle of about 30° from the galactic plane. This indicates that the heliosphere is in one hemisphere to a larger extent 'magnetically squeezed' than in the other half, causing an asymmetry in the north-south directions. The width of the inner heliosheath may therefore be wider in the northern regions than in the southern regions of the heliosphere. Stone et al. [5] also reported that the intensity of 4-5 MeV protons, accelerated by the TS near Voyager 2, was three times that observed concurrently by Voyager 1, indicating possible differences in particle populations at the two locations which could also be temporal effects. Voyager 1 is approaching the HP at about 3 AU per year so the anticipation is that the spacecraft, now at ~113 AU, may cross the HP within the next two years. It will then move into interstellar space!

## 2. Cosmic Rays in the Heliosphere

Charged particles with energies larger than 1 MeV, and present in the heliosphere, can be classified in four main populations: (1) Galactic cosmic rays which clearly originated outside the heliosphere. When arriving at Earth, these particles are composed of ~98% nuclei (mostly protons), fully stripped of all their electrons, and ~2% electrons, fewer positrons and anti-protons. (2) Solar energetic particles which originated mainly from solar flares, coronal mass ejections and shocks in the interplanetary medium. They occur sporadically, observed in the inner heliosphere, usually only for several hours mainly during solar maximum activity, and may have energies up to hundreds of MeV. These events are directly linked to what is called space weather. (3) The anomalous components (ACRs) which originate as interstellar neutral atoms that become mainly singly ionized in the heliosphere. They get transported by the solar wind, as so-called pick-up ions, in the heliosheath where they get accelerated up to a ~100 MeV through various processes which are presently heavily debated. (4) The Jovian electrons which dominate the observed low energy electron spectrum, up to 30 MeV, within the first 10 AU from the Sun.

The best illustration of GCR modulation over a long-term is the records of the intensity of GCRs observed with so-called neutron monitors (NMs) on ground level. This is shown in Figure 3. The most important variability time scale related to solar activity is the 11-year cycle. This quasi-periodicity is convincingly reflected in records of sunspots since the early 1600's, and in other solar activity indices. The next important and interesting phenomenon shown in Figure 3 is the 22-year cycle caused by the gradient, curvature and current sheet drifts of GCRs which are directly related to the global gradients and curvatures of the HMF and the fact that it reverses polarity every 11 years. Because drift patterns are reversed for oppositely charged particles, electrons (or protons) and positrons (or anti-protons) behave differently inside the heliosphere during a given solar magnetic polarity cycle. They therefore probe different regions of the heliosphere during the same cycle. Simultaneous measurements of GCR electrons and positrons (protons and anti-protons) are thus a crucial test for our understanding of charge-sign dependent modulation in the heliosphere, in particular how the drifts that they experience are affected by turbulence as a function of decreasing energy, position in the heliosphere and over a complete solar activity cycle [8,9,10]. Observations of these GCR anti-particles are presently made by the PAMELA mission [11].

In this context a very useful proxy for solar activity is the so-called tilt angle ($\alpha$) of the wavy heliospheric current sheet (HCS), which turned out to be one of the most successful modulation parameters after it was realized that the gradient and curvature drifts of GCR particles should play an important role. The tilt angles of the HCS are shown in Figure 4. Tilt angles between 5-25° are indicative of minimum solar activity, between 30-50° of moderate solar activity

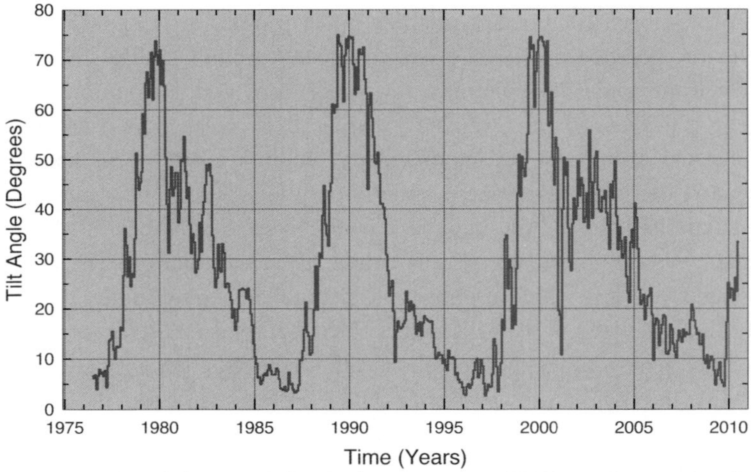


Figure 4. Tilt angles of the wavy heliospheric current sheet (HCS); maximum values indicate extreme solar activity when the polarity of the solar magnetic field changes; minimum values indicate when solar modulation is at its minimum such as in 2009. (With the courtesy of Todd Hoekseman; see http://wso.stanford.edu for the two approaches in calculating these values).

and $\alpha$ >60° of maximum solar activity. At extreme solar maximum activity the HMF reverses its polarity and then the tilt angle begin to decrease again. These values are widely used in GCR data interpretation and in numerical modeling. The modulation effects of the wavy HCS and global drifts, the subsequent 22-year cycle and corresponding charge-sign dependence have been studied in detail [12,13,14]. Periods of maximum GCR modulation are more complex; they may last only three years (e.g., 1969–1971), or up to six years (e.g., 1979–1984), or may temporarily be dominated by a massive CR decrease as in 1991.

The recent solar minimum activity period was extra-ordinary long, lasting almost two years longer than the previous ones. The HMF (measured and estimated) at Earth was at its lowest value since 100 years ago [15]; this may be the first indication that we are about to observe an additional solar activity cycle Consequently, GCRs reached their highest intensity level, over five solar cycles, at the end of 2009 (see Figure 3). The next solar activity cycle is predicted to have significantly lower sunspots than the previous ones.

## 2.1. *Theoretical Background and Numerical Models*

Numerical models for the global and long-term modulation of CRs in the heliosphere are based on solutions of Parker's [16] time-dependent transport equation:

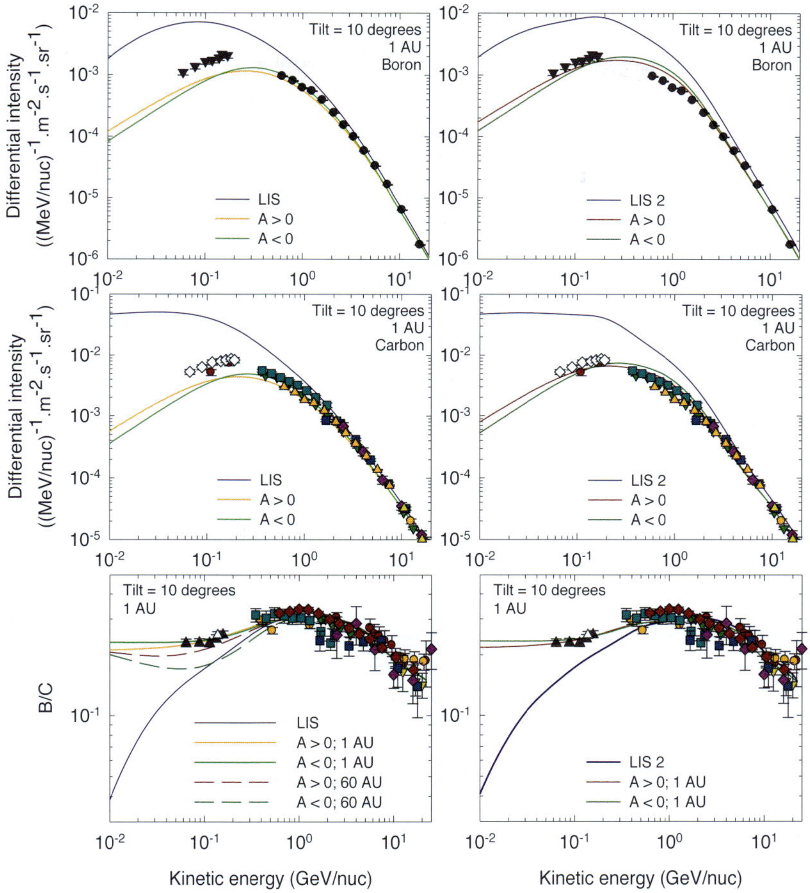


Figure 5. Top and middle panels: Computed spectra for Boron (top) and Carbon (middle) at Earth compared to the corresponding LIS, at 120 AU, for both polarity cycles compared to B and C observations. Bottom panel: The ratio B/C as a function of kinetic energy for both polarity cycles with $\alpha = 10°$ (solar minimum conditions) compared to observations. LIS values are from [19,20]. Compilation of observational data is from [17,18,19,20] and the model is described by [21].

$$\frac{\partial f}{\partial t} = -(\mathbf{V} + \langle \mathbf{v_D} \rangle) \cdot \nabla f + \nabla \cdot (\boldsymbol{K}_s \cdot \nabla f) + \frac{p}{3}(\nabla \cdot \mathbf{V}) \frac{\partial f}{\partial p} + J_{source}, \qquad (1)$$

where $f(\mathbf{r},p,t)$ is the CR distribution function; $p$ is momentum, $\mathbf{r}$ is position, $\mathbf{V}$ is the solar wind velocity and $t$ is time. Terms on the right-hand side represent respectively convection, gradient and curvature drifts, diffusion, adiabatic energy changes, and a source function e.g., for the anomalous CR component. The tensor $\boldsymbol{K}_s$ consists of a parallel diffusion coefficient ($K_{||}$) and two perpendicular diffusion coefficients, one in the radial direction ($K_{\perp r}$) and one in the polar direction ($K_{\perp\theta}$). The pitch angle averaged guiding centre drift velocity is given by $\langle \mathbf{v}_D \rangle = \nabla \times (K_A \mathbf{e}_B)$, with $\mathbf{e}_B = \boldsymbol{B}/B_m$ and $B_m$ the magnitude of the modified background heliospheric magnetic field (HMF), with $K_A$ the off-diagonal element of the full diffusion tensor. The present understanding of the mechanisms of global modulation in the heliosphere, as described above, is considered essentially correct. However, the main obstacle in solving Eq. (1) is insufficient knowledge of the spatial, rigidity and especially the temporal dependence of the diffusion coefficients including the underlying features of the magnetic field turbulence, the size and geometry of heliosphere. Serious efforts are being made to improve the situation to come to a formulation in which the diffusion coefficients are calculated from basic diffusion theory and from the underlying fluctuating parameters, based on plasma and turbulence theories, using known features of the solar wind and the HMF. These approaches must be tested against CR observations. Diffusion theory involves several turbulence parameters so that one needs to understand how solar wind turbulence evolves throughout the heliosphere, also at high heliolatitudes, and now also in the heliosheath.

All GCR transport models require knowledge of the local interstellar intensities, the global structure and geometry of heliosphere, the solar wind velocity profile and the HMF including the heliospheric current sheet. An example of the application of the numerical solution of Eq. (1) using all available knowledge is shown in Figure 5. The top and middle panels show the computed differential intensities for Boron and Carbon at Earth for both solar magnetic polarity cycles( $A > 0$; $A < 0$) compared to B and C observations [17, 18; see 19 and 20 for full references]. Computations were done with the LIS for B and C by [19,20]. The model is described by [21]. The bottom panel shows B/C as a function of kinetic energy for both polarity cycles, with a HCS tilt angle $\alpha = 10°$, representing solar minimum conditions, compared to observations. The computations were done with a model that included a TS at 90 AU and a HP at 120 AU where the interstellar B/C is shown as a reference.

## 2.2. *Modulation Models, Solar Cycles and Charge-sign Dependence*

A departure point for the time-dependence of global long-term modulation is that propagating barriers (solar wind and magnetic field structures inhibiting the easy access of CRs) are formed (and later dissipate) in the heliosphere during the 11 year activity cycle. This is especially applicable to the phase of the solar activity cycle before and after solar maximum conditions when large steps in the

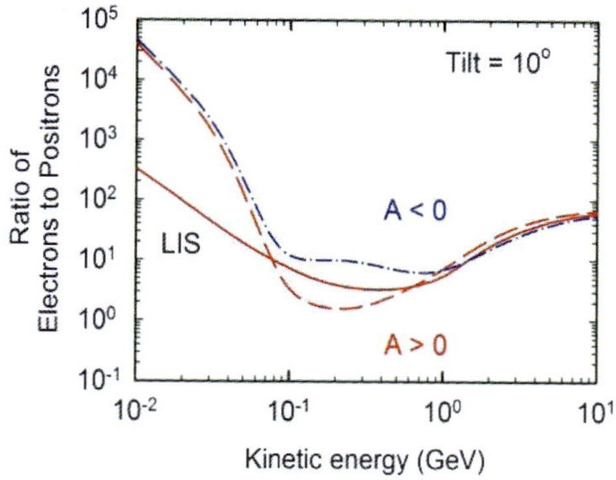


Figure 6. Computed electron to positron ratios at Earth for the two polarity cycles of the Sun compared to the ratio of the LIS at 120 AU. Differences above about 80 MeV are caused by gradient, curvature and current sheet drifts in the heliosphere during solar minimum activity [26].

particle intensities are observed. A wide range of interaction regions occur in the heliosphere, with global merged interaction regions (GMIRs) the largest, introduced by [22]. They observed that a clear relation exists between GCR decreases (and recoveries) and the time dependent decrease (and recovery) of the HMF magnitude local to the observation point. The paradigm on which these modulation barriers is based is that interaction (and rarefaction) regions form with increasing radial distance from the Sun. These relatively narrow interaction regions get azimuthally extended to become CIRs. They spread, merge and interact to form eventually GMIRs that can become large in extent and capable of causing the large step-like changes in CRs. It was shown by [23] that the affects on long-term modulation depend on their rate of occurrence, the radius of the heliosphere (i.e. how long they stay inside the modulation volume), the speed with which they propagate, their spatial extent and amplitude, especially their latitudinal extent (to disturb drifts), and the background modulation conditions they encounter. Drifts on the other hand, dominate the solar minimum modulation periods up to four years so that during an 11 year cycle a transition must occur (depending how solar activity develops) from a period dominated by drifts to a period dominated by these propagating structures. The largest of the step decreases and recoveries shown in Figure 3 are caused by these GMIRs.

The 11 and 22 year cycles together with the step-like modulation are good examples of the interplay of the main modulation mechanisms as shown by [24]. They illustrated that it was possible to simulate a complete 22 year modulation cycle by including a combination of drifts, determined by time-dependent tilt

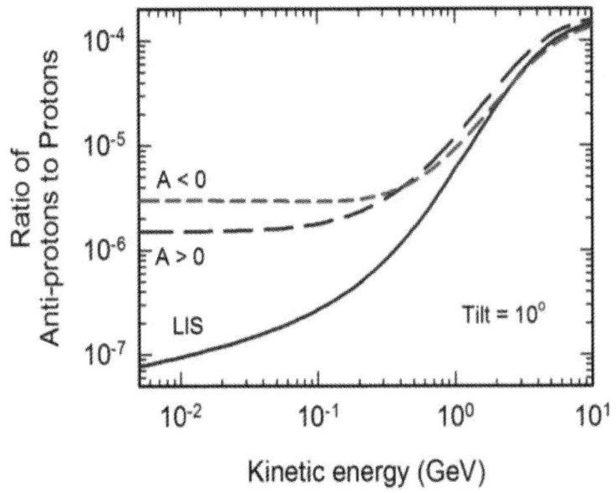


Figure 7. Similar to Figure 6 but for the computed ratio of galactic protons to anti-protons, during solar minimum modulation, compared with the ratio of the corresponding LIS at the heliopause, here placed at 120 AU [27].

angles, and GMIRs in a time-dependent modulation model. The compound modeling approach was subsequently introduced by [25] who showed that the concept of propagating modulation barriers, together with changes in the HCS and drifts, could reproduce observations remarkably well over a period of 22 years. This applies to observations at Earth, along the Ulysses and Voyager 1 and 2 trajectories. An important accomplishment of this compound approach is that it also produces the observed charge-sign dependent GCR modulation from solar minimum to maximum activity. As mentioned above, charge-sign dependent modulation is a direct indication of particle drifts in the heliosphere, illustrated in Figures 6 and 7 for the intensity ratios of cosmic ray electrons, positrons, protons and anti-protons [26,27] respectively.

Considering the modulation shown in Figure 3, the relevant question within the context of present Voyager 1 and 2 observations is how much modulation occurs inside the heliosheath and where does CR modulation actually start? Can the heliosheath consider being a modulation 'barrier'? The process is of course highly energy dependent. An illustrative example of the amount of modulation that GCR protons may experience in the heliosheath, in the nose direction of the heliosphere, was computed by [28] and shown in Figure 8. At energies $E <$ ~0.02 GeV, GCRs may experience > 80% modulation in the heliosheath for both polarity cycles. For all four the modulation conditions ($\alpha$ = 10° as solar minimum; $\alpha$ = 75° as solar maximum activity), the heliosheath modulation will eventually reach 0% but at different energies, indicating that it differs significantly with energy as well as with drift cycles. The negative values indicate that at these energies GCRs may get re-accelerated at the heliospheric

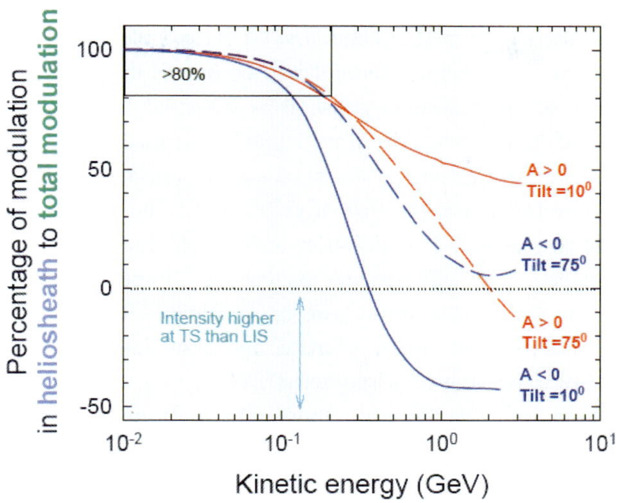


Figure 8. Computed percentage of GCR modulation in the heliosheath with respect to the total modulation (between 120 AU and 1 AU) for the two magnetic polarity cycles ($A > 0$ and $A < 0$), for solar minimum ($\alpha = 10°$) and for moderate maximum ($\alpha = 75°$) conditions, in the equatorial plane in the nose direction of the heliosphere. Negative percentages mean that GCRs are re-accelerated at the TS under these conditions [28].

TS so that the intensity actually becomes higher there than the corresponding values of the LIS at the HP. How much gradient and curvature drifts actually may occur in the heliosheath is still unanswered.

## 3. Discussion and Conclusions

Heliospace physics forms a part of the universal physical processes that can be used to gain better understand of the features and characteristics of space, from geospace to heliospace into interstellar space, and finally into galactic space. Heliospheric physics has become most relevant in order to understand processes occurring in astrophysics and is actively studied. The recent crossings of the heliospheric termination shock by the two Voyager spacecraft have been a major accomplishment that has renewed the interest in GCR modulation and the physics of the heliosheath, and what is to be expected beyond the HP when these spacecraft move into the local interstellar medium. An important issue is what are the values of the very local interstellar spectra for the different GCR species at energies below a few hundred MeV?

The study of the heliosheath, the heliopause and the heliospheric interface with the local interstellar medium and how GCRs and the anomalous component respond to the global dynamics thereof will be one of the prominent heliospace research topics for the coming years. It is essential to fully understand the dynamics of the heliosphere, the time variation of local ISM, and beyond, in order to appreciate the variations on the scale of thousands of years, and much longer, that seem to exist in the flux of GCRs [29]. The effects of the dynamics of the heliosphere on GCR modulation have been studied quantitatively only recently, an aspect that will become increasingly important. Variation in the GCR intensity in the galactic spiral arms is significant [30] so that it should have an effect on heliospheric space climate and thus have a clear imprint on terrestrial archives over the very long term [29].

Observations of galactic and anomalous CRs in the outer and inner heliosphere, together with the solar wind and magnetic field, have also caused new controversies and scientific issues. Several challenges need to be addressed, examples are: What is the spatial-time-dependence of the global strength and structure of the TS? How are energetic particles, in particular the anomalous component [31], accelerated beyond the TS? What are the global properties of the plasmatic flow beyond the TS and what is happening to it in the heliotail? How does the interstellar flow interact with the heliosphere beyond the HP? Understanding both the macro and micro physics of the dynamic heliosphere will give the theoretical and modeling tools to study broader issues in astrophysics.

## Acknowledgments

MSP thanks the organizers for the invitation to participate in this workshop. Financial support of the South African National Research Foundation (NRF) and the SA Centre for High Performance Computing (CHPC) is appreciated. The authors are grateful for informative discussions on cosmic rays and the heliosphere with Mirko Boezio, Bernd Heber, Ingo Büsching, Horst Fichtner, Klaus Scherer and Bill Webber.

## References

1. H-J. Fahr, *Adv. Space Res.* **34**, 3 (2004).
2. K. Scherer and S. E. S Ferreira, *ASTRA* **1**, 17 (2005).
3. U. W. Langner, M. S. Potgieter, H. Fichtner and T. Borrmann, *Astrophys. J.* **640**, 1119 (2006).

4. J. D. Richardson, E. C. Stone, J. C. Kasper, J. W. Belcher and R. B. Decker, *Geophys. Res. Lett.* **36**, L10102 (2009).
5. E. C. Stone, A. C. Cummings, F. B. McDonald, B. C. Heikkila, N. Lal and W. R. Webber, *Nature* **454**, 71 (2008).
6. J. L. Snyman, Master's thesis, North-West University, Potchefstroom, South Africa (2007).
7. M. Opher, F. Alouani Bibi, G. Toth, J. D. Richardson, V. V. Izmodenov and T. I. Gombosi, *Nature* **462**, 1036 (2009).
8. M. S. Potgieter, *Adv. Space Res.* **41**, 245 (2008).
9. M. S. Potgieter, *J. Atmospheric & Solar-Terres. Phys.* **70**, 207 (2008).
10. B. Heber and M. S. Potgieter, *Space Sci. Rev.* **127**, 117 (2006).
11. O. Adriani, G. C. Barbarino, G. A. Bazilevskaya, R. Bellotti, et al., *Nature* **458**, 407 (2009).
12. E. J. Smith, *J. Geophys. Res.* **106**, 15819 (2001).
13. J. Jiang, R. Cameron, D. Schmitt and M. Schüssler, *Astrophys. J.* **709**, 301 (2010).
14. S. E. S. Ferreira and M. S. Potgieter, *Astrophys. J.* **603**, 744 (2004).
15. L. Svalgaard and E. W. Cliver, *J. Geophys. Res.* A09111, 115 (2010).
16. E. N. Parker, *Planet. Space Sci.* **13,** 9 (1965).
17. J. J. Engelmann, P. Ferrando, A. Soutoul, P. Goret and E. Juliusson, *Astron. Astrophys.* **233**, 96 (1990).
18. S. A. Stephens and R. E. Streitmatter, *Astrophys. J.* **505**, 266 (1998).
19. I. V. Moskalenko, A. W. Strong, J. F. Ormes and M. S. Potgieter, *Astrophys. J.* **565**, 280 (2002).
20. I. V. Moskalenko, A. W. Strong, S. G. Mashnik and J. F. Ormes, *Astrophys. J.* **586**, 1050 (2003).
21. M. S. Potgieter and U. W. Langner, *Ann. Geophys.* **22**, 3729 (2004).
22. L. F. Burlaga, F. B. McDonald and N. F. Ness, *J. Geophys. Res.* **98**, 1 (1993).
23. M. S. Potgieter, J. A. le Roux, L. Burlaga and F. B. McDonald, *Astrophys. J.* **403**, 760 (1993).
24. J. A. le Roux and M. S. Potgieter, *Astrophys. J.* **442**, 847 (1995).
25. S. E. S. Ferreira and M. S. Potgieter, *Astrophys. J.* **603**, 744 (2004).
26. M. S. Potgieter and U. W. Langner, *Astrophys. J.* **602**, 993 (2004).
27. U. W. Langner and M. S. Potgieter, *Adv. Space Res.* **34**, 144 (2004).
28. U. W. Langner, M. S. Potgieter and W. R. Webber, *J. Geophys. Res.* **108**, 8039, LIS14 1-9 (2003).
29. K. Scherer, et al., *Space Sci. Rev.* **127**, 327 (2006).
30. I. Büsching and M. S. Potgieter, *Adv. Space Res.* **42**, 504 (2008).
31. R. D. Strauss, M. S. Potgieter and S. E. S. Ferreira, *Astron. Astrophys.* **513**, A24 (2010).

## A Markov Chain Monte Carlo technique to sample transport and source parameters of Galactic cosmic rays

A.Putze[1], L. Derome[2], F. Donato[3], and D. Maurin[2]

[1] *The Oskar Klein Centre for Cosmoparticle Physics, Department of Physics, Stockholm University, AlbaNova University Center, SE-10691 Stockholm, Sweden*
*antje@fysik.su.se*

[2] *Laboratoire de Physique Subatomique et de Cosmologie, Université Joseph Fourier Grenoble 1, CNRS/IN2P3, Institut Polytechnique de Grenoble, 53 avenue des Martyrs, Grenoble, 38026, France*

[3] *Dept. of Theoretical Physics and INFN, via Giuria 1, 10125 Torino, Italy*

We implemented a Markov Chain Monte Carlo technique to estimate the probability-density functions of the cosmic-ray transport and source parameters in a diffusion model. From the measurement of the B/C ratio and radioactive cosmic-ray clocks, we calculate their probability density functions, with a special emphasis on the halo size $L$ of the Galaxy and the local underdense bubble of size $r_h$. Finally, we check the compatibility of the primary fluxes with the transport parameters derived from the B/C analysis and then derive the source parameters (slope, abundance, and low-energy shape). We conclude that the size of the diffusive halo depends on the presence/absence of the local underdensity damping effect on radioactive nuclei. Models based on fitting B/C are compatible with primary fluxes. The different spectral indices for the propagated primary fluxes up to a few TeV/n can be naturally ascribed to transport effects only, implying universality of elemental source spectra.

*Keywords*: Cosmic rays; Propagation; Diffusion model; Markov Chain Monte Carlo; Local Bubble

### 1. Introduction

One issue of cosmic-ray physics is the determination of the transport parameters in the Galaxy. Such a determination is based on the analysis of the secondary-to-primary ratio (e.g., B/C). We proposed in Ref. 1 a new method to extract these parameters using a Markov Chain Monte Carlo (MCMC) algorithm (see e.g. Refs. 2,3 and references therein), an efficient and sound numerical tool which covers efficiently the parameter space and enables the enlargement of the parameter space at a minimal computing time cost. The MCMC algorithm provides the probability density function (PDF) of the model parameters, based on selected experimental data. A detailed description of the approach used is given in Ref. 1.

The model parameter values presented here correspond to the most probable values, i.e., the peak value of the marginalised parameter PDFs, whereas the $\chi^2$ values represent the best-fit model for which the parameter values are not given.

## 2. Model

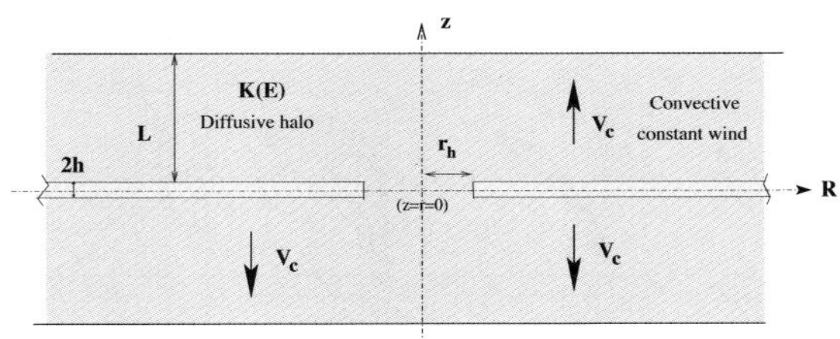


Fig. 1. Sketch of the model: sources and interactions are restricted to the thin-disc. Diffusion and convection transport nuclei in the disc and in the halo. The local bubble is featured as a hole in the disc.

We consider the 1D thin disc Diffusion Model[4] with constant Galactic wind $V_c$ and minimal reacceleration. This model is shown in Fig. 1. The diffusion coefficient and the injection spectrum are given by

$$K(E) = K_0 \beta R^{\delta}, \qquad \text{and} \qquad Q(E) = q\beta^{\eta_S} R^{-\alpha},$$

respectively. The standard diffusion model can be modified in order to take into account the under-dense local interstellar medium[5] (LISM). It is modelled as a hole in the Galactic disc.[6]

The free parameters of this model are the normalisation $K_0$ and the slope $\delta$ of the diffusion coefficient, the constant Galactic wind velocity $V_c$, the Alfvén velocity $V_a$ related to the reacceleration, the halo size $L$ of the Galaxy, the radius of the local bubble $r_h$, and finally, the normalisation $q$, the low-energy shape $\eta_S$, and the slope $\alpha$ of the injection spectrum.

## 3. Results

We start with the standard diffusion model ($r_h = 0$). There is a well-known degeneracy between the parameters $K_0$ and $L$ when stable species are considered: all models with the same ratio $K_0/L$ lead to the same grammage, $\lambda_{esc} = n\bar{m}vhL/K(E)$, hence to the same secondary-to-primary ratio.

Here we used B/C data only and we fixed $L = 4\,\mathrm{kpc}$. As found in previous analyses,[7,8] the diffusion model with both reacceleration and convection reproduces the B/C data more accurately than without. A more detailed description of the study and of the results obtained can be found in Ref. 9.

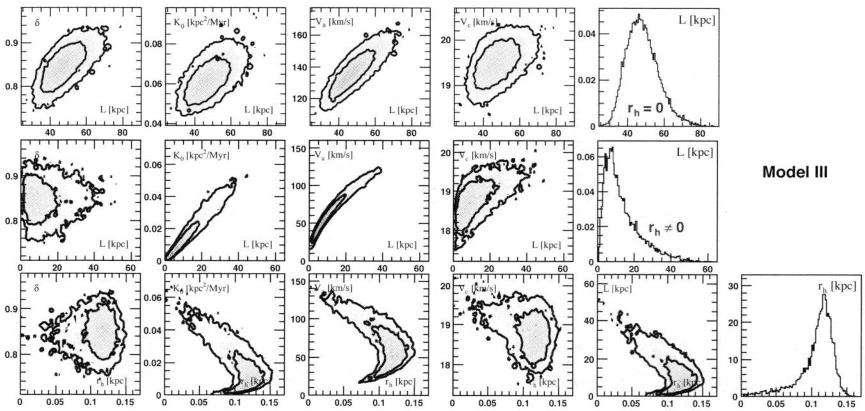


Fig. 2. Marginalised posterior PDFs of the diffusive halo size $L$ (right panels of the 1st and 2nd row) and the local bubble radius $r_h$ (right panel of the 3rd row) for the standard ($r_h = 0$, 1st row) and the modified diffusion model ($r_h \neq 0$, 2nd and 3rd row), as constrained by using B/C and $^{10}$Be/$^{9}$Be data. The correlations between the geometrical parameters $L$ and $r_h$ and the transport parameters $\delta$, $K_0$, and $V_a$, are shown in the 2D histograms. The colour code corresponds to the regions of increasing probability (from paler to darker shades), and the two contours delimit regions containing respectively 68% and 95% (inner and outer contour) of the PDF.

The degeneracy between $K_0$ and $L$ is lifted when radioactive secondary nuclei are considered. Indeed, a nucleus with a decay time smaller than $L^2/K_0$ doesn't reach the border of the Galaxy and is hence only sensitive to the diffusion coefficient $K_0$. We now take $L$ as a free parameter and use $^{10}$Be/$^{9}$Be (additionally to the B/C data) to constrain it. This radioactive ratio has no impact on the transport parameter values. The PDF of $L$ and its correlation with the transport parameters are shown in the first row of Fig. 2. We find $L = 46^{+9}_{-8}\,\mathrm{kpc}$, i.e., values significantly larger than those quoted in the literature.

In the last step, the radius of the under-density in the LISM is added as a free parameter. The net effect of the local bubble is an exponential decrease of the radioactive flux (see Ref. 6 for the detailed calculation). The resulting PDFs for $L$ and $r_h$ are shown in the two bottom rows of Fig. 2. The most likely values and uncertainties at 68% CL of the two parameters are $L = 8^{+8}_{-7}$ kpc and $r_h = 120^{+20}_{-20}$ pc. The local bubble parameter is correlated to the halo size of the Galaxy: the halo size $L$ is lowered to the range $1 - 50$ kpc when $r_h \neq 0$. The reduced $\chi^2$ value of the best fit for this model is 1.28, which is better than 1.41 for the model with no hole, a statistically significant improvement. This confirms the findings of Ref. 6: The standard model $r_h = 0$ is disfavoured. The bubble radius is consistent with the estimations from direct measurements. A more detailed study of the diffusion model is presented in Ref. 9.

Present data on primary fluxes alone cannot constrain significantly the transport parameters. Therefore we fix the transport parameters, and then constrain the source parameters. Below, $\alpha_i$, $q_i$ and $\eta^i_S$ are free parameters for each primary species considered from C to Fe, which allows us to i) test the universality of the source spectra, ii) take into account the correlations between the normalisation $q_i$ and the spectral index $\alpha_i$,[1] and iii) inspect the systematic spread on the source parameters as several configurations of the diffusion model, consistent with B/C data,[10] are taken. The resulting most probable values as well as the 68% and 95% error bars for the relative abundances (normalised to Silicon = 100) and the source spectral index are shown in Fig. 3 (for the full analysis, and also the results on p and He fluxes, see Ref. 11).

We first underline that the 95% CL relative uncertainty for the parameters ranges from 5% on $\alpha_i$ to 20% on $q_i$. CREAM data cover too narrow an energy range to give stringent constraints, so we do not comment on them further below. Independent of the model and data considered, the source slope $\alpha$ is constrained to fall in the range $2.2 - 2.5$, which is a robust prediction considering the wide range of $\delta$ ($0.23 - 0.86$) covered by the models. It also means that the asymptotic value for the propagated spectra ($\gamma_{\rm asymp.} \equiv \alpha + \delta$), which falls in the range $2.7 - 3.0$, is not reached in the GeV/n to TeV/n regime: residual propagation effects (reacceleration, convection, spallations) are still active. Finally, most of the derived relative source abundances are in agreement with those derived by earlier groups[1,12,13] using a Leaky-Box model. The difference for Ar and Ca may be related to the fact these elements were assumed to be pure primary species in this analysis. The lack of the secondary contribution translates in a higher primary flux required to match the data. Part of the discrepancy could also be related to the fact that the highest energy data point is better

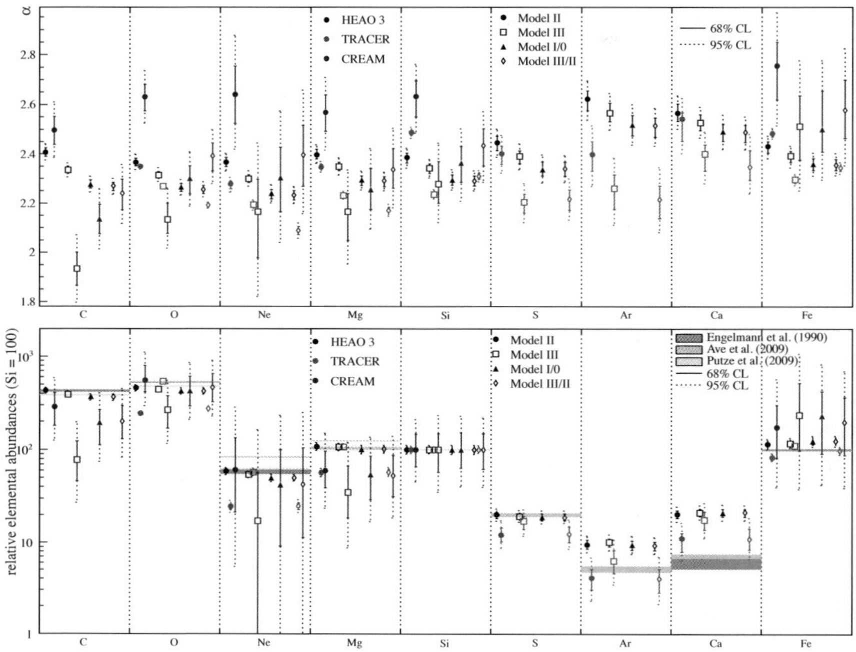


Fig. 3. Source spectral index $\alpha$ (top figure) and relative elemental abundances normalised to Silicon = 100 (bottom figure) of primary nuclei from C to Fe. The source parameters were fitted to the HEAO-3 (black symbols), TRACER (red symbols), and CREAM (blue symbols) data for 4 different sets of transport parameters (Model II: filled circle, III: open square, I/0: filled triangle, III/II: open diamond). The relative elemental abundances are compared to those resulting from studies using a Leaky-Box propagation model by Engelmann et al.[12] (green band), Ave et al.[13] (orange band), and Putze et al.[1] (yellow band).

fitted than for the other elements (not shown). The consistently low values for all elements obtained from the TRACER data for Models II and III/II can be simply explained in terms of a too large Si abundance obtained from the TRACER data for these models. This explanation is supported by the agreement between the abundances obtained for Model III, which gives a good fit to Si data, and those derived from HEAO-3 data. A more detailed study of the source parameters is given in Ref. 11.

## 4. Conclusion

We have implemented a Markov Chain Monte Carlo to extract the posterior distribution functions of the propagation and source parameters. Taking advantage of its sound statistical properties, confidence intervals for the propagation and source parameters can be given.

The analysis was performed for the one-dimensional diffusion model. The best-fit model includes both convection and reacceleration for the transport of cosmic rays. We use the $^{10}$Be/$^{9}$Be ratio to lift the degeneracy between $K_0$ and $L$. The uncertainty on $L$ remains large due to the limited energy range covered by the radioactive nuclear data. We note that, for the standard diffusion model (no hole), the value found for $L$ is larger than that claimed in other studies. We also present the PDF of the local bubble size parameter, $r_h$, pointing to a value $\sim 120$ pc. The presence of an underdensity in the LISM is favoured by cosmic-ray data. The effect of including a bubble is a reduction of the Galactic halo size $L$ to a range of $1 - 50$ kpc when $r_h \neq 0$. We have also studied primary nuclei spectra from C to Fe using models based on fitting B/C, which are well reproduced without a need for an artificial break in the injection slope. These nuclei point to $\alpha$ around 2.2–2.5 regardless of the propagation configuration used.

## References


1. A. Putze, L. Derome, D. Maurin, L. Perotto and R. Taillet, *Astronomy and Astrophysics* **497**, 991(April 2009).
2. A. Lewis and S. Bridle, *Physical Review D* **66**, p. 103511(November 2002).
3. J. Dunkley, M. Bucher, P. G. Ferreira, K. Moodley and C. Skordis, *Monthly Notices of the Royal Astronomical Society* **356**, 925(January 2005).
4. F. C. Jones, A. Lukasiak, V. Ptuskin and W. Webber, *The Astrophysical Journal* **547**, 264(January 2001).
5. R. Lallement, B. Y. Welsh, J. L. Vergely, F. Crifo and D. Sfeir, *Astronomy and Astrophysics* **411**, 447(December 2003).
6. F. Donato, D. Maurin and R. Taillet, *Astronomy and Astrophysics* **381**, 539(January 2002).
7. D. Maurin, F. Donato, R. Taillet and P. Salati, *The Astrophysical Journal* **555**, 585(July 2001).
8. D. Maurin, R. Taillet and F. Donato, *Astronomy and Astrophysics* **394**, 1039(November 2002).
9. A. Putze, L. Derome and D. Maurin, *Astronomy and Astrophysics* **516**, p. A66(June 2010).
10. D. Maurin, A. Putze and L. Derome, *Astronomy and Astrophysics* **516**, p. A67(June 2010).
11. A. Putze, D. Maurin and F. Donato, *ArXiv e-prints* (November 2010).
12. J. J. Engelmann, P. Ferrando, A. Soutoul, P. Goret and E. Juliusson, *Astronomy and Astrophysics* **233**, 96(July 1990).
13. M. Ave, P. J. Boyle, C. Höppner, J. Marshall and D. Müller, *The Astrophysical Journal* **697**, 106(May 2009).

# PAMELA Through a Magnetic Lens

J. P. ROBERTS*

*Center for Cosmology and Particle Physics, New York University,*
*New York, NY 10003, USA*
**E-mail: jr198@physics.nyu.edu*

The PAMELA satellite has observed an excess of positrons over electrons in the energy range $1 - 100$ *GeV* that increases with energy. We propose that the excess is due to heliospheric modulation. We motivate this from the known form of the heliospheric magnetic field and predict that the excess will disappear as we enter a period of solar maximum activity.

*Keywords*: PAMELA, positron excess, heliospheric modulation, AMS-02.

## 1. Introduction

The PAMELA satellite[1] has observed a rise in the positron to electron ratio that increases with energy above 10 *GeV*. This has been interpreted as a rise in the positron fraction in the local interstellar spectrum of cosmic rays. We propose[2] that the rise in the positron fraction is not present in the interstellar spectrum, but is a result of the passage of cosmic rays through the heliospheric magnetic field. As we transition from the current solar minimum into a solar maximum and the magnetic field becomes chaotic the charge asymmetry will disappear, followed by an electron excess in the following solar minimum. AMS-2 will be perfectly positioned to observe these effects. If such a change in the positron to electron ratio is observed it will be unequivocal proof that the positron excess was a heliospheric effect.

## 2. The heliospheric magnetic field

During solar minimum the heliospheric magnetic field is well described by a Parker spiral.[3]

$$\mathbf{B} = B_0 \frac{r_0^2}{r^2} \left[ \hat{\mathbf{e}}_r - \frac{\Omega_\odot (r - b) \sin\theta}{V} \hat{\mathbf{e}}_\phi \right], \tag{1}$$

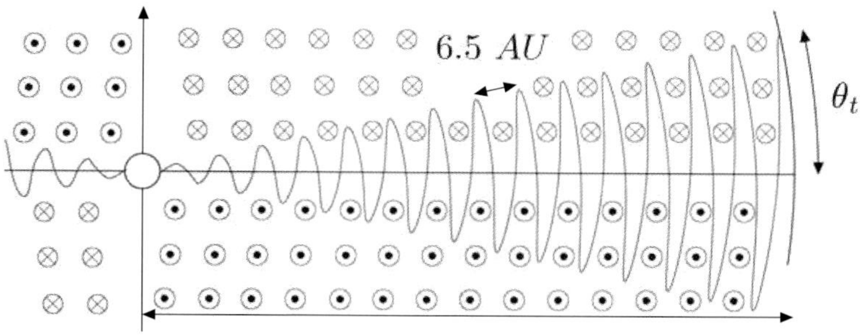


Fig. 1. A 2D schematic of the form of the magnetic field inside the heliosphere.

where $(r, \theta, \phi)$ are heliocentric spherical coordinates, $B_0$ is the magnetic field at $r_0$, V is the solar wind speed, b is the radius at which the field is purely radial and $\Omega_\odot$ is the equatorial rotation of the sun. We take the magnetic field at earth during the recent solar minimum to be 3.3 $nT$.

The magnetic field is oriented oppositely above and below the plane separated by a zero field current sheet. The current sheet oscillates with the rotation of the sun causing a 'ballerina skirt' effect with tilt angle $\theta_t$. A schematic of the heliospheric magnetic field is shown in Fig. 1. Beyond a few AU the magnetic field is predominantly azimuthal with a strength that falls as $1/R$. This acts as a barrier to the radial motion of cosmic rays and, combined with the convection of the solar wind, shields the center of the solar system from low energy interstellar cosmic rays.

## 3. The magnetic lens

The current sheet creates a charge asymmetry in the transport of cosmic rays, first noted by Levy.[4] A particle of one charge entering the heliosphere will be deflected towards the ecliptic plane. Once it crosses the current sheet the direction of the magnetic field reverses causing it to be deflected back towards the ecliptic. We show the resulting path in Fig. 2 for a flat current sheet. Oppositely charged particles are deflected away. These paths allows a fraction of the unsuppressed interstellar flux of positrons to penetrate the center of the solar system and this fraction grows with momentum:

$$f(p) = \frac{r_g}{r} = \sin\theta = \frac{2.2 \times 10^{-2} p_\perp (GeV/c)}{|q| B_E(nT)}. \tag{2}$$

where $B_E$ is the magnetic field strength at earth.

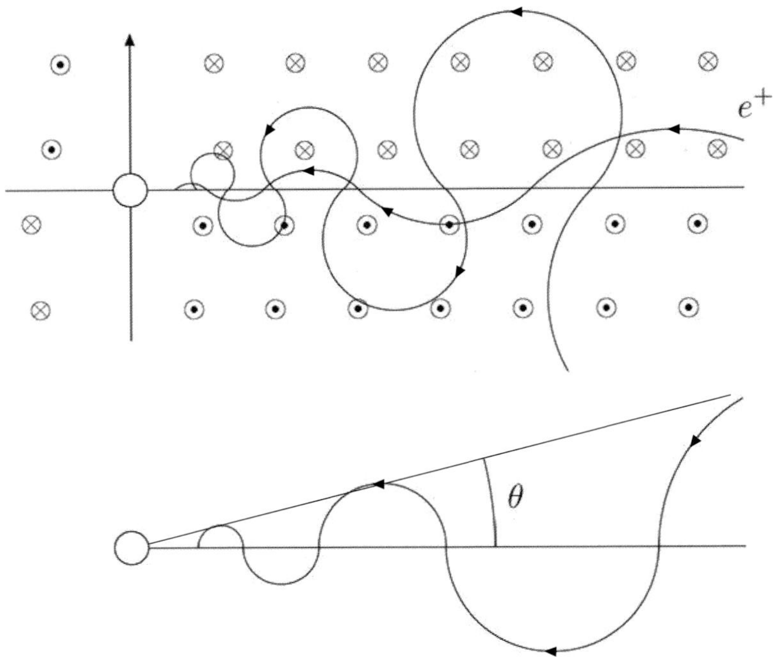


Fig. 2. Particle paths with a flat current sheet.

If the local flux of cosmic rays (before the lens is considered) is suppressed with respect to the interstellar flux, then the total flux of both electrons and positrons at earth is:

$$F_{1\mathrm{AU}}(e^+) = (A + f(p))F_{80\mathrm{AU}}(e^+)$$
$$F_{1\mathrm{AU}}(e^-) = AF_{80\mathrm{AU}}(e^-) \tag{3}$$

where $A$ is the local suppression factor.

In Fig. 3 we use Eq. (3) to fit the PAMELA data using a Galprop[5] model to provide input positron and electron fluxes (dashed red line). The general form of the energy dependence is promising, but the quality of the fit is poor. In generating the blue line we require $A = 0.23$.

The poor fit is not surprising given the rough nature of our model. As a first step we must consider the effect of the oscillations of the current sheet. At low energies ($< 10\ GeV$) particles travel along the current sheet as for a flat current sheet. The oscillations of the current sheet result in a much longer path length for the particle and therefore reduces the inward radial velocity. This radial velocity is below the velocity of the solar wind,

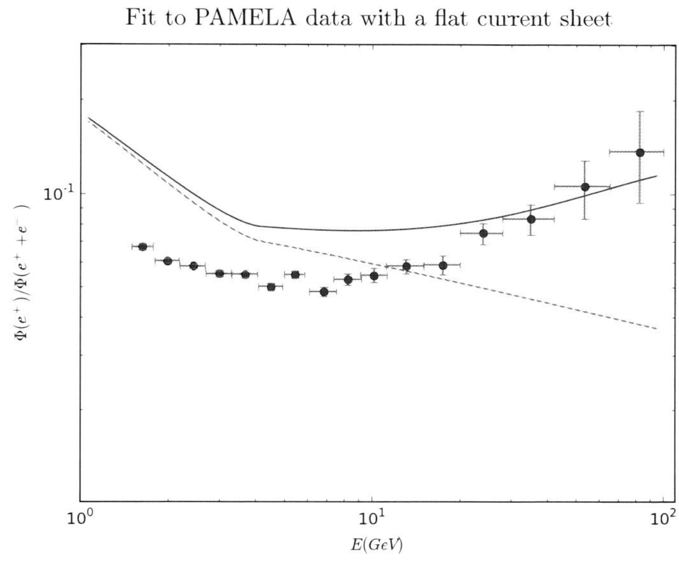


Fig. 3. We use the flat current sheet to fit to the PAMELA data.

the inward drift is overwhelmed by the outward convection and there will be no positron excess.

At larger energies particles do not follow the current sheet. If the Larmor radius of the particle is larger than the amplitude of the current sheet the particle can cross the sheet once per rotation and the particle effectively sees a flat current sheet. This is only true for those particles that travel outside the volume traversed by the current sheet. There are also particles that have a Larmor radius larger than the wavelength of the current sheet and that stay within the volume occupied by the current sheet. They can cross between oscillations of the current sheet rather than travelling along it - allowing them to penetrate deep into the solar system until their Larmor radius becomes smaller than the wavelength of the current sheet. This process is charge symmetric as shown in the near horizontal trajectories of an electron and a positron in Fig. 4.

In our crude approximation we don't allow any particle that remains within the volume of the current sheet to contribute to any charge asymmetry. The tilt of the current sheet defines the volume and pushes the appearance of the effect of the magnetic lens to higher energies.

$$f'(p) = \sin\theta(p) - \delta, \quad f' \geq 0 \tag{4}$$

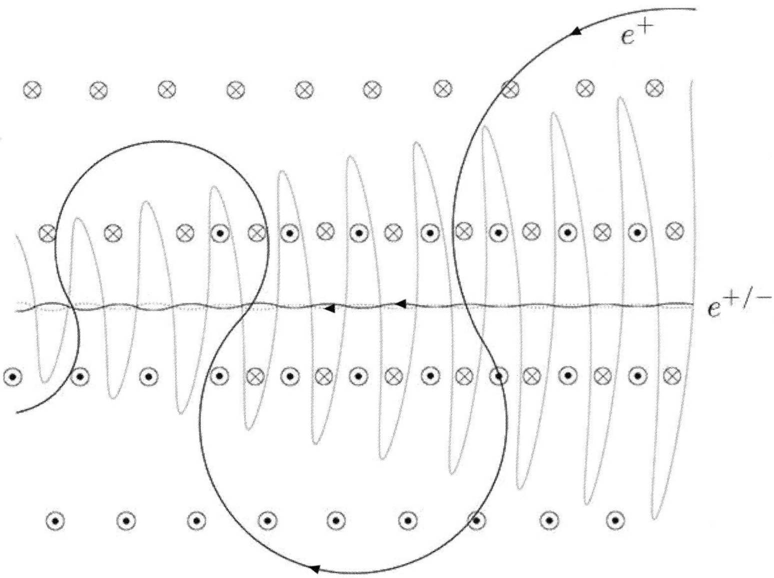


Fig. 4. Particle paths taking into account the wavy current sheet.

where $\delta$ is proportional to the volume occupied by the current sheet.

In Fig. 5 we fit to the data using Eq. (4). The fit requires $A = 0.15$, $\delta = 0.09$. We only fit to points above 10 $GeV$ where we know that conventional solar modulation is negligible. The fit is good and the required parameters are not unreasonable, but our analytic approximation is rough and a full numerical simulation (currently underway) is required to obtain a quantitative rather than a qualitative understanding of high energy solar modulation.

## 4. Conclusions

The paths that interstellar cosmic rays take to the center of the solar system change from diffusion, dispersion and convection at 1 $GeV$ to direct transport with little deviation at 1 $TeV$. This transition occurs in a charge asymmetric manner with positrons penetrating the central solar system at lower energies than electrons due to the magnetic lens created by the heliospheric magnetic field. This introduces a charge asymmetry in the local flux of cosmic rays as long as there is a local deficit in comparison to the interstellar spectrum. The asymmetry increases with energy.

If the origin of the measured excess is due to the solar magnetic field

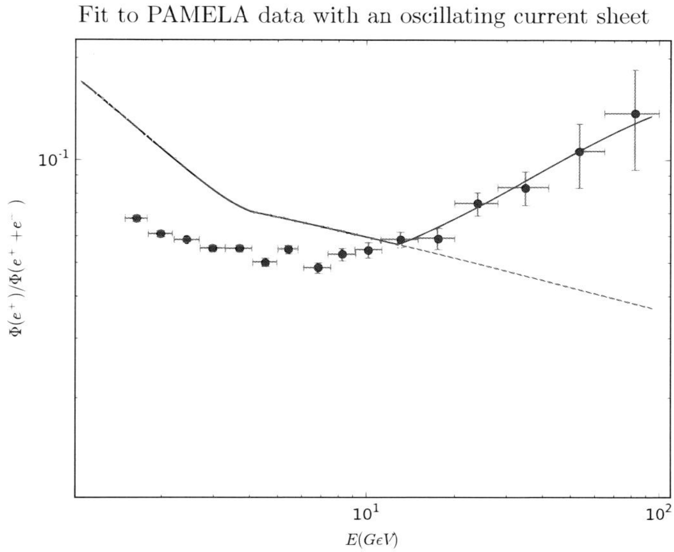


Fig. 5. The fit to the PAMELA data with an oscillating current sheet.

it will be easy to test. The sun is entering a period of solar maximum activity which will disrupt the heliospheric magnetic field. We predict that when AMS-2 begins to take data in early 2011 it will observe little or no positron excess . The excess will disappear throughout 2011 and become an electron excess during the following solar minimum. Such an observation would be incontrovertible evidence that the excess observed by PAMELA was of heliospheric origin.

## Acknowledgements

JPR's work was supported by NSF Awards PHY-0758032 and PHY-0449818 and DoE Award No. DE-FG02-06ER41417.

## References

1. O Adriani, G Barbarino, and G Bazilevskaya. *Nature*, Jan 2009.
2. J P Roberts. *arXiv*, [1005.4668].
3. E. N Parker. *Astrophysical Journal*, 128:664, Nov 1958.
4. E H Levy. *Nature*, 261(5559):394, Jun 1976.
5. A. W Strong et al. *arXiv*, astro-ph.HE, [0907.0559] 2009.

# ANALYSIS OF POSSIBILITY OF COSMIC RAYS PROTON ANISOTROPY PHASE AND AMPLITUDE AND ELECTRON SPECTRA DESCRIPTION AT TEV-REGION WITHIN THE BOUNDS OF THE SAME SET OF SOURCES

OLGA STRELNIKOVA

*e-mail: olga.n.strelnikova@gmail.com*
*Skobeltsyn Institute of Nuclear Physics, Moscow State University, Leninskie Gory, 1, Moscow, 119192, Russia*

VLADIMIR PTUSKIN

*Pushkov Institute of Terrestrial Magnetism, Ionosphere and Radio Wave Propogation (IZMIRAN), Russian Academy of Sciences, Troitsk, Moscow region 142092, Russia*

LYUBOV SVESHNIKOVA

*Skobeltsyn Institute of Nuclear Physics, Moscow State University, Leninskie Gory, 1, Moscow, 119192, Russia*

The contributions of nearby sources registered in radio, X-ray and gamma region in last experiments FERMI, Chandra, HESS to electron and positron spectra near the Earth and to proton anisotropy are investigated in this paper. Also the possible contribution to lepton spectra of pulsars' magnetospheres is evaluated. Unlike others we take into account the arm structure in source distribution through the galactic disc, escape time dependence of accelerated in SNR electrons and calculate the cosmic rays proton anisotropy simultaneously. Since the proton anisotropy is very sensitive to the source distribution and SNR spectra it restricts the choice of these parameters. We present the results of electron spectra and proton anisotropy calculations which fit the experimental results in TeV region.

## Introduction

After the appearance of new experimental data in TeV-region of cosmic ray (CR) electron spectra [1,2,3,4,5,6] and proton anisotropy [7,8,9,10] a new possibility to conclude about contributions of nearby sources into CR flux near the Earth appeared. One can solve this question by simultaneous investigation of electron flux and proton anisotropy within the bounds of the same set of sources. The second problem is whether electron amount from SNR is enough for description of the experimental data at GeV-energy region. Also we try to estimate the possible contribution from pulsars' magnetospheres to electron and positron flux near the Earth.

## 1. Method of Calculations and Approximations

Usually the propagation of CR in the Galaxy is described in the diffusion approximation. The most adequate is the model with flat extended halo and continuous source distribution within the galactic disk [11].

However this method doesn't account for fluctuations in CR concentration and anisotropy existing due to the statistical nature of sources [12]. That's why we use statistical approach of estimation of CR characteristics and their fluctuations which consists in consideration of the statistical ensemble of the galactic sources and summation of contributions from all sources, each of those was located at distance R from the Earth and emitted spectra *Q(E) t* years ago.

The particle propagation on the way to the Earth is described in terms of Green function, which depends on diffusion coefficient, distance of the source and its age [12,13,14]. According to GAPLROP [11] under 10 GeV the diffusion coefficient *D(E)* is power function of energy $D \sim E^{-\alpha}$ with $\alpha$=0.3-0.6, so at 1 TeV $D=(4\text{-}7)\text{x}10^{29}$ $\text{sm}^2\text{s}^{-1}$ and spectrum at source is also power function of energy $Q \sim E^{-g}$ with $g$=-2.4.

In our calculation we divided all the sources into 2 groups. The first group contains actual nearby young sources registered in radio, TeV, X-ray and gamma-emission, which are located within the distance $R_{near}$ and are not older than $T_{near}$ years. The second group consists of all other sources which are pointed randomly. Values $R_{near}$ и $T_{near}$ were chosen to satisfy the condition that fluctuations of CR flux near the Earth must not exceed 20%.

## 2. Selection of Sources

In case of common values of Galactic radius (15 kps), halo width (4 kps) and CR lifetime ($1.4 \times 10^8$ years at 1 GeV) the total quantity of sources makes up 2.9 million. In case of the continuous source distribution one may expect for about 20 sources with distance $R < 1.5$ kps and age $T < 10^5$ years and about 10 sources with $R < 1$ kps and $T < 10^5$ years.

For source distribution of core collapse supernovae (SNR type II, Ibc) we choose the spiral structure and for supernovae of Ia type we choose the uniform distribution. In total such distribution of SNR doesn't contradict the experimental results [15].

We analyzed SNRs [16], pulsar wind nebulae (PWN) [17], TeV-sources [18], gamma pulsars [19] and radio pulsars [20] catalogs. There are 6 SNRs of Ia type, 19 PWNe and 11 gamma-pulsars with distance $R < 1.5$ kps and age $T<10^5$ years thereby the amount of actual nearby young sources coincides with expected statistics in assumption of SNR rate 1/50.

## 3. CR Anisotropy

The main causes of CR anisotropy are the movement of the Earth through the space and the nearby sources. The nearby sources make the main contribution to CR anisotropy and lead to divergence in distribution function of CR concentration [13,15,12].

In the model of isotropic diffusion the anisotropy $\boldsymbol{A_r}$ at direction $\boldsymbol{r}$:

$$\mathbf{A}_r = 3\frac{D}{c}\frac{1}{N}\frac{\partial N}{\partial \mathbf{r}}. \quad \text{(Eq.1)}$$

In our calculations the total anisotropy from ensemble of sources is the sum of contributions of nearby young sources and all others.

Commonly the harmonic analysis of CR flux intensity $I$ as a function of sidereal time $t$

$$I = I_0 + A\cos(t - \varphi_n) \quad \text{(Eq.2)}$$

is considered in experiments [7-10]. In recent years the two-dimensional variations of CR flux as a function of sidereal time were measured at fixed declinations δ in bands with different Δδ by Super Kamiokande[8], TIBET [9], MILAGRO [10].

In our work we fixed the direction with right ascension $\alpha_k$ and declination $\delta_k$, calculated the projection of concentration gradient to this direction and then summed contributions from all sources. Then we averaged the result on fixed declination, found its deviation from the average value, found the sidereal time dependence and calculated the anisotropy amplitude and direction in equatorial coordinates. By this way we attempted to reproduce the experimental measurement procedure.

As a result of our calculations we came to the following conclusions.

1. The proton anisotropy has maximum along the celestial equator (at $\delta_k = 0$), at higher latitudes the anisotropy amplitude decreases as $\cos\delta_k$ .
2. For all latitudes the sinusoid maximum at each declination is located at the same right ascension $\alpha_{max}$. That's why one can determine only the right ascension α of anisotropy, but can't determine the true declination δ of a source.
3. If the majority of CR sources are distributed in the galactic disk then according to our calculations the most likely anisotropy maximum must be at α~3-7 hours (direction to Orion Spur) or at 18-22 hours (direction to the center of the Galaxy) and minimum must be at 12-13 hours. Predominance

of any directions of maximum depends on the source distribution along the galactic radius and on location of nearest arms.

Our conclusions coincided with recent assertions of Baksan group [7] about the imitation of the anisotropy maximum at δ=0.

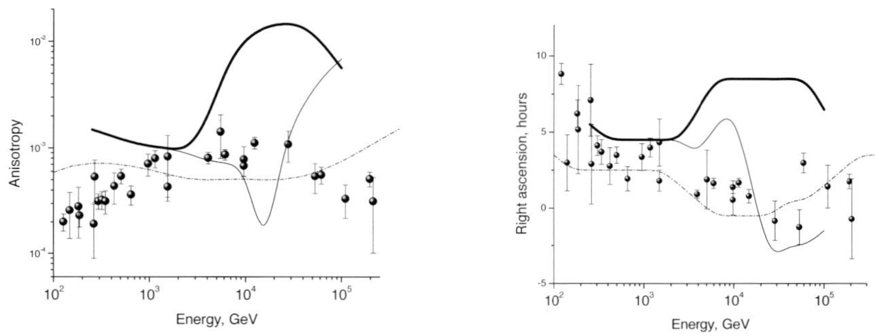


Figure 1. The amplitude of the anisotropy (left panel) and the right ascension (right panel) in comparison with experimental data from [9]. Dash-dotted line – without nearby sources, thick line – with all nearby sources including Vela X, thin line with all nearby sources including Vela X but Vela's contribution equals 20% from contribution of standard SNR.

Calculated magnitudes of the anisotropy direction and amplitude for δ=0 for diffusion coefficient $D \sim E^{0.3}$ and arm structure [11] in comparison with [9] are shown in Figure 1. The calculated curve fits the experimental results best of all in absence of nearby sources (dash-dotted curve). If we include all nearby actual sources with Vela X contribution as "standard" SNR (thick line) the theoretical curve is in obvious contradiction with the experiment. The reduction of Vela X's contribution to 20% allows for much better agreement between experimental and calculated data. This suggestion is confirmed by HESS' results [18].

## 4. CR Electron Spectra

We calculated the electron spectra near the Earth from the same set of sources in the same assumptions. Also we took into account the dependence of running-away particles maximum energy on SNR age according [21]. This dependence leads to narrowing of the spectrum of young sources due to absence of low-energy particles which can be emitted by only old source.

We found that one can satisfactory describe the experimental electron spectra in assumption of equal amount of electrons from the two groups of sources. Vela X contributes mostly in high energy region, but it contradicts the proton anisotropy data. The dashed line in Figure 2 shows the electron spectra from SNR with Vela X, but its contribution is 20% from "standard" SNR contribution. There is some lack of electrons between our and experimental results in this case. We decided to estimate the possible electron-positron flux from pulsars' magnetospheres to fill this lack following [22] and [19].

Our results are shown in Figure 2. We found the contribution from nearby young pulsars to be essential only in high-energy region (dotted line).

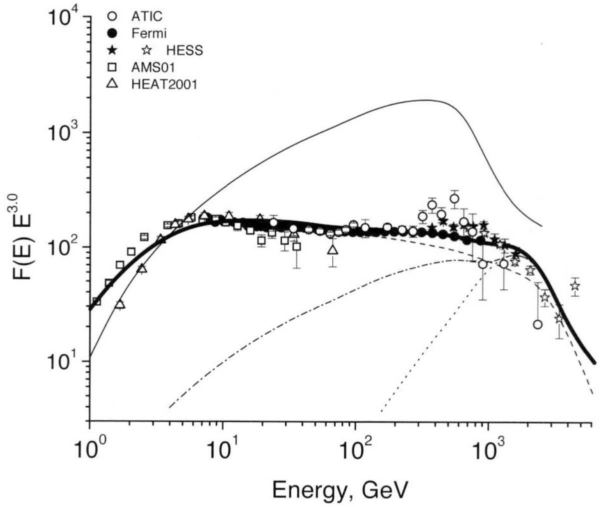


Figure 2. Experimental electron spectra with comparison with calculated results. Dashed line – spectra from all SNRs with Vela's contribution about 20% from standard SNR contribution, dotted line – electrons from nearby young gamma-pulsars with $E_{out}=10^{48}$ ergs, thin line – electrons from all pulsars with $E_{out}=10^{48}$ ergs, dash-dotted line – electrons from all pulsars with $E_{out}=10^{48}$ ergs for nearby pulsars and average $E_{out}=10^{46}$ ergs for others, thick line – total flux from SNRs and pulsars.

Total energy $E_{out}$ that nearby pulsars convert into pair energy according [22] and [19] is about $10^{48}$ ergs. If we account for nearby old and far pulsars with the same $E_{out}$ then results of calculations will contradict the experiment (thin line). The summary electron spectra from SNR and pulsars (thick line) fit experimental data if average $E_{out}$ for random gamma-pulsars is $3\text{x}10^{46}$ ergs and $E_{out}=10^{48}$ ergs for nearby pulsars (dash-dotted line). The reducing of $E_{out}$ is equivalent to assumption that only few percents of pulsars emit high energy electron-positron pairs.

We calculated the positron fraction at the Earth from SNRs and pulsars in assumption of equivalence of electron and positron amount in flux from pulsars.

The secondary positron flux was calculated using GALPROP. We found that if we take into account the pairs from pulsars than it becomes possible not only to fit the experimental electron spectra but also to explain the growth of positron fraction as shown by PAMELA.

**Discussion**

We calculated the CR electron spectra and proton anisotropy from actual nearby young sources together with randomly distributed old and distant sources: SRNs, PWNe and gamma-pulsars. It is shown that the model with considered actual sources can reproduce the experimental data on CR electron spectra and proton anisotropy in assumptions of electron acceleration in SNR shocks and equal average emission intensity of nearby and far sources through entire Galaxy. There is no need to use additional coefficients for correlations of the two source groups: actual and random ones. There is no indication on presence of nearby unidentified sources. However we are to suppose Vela X to produce less than 20% CR in comparison with standard SNR.

We simulated the experimental procedure of anisotropy measurement. Our conclusions coincided with recent assertions of Baksan group [7] about the two dimensional anisotropy: the flux excess in Taurus direction [8-10] most probably is a result of anisotropy measurement method.

The addition of pair contribution from pulsars' magnetospheres to electron flux from SNRs allows for much better agreement between the calculated electron spectra and experimental one if average $E_{out}$ for gamma-pulsars is $3x10^{46}$ ergs. Besides, this suggestion allows for the explanation of the growth of positron fraction in PAMELA experiment.

**References**


1. Abdo A.A. et al., *Phys. Rev. Lett.***102** (2009) 181101
2. Aharonian F. et al., *Phys. Rev. Lett.* **101** (2008) 261104
3. Aharonian F. et al., *Astro-ph/*0905.0105 (2009)
4. Chang et al., *Nature* **456** (2008) 362
5. Grasso. D., *Astro-ph/*0905.0636v3
6. Adriani O. et al., *Nature* **458** (2009b) 607
7. Kozyarivsky V.A. et al., *Astro-ph/*0406059, 2004
8. Guillian G.et al., *Astro-ph*/0508468v4, 2008
9. Amenomori M et al., *Astro-ph/* 0505114v1. 2005
10. Kolterman B.E., *30th Intern. Cosm. Ray Conf.*, Merida, Mexico, 2007, icrc0722
11. Strong W., Moskalenko I. V., Ptuskin V. S., *Ann. Rev. Nucl. Part. Sci.* **57** (2007) 285
12. Lagutin, A.A., Nikulin, Y.A., *JETP* **81**, 825–830, 1995

13. Gendelev L., *Astro-ph*/1001.4540v1
14. Lee, M.A., *ApJ* **229**, 424–431, 1979. 298
15. Case G., Bhattacharya., *A&A Supplement.* **Ser.120**, p.437
16. Green. D. A., *Astroph*/0905.3699v2
17. Kargaltsev O., Pavlov G. G., *Astro-ph*/0801.2602v2
18. H.E.S.S. http://www.mpi-hd.mpg.de/hfm/HESS/pages/home/sources
19. Abdo A., *Astro-ph*/0910.1608v1
20. Manchester R. N. et al., *Astro-ph*/0412641v1
21. Ptuskin V., Zirakashvili V., *Astro-ph*/0302053v1
22. Gendelev L. *Astro-ph*/1001.4540v1

# INTERSTELLAR GAMMA RAYS AND COSMIC RAYS: NEW INSIGHTS FROM FERMI-LAT AND INTEGRAL

A. W. STRONG*, on behalf of the Fermi-LAT Collaboration

*Max-Planck-Institute für extraterrestrische Physik*
*Garching, Germany*
**E-mail: aws@mpe.mpg.de*

In over two years of operation Fermi-LAT has revolutionized our knowledge of the gamma-ray sky. Interstellar gamma rays are part of this new era and allow unprecedented tests for models of cosmic rays in the Galaxy. The extension to lower energies with INTEGRAL/SPI data is also evolving. The global multi-wavelength luminosity of the Milky Way has been derived, with implications for the Galactic energy balance and the radio-FIR correlation.

*Keywords*: cosmic rays, gamma rays, interstellar medium, galaxies, Milky Way, high-energy astrophysics

## 1. Introduction

In more than two years of operation so far, Fermi-LAT has revolutionized our knowledge of the gamma-ray sky. Interstellar gamma rays are part of this new era and allow unprecedented tests for models of cosmic rays (CR) in the Galaxy. The extension to lower energies with INTEGRAL/SPI data is also making significant advances. There are also new results on the global multiwavelength luminosity of the Galaxy, with implications for the energy balance and the radio-FIR correlation. For a pre-Fermi review of CR and gamma rays see Ref. 1.

## 2. Fermi results on interstellar emission

First I briefly summarize the main Ferm-LAT results on interstellar gamma rays so far. The first major result[2] was to show that the 'GeV excess' above the expected interstellar emission was evidently an EGRET instrumental effect; instead the Fermi-LAT spectrum at intermediate latitudes (i.e. emission in the our local neighbourhood of the Galaxy) agrees rather well with that expected from locally-measured CR interacting with atomic and molecular hydrogen, plus inverse Compton emission. A second result[3] was a detailed study of the local gamma-ray emission from CR interacting with atomic hydrogen using a correlation analysis ; again good agreement with the expected spectrum from CR interactions was found. Thirdly[4] a study of the 2nd Galactic quadrant - probing the outer Galaxy - allowed determination of the $H_2$-to-CO factor ($X_{CO}$), confirming the increase with Galactocentric radius[13] and finding a rather lower local value than previously obtained with EGRET data. This analysis showed a surprisingly large level of emission in the outer Galaxy. This has been confirmed by a new study of the 3rd quadrant[5] . The outer galaxy emission is greater than expected if CR sources follow supernova remnants as traced by e.g. pulsars. CR propagation in a larger halo (10 kpc height) can help to reproduce this, but possibly large amounts of gas not traced by 21-cm atomic hydrogen and CO (tracer of $H_2$) surveys are present in the outer Galaxy[6] , and this could help explain this emission. Another aspect which requires good modelling of the Galactic emission is the determination of the extragalactic gamma-ray background[7] .

Meanwhile gamma-ray emission has been detected in five (non-AGN) extragalactic systems : LMC[8] , SMC[9] , M31 (with an upper limit for M33)[10] and the starburst galaxies M82 and NGC253[11] . Since this emission is presumably mainly interstellar as in the Galaxy, the combination of these results throws light on CR in a wider context. These results are discussed by P. Martin (this conference).

## 3. Comparison of large-scale Galactic emission models with Fermi-LAT data

The data are from the first year of Fermi-LAT observations, using a gamma-ray event class optimized for diffuse emission with low background[7] . Counts and exposure were converted to HEALPIX for processing.

The model was generated with the GALPROP[a] code (see Ref. 1 and references therein). The model includes diffusive reacceleration with a 4 kpc halo height, and details of the parameters are given in a related paper[12] . It uses the locally-measured CR proton and Helium spectra, and the Fermi-LAT measurements of the electron (plus positron) spectrum, for normalization at the Sun. For gamma-ray predictions, gas column densities from HI, CO surveys, far infrared dust emission as a tracer of additional gas, and a model for HII, are used. The cosmic-ray source distribution follows pulsars,

[a]http://galprop.stanford.edu

which are taken as a tracer of supernova remnants, the likely source of CR. The radial variation of $X_{CO}$ is taken from earlier work[13] . For inverse Compton, the interstellar radiation field of GALPROP[14] is used. The extragalactic (or at least isotropic) background is from the Fermi-LAT analysis[7] . Known gamma-ray sources are taken from the Fermi 1FGL catalogue[15] and included in the model. The model skymaps for each physical component were convolved to Fermi-LAT with the appropriate energy-dependent point-spread-function.

It is important to note that the chosen model is just illustrative, makes no claim to be unique, nor has it been specially fitted to Fermi data (this is ongoing work to be reported elsewhere); for this reason it is referred to as an '*a priori* model'. Even without such details, the quality of Fermi-LAT data allows interesting comparisons.

Figures 1 and 2 show the spectrum of the inner Galaxy, and longitude and latitude profiles at 1.4 GeV. Such a basic model reproduces the data over much of the sky within $\approx$ 20% and, remarkably, from the Galactic plane to the poles in latitude over 2 decades dynamic range. The latter shows the importance of inverse Compton emission which dominates at high latitudes. The spectrum of the inner Galaxy is in reasonable agreement with the model, while there is much room for adjustment of the parameters which could improve the fit; this is the subject of forthcoming work by the Fermi-LAT collaboration. The latitude profile in particular can be improved with a larger halo height of about 10 kpc instead of 4 kpc (also shown in Figure 2), and this is consistent with the higher outer Galaxy emission mentioned in Section 2. While the global agreement is good, there are significant residuals over the sky, again a subject for future detailed studies[b].

## 4. Contribution from unresolved sources

While Fermi-LAT has detected 1451 gamma-ray sources[15] which can be accounted for in comparisons of the diffuse emission with models, the underlying source populations will certainly contribute to the remaining emission. Current estimates put the level of this emission at 10-20% of the total emission, depending on energy and direction, which corresponds to at least as much as the detected sources. This is therefore a major uncertainty affecting any comparison of diffuse models with data, especially towards the

[b]Other studies have given indications for large-scale features not predicted by standard CR propagation models, such as the 'Fermi bubbles'[16] , but these will not be discussed here

inner Galaxy. In fact, the spectrum of the inner Galaxy (Figure 1) reveals an excess at GeV energies over that expected, and this could plausibly be attributed to unresolved source populations, in particular pulsars. A detailed study of this aspect is in preparation.

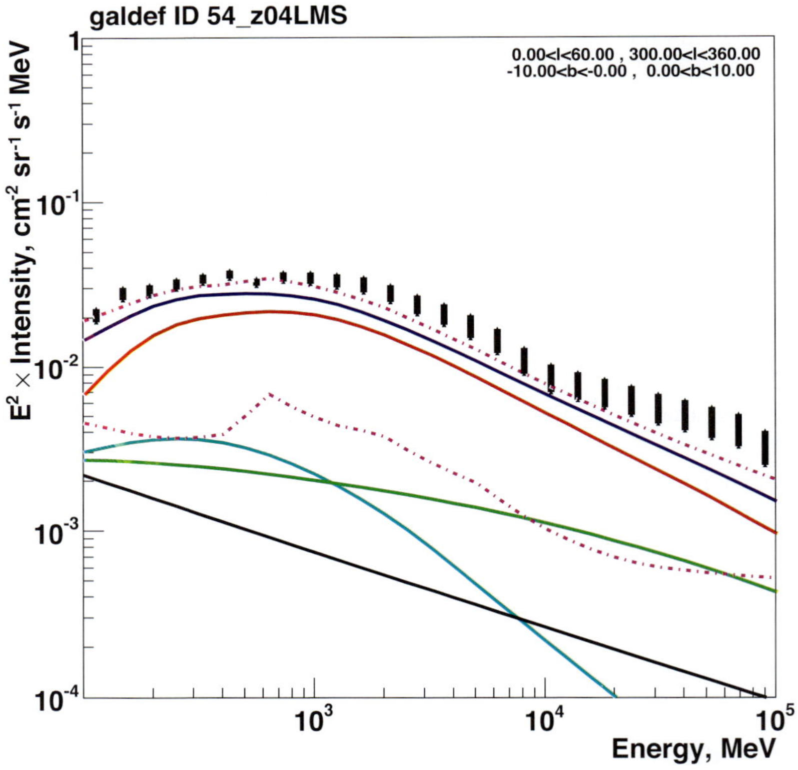


Fig. 1. Spectrum of inner Galaxy ( $300^o < l < 60^o$, $|b| < 10^o$). Model from GALPROP, based on locally-measured cosmic ray spectra and a halo height 4 kpc. Lines show the components of model; red: pion-decay, green: inverse Compton, black: extragalactic/isotropic, blue: total without sources, magenta dashed: Fermi-detected sources and total including sources. Data: black vertical bars: Fermi-LAT (PRELIMINARY).

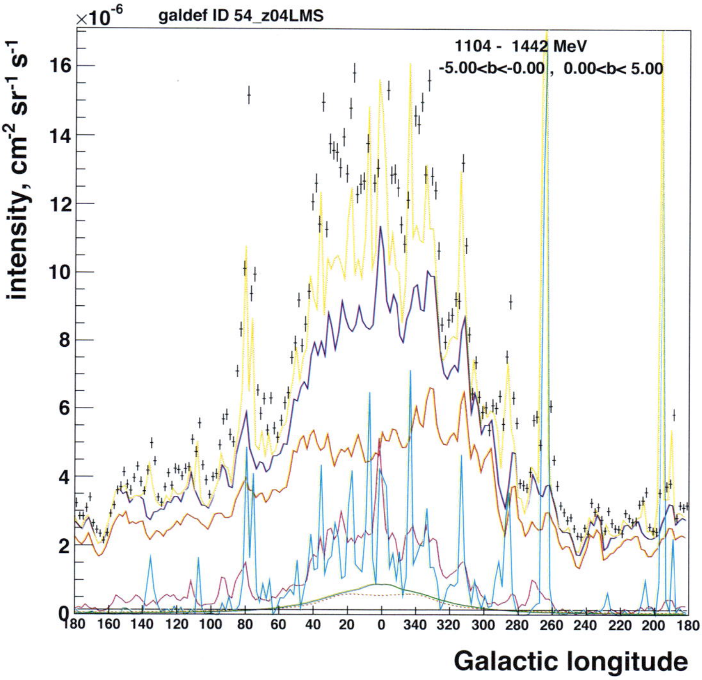


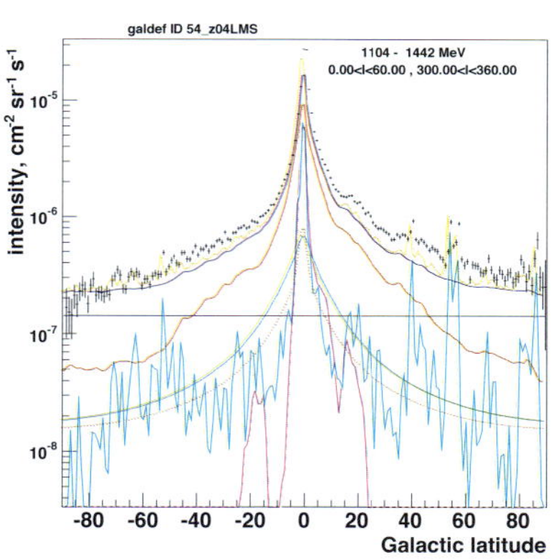


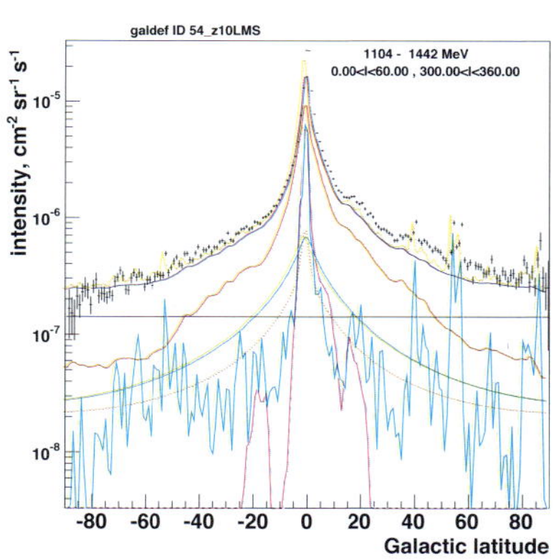


Fig. 2. Longitude ($|b| < 5^o$) and latitude ($300^o < l < 60^o$) profiles using Fermi-LAT (PRELIMINARY) data for 1.1–1.4 GeV. Model as in Figure 1, except right latitude profile has halo height 10 kpc. Lines show components of the model; red solid: HI, red dashed: HII, magenta: $H_2$, green: inverse Compton, cyan: Fermi-detected sources, black: extragalactic/isotropic, blue : total without sources, orange : total including sources.

## 5. Multiwavelength luminosity of the Galaxy

Although spectral energy distributions (SEDs) are plentifully available for AGN, SNRs, pulsars and many other objects, for the Milky Way it is more difficult because of our postion inside the Galaxy. Recently[12] it has been possible to synthesize the multiwavelength spectrum of the Galaxy from radio to gamma rays, based on the GALPROP model which includes all CR-related processes (although not compact sources, lines etc). The model is compatible with radio, gamma-ray and CR data, but is obviously not unique; discussion of the possible range of models is given in that paper. Figure 3 shows such an SED, including the optical and infrared components as well. This is useful as the basis for estimating the gamma-ray background from normal galaxies, and for comparison with external galaxies such as M31[10] which are similiar to the Milky Way (see also P. Martin, these proceedings). In addition it allows us to address the question of whether the Galaxy is a lepton calorimeter: the answer is affirmative provided both synchroton and inverse Compton emission are included. This is useful information for the interpretation of the radio-far-infrared correlation observed for a wide range of galaxies: it provides the only case where the lepton calorimetry can be directly checked via gamma rays, synchrotron and direct measurements of cosmic ray leptons. Normally only the synchrotron component is considered in assessing calorimetry. A spin-off of this study was to show that the Milky Way lies on the radio-FIR relation defined by other galaxies.

## 6. Extension to hard X-rays and low-energy gamma rays

While Fermi-LAT has revolutionized our knowledge of the gamma-ray sky above 100 MeV, progress at lower energies is slower. In the MeV range no new data since GRO/COMPTEL have been available[c]. The SPI instrument on the INTEGRAL satellite has improved the situation with a spectrum of diffuse emission from 20 keV to 1 MeV[14,17] and more recently to 2 MeV (Bouchet et al. in preparation). Figure 4 shows a combination of Fermi-LAT, COMPTEL and INTEGRAL/SPI data for the inner Galaxy. Of particular interest is the fact that inverse Compton emission is the main process in the range 100 keV to a few MeV, and accounts well for the

[c]The sensitivity contrast between 1–30 MeV and >100 MeV is now enormous, comparing COMPTEL with Fermi-LAT: about a factor of 50. This makes it all the more regrettable that no new MeV instrument is confirmed to fly in the coming decade. The MeV sky will remain *cielo incognito* for some time to come

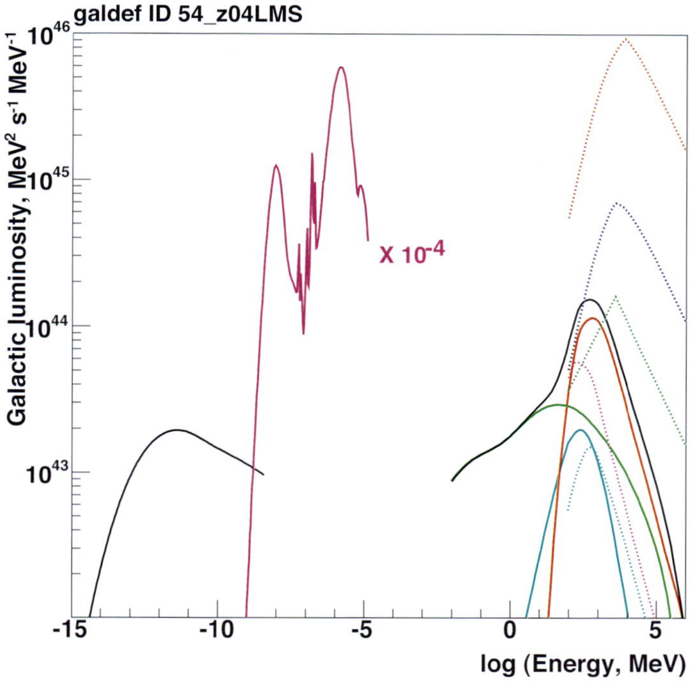


Fig. 3. Luminosity spectrum of the Galaxy, from Ref. 12. Components: black lines: synchrotron and total gamma rays; continuous lines red: pion-decay; cyan: bremsstrahlung; green: inverse Compton; dotted lines: cosmic rays: red: protons, blue: Helium, green: electrons, magenta: secondary positrons, cyan: secondary electrons. Magenta lines: optical and infrared, scaled by $10^{-4}$ for clarity. From Ref 12. Reproduced by permission of the AAS.

observed power-law continuum spectrum there. Note that the data shown includes other diffuse components like annihilation positronium and 511 keV, and $^{26}$Al which are not included in the model, while below 100 keV, hard X-ray sources dominate the emission[18] . This figure illustrates the importance of extending the observations to lower as well as higher energies. The excess in the MeV range observed by COMPTEL has been known for a long time and is plausibly attributable to discrete sources - however our lack of knowledge of the MeV sky impedes a more detailed discussion.

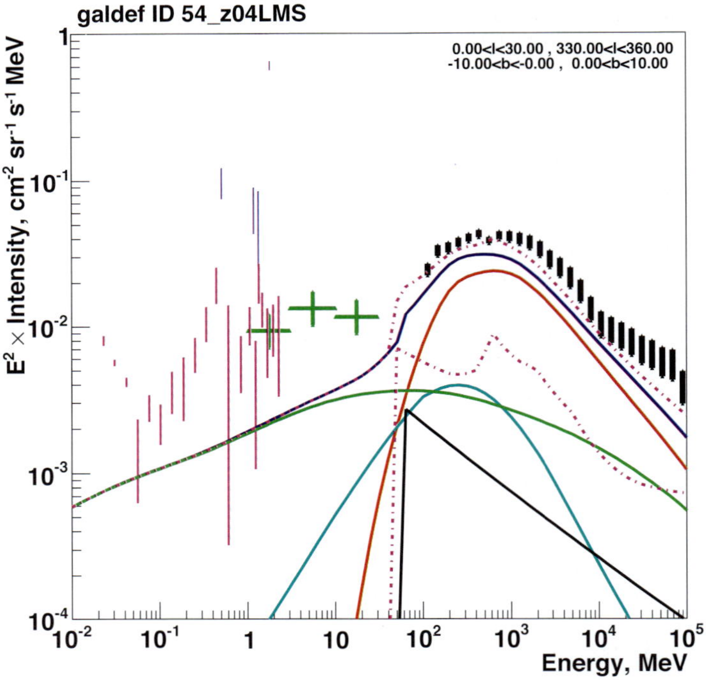


Fig. 4. Spectrum of inner Galaxy ( $330^o < l < 30^o$, $|b| < 10^o$). Data from INTEGRAL/SPI (L. Bouchet, private communication), COMPTEL and Fermi-LAT. Model and components coding as in Figure 1. Data: black vertical bars: Fermi-LAT (PRELIMINARY), green crosses: COMPTEL, magenta and blue vertical bars: INTEGRAL/SPI (magenta: continuum, blue: line emission ). The longitude range is less than in Figure 1 to acommodate the INTEGRAL/SPI measurement. The model below 100 MeV does not include positronium, line emission or point sources.

## 7. Acknowledgements

The *Fermi* LAT Collaboration acknowledges support from a number of agencies and institutes for both development and the operation of the LAT as well as scientific data analysis. These include NASA and DOE in the United States, CEA/Irfu and IN2P3/CNRS in France, ASI and INFN in Italy, MEXT, KEK, and JAXA in Japan, and the K. A. Wallenberg Foundation, the Swedish Research Council and the National Space Board in Sweden. Additional support from INAF in Italy and CNES in France for science analysis during the operations phase is also gratefully acknowledged.

## References


1. A. W. Strong, I. V. Moskalenko and V. S. Ptuskin, *Annual Review of Nuclear and Particle Science* **57**, 285 (2007).
2. A. A. Abdo *et al.*, *Physical Review Letters* **103**, 251101 (2009).
3. A. A. Abdo *et al.*, *The Astrophysical Journal* **703**, 1249 (2009).
4. A. A. Abdo *et al.*, *The Astrophysical Journal* **710**, 133 (2010).
5. M. Ackermann *et al.*, *The Astrophysical Journal* **726**, 81 (2011).
6. P. P. Papadopoulos, W. Thi and S. Viti, *The Astrophysical Journal* **579**, 270 (2002).
7. A. A. Abdo *et al.*, *Physical Review Letters* **104**, 101101 (2010).
8. A. A. Abdo *et al.*, *Astronomy and Astrophysics* **512**, A7 (2010).
9. A. A. Abdo *et al.*, *Astronomy and Astrophysics* **523**, A46 (2010).
10. A. A. Abdo *et al.*, *Astronomy and Astrophysics* **523**, L2 (2010).
11. A. A. Abdo *et al.*, *The Astrophysical Journal Letters* **709**, L152 (2010).
12. A. W. Strong, T. A. Porter, S. W. Digel, G. Jóhannesson, P. Martin, I. V. Moskalenko, E. J. Murphy and E. Orlando, *The Astrophysical Journal Letters* **722**, p. L58 (2010).
13. A. W. Strong, I. V. Moskalenko, O. Reimer, S. Digel and R. Diehl, *Astronomy and Astrophysics* **422**, L47 (2004).
14. T. A. Porter, I. V. Moskalenko, A. W. Strong, E. Orlando and L. Bouchet, *The Astrophysical Journal* **682**, 400 (2008).
15. A. A. Abdo *et al.*, *The Astrophysical Journal Supplement Series* **188**, 405 (2010).
16. M. Su, T. R. Slatyer and D. P. Finkbeiner, *The Astrophysical Journal* **724**, 1044 (2010)
17. L. Bouchet, E. Jourdain, J. Roques, A. Strong, R. Diehl, F. Lebrun and R. Terrier, *The Astrophysical Journal* **679**, 1315 (2008).
18. M. Revnivtsev, S. Sazonov, E. Churazov, W. Forman, A. Vikhlinin and R. Sunyaev, *Nature* **458**, 1142 (2009).

# ENERGY LOSS FOR ELECTRONS IN THE HELIOSPHERE AND LOCAL INTERSTELLAR SPECTRUM FOR SOLAR MODULATION

P. Bobik[1], G. Boella[2,3], M.J. Boschini[2,4], C.Consolandi[2], S. Della Torre[2,5], M. Gervasi[2,3], D. Grandi[2], M. Elmo[3], K. Kudela[1], E. Memola[2], S. Pensotti[2,3], P.G. Rancoita[2], D. Rozza[3] and M. Tacconi[2]*

[1] *Institute of Experimental Physics, Kosice (Slovak Republic)*
[2] *Istituto Nazionale di Fisica Nucleare, INFN Milano-Bicocca, Milano (Italy)*
[3] *Department of Physics, University of Milano Bicocca, Milano (Italy)*
[4] *CILEA, Segrate (MI) (Italy)*
[5] *Department of Physics and Maths, University of Insubria, Como (Italy)*

***E-mail: mauro.tacconi@mib.infn.it***

Galactic Cosmic Rays (GCR) entering the Heliosphere are affected by the solar modulation, a combination of diffusion, convection, magnetic drift and adiabatic energy loss usually seen as a decrease in the flux at low energy (less than $\sim$ 10 GeV). We have improved a quasi time-dependent 2D Stochastic Simulation code describing this effects. We focused our attention on the electron modulation, adding energy losses in the Heliosphere that can be neglected for protons and ions: inverse Compton, ionization, synchrotron and bremsstrahlung. These effects have been evaluated in the region affected by the solar magnetic field, up to 100 AU, where the environment conditions are not constant, especially the magnetic field intensity and the photon density. In our calculation the inverse compton energy losses are dominant, but they contribute only a few percent in comparison with the adiabatic losses. We also compared the Local Interstellar Spectrum (LIS) of primary electrons with experimental data collected in the past years at energies $\geq$ 20 GeV. We found that, inside one standard deviation, LIS fits the data and can be used in a Monte carlo code reproducing CR propagation in the Heliosphere.

*Keywords*: cosmic rays; heliosphere; solar modulation; energy losses.

## 1. Introduction

Electrons and positrons are a small fraction (1%) of the Galactic Cosmic Rays at the Earth in the GeV-TeV region of the spectrum. They undergo the same mechanisms of cosmic rays proton component. Compared to protons, they have a lower mass, a relativistic velocity and are affected by efficient electromagnetic energy losses. These losses are the results of inverse Compton scattering, ionization, synchrotron and bremsstrahlung processes inside the radiation and magnetic fields of the astrophysical medium they are moving. These processes have been already described in literature and applied to the interstellar environment. Due to these energy losses electrons can not diffuse for a long distance. Therefore the sources of Cosmic Rays electrons measured at the Earth must be placed close to us ($\lesssim$ 1 Kpc). Moreover the timelife of Cosmic Rays electrons moving in the interstellar medium is relatively short. The study of these effects offers a tool to better understand the recent history of the solar system neighborhood. In this paper we concentrate on the calculation of these processes inside the solar cavity, in order to implement them inside a Montecarlo code of Cosmic Ray propagation in the heliosphere.

## 2. Local Interstellar Spectrum

Many authors described the high energy part of the electron spectrum in cosmic rays. Starting from some assumptions on galactic sources of electron and positron, particles are propagated in the interstellar medium using a diffusion model[1]. We then obtain primary and secondary fluxes of these Cosmic Rays at the border of the heliosphere: this is what we call Local Interstellar Spectrum (LIS). This diffusion approach allows the use of more realistic models for diffusion constants, still permitting an analytical solution. We focused our attention on the work of Zhang & Cheng[2] that published the following analytical expression for the GCR electron LIS:

$$J_{lis}(E) = \frac{1600\ \epsilon^{-1.1}}{1 + 11\ \epsilon^{0.9} + 3.2\ \epsilon^{2.15}}\ \text{GeV}^{-1}\ \text{m}^{-2}\ \text{sr}^{-1}\ \text{s}^{-1} \qquad (1)$$

We compared this LIS with the data of 6 experiments that measured CR fluxes in the past years. For this analysis we selected data at energy $\geq$ 20 GeV, because in this range the effect of solar modulation can be considered negligible. We fitted data with the expression:

$$F_{lis}(E) = \frac{C\ \epsilon^{\gamma}}{1 + 11\ \epsilon^{0.9} + 3.2\ \epsilon^{2.15}}\ \text{GeV}^{-1}\ \text{m}^{-2}\ \text{sr}^{-1}\ \text{s}^{-1} \qquad (2)$$

where we introduced the two parameters $C$ and $\gamma$ to be best-fitted using the experimental data. In Table 1 we report the two parameters calculated for each experiment with their error bar (1 standard deviation). We show the fitted value of the two parameters $\gamma$ and $C$ for the various experiments in

Fig. 1. In the same figure we show also the average value in comparison with the values reported in Eq. 1. We found an agreement between experimental and theoretical values, inside the uncertainty. In Fig. 1 we can also have a rough estimation of the amount of statistical and systematic uncertainty for each experiment in comparison to the others.

Table 1. The two parameters $C$ and $\gamma$ fitted using data by several experiments.

| Experiment | Year | ΔE (GeV) | C | δC | γ | δγ |
|---|---|---|---|---|---|---|
| PRINCE[3] | 1975 | 5-200 | 2094,990 | 440,796 | -1,336 | 0,182 |
| MASS[4] | 1991 | 7.5-45 | 1314,260 | 353,374 | -1,025 | 0,177 |
| CAPRICE[5] | 1994 | 0.46-43.6 | 2304,000 | 757,602 | -1,411 | 0,252 |
| BETS[6] | 1997-8 | 10-100 | 1591,800 | 739,986 | -1,064 | 0,151 |
| NISHIMURA[7] | 1996-8 | 30-300 | 1536,000 | 440,769 | -1,158 | 0,265 |
| AMS-01[8] | 1998 | 0.1-40 | 2304,000 | 670,023 | -1,343 | 0.127 |

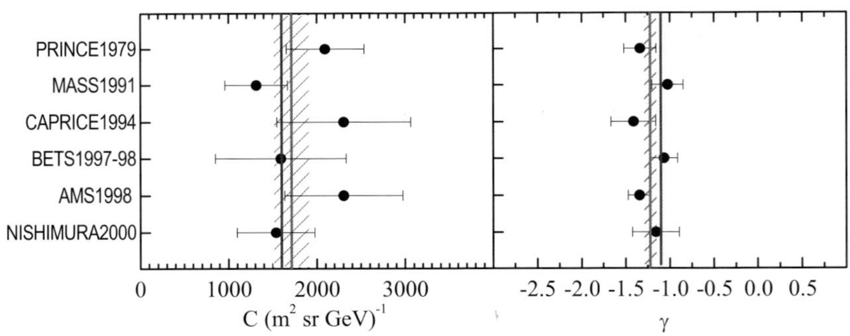


Fig. 1. The two parameters $C$ and $\gamma$ fitted using data by experiments reported in Table 1. We show also the average value (red line) to be compared with the theoretical value reported in Eq. 1 (blue line).

## 3. Propagation and energy losses

In the last years the AMS-Milano group developed a quasi time-dependent 2D stochastic simulation code able to describe the propagation of cosmic rays inside the Heliosphere. We have improved the code taking into account also the additional energy losses experienced by Cosmic Rays during the propagation in Heliosphere. In the case of protons (or nuclei) these energy losses can be neglected, while for electrons this assumption is non more valid, due to their small mass. In the following sections we describe the several processes producing energy losses in Cosmic Rays.

### 3.1. *Ionization*

Electrons moving in a region containing gas or plasma lose their energy by means of the ionization precesses. This appens wen electrons move across the interstellar medium (ISM) as well as inside the solar cavity. Ionization energy losses are proportional to the logarithm of the electron energy and, in the case of neutral plasma ($n_{ion} \sim n_e$) can be described by the following relation:

$$\left.\frac{dE}{dt}\right|_{ion} = -\frac{3}{4}\sigma_T c(mc^2)n_e\left[\ln\left(\frac{E}{mc^2}\right) + 2\ln\left(\frac{mc^2}{h\nu_{pl}}\right)\right] \tag{3}$$

where $E$ is the energy of electrons, $n_e$ the atoms density in the ISM (or in the Heliosphere), $\sigma_T = \dfrac{e^4}{6\pi\epsilon_0^2 m^2 c^4}$ the cross section and $\nu_{pl} = \dfrac{e}{2\pi}\sqrt{\dfrac{n_e}{\epsilon_0 m}}$ the plasma frequency.

### 3.2. *Bremsstrahlung*

Bremsstrahlung energy loss is due to acceleration (or actually deceleration) of electrons interacting with charged particles. It's proportional to the particle's energy and, in the case of neutral plasma could be represented by:

$$\left.\frac{dE}{dt}\right|_{brems} = -\frac{3}{8\pi}\sigma_T c\alpha E n_H \phi_H \tag{4}$$

where $\sigma_T$ is the Thompson cross section, $c$ the speed of light, $\alpha$ is the fine-structure constant, $n_H$ is the space medium density and $\phi_H = 29.97$ for hydrogen plasma.

### 3.3. *Syncrotron*

Syncrotron energy loss is proportional to $E^2$ and it is related to interaction (and acceleration) of cosmic particles with magnetic fields:

$$\left.\frac{dE}{dt}\right|_{sinc} = -\frac{4\sigma_T c}{3(mc^2)^2}U_{mag}E^2 \tag{5}$$

where $U_{mag} = \frac{B^2}{2\mu_0}\left[\frac{eV}{m^3}\right]$ is the magnetic field density.

### 3.4. *Inverse-Compton*

The Inverse-Compton energy loss is due to interaction of an ultra-relativistic electron with the soft background radiation. This process could be described by two case: Thomson approximation and Klein-Nishina case. Thomson approximation is valid at low energy. In this case the Compton cross section is used and the energy loss rate is proportional to $E^2$. Klein-Nishina is an extreme case where both photon and particle's energy are very high (e.g. $\frac{4\varepsilon\gamma}{m_e c^2} \gg 1$, where $\varepsilon$ is photon energy, $\gamma$ is Lorentz relativistic factor and $m_e c^2$ is the rest mass of particle). Defining $\xi = \frac{\gamma k_b T}{m_e c^2}$ (with $k_b$ Boltzmann constant) the Inverse-Compton energy loss in presence of a black body radiation is[9]:

$$-\left.\frac{dE}{dt}\right|_{IC} = \begin{cases} \frac{4}{3}\sigma_T c U_{ph}\left[\left(\frac{E}{mc^2}\right)^2 - 1\right] & \xi < 3.8\cdot 10^{-4} \\ 10^{-45}\frac{E^2(k_b T)^4}{\xi} e^{\sum_{i=0}^{6} c_i (\ln\xi)^i} & 3.8\cdot 10^{-4} \le \xi \le 1.8\cdot 10^3 \\ \frac{\sigma_T}{16}\frac{(m_e c k_b T)^2}{\hbar^3} \ln(4\xi) - 1.9805 & \xi > 1.8\cdot 10^3 \end{cases} \tag{6}$$

where $c_i = \{74.77, -0.1953, -0.0997, 0.004352, 0.0003546, -0.0000301\}$ and $U_{ph}$ is the photon energy density.

### 3.5. *Energy loss in the heliosphere*

The heliosphere is a dynamic place strongly related to the presence of the Sun and to its activity. While in the intergaliactic medium it is possible to account for average quantities, in the solar cavity we need to evaluate them locally, as a function of the distance from the Sun. The plasma density is measured at $r_0 = 1$ AU and is $n_{pl}(1\text{AU}) \approx 8 - 9$ protons/cm$^3$. Using the continuity equation it is possible to estimate the plasma density in the eliosphere at every distance from the Sun:

$$n_{pl} = n_{pl}(1\text{AU})\left(\frac{r_0}{r}\right)^2 \tag{7}$$

Moreover, assuming that the electromagnetic emission of the Sun is a black body, we can estimate the photon density inside the heliosphere:

$$U_{ph} = \frac{L_\theta}{4\pi c}\left(\frac{r_0}{r}\right)^2 \tag{8}$$

where $L_\theta = 3.846 \cdot 10^{26}$ W.

As heliospheric magnetic field (HMF) we used a modified Parker field, according to Jokipii & Kota[10], that enhanced the magnitude of HMF in the polar regions. A comparison among the several energy loss processes inside the eliosphere at 1 AU is shown in Figure 2. We found that Inverse-Compton is the dominant process, due to the large quantity of photon emitted by the Sun.

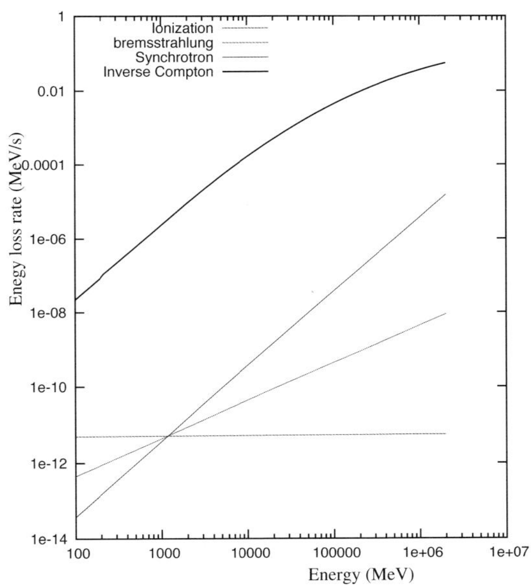


Fig. 2. Energy loss rate of electrons inside the heliosphere at 1 AU, due to several processes.

We estimated the importance of these energy loss processes inside the heliosphere, in comparison with the adiabatic energy losses. We used our MonteCarlo simulation code of the heliosphere[11] and adapted it to take into account these processes[12]. We estimated first the average time of permanence of GCR leptons inside the heliosphere, considering only particles reaching the Earth. From the time spent in the heliosphere by these particles we estimated the total energy loss due to Inverse-Compton process ($\Delta E_{IC}$), neglecting other losses. We compare this $\Delta E_{IC}$ with the total energy lost by the same particles in the heliopshere. We estimated the adiabatic energy loss ($\Delta E_{FF}$) by using a Force Field approach[13]. Using a typical modulation potential occurring in periods of medium solar activity ($\Phi = 550$ MV), we estimate an adiabatic energy loss $\Delta E_{FF} = 0.55$ GeV, roughly constant at all energies. In Table 2 we compare the two energy losses. Inverse-Compton is always several order of magnitude lower than the total energy loss. At energies around 10 GeV Inverse-Compton energy loss is 5% of the adiabatic loss.

Table 2. Energy loss estimation for different initial particle's energy.

| $E_0$ (GeV) | $\Delta E_{IC}$ (GeV) $\times 10^{-3}$ | $\Delta E_{IC}/E_0$ % | $\Delta E_{FF}$ (GeV) | $\Delta E_{FF}/E_0$ % | $\Delta E_{IC}/\Delta E_{FF}$ % |
|---|---|---|---|---|---|
| 1 | 0.145 | 0.014 | 0.55 | 55 | 0.025 |
| 10 | 1.056 | 0.010 | 0.55 | 5.5 | 5.5 |
| $10^2$ | 2.335 | 0.0023 | 0.55 | 0.55 | 0.42 |
| $10^3$ | 1.815 | 0.00018 | 0.55 | 0.055 | 0.33 |

## 4. Conclusions

In this contribution we have compared theoretical LIS for electrons with experimental data. We used in our analysis datasets of several experiments operated in different periods. We considered only data up to 20 GeV where solar modulation is negligible. Moreover we improved the our MonteCarlo model of Cosmic Rays electrons propagation, introducing additional energy loss mechanisms, such as inverse Compton scattering, ionization, synchrotron and bremsstrahlung. We found that the major contribution is due to Inverse-Compton process. But anyway this is not exceeding a few percent of the total energy loss in the heliosphere, always dominated by the adiabatic energy loss process.

## References


1. I.V. Moskalenko, and A.W. Strong, *Astrophys. J.*, **493**, 694 (1998).
2. L. Zhang & K.S. Cheng, *Astron. & Astrophys.*, **368**, 1063 (2001).
3. T.A. Prince, *Astrophys. J.*, **227**, 676 (1979).
4. C. Grimani, et al., *Astron. & Astrophys.*, **392**, 287 (2002).
5. M. Boezio, et al., *Astrophys. J.*, **532**, 653 (2000).
6. S. Torii, et al., *Astrophys. J.*, **559**, 973 (2001).
7. J. Nishimura, et al., *Adv. Space Res*, **26**, 1827 (2000).
8. J. Alcaraz, et al., *Phys. Lett. B*, **484**, 10 (2000).

9. T. Delahaye, et al., *Astron. & Astrophys.*, in press, arXiv: 1002.1910v1 (2010).
10. J.R. Jokipii, and J. Kota, *Geophys. Res. Lett.*, **16**, 1 (1989).
11. P. Bobik, M.J. Boschini, C. Consolandi, S. Della Torre, M. Gervasi, D. Grandi, K. Kudela, S. Pensotti, and P.G. Rancoita, *Proton and antiproton modulation in the heliosphere for different solar conditions and AMS-02 measurements prediction*, ICATPP Conference on Cosmic Rays for Particle and Astroparticle Physics, Como 7-8/10/2010, this Proceedings.
12. P. Bobik, M.J. Boschini, C.Consolandi, S. Della Torre, M. Gervasi, D. Grandi, K. Kudela, S. Pensotti, and P.G. Rancoita, *Electron and Positron solar modulation and prediction for AMS-02*, ICATPP Conference on Cosmic Rays for Particle and Astroparticle Physics, Como 7-8/10/2010, this Proceedings.
13. L.J. Gleeson, and W.I. Axford, *Astrophys. J.*, **154**, 1011 (1968).

# Cosmic Rays from Astrophysical Sources

| | |
|---|---|
| P. Blasi | Cosmic ray acceleration in supernova remnants |
| D. Caprioli | $\gamma$-rays from heavy nuclei accelerated in supernova remnants |
| I. Cernuda | Anisotropies in the Cosmic-ray electron spectrum: A way to discriminate between exotic and astrophysical sources? |
| M. Dormody | Cosmic-ray electrons and positrons from gamma-ray pulsars |
| J. Lavalle | Galactic electrons and positrons at the Earth: new estimate of the primary and secondary fluxes |
| P. Mertsch | The PAMELA anomaly indicates a nearby cosmic ray accelerator |
| C. Monte | Observations of Intermediate Synchrotron Peaked blazars with the Fermi-LAT |
| G. Morlino | Shock acceleration in partially neutral plasmas |
| J. Pochon | Pulsar electrons detection in AMS-02 experiment. Model status and discovery potential |
| A.M. Taylor | The CR Connection: UHE Primaries and Secondaries from UHECR Sources |
| R. Walter | $\eta$ Carinae: a very Large Hadron Collider |
| M. Paganoni | Readiness of the CMS computing fot the data taking |
| A. Papa | The $\mu + \rightarrow e + \gamma$ decay from the MEG experiment: results from the first physics run |
| J. Pinfold | QUARTIC - a Precise Time-of-Flight Counter for the ATLAS Forward Physics Project |
| A. Salamon | Readout of the LHCb muon system with the TELL1 board |
| L. Sartori | Associative Memory design for Fast Tracker at LHC |
| C. Schmitt | ATLAS cosmic ray results |
| M. Shamim | Optimization and Performance of the ATLAS Tau Trigger with Cosmics Data |
| A. Sidoti | Trigger monitoring in ATLAS |

# COSMIC RAY ACCELERATION IN SUPERNOVA REMNANTS

P. BLASI

*INAF/Osservatorio Astrofisico di Arcetri*
*Largo E. Fermi, 5 Firenze, 50126 Italy*
**E-mail: blasi@arcetri.astro.it*

We review the main observational and theoretical facts about acceleration of Galactic cosmic rays in supernova remnants, discussing the arguments in favor and against a connection between cosmic rays and supernova remnants, the so-called supernova remnant paradigm for the origin of Galactic cosmic rays. Recent developments in the modeling of the mechanism of diffusive shock acceleration are discussed, with emphasis on the role of 1) magnetic field amplification, 2) acceleration of nuclei heavier than hydrogen, 3) presence of neutrals in the circumstellar environment. The status of the supernova-cosmic ray connection in the time of Fermi-LAT and Cherenkov telescopes is also discussed.

*Keywords*: Style file; LaTeX; Proceedings; World Scientific Publishing.

## 1. Introduction

The possibility that the bulk of cosmic rays (CRs) observed at Earth may be generated in supernova remnants (SNRs) dates back to the '30s[1] and with time it has acquired the rank of a paradigm, mainly because of the fact that on energetic grounds[2] supernovae are the only class of sources in the Galaxy that can provide enough energy as to explain the CR flux observed at Earth. The basic requirements for the SNR paradigm are: 1) that SNRs may accelerate with typical efficiency of $\sim 10-20\%$; 2) that the spectrum of individual elements and the consequent all-particle spectrum are reproduced, including the presence of a knee at $\sim 3\times 10^{15}$ eV; 3) that the chemical abundances of nuclei are well described; 4) that the multifrequency observations of individual remnants, from the radio to the gamma ray band, are well described; 4) that the anisotropy induced by the spatial distribution of SNRs in the Galaxy is compatible with observations.

The acceleration mechanism that is usually assumed to work in SNRs is diffusive shock acceleration (DSA),[3,4] but the energetic requirement that at least $\sim 10-20\%$ of the kinetic energy of the supernova shell is converted to CRs leads to realize immediately that the standard test-particle version of the theory describing this process is not applicable to the description of CR acceleration: the reaction of accelerated particles onto the accelerator cannot be neglected and in fact it is responsible for spectral features (such as spectral concavity) that may represent potential signatures of CR acceleration. There is another, possibly more important reason why DSA must include the reaction of accelerated particles: the standard diffusion coefficient typical of the interstellar medium (ISM) only leads to maximum energies of CRs in the range of $\sim$ GeV, rather than $\sim 10^6$ GeV (around the knee) required by observations. The only way that the mechanism can play a role for CR acceleration is if the accelerated particles generate the magnetic field structure on which they may scatter,[5] thereby reducing the acceleration time and reach larger values of the maximum energy. A non linear version of DSA including the dynamical reaction of accelerated particles on the shock was developed by many authors (see Ref. 6 for a review) and more recently completed with the inclusion of self-generation of magnetic field[7–9] and acceleration of nuclei other than Hydrogen.[10]

From the observational point of view the detection of narrow rims in the X-ray emission of several SNRs (Ref. 11 and references therein) has provided an important support to the idea that CRs may amplify the magnetic field close to the shock surface, thereby leading to reach higher energies: these rims are in fact most easily interpreted as the result of the synchrotron emission on a time scale comparable with the loss length of the highest energy electrons.[12] A simple estimate leads to magnetic fields of order $\sim 100-1000\mu G$ downstream of the shock, which are hard to interpret as the result of solely the compression of the field at the shock surface since the ISM magnetic field is typically $\sim 1-10\mu G$ and the compression at a strong shock is a factor $\sim 4$. It is important to realize that the CR induced magnetic field amplification occurs upstream of the shock; the perpendicular components of the field are further compressed at the shock surface. Moreover the overall structure of the rims, at least in the case of SN1006 appears to be inconsistent with the absence of magnetic field amplification upstream of the shock.[13] However alternative explanations, involving acceleration at perpendicular shocks and magnetic field amplification due to turbulent eddies downstream of the shock[14] may still beb plausible explanations of the data.

The rigidity dependent nature of DSA leads to predict higher energies for accelerated nuclei (if they get fully ionized), so that the knee may result from the superposition of the spectra of accelerated nuclei, if the magnetic field is amplified to sufficiently high levels. Unfortunately the problem of injection, hard enough for protons and electrons, becomes even harder for nuclei especially for those that may result from sputtering of dust grains.[15] Some phenomenological attempts to calculating the contribution of nuclei in the context of non linear theory of DSA to the all-particle spectrum observed at Earth have recently been carried out.[10,16,17]

A crucial step towards confirming or rejecting the SNR paradigm might be made through gamma ray observations both in the TeV energy range, by using Cherenkov telescopes, and in the GeV energy range accessible to the Fermi gamma ray telescope. Gamma radiation can be produced mainly as a result of inverse Compton scattering (ICS) of relativistic electrons on the photon background and in inelastic proton-proton scatterings with production and decay of neutral pions. The present observational situation is rather puzzling and deserves some special discussion (see §4.1): most gamma ray spectra observed by Fermi (see[18,19] for reviews) (with some important exceptions, e.g. RX J1713-3946) hint to rather steep spectra of accelerated particles ($\propto E^{-\gamma}$, with $\gamma \sim 2.4 - 3$), which are not easy to accomodate in the context of NLDSA that predicts flat spectra, possibly even flatter than $E^{-2}$ at high enough energy.[20] Flat injection spectra would also lead to require a steep dependence of the Galactic diffusion coefficient on energy, which in turn is known to result in exceedingly large anisotropy.[21] The most likely explanation for this discrepancy might lie in a rather subtle detail of the DSA theory, namely that the velocity relevant for particle acceleration is the velocity of waves with respect to the plasma. Usually the wave speed is negligible compared with the plasma velocity in the shock frame, but in the presence of magnetic field amplification this condition might be weakly violated. This is very bad news in that the spectral changes induced by this effect depend not only on the wave speed but on the wave polarization as well. In[10,17] the authors show that there are situations in which the spectral steepening can indeed be sufficient to explain the observed spectrum of CRs and required by Fermi data on some SNRs.

## 2. Efficient CR acceleration in SNRs

We first summarize the energetic requirements imposed by the SNR paradigm together with the diffusive propagation of CRs in the Galaxy. In a simple model in which the sources are all located in an infinitely thin disc with radius $R_d$ and the diffusion coefficient is constant within a halo of size $H$, the density of CRs is easily shown to be

$$n_{CR}(E) \approx \frac{N(E)\mathcal{R}_{SN}}{2\pi R_d^2}\frac{H}{D(E)}, \qquad (1)$$

where $\mathcal{R}_{SN}$ is the rate of supernova explosions in the Galaxy, assumed to be spatially constant, $N(E) \propto E^{-\gamma}$ is the spectrum of CRs produced by an individual SNR (assumed to occur instantaneously) and $D(E)$ is the diffusion coefficient, for which we adopt a typical functional form $D(E) = 10^{28}D_{28}(\rho/3GV)^{\delta}cm^2/s$, where $\rho = E/Z$ is particle rigidity. In the following we concentrate our attention on ultrarelativistic protons, therefore $Z = 1$ and their speed is the speed of light, $c$. The flux observed at Earth is $\phi(E) = cn_{CR}(E)/(4\pi)$. In Refs. 22,23 the authors find that in order to fit at the same time the B/C ratio and the antiproton data one has to use $D_{28}/H_{kpc} \sim 1$, therefore one can write the CR flux as

$$\phi_{CR}(E) \approx 2.4E_{51}\xi_{CR}R_{d,15}^{-2}\mathcal{R}_{SN,30}(\gamma-2)3^{\delta}E_{TeV}^{-2.73}TeV^{-1}m^{-2}s^{-1}sr^{-1}, \qquad (2)$$

where $E_{51} = E_{SN}/10^{51}erg$, $\xi_{CR}$ is the CR acceleration efficiency, $\mathcal{R}_{SN,30}$ is the SN rate in units of one every 30 years and $R_{d,10} = R_d/15kpc$. It is easy to see that in order to fit the observed proton spectrum, $8.7 \times 10^{-2}E_{TeV}^{-2.73}TeV^{-1}m^{-2}s^{-1}sr^{-1}$ one has to assume an acceleration efficiency $\xi_{CR} \sim 7\%$ for $\delta = 1/3$, $\xi_{CR} \sim 11\%$ for $\delta = 0.54$ and $\xi_{CR} \sim 58\%$ for $\delta = 0.7$ for the reference values of the parameters. These should be considered as lower limits to the required efficiencies. These numbers immediately stress the need for a theory of particle acceleration that takes into account the possible dynamical reaction of the accelerated particles on the accelerator.

3 another reason was discussed for having a reaction of CRs on the acceleration process: streaming instability induced by the accelerated particles leads to magnetic field amplification upstream of the shock (where a spatial gradient exists). In the absence of damping and if only resonant streaming instability is excited, the strength of the amplified magnetic field can be estimated as a function of the initial field $B_0$ (by using an extrapolation of quasi-linear theory) as:[8]

$$\delta B \sim B_0\sqrt{2M_A\xi_{CR}}, \qquad (3)$$

where $M_A \gg 1$ is the Alfvenic Mach number. For typical values of $M_A \sim 1000$ one obtains $\delta B/B_0 \sim 30$ and the field is further amplified by compression when the fluid element crosses the shock from upstream to downstream. Larger values of the amplification factor can, at least in principle, be reached if non-resonant streaming instabilities are at work.[24] It is however not clear which role the field can play in this case in terms of scattering particles up to the knee energy since the unstable waves are produced at wavelengths much shorter than the Larmor radius of the particles generating them. Non linear effects in wave evolution can lead however to transporting energy towards longer wavelengths, though it is not clear whether this process is fast enough to occur within the precursor and thereby lead to efficient particle scattering, essence of the acceleration process.

These two processes, dynamical reaction of accelerated particles and magnetic field amplification, are the two most important ones to take into account when describing non-linear particle acceleration in SNR shocks. The main predictions of the theory of non-linear DSA are the following: 1) spectral concavity induced by the momentum dependence of the diffusion coefficient in the upstream precursor; 2) effects related to magnetic field amplification (X-ray rims downstream, spatial profile of the upstream X-ray brightness, maximum energy of accelerated particles, possibly accessible through the gamma ray range); 3) lower temperatures of the thermal plasma downstream, induced by the simple fact that less energy is left for plasma heating after particle acceleration.

## 3. Observational signatures of efficient CR acceleration

In this section we briefly discuss some of the observational evidence pointing towards efficient CR acceleration in individual SNRs. It is worth keeping in mind that while the SNR paradigm for the origin of CRs is usually quoted without any specific reference to the type of supernovae involved, the particle acceleration process is strongly dependent upon the environment in which the supernova goes off. For instance we can expect that the blast wave of a SN-type Ia has a larger Mach number than that of a type II SN, mainly because of the lower temperature of the ordinary ISM compared with that of the material around a massive star with powerful winds. For the same reason we can expect the gas density to be lower for type-II SNe. While type-II SNe are more frequent than type Ia, it is probably more difficult to observe a gamma ray emission of hadronic origin from these objects because of the lower gas density. This short warning is meant to stress that the constraints obtained on the particle acceleration efficiency of a few SNRs are hard to generalize to the whole zoo of SNRs.

### 3.1. *X-ray emission from rims and precursor*

Non-thermal X-ray radiation is produced in SNRs through synchrotron emission of high energy electrons in the magnetic field around the shock. The emission is dominated by the region downstream of the shock where the magnetic field is stronger and is cut off at a frequency that, in the case of Bohm diffusion, is independent of the strength of the magnetic field:

$$\nu_{max} \approx 0.2\ u_8^2\ \mathrm{keV}, \tag{4}$$

where $u_8 = u_{sh}/(10^8 cm/s)$ is the shock velocity in units of 1000 km/s. The maximum energy of accelerated electrons depends on the strength of the amplified magnetic field and can be estimated as

$$E_{max} \approx 10\ B_{100}^{-1/2}\ u_8\ \mathrm{TeV}, \tag{5}$$

where $B_{100} = B/100\mu G$ is the magnetic field in units of $100\mu G$.

The emission region has a spatial extent at $\nu \sim \nu_{max}$ which is determined by diffusion and can be written as

$$\Delta x \approx \sqrt{D(E_{max})\tau_{loss}(E_{max})} \approx 0.04\ B_{100}^{-3/2}\ \mathrm{pc}. \tag{6}$$

The typical thickness of the X-ray rims found through high resolution observations is of order $\sim 10^{-2}$ pc, thereby leading to predicting values of the magnetic field of $100-300\mu G$ downstream of the shock (or $\sim 20-80\mu G$ upstream of the shock, if the field is amplified by the streaming of accelerated particles and further compressed at the shock).

In Fig. 1 we show the X-ray brightness profile (histogram) of part of the rim of SN 1006.[25] The thick lines refer to the predictions of NLDSA[13] for three values of the injection efficiency (larger values of $\xi$ correspond to lower efficiencies). The thin lines are the predicted radial profiles in test-particle theory for two values of the upstream magnetic field (as indicated). Two pieces of information arise from this figure: 1) the narrow rims ($\sim 10-20$ arcsec) downstream can only be reproduced for efficient particle acceleration scenarios; 2) the predicted X-ray emission from the precursor region drops below the background for the same cases in which the rims are present. In other words, the non-detection of the precursor X-ray emission might be additional evidence for efficient acceleration. This is due to the fact that the spatial extent of the precursor is reduced when magnetic fields are amplified by CR streaming.

Although very interesting, this interpretation is not totally unique: it could be that the magnetic field is not amplified upstream by CRs,

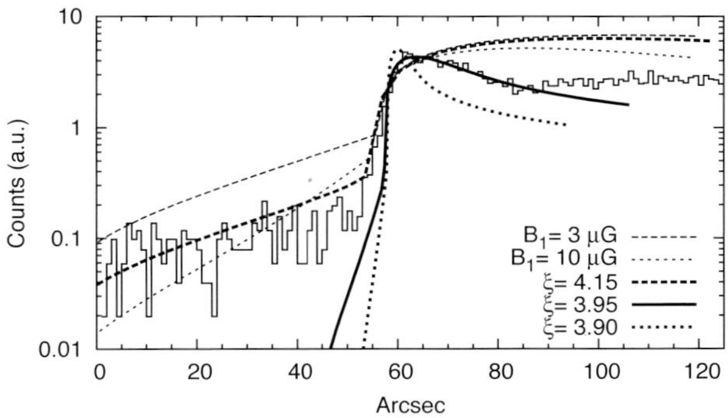


Fig. 1. Radial profile of the emission in the 0.8-2.0 keV band, extracted from a narrow strip around the rim of SN 1006 (thin black line). Overplotted are the theoretical predictions of NLDSA for 3 different injection efficiencies (thick lines) and those of test-particle theory for two different values of the pre-existing turbulent magnetic field upstream (thin lines).

but rather enhanced due to fluid instabilities downstream in quasi-perpendicular shocks.[14] In this case the non-detection of the upstream emission could be due to a dominantly perpendicular topology of the magnetic field lines.

### 3.2. *Collisionless shocks in partially ionized plasmas*

The blast waves produced by supernovae explosions give rise to collisionless shocks. The formation of the shock, as well as of the precursor is due to plasma processes which affect the ionized component of the plasmas involved. The neutral component, if present, is affected by the presence of the shock and by particle acceleration only through the process of charge exchange and, to a lesser extent, through ionization. A neutral atom can exchange an electron with an ion if the two are moving with different speeds. This can happen in three situations: 1) if the temperatures of ions and neutrals are different; 2) if ions and neutrals share the same temperature but two atoms lie on different parts of the thermal velocity distribution; 3) If the ions and neutrals have different bulk velocities.

In the absence of CR induced precursor, the effect of charge exchange appears after shock crossing: downstream of the shock the ionized plasma is compressed and shock heated, while the neutral gas keeps moving with the upstream speed $u_0$ (neutrals might also have an intrinsic thermal motion, but for the present discussion we can neglect this point). At this point, charge exchange may take place in that cold neutral can lose an electron to a hot ion. As a result, a population of colder ions and one of warmer neutrals are produced.

The observational signature of this phenomenon is clear in the so-called *Balmer dominated shocks*, where the Balmer lines are clearly detected. Charge exchange leads to the formation of a narrow line, corresponding to the neutrals that keep the same 'temperature' as they had upstream and did not suffer charge exchange, and a broad line corresponding to the neutrals that have been produced through charge exchange and were hot ions before the process. The measurement of the width of the narrow and broad Balmer lines allows one to measure the temperature of the two components.

The situation becomes more interesting in the presence of cosmic rays: when the pressure of accelerated particles upstream of the shock generates the precursor, in the shock frame the bulk motion of ions is slowed down with respect to neutrals. Neutral atoms keep moving with the velocity that ions have at upstream infinity, $u_0$, therefore a difference in bulk velocity between the two species arises. At the shock crossing, the ionized plasma suffers shock heating and compression while the neutral component does not feel the shock and keeps moving with speed $u_0$. In the presence of CR acceleration, the temperature of the ions downstream is clearly lower that in the absence of CRs, simply because of energy conservation (there is now another component, CRs, into which the ram pressure $\rho_0 u_0^2$ can be channelled). Therefore one can expect that the width of the broad Balmer line is somewhat smaller than in the absence of CRs. This phenomenon has been recently observed[26] in the SNR RCW86.

Measurements of the proper motion of the shock lead to a shock velocity of $u_0 = 6000 \pm 2800$ km/s, which in turn should imply a downstream temperature of $T_2 = 20-150$ keV or $T_2 = 12-90$ keV depending on whether protons and electrons reach or not thermal equilibrium downstream, and assuming that no CR acceleration is taking place. The temperature inferred from the width of the broad Balmer line is $T_2 = 2.3 \pm 0.3$ keV. The authors interpreted this discrepancy as the result of effective CR acceleration at the SNR shock, with an estimated efficiency of $\sim 50-60\%$ (this numerical value is however rather model dependent).

As pointed out above, in CR modified shocks some level of charge exchange is expected to take place in the precursor because of the difference in bulk velocity between the ionized plasma and the neutrals. The process is responsible for producing a population of warmer neutrals upstream. After shock crossing this leads to a narrow Balmer line which is broader than in the absence of a CR induced precursor. This effect was observed in several

SNRs.[27] An independent signature of the same phenomenon was recently found in the Tycho SNR:[28] the authors used the Hubble Space Telescope to measure the intensity of the broadened narrow Balmer line as a function of position around the shock and claim that there is substantial emission from the region in front of the shock. This finding could be either the signature of different bulk velocities between ions and neutrals (precursor) or of a different temperature of the two components. In this latter case, the ions could have been heated up due to the action of turbulent heating in the precursor.

In the future the importance of the detection of anomalous widths of Balmer lines in SNR shocks will plausibly increase especially if a clear correlation with the detection of narrow non-thermal X-ray rims could be established. This correlation would represent a crucial confirmation that magnetic field amplification and efficient CR acceleration are related to each other.

### 3.3. *Maximum energy and the knee*

The spectrum of individual elements in CRs carries more information than the all-particle spectrum in terms of understanding particle acceleration in the sources. The spectrum of protons measured by KASCADE[29] shows a pronounced decline at $\sim few \times 10^6$ GeV, which is likely to be produced by the acceleration process running out of steam. Although other experiments appear to disagree on the details of the measurement, they also show a decline in the same energy range. The situation for He nuclei is similar but it becomes rapidly worse for heavier nuclei. If the interpretation of this decline in terms of a maximum energy at the sources is correct, this implies that CR protons need be accelerated at least to $\sim few \times 10^6$ GeV in SNRs, in they are the sources.

For Bohm diffusion, this result can be achieved in SNRs only if the magnetic field is amplified by $\sim 100$, a number surprisingly close to that inferred from the width of X-ray rims. This conclusion was also confirmed in the context of NLDSA.[30] Since the maximum energy reached through diffusive shock acceleration scales with the charge of the nucleus, another consequence of this line of thought is that at energies above $\sim few \times 10^6$ GeV the chemical composition should become dominated by heavier nuclei and that the spectrum of Galactic CRs should eventually end at energies of order $\sim 10^8$ GeV with an iron dominated composition. At this point a transition to extragalactic CRs is expected.

## 4. Problematic aspects of the SNR paradigm

Can we consider the SNR paradigm as proven right? As discussed in the previous section there are plenty of evidences that at least some of the SNRs we observe are accelerating CRs efficiently. What does not allow us to claim a full confirmation is the existence of several loose ends as well as the fact that other SNRs do not seem to be accelerating (hadronic) CRs to the level required by the paradigm. It is however worth stressing here again a crucial point. The simplicity of the SNR paradigm contrasts with the complexity of Nature in many different ways and while the general theory makes some very general predictions on a "typical" SNR, the behaviour of an individual SNR depends on many physical phenomena which are specific of the environment where the supernova explodes and are much harder to predict. Below we list some of the aspects that represent possible challenges for the SNR paradigm.

### 4.1. *Gamma ray observations*

One of the observational signatures of SNRs as sources of CRs that has been long waited for is the detection of gamma radiation from the production and decay of neutral pions. Aside from the experimental difficulties in detecting the gamma rays from SNRs, it may be even more problematic to establish whether a detected flux can be unambiguously attributed to $\pi^0$ decay or rather to ICS of accelerated electrons. In the last few years the gamma ray detections of SNRs have proliferated both with Cherenkov telescopes[31–33] and with the Fermi-LAT telescope.[34]

The first case that was strongly suggestive of a hadronic origin of CRs was SNR RX J1713-3946, detected by HESS in the TeV energy region and later by Fermi[18] at lower energies. The combined spectrum appears to be relatively flat and probably of hadronic origin (see §4.2 for more discussion on this point). Rather unexpectedly most of the SNRs detected by Fermi have a rather steep gamma ray spectrum, with slope $\sim 2.4 - 3$, quite unlike the flat or even concave spectra predicted in the context of NLDSA. One might argue that the steep spectra are mostly observed in old SNRs with a nearby molecular cloud that serves as a target for inelastic CR collisions. This possibility is certainly to be kept in mind.

Whether such steep spectra may be compatible with having efficient CR acceleration and magnetic field amplification still remains to be investigated, but it is interesting to notice that, as we discuss in §4.3, the observed spectrum of CRs at Earth also leads to a preference of relatively steep injection spectrum. Flat injection implies in fact a too large CR anisotropy.[21]

### 4.2. *X-ray lines*

As mentioned above, gamma rays were detected from SNR RX J1713-3946. In Ref. 35 a detailed analysis of the arguments in favor and against a hadronic origin of this radiation was carried out. The authors conclude that a hadronic origin is favored but that there are some problems with it, the most important of which is that the predicted thermal X-ray bremsstrahlung emission exceeds observations unless the temperature of electrons is much smaller than that of protons. Indeed it can be expected that electrons are thermalized to a temperature $\sim m_e/m_p$ lower than for protons and that they gradually thermalize with protons on time scales as large as the time scale for Coulomb scattering, although other collisionless processes may lead to faster thermalization. In Ref. 36 it was argued that Coulomb scattering is sufficient to heat up the electrons to a temperature of $\sim 1$ keV during the expansion of the remnant, enough to excite emission lines whose intensity should have been detected. From the non-detection the authors conclude that the observed gamma rays cannot be of hadronic origin.

### 4.3. *Spectra at Earth*

In the simple case of a power law injection spectrum of CRs $\propto E^{-\gamma}$, the spectrum observed at Earth is $n(E) \sim E^{-\gamma-\delta}$, where the diffusion coefficient is assumed to be $D(E) \propto E^{\delta}$. This simple argument implies that the flatter the injection spectrum the larger must be the value of $\delta$. However $\delta > 0.5 - 0.6$ leads to excessive anisotropy of the CRs observed at Earth, thereby imposing a preference for relatively steep injection spectra. For instance for a Kolmogorov diffusion coefficient $D(E) \propto E^{1/3}$ the required injection spectrum must be $\sim E^{-2.4}$. As discussed above, such steep spectra are not easy to obtain in NLDSA with efficient acceleration and magnetic field amplification. The only possibility, recently suggested in Refs. 10,17 is that the magnetic field amplification also leads to introduce a relatively large velocity of the scattering centers responsible for particle diffusion in the shock region. This phenomenon leads to steeper spectra even in the context of NLDSA. The net effect is however to have lower magnetic fields and lower efficiencies of particle acceleration.[10] In Ref. 10 the authors calculate the spectra of different nuclei at the source and after propagation: the injected spectra are steeper than in standard NLDSA, but they still require $D(E) \propto E^{0.54}$, in agreement with previous calculations[17] and sort of borderline from the point of view of anisotropy.

It is important to keep in mind that calculations of the spectra of CRs released by SNRs into the ISM are highly uncertain because of the lack of a proper understanding of the processes responsible for the escape of CRs from the sources.[20,37,38]

## 5. Conclusions

We provided a short review of the main arguments in favor and against the so called SNR paradigm for the origin of Galactic CRs. There is a train of recent evidence that efficient CR acceleration takes place in several SNRs. Probably the most striking findings are the detection of narrow X-ray rims interpreted as evidence for magnetic field amplification, and numerous observational results on Balmer dominated shocks, which also lead to conclude that CRs are being accelerated effectively.

On the other hand, there are a few pieces of the puzzle that do not appear to be in place: the non-detection of X-ray lines from SNR RX J1713-3946 is certainly problematic for a scenario in which the detected gamma rays are of hadronic origin. Fermi-LAT gamma ray observations of several SNRs with steep spectra also seems to be at odds with the standard predictions of NLDSA which is expected to describe efficient CR acceleration in SNR shocks. Steep spectra are also suggested by observations of CR anisotropy. We have discussed how the last two issues mentioned above could be understood by taking into account the finite velocity of the waves scattering particles close to the shock.

## Acknowledgments

The author is grateful to his collaborators, E. Amato, D. Caprioli, G. Morlino, for precious scientific collaboration on the topics discussed here.

## References

1. W. Baade and F. Zwicky, *Physical Review* **46**, p. 76(Jul 1934).
2. V. L. Ginzburg and S. I. Syrovatskii, *The Origin of Cosmic Rays* (Jan 1964).
3. A. R. Bell, *Royal Astronomical Society* **182**, p. 147(Jan 1978), A&AA ID. AAA021.143.001.
4. R. Blandford and D. Eichler, *Physics Reports* **154**, p. 1(Oct 1987).
5. P. O. Lagage and C. J. Cesarsky, *Astronomy and Astrophysics (ISSN 0004-6361)* **125**, p. 249(Sep 1983).

6. M. Malkov and L. Drury, *Reports on Progress in Physics* **64**, p. 429 (2001).
7. E. Amato and P. Blasi, *Monthly Notices of the Royal Astronomical Society: Letters* **364**, p. L76(Nov 2005).
8. E. Amato and P. Blasi, *Monthly Notices of the Royal Astronomical Society* **371**, 1251(Sep 2006).
9. D. Caprioli, P. Blasi, E. Amato and M. Vietri, *The Astrophysical Journal* **679**, p. L139(Jun 2008).
10. D. Caprioli, P. Blasi and E. Amato, *arXiv* **astro-ph.HE**(Jul 2010).
11. E. Parizot, A. Marcowith, J. Ballet and Y. A. Gallant, *Astronomy and Astrophysics* **453**, p. 387(Jul 2006).
12. H. J. Völk, E. G. Berezhko and L. T. Ksenofontov, *Astronomy and Astrophysics* **433**, p. 229(Apr 2005).
13. G. Morlino, E. Amato, P. Blasi and D. Caprioli, *Monthly Notices of the Royal Astronomical Society: Letters* **405**, L21(Apr 2010).
14. J. Giacalone and J. R. Jokipii, *The Astrophysical Journal* **663**, p. L41(Jul 2007).
15. D. C. Ellison, L. O. Drury and J.-P. Meyer, *Astrophysical Journal v.487* **487**, p. 197(Sep 1997).
16. E. G. Berezhko and H. J. Völk, *The Astrophysical Journal* **661**, p. L175(Jun 2007).
17. V. Ptuskin, V. Zirakashvili and E.-S. Seo, *The Astrophysical Journal* **718**, p. 31(Jul 2010).
18. S. Funk, Fermi-lat observations of young supernova remnants (2009).
19. T. Tanaka, Fermi lat observations of supernova remnants interacting with molecular clouds (2009).
20. D. Caprioli, P. Blasi and E. Amato, *Monthly Notices of the Royal Astronomical Society* **396**, 2065(Jul 2009).
21. V. S. Ptuskin, F. C. Jones, E. S. Seo and R. Sina, *Advances in Space Research* **37**, p. 1909(Jan 2006).
22. G. D. Bernardo, C. Evoli, D. Gaggero, D. Grasso and L. Maccione, *eprint arXiv* **0909**, p. 4548(Sep 2009).
23. A. W. Strong and I. V. Moskalenko, *Advances in Space Research* **27**, p. 717(Jan 2001).
24. A. R. Bell, *Monthly Notices of the Royal Astronomical Society* **353**, p. 550(Sep 2004).
25. S. Katsuda, R. Petre, K. S. Long, S. P. Reynolds, P. F. Winkler, K. Mori and H. Tsunemi, *The Astrophysical Journal Letters* **692**, p. L105(Feb 2009).
26. E. A. Helder, J. Vink, C. G. Bassa, A. Bamba, J. A. M. Bleeker, S. Funk, P. Ghavamian, K. J. V. D. Heyden, F. Verbunt and R. Yamazaki, *Science* **325**, p. 719(Aug 2009).
27. J. Sollerman, P. Ghavamian, P. Lundqvist and R. C. Smith, *Astronomy and Astrophysics* **407**, p. 249(Aug 2003).
28. J.-J. Lee, J. C. Raymond, S. Park, W. P. Blair, P. Ghavamian, P. F. Winkler and K. Korreck, *arXiv* **astro-ph.HE**(May 2010).
29. W. D. Apel, J. C. Arteaga, A. F. Badea, K. Bekk, J. Blümer, H. Bozdog, I. M. Brancus, M. Brüggemann, P. Buchholz, F. Cossavella, K. Dau-

miller, V. de Souza, P. Doll, R. Engel, J. Engler, M. Finger, D. Fuhrmann, H. J. Gils, R. Glasstetter, C. Grupen, A. Haungs, D. Heck, J. R. Hörandel, T. Huege, P. G. Isar, K.-H. Kampert, D. Kang, D. Kickelbick, H. O. Klages, Y. Kolotaev, P. uczak, H. J. Mathes, H. J. Mayer, J. Milke, B. Mitrica, S. Nehls, J. Oehlschläger, S. Ostapchenko, S. Over, M. Petcu, T. Pierog, H. Rebel, M. Roth, G. Schatz, H. Schieler, F. Schröder, O. Sima, M. Stümpert, G. Toma, H. Ulrich, J. van Buren, W. Walkowiak, A. Weindl, J. Wochele, M. Wommer and J. Zabierowski, *Astroparticle Physics* **31**, p. 86(Mar 2009).
30. P. Blasi, E. Amato and D. Caprioli, *Monthly Notices of the Royal Astronomical Society* **375**, 1471(Feb 2007).
31. Y. A. Gallant and H. E. S. S. Collaboration, *American Astronomical Society* **11**, p. 738(Feb 2010).
32. E. Carmona, M. T. Costado, L. Font, J. Zapatero and for the MAGIC Collaboration, *eprint arXiv* **0907**, p. 1009(Jul 2009).
33. J. H. Buckley and V. Collaboration, *American Astronomical Society* **11**, p. 738(Feb 2010).
34. S. Funk and F.-L. collaboration, *American Astronomical Society* **11**, p. 655(Feb 2010).
35. G. Morlino, E. Amato and P. Blasi, *Monthly Notices of the Royal Astronomical Society* **392**, 240(Jan 2009).
36. D. C. Ellison, D. J. Patnaude, P. Slane and J. Raymond, *The Astrophysical Journal* **712**, p. 287(Mar 2010).
37. D. Caprioli, E. Amato and P. Blasi, *Astroparticle Physics* **33**, 160(Apr 2010).
38. L. O. Drury, *eprint arXiv* **0907**, p. 4799(Sep 2010).

# $\gamma$-RAYS FROM HEAVY NUCLEI ACCELERATED IN SUPERNOVA REMNANTS

D. CAPRIOLI*, P. BLASI and E. AMATO

*INAF/Osservatorio Astrofisico di Arcetri,*
*50125 Firenze, Italy*
**E-mail: caprioli@arcetri.astro.it*

We investigate the theoretical and observational implications of the acceleration of protons and heavier nuclei in supernova remnants (SNRs). By adopting a semi-analytical technique, we study the non-linear interplay among particle acceleration, magnetic field generation and shock dynamics, outlining a self-consistent scenario for the origin of the spectrum of Galactic cosmic rays as produced in this class of sources. Moreover, the inferred chemical abundances suggest nuclei heavier than Hydrogen to be relevant not only in the shock dynamics but also in the calculation of the $\gamma$-ray emission from SNRs due to the decay of neutral pions produced in nuclear interactions.

*Keywords*: cosmic rays, acceleration of particles, SNR, $\gamma$-rays

## 1. Cosmic rays and supernova remnants

For more than 70 years scientists have been regarding supernova remnants (SNRs) as the most plausible sources for Galactic cosmic rays (GCRs), but only in recent times the theoretical comprehension of the ongoing processes have made important steps forward (see also P. Blasi's contribution in this volume). Once a model for CR propagation in the Milky Way is assumed, it is possible to infer from observations at Earth the spectra of single chemical species expected at the sources. In this work we investigate a scenario in which Galactic SNRs are responsible for the acceleration of protons and heavier nuclei (hereafter HN), trying to disentangle the solid, physical ingredients and the phenomenological recipes which have to be included in order to account for many observational constraints. In particular we show that, according to this *SNR paradigm* for the origin of GCRs, HN may play a fundamental role in the shock dynamics and also contribute in a non-negligible way to the $\gamma$-ray emission from SNRs.

We consider here a semi-analytical approach to the problem of non-linear diffusive acceleration of particles at shocks (NLDSA) following the basic implementation put forward in Ref. 1–3, also including the effects of the magnetic field amplification on the shock dynamics[4] and the pressure of the most abundant CR species.[5] The evolution of a remnant in a homogeneous circumstellar medium is followed in a quasi-stationary way as in Ref. 6 and coupled with the acceleration of particles at the forward shock as in Refs. 5,7. For the details of the computational apparatus, the reader may refer to the papers above. Here, we only would like to highlight how such a semi-analytical approach to NLDSA is typically much faster than, and as much rigorous as, other fully numerical or Monte Carlo methods for non-relativistic shocks,[8] and therefore it is very useful for including multi-specie CRs or, in general, for studying problems with a wide range of environmental parameters.

In our calculations protons are injected into the acceleration process from the thermal bath through a "thermal-leakage" mechanism,[9] which is sensitive to the shock dynamics and tends to suppress the number of injected particles when the the acceleration becomes more and more efficient. The amounts of injected HN are tuned by hand relatively to protons to reproduce the abundances measured at Earth. This recipe, while including a reasonable feedback which self-regulates the efficiency of the injection, has to be regarded as a necessary phenomenological description since, unfortunately, the HN injection at SNR shocks is still poorly understood for two main reasons. First, injection strongly depends on the charge/mass ratio and it is very difficult to follow the degree of ionization of heavy atoms during their acceleration; second, refractory elements (Mg, Si, Fe,...) are thought to be injected as a result of sputtering of accelerating dust grains.[10,11] The latter phenomenon, though very hard to deal with quantitatively, is however expected to produce largely suprathermal (but still non-relativistic) ions able to cross the shock from downstream to upstream because of their large gyroradii. For the same reason, partially ionized heavy atoms are expected to be preferentially injected, in agreement with the relative abundances measured in GCRs.[10,11] Nevertheless, since our abundances are tuned on the relativistic region of the GCR spectrum, we can bypass the problem above and make safe predictions about the role of HN in the shock dynamics and in the SNR $\gamma$-ray emission.

At any given time the shock dynamics is regulated by the non-linear interplay between particle acceleration, occurring via first-order Fermi mechanism, and magnetic field amplification, which we model as due to resonant streaming instability excited by all the accelerated particles.[12] On one hand, the pressure in CRs diffusing around the shock leads to the formation of a precursor which slows down the incoming fluid and tends to make the shock weaker while, on the other hand, the pressure in the shape of self-generated magnetic turbulence may become comparable to, or even larger than, the gas pressure upstream, preventing an excessive modification of the velocity profile. This *magnetic feedback*[13] has proved itself to be a very common mechanism able to account, at the same time, for both the level of magnetization and the hydrodynamics inferred by multi-wavelength studies of young SNRs.[4] In addition, when the velocity of the scattering centres, which we assume to be transverse Alfvén waves as predicted by the quasi-linear theory of resonant streaming instability, becomes a non-negligible fraction of the fluid velocity, the compression ratios actually felt by the fluid and by the accelerated particles are no longer the same. A physically motivated account for the relative velocity between the fluid and the waves leads to conclude that the more efficient the magnetic field amplification (the larger the Alfvén velocity) is, the steeper the spectra of the accelerated particles are (see e.g. §7 of Ref. 5).

## 2. Particle escape from SNRs

In order to explain the observations at Earth, one has to account for *when* and *how* accelerated particles leave the source and become CRs. There are three main mechanisms playing a potential role in the problem: escape from upstream, escape from downstream and release at the "death" of the remnant (after having undergone adiabatic losses). In general, the total CR flux provided by a single source is given by the convolution over its life of these three contributions. Let us discuss them in some more detail.

During the ejecta-dominated stage the shock velocity $V_{sh}$ is roughly constant: the magnetic field is expected to increase, in turn leading to more and more energetic particles. In such a situation no particle can diffuse away from upstream. During the Sedov-Taylor stage, instead, both the shock velocity and the magnetization level decrease: the diffusion length increases as $1/V_{sh} \propto t^{3/5}$ and hence the highest-energy particles are no longer able to make it back to the shock, since it moves as $R_{sh} \propto t^{2/5}$.[14] Quantitatively speaking, we assume that at any moment the system can confine only particles with diffusion length $D(E, \delta B)/V_{sh} \leq \chi R_{sh}$, where $D(E, \delta B)$ is the Bohm diffusion coefficient in the amplified field $\delta B$, by imposing the distribution function of the accelerated particles to vanish for radii larger than $(1 + \chi)R_{sh}$ (see also Ref. 3).

Clearly, most of accelerated particles are advected behind the shock and, trapped in the expanding shell, cannot do but losing energy adiabatically. If, for any reason (like, for instance, the presence of strong density or magnetic patterns in the circumstellar medium), the SNR shell were "broken", preferential conduits might channel particles from downstream into the Galaxy. Such a contribution is in principle very hard to model, thus we assume that, at any given time, a 10% of the downstream accelerated particles is injected into the Galaxy, contributing to the diffuse GCRs.[7] Finally, the remainder of the advected particles should be released at the "death" of the SNRs, which is also quite difficult to define: it might coincide with the beginning of the radiative stage, but actually it should be more closely related to the damping of the magnetic field and hence to the transition from a SNR-like to a Galactic-like regime for the particle diffusion.

We consider here a "benchmark" SN explosion releasing 1.4 solar masses with total kinetic energy $10^{51}$erg in the homogeneous circumstellar environment, taken with temperature $10^6$K and particle density $\rho_0 = 0.01 m_{\rm H}{\rm cm}^{-3}$. The remnant evolution is thus followed as in Ref. 6, Tab. 7, and the SNR is imposed to die at the end of the Sedov stage. The transport in the Galaxy is then calculated within a simple *leaky-box* model, assuming a halo height of 3.5 kpc, a Galaxy radius of 10 kpc and a residence time $\tau_{esc}(R) = 20(R/10{\rm GV})^{-0.55}$ Myr, where $R$ is the nucleus rigidity. When spallation (mainly relevant at low energies and for HN) is also included,[15] the spectra of the GCRs is recovered in terms of slope, normalization and cut-off (see Figs. 3,4 in Ref. 5).

It is however worth keeping in mind that the recipe linking magnetic field amplification and the velocity of the scattering centers is rather phenomenological and it has not been thoroughly calculated from first principles, yet. Remarkably, at the moment it represents the only proposed effect able to produce source spectra steeper than $E^{-2}$ (see also Ref. 16) and therefore consistent with the inferred value of $\delta = 0.3 - 0.6$ for the scaling of the Galactic residence time $\tau_{esc} \propto E^{-\delta}$. In fact, in Ref. 17 such an effect is not included and $\delta = 0.75$ is required to recover the GCR spectrum.

## 3. The role of heavy nuclei

The chemical abundances fitting the GCR data at Earth are self-consistently put in the computational apparatus in a recursive way until consistency is reached, since the non-linear nature of the acceleration process does not allow an *a priori* exact determination of the proton/HN ratios. Provided that the relative abundances do not depend on time, such a process allow us to investigate the role of HN at any given time of the

SNR evolution.

In the left panel of Fig. 1 a snapshot of the shock profile at the age $t = 2000$ yr (beginning of the Sedov phase) is shown, in units of $x_0 = \chi R_{sh} \simeq 0.2 R_{sh}$, i.e. the distance of the free-escape boundary from the shock. $U(x)$ is the local velocity in units of $V_{sh}$, $P_g(P_w)$ is the gas (magnetic field) pressure and the other curves correspond to the contributions to the pressure by accelerated particles of different species, as in the legend; all pressures are normalized to $\rho_0 V_{sh}^2$.

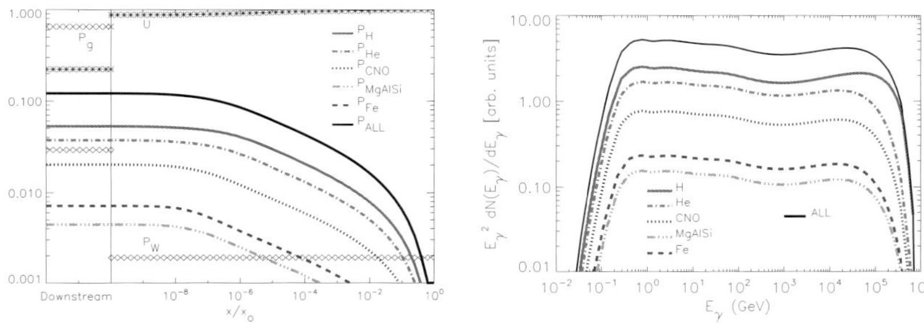


Fig. 1. Left panel: spatial profile of the modified shock. See text for the details of the different curves. Right panel: $\gamma$-ray spectrum from $\pi^0$ decay produced in nuclear collisions, in arbitrary units. Both panels correspond to a SNR age of about 2000 yr.

It is easy to see that more than 10% of the bulk pressure is channelled into accelerated particles, and that HN contribute as much as the protons to the shock dynamics. Most of their contribution comes from He nuclei, but heavier elements still account for about 25% of $P_{\mathrm{ALL}}$. As outlined above, the pressure in magnetic turbulence $P_w$ turns out to be larger than the gas pressure $P_g < 10^{-4}$, and thus the precursor is dynamically dominated by the interplay between CRs and magnetic field.[4,13] Moreover, since $P_w \propto P_{\mathrm{ALL}}$, we can postulate that HN play a non-negligible role also in the amplification of the background magnetic field via streaming instability.

Another interesting consequence of the acceleration of HN is that they may make an impact on the hadronic $\gamma$-ray emission from SNRs, i.e. the emission due to the decay of neutral pions produced in nuclear collisions between relativistic particles and background gas.[18] For distributions of accelerated particles $\propto E^{-2}$, the contribution to the $\gamma$-ray flux by different chemicals is simply proportional to their relative abundances (see also §5 in Ref. 5). This means that, for the chemical composition determined above, HN contribute in a substantial way to the $\gamma$-ray flux from a SNR (more than a factor 2 for the case depicted in Fig. 1). The maximum $\gamma$-ray energy produced by all the HN is typically half the one produced by protons, since the acceleration is rigidity-dependent while the energy of the secondaries scales with the energy per nucleon. This effect leads to an alteration in the shape of the cut-off, a signature probably too hard to be discriminated by observations (ALL vs H curve in the right panel of Fig. 1).

Very generally, the inclusion of HN implies that the observed $\gamma$-ray flux from a given SNR may be achieved with, say, half the density of the target gas with respect to a case accounting for accelerated protons only. As a consequence, the HN contribution is of great importance when the density of the circumstellar medium is deduced by the level of $\gamma$-rays of hadronic origin. A different estimate of the circumstellar density, in principle, leads also to a different modelling of the SNR evolution, affecting the inferred age, radius, expansion rate and, eventually, distance of the object. Even more interestingly, a reduced circumstellar density would also imply a strong suppression ($\propto \rho^2$) of the expected thermal emission of the shocked plasma, both in terms of continuum and lines. Possible HN overabundances, with respect to the "standard" values above, might as well account for the lack of detection of thermal emission in single $\gamma$-ray-bright SNRs as, for instance, RX J1713.7-3946.

## References

1. E. Amato, P. Blasi, *MNRAS Lett.* **364**, 76 (2005)
2. E. Amato, P. Blasi,*MNRAS* **371**, 1251 (2006)
3. D. Caprioli, E. Amato, P. Blasi, *APh* **33**, 307 (2010)
4. D. Caprioli, P. Blasi, E. Amato, M. Vietri,*MNRAS* **395**, 895 (2009)
5. D. Caprioli, P. Blasi, E. Amato, *Accepted in APh*, astro-ph:1007.1925
6. J.K. Truelove, C.F. Mc Kee, *Apj Supplement Series* **120**, 299 (1999)
7. D. Caprioli, E. Amato, P. Blasi, *APh* **33**, 160 (2010)
8. D. Caprioli, H. Kang, A. Vladimirov, T.W. Jones, *MNRAS* **417**, 1773 (2010)
9. P. Blasi, S. Gabici, G. Vannoni, *MNRAS* **361**, 907 (2005)
10. J-P. Meyer, L.O'C. Drury, D.C. Ellison, *ApJ* **487**, 182 (1997)
11. D.C. Ellison, L.O'C. Drury, J-P. Meyer, *ApJ* **487**, 197 (1997)
12. A.R. Bell,*MNRAS* **182**, 147 (1978)
13. D. Caprioli, P. Blasi, E. Amato, M. Vietri, *ApJ Letters* **679**, 139 (2008)
14. D. Caprioli, P. Blasi, E. Amato, *MNRAS* **396**, 2065 (2009)
15. J.R. Hörandel, N.N. Kalmykov, A.V. Timokhin, *APh* **27**, 119 (2007)
16. V.S. Ptuskin, V.N. Zirakashvili, E.S. Seo, *ApJ* **718**, 31 (2010)
17. E.G. Berezhko, H.J. Völk, *ApJ Letters* **661**, 175 (2007)
18. L.O'C. Drury, F. Aharonian, H.J. Völk, *A&A* **287**, 959 (2004)

# Anisotropies in the Cosmic-ray electron spectrum: A way to discriminate between exotic and astrophysical sources?

I. CERNUDA

*Basic Research Division, CIEMAT, Madrid, 28008, Spain*
*E-mail: ignacio.cernuda@ciemat.es*
*www.ciemat.es*

Over the last two years, the release of new data from Cosmic-ray (CR) experiments such as PAMELA[1] and Fermi,[2] has brought a lot of excitement to the astroparticle physics community over the so-called positron excess. Indeed, these experiments have revealed a surprisingly large flux of high energy electrons and positrons that strongly suggest the existence of a new source of primary CR. A considerably effort has been devoted to explain this excess from already known astrophysical sources (pulsars,SNRs) and from exotic ones (Dark Matter (DM) annihilation/decay), but not an absolute proof to discriminate between both scenarios has been found yet. In this work we propose the study of CR electron (CRe) anisotropies as a tool to sort the puzzle out. We address the problem of the electron anisotropy in a pulsar and a DM annihilation scenario and estimate the prospect that a small dipole anisotropy might be found by the Fermi observatory.

*Keywords*: Dark-Matter;Pulsar;Anisotropy;Galactic Cosmic-Rays.

## 1. Introduction

The study of Cosmic-rays has proved to be of utmost importance in the understanding of the processes that govern our galaxy and has became a privileged field for the discovery of new physics. The current availability of precision measurements in a number of recent experiments such as PAMELA and Fermi, or the impending AMS-02, has provided a unique opportunity to challenge the theoretical framework that builds our comprehension of Nature. Certainly, the recent advent of new data on CR electrons and positrons has raised disagreements with our current knowledge of CR production and propagation. The observation of an excess in the CRe spectrum has been interpreted as the manifestation of a new CR primary source, either exotic (DM) or conventional (e.g. pulsars, SNRs). However,

although unlikely, a secondary origin cannot be ruled out yet, in view of the high theoretical uncertainties.
It has been shown in previous works the plausibility of these scenarios, but in the absence of multimessenger and multiwavelength studies, the models fail to out-stand from each other. In this proceeding we approach the problem of the CR electron anisotropy in two competing scenarios as a strategy to distinguish the proposed models, and discuss the implications derived from the study.
In the first section, we describe the propagation framework we will use in this work. In the second section we review the CRe production in pulsars and estimate the contribution to the local spectra from all known gamma-ray pulsars as listed in the ATNF catalogue.[3] In the third section we consider the possible contribution from DM annihilation, in view of the more recent N-body simulations of Milky Way sized DM halos.
Finally, we calculate the predicted dipole anisotropy produced in the considered scenarios and discuss its measurement prospects.

## 2. Propagation

The processes to what Galactic CR are subjected following their production at the sources, are well established since the early work from Ginzburg.[4] For CR leptons, the relevant processes are diffusion, as a result of the scattering off of the random galactic magnetic fields, and large energy losses due to synchrotron radiation and Inverse Compton (IC) off the background radiation (CMB and starlight). The diffusion-loss equation describing the evolution of the number density $n(E, \vec{x}, t)$ that takes into account these processes can be expressed as:

$$\frac{\partial}{\partial t}n(E,\vec{x},t) = D(E)\cdot\nabla^2 n(E,\vec{x},t) - \frac{\partial}{\partial E}(b(E)n(E,\vec{x},t)) + Q(E,\vec{x},t) \quad (1)$$

Where $b(E) \simeq E^2/(\tau_E \cdot \mathrm{GeV})$ codifies the energy losses, with $\tau_E = 10^{16}$s and $D(E) = D_0 E^\gamma$ the diffusion term. We adopt a cylindrical diffusion zone of radius R=20Kpc and half-width L=4Kpc. For the diffusion coefficient, we consider a *MED* setup which is consistent with the boron to carbon ratio B/C. Eq. 1 can be solved for non stationary sources (e.g. pulsars or SNRs) or for the steady state which is the case of a DM source. For the former case, the solution reads[5] :

$$\phi(E,r,t) = \frac{\beta c}{4\pi}\frac{Q_0}{\pi^{3/2}r^3}\left(\frac{r}{D_{diff}}\right)^3 (1-atE)^{\alpha-2}E^{-\alpha}e^{-\frac{E}{(1-atE)E_c}}e^{-\left(\frac{r}{D_{diff}}\right)^2} \quad (2)$$

where the distance scale is $D_{diff}(E,t) \simeq 2\sqrt{D(E)t\frac{1-(1-E/E_{max})^{1-\gamma}}{(1-\gamma)E/E_{max}}}$

For the latter case, the flux $\phi_{\odot}(E) = \frac{\beta c}{4\pi} n(\vec{x}_{\odot}, E)$ is calculated as in Ref. 6 with the number density n:

$$n(\vec{x}, E) = \int_{E}^{\infty} dE_s \int_{Halo} d^3\vec{x}_s G(\vec{x}, E \leftarrow \vec{x}_s, E_s) Q(\vec{x}_s, E_s) \tag{3}$$

where the green function is:

$$G(\vec{x}, E \leftarrow \vec{x}_s, E_s) = \frac{\tau_E}{E^2} \left\{ \frac{1}{\pi D_{diff}^2} \right\}^{3/2} \exp\left\{ -\frac{(\vec{x} - \vec{x}_s)^2}{D_{diff}^2} \right\} \tag{4}$$

and the diffusion length is $D_{diff} = \sqrt{4D_0(\tilde{t} - \tilde{t}_s)}$, where $\tilde{t} = \tau_E \left\{ \frac{E^{\gamma-1}}{1-\gamma} \right\}$ Finally, the background has been calculated with a standard version of the public GALPROP code for the relevant species.

## 3. Astrophysical sources

Pulsars have been for a long time known to produce electromagnetic cascades in their magnetosphere. It is only natural to think that they can operate as $e^{\pm}$ sources. However, it has not been until recently, with the release of high energy precision data, that the possibility of pulsars being an actual primary source of CR electrons has been revisited.

### 3.1. *The Source Term*

Gamma-ray pulsars are indeed expected to produce pairs of electrons and positrons as a result of electromagnetic cascades produced by the acceleration of electrons in their magnetosphere. The simplest way to model the $e^{\pm}$ production is the so called magnetic dipole rotator. In this schema, it is assumed that the pulsar losses all the rotational energy through magnetic dipole radiation, resulting in a breaking $\Omega(t) = \frac{\Omega_0}{(1+t/\tau_0)^{1/2}}$. The spin-down power $\dot{E}$ of the pulsar can be then easily obtained: $\dot{E} = I\Omega\dot{\Omega} = \frac{1}{2}I\Omega_0^2 \frac{1}{\tau_0} \frac{1}{(1+t/\tau_0)^2}$, from which a fraction $f_{e^{\pm}}$ will be translated into pair production. We can further calculate the total released energy just integrating the spin-down power over the pulsar age $T$ to obtain $E_{out}[ST] = f_{e^{\pm}} \dot{E} \frac{T^2}{\tau_0}$. If we assume that the electron injection spectrum can be described by a power-law, then the source Q(E,r,t) from a single pulsar located at r, injecting electrons with energy E at time t can be expressed by: $Q(E,r,t) = Q_0 \cdot \mathcal{L}_{e^{\pm}}(t) E^{-\alpha} \exp(-E/E_c) \delta(r)$ where $Q_0$ must satisfy the total energy release constraint, and $\mathcal{L}_{e^{\pm}}(t) = f_{e^{\pm}} \dot{E}$. A spectral cutoff has been considered to take into account the drop observed by Hess at E $\sim$ 1 TeV. To estimate the potential sources that can contribute to the local electron spectrum, we make use of the Australian Telescope National Facility (ATNF) pulsar catalogue, from which we take all the sources labelled as gamma-ray pulsars according to the condition $g = 5.5P^{26/21}B_{12}^{-4/7} < 1$ in terms of the pulsar period P and the pulsar surface magnetic field $B_{12}$. Furthermore, from this collection we must subtract those with ages inferior to $10^4 - 10^5$ years which are considered to be surrounded by a pulsar wind nebula (PWN), thus, restricting the free diffusion of the electrons to the interstellar medium (ISM).

## 4. Dark-Matter sources

Dark matter annihilation or decay can, in fact, produce a large amount of positrons and electrons, hence, it constitutes a very attractive candidate to explain the reported excesses. Nevertheless, conventional WIMP's annihilation rates with canonical cross sections $\langle\sigma v\rangle = 3\text{x}10^{-26}\text{cm}^3/\text{s}$ are too little to make up the observed abundances. Moreover, the absence of an excess in antiprotons indicates that the annihilation through hadronic channels must be strongly suppressed. There are, however, a few ways to override these constraints (for instance we can consider a new light force carrier into which the DM particle annihilates, leading to a Sommerfeld enhancement at low velocities). In this section we will not detail a specific model but instead, assume that the required normalisation is met by means of any of the mechanisms considered in related works.

### 4.1. *The Source Term*

We will focus our discussion on DM annihilation rather than decaying DM. The results for the latter can be found elsewhere. The signal that results from DM annihilation, depends essentially on the squared DM density from the astrophysics point of view, and the annihilation cross section from the particle physics side: $Q(E,\vec{r}) = \frac{1}{2}\frac{\rho^2(\vec{r})}{m_{DM}^2}\Sigma\langle\sigma v\rangle_i\frac{dN_i(E)}{dE}$. The mass distribution, abundance, and internal structure of DM halos is investigated with the use of high-resolution numerical simulations such as the Via Lactea II (VLII) and Aquarius. Both simulations agree in the existence of a smooth spherical halo profile and a clumpy component with a virialised subhalo abundance $\propto M_{sub}^{[-1.9,-2.0]}$. In this work, we will consider a NFW/Einasto profile for the smooth component in the VLII/Aquarius cases, and model the substructure as in Pieri:[7] $\rho(M_{sub},\vec{r}) = \frac{dP_V(\vec{r})}{dV}\frac{dP_M(m,\vec{r})}{dm}$

where $\frac{dP_M^{\rm Aq|VLII}(m,\vec{r})}{dm} \propto (\frac{M_{sub}}{M_\odot})^{1.9|2.0}$ and the spatial distribution: $\frac{dP_V^{\rm Aq}(\vec{r})}{dV} \propto \exp\left[-\frac{2}{\alpha}((\frac{r}{r_s})^\alpha - 1)\right]$ and $\frac{dP_V^{\rm VLII}(\vec{r})}{dV} \propto \frac{1}{(1+\frac{r}{r_s})^2}$.
We describe the subhalos by their concentration parameter $c_{200}(M_{sub}, \vec{r}) = r_{200}/r_s$ as in Ref. 7 and we account for the tidal effects, imposing that the scale radius be inferior to the tidal radius, according to the Roche criterion.

## 5. Anisotropies

High energy charged galactic CRs diffuse in the galactic magnetic fields arriving Earth with highly isotropic direction distributions. Despite that, a number of ground-based experiments have reported the existence of intermediate to large scale anisotropies that could provide useful information on the CR production sites. In the case of CR electrons, it has been argued[8] that anisotropies can be used to disentangle the possible contributions to the electron spectrum of DM and conventional astrophysical sources. Actually, two of the more luminous objects that could contribute to the primary electrons, namely Monogem and Geminga, lay roughly in the anti-center of the galaxy with respect Earth's location. In contrast, DM halos are characterised by spherically symmetric density distributions that falls with the distance, suggesting a potential CRe dipole anisotropy towards the galactic center. In this section we estimate the predicted degree of anisotropy that can be expected from the scenarios we have reviewed and compare it with the potential of Fermi to detect it in a 5-years survey.

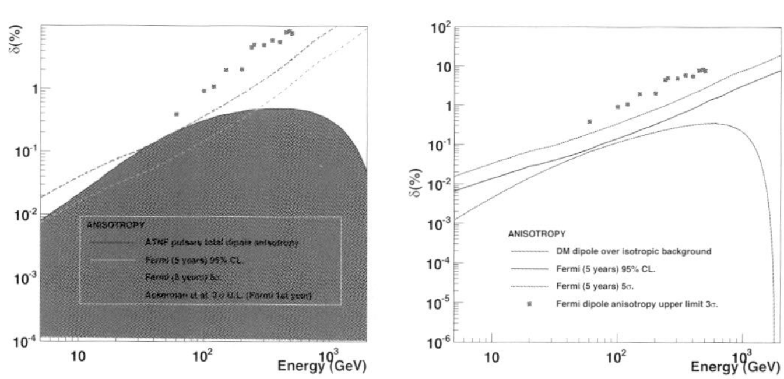


(a) Pulsar Scenario (b) DM Aquarius Scenario

Fig. 1. Dipole anisotropy

In the diffusion model of CR propagation, the level of dipolar anisotropy from a single source is calculated to be[9] $\delta_i = \frac{3D}{c}\frac{|\nabla N|}{N}$. In the case of multiple sources, it can be generalised as $\delta = \frac{\sum \phi_i(E,r,t)\langle\delta_i \hat{r}_i \hat{n}_i\rangle}{\phi_T}$, where the product $\langle\delta_i \hat{r}_i \hat{n}_i\rangle$ is the projection of the individual dipole over the direction of maximum intensity. In the case of pulsars, the sum ranges for the collection of pulsars considered. In the case of DM, we make use of the framework sketched in section 3 and switch the sum conveniently to an integral over the DM distribution. Interestingly, Fermi has recently published the first results in anisotropy searches for $\sim$1 year data taking.[10] The results point out that for the collected statistics, no anisotropic features have been found, and the CRe dipole anisotropy upper limits (U.L). are above the expected levels derived in this study (Fig. 1). However, we estimate that for these models, an anisotropy measurement could be within Fermi's reach in a few years.

## 6. Discussion and Conclusions

We have shown that the positron excess can be attributed both to astrophysical sources and to DM annihilation, although the latter is already in tension with the continuously improving new data. Moreover, in order to distinguish the different scenarios, the information encoded in the electron spectrum lies powerless to disentangle the issue. In addition to multimessenger and multiwavelength searches, the study of CRe anisotropies is proposed to address the problem. The recent results from the Fermi collaboration on CRe anisotropies, suggest that no anisotropy is present at certain level, but the U.L. are still above the expectations for the considered sources. We still should wait for Fermi to accumulate more statistics to be able to put severer constraints and, hopefully, provide hints on the origin of the positron excess.

## References

1. O. Adriani *et al.*, *Nature* **458**, 607 (2009).
2. FERMI/LAT collaboration, *Phys. Rev. Lett.* **102** (2009).
3. e. a. Manchester, *Astron. J.* **129**, 1993 (2005).
4. V. Ginzburg and S. Syrovatskii (Pergaemon, Oxford, 1964).
5. A. M. Atoyan, F. A. Aharonian and H. J. Völk, *Phys. Rev. D* **52**, 3265 (1995).
6. E. Baltz and A. Edsjo, *Phys. Rev. D* **59**, p. 023511 (1998).
7. L. P. et al., *arXiv:0908.0195v1[astro-ph.HE]* (2009).
8. I. Cernuda, *Astroparticle Physics* **43**, 59 (2010).
9. C. Mao and C. Shen, *Chinese Journal of Physics* **10**, 16 (1971).
10. A. et al., *to appear in Physical Review D* (2010).

# Cosmic-ray electrons and positrons from gamma-ray pulsars

M. DORMODY*

*Department of Physics, University of California, Santa Cruz,*
*Santa Cruz, CA 95064, USA*
**E-mail: dormody@scipp.ucsc.edu*

We analyze the contribution of gamma-ray pulsars from the first Fermi-Large Area Telescope (LAT) catalogue to the local flux of cosmic-ray electrons and positrons ($e^+e^-$). We present new distance estimates for all Fermi gamma-ray pulsars and estimate the contribution of gamma-ray pulsars to the local $e^+e^-$ flux, in the context of a simple model for the pulsar $e^+e^-$ emission. We find that 10 of the Fermi pulsars potentially contribute significantly to the measured $e^+e^-$ flux in the energy range between 100 GeV and 1 TeV. We argue that known radio pulsars fall in regions of parameter space where the $e^+e^-$ contribution is predicted to be typically much smaller than from those regions where Fermi-LAT pulsars exist. However, we find that a few known radio pulsars that have not yet been detected by Fermi can also significantly contribute to the local $e^+e^-$ flux if they are closer than 2 kpc, and if they have a characteristic age on the order of one mega-year.

*Keywords*: cosmic ray theory; gamma ray theory; radio pulsars; gamma-ray pulsars.

## 1. Electron-Positron Increase

PAMELA measured an $\frac{e^+}{e^+ + e^-}$ increase above 10 GeV,[1] which can be attributed to Weakly Interacting Massive Particles (WIMPs),[2] or possibly gamma-rays interacting with optical-UV sources.[3] A simpler approach uses known astrophysical sources as the main contributers of the positron flux such as pulsars.[4] A two component model that can account for this excess is:

(a) a distant source of $e^-$, spaced uniformly
(b) nearby sources of very high energy $e^-$ with a spectrum extending beyond 1 TeV ($\sim$ 100 pc, $\sim$ 100 kyr).[5]

The goal of this study is to determine which gamma-ray pulsars contribute the most to the observed $e^+e^-$. These proceedings are a summary of the previously published study.[6]

## 2. Solving the diffusion equation

The spherically symmetric diffusion equation is given as

$$\frac{\partial f}{\partial t} = \frac{D(\gamma)}{r^2}\frac{\partial}{\partial r}r^2\frac{\partial f}{\partial r} + \frac{\partial}{\partial \gamma}(P(\gamma)f) + Q \tag{1}$$

with an injection spectrum for $e^+e^-$ produced from pulsars

$$Q(E,t,\vec{r}) = Q_0\left(\frac{E}{1\text{ GeV}}\right)^{-\Gamma} e^{-\frac{E}{E_{\text{cut}}}}\ \delta(t-t_0)\ \delta(\vec{r}) \tag{2}$$

where $f(r,t,\gamma)$ is the distribution function, $\gamma = \frac{E_{e\pm}}{m_e c^2}$, $D(\gamma) \sim \gamma^\delta$, $E_{\text{cut}} = 1$ TeV, and $\Gamma = 1.7$. The energy loss $P(\gamma) = -\frac{dE}{dt} = p_0 + p_1\gamma + p_2\gamma^2$ where $p_0$ is Coulomb energy loss, $p_1$ is Bremmstrahlung energy loss, and $p_2$ is Inverse Compton scattering and Synchrotron radiation.[5] For shorter time scales ($t \leq 10^7$ yr), Inverse Compton and synchrotron energy losses dominate and the distribution function is:[5]

$$f(r,t,\gamma) = \frac{N_0\gamma^{-\alpha}}{\pi^{3/2}r^3}(1-p_2 t\gamma)^{\alpha-2}\left(\frac{r}{r_{\text{dif}}}\right)^3 e^{-(\frac{r}{r_{\text{dif}}})^2} \tag{3}$$

where

$$r_{\text{dif}} \sim 2\sqrt{D(\gamma)t\frac{1-(1-\frac{\gamma}{\gamma_{\text{cut}}})^{1-\delta}}{(1-\delta)\frac{\gamma}{\gamma_{\text{cut}}}}} \tag{4}$$

Such astrophysical sources must be constrained by age and distance from the sun. Relativistic $e^-$ diffusing in the ISM against synchrotron and Inverse Compton energy losses constrain $t \sim 10^5$ yr, and the source distance is constrained by the diffusion radius $r \leq 2\sqrt{Dt} \approx 100 - 300$ pc where $D_0 \sim 10^{28}$ cm$^2$ s$^{-1}$.

Gamma-ray pulsars can satisfy this requirement. A pulsar is a rapidly rotating neutron star that emits radiation across a broad frequency spectrum. By definition, gamma-ray pulsars have observed emission in gamma-rays, and radio pulsars have observed emission in radio. Due to beaming geometry, we might only detect pulsars in one wavelength. Radio observations are more sensitive to pulsar detections than gamma-ray observations ($>$ 1800 observed radio pulsars and $\sim$ 60 observed gamma-ray pulsars). We used the Fermi Pulsar Catalog,[7] which is comprised of 24 radio-loud pulsars (including 8 MSPs), 16 blind search pulsars, and 6 EGRET pulsars. We ignore recycled pulsars (MSPs) in this study.

## 3. Estimating Pulsar distances

Measuring the distance to the pulsars is important in establishing the $e^+e^-$ spectrum, but distances to gamma-ray pulsars is non-trivial. Pulsar distances are determined using parallax, radio Dispersion Measure, X-ray measurements, or absorption/emission lines in HI. However, gamma-rays do not reflect or refract, and there is no way to determine distance using gamma-ray data alone. Instead, we adopt a phenomenological approach. We model the distance using an assumption on gamma-ray luminosity and spin-down energy:

$$L_\gamma \approx w\dot{E} \approx C \times \left(\frac{\dot{E}}{10^{33}\ \mathrm{erg\ s^{-1}}}\right)^{1/2} \times 10^{33}\ \mathrm{erg\ s^{-1}} \tag{5}$$

where, $L_\gamma = f_\Omega(4\pi D^2)G_\gamma$, $f_\Omega$ is model-dependent flux correction factor taken from the Pulsar Atlas,[8] $w$ is the gamma-ray efficiency, and $G_\gamma$ is the gamma-ray energy flux.

We calibrated this equation with the gamma-ray pulsars with distances measured from other means, and obtained a value of $C = 6.04 \pm 0.75$. The plots can be found in figure 1. There are 2 cases of large over/under estimation. PSR J2021+4026 distance is underestimated possibly due to large unpulsed emission. PSR J0659+1414 is overestimated due to inferred gamma-ray luminosity being 30 times lower than the general trend for gamma-ray pulsars.

## 4. Pulsar emission model

The cosmic-ray diffusion-loss transport equation is given as:

$$\frac{\partial N_e(E,t,\vec{r})}{\partial t} - D(E)\nabla^2 N_e - \frac{\partial}{\partial E}(b(E)N_e) = Q(E,t,\vec{r}) \tag{6}$$

The injection spectrum is given in equation 2. The solution is

$$\begin{aligned} N_e(E,t,\vec{r}) = {} & \frac{Q_0}{\pi^{3/2}R^3_{\mathrm{diff}}(E,t)}\left(1-\frac{E}{E_{\max}(t)}\right)^{\Gamma-2}\left(\frac{E}{1\ \mathrm{GeV}}\right)^{-\Gamma} \\ & \times \exp\left[-\frac{E}{E_{\mathrm{cut}}}\frac{1}{1-\frac{E}{E_{\max}}} - \left(\frac{r}{R_{\mathrm{diff}}}\right)^2\right] \end{aligned} \tag{7}$$

with

$$R_{\mathrm{diff}}(E,t) \simeq 2\sqrt{D(E)t\frac{1-(1-\frac{E}{E_{\max}})^{1-\delta}}{(1-\delta)\frac{E}{E_{\max}}}} \tag{8}$$

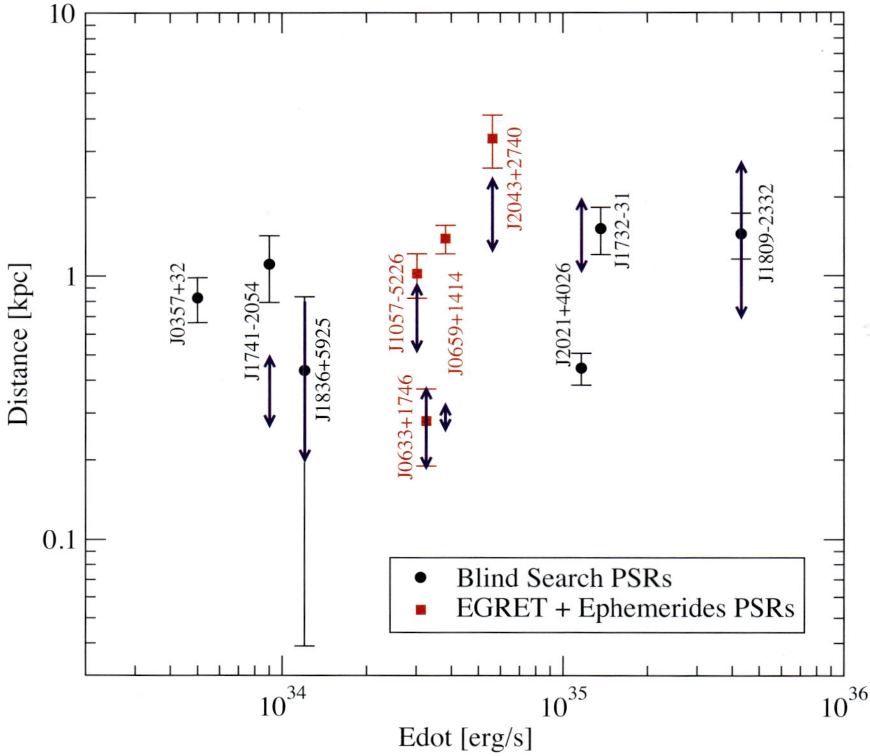


Fig. 1. Distance estimations for the gamma-ray pulsars that contribute the most to the $e^+e^-$ flux. Blue arrows indicate pulsar distances, when available. The slight over/under estimation is due to error in $w$, in the universal model assumption.

using the normalization requirement:

$$\int_{m_e}^{\infty} E \times Q(E) dE = E_{\text{out}} = \eta \frac{\dot{E} t_{ch}^2}{\tau} \quad (9)$$

The individual spectra for the contributing pulsars can be found in figure 2.

## 5. Parameter spaces of detected sources

The distribution of the contribution at 100 GeV to the total $e^+e^-$ flux is narrow in age ($10^5 - 10^6$ yr) and spin-down ranges ($10^{34} - 10^{35}$ erg s$^{-1}$. This distribution is relatively narrow, and has a nearly uniform contribution. The % contribution at the peak energy to the total $e^+e^-$ flux is steeply declining with distance, and the peak energy is between 100 GeV and 1 TeV.

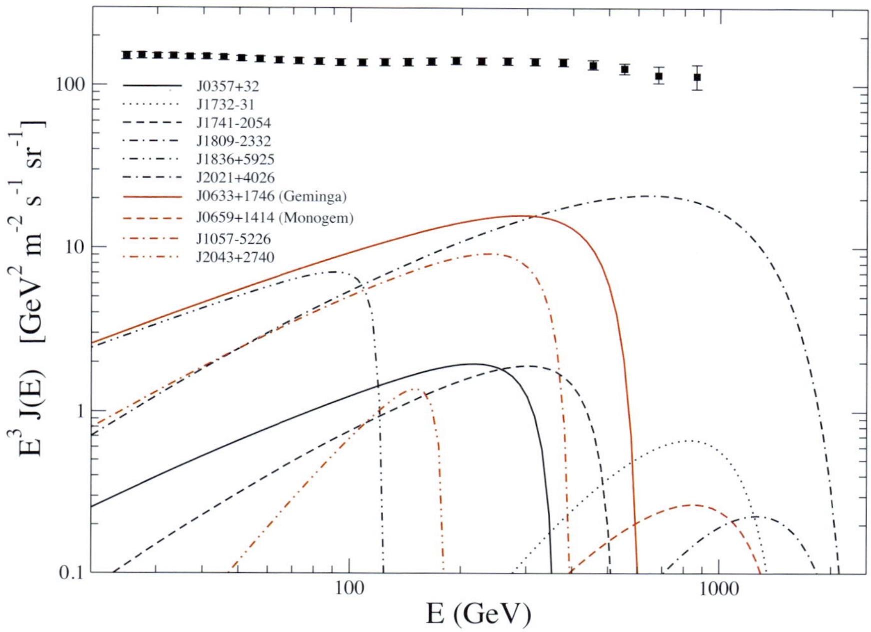


Fig. 2. Individual spectra of the 10 contributing pulsars. We assume injection efficiency of $\eta = 0.4$, $\Gamma = 1.7$, and $E_{\rm cut} = 1$ TeV. We ignore time between pulsar birth and injection, as this time delay is insensitive for contributing pulsars. The top data points are the $e^+e^-$ spectrum as seen by Fermi.[9]

What is the parameter space of these contributing, detected pulsars? The age range is limited by enough time for the charged particles to propagate through to the observer, or $t_{\rm min}$ set by $r_{\rm dif} \sim \sqrt{Dt}$. In addition, the maximum age is set by the cutoff energy $t_{\rm max} \leq E^{-1}$. Other pulsars in the ATNF catalog within this distance range do not contribute due to their age being too young and their luminosity being too low.

## 6. Conclusions and Acknowledgements

Our study has shown that gamma-ray pulsars can contribute significantly ($> 1\%$) ($\eta/0.4$) to the $e^+e^-$ flux. Ten pulsars from the Fermi Pulsar Catalog contribute, and of those, 6/10 are blind-search gamma-ray selected pulsars, and 4/10 are known radio-loud pulsars (including Geminga). Any undetected (so far) gamma-ray pulsars can contribute if the satisfy two conditions:

(a) they have a large $L_\gamma$

(b) $r < 3$ kpc for $\tau \sim 1$ Myr.

The author graciously acknowledges the organizing committee for the invitation to the conference. In addition, the author acknowledges travel funding from the AAS International Travel Grant committee and the Deparment of Energy.

## References

1. O. Adriani, G. C. Barbarino, G. A. Bazilevskaya, R. Bellotti, M. Boezio, E. A. Bogomolov, L. Bonechi, M. Bongi, V. Bonvicini, S. Bottai, A. Bruno, F. Cafagna, D. Campana, P. Carlson, M. Casolino, G. Castellini, M. P. de Pascale, G. de Rosa, N. de Simone, V. di Felice, A. M. Galper, L. Grishantseva, P. Hofverberg, S. V. Koldashov, S. Y. Krutkov, A. N. Kvashnin, A. Leonov, V. Malvezzi, L. Marcelli, W. Menn, V. V. Mikhailov, E. Mocchiutti, S. Orsi, G. Osteria, P. Papini, M. Pearce, P. Picozza, M. Ricci, S. B. Ricciarini, M. Simon, R. Sparvoli, P. Spillantini, Y. I. Stozhkov, A. Vacchi, E. Vannuccini, G. Vasilyev, S. A. Voronov, Y. T. Yurkin, G. Zampa, N. Zampa and V. G. Zverev, *Nature* **458**, 607(April 2009).
2. A. J. Tylka, *Physical Review Letters* **63**, 840(August 1989).
3. F. A. Aharonian and A. M. Atoyan, *Journal of Physics G Nuclear Physics* **17**, 1769(November 1991).
4. S. Profumo, *ArXiv e-prints* (December 2008).
5. F. A. Aharonian, A. M. Atoyan and H. J. Voelk, *A&A* **294**, L41(February 1995).
6. L. Gendelev, S. Profumo and M. Dormody, *Journal of Cosmology and Astro-Particle Physics* **2**, 16(February 2010).
7. A. A. Abdo and et. al., *ApJS* **187**, 460(April 2010).
8. K. P. Watters, R. W. Romani, P. Weltevrede and S. Johnston, *ApJ* **695**, 1289(April 2009).
9. A. A. Abdo and et. al., *Physical Review Letters* **102**, 181101(May 2009).

# GALACTIC ELECTRONS AND POSITRONS AT THE EARTH: NEW ESTIMATE OF THE PRIMARY AND SECONDARY FLUXES

J. LAVALLE*

*Departamento de Física Teórica (UAM) & Instituto de Física Teórica (UAM/CSIC) Universidad Autónoma de Madrid, Cantoblanco, 28049 Madrid — Spain*
**E-mail: julien.lavalle@uam.es*

*Dipartimento di Fisica Teorica, Università di Torino & INFN, via Giuria 1, 10125 Torino — Italy*

Measurements of a rise in the cosmic positron fraction up to $\sim 100$ GeV by the PAMELA satellite have triggered putative interpretation attempts in terms of dark matter (DM) annihilation or decay, even though generic DM particle models are not expected to significantly contribute to this signal. Such an exotic abundance has almost made (part of) the scientific community forget that conventional astrophysical cosmic-ray (CR) sources, like pulsars, have long been proven to provide natural explanations to this observation, while even other astrophysical mechanisms were recently demonstrated to potentially feed the high energy positron budget. In this proceeding, we show how considering conventional astrophysical sources — supernova remnants (SNRs) and pulsars — in a consistent CR propagation framework can easily fit the CR electron and positron data.

*Keywords*: Galactic cosmic rays; cosmic-ray sources; cosmic-ray propagation.

## Introduction

Indications of an increase in the CR positron fraction above a few GeV have been collected for quite a long time,[1–4] but the recent PAMELA measurements have reached an unprecedented statistics allowing much more detailed analyses.[5] This increase can hardly be explained in terms of secondary positrons[a],[6,7] though peculiar spatial effects are still worth being investigated into more details[8] — see also an original proposal about a possible solar magnetic lens effects.[9]

Antimatter CRs as potential probes of DM annihilation were first proposed in Ref. 10, the main argument being that the astrophysical background is much more rare than its matter counterpart, and is mostly expected to be of secondary origin. Later, as already mentioned in Ref. 11, it was shown that the annihilation of supersymmetric neutralinos, the most popular DM species candidate so far, could hardly generate observable features in the local positron spectrum unless the annihilation rate is substantially boosted,[12] but DM substructures were expected to play such an amplifier role, as suggested in Ref. 13. This conclusion was also derived for other theories, like extra-dimensional DM models (*e.g.* Ref. 14), while it was recently demonstrated that substructures can actually hardly be at the origin of the large required enhancement.[15,16] Still, PAMELA measurements have generated a lot of activity in this field (see *e.g.* Ref. 17).

Although interpretations in terms of standard astrophysical sources were only a few in the past decades,[18–22] they all relied on quite sound physical arguments: pulsars were already predicted to produce a significant amount of $e^+e^-$ pairs only few years after their discovery,[23] which is now indirectly proven from the more recent observations of gamma-rays originating from pulsar magnetospheres. Another way to generate primary-like positrons, also relying on sound astrophysical mechanisms and recently widely surveyed, is to invoke the diffusive acceleration of those secondary positrons created at SNR shocks.[24,25] It therefore seems clear that standard astrophysics can provide natural explanations to what is called the positron *excess*, which strongly lowers the relevance of any other exotic and ad-hoc solution. Nevertheless, we note that fully consistent and complete predictions of primary and secondary CR leptons were barely available until recently. For instance, the popular `GALPROP` numerical code for CR propagation did not include any primary positron source,[6,26] but we stress that the current data on the local positron flux hardly exhibits any excess with respect to a single secondary component,[7,27] only the positron fraction does.

Connected to the positron issue, claims for an excess in the $e^+e^-$ flux followed the release of Fermi data,[28] while it is known for long that the local high energy electron and positron budget must be affected by fluctuations due to local sources.[29,30] In most of cases, such claims relied on considering

[a]For the non-expert reader, a primary cosmic ray is produced — or/and accelerated — at sources, whereas a secondary cosmic ray originates from nuclear interactions between primary CR nuclei and the interstellar gas (ISG).

GALPROP as a standard model, despite the many cautions provided by the authors themselves .[31] Hence, trying to sketch a self-consistent method to derive trustworthy predictions for CR electrons and positrons is likely valuable. In this proceeding, we summarize the method we recently developed in Ref. 32. We first recall some bases of CR propagation, and then review the main aspects of the proposed model before concluding.

**Bases for cosmic-ray propagation**

Reviews and books on CR propagation are numerous, and we refer the reader to *e.g.* Ref. 31 for a recent review, and to Refs. 33 and 34 for valuable books. A detailed description of CR electron[b] propagation can be found in Ref. 32.

Once produced, stable charged CRs in the GeV-TeV energy range may experience different processes. The dominant ones are diffusion in space (due to scattering with magnetic turbulences and to convective winds) and diffusion in momentum (energy losses — negligible for CR nuclei — and diffusive reacceleration — negligible above a few GeV). CR electron propagation in this energy range is almost completely set by spatial diffusion and energy losses, the latter being mainly due to inverse Compton interactions with the interstellar radiation field (ISRF, including the CMB) and the Galactic magnetic field (synchrotron). In steady state and when convection is neglected, the master propagation equation associated with an electron density $\mathcal{N} \equiv dn/dE$ reads

$$-\vec{\nabla}\left\{K(E,\vec{x})\,\vec{\nabla}\mathcal{N}(E,\vec{x})\right\} - \partial_E\left\{b(E)\,\mathcal{N}(E,\vec{x})\right\} = \mathcal{Q}(E,\vec{x})\,, \tag{1}$$

where $\mathcal{Q}$ is the source term, $K(E,\vec{x})$ is the diffusion coefficient, and $b(E) = -dE/dt$ is the energy loss rate; being set by Compton interactions, the latter strongly increases with energy ($b(E) \propto E^2$ in the Thomson approximation). The typical energy loss timescale is $\tau_l \approx 10^{16}$s at 1 GeV for Galactic electrons. Besides, the diffusion coefficient is usually considered as homogeneous and modeled as a power law, $K(E) \simeq K_0(E/E_0)^\delta$. The normalization and index can be determined locally from observed secondary-to-primary ratios of stable nuclei species which mostly depend on $K_0/L$, where $L$ is the half-thickness of the diffusion zone[36,37] — values are found close to $K_0(E_0 = 1\,\mathrm{GeV}) \approx 3.4 \times 10^{27}\mathrm{cm}^2/\mathrm{s}$, $\delta \approx 0.7$ and $L \approx 4$ kpc. The above equation (without spatial boundaries) admits a solution in terms of

[b]The term *electrons* will refer to both electrons and positrons when discussing general propagation features.

a Green function

$$\mathcal{G}(E, \vec{x} \leftarrow E_s, \vec{x}_s) = \frac{\exp\left\{-\frac{(\vec{x}_s - \vec{x})^2}{\lambda^2}\right\}}{b(E)\,(\pi\,\lambda^2)^{\frac{3}{2}}} \text{ with } \lambda^2 \equiv 4\int_E^{E_s} dE'\,\frac{K(E')}{b(E')}\,, \quad (2)$$

where $\lambda$ is the mean propagation scale, of the order of $\sqrt{K_0\,\tau_l} \approx 2$ kpc, *i.e.* usually less than $L$ (spatial boundaries can be neglected to a good approximation): due to efficient energy losses, electrons with energies above a few tens of GeV are mostly local. This Green function allows the computation of the electron density at the Earth given a source term $\mathcal{Q}(E_s, \vec{x}_s)$ which features the spatial and energy distributions before propagation.

Primary and secondary CRs of astrophysical origin are injected in the Galactic plane, where the standard sources (SNRs and pulsars) and the ISG are located. For secondary positrons, one can approximate the source term as a power law spectrum of index $\gamma_s$, reminiscent from the CR nuclei index, and homogeneously distributed in the Galactic plane, so that $\mathcal{Q}_s \propto E^{-\gamma_s}\delta(z)$. Thus (in the Thomson approximation), the secondary positron flux can be derived analytically: $\phi_s \propto \sqrt{\tau_l/K_0}\,E^{-\tilde{\gamma}_s}$, where $\tilde{\gamma}_s = \gamma_s + 0.5(\alpha + \delta - 1)$ — $\alpha \equiv \ln(b(E)/b(E_0)/\ln(E/E_0)$, *i.e.* $\alpha = 2$ in the Thomson approximation. This reasoning also holds for high energy primaries when neglecting the discreteness of the local source distribution. By using the index inferred from CR nuclei interactions with the ISG, *i.e.* $\gamma_s \simeq 2.75$, one readily gets $\tilde{\gamma}_s \simeq 3.45$, close to the observed positron-only index.[7,32] Nevertheless, predictions for the secondary positron flux fail to fit the positron fraction data — $f_{e^+} \equiv \phi_{e^+}/(\phi_{e^-} + \phi_{e^+})$ — leading instead to a fraction decreasing with energy.[7,32] Further taking into account the hardening in the proton spectrum above $\sim 2$ TeV recently observed by the CREAM experiment[38] results in a slightly harder secondary positron spectrum, but is obviously not enough to explain the measured increase.[27] Predictions for the secondary positron fraction are displayed in Fig. 1, with a conservative range for theoretical uncertainties in the propagation parameters — these predictions can be reproduced from the fitting functions provided in Ref. 27.

**Towards self-consistent predictions for the electron and positron fluxes**

Pulsars have long been proposed as sources of CR positrons, and were even demonstrated, more than 15 years ago, to provide a good fit to the observed positron fraction above 5 GeV.[18–22] Since there are also other astrophysical mechanisms beside pulsars that could feed the positron budget, the question is not really whether standard astrophysics can explain the data, but what

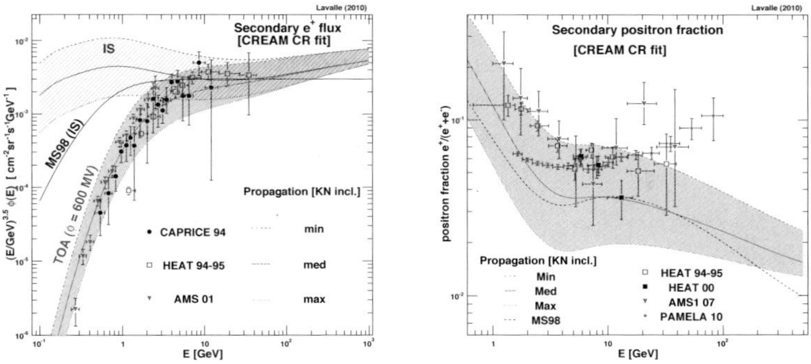


Fig. 1. Left: Predictions for the secondary positron flux with a conservative theoretical uncertainty band for the propagation parameters. Right: corresponding positron fraction.[32] MS98 corresponds to the prediction calculated in Ref. 6.

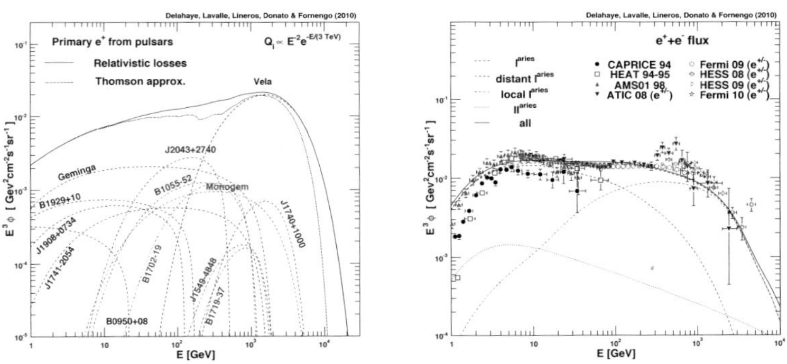


Fig. 2. Left: contribution of local pulsars to the positron flux. Right: example of a self-consistent prediction for both electrons and positrons. These plots are borrowed from Ref. 32.

a consistent CR model should look like. In particular, we can first note that the whole set of data such a model has to fit must also include, beside the positron fraction, the individual spectra of electrons and positrons, and their sum as recently measured by Fermi and HESS up to TeV energies.[28,39]

The lack of primary positrons in the popular GALPROP model illustrates by itself that although a standard paradigm for CR sources and CR propagation is well established,[33,35,40] a *standard model* of Galactic CRs is still far from reached. In fact, if the discussion about the PAMELA data finally pointed out this defect, this is certainly not the only one. For instance, it has long been known that considering a smooth spatial distribution of sources, as it was done in GALPROP for high energy electrons, failed for local predictions, just because the discreteness of local sources has to manifest itself at in the form of spectral fluctuations.[29,30] This other defect is at the origin of many discussions about what was too naively quoted the *Fermi electron excess* by many authors. Yet, this confusion cannot be attributed to the GALPROP model itself, since it was dedicated more to studies on diffuse multi-wavelength Galactic emissions originating from CR interactions with the interstellar medium (ISM), either hadronic and electromagnetic, than to the local electron budget.

The presence of such high energy spectral fluctuations can be understood as follows. The typical energy loss rate for electrons (inverse Compton, synchrotron) is locally of order $b(E) = -dE/dt = E_0/\tau_l \approx 1/300\,(E/1\,\text{GeV})^2$ GeV/Myr in the Thomson approximation.[6,32,35] For a 1 TeV electron, the energy loss timescale is $\tau_{\text{TeV}} = (E = 1\,\text{TeV})/b(E = 1\,\text{TeV}) \approx 0.3$ Myr, which, combined with a diffusion coefficient $K(E) = \beta K_0 (E/E_0)^\delta$, with a normalization of order $K_0 = K(E = E_0 = 1\,\text{GeV}) \approx 10^{-2}\text{kpc}^2/\text{Myr}$,[26,36,37] gives a propagation scale of order $\lambda \approx \sqrt{K(E = 1\,\text{TeV})\,\tau_{\text{TeV}}} \approx 1$ kpc. If we further assume a supernova explosion rate of $\Gamma_\star = 5/100\,\text{yr}$ homogeneously distributed in a thin Galactic disk of radius $R = 20$ kpc, then the typical number of sources expected to contribute to the TeV electron flux is of order $N_\star = (\lambda/R)^2\,\Gamma_\star\,\tau_{\text{TeV}} = \mathcal{O}(10)$, the poissonian fluctuations of which is therefore of the same order.

A consistent prediction of the CR $e^+e^-$ fluxes at the Earth (or at any point in the Galaxy) should at least include the contributions of (i) secondaries, (ii) a smooth distribution of sources for the distant primary component (relevant to the low energy flux), with a radial cut-off set to a distance from the Earth for which the source discreteness effects can safely be averaged out ($\sim 2$ kpc), and (iii) the discrete sources located inside this cut-off radius (relevant to the high energy flux). For the latter contribution, different catalogs are available to constrain the relevant parameters, *e.g.* the position and age estimates.[41,42] We emphasize that for point sources, Klein-Nishina effects must be included in the treatment of energy losses to derive reliable predictions.[30,32]

Beside consistency in the spatial distribution of sources and propagation, there is still a major issue to mention: the injection of electrons from SNRs and pulsars. Indeed, though we can reasonably motivate an averaged power law spectrum associated with an injection rate set by the explosion rate of supernovæ and some energetics considerations for the distant component,[40] an accurate modeling of local sources is of paramount importance, since few of them might dominate the overall flux above $\sim 100$ GeV: the overall spectrum piles up all individual contributions and likely departs from a simple power law at high energy. Individual amplitudes and spectral shapes strongly depend not only on the injected spectra, but also on the source distances and ages.[32] A big uncertainty concerns the dynamics of CR injection into the ISM.[43,44] Nevertheless, the big advantage of local sources is that we have at hand a numerous multi-wavelength observations to build a rather constrained model for each of them.[45] Although a thorough study involving sophisticated modelings for both local pulsars and SNRs remains to be achieved, some preliminary efforts have already been undergone in this direction. In Ref. 32, we have notably demonstrated that considering the observational properties of local sources is sufficient to explain the whole set of data on CR electrons and positrons despite the large theoretical uncertainties, without overtuning the parameters (*e.g.* invoking very hard spectral indices, *etc.*) — see Fig. 2. In this study, few sources are shown to dominate the high energy $e^+e^-$ budget, as expected. This means that purely stochastic approaches which do not include the known local sources are likely strongly overestimating the actual uncertainties in the expected local lepton flux[c], and that, in contrast, other approaches based on refining the observed sources modeling are certainly more promising. This paves the road to further developments.

**Conclusion**

In this contribution, we have discussed the interpretation of the recent data collected on CR electrons and positrons. We have shown that a propagation model self-consistently including (i) a smooth spatial distribution and averaged spectral properties for distant sources (relevant to the low energy flux), (ii) discrete contributions from known local sources with observationally constrained properties (relevant to the high energy flux), (iii) with relativistic energy losses for both components, naturally provides a fairly good fit to all data separately. In particular, few sources may dominate the high energy budget, which is worth to be investigated into more details in the future. Nevertheless, we emphasize that though a standard paradigm

[c]Some Monte Carlo or stochastic approaches are only based on a SN explosion rate and energetic considerations, whereas any random source should be tested as potential observable multi-wavelength emitters, in which case some of them would fall in the contingent of known sources, or would be excluded. Part of the available phase space is therefore already known (observed local sources), an information which is usually not included in these probabilistic treatments.

is now well established for CR propagation and sources, a standard model is still far from reached, notably for Galactic CR electrons for which local effects are very important at high energy. This will have impact in our interpretation of the data on diffuse emissions, for which claims of a few tens percent excesses may have poor physical relevance. Searches for exotic signatures in the CR data are therefore more challenging than expected, and great theoretical and observational efforts will be necessary to really to hope to control the astrophysical backgrounds in the near future. Current fundamental issues such as the dark matter enigma are worth these efforts.

**Acknowledgments:**

We thank the organizers for their warming invitation and welcome. We also thank all the co-authors of Ref. 32, T. Delahaye, R. Lineros, F. Donato and N. Fornengo. We acknowledge support from M. Cirelli and the IPhT CEA Saclay for the participation to the conference. This work was partly supported by the Spanish MICINNs Consolider-Ingenio 2010 Program under grant MultiDark CSD2009-00064, the MICINN under grant FPA2009-08958, the Community of Madrid under grant HEPHACOS S2009/ESP-1473, and the European Union under the Marie Curie-ITN program PITN-GA-2009-237920.

## References

1. J. L. Fanselow, R. C. Hartman, R. H. Hildebrad and P. Meyer, *ApJ* **158**, 771(November 1969).
2. R. L. Golden, B. G. Mauger, S. Horan, S. A. Stephens, R. R. Daniel, G. D. Badhwar, J. L. Lacy and J. E. Zipse, *A&A* **188**, 145(December 1987).
3. S. W. Barwick, J. J. Beatty, A. Bhattacharyya *et al.*, *ApJL* **482**, L191+(June 1997).
4. J. Alcaraz, B. Alpat, G. Ambrosi *et al.*, *Physics Letters B* **484**, 10(June 2000).
5. O. Adriani, G. C. Barbarino, G. A. Bazilevskaya *et al.*, *Nature* **458**, 607(April 2009).
6. I. V. Moskalenko and A. W. Strong, *ApJ* **493**, 694(January 1998).
7. T. Delahaye, F. Donato, N. Fornengo, J. Lavalle, R. Lineros, P. Salati and R. Taillet, *A&A* **501**, 821(July 2009).
8. N. J. Shaviv, E. Nakar and T. Piran, *Physical Review Letters* **103**, 111302(September 2009).
9. J. P. Roberts, *ArXiv e-prints* [arXiv:1005.4668] (May 2010).
10. J. Silk and M. Srednicki, *Physical Review Letters* **53**, 624(August 1984).
11. A. J. Tylka, *Physical Review Letters* **63**, 840(August 1989).
12. E. A. Baltz and J. Edsjö, *Phys. Rev. D* **59**, p. 023511(January 1998).
13. J. Silk and A. Stebbins, *ApJ* **411**, 439(July 1993).

14. J. Lavalle, E. Nezri, E. Athanassoula, F.-S. Ling and R. Teyssier, *Phys. Rev. D* **78**, 103526(November 2008).
15. J. Lavalle, J. Pochon, P. Salati and R. Taillet, *A&A* **462**, 827(February 2007).
16. J. Lavalle, Q. Yuan, D. Maurin and X.-J. Bi, *A&A* **479**, 427(February 2008).
17. M. Cirelli, M. Kadastik, M. Raidal and A. Strumia, *Nuclear Physics B* **813**, 1(May 2009).
18. J. Arons, Particle acceleration by pulsars, in *Origin of Cosmic Rays*, ed. G. Setti, G. Spada, & A. W. Wolfendale, IAU Symposium, Vol. 941981.
19. A. K. Harding and R. Ramaty, The Pulsar Contribution to Galactic Cosmic Ray Positrons, in *International Cosmic Ray Conference*, , International Cosmic Ray Conference Vol. 21987.
20. A. Boulares, *ApJ* **342**, 807(July 1989).
21. F. A. Aharonian, A. M. Atoyan and H. J. Voelk, *A&A* **294**, L41(February 1995).
22. X. Chi, K. S. Cheng and E. C. M. Young, *ApJL* **459**, L83+(March 1996).
23. P. A. Sturrock, *Nature* **227**, 465(August 1970).
24. E. G. Berezhko, L. T. Ksenofontov, V. S. Ptuskin, V. N. Zirakashvili and H. J. Völk, *A&A* **410**, 189(Octtober 2003).
25. P. Blasi, *Physical Review Letters* **103**, 051104(July 2009).
26. A. W. Strong and I. V. Moskalenko, *ApJ* **509**, 212(December 1998).
27. J. Lavalle, *ArXiv e-prints* [arXiv:1011.3063] (November 2010).
28. A. A. Abdo, M. Ackermann, M. Ajello *et al.*, *Physical Review Letters* **102**, 181101(May 2009).
29. C. S. Shen, *ApJL* **162**, L181+(December 1970).
30. T. Kobayashi, Y. Komori, K. Yoshida and J. Nishimura, *ApJ* **601**, 340(January 2004).
31. A. W. Strong, I. V. Moskalenko and V. S. Ptuskin, *Annual Review of Nuclear and Particle Science* **57**, 285(November 2007).
32. T. Delahaye, J. Lavalle, R. Lineros, F. Donato and N. Fornengo, *A&A* **524**, A51+(December 2010).
33. V. S. Berezinskii, S. V. Bulanov, V. A. Dogiel and V. S. Ptuskin, *Astrophysics of cosmic rays* (Amsterdam: North-Holland, 1990, edited by Ginzburg, V.L., 1990).
34. M. S. Longair, *High energy astrophysics. Vol.1: Particles, photons and their detection* (High Energy Astrophysics, by Malcolm S. Longair, pp. 436. ISBN 0521387736. Cambridge, UK: Cambridge University Press, March 1992., March 1992).
35. M. S. Longair, *High energy astrophysics. Vol.2: Stars, the galaxy and the interstellar medium* (Cambridge: Cambridge University Press, —c1994, 2nd ed., 1994).
36. D. Maurin, F. Donato, R. Taillet and P. Salati, *ApJ* **555**, 585(July 2001).
37. A. Putze, L. Derome and D. Maurin, *A&A* **516**, A66+(June 2010).
38. H. S. Ahn, P. Allison, M. G. Bagliesi *et al.*, *ApJL* **714**, L89(May 2010).
39. F. Aharonian, A. G. Akhperjanian, U. Barres de Almeida *et al.*, *Physical Review Letters* **101**, 261104(December 2008).
40. M. A. Malkov and L. O'C Drury, *Reports on Progress in Physics* **64**,

429(April 2001).
41. R. N. Manchester, G. B. Hobbs, A. Teoh and M. Hobbs, *AJ* **129**, 1993(April 2005).
42. D. A. Green, *Bull. Astron. Soc. Ind.* **37**, 45(March 2009).
43. L. O'C. Drury, *ArXiv e-prints* [arXiv:1009.4799] (September 2010).
44. P. Blasi and E. Amato, *ArXiv e-prints* [arXiv:1007.4745] (July 2010).
45. V. Tatischeff, *A&A* **499**, 191(May 2009).

# The 'PAMELA anomaly' indicates a nearby cosmic ray accelerator

P. MERTSCH* and S. SARKAR

*Rudolf Peierls Centre for Theoretical Physics, University of Oxford,*
*Oxford OX1 3NP, UK*
**E-mail: p.mertsch1@physics.ox.ac.uk*

We discuss the recently observed 'excesses' in cosmic ray electron and positron fluxes which have been widely interpreted as signals of dark matter. By considering the production and acceleration of secondary electrons and positrons in nearby supernova remnants, we predict an additional, harder component that becomes dominant at high energies. The unknown spatial distribution of the supernova remnants introduces a stochastic uncertainty which we estimate analytically. Fitting the prediction for different source distributions to the total electron + positron flux measured by Fermi–LAT fixes all free parameters and allows us to 'postdict' the rise in the positron fraction seen by PAMELA. A similar rise in the B/C ratio is predicted at high energies.

*Keywords*: dark matter indirect detection, galactic cosmic rays

## 1. Introduction

The recent findings of a rise in the positron fraction by PAMELA[1] and an excess in the total $e^+ + e^-$ flux by Fermi–LAT[2] have generated a lot of interest because these might be signals of dark matter annihilation or decay. However the expected signals are much smaller than those observed and models which attempt to circumvent this by invoking exotic new physics are increasingly constrained by $\gamma$–ray and radio observations. It is thus important to investigate more prosaic astrophysical explanations, e.g. nearby pulsars[3,4] or supernova remnants (SNRs). In this context we wish to revisit the predictions of the standard SNR origin model of galactic cosmic rays (GCRs) wherein only primary electrons and nuclei are accelerated in SNRs by diffusive shock acceleration (DSA).[5]

We consider the recent proposal[6] that some of the secondary $e^+$ made *in* the SNRs are also accelerated by DSA. By accounting for the discrete distribution of the SNRs, we are then able to match the predicted fluxes to the measured $e^+ + e^-$ flux. Having thus fixed all the free parameters of our model we can make an independent 'postdiction' for the positron fraction which agrees remarkably well with the PAMELA data. We also predict a similar rise for the B/C ratio as a definitive test of our model.

## 2. The discreteness of sources

Although SNRs can grow to become quite large (over $\sim 100\,\text{pc}$ in diameter) towards the end of their lifetime, they are still much smaller than the Kpc scales relevant for GCR propagation. Furthermore, the bulk of the particles accelerated by a SNR will be released towards the end of the Sedov–Taylor phase when they can no longer be confined by magnetic turbulence. We are therefore not dealing with a continuous distribution of sources within the galactic disk but with a large number of *discrete* sources, both in space and time. However, as we do not know the exact distribution, this discreteness introduces an uncertainty in the predicted total $e^+ + e^-$ flux. However, a generalised central limit theorem and the analytical form of the Green's function for the diffusion–energy loss problem can be used to infer the statistical properties of the flux, i.e. its average and quantiles. Our results are in accordance with earlier studies using the Monte Carlo approach.[7–9]

The flux from a source $i$ that injected $e^-$ or $e^+$ a time $t_i$ ago at a distance $L_i$ from the observer is given by the Green's function[10] $G_{\text{disk}}(E, L_i, t_i)$ of the diffusion equation: $J_i(E) = c/(4\pi) G_{\text{disk}}(E, L_i, t_i)$. The flux $J(E)$ of $N$ identical sources at distances $\{L_i\}$ and times $\{t_i\}$ is just the sum of the individual fluxes,

$$J = \sum_{i=1}^{N} J_i(E) = \frac{c}{4\pi} \sum_{i=1}^{N} G_{\text{disk}}(E, L_i, t_i)\,. \tag{1}$$

As a function of the random variables $L$ and $t$, the Green's function $Z \equiv G_{\text{disk}}(E, L, t)$ is itself a random variable with probability density $f_Z$, expectation value $\mu_Z$ and standard deviation $\sigma_Z$. If the central limit theorem were applicable, the fluxes $J$ for different realisations of the same source density at a fixed energy $E$ would follow a normal distribution with mean $\mu_J = c/(4\pi) N \mu_Z$ and standard deviation $\sigma_J = c/(4\pi)\sqrt{N}\sigma_Z$.

In the case under consideration, $L$ and $t$ are assumed to be independent random variables with probability densities $f_L$ and $f_t$, respectively. We choose the sources to be distributed homogeneously in a ring with inner and outer radius $r_1$ and $r_2$, so $f_L = 2L/(r_2^2 - r_1^2)$ for $r_1 \leq r < r_2$ and 0 otherwise. For simplicity we also assume the source rate to be constant in time, i.e. $f_t = 1/t_{\text{max}}$ for $0 \leq t \leq t_{\text{max}}$ and 0 otherwise, up to a maximum time set by the minimum energy $E_{\rm min}$ under consideration: $t_{\rm max} = (b_0 E_{\rm min})^{-1}$. Here, $b_0$ is the normalisation of the energy loss rate of the $e^-/e^+$: $b(E) = b_0 E^2$.

For $E > 10\,\mathrm{GeV}$ and $r_1 \to 0$ (the distance to the nearest source is not limited physically) we find the expectation value:

$$\mu_J = \frac{c}{4\pi} \frac{1}{\sqrt{4D_0 b_0 (1-\delta)}} \frac{N}{\pi r_2^2} \frac{Q_0}{t_{\rm max}} E^{-\gamma-1+(1-\delta)/2} \frac{\Gamma\left(\frac{\gamma-1}{1-\delta}\right)}{\Gamma\left(\frac{2\gamma-\delta-1}{2(1-\delta)}\right)}\,. \tag{2}$$

where the interstellar diffusion coefficient $D_0 = 10^{28}\,\mathrm{cm^2\,s^{-1}}$ at $E = 1\,\mathrm{GeV}$, and $\delta \simeq 0.6$ is its spectral index ($D \propto E^\delta$). We set $b_0 = 10^{-16}\,\mathrm{GeV^{-1}\,s^{-1}}$ and $r_2 = 15\,\mathrm{kpc}$ as is standard. The total number $N$ of sources that are needed to reproduce the (observed) number $\mathcal{N} \simeq 300$ of SNRs active in the Galaxy at any given time depends on the average lifetime of a SNR, $\tau_{\rm SNR}$, which is suggested to be $\sim 10^4\,\mathrm{yr}$,[11] hence $N = \mathcal{N} t_{\rm max}/\tau_{\rm SNR} = 10^6$. With $Q_0 = 8.4 \times 10^{49}\,\mathrm{GeV^{-1}}$, the expectation value,

$$E^3 \mu_J \simeq 150\,\mathrm{GeV^{-1}\,cm^{-2}\,s^{-1}\,sr^{-1}}\,, \tag{3}$$

closely matches the featureless $E^{-3}$ spectrum measured by Fermi–LAT.[2]

Whereas the expectation value $\mu_J$ is well-defined, the variance $\sigma_J^2$ diverges because $f_Z(z)$ has a long power law tail. Nevertheless, instead of the central limit theorem, one can apply a generalised version[12] which allows determination of the distribution of $J$ from the asymptotic behaviour of $f_Z(z)$ for $z \to \infty$. The distribution of $J$ is a so-called stable distribution[13] — an asymmetric generalisation of a Gaussian which also exhibits a power law behaviour for large values of $J$. The uncertainty interval around the expectation value $\mu_J = c/(4\pi) N \mu_Z$ can then be defined by quantiles of the stable distribution. The energy dependence of the quantiles turns out to be always harder than that of the expectation value, so the fluctuations are growing with energy. Figure 1 shows the uncertainty bands around the expectation value for the flux $\mu_J$, together with the expected fluxes from 50 random realisations of the above source distribution.

To calculate the fluxes of $e^-$ and $e^+$ at Earth, we perform a Monte Carlo calculation by considering a large number of realisations of randomly distributed sources according to a probability density function which reflects our present knowledge concerning the distribution of SNRs in the Galaxy. We emphasise that the better the expected flux of $e^-$ and $e^+$ from such a realisation of the source density matches the measured fluxes, the closer is the underlying distribution of sources likely to be to the actual one. We do not consider any variations between the SNRs but assume a prototypical

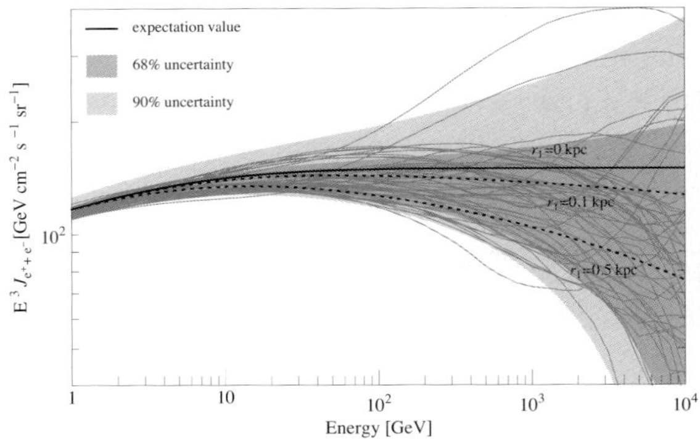


Fig. 1. Fluxes of high energy $e^-$ from ensembles of sources uniformly distributed in a disk around the observer. The solid line denotes the expectation value for the sum of fluxes from $N$ discrete, transient sources while the dashed lines show the expectation values if the sources are limited to a ring with inner radius $r_1$ (normalised to the expectation value for $r_1 = 0$ at 1 GeV). The coloured bands quantify the fluctuations and contain, respectively, 68 % and 90 % of the calculated fluxes. The fluxes from 50 random realisations of an ensemble of $N$ individual sources are shown by the thin grey lines.

set of source parameters which we determine from a compilation of $\gamma$-ray SNRs, see Sec. 3. Of course all SNRs are not the same, however variations of the source parameters would only introduce additional fluctuations in the fluxes without altering their average.

For a more realistic distribution of source distances, we model the SNR density by a logarithmic spiral tracing the gas density (see Fig. 2) and also include the known radial distribution of SNRs in the Galaxy from radio surveys. We transform to a coordinate system centred on the Sun and integrate over azimuth such that our calculated probability density (Fig. 3) encodes the average surface density of SNRs at a particular distance.

## 3. Primary electrons

The injection of primary $e^-$ by SNRs is parametrised as:

$$R_{e^-} = R^0_{e^-} \left(\frac{E}{\text{GeV}}\right)^{-\gamma} \mathrm{e}^{-E/E_{\text{cut}}} . \tag{4}$$

The spectral index $\gamma$ and the cut-off energy $E_{\text{cut}}$ can be obtained from the spectral indices of SNRs in $\gamma$-rays as measured by Imaging Air Cerenkov Telescopes (IACTs), like HESS, MAGIC and VERITAS. (We assume the same spectral index for the hadronic and the electronic components, as predicted for DSA.) We have compiled[14] a list of all SNRs detected by IACTs

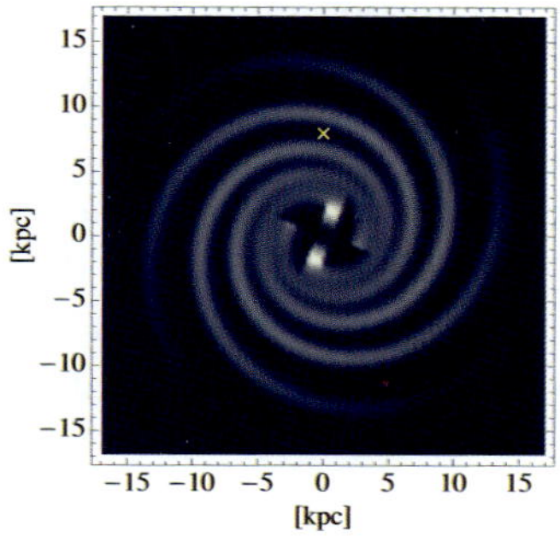


Fig. 2. The adopted distribution of SNRs in the Galaxy — the cross denotes the position of the Sun.

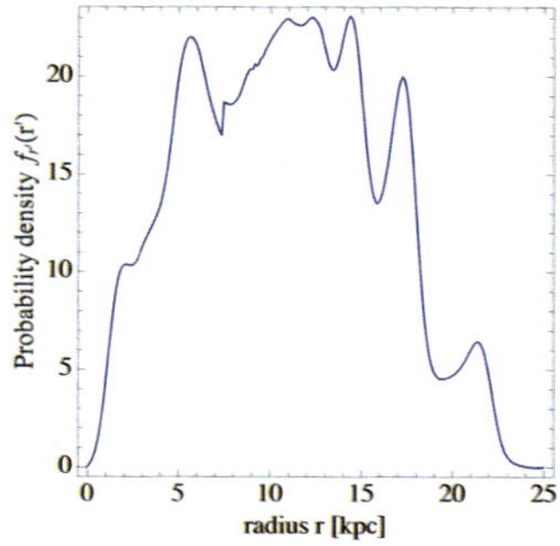


Fig. 3. The probability density for the distance of a SNR from the Sun.

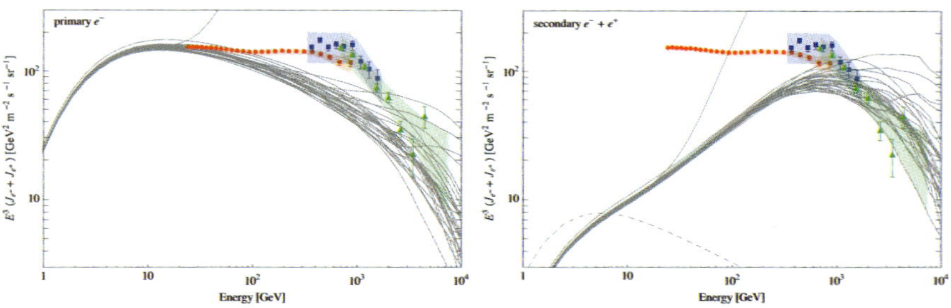


Fig. 4. Predicted spectra of $e^-$ and $e^+$ with data from Fermi–LAT[2] (red circles) and HESS[17,18] (blue squares & green triangles). **Left:** Primary $e^-$ after propagation to Earth. **Right:** Secondary $e^-$ and $e^+$ from cosmic ray interactions, created during propagation (dashed line) and created during acceleration in SNRs (full lines).

and find the typical values to be $\gamma = 2.4$ and $E_{\text{cut}} = 20\,\text{TeV}$. The normalisation $R^0_{e^-}$ is determined by fitting the electron flux at Earth from our Monte Carlo computation to the preliminary measurement by PAMELA at 10 GeV;[15] the secondary fluxes can be neglected for this purpose. We find $R^0_{e^-} = 1.8 \times 10^{50}\,\text{GeV}^{-1}$ for $\gamma = 2.4$ which corresponds to a total injection energy of $\int \mathrm{d}E\, E\, R_{e^-}(E) \simeq 7 \times 10^{47}\,\text{erg}$. Solar modulation is treated in the spherical force–field approximation[16] with a potential of $\phi = 600\,\text{MV}$.

The primary electron fluxes on Earth from a large number of realisations of the source distribution are shown in the left panel of Fig. 4 and clearly exhibit a deficit with respect to the measurements by Fermi–LAT and HESS.

## 4. Additional positrons from the acceleration of secondaries

It is usually assumed that secondary $e^-$ and $e^+$ are produced only through spallation (mainly on interstellar hydrogen and helium) during the galactic propagation of hadronic cosmic rays. At energies above 10 GeV, the spectrum of such secondaries after propagation should be softer than that of the primaries (by the power $\delta$, the spectral index of the diffusion coefficient), and the positron fraction should therefore be falling. We calculate secondary $e^+$ from propagation of protons and nuclei following Ref. 19, using however the propagation parameters specified above.

It was recently suggested[6] that the acceleration of secondary $e^-$ and $e^+$ in the cosmic ray sources, i.e. SNRs, could lead to a harder source spectrum of $e^+$ that can explain the observed rise in the positron fraction. As the volume in which particles that participate in DSA grows with the diffusion coefficient $D_{\text{Bohm}}(p) \propto p$, the fraction of the secondaries which can participate in DSA also increases linearly with momentum. The resulting source spectrum $R_\pm$ of secondary $e^-$ and $e^+$ is then a sum of two power laws, corresponding to the unaccelerated and the accelerated secondaries:

$$R_\pm \simeq R^0_\pm \, p^{-\gamma} \left[ 1 + \left( \frac{p}{p_{\text{cross}}} \right) \right] , \tag{5}$$

where the "cross-over" momentum, $p_{\text{cross}}$, satisfies

$$D(p_{\text{cross}}) = \frac{3}{4} \frac{r u_1^2 \tau_{\text{SNR}}}{(\gamma + 2)(1/\xi + r^2)} . \tag{6}$$

As has been noted,[6] this mechanism is most efficient for *old* SNRs where field amplification by the shock wave is not very effective anymore. We therfore introduce a fudge factor $K_{\text{B}}$ that parameterises the effect of the smaller field amplification on the otherwise Bohm-like diffusion coefficient in the SNR,

$$D(E) = 3.3 \times 10^{22} K_{\text{B}} \left( \frac{B}{\mu\text{G}} \right)^{-1} \left( \frac{E}{\text{GeV}} \right) \text{cm}^2\text{s}^{-1} . \tag{7}$$

A break in the source spectrum occurs at $p_{\text{break}}$ because the growth of the acceleration zone is bounded by the physics size of the SNR. The source spectrum $R_\pm$ thus returns to a $p^{-\gamma}$ dependence around $p = p_{\text{break}}$. At even higher energies the secondary spectrum cuts off at the same $E_{\text{cut}}$ as the primary $e^-$ (see Sec. 3).

Following Refs. 6,20, the parameters are chosen to be: $u_1 = 0.5 \times 10^8\,\text{cm}\,\text{s}^{-1}$, $n_{\text{gas},1} = 2\,\text{cm}^{-3}$, $B = 1\,\mu\text{G}$. Instead of fixing the normalisation of the injection spectrum *ad hoc*, we determine it from $\gamma$–ray observations

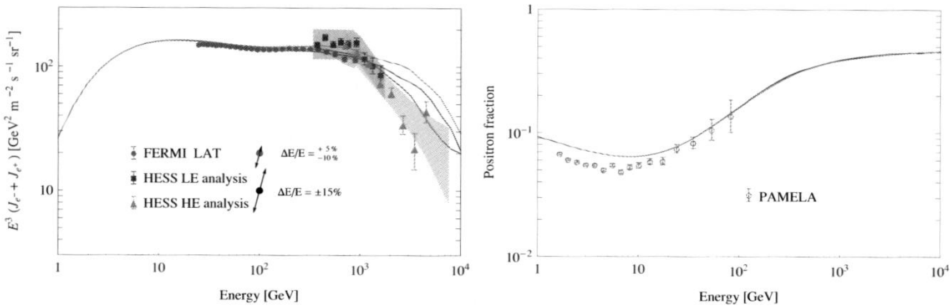


Fig. 5. **Left:** Predicted spectra of the total flux of primary and secondary $e^-$ and $e^+$. **Right:** The predicted positron fraction after fixing all free parameters using the total $(e^+ + e^-)$ flux, compared with data from PAMELA.[1]

of SNRs. Knowing the cross-sections for the production of $\gamma$-rays (from the decay of neutral pions) and the production of secondary $e^-$ and $e^+$ (from the decay of charged pions), we determine the normalisation

$$R_+^0 = 7.4 \times 10^{48} \left(\frac{\tau_{\rm SNR}}{10^4 {\rm yr}}\right)\left(\frac{Q_\gamma^0}{5.7 \times 10^{33} {\rm s}^{-1} {\rm TeV}^{-1}}\right) {\rm GeV}^{-1}. \tag{8}$$

where $Q_\gamma^0$ is the observed typical luminosity of SNR in $\gamma$-rays as determined from the same compilation of data as above.

The right panel of Fig. 4 shows the expected flux of secondary $e^-$ and $e^+$ for 30 realizations of the possible distribution of SNRs in our Galaxy. Clearly this component can potentially match the high energy Fermi–LAT and HESS data. We also show the secondary $e^-$ and $e^+$ from the propagation of protons and nuclei, which is subdominant.

## 5. Total electron–positron flux and positron fraction

Figure 5 shows the total $(e^+ + e^-)$ flux obtained by adding the primary $e^-$ and the secondary $e^-$ and $e^+$ (see. Fig. 4). As explained above, we selected just those (3) realisations of the source distribution which give the best fit to the measurements by Fermi–LAT and HESS. By fitting to the $(e^+ + e^-)$ flux we have also fixed the value for the diffusion coefficient near the SNR shock wave (which determines the ratio of the accelerated to the unaccelerated secondaries) to about 15 times the Bohm value i.e. $K_B \simeq 15$.

Having thus fixed all free parameters we can make an independent prediction for the positron fraction which is shown in the right panel of Fig. 5. It is in very good agreement with the PAMELA measurements above $\gtrsim 10\,$GeV; we do not expect to match the data below 10 GeV be-

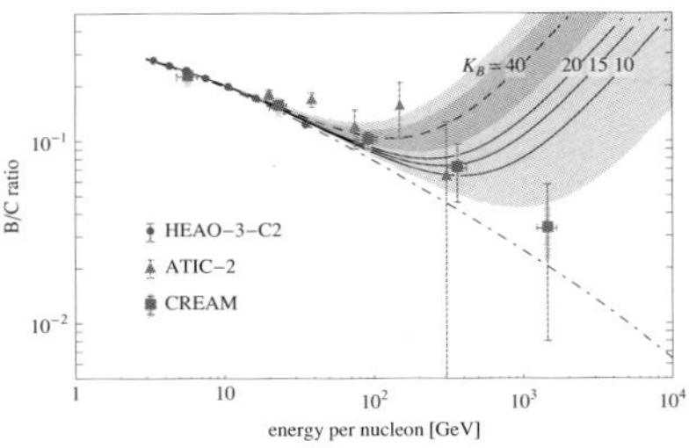


Fig. 6. The predicted B/C ratio in cosmic rays (from Ref. 22) for the 'leaky box model' with production of secondaries during propagation only (dot-dashed line), and including production and acceleration of secondaries in nearby SNRs (solid lines) for values of the diffusion coefficient near the shock wave which best fit the $e^{\pm}$ spectrum (see Sec. 5). The dashed line corresponds to the value of the diffusion co-efficient required to fit ATIC-2 data on Ti/Fe (from Ref. 22), along with its $1\sigma$ and $2\sigma$ error bands. The data points are from HEAO-3-C2 (circles),[24] ATIC-2 (triangles)[25] and CREAM (squares).[26]

cause our analytic approach neglects complications such as convection and reacceleration in the ISM which can be important at these energies.

## 6. Nuclear secondary-to-primary ratios

The acceleration of secondaries described above modifes not only the spectra of secondary $e^-$ & $e^+$ but also of other charged secondaries. This provides a direct test of this explanation for the rise in the positron fraction, e.g. a rise in the antiproton-to-proton ratio is also predicted[20] (consistent with the measurements by PAMELA[21] so far). We have argued[22] that *nuclear* secondary-to-primary ratios can be used to not only test this model but also to discriminate against alternative explanations such as dark matter annihilation/decay or nearby pulsars which cannot affect nuclear abundances.

We have calculated the rise in nuclear secondary-to-primary ratios expected from the acceleration of secondaries, in particular, we have used recent ATIC-2 data on the titanium-to-iron ratio which exhibit a *rise* around 100 GeV/n to fix the diffusion coefficient. We find a 'fudge factor' of $K_B \simeq 40$ is necessary, albeit with large error bars due to the limited statistics of the ATIC-2 data. Again, we can make an independent prediction for a different secondary-to-primary ratio, namely boron-to-carbon (B/C). Our prediction, together with the canonical expectation for a purely secondary origin through spallation in the ISM, are shown in Fig. 6 with a selection of data. Currently, B/C is being measured by PAMELA, and it is also a prime goal of the upcoming AMS-02 mission.[23]

## 7. Conclusion

We have discussed the recently measured excesses in the $e^+$ fraction and in the total $e^+ + e^-$ flux in galactic cosmic rays. The fluctuations induced by the discreteness of the cosmic ray sources has been estimated assuming these to be SNRs. However when the source spectral index is estimated from $\gamma$-ray observations of SNRs, the expected total $e^+ + e^-$ flux at Earth is deficient. Adding the flux of secondary $e^-$ and $e^+$ produced and accelerated in nearby SNRs brings the model prediction back in agreement with the measurements and naturally implies a rise in the $e^+$ fraction. A crucial test of this idea is to determine if nuclear secondary-to-primary ratios, e.g. B/C, also increase with energy. Such nearby SNRs ought also to be detected by the IceCube neutrino observatory in a few years.[22] This would be the first direct astronomical identification of the sources of galactic cosmic rays.

## References

1. O. Adriani *et al.*, *Nature* **458**, 607 (2009).
2. A. A. Abdo *et al.*, *Phys. Rev. Lett.* **102**, p. 181101 (2009).
3. F. Aharonian, A. Atoyan and H. Volk, *Astron. Astrophys.* **294**, L41 (1995).
4. D. Hooper, P. Blasi and P. D. Serpico, *JCAP* **0901**, p. 025 (2009).
5. M. A. Malkov and L. O'C Drury, *Rept. Prog. Phys.* **64**, 429 (2001).
6. P. Blasi, *Phys. Rev. Lett.* **103**, p. 051104 (2009).
7. M. Pohl and J. A. Esposito, *Astrophys. J.* **507**, p. 327 (1998).
8. A. W. Strong and I. V. Moskalenko (2001).
9. S. P. Swordy, in *Proceedings of the 28th ICRC, Tsukuba, Japan*, 2003.
10. V. L. Ginzburg *et al.* Amsterdam, Netherlands: North-Holland (1990) 534 p.
11. S. P. Reynolds, *Annu. Rev. Astron. Astrophys.* **46**, p. 89 (2008).
12. B. V. Gendenko and A. N. Kolmogorov, *Limit Distributions for Sums of Independent Random Variables* (Addison-Wesley, Cambridge, MA, USA, 1954).
13. J. P. Nolan, *Stable Distributions - Models for Heavy Tailed Data* (Birkhäuser, Boston, 2010).
14. M. Ahlers, P. Mertsch and S. Sarkar, *Phys. Rev.* **D80**, p. 123017 (2009).
15. E. Mocchiutti, (2009), Talk at 2nd RICAP, Rome, 13-15 May 2009.
16. L. J. Gleeson and W. I. Axford, *Astrophys. J.* **154**, p. 1011 (1968).
17. F. Aharonian *et al.*, *Phys. Rev. Lett.* **101**, p. 261104 (2008).
18. F. Aharonian *et al.*, *Astron. Astrophys.* **508**, p. 561 (2009).
19. T. Delahaye *et al.*, *Astron. Astrophys.* **501**, 821 (2009).
20. P. Blasi and P. D. Serpico, *Phys. Rev. Lett.* **103**, p. 081103 (2009).
21. O. Adriani *et al.*, *Phys. Rev. Lett.* **102**, p. 051101 (2009).
22. P. Mertsch and S. Sarkar, *Phys. Rev. Lett.* **103**, p. 081104 (2009).
23. AMS Webpage, `http://ams.cern.ch/`.
24. J. J. Engelmann *et al.*, *Astron. Astrophys.* **233**, 96(July 1990).
25. A. D. Panov *et al.* (2007), Proc. 30th ICRC, Merida.
26. H. S. Ahn *et al.*, *Astropart. Phys.* **30**, 133 (2008).

# Observations of Intermediate Synchrotron Peaked blazars with the Fermi-LAT

C. MONTE * on behalf of the Fermi-LAT collaboration

*Dipartimento Interateneo di Fisica "M.Merlin", Università e Politecnico di Bari and Istituto Nazionale di Fisica Nucleare, Sezione di Bari, Via Amendola 173, 70126 Bari, Italy*

After the first year of scientific activity with the Large Area Telescope (LAT), the primary instrument onboard *Fermi*, a catalog of Active Galactic Nuclei has been published. The 1st LAT AGN Catalog (1LAC) includes 671 gamma-ray sources located at high Galactic latitudes, detected with a test statistic greater than 25 and associated statistically with 709 AGNs (due to some multiple associations). The multi-wavelength quasi-simultaneous Spectral Energy Distributions (SEDs) of these blazars show the typical two-bump signature attributed to synchrotron and inverse Compton emission. The study of the position of the first peak led to a new SED-based classification of blazars as Low Synchrotron Peaked (LSP), Intermediate Synchrotron Peaked (ISP) and High Synchrotron Peaked (HSP) blazars.

A particular and interesting group of objects is composed of ISP blazars, i.e. sources for which the synchrotron emission peaks at intermediate frequencies (from $10^{14}$ Hz to $10^{15}$ Hz). These objects are expected to have their SED high-energy peak centered on the *Fermi*-LAT band (from 20 MeV to 300 GeV). The brightest 1LAC ISP sources have been selected and they have been analyzed covering a period of 22 months of *Fermi*-LAT gamma-ray data in order to investigate their spectral features and to characterize the temporal evolution of their gamma-ray spectra.

*Keywords*: Gamma rays astronomical observations, Active Galactic Nuclei

## 1. Introduction

The Gamma-ray Large Area Space Telescope was launched on June 11, 2008 into a low Earth circular orbit at an altitude of 550 km and an inclination of 25.6°. It began its scientific operations two months later, and shortly thereafter, it was renamed the *Fermi Gamma-ray Space Telescope*.

The Large Area Telescope (LAT), the main instrument onboard *Fermi*,

*claudia.monte@ba.infn.it

is a pair-production telescope with large effective area ($\sim 8000\,\text{cm}^2$, on axis for $E > 1\,\text{GeV}$) and field of view ($\sim 2.4\,\text{sr}$ at $1\,\text{GeV}$), sensitive to gamma-rays in the energy range from 20 MeV to more than 300 GeV.[1] Thanks to its major sensitivity with respect to its precedessor EGRET and the Italian Space Agency′s Astro-rivelatore Gamma a Immagini Leggero (AGILE), the LAT is able to study the gamma-ray sky in unprecedented detail. In sky survey mode (*Fermi*′s main observing mode), the LAT observes the entire sky every 3 hours, providing uniform exposure on the timescale of days.

One of the major scientific goals of the *Fermi* mission is to investigate high-energy emission in active galactic nuclei (AGNs). Although AGNs are probably one of the better known class of gamma-ray emitting objects, there are many questions that are yet unanswered, as the mechanisms by which the particles are accelerated, the precise site of the gamma-ray emission and the origin of AGN variability.[2]

## 2. The First Large Area Telescope AGN Catalog (1LAC)

Of the 1043 high-latitude ($|b| > 10°$), gamma-ray sources in the *Fermi* Large Area Telescope First Source Catalog (1FGL),[3] detected in 11 months (from August 4, 2008 to July 4, 2009) of scientific data collected with standard sky-survey observation, the 1st LAT Catalog of Active Galactic Nuclei (1LAC),[2] includes 671 sources detected with a test statistic greater than 25 and statistically associated with 709 AGNs (due to some multiple associations).

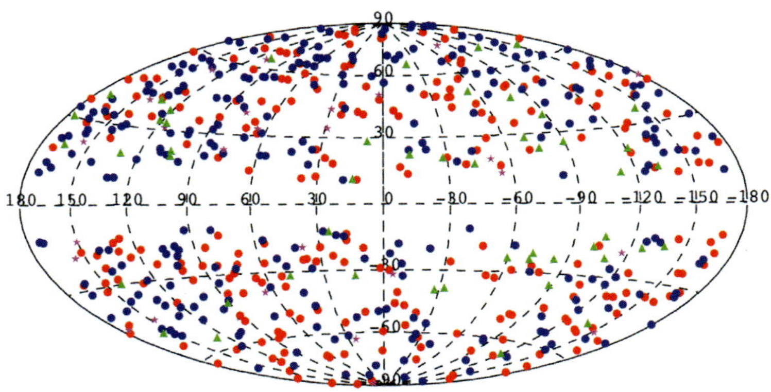


Fig. 1. Locations of the associated sources in the 1LAC. Red full circles: FSRQs, blue full circles: BL Lacs, magenta stars: radio galaxies, green triangles: AGNs of unknown type.[2]

The 1LAC includes 300 BL Lacertae (BL Lacs) blazars, 296 Flat Spectrum Radio Quasar (FSRQs) blazars, 41 AGNs of other types (including radio galaxies) and 72 AGNs of unknown type, distributed in the sky as shown in figure 1. It is clear that the distribution is not isotropic: there are more sources in the northern galactic hemisphere than in the southern one.[2]

## 3. The SED based classification of blazars

AGNs are highly variable objects, therefore simultaneous observations in different energy bands are essential to understand the emission processes that take place in these objects. For 48 sources of the LBAS (LAT Bright AGN Sample)[?] it has been possible to combine *Fermi*-LAT three months (August 4, 2008 to October 31, 2008) gamma-ray data with simultaneous (or quasi-simultaneous) observations at other wavelengths, in order to assemble high-quality and quasi-simultaneous Spectral Energy Distributions (SEDs).[4]

All the SEDs clearly show the typical two-bump signature attributed to synchrotron and inverse Compton emission and the study of this unprecedented collection of blazars has allowed one to estimate a certain number of parameters characterizing the SED of blazars, as the frequency of the synchrotron peak ($\nu_S^{peak}$) and of the inverse Compton peak ($\nu_{IC}^{peak}$). The determination of $\nu_S^{peak}$ for a large number of blazars has permitted one to extend the usual BL Lacs classification[5] (Low Energy peaked BL Lacs and High energy peaked BL Lacs) to all blazars, introducing a a new SED-based classification scheme:

- *Low Synchrotron Peaked (LSP)* blazars, that are sources with the synchrotron peak at low energy i.e. in the far IR or IR band (or $\nu_S^{peak} \leq 10^{14}$ Hz).
- *Intermediate Synchrotron Peaked (ISP)* blazars, i.e. sources for which the synchrotron emission peaks at intermediate frequencies (from $10^{14}$ Hz to $10^{15}$ Hz).
- *High Synchrotron Peaked (HSP)* blazars that have the synchrotron peak at UV or higher energies ($\nu_S^{peak} \geq 10^{15}$ Hz).

## 4. The Intermediate Synchrotron Peaked (ISP) blazars analysis

For ISP blazars, the inverse Compton (high energy) peak in the SED is expected to be exactly within the *Fermi*-LAT energy band. Consequently, looking at the gamma-ray spectra of 1LAC blazars,[2] a preliminary list of about 30 sources showing the SED high energy peak centered in the *Fermi*-LAT energy band has been built. For these sources, that are ISP candidates, LAT 22 months data (from August 4, 2008 to June 4, 2010) have been analyzed in the energy range from 100 MeV to 300 GeV, in order to obtain the spectral energy distributions in the gamma-ray band and the light curves.

### 4.1. *Spectral Energy Distributions*

For the spectral analysis, different methods have been used. In the first case (unfolding method) the spectra have been evaluated using a deconvolution technique[6] based on the Bayes theorem, that allows one to take into account the energy dispersion.

In the second case, spectra have been reconstructed using *Fermi* science tool gtlike,[7] where a parametric model (Power Law, PL) is assumed in each individual energy bin for the source spectrum and for the background components and a maximum likelihood fit is performed for each energy bin.

A very simple analytical function that can model the shape of the peak in the gamma-ray band with nearly symmetric curvature around the maximum is the Log-Parabola (LP).[8] In the third method of analysis, we modeled the source under investigation using the log-parabola function in the whole energy band (from 100 MeV to 300 GeV) and than a maximum likelihood fit has been performed.

As an example, the results of the analysis for two sources are shown in

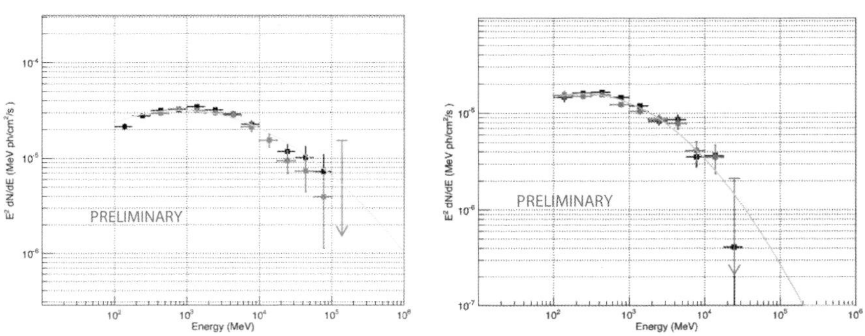


Fig. 2. PKS 0426-380 (left) and B3 0917+449 (right) SED: unfolding method (black points), gtlike with the source modeled with a PL in each energy bin (magenta points) and gtlike with the source modeled with a LP in the whole energy band (cyan line). The upper limits cut has been fixed at a value of $TS = 10$.

figure 2. The plot on the left and the plot on the right show the gamma-rays SED respectively for the PKS 0426-380 (1FGL J0428.6-3756, celestial coordinates: RA = 67.1567°, DEC = −37.94121°) and for the B3 0917+449 (1FGL J0920.9+4441, celestial coordinates RA = 140.2412°, DEC = 44.6888°), obtained with the different methods of analysis: the unfolding method (black points), gtlike with the source modeled with a PL in each energy bin (magenta points) and gtlike with the source modeled with a LP in the whole energy band (cyan line).

It is evident that the results of the different methods are consistent and that the sources can be modeled with a log-parabola function with the peak at about 1 GeV, exactly in the middle of the *Fermi*-LAT energy band.

## 4.2. *Light Curves*

The plots in figure 3 show the light curve for the integral flux greater than 100 MeV for the PKS 0426-380 obtained dividing the entire 22 month data set in time intervals respectively of 1 month each (upper plot) and 15 days each (lower plot).

For each time interval, a maximum likelihood fit has been performed considering a region of interest of 15 degrees around the source under in-

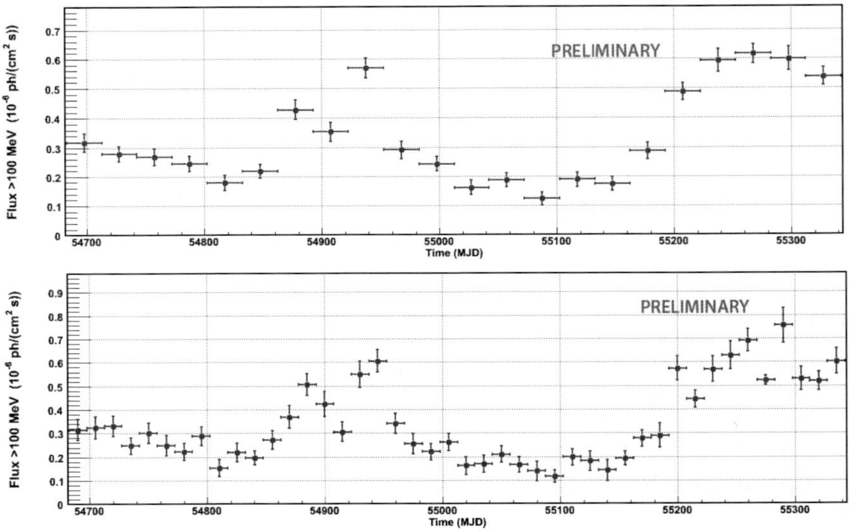


Fig. 3. Light curve for the integral flux greater than 100 MeV for the PKS 0426-380 obtained dividing the entire 22 month data set in time intervals of 1 month each (upper plot) and 15 days each (lower plot).

vestigation and modeling the source under study with a power law with all spectral parameters free, all sources within 10 degrees with a PL with spectral index fixed at the 1FGL catalog value and the normalization parameter free, all sources between 10-15 degrees with a PL with all parameters fixed at the 1FGL catalog values. The light curves shown here are representative of the work done for all sources in the sample of ISP candidates.

## 5. Conclusions

A sample of about 30 ISP candidates has been selected from the 1LAC. Preliminary spectral analysis performed over a period of 22 months seems to show that the log-parabola function can be used to model the shape of the SED high-energy peak in the *Fermi*-LAT gamma-ray energy band for these sources. The study of the temporal evolution has started with the construction of the light curves with 1 month and 15 days time intervals. The analysis results shown here are still preliminary: a further analysis will be performed, especially to investigate any spectral features of these new class of blazars.

## Acknowledgments

The *Fermi* LAT Collaboration acknowledges support from a number of agencies and institutes for both development and the operation of the LAT as well as scientific data analysis. These include NASA and DOE in the United States, CEA/Irfu and IN2P3/CNRS in France, ASI and INFN in Italy, MEXT, KEK, and JAXA in Japan, and the K. A. Wallenberg Foundation, the Swedish Research Council and the National Space Board in Sweden. Additional support from INAF in Italy and CNES in France for science analysis during the operations phase is also gratefully acknowledged.

## References

1. W.B. Atwood et al., *ApJ*, **697**, 1071 (2009)
2. A.A. Abdo et al., *ApJ*, **715**, 429 (2010)
3. A.A. Abdo et al., *ApJS*, **188**, 405 (2010)
4. A.A. Abdo et al., *ApJ* **716**, 30 (2010)
5. P.Padovani & P. Giommi, *ApJ*, **444**, (1995)
6. M.N. Mazziotta, *Proc. of the 31 ICRC*, LODZ (2009)
7. http://fermi.gsfc.nasa.gov/ssc/data/analysis/documentation/Cicerone/
8. E. Massaro et al., *A&A* **448**, 861 (2006)

# Shock acceleration in partially neutral plasmas

G. MORLINO*, E. AMATO, P. BLASI and D. CAPRIOLI

*INAF, Osservatorio Astrofisico di Arcetri,*
*L.go E. Fermi-5, Firenze, 50125, Italy*
**E-mail: morlino@arcetri.astro.it*

We present the non-linear theory of shock acceleration applied to SNRs expanding into partially neutral plasma. Using this theory we show how the Balmer lines detected from young SNRs can be used to test the efficiency of shocks in the production of cosmic rays. In particular we investigate the effect of charge-exchange process between protons and neutral hydrogen occurring in the precursor formed ahead of the shock. In this precursor the CR pressure accelerate the ionized component of the plasma and a relative velocity between protons and neutral hydrogen is established. On the other hand the charge-exchange process tends to equilibrate ions and neutrals resulting in the heating of both components. We show that even when the shock convert only a few per cent of the total bulk kinetic energy into CRs, the heating is efficient enough to produce a detectable broadening of the narrow Balmer lines emitted by the neutral hydrogen.

*Keywords*: Particle acceleration; Supernova Remnant; Interstellar medium.

## 1. Introduction

Supernova remnants (SNRs) are thought to be the primary source of Galactic cosmic rays (CRs). The diffusive shock acceleration (DSA) occurring around the forward shock waves is considered the most promising mechanism for the acceleration of particles up to very high energies. In order to explain the flux of CRs observed at the Earth, the DSA should convert a fraction $\sim 10-30\%$ of the shock kinetic energy into non thermal particles. Indeed, from the theoretical point of view, the DSA can be very efficient in producing accelerated particles, but a clear confirmation of this prediction still lacks, even if many circumstantial evidences have been collected (see e.g. Ref. 1). A possible way to investigate the efficiency of CRs production is through the analysis of the Balmer lines associated with shocks of SNRs which propagate in a partially ionized plasma. The hydrogen lines of a Balmer shocks consist of two superimposed components: the narrow component is emitted by neutral hydrogen after entering the shock front and the broad component by hot protons after undergoing charge exchange with incoming neutral hydrogen atoms. The width of the broad component reflects the proton temperature behind the shock front, while the width of the narrow component provide a direct measurement of the temperature upstream of the shock.[2,3]

We stress that, when the shock is not modified by the presence of CRs, the charge exchange process and the Balmer emission both occurs only downstream of the shock. On the other hand, when CRs are accelerate efficiently, the shock structure is modified and a precursor is generated upstream: the CRs pressure slows down and compresses the ionized plasma. Even if the neutral component is not directly affected by the CRs pressure, a relative velocity between ions and neutral is now established. Hence the charge exchange can also occurs upstream of the shock, resulting in the pre-heating of the neutral hydrogen. We do expect two remarkable consequences: 1) the narrow Balmer lines can be emitted also from the upstream region and 2) the typical width of these lines becomes larger with respect to the case without CRs, because the hydrogen temperature increases.

A further consequence of efficient acceleration is that the temperature of the shocked ions downstream is lower with respect to the case with no acceleration. This occurs because a non negligible fraction of the shock kinetic energy is converted into CRs, rather than into thermal energy. As a consequence also the width of the broad Balmer lines is affected, being reduced with respect to the case with no acceleration.

Remarkably all these aspects have been observed in some Balmer-dominated shocks:1) from a region of the Tycho remnant, the emission of narrow Balmer lines has been detected upstream of the shock;[4] 2) in several SNRs the narrow Balmer lines present a width incompatible with the typical temperature of the interstellar medium;[5] 3) in two different cases where the Balmer emission has been detected, i.e. RCW 86[6] and SNR 0509-67.5,[7] the width of the broad lines led to a downstream temperature which is lower than that estimated form the measurement of the shock proper motion, suggesting that a fraction of the total energy is converted in CRs.

In order to quantify all these effects simultaneously, in this paper we present the solution for the stationary case of DSA when the shock propagates into a partially ionized plasma. We use a semi analytical method, similar to those used in Ref. 8, specialized for plane shock geometry.

## 2. Interaction between ions and neutrals in the precursor

The process of charge exchange can occurs in the precursor only if its typical length scale is smaller than the precursor length. The length of the precursor created by CR pressure is of the order of the diffusion length of the particles with maximum energy, i.e.:

$$L_{prec} \simeq D(E_{\max})/u_0 = 3 \cdot 10^{17} E_{\rm TeV} B_{\mu G}^{-1} u_8^{-1}\,{\rm cm}\,, \quad (1)$$

where the maximum energy is expressed in TeV, the magnetic field in $\mu$G and the shock speed is $u_0 = u_8 10^8$cm/s. We assume that particles diffuse with a Bohm-like diffusion coefficient, $D(p) = (c/3) r_L(p)$, where $r_L = pc/eB$ is the Larmor radius and the magnetic field strength is assumed constant in the precursor. Let call $n_H$ the neutral hydrogen density, the length scale for a proton to make charge exchange upstream is:

$$L_{ce} = u_0 / \left(n_H\, \sigma_{ce}(v_r)\, v_r\right) \simeq 10^{15} (u_0/v_r)(n_H/1{\rm cm}^{-3})^{-1}\,{\rm cm}\,. \quad (2)$$

We note that when the relative speed between ions and neutral hydrogen is $v_r \lesssim 2000$ km/s, the charge exchange cross section is roughly constant.[3] Comparing $L_{ce}$ and $L_{prec}$ we see that a single proton can charge exchange several times in the precursor before crossing the shock. In this work we neglect the role of ionization because the ionization cross section is much smaller than the charge exchange one when $v_r \lesssim 1000$ km/s, but we will deserve a more extensive treatment of ionization in a future paper.

## 3. Hydrodynamics with charge-exchange

The kinetics of a system of neutral hydrogen and protons interacting via charge exchange are described by the kinetic equations for each species:

$$\frac{\partial f_i}{\partial t} + v_x \frac{\partial f_i}{\partial x} = \beta_i(x, \mathbf{v}) f_j(x, \mathbf{v}) - \beta_j(x, \mathbf{v}) f_i(x, \mathbf{v})\,, \quad (3)$$

where $i, j = H, p$ and $f_i$ is the distribution function. Eq. (3) is written in one spatial dimension because we are considering plane shocks propagating along the $x$ direction, as a consequence $f_i$ is only a function of the position $x$ and of the particle velocity. The change exchange frequency is defined by:

$$\beta_i(x, \mathbf{v}) = \int d\mathbf{w}\, |\mathbf{v} - \mathbf{w}| \sigma_{ce}(|\mathbf{v} - \mathbf{w}|) f_i(x, \mathbf{w})\,. \quad (4)$$

Notice that the charge exchange cross section is only a function of the relative speed $v_r \equiv |\mathbf{v}_H - \mathbf{v}_p|$. Now we consider only the stationary solution, hence $\partial f_i/\partial t = 0$. From the integration of first, second and third momenta of Eq. (3) we get the fluid equations for the neutral hydrogen:

$$\frac{\partial}{\partial x}\left[\rho_H u_H\right] = 0\,, \tag{5}$$

$$\frac{\partial}{\partial x}\left[\rho_H u_H^2 + P_H\right] = -q_m\,, \tag{6}$$

$$\frac{\partial}{\partial x}\left[\frac{1}{2}\rho_H u_H^3 + \frac{\gamma}{\gamma-1}P_H u_H\right] = -q_e\,. \tag{7}$$

The term $q_m$ and $q_e$ are the flux of momentum and energy which are transferred from the neutral hydrogen to the protons, i.e:

$$q_m = \int dv_H dv_p\,(v_H - v_p)\,\sigma_{ce}(v_r)v_r f_H(v_H)f_p(v_p) \tag{8}$$

$$q_e = \int dv_H dv_p\,\frac{1}{2}(v_H^2 - v_p^2)\,\sigma_{ce}(v_r)v_r f_H(v_H)f_p(v_p)\,. \tag{9}$$

In order to get an analytic expression for $q_m$ and $q_e$ we follow Ref. 9. First of all we assume that both the hydrogen and protons distribution are Maxwellian, with a bulk velocity $\mathbf{u}_i(x)$, along the $x$ direction, and a temperature $T_i(x)$, i.e. $f_i(x,\mathbf{v}) = \rho_i\,(\pi\,v_{T_i}^2)^{-3/2}e^{-(\mathbf{v}-\mathbf{u}_i)^2/v_{T_i}^2}$, where the thermal speed is defined as $v_{T_i} = \sqrt{2P_i/\rho_i}$. Moreover we note that for $v_r \lesssim 2000$ km/s, the charge exchange cross section decreases only logarithmically with increasing $v_r$, hence we can approximate $\sigma_{ce}$ as a constant evaluated at the average relative speed and pull it out of the integrals. Under this assumptions momentum and energy transfer can be expressed as follows:

$$q_m = \sigma_{ce}(\bar{u})\rho_p\rho_H(u_H - u_p)\left[\bar{u} + v_{T_H}/h_1(H,p) + v_{T_p}/h_1(H,p)\right]\,; \tag{10}$$

$$q_e = \sigma_{ce}(\bar{u})\rho_p\rho_H\frac{1}{2}\left\{\bar{u}(u_H^2 - u_p^2) + \frac{3}{2}\left[v_{T_H}^2 h_2(p,H) - v_{T_p}^2 h_2(H,p)\right] + \right.$$
$$\left. +2(u_H - u_p)\left[\frac{u_H v_{T_H}^2}{h_1(p,H)} + \frac{u_p v_{T_p}^2}{h_1(H,p)}\right]\right\} \tag{11}$$

where the average relative speed $\bar{u}$, and the functions $h_{1,2}$ are defined as:

$$\bar{u} = \left[\frac{4}{\pi}(v_{T_i}^2 + v_{T_j}^2) + (u_i - u_j)^2\right]^{1/2}\,; \tag{12}$$

$$h_1(i,j) = \left[4\left(\frac{4}{\pi}v_{T_i}^2 + (u_i - u_j)^2\right) + \frac{9\pi}{4}v_{T_n}^2\right]^{1/2}\,; \tag{13}$$

$$h_2(i,j) = \left[\frac{4}{\pi}v_{T_i}^2 + \frac{64}{9\pi}v_{T_j}^2 + (u_i - u_j)^2\right]^{1/2}\,. \tag{14}$$

The fluid equations for protons are similar to those for hydrogen, but include the contribution due to CRs. Following Ref. 8 it is easy to show that the flux conservation of mass, momentum and energy are:

$$\frac{\partial}{\partial x}\left[\rho_p u_p\right] = 0 \tag{15}$$

$$\frac{\partial}{\partial x}\left[\rho_p u_p^2 + P_p + P_c\right] = q_m \tag{16}$$

$$\frac{\partial}{\partial x}\left[\frac{1}{2}\rho_p u_p^3 + \frac{\gamma}{\gamma-1}P_p u_p\right] = -u_p\frac{\partial P_c}{\partial x} + q_e\,. \tag{17}$$

where $P_c$ is the CR pressure which is connected to the CR distribution function through $P_c(x) = 4\pi/3\int p^2 dp\, pv f_c(x,p)$. Equations (5)-(7) and (15)-(17) are coupled through $q_m$ and $q_e$ and can be solved numerically to get the pressures and the bulk velocities of both protons and hydrogen as implicit functions of $P_c$. In order to close the system of equations we need to solve also the transport equation for the CR distribution function, $f_c$. All the system can be solved using iterative techniques similar to those used in Ref. 8. We will describe the full procedure in a more detailed work.

It is worth stressing that the technique used allows one to take into account other important effects in the shock acceleration theory, like the magnetic field amplification and the turbulent heating. For the sake of simplicity here we neglect such complications in order to isolate the role of charge exchange.

## 4. Discussion and Conclusions

Now we consider a typical case of shock with the following parameters: shock speed $u_0 = 2000$ km/s, total upstream density $\rho_0 = 1\text{cm}^{-3}$, neutral fraction 50%, upstream temperature $T_0 = 10^4$ K, upstream magnetic field strength $B_1 = 10\mu$G and maximum momentum of accelerated protons is fixed to $p_{\max} = 10^4 m_p c$. The parameter which regulate the injection efficiency is taken $\xi_{\text{inj}} = 3.95$.[8] For these values we get $L_{prec}/L_{ce} \simeq 150$, hence we do expect an efficient role of the charge exchange, while $L_{prec}/L_{ion} \simeq 10^{-4}$ hence the ionization can be neglected as stated above. In Fig. 1 we show the velocity and the temperature profiles in the precursor for both protons and hydrogen. The dotted line shows the normalized CR pressure, which at the sub-shock position reaches the value $P_c/\rho_0 u_0^2 = 0.14$, hence we are dealing with a mildly efficient shock. The upstream temperatures of protons and hydrogen reach the values of $5\cdot 10^5$ K and $4.7\cdot 10^5$ K, respectively. Protons are slightly wormer than hydrogen because they also suffer the adiabatic compression due to the CR pressure. The narrow Balmer line width corresponding to the hydrogen temperature at ths subshock is 46 km/s, to be compared with 21 km/s resulting in the absence af a CR precursor. The downstream temperature of shocked ions is $6.7 \cdot 10^7$ K, which produces a broad Balmer line with a width of 1774 km/s; in the absence of acceleration the same width would be 2039 km/s. As already pointed out, this difference arises because when the acceleration is efficient a non negligible fraction of the total shock kinetic energy is channelled into CRs rather than into thermal energy.

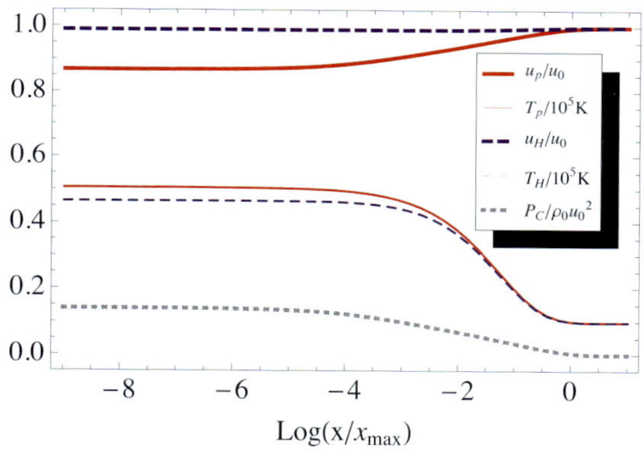


Fig. 1. Profiles of velocities and temperatures of hydrogen and protons in the precursor upstream of the shock. The thick-dotted line shows the CR pressure. All quantities are normalized as shown in the legend. The typical precursor length is $x_{\max} = D(p_{\max})/u_0$.

## References

1. G. Morlino, E. Amato, and P. Blasi, *MNRAS*, **392**, 240 (2009)
2. R. A. Chevalier, R. P. Kirshner, and J. C. Raymond, *ApJ*, **235**, 186 (1980)
3. K. Heng, *PASA*, **27**, 23 (2009)
4. J. J. Lee, J. C. Raymond, S. Park, W. P. Blair, P. Ghavamian, P. F. Winkler, and K. Korreck, *ApJ*, **715**, L146 (2010).
5. J. Sollerman, P. Ghavamian, P. Lundqvist, R. C. Smith, *A&A*, **407**, 249 (2003)
6. E. A. Helder, J. Vink, C. G. Bassa, A. Bamba, J. A. M. Bleeker, S. Funk, P. Ghavamian, K. J. van der Heyden, F. Verbunt, R. Yamazaki, *Science*, **325**, 719 (2009)
7. E. A. Helder, D. Kosenko, J. Vink, *ApJ*, **719L**, 140 (2010)
8. E. Amato, and P. Blasi, *MNRAS*, **371**, 1251 (2006)
9. H. L. Pauls, G. P. Zank, and L. L.Williams, *Jour. of Geophys. Research*, **100**, 21 (1995)

# PULSAR ELECTRONS DETECTION IN AMS-02 EXPERIMENT. MODEL STATUS AND DISCOVERY POTENTIAL.

Jonathan Pochon*

*Instituto de Astrofísica de Canarias,*
*C. Vía Láctea, 38200, La Laguna, Tenerife, Spain*
**E-mail: jgpochon@iac.es*

The measurements of cosmic rays electrons ($e^{\pm}$) have begun a new era a few years ago with high precision experiments like PAMELA and Fermi-LAT. Both experiments show an excess above 10 GeV. Studying Fermi-LAT data, the total electron flux seems to be steeper than expected. While these new measurements have not closed the debate, results from AMS-02 are expected to reach the accuracy needed to determine a full description of these excesses, and hopefully give some evidence on the possible source. We will present in this paper, the AMS-02 capacity in the case of positrons produced by pulsars.

*Keywords*: AMS-02 detector, electrons pulsar, anisotropy, inverse Compton.

## Introduction

Since the 70s most of the experiments, namely TS93[1], HEAT[2], Caprice[3], AMS-01[4], PAMELA[5], exhibit a deviation from the standard framework in the positron fraction ($\frac{e^+}{e^+ + e^-}$), summarized in Ref. **6**. Same disagreement appears in the total flux of $e^{\pm}$ measured by Fermi-LAT[7], balloon experiments like ATIC[8] and ground ones like HESS[9] which present a steeper flux, although Fermi-LAT measurements do not agree with the ATIC bump result[10,11]. The standard propagation model[12], still suffering uncertainties, seems unlikely able to reproduce such excess. A possible explanation could be a primary source, astrophysical objects can be involved like pulsars[13] or supernovæand/or dark matter particles, which can produce $e^{\pm}$ through annihilation[14] or decaying[15]. These uncertainties about the excess origin makes AMS-02, which be launched by February 2011 for 10 operation years, a potential crucial detector able to bring more precise measurements. This paper will discuss the electrons produced by pulsars using previous studies context[6,16,17].

## 1. Electrons flux from pulsars

P. Goldreich and W.H. Julian[18] have demonstrated that magnetic field tears away electrons from the pulsar surface. EGRET[19] has observed a pulsed gamma-ray emission at GeV scale, interpreted as a proof of magnetosphere implication for particle acceleration. A pulsar can be modeled like a rotating neutron star, characterized by $\Omega = 2\pi/P$ [Hz], with a dipolar magnetic field misaligned with its rotation axis, producing pulsed radiation from the magnetic poles. Mature pulsars could contribute to $e^{\pm}$ flux because produced particles are no longer trapped[20]. The total energy that a mature pulsar can inject in magnetic dipole radiation is[16,17] $E_{tot} \approx f_{e^{\pm}}\dot{E}T^2$ where $T$ is the typical age, $\dot{E}$ the spin-down power and assuming an efficiency factor $f_{e^{\pm}}$ for $e^{\pm}$ pair production, $f_{e^{\pm}} \sim$ few %[16]. The spectral shape is limited at high energy by $E_{max} = 3.10^3/t_5$ GeV, as a consequence of energy losses and a cut-off is expected $E_{cut}$ around the TeV scale. Below that, we will assume a power law $Q(E_e) \propto E^{-\alpha}$ with $\alpha$ between 1.4 and 2.2. After production, $e^{\pm}$ mostly loose energy by synchrotron radiation and inverse Compton scattering. These processes are calculated by solving the transport equation in the standard diffusion approximation (neglecting convection), which for local sources can be expressed with a spherical symmetry, and reduced to the form[21]

$$\frac{\partial}{\partial t}\frac{dn_e}{dE_e} = \frac{K(E_e)}{r^2}\frac{\partial}{\partial r}\left[r^2\frac{\partial}{\partial r}\frac{dn_e}{dE_e}\right] + \frac{\partial}{\partial E_e}\left[b(E_e)\frac{dn_e}{dE_e}\right] + Q(E_e) \quad (1)$$

Here, $dn_e/dE_e$ is the number density of $e^{\pm}$ per unit energy, $K(E_e)$ is the diffusion parameter, $b(E_e)$ is the rate of energy loss, and $Q(E_e)$ the source term. We assume $K(E_e) \equiv K_0(E_e/1\mathrm{GeV})^{\gamma}$, with $K_0$ and $\gamma$ specified in Ref. **22**, where scenario MAX maximizes the electron flux and MIN minimizes it, and $b(E_e) = -bE_e^2$ with $b = 10^{-16}\mathrm{GeV}^{-1}\mathrm{s}^{-1}$. Assuming a power law injection with a cut-off $E_{cut}$, the solution of equation **1** is given by[21,10]

$$\frac{dn_e}{dE_e} = \frac{Q(E_e)}{\pi^{3/2}r^3}(1-btE_e)^{\alpha-2}\left(\frac{r}{r_{diff}}\right)^3 e^{-\frac{E_e}{(1-E/E_{max})E_{cut}}} e^{(r/r_{diff})^2} \quad (2)$$

where $E < E_{max} \approx 1/(bt)$, and by $dn_e/dE_e = 0$ otherwise, with $r_{diff}$ being

$$r_{diff} \approx 2\sqrt{K(E_e)t\frac{1-(1-E/E_{max})^{1-\gamma}}{(1-\gamma)E/E_{max}}} \quad (3)$$

Table 1. Data for selected nearby pulsars with $f_{e^\pm}$ calculated for different propagation set (see Ref. **22**).

| Object Name | $E_{tot}^{out}$ (GeV) | $f_{e^\pm}^{MAX}$ | $f_{e^\pm}^{MIN}$ |
|---|---|---|---|
| Geminga [J0633+1746] | $74.10^{50}$ | 0.04 | 0.33 |
| Monogem [B0656+14] | $9.10^{50}$ | 0.03 | 0.25 |
| Vela [B0833-45] | $17.10^{50}$ | 0.03 | 0.25 |
| B0355+54 | $281.10^{50}$ | 0.30 | 0.40 |
| Loop I [SNR] | $43.10^{50}$ | 0.04 | 0.33 |

The local $e^\pm$ flux from pulsars is fixed by pulsar distance $r$, the time of injection $t$ which is the pulsar age $t = T$ for mature ones, and normalized with the $E_{tot}$. One can calculate that very old pulsars contribute less than 10 or 100 times than young ones. Secondly, propagation set (see Ref. **22**) shapes energy spectra, indeed MAX scenario maximizes electron flux and gives a bump around $E_{max}$ scale. From there, closest pulsars, described in table **1** (more details in Ref. **6**), can be studied as a primary source to explain the positron excess. Figure **1** presents positron fraction measured by AMS-02[23,24] during 3 years, produced by closest pulsars with power law injection $\alpha = 2$ and $E_{cut} = 1$ TeV for propagation set MAX. Each pair production efficiency $f_{e^\pm}$ pulsar is adjusted arbitrary to reproduce PAMELA positron. As expected, MAX scenario allows the lowest $f_{e^\pm}$ around some %, which is in agreement with expectations[16] and allows one to distinguish structure in the flux. On the other hand, MIN one needs too high $f_{e^\pm}$ becoming unlikely, and the distribution is too smooth to be interpreted.

Two aspects can be discussed, first the furthest pulsars could be considered as a continuum[25,16,6]. This contribution must reduce positron fraction structure leading to observe at least a plateau above 500 GeV with a slow decreasing[6]. Second for this study, electrons background was assumed for a mean value, Delahaye *et al.*[26] presented uncertainties for leptons background where background can be higher and, in this case pulsars flux may be lower.

## 2. Anisotropy: strong evidence for pulsar contribution?

The propagation of positrons-electrons does not allow one to pinpoint each possible $e^\pm$ source, reducing the chance of constraining models. Mao & Shen[27] have proposed that the closest pulsars can induce an anisotropy in the pulsars direction. Although propagation affects electron trajectory, close-by pulsars can induce a small dipole anisotropy for electrons-positrons and positrons case, which should be present at sufficiently high energy[27]. In

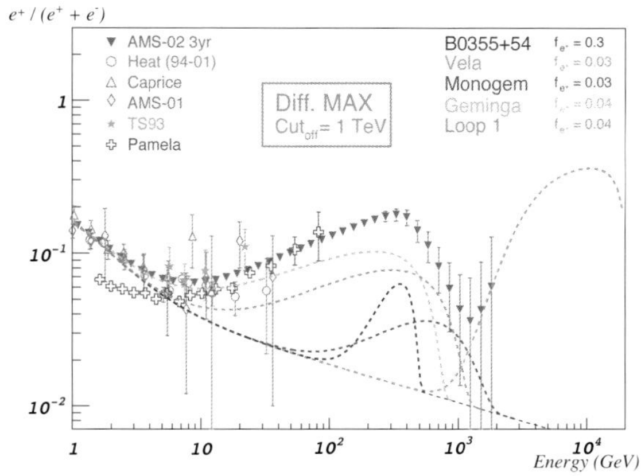


Fig. 1. *Positron fraction reproduced by pulsar contributions, $\alpha = 2$ and $E_{cut} = 1$ TeV, for propagation scenario MAX (see Ref. 22) with AMS-02 capacity.*

a very general way, the anisotropy of the $e^{\pm}$ flux associated with diffusive propagation can be calculated as[28]

$$\delta = \frac{I_{max} - I_{min}}{I_{max} + I_{min}} = \frac{3K|\nabla(dN_e/dE_e)|}{c(dN_e/dE_e)} = \frac{3}{2c}\frac{r}{t}\frac{(1-\gamma)E/E_{max}}{1-(1-E/E_{max})^{1-\gamma}}\frac{N_e^{Puls}}{N_e^{tot}} \tag{4}$$

where $\nabla(dN_e/dE_e)$ is the gradient of $e^{\pm}$ density, $r$ and $t$ are respectively the distance and age of the pulsar, $N_e^{Puls}$ and $N_e^{tot}$ are $e^{\pm}$ contribution from pulsars and from the electron background, taken from Ref. **29**. From the detection point of view, to observe an anisotropy at the $2\sigma$ level, the observations must satisfy the condition $\delta \geq 2\sqrt{2}/\sqrt{N_{evts}}$ where $N_{evts}$ is the number of events collected above an energy threshold.

Studying positron is a direct way to connect to the primary $e^{\pm}$ source. Therefore, anisotropy in $e^{+}$ flux is the best way to characterize a close-by source and to constrain it, this study was well presented by Busching *et al.*[30]. Figure **2** shows $e^{+}$ anisotropy expected for the two closest pulsars: Geminga and Monogem, where each pulsars flux is tuned to reproduce positron excess alone being the most favorable case. After five years, PAMELA wouldn't get enough statistics to detect any $e^{+}$ anisotropy. AMS-02 will reach Geminga anisotropy with 10 years of data. About $e^{\pm}$ anisotropy, 10 years will be also necessary for AMS-02 to observe Geminga anisotropy under 10 GeV when Fermi-LAT must be able to detect it above

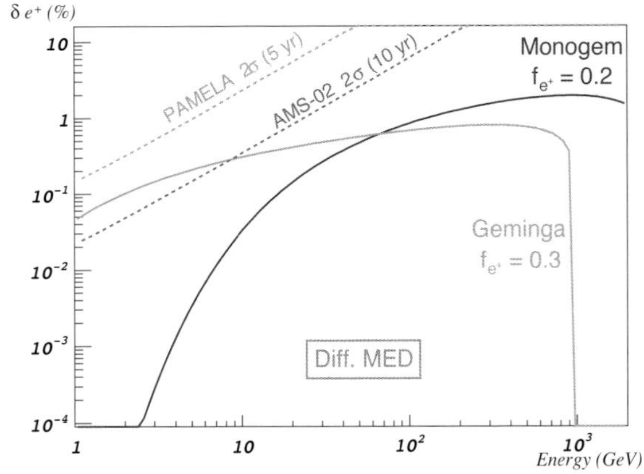


Fig. 2. *$e^+$ anisotropy for Geminga and Monogem, set to reproduce alone positron excess, with sensitivity for AMS-02, PAMELA and Fermi-LAT for MED propagation scenario*

10 GeV, away from solar modulation. Geminga and Monogem were supposed to be sufficiently distant from each other to induce an individual dipole. Nevertheless, according particles propagation and respective pulsar galactic coordinates, one can assume that total measured anisotropy could be sum of the two. This sum will help to detection but without sufficient enhancement above 10 GeV.

## 3. Constraints from $\gamma$ emission

Looking the electron pulsar, two assumptions are needed to evaluate electron injection flux $Q(E) \propto K_e E^{-\alpha}$. First, the normalisation parameter $K_e$ depends on the pair production efficiency $f_{e^\pm}$ which leads to how many electrons can we expect from the whole energy released by pulsar. The second is the power-law represented by $\alpha$ assuming to be around the mean value $\alpha = 2$. One can imagine to constrain mature pulsar studying present gamma-ray emission to determine the correlation between present $e^\pm$ production and $e^\pm$ contribution to positron excess. Electron from pulsars interact with interstellar medium producing synchrotron and inverse Compton flux. Y. Hasan *et al.* in Ref. **31** used a minimal model, where electrons interact with the CMB, to produce an inverse Compton spectra able to adjust Milagro observations for Geminga.

For Geminga, an inverse Compton flux is calculated on Fig. **3** for different injection energies interacting CMB only. Observing for one year Fermi-

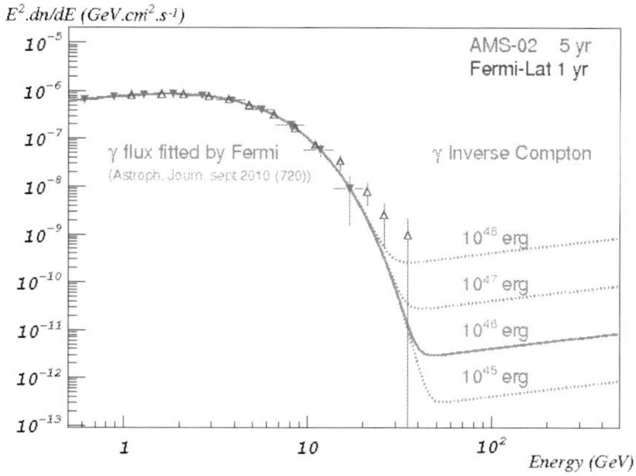


Fig. 3. *Geminga $\gamma$ emission with Fermi-LAT data*[32] *and AMS-02 capacity.*

LAT has obtained synchrotron flux measurement and would be able to discover the inverse Compton part, and from there actual electron flux should be constrained. Unfortunately, AMS-02 $\gamma$ sensitivity won't be enough to detect such radiation.

## 4. Conclusion

In this article, we considered that positron excess is the result of $e^{\pm}$ pairs production in nearby pulsars. AMS-02 will the only experiment able to investigate positrons at high energy, and therefore to constrain more efficiently $e^{+}$ sources. The closest pulsars can contribute to positron fraction and propagation should modify contributions, in two extremes ways: propagation scenario which maximizes electrons flux should allow each pulsar to be noticed, and the other hand the one minimizing it will smooth all contributions giving a continuum flux. The MAX scenario is preferred from experimental point of view, and it is in agreement with pulsars models implying low pair production parameters, around %. Without direct detection of pulsars, anisotropy studies in the positron and positron-electron flux will be powerful test to establish pulsar production proof. If Geminga and Monogem are the only contribution to positron excess, they should induce enough $e^{\pm}$ anisotropy to be detected in 5 years with $2\sigma$ significance by Fermi-LAT, and AMS-02 will be able to detect an $e^{+}$ anisotropy after 10 years. A new possibility to evaluate electrons contribution is to consider inverse Compton emission, assuming a correlation with present electrons production. Fermi-LAT would detect this radiation for Geminga in few years. To conclude, AMS-02 will be a central detector studying from leptons to nuclei able to understand propagation and source candidates.

## References


1. R.L. Golden *et al.*, ApJ, 457 (1996), L103.
2. Ma Duvernois *et al.*, ApJ, 559 (2001), 296-303; S. Barwick *et al.*, Phys. Rev. Lett., 75 (1995) 390; ApJ, 482 (1997) L191.
3. M. Boezio *et al.*, ApJ, 532 (2000), 653.
4. J. Alcaraz *et al.*, Phys. Lett. B484 (2000) 10.
5. O. Adriani *et al.*, Nature 458, (2009) 607-609.
6. J. Pochon, arXiv:1002.4213 (2010).
7. A.A. Abdo *et al.*, Phys. Rev. Lett. 102 (2009) 181101.
8. J. Chang *et al.*, Nature 456, (2008) 362-365.
9. H.E.S.S collaboration, Phys. Rev. Lett. 101 (2008) 261104.
10. D. Grasso *et àl.*, arXiv:0905.0636 (2009).
11. Fermi-LAT collaboartion, arXiv:1008.3999 (2010).
12. I.V. Moskalenko and A.W. Strong, ApJ, 493 (1998a) 694; D. Maurin *et al.*, arXiv:astro-ph/0212111.
13. J.H. Taylor and D.R. Stinebring, A. Rev. Astr. Astrophys 24 (1986) 285
14. M. Cirelli and A. Strumia, arXiv:0808.3867 (2008).
15. J. Zhang *et al.*, arXiv:0812.0522 (2008).
16. D. Hooper *et al.*, JCAP 0901 (2009) 025.
17. S. Profumo, arXiv:0812.4457 (2008).
18. P. Goldreich and W.H. Julian, ApJ, 157 (1969) 869.
19. D.J. Thompson *et al.*, ApJ, 436 (1994) 229.
20. X. Chi *et al.*, ApJ, 459 (1996) L83.
21. A.M. Atoyan *et al.*, Phys. Rev., D52 (1995) num. 6.
22. T. Delahaye *et al.*, Phys. Rev., D77 (2008) 063527; arXiv:0712.2312.
23. J. Pochon, PhD. thesis (2005).
24. P. Maestro, PhD. thesis (2003).
25. L. Zhang & K.S. Cheng, A&A, 368 (2001) 1063.
26. T. Delahaye et al., A&A.501,(2009) 821-833.
27. C.Y. Mao et C.S. Shen, Chinese Journal Phys. vol. 10 no. 1 (1972).
28. P. Ginzburg & S.I. Syrovatskii, The Origin of Cosmic Rays (1964).
29. S.W. Barwick *et al.*, ApJ, 498 (1998) 779.
30. I. Busching *et al.*, ApJ, 678 (2008) L39, 0804.0220.
31. Y. Hasan *et al.*, arXiv:0810.2784 (2008).
32. Fermi-LAT collaboration, arXiv:1007.1142 (2010).

# THE CR CONNECTION: UHE PRIMARIES AND SECONDARIES FROM UHECR SOURCES

A. M. Taylor*

*ISDC, University of Geneva,*
*Geneva, 1290, Switzerland*
**andrew.taylor@unige.ch*

During their acceleration, both UHECR nuclei and protons may interact with the radiation field present in the accelerator region. At UHE, such interactions lead to photo-disintegration and photo-pion production respectively. Thus, a close connection exists between the disintegration rates of UHECR nuclei in the source and the secondary UHE neutrino flux producible by the source. Upon escaping the accelerator, UHECR go on to generate a secondary UHE photon flux through photo-pion interactions with the cosmic microwave background (CMB) photons, These secondary UHE photons produced (via neutral pion decay) will themselves also interact with the cosmic radio background (CRB) photons. This leads to a second connection between UHE photons and UHECR composition. We here investigate both these connections to draw conclusions on what the presence of UHECR nuclei can tell us about the UHECR composition and their sources.

*Keywords*: UHECR nuclei; UHE neutrinos; UHE photons; UHECR sources

## 1. Multi-Messenger Approaches to Problems

Our multi-messenger approach is here divided into two sections. In the first of these the UHECR nuclei-proton connection is investigated. A general relationship between the interaction rates of both UHECR nuclei and protons with a general radiation field is determined. This result is used to demonstrate the strong constraints that may be placed on the source opacity to the UHECR they accelerate. In the second section, the UHECR-UHE photon connection is investigated. Motivated by the extreme cases of UHECR proton and UHECR iron nuclei fits to recent Auger results,[1] the subsequent photon flux obtained for these two cases is derived. This photon fraction is used as an indirect measure of the UHECR composition. We conclude with a summary on what we can learn through these two multi-messenger approaches.

## 2. Why Consider UHECR Nuclei?

Recently, the Auger collaboration has released its results on the measurements of both $\langle X_{\max}\rangle$ and RMS($X_{\max}$) using almost 4000 CR events detected by the fluorescence telescopes. These results are shown in fig. 1. For comparison, the theoretical hadronic model expectations for different nuclear species are also shown. Under the assumption that these hadronic models are representative of the present uncertainty that exists in our description of UHECR atmospheric cascades, these results suggest that UHECR become significantly heavier at energies above $10^{19}$ eV. Interestingly, preliminary results from LHC measurements suggest that the range of hadronic models considered by the UHECR community do indeed span the present range of uncertainties in the high energy physics of UHECR cascades.[2]

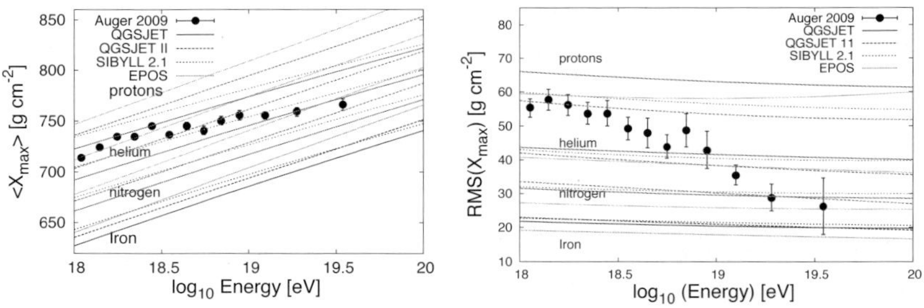


Fig. 1. **Left-Panel:** The $\langle X_{\max}\rangle$ distribution of arriving UHECR measured by the Auger observatory.[1] **Right-Panel:** The RMS($X_{\max}$) distribution of arriving UHECR measured by the Auger observatory.[1]

## 3. The UHECR nuclei-proton Connection

The interaction rates of UHECR with background photons follow the general expression,

$$R = \frac{1}{2\Gamma_p^2}\int_0^{\infty}\frac{1}{\epsilon_\gamma^2}\frac{dn_\gamma}{d\epsilon_\gamma}d\epsilon_\gamma\int_0^{2\Gamma_p\epsilon_\gamma}\epsilon'_\gamma\sigma_{p\gamma}(\epsilon'_\gamma)d\epsilon'_\gamma \qquad (1)$$

With both nuclei-photon and proton-photon interaction cross-sections possessing an initial resonance, which dominates the interaction rate integral given above in eqn 1, an approximation for the result may be found for these cases. Applying this to UHECR protons, gives interaction rates,

$$R_{p\gamma} \approx \sigma_{p\gamma} \int_{(\xi_{p\gamma}-\Delta_{p\gamma})/2\Gamma_p}^{(\xi_{p\gamma}+\Delta_{p\gamma})/2\Gamma_p} d\epsilon_\gamma \frac{dn_\gamma(\epsilon_\gamma)}{d\epsilon_\gamma}, \tag{2}$$

where $\Gamma_p$ is the proton's Lorentz factor, $\sigma_{p\gamma}$ is the mean proton-photon interaction cross-section, $dn_\gamma/d\epsilon_\gamma$ is the radiation field distribution, $\xi_{p\gamma}$ is the mean energy of the cross-section's resonance, and $\Delta_{p\gamma}$ is the width of the cross-section's resonance.

Similarly, applying 1 to photo-disintegration reactions,

$$R_{Fe\gamma} \approx \sigma_{Fe\gamma} \int_{(\xi_{Fe\gamma}-\Delta_{Fe\gamma})/2\Gamma_p}^{(\xi_{Fe\gamma}+\Delta_{Fe\gamma})/2\Gamma_p} d\epsilon_\gamma \frac{dn_\gamma(\epsilon_\gamma)}{d\epsilon_\gamma}. \tag{3}$$

The delta-resonance and giant-dipole-resonances are describable by the following parameters, $\sigma_{p\gamma} \approx 0.5$ mb, $\xi_{p\gamma} \approx 310$ MeV, $\Delta_{p\gamma} \approx 100$ MeV, $\sigma_{Fe\gamma} \approx 80$ mb, $\xi_{Fe\gamma} \approx 18$ MeV and $\Delta_{Fe\gamma} \approx 5$ MeV. The incorporation of these parameters into 2 and 3, gives the ratio between photo-pion and photo-disintegration rates,

$$R_{Fe\gamma}(\Gamma_p) \approx \frac{\sigma_{Fe\gamma}}{\sigma_{p\gamma}} R_{p\gamma}(15\ \Gamma_p) = 160 R_{p\gamma}(15\ \Gamma_p)\,. \tag{4}$$

Thus, the close connection of the photo-pion and photo-disintegration rates is unavoidable, and to first order is independent of the spectral shape of the target radiation field in the source.

With the recent Auger results suggesting that nuclei are significantly present in the UHECR flux produced from the source at energies above $10^{19}$ eV, *fewer than 1%* of UHECR protons can be expected to interact in the source at energies above $10^{18}$ eV.

## 4. The UHECR-UHE-photon Connection

During their propagation though extragalactic space, UHECR protons lose a significant fraction of their energy through interactions with the CMB via pair creation and photo-pion production interactions. Thus an associated UHE-photon flux arises from the decay of the neutral pions, $\pi^0 \rightarrow \gamma\gamma$, created through the $p + \gamma \rightarrow p + \pi^0$ interaction channel. These high energy photons also have their propagation range limited by background radiation fields (CMB and radio background- see left panel of fig. 2). If, besides the CMB, an additional radio background contribution is also assumed,[3,4] the interaction length for $10^{19}$ eV photons is approximately a few Mpc (see right panel of fig. 2). Indeed, as seen in fig 2, even if the radio background component is reduced to the CMB only component, the interaction length at $10^{19}$ eV increases by only a factor of 2. Thus, the uncertainties in these photon fraction calculations, due to the uncertainty in the radio background component, is small and may be considered negligible.

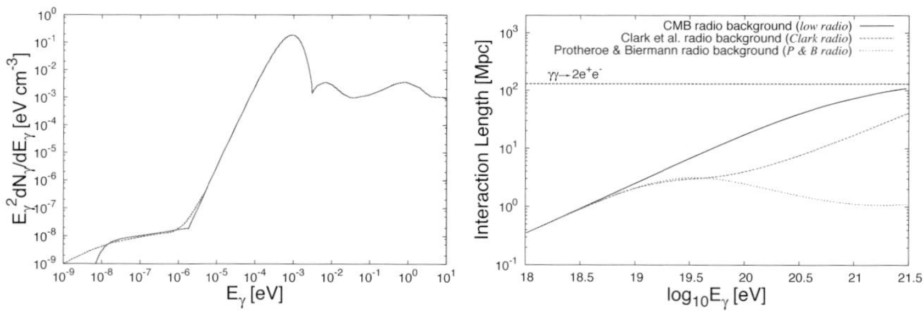


Fig. 2. **Left-Panel:** The background radiation fields impeding both UHECR proton and UHE photon propagation. The different lines in the figure represent the present uncertainty in the radio background. **Right-Panel:** The interaction rates of UHE photons in the background radiation fields. The uncertainties in the rates, due to the uncertainties in the radio background, are shown to be small at the energy of interest ($10^{19}$ eV).

To calculate the UHECR flux at Earth, a CR source spectrum of the form $dN_{CR}/dE_{CR} \sim E_{CR}^{-\alpha} \exp(-E_{CR}/E_{\max})$ is adopted. The value of the spectral index used in this study, $\alpha = 2$, is motivated by non-relativistic first order Fermi acceleration. A best-fit value for $E_{\max}$ was found for both the "100% protons" and "100% nuclei" cases by allowing a scan over a range of $E_{\max}$ from $(Z/26) \times 10^{20}$ to $(Z/26) \times 10^{22}$ eV. The assumed spatial distribution of CR sources is $dN/dV \sim (1+z)^3$, where $z$ is the redshift of the source. A maximum redshift of $z_{\max} = 1$ for the sources has been used for these results. Note that for the cosmologically nearby sources, to which the photon fraction is sensitive,[5] the $z$ dependence of the cosmological source distribution is of little relevance.

In fig. 3, we show two (extreme) cases for the UHECR composition: a proton only composition and an iron nuclei only composition. As can be seen from these best fit results, a considerably larger cut-off (and subsequently a larger energy budget) is required to best-fit the data for the "100% proton" case than for the "100% iron" nuclei case. To obtain these results, a Monte Carlo simulation description of both UHECR proton and nuclei propagation have been employed, as described in ref.[6]

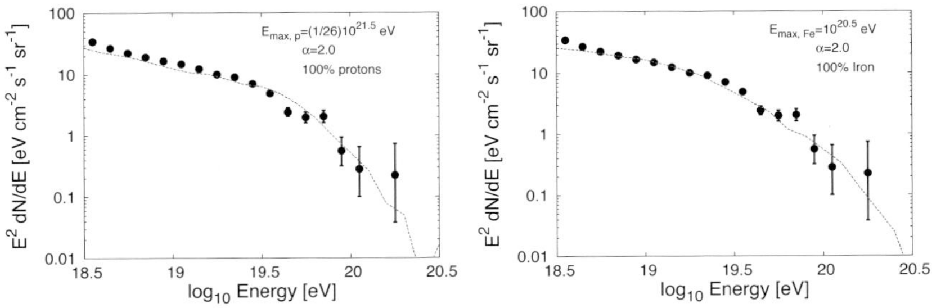


Fig. 3. **Left-Panel:** A best-fit proton only recent to recent Auger results[7] with a source injection spectrum of $\alpha = 2$ and the cut-off free to vary. **Right-Panel:** A best-fit iron only recent to recent Auger results[7] with a source injection spectrum of $\alpha = 2$ and the cut-off free to vary.

For these extreme cases, we calculate the subsequent UHE photon fluxes expected to contaminate these UHECR fluxes, as shown in fig. 4. Once again, to obtain these results, a Monte Carlo description of the UHE-photons, injected from photo-pions ($\pi^0$) produced through UHECR protons interactions with background radiation.[8] These results highlight the fact that the heavy nuclei model of UHECR can lead to a dramatic reduction in the photon fraction to levels unobservable by present UHECR detectors such as the Auger observatory.[9]

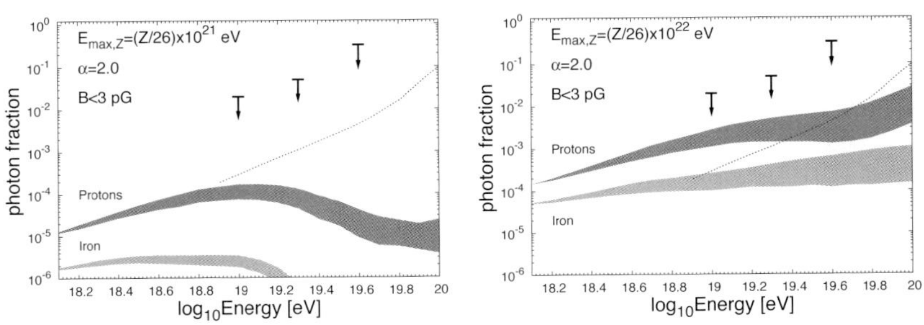


Fig. 4. **Left-Panel:** The photon fraction results obtained for a maximum energy of $E_{\rm max} = 10^{21}$ eV and a source spectral index of $\alpha = 2$. **Right-Panel:** The photon fraction results obtained for a maximum energy of $E_{\rm max} = 10^{22}$ eV and a source spectral index of $\alpha = 2$.

## 5. Conclusions

The first of our multi-messenger approaches discussed the relationship between UHECR nuclei and proton interaction rates with background radiation. Using these results, the presence of Fe in the arriving UHECR was used to demonstrate that these results demand for much smaller neutrino fluxes to be expected from these sources than obtained from calculations which ignore the strong constraints placed by the existence of UHECR nuclei. If iron nuclei are indeed being observed in the UHECR spectrum at energies above $10^{19}$ eV, *fewer than 1%* of UHECR protons can be interacting with the source's radiation field at energies above $10^{18}$ eV.

The second of our multi-messsenger approaches utilised the potential detection of the photon fraction component of UHECR spectrum. The photon fraction was demonstrated to be a powerful diagnostic tool for the UHECR composition. This followed from best-fit results for both proton and iron primary models. These fits were made to the most recent UHECR flux results[7] under the assumption that the source injection spectrum takes the form $dN/dE \propto E^{-2}e^{-E/E_{\rm max}}$. Through a comparison of these photon fraction results to predicted 20 year Auger predicted limits,[9] the predictions of these two extreme models is shown to span the present detection range. Thus, the photon fraction has some potential as a discriminator between different UHECR compositions.

## 6. Acknowledgments

A .M. Taylor would like to thank the organisers for inviting him to present his work at the 2010 Villa Como conference.

## References

1. J. Abraham *et al.* [Pierre Auger Observatory Collaboration], Phys. Rev. Lett. **104** (2010) 091101 [arXiv:1002.0699 [astro-ph.HE]].
2. S. Ostapchenko, arXiv:1010.0137 [astro-ph.HE].
3. T. A. Clark, L. W. Brown and J. K. Alexander, Nature 228, 847-849
4. R. J. Protheroe and P. L. Biermann, Astropart. Phys. **6** (1996) 45 [Erratum-ibid. **7** (1997) 181] [arXiv:astro-ph/9605119].
5. A. M. Taylor and F. A. Aharonian, Phys. Rev. D **79** (2009) 083010 [arXiv:0811.0396 [astro-ph]].
6. D. Hooper and A. M. Taylor, Astropart. Phys. **33** (2010) 151 [arXiv:0910.1842 [astro-ph.HE]].
7. J. Abraham *et al.* [The Pierre Auger Collaboration], Phys. Lett. B **685** (2010) 239 [arXiv:1002.1975 [astro-ph.HE]].
8. D. Hooper, A. M. Taylor and S. Sarkar, arXiv:1007.1306 [astro-ph.HE].
9. M. Risse and P. Homola, Mod. Phys. Lett. A **22**, 749 (2007).

# $\eta$ Carinae: a very Large Hadron Collider

R. WALTER*, C. FARNIER, J.-C. LEYDER

*INTEGRAL Science Data Centre, Université de Genève*
*Chemin d'Ecogia 16, CH-1290 Versoix, Switzerland*
**E-mail: Roland.Walter@unige.ch*

Eta Carinae is the colliding wind binary with the largest mass loss rate in our Galaxy and the only one in which hard X-ray emission has been detected. Eta Carinae is therefore a primary candidate to search for particle acceleration by probing its gamma-ray emission. We used the first 21 months of Fermi/LAT data to extract gamma-ray (0.2-100 GeV) images, spectra and light-curves, then combined them with multi-wavelength observations to model the non-thermal spectral energy distribution. A bright gamma-ray source is detected at the position of eta Carinae. Its flux at a few 100 MeV corresponds very well to the extrapolation of the hard X-ray spectrum towards higher energies. The spectral energy distribution features two distinct components. The first one extends over the keV to GeV energy range, and features an exponential cutoff at $\sim$ 1 GeV. It can be understood as inverse Compton scattering of ultraviolet photons by electrons accelerated up to $\gamma \sim 10^4$ in the colliding wind region. The second component is a hard gamma-ray tail detected above 20 GeV. It could be explained by $\pi^0$-decay of accelerated hadrons interacting with the dense stellar wind. The ratio between the fluxes of the $\pi^0$ and inverse Compton components is roughly as predicted by simulations of colliding wind binaries. This hard gamma-ray tail can only be understood if emitted close to the wind collision region. The energy transferred to the accelerated hadrons ($\sim$ 5% of the collision mechanical energy) is comparable to that of the thermal X-ray emission.

## 1. Introduction

About 30 early-type stellar systems feature synchrotron radiation in the radio domain, a signature of electron acceleration.[1] Diffusive shock acceleration in stellar wind collisions,[2] either in colliding wind binaries or OB associations, is the most likely acceleration process and a candidate for cosmic ray acceleration.[3,4] Gamma-rays, emitted by hadrons accelerated in stellar wind collisions, have however not yet been identified. Located at a distance of 2.3 kpc,[5] $\eta$ Carinae is one of the most massive (80–120 $M_\odot$[6,7]) and brightest stellar system in the Galaxy and features the strongest mass loss rate known. As, in addition, its colliding wind region is relatively wide, $\eta$ Carinae is a primary candidate for gamma-ray detection.

A hard X-ray tail had first been observed towards $\eta$ Carinae by *BeppoSAX*,[8] and has subsequently been unambiguously confirmed by both *INTEGRAL*[9,10] and *Suzaku*[11] observations. This gives a strong evidence that the wind collision leads to non-thermal electron acceleration, and suggests the possibility of a gamma-ray detection.

## 2. *Fermi/LAT* observations

We have used *Fermi/LAT* data accumulated in the Carina region during 21 months, from August 4, 2008 to April 3, 2010. The Fermi source FGL J1045.0-5942 with $\alpha$=161.265 and $\delta$=–59.7015 (95% confidence radius of 1.18 arcmin) is perfectly compatible with the position of $\eta$ Carinae. The presence of the source is indubitable, with a test statistic TS $> 2800$ ($\approx 53\ \sigma$) for the 200 MeV to 100 GeV energy range. Given the number of pulsars detected within $2^o$ of the Galactic plane the probability to find one of them in the 95% confidence region of $\eta$ Carinae is $2.5\times10^{-5}$. Moreover, we conducted a blind search frequency analysis and did not find any evidence for a pulsation in FGL J1045.0-5942.

The spectral analysis of FGL J1045.0-5942 was performed using the maximum likelihood method. The region modeling includes two diffuse emissions (Galactic plane and isotropic) and 35 point-like sources listed in the 1FGL catalogue. The resulting spectral energy distribution are presented in Fig. 1. A curvature at low energy and a modification of the spectral slope at high-energy are clearly detected.

## 3. Particle acceleration in $\eta$ Carinae

In the shock region, particle acceleration is counterbalanced by two main cooling processes. **(A) Inverse Compton scattering** of electrons in the intense ultraviolet radiation field of the stars. The cooling time scale is

$$t_{IC} = \frac{3\gamma m_e c^2}{4\sigma_T c \gamma^2 \beta^2 U_{rad}} = \frac{3\pi R^2 m_e c^2}{\sigma_T \gamma \beta^2 L} \approx \frac{R^2_{10^{14}\mathrm{cm}}}{\gamma_{10^4} L_{5\cdot10^6\mathrm{L}_\odot}} \times 6 \cdot 10^2 \text{ s},$$

where $L$ is the stellar ultraviolet luminosity of the primary star and $R$ is its distance to the colliding wind region. **(B) Proton-proton interactions** and subsequent pion decay. The interaction timescale for protons is inversely proportional to the density of matter in the post-shock region, where the particles are trapped by the magnetic field. Hydrodynamic simulations[12] indicate that the density in the shock region is larger than the unperturbed wind density, by a factor of $\delta$ ~1–100. The p-p cooling time scale can therefore be written :

$$t_{pp} = \frac{1}{\sigma_{pp}\delta nc} = \frac{4\pi R^2 m_p V_w}{\sigma_{pp}\delta \dot{M} c} \approx \frac{R^2_{10^{14}\mathrm{cm}}\ V_{10^3\mathrm{km/s}}}{\delta_{10}\ \dot{M}_{10^{-4}\mathrm{M}_\odot/\mathrm{yr}}} \times 4 \cdot 10^5\ \mathrm{s},$$

where $V$ is the typical wind velocity and $\dot{M}$ is the mass loss rate of the primary star. Equating the acceleration time $t_{acc} = \frac{R_L}{c}\left(\frac{c}{V}\right)^2$ respectively with $t_{IC}$ and $t_{pp}$ for electrons and protons ($R_L$ is the Larmor radius) provides the maximum characteristic energy of the particle distribution. For electrons, a powerlaw spectrum is expected, with an exponential cutoff at

$$\gamma_{max,e} = \sqrt{\frac{3\pi e c^2}{\sigma_T \beta^2}}\sqrt{\frac{B \cdot R^2}{L}\frac{V}{c}} \approx \sqrt{\frac{B_{1\mathrm{G}} \cdot R^2_{10^{14}\mathrm{cm}}}{L_{5\cdot 10^6 \mathrm{L}_\odot}}V_{10^3\mathrm{km/s}}} \times 3 \cdot 10^4,$$

where $B$ is the magnetic field in the shock region. $\gamma_{max,e}$ should be fairly independent of the orbital position in the dipole approximation for a magnetic field varying as $R^{-2}$. The gamma-ray spectrum will therefore feature an exponential cutoff at the maximal energy.[13] A cutoff energy of $\sim$ 1 GeV corresponds to a magnetic field of $\sim$ 50 G at the stellar surface.

For protons, the maximum characteristic energy is limited by proton-proton interactions to

$$\gamma_{max,p} = \frac{4\pi R^2 e B}{\sigma_{pp}\delta \dot{M}}\left(\frac{V}{c}\right)^3 \approx \frac{R^2_{10^{14}\mathrm{cm}} B_{1\mathrm{G}} V^3_{10^3\ \mathrm{km/s}}}{\delta_{10}\dot{M}_{10^{-4}\mathrm{M}_\odot/\mathrm{yr}}} \times 4 \cdot 10^4.$$

The non-thermal spectral energy distribution of $\eta$ Carinae has been represented with a model consisting of two cutoff powerlaw distributions for the electrons and for the interacting protons. The model and the data are shown together in Fig. 1. The magnetic field and the electron energy distribution were adjusted to match the upper limit on the radio synchrotron emission, derived from the minimal thermal emission detected with *ACTA*,[14] and to match the inverse Compton continuum determined by *INTEGRAL* and *Fermi*. The slope and cutoff energy of the interacting proton energy distribution were fixed to 2.25 and $10^4$, respectively, the same values as found for electrons. and its normalization was fitted to match the high-energy gamma-ray tail using $\pi^0$-decay.[15] The proton and ultraviolet photon energy densities in the shock region were fixed to the values expected for a representative distance of $R = 10^{14}$ cm.

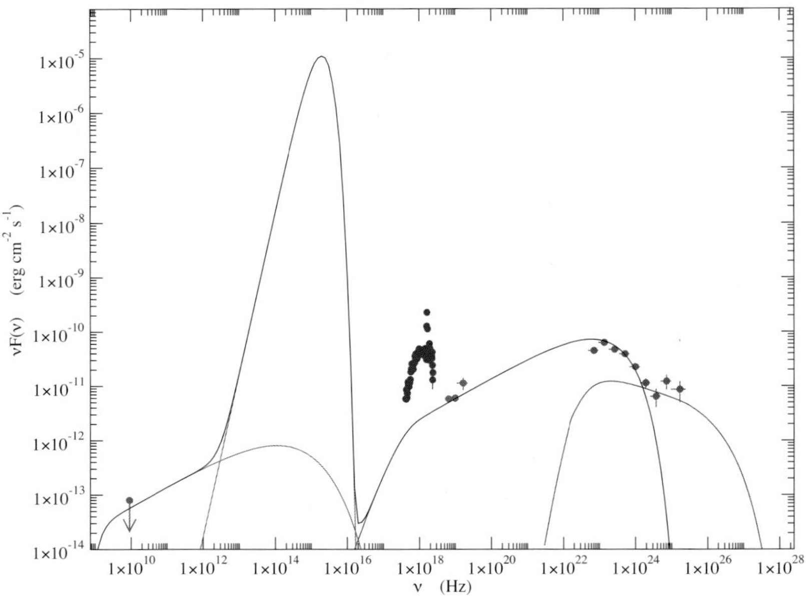


Fig. 1. Spectral energy distribution of $\eta$ Carinae including *BeppoSAX/MECS* (blue), *INTEGRAL/ISGRI* (red) and *Fermi/LAT* (purple) data, and a radio upper limit to the synchrotron emission (green). From low to high energies are shown the synchrotron, stellar emission, inverse Compton and $\pi^0$-decay spectral components.

## 4. Energetic

The wind momentum ratio of $\eta$ Carinae is $\eta = (\dot{M}_2 V_2)/(\dot{M}_1 V_1) \approx 0.2$.[16] The half-opening angle of the shock region is therefore $\sim 1$ rad, the fraction of the wind involved in the wind collision region is $\sim 10\%$, and the mechanical energy available in that region is $\sim 200$ $L_\odot$. For a density of cold protons of $3 \cdot 10^9$ cm$^{-3}$ in the shock region, the normalization of the $\pi^0$-decay spectrum requires a total interacting proton energy $E_p \sim 10^{40}$ erg. The energy injected in the shock to sustain the observed proton distribution is of the order of $E_p/t_{pp} \sim 10$ $L_\odot$. The gamma-ray spectrum of $\eta$ Carinae thus indicates that $\sim 5\%$ of the shock mechanical energy (or less than 1% of the total wind mechanical luminosity) is transferred to accelerated protons downstream. This is in agreement with recent numerical simulations of relativistic collisionless shocks,[17] if a significant fraction of the hadrons interacts and generates $\pi^0$.

Modeling the thermal X-ray emission depends on detailed hydrodynamical simulations[16] and could be affected by many phenomena.[18] Such simulations can explain an X-ray luminosity of $\approx 20\ L_\odot$ above a few keV. Our analysis hence indicates that the fraction of the shock energy accelerating protons is similar to that emitting observable X-rays in $\eta$ Carinae.

## 5. Conclusions

We have detected a bright gamma-ray source at the position of $\eta$ Carinae. Its flux at a few 100 MeV corresponds very well to the extrapolation of the hard X-ray spectrum of $\eta$ Carinae (as measured by *INTEGRAL* and *Suzaku*) towards higher energies. The spectral energy distribution, which corresponds to an average over almost half of the orbit of $\eta$ Carinae, features two spectral components.

The first one is a powerlaw extending from keV to GeV energies, with an exponential cutoff at $\sim 1$ GeV. This component can be understood assuming inverse Compton scattering of stellar photons by electrons accelerated up to $\gamma \sim 10^4$ in the wind collision region. The observed cutoff energy implies a magnetic field of $\sim 50$ G at the stellar surface.

The second component, a hard gamma-ray tail, is detected above 20 GeV. This bright component could be explained by $\pi^0$-decay of accelerated hadrons interacting with the dense stellar wind in the shock region. The ratio between the fluxes of the $\pi^0$ and inverse Compton components is roughly as predicted by simulations.[19,20] Bremsstrahlung emission is expected at a much lower level and with a cutoff energy similar to that of the inverse Compton component.

The hard gamma-ray tail can only be understood if emitted close to the wind collision region. Indeed, the external shock between the *Homunculus* nebula and the interstellar medium occurs at a density and magnetic field strength that are much too small to explain the observed emission, even if a small fraction of the hard X-ray emission could come from inverse Compton scattering of infrared photons in the external shock.[10,21]

The energy transmitted to the accelerated particles ($\sim 5\%$ of the wind collision mechanical energy) is of the same order as that of the observed thermal X-ray emission, providing an important observational constraint for future numerical hydrodynamical models of the colliding wind region in $\eta$ Carinae.

## References

1. M. De Becker, *A&A Rev.* **14**, 171(November 2007).
2. P. Benaglia and G. E. Romero, *A&A* **399**, 1121(March 2003).

3. W. I. Axford, *Annals of the New York Academy of Sciences* **375**, 297(December 1981).
4. M. Casse and J. A. Paul, *ApJ* **258**, 860(July 1982).
5. N. Smith, *ApJ* **644**, 1151(June 2006).
6. K. Davidson and R. M. Humphreys, *ARA&A* **35**, 1 (1997).
7. D. J. Hillier, K. Davidson, K. Ishibashi and T. Gull, *ApJ* **553**, 837(June 2001).
8. R. F. Viotti, L. A. Antonelli, C. Rossi and S. Rebecchi, *A&A* **420**, 527(June 2004).
9. J. Leyder, R. Walter and G. Rauw, *A&A* **477**, L29(January 2008).
10. J. Leyder, R. Walter and G. Rauw, *A&A* , p. in press (2010).
11. A. Sekiguchi, M. Tsujimoto, S. Kitamoto, M. Ishida, K. Hamaguchi, H. Mori and Y. Tsuboi, *ArXiv e-prints* (March 2009).
12. J. M. Pittard, *MNRAS* **396**, 1743(July 2009).
13. D. Eichler and V. Usov, *ApJ* **402**, 271(January 1993).
14. R. A. Duncan and S. M. White, *MNRAS* **338**, 425(January 2003).
15. S. R. Kelner, F. A. Aharonian and V. V. Bugayov, *Phys. Rev. D* **74**, 034018(August 2006).
16. J. M. Pittard and M. F. Corcoran, *A&A* **383**, 636(February 2002).
17. A. Spitkovsky, *ApJ* **682**, L5(July 2008).
18. E. R. Parkin, J. M. Pittard, M. F. Corcoran, K. Hamaguchi and I. R. Stevens, *MNRAS* **394**, 1758(April 2009).
19. J. M. Pittard and S. M. Dougherty, *MNRAS* **372**, 801(Octtober 2006).
20. A. Reimer, M. Pohl and O. Reimer, *ApJ* **644**, 1118(June 2006).
21. S. Ohm, J. A. Hinton and W. Domainko, *ArXiv e-prints* (June 2010).

# Cosmic Rays from Exotic Sources

| | |
|---|---|
| T. Bringmann | Gamma rays from dark matter |
| C. Combet | Introducing CLUMPY: a public code for gamma-ray emission from Dark Matter annihilation in the Galaxy |
| T. Delahaye | Cosmic rays and dark matter indirect detection |
| M.H. Reno | Neutrinos from dark matter |
| P. Salati | Charged cosmic rays from dark matter |
| D. Tran | Gamma-ray and neutrino signatures of unstable dark matter |
| C. Weniger | Gamma-ray anisotropies from decaying dark matter |

# GAMMA RAYS FROM DARK MATTER

T. BRINGMANN

*II. Institute for Theoretical Physics, University of Hamburg,*
*Luruper Chaussee 149, D-22761 Hamburg, Germany*
*E-mail: torsten.bringmann@desy.de*
*http://unith.desy.de*

A leading hypothesis for the nature of the elusive dark matter are thermally produced, weakly interacting massive particles that arise in many theories beyond the standard model of particle physics. Their self-annihilation in astrophysical regions of high density provides a potential means of indirectly detecting dark matter through the annihilation products, which nicely complements direct and collider searches. Here, I review the case of gamma rays which are particularly promising in this respect: distinct and unambiguous spectral signatures would not only allow a clear discrimination from astrophysical backgrounds but also to extract important properties of the dark matter particles; powerful observational facilities like the Fermi Gamma-ray Space Telescope or upcoming large, ground-based Cherenkov telescope arrays will be able to probe a considerable part of the underlying, e.g. supersymmetric, parameter space. I conclude with a more detailed comparison of indirect and direct dark matter searches, showing that these two approaches are, indeed, complementary.

*Keywords*: dark matter, indirect detection, gamma rays

## 1. Introduction

The existence of a sizable non-baryonic, cold dark matter (DM) component in the universe is supported by a wealth of observations on distance scales that range from individual galaxies over clusters of galaxies all the way out to fluctuations of the cosmic microwave background, spanning thus a range from tens of kpc to several Gpc. The *amount* of DM, $\Omega_\chi = 0.227 \pm 0.014$ according to the most recent estimates[1], can now be determined much more precisely than in the times of Zwicky who who first formulated the DM problem almost 80 years ago;[2] as of today, however, the *nature* of DM still remains an open question[a].

[a]Actually, the situation has recently become quite interesting as both direct[3] and cosmic ray[4] experiments have reported data that could be interpreted in terms of DM; at

A theoretically particularly appealing solution is a class of DM candidates that go under the name of Weakly Interacting Massive Particles (WIMPs): new effects related to electroweak symmetry breaking (currently being searched for at the CERN LHC) generically predict such particles which would be produced as thermal relics with the right density today.[5] While the lightest supersymmetric neutralino is often taken as a standard template for such a WIMP, alternative candidates like the lightest Kaluza-Klein photon in theories with universal extra dimensions[6] or the lightest $T$-odd particle in little Higgs models[7] provide interesting alternatives. Apart from being well-motivated from particle physics, WIMPs have the advantage of being, at least in principle, detectable by means other than gravitational: at colliders (where the main signature would be missing transverse energy), in direct detection experiments (where one hopes to measure the recoil energy of WIMPs off the nuclei of terrestrial detectors) or indirectly through the observation of WIMP annihilation products in cosmic rays.

Among possible messengers for indirect DM searches, there are several advantages connected to gamma rays: usually produced at rather high rates, they propagate unhindered through the galactic halo; this means that they point directly back to the source and no assumptions about the diffusive halo are necessary like in the case of charged particles. What is maybe even more important is that they provide very clear and distinct spectral signatures to look for – a point which will now be taken up in more detail.

## 2. Spectral signatures

The differential gamma-ray flux from a source with DM density $\rho$ in the direction $\psi$ is given by

$$\frac{d\Phi_\gamma}{dE_\gamma}(E_\gamma,\psi) = \frac{\langle\sigma v\rangle_{\rm ann}}{4\pi m_\chi^2}\sum_f B_f\frac{dN_\gamma^f}{dE_\gamma} \times \frac{1}{2}\int_{\Delta\psi}\frac{d\Omega}{\Delta\psi}\int_{\rm l.o.s} d\ell(\psi)\rho^2(\mathbf{r})\,, \quad (1)$$

where $\langle\sigma v\rangle_{\rm ann}$ is the total annihilation cross section, $m_\chi$ the mass of the DM particle, $B_f$ the branching ratio into channel $f$ and $N_\gamma^f$ the number of resulting photons; these quantities depend only on the underlying particle physics model and can thus be determined to a rather good accuracy for any given DM model. The right part of the above expression is a measure for the number of DM particle pairs along the line of sight of a detector with an opening angle $\Delta\psi$; since it depends strongly on the largely unknown

present, however, no compelling evidence for the DM and against competing, conventional explanations has been presented.

distribution of DM, there is usually a considerable uncertainty connected to the overall *normalization* of the expected annihilation flux.[b]

In order to claim a DM detection in gamma rays, i.e. a convincing discrimination of the signal from more conventional astrophysical sources, it is thus important to rely on clear *spectral signatures*. In general, one can distinguish four different types of contributions to the gamma-ray spectrum:

(1) At tree-level, WIMPs annihilate into pairs of quarks, leptons, Higgs and weak gauge bosons. *Secondary photons* are produced in the hadronization and further decay of these primary annihilation products, mainly through $\pi_0 \to \gamma\gamma$. The result is a featureless spectrum with a rather soft cutoff at $m_\chi$, almost indistinguishable for the various possible annihilation channels (with the exception of $\tau$-lepton final states).[8]
(2) Whenever charged annihilation products are present, additional *internal bremsstrahlung* (IB) photons appear at $\mathcal{O}\left(\alpha_{\rm em}\right)$. They generically dominate at the highest energies that are kinematically accessible, and thereby add pronounced signatures to the spectrum; viz. a very sharp cutoff at $m_\chi$ and bump-like features at slightly smaller energies.[9,10] More recently, it has been pointed out that also electroweak corrections, with a final state $W$ or $Z$ boson instead of a photon, can give sizable corrections that could visibly effect the spectrum.[11]
(3) Necessarily loop-suppressed, and thus only at $\mathcal{O}\left(\alpha_{\rm em}^2\right)$, *monochromatic* $\gamma$ *lines* result from the annihilation of DM particles into two-body final states containing a photon.[12] While in principle providing a striking signature, these processes are usually subdominant and thus not actually visible when taking into account realistic detector resolutions;[10,13] examples of particularly strong line signals, however, exist.[14]
(4) In models with large branching fractions into $e^+/e^-$ pairs, finally, there is also a contribution from *inverse Compton* scattering of highly energetic $e^\pm$ on starlight and the cosmic microwave background. It only appears at energies considerably below $m_\chi$, however, and does not result in pronounced spectral signatures like in the previous two cases.[15]

The distinct spectral features connected to IB even remain visible when taking into account realistic energy resolutions for the detector. As an ex-

[b]The gamma-ray flux only factorizes as indicated in Eq. (1) if $\langle\sigma v\rangle_{\rm ann}$ is largely independent of the relative velocity $v$ of the annihilating particles; while this is often the case (valid to a very high accuracy for $s$-wave annihilations, in particular), there are situations (e.g. pure $p$-wave or Sommerfeld enhanced annihilations) where this is not the case and one has to perform the angular average and line integral over the full expression.

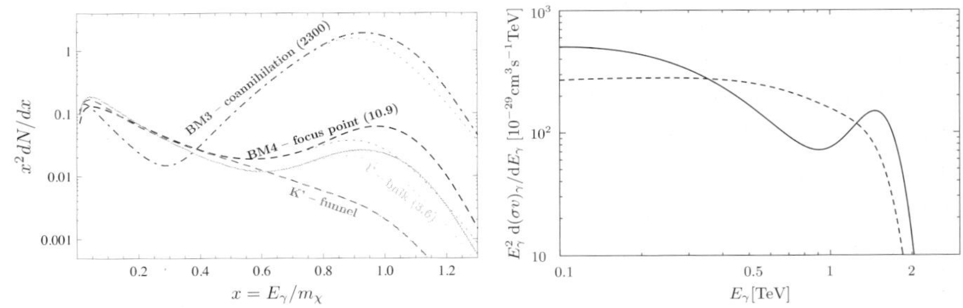


Fig. 1. This figure shows typical DM annihilation spectra: in the left panel, four benchmark models representative of the four cosmologically relevant regions in the mSUGRA parameter space are plotted, as roughly seen by a detector with an energy resolution of 10% (from Ref. 18). For each of these models, the number of IB over secondary photons (at energies $E_\gamma > 0.6\, m_\chi$) is indicated in parenthesis and the dotted line shows the same spectrum without including the line signal. The right panel (from Ref. 19) compares the spectra of Wino (solid line) and Kaluza-Klein DM (dashed line), assuming an energy resolution of 15% and, for the sake of comparison, a common mass of $m_\chi = 1.5\,\mathrm{TeV}$.

ample, consider the case of minimal supergravity (mSUGRA) where one can single out four regions in the underlying parameter space that give the correct DM relic density (see, e.g., Ref. 16 for a discussion). The left panel of Fig. 1 shows the type of annihilation spectra typically encountered in these cases, smeared with a Gaussian to simulate the effect of an energy resolution of 10%. Clearly, one could use these spectra – even in their smoothed versions – to discriminate between the cosmologically interesting regions in the mSUGRA parameter space, and thus obtaining important information about the nature of the annihilating DM. The same figure also indicates the comparably small contribution from line signals and states the ratio of IB over secondary photons at high energies. The latter can be sizable especially in the $\tilde{\tau}$-coannihilation region, thereby significantly improving the detectability of such DM candidates in future gamma-ray experiments.[17]

As a further example, the right panel of Fig. 1 compares the spectra of annihilating Wino DM[20] and Kaluza-Klein DM in theories of universal extra dimensions;[21] to distinguish these two cases, an energy resolution of 15% is clearly sufficient. Yet another example of the discrimination power of IB-dominated signals is given in Ref. 22 for DM models that have been invoked to explain the already mentioned anomalies[4] in cosmic rays.

## 3. Dark matter sources

When looking for suitable targets for indirect DM searches in gamma rays, one obviously will prefer places with a high (dark) matter density. At least as important, however, is a low astrophysical background or, rather, a high signal to background ratio. Since it is quite difficult to constrain the DM distribution from kinematic observations – especially in the innermost part of the Milky Way, which is *not* DM-dominated – one usually has to infer the DM density profile from $N$-body simulations. For non-interacting cold DM (like WIMPs), these consistently find cuspy DM distributions in virialized halos; the most recent, highest-resolution simulations[23] tend to somewhat favor the Einasto profile[24]

$$\rho_\chi(r) = \rho_s \exp\left[-\frac{2}{\alpha}\left(\left(\frac{r}{r_s}\right)^\alpha - 1\right)\right], \tag{2}$$

at the expense of introducing one more free parameter, over the slightly steeper NFW profile[25] which simply scales like $\rho_\chi \propto r^{-1}$ at small distances $r$ from the center. Such simulations, however, have to be taken with a grain of salt since they do not (fully) include the effect of baryons; another point to keep in mind is that one often has to extrapolate the above scaling behavior several orders of magnitude beyond the actual numerical resolution in order to make quantitative predictions for the expected annihilation flux.

Observations of external galaxies, especially those small in mass and dominated by DM, are sometimes quoted as evidence for shallow DM profiles with a central core.[26] While this evidence is not necessarily conclusive given that at least dwarf galaxies are also consistent with NFW profiles,[27] such a potential discrepancy with the results from $N$-body simulations could be attributed to the effect of baryons when including dynamical friction and stellar feedback processes.[28] For *baryon-dominated* systems like our own galaxy, the presence of baryons could actually also lead to a *steepening* of the DM profile in a process known as *adiabatic contraction*.[29] This is an issue of ongoing debate and there is evidence both in favour of[30] and against[31] such an additional steepening of the DM profile (for a discussion, see Ref. 32). The infall of (dark) matter onto a black hole at the center of the halo, like the supermassive black hole at the center of the Milky Way, could lead to an even more spiky DM profile,[33] though this strongly depends on the halo formation history.[34] The upshot is that the NFW (or Einasto) profile is a natural assumption for WIMP DM – as long as one keeps in mind that the currently not very well understood role of baryons may somewhat change this picture.

Given this assumption, the *galactic center* is the single-most brightest source of DM annihilation in the sky. At the same time, however, it is also an astrophysically very rich and complex environment, which inevitably means that it will generally be quite difficult to unambiguously distentangle a DM signal from the background.[35] The situation becomes further complicated by the fact that adopting other DM profiles would change the expected gamma-ray flux by several orders of magnitude.[36]

For that reason, it may be advantageous to broaden the view and include the diffuse gamma-ray emission from the whole Milky Way halo, in which case the expected *angular distribution* of the DM signal would be a nice signature to look for (especially when combined with the observation of spectral signatures).[37] Indeed, when focussing on the signal to background ratio, there are indications that the optimal angular window of observation is an annulus of a few degrees around the galactic center.[38] There is also a DM-induced contribution to the extra-galactic flux, stemming from annihilations in external halos at higher redshift and then integrated over the cosmological evolution.[39] While the challenge lies in the large extragalactic background, which is even less understood than its galactic counterpart, there could be prominent spectral signatures to look for, like the $\pi^0$ peak or line signals which would be modified in a characteristic way due to the cosmological redshift.

As for discrete sources, *dwarf galaxies* are the most DM dominated objects observed so far, with mass-to-light ratios of up to $\sim 1000$, and therefore represent in some sense the opposite extreme of the galactic center: while the DM-signal in gamma rays is expected to be rather faint, it should by far outshine any astrophysical backgrounds. Consequently, these objects have a long history as targets for DM searches,[40] which has been triggered anew[41,42] with the detection[43] of a new class of ultra-faint galaxies. Kinematic observations of the stars in these galaxies allow an ever better determination of their density profile and the resulting uncertainty in the annihilation flux is considerably less than for the galactic center.[41]

Numerical simulations confirm the theoretical expectation that DM is not distributed smoothly in the galactic halos but forms a lot of clumps and *substructures*. Large DM clumps, if not massive enough to trigger star formation and thus act as a possible seed for a dwarf galaxy,[44] would thus be ideal targets for the purpose of DM detection;[45] if discovered as unidentified gamma-ray sources in all-sky surveys, a follow-up observation with the much better sensitivity of ground-based telescopes could verify their dark nature. With reasonably optimistic assumptions about the particle nature of the DM, the results of simulations seem to indicate that the discovery of at least a few DM clumps could soon be within reach.[46] The overall signal from very small, unresolved clumps is, in fact, already constrained by observations of the diffuse background.[47]

Another interesting target for indirect searches are galaxy clusters which are very massive, DM dominated systems; the gamma-ray background from cosmic rays may, however, complicate the discrimination of an annihilation signal[48] and a multi-wavelength oriented approach could therefore be more appropriate.[49] Finally, an interesting possibility might be the existence of DM mini-spikes around intermediate mass black holes[50] which, however, already seems to be in some tension with current observational data.[51]

The annihilation signal in all these cases could be enhanced[52] by the existence of halo sub-(sub-)structure which is predicted to extend down to subhalo mass scales of $M_{\rm cut} \sim 10^{-11}$–$10^{-3} M_\odot$,[53] where the numerical value of the cutoff depends on the DM particle properties and is set by the kinetic decoupling of DM in the early universe.[54] The reason for this is simply that the line-of sight integral in Eq. (1) effectively constitutes an average and one always has $\langle \rho^2 \rangle > \langle \rho \rangle^2$ for inhomogeneous distributions. Roughly speaking, the boost factor receives about the same contribution from each logarithmic interval in the subhalo mass, though the usual cautionary remarks on the underlying extrapolation of results from numerical simulations apply here as well. Near the galactic center, the effect should be rather small since most clumps have likely been destroyed due to tidal disruption. One should note that the term "boost-factor" is not used uniquely in the literature and in general depends both on the particle species and the energy (at least for charged particles);[55] sometimes, it is even used to include particle-physics effects like the recently much discussed Sommerfeld enhancement.[56]

## 4. Experiments and observational status

Gamma-ray observations can be performed either in space or on earth. The sensitivity of space-based telescopes is generally limited by rather small effective areas and there is an upper bound on the photon energy that can be resolved ($\sim 300$ GeV for the already operating Fermi[57] satellite, $\sim 1$ TeV for AMS-02[58] which will be installed on the international space station in early 2011). A great field of view, on the other hand, makes the discovery of previously unknown sources a relatively straight-forward task. Earth-based telescopes like HESS,[59] VERITAS,[60] MAGIC[61] or the planned AGIS[62] and Air Cherenkov Telescope Array (CTA)[63] are complementary in that they have to face a lower threshold for the observable energy since the discrimination of electromagnetic from hadronic showers becomes increasingly difficult and below about 10 GeV electromagnetic showers produced in the atmosphere do not reach the ground anymore; due to an effective area of up to several km$^2$ they can, however, achieve much better sensitivities than space-based telescopes. Their rather small feld of view makes them ideal for pointed observations.

A DM induced gamma-ray signal from the galactic center has been claimed almost every time new observational data from this region became available,[21,64] but so far more refined analyses and new data always tended to disfavor the DM hypotheses previously put forward.[65] The ongoing data taking and analysis of the Fermi collaboration will in any case be a crucial step in determining the background sufficiently well so as to eventually be able to discriminate even a subdominant DM component. In passing, we note that in particular the contribution of light DM particles to the observed flux is severely bound by observations in other cosmic rays species.[66]

In the recent past, there have been a number of null searches for DM signals which place limits on the relevant DM properties (its mass and annihilation cross section, in particular) that start to actually touch the theoretically favored parameter space of standard WIMP candidates. Among these new limits many result from the first year of Fermi data (which could, consequently, be improved considerably during the remaining years of operation) – including an all-sky search for line-signals,[67] cosmological DM annihilations[68] in the extragalactic background and detailed observations of galaxy clusters[69] as well as dwarf galaxies;[70] an analysis of unidentified point sources, among which there could be DM clumps, is underway. The major operating Air Cherenkov Telescopes have all identified dwarf galaxies as interesting targets, too;[71] given the relatively short observation times of only 20-50 h, however, the resulting limits on the annihilation cross section can't compete with the ones derived in Ref. 70. Taken at face value, the latter actually represent the overall strongest of such limits; in fact, these results will likely improve even further by a "stacking" of all the dwarf galaxies, including Segue 1 for which the analysis is still ongoing, i.e. by using the fact that the DM spectrum should be the same in each case and that the luminosity *ratios* of these objects are much better constrained than the individual flux from any given dwarf.[72] One should of course keep in mind, however, that a comparison of limits derived from different sources is always a delicate issue – both since the underlying assumptions are often difficult to compare and due to the already discussed, large astrophysical uncertainties involved.

## 5. Future prospects and comparison with direct searches

As already alluded to, the effective area will always be a limiting factor for space-based experiments. Ground-based detectors, on the other hand,

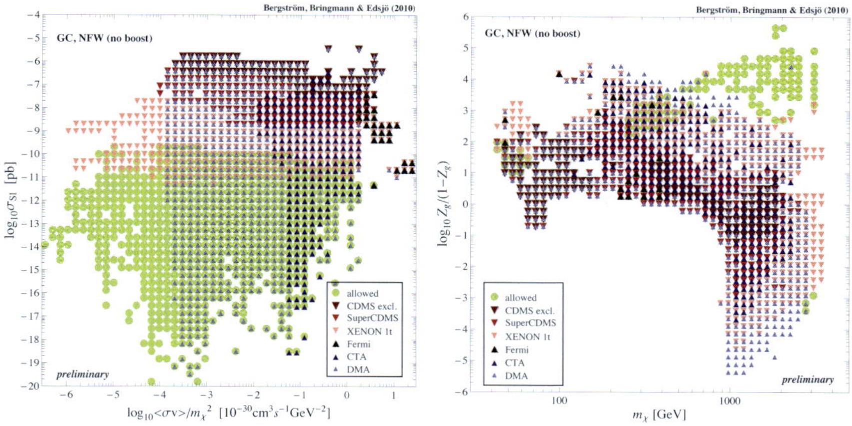


Fig. 2. The reach of direct and indirect detection experiments is shown for the case of neutralino DM, both for presently operating and for future experiments. For this purpose,the left panel shows the spin-independent cross section relevant for direct searches versus a quantity directly relevant for indirect searches, $\langle\sigma v\rangle/m_\chi^2$. The left panel shows the gaugino fraction $Z_g$ of the neutralino versus its mass $m_\chi$. Further details are provided in the text; for even more, including an updated version of the figures, see Ref. 73.

are multi-purpose experiments dedicated to the physics of extreme objects, with DM often considered a mere side-aspect of the overall science goals; even for CTA one can probably not realistically expect to allocate more than about 5% of the available observation time for DM searches.

In order to assess the real potential of indirect searches, it is therefore timely to think about the possibility of a *dedicated* DM experiment[73] – a concept that is, indeed, very familiar from direct searches but has not been explored so far for indirect searches. To see how far one could possibly get with such an approach, Ref. 73 introduces the idea of a "Dark Matter Array" (DMA) with a CTA-like setup, but optimized for DM searches. Compared to CTA, DMA could gain roughly two orders of magnitude in sensitivity by allowing for an energy threshold of 10 GeV (note that at high altitudes above sea-level, even 5 GeV seems to be possible with current technology[74]), an effective area of about 10 km$^2$ and 5000 h observation time instead of the usually considered 50 h, implementing thus the idea of a dedicated experiment that only would observe a selected, very small number of targets highly relevant for DM.

In Fig. 2, the reach of indirect searches in gamma rays is compared to current and planned direct detection experiments like (Super)CDMS and XENON.[75] For the indirect searches, the background model[76] of the Fermi

LAT group was employed and it was assumed that the angular resolution is good enough to distinguish the contribution from the galactic center point source[77] observed by HESS. The figure shows the result of large scans over the parameter space for the supersymmetric neutralino using DarkSUSY,[78] where only models were included that satisfy all current collider bounds and give the correct relic density. Clearly, the parameter space of well-motivated DM candidates extends to much smaller direct detection cross sections than what is usually shown in corresponding exclusion plots; such models would only be accessible by indirect searches and the DMA would be ideal for such a task. Similar conclusions can be drawn for the observation of dwarf galaxies;[73] for the galactic center, observational prospects of course improve significantly for clumpy or more spiky profiles than NFW. From the right panel of Fig. 2, one can see that direct searches are particularly well suited for mixed neutralinos, while very massive Higgsinos could only be probed by a setup like DMA; very massive Gauginos are more difficult to probe and would require rather favorable astrophysical conditions. Finally, it is worth stressing that the most massive models in reach of XENON1t and/or DMA would be out of reach for the LHC, demonstrating once again the complementarity of all these approaches.

## 6. Conclusions

While we have (probably) not seen a DM signal so far, such a detection could well be just around the corner: indirect searches place ever more stringent constraints that start to touch the parameter region of interest for WIMPs – which one can argue to be the *a priori* best motivated DM candidates from the point of view of particle physics. Gamma rays are particularly interesting in this respect since they carry *spectral information* which could be used to infer detailed properties of the particle nature of DM, especially if the observed signal extends to the highest kinematically accessible energies. If one could correlate such results with findings from indirect searches at other wavelengths or from other messengers, this would of course allow an even better determination of the DM properties.

However, one should keep in mind that the parameter space of well-motivated DM candidates extends well beyond the reach of any currently planned indirect detection experiment. In order to fully make use of the potential of indirect searches, one would need a *dedicated DM experiment* like the Dark Matter Array (DMA) which would prove truly complementary to other approaches and could, even without very optimistic assumptions about the astrophysical distribution of DM, constrain models that are completely out of reach for direct detection experiments or the LHC.

**Acknowledgments**

I would like to thank the organizers for inviting me to this conference. As an Emmy Noether fellow, I gratefully acknowledge support from the German Research Foundation (DFG) under grant BR 3954/1-1.

**References**

1. E. Komatsu *et al.*, arXiv:1001.4538.
2. F. Zwicky, *Helv. Phys. Acta* **6**, 110 (1933).
3. R. Bernabei *et al.*, *Eur. Phys. J. C* **56**, 333 (2008); C. E. Aalseth *et al.*, arXiv:1002.4703; Z. Ahmed *et al.*, *Science* **327**, 1619 (2010).
4. O. Adriani *et al.*, *Nature* **458**, 607 (2009); J. Chang *et al.*, *Nature* **456**, 362 (2008); A. A. Abdo *et al.*, *Phys. Rev. Lett.* **102**, 181101 (2009).
5. for reviews see, e.g., G. Jungman, M. Kamionkowski and K. Griest, *Phys. Rept.* **267**, 195 (1996); L. Bergström, *Rept. Prog. Phys.* **63**, 793 (2000); G. Bertone, D. Hooper and J. Silk, *Phys. Rept.* **405**, 279 (2005).
6. D. Hooper and S. Profumo, *Phys. Rept.* **453**, 29 (2007).
7. A. Birkedal *et al.*, *Phys. Rev. D* **74**, 035002 (2006).
8. see, e.g., G. Bertone *et al.*, arXiv:astro-ph/0612387; J. A. R. Cembranos *et al.*, arXiv:1009.4936.
9. A. Birkedal, K. T. Matchev, M. Perelstein and A. Spray, hep-ph/0507194.
10. T. Bringmann, L. Bergström and J. Edsjö, *JHEP* **0801**, 049 (2008).
11. N. F. Bell *et al.*, *Phys. Rev. D* **78**, 083540 (2008). M. Kachelriess, P. D. Serpico and M. A. Solberg, *Phys. Rev. D* **80**, 123533 (2009); P. Ciafaloni and A. Urbano, *Phys. Rev. D* **82**, 043512 (2010); C. E. Yaguna, *Phys. Rev. D* **81**, 075024 (2010); P. Ciafaloni *et al.*, arXiv:1009.0224 [hep-ph]. N. F. Bell, J. B. Dent, T. D. Jacques and T. J. Weiler, arXiv:1009.2584.
12. L. Bergström, P. Ullio and J. H. Buckley, *Astropart. Phys.* **9**, 137 (1998).
13. L. Bergström, T. Bringmann, M. Eriksson and M. Gustafsson, *JCAP* **0504**, 004 (2005); G. Bertone *et al.*, arXiv:1009.5107.
14. see, e.g., M. Gustafsson *et al.*, *Phys. Rev. Lett.* **99**, 041301 (2007).; E. Dudas *et al.*, JHEP **0908**, 014 (2009); Y. Mambrini, JCAP **0912**, 005 (2009).
15. see, e.g., J. Zhang *et al.*, *Phys. Rev. D* **80**, 023007 (2009); E. Borriello, A. Cuoco and G. Miele, *Astrophys. J.* **699**, L59 (2009); M. Cirelli and P. Panci, *Nucl. Phys. B* **821**, 399 (2009); L. Zhang, F. Miniati and G. Sigl, arXiv:1008.1801.
16. M. Battaglia *et al.*, *Eur. Phys. J. C* **33**, 273 (2004).
17. T. Bringmann, M. Doro and M. Fornasa, *JCAP* **0901**, 016 (2009); M. Cannoni *et al.*, *Phys. Rev. D* **81**, 107303 (2010).
18. T. Bringmann, *PoS* **IDM2008**, 026 (2008).
19. L. Bergström *et al.*, *AIP Conf. Proc.* **861**, 814 (2006).
20. L. Bergström, T. Bringmann, M. Eriksson and M. Gustafsson, *Phys. Rev. Lett.* **95**, 241301 (2005).

21. L. Bergström, T. Bringmann, M. Eriksson and M. Gustafsson, *Phys. Rev. Lett.* **94**, 131301 (2005).
22. M. Perelstein and B. Shakya, arXiv:1007.0018.
23. J. Diemand *et al.*, *Nature* **454**, 735 (2008); V. Springel *et al.*, *MNRAS* **391**, 1685 (2008).
24. J. Einasto, *Trudy Inst. Astroz. Alma-Ata* **51**, 87 (1965).
25. J. F. Navarro, C. S. Frenk and S. D. M. White, *Astrophys. J.* **462**, 563 (1996).
26. G. Gilmore *et al.*, *Astrophys. J.* **663**, 948 (2007); S.-H. Oh *et al.*., *Astronom. J.* **136**, 2761 (2008); M. Spano *et al.*., *MNRAS* **383**, 297 (2008).
27. L. E. Strigari, C. S. Frenk and S. D. M. White, arXiv:1003.4268.
28. S. Mashchenko, J. Wadsley and H. M. P. Couchman, *Science* **319**, 174 (2008); S. Peirani, S. Kay and J. Silk, *Astron. Astrophys.* **479**, 123 (2008); T. Goerdt *et al.*, *MNRAS* **368**, 1073 (2006); T. Goerdt, J. I. Read, B. Moore and J. Stadel, arXiv:0806.1951.
29. G. R. Blumenthal, S. M. Faber, R. Flores and J. R. Primack, *Astrophys. J.* **301**, 27 (1986).
30. see, e.g., M. Gustafsson, M. Fairbairn and J. Sommer-Larsen, *Phys. Rev. D* **74**, 123522 (2006); J. R. Jardel and J. A. Sellwood, *Astrophys. J.* **691**, 1300 (2009); D. Ceverino and A. Klypin, *Astrophys. J.* **695**, 292 (2009).
31. see, e.g., A. El-Zant, I. Shlosman and Y. Hoffman, *Astrophys. J.* **560**, 636 (2001); C. Tonini, A. Lapi, F. Shankar and P. Salucci, Astrophys. J. **638**, L13 (2006); E. Romano-Diaz, I. Shlosman, Y. Hoffman and C. Heller, *Astrophys. J.* **685**, L105; Astrophys. J. **702**, 1250 (2009).
32. W. J. G. de Blok, arXiv:0910.3538; J. R. Primack, arXiv:0909.2247.
33. S. Colafrancesco, S. Profumo and P. Ullio, *Phys. Rev. D* **75**, 023513 (2007).
34. P. Ullio, H. Zhao and M. Kamionkowski, *Phys. Rev. D* **64**, 043504 (2001); D. Merritt, M. Milosavljevic, L. Verde and R. Jimenez, *Phys. Rev. Lett.* **88**, 191301 (2002).
35. G. Zaharijas and D. Hooper, *Phys. Rev. D* **73**, 103501 (2006).
36. N. Fornengo, L. Pieri and S. Scopel, *Phys. Rev.* D **70**, 103529 (2004).
37. S. Dodelson, D. Hooper and P. D. Serpico, *Phys. Rev. D* **77**, 063512 (2008).
38. P. D. Serpico and G. Zaharijas, *Astropart. Phys.* **29**, 380 (2008).
39. P. Ullio, L. Bergström, J. Edsjö and C. G. Lacey, *Phys. Rev. D* **66**, 123502 (2002).
40. E. A. Baltz *et al.*, *Phys. Rev. D* **61**, 023514 (2000); C. Tyler, *Phys. Rev. D* **66**, 023509 (2002); N. W. Evans, F. Ferrer and S. Sarkar, *Phys. Rev. D* **69**, 123501 (2004).
41. L. E. Strigari *et al.*, arXiv:0709.1510.
42. see, e.g., T. Bringmann, M. Doro and M. Fornasa, *JCAP* **0901**, 016 (2009); L. Pieri *et al.*, *Astron. Astrophys.* **496**, 351 (2009); G. D. Martinez *et al.*, *JCAP* **0906**, 014 (2009); R. Essig, N. Sehgal and L. E. Strigari, *Phys. Rev. D* **80**, 023506 (2009); P. Scott *et al.*, *JCAP* **1001**, 031 (2010).
43. V. Belokurov *et al.*, *Astrophys. J.* **654**, 897 (2007); M. Geha, M. R. Blanton, M. Masjedi and A. A. West, *Astrophys. J.* **653**, 240 (2006).
44. L. E. Strigari *et al.*, *Astrophys. J.* **669**, 676 (2007).
45. G. Lake, *Nature* **346**, 39 (1990).

46. L. Pieri, J. Lavalle, G. Bertone and E. Branchini, arXiv:0908.0195; B. Anderson *et al.*, *Astrophys. J.* **718**, 899 (2010).
47. L. Pieri, G. Bertone and E. Branchini, *MNRAS* **384**, 1627 (2008).
48. T. E. Jeltema, J. Kehayias and S. Profumo, *Phys. Rev. D* **80**, 023005 (2009).
49. S. Colafrancesco, S. Profumo and P. Ullio, *Astron. Astrophys.* **455**, 21 (2006).
50. G. Bertone, A. R. Zentner and J. Silk, *Phys. Rev. D* **72**, 103517 (2005).
51. T. Bringmann, J. Lavalle and P. Salati, *Phys. Rev. Lett.* **103** (2009) 161301.
52. J. Silk and A. Stebbins, *Astrophys. J.* **411**, 439 (1993).
53. T. Bringmann, *New J. Phys.* **11**, 105027 (2009).
54. A. M. Green, S. Hofmann and D. J. Schwarz, *JCAP* **0508**, 003 (2005).
55. see, e.g., T. Bringmann, *JCAP* **0508**, 006 (2005); J. Lavalle, J. Pochon, P. Salati and R. Taillet, *Astron. Astrophys.* **462**, 827 (2007).
56. J. Hisano, S. Matsumoto and M. M. Nojiri, *Phys. Rev. Lett.* **92**, 031303 (2004); J. Hisano, S. Matsumoto, M. M. Nojiri and O. Saito, *Phys. Rev. D* **71**, 063528 (2005); N. Arkani-Hamed, D. P. Finkbeiner, T. R. Slatyer and N. Weiner, *Phys. Rev. D* **79**, 015014 (2009).
57. http://fermi.gsfc.nasa.gov/
58. http://www.ams02.org/
59. http://www.mpi-hd.mpg.de/hfm/HESS/
60. http://veritas.sao.arizona.edu/
61. http://wwwmagic.mppmu.mpg.de/
62. http://gamma1.astro.ucla.edu/joomla/
63. http://www.cta-observatory.org/
64. see, e.g., A. Cesarini *et al.*, *Astropart. Phys.* **21**, 267 (2004); D. Horns, *Phys. Lett. B* **607**, 225 (2005) [Erratum-ibid. B **611**, 297 (2005)].
65. F. W. Stecker, S. D. Hunter and D. A. Kniffen, *Astropart. Phys.* **29**, 25 (2008); F. Aharonian *et al.*, *Phys. Rev. Lett.* **97**, 221102 (2006) [Erratum-ibid. **97**, 249901 (2006)].
66. T. Bringmann, arXiv:0911.1124; J. Lavalle, arXiv:1007.5253.
67. A. A. Abdo *et al.*, *Phys. Rev. Lett.* **104**, 091302 (2010).
68. A. A. Abdo *et al.*, *JCAP* **1004**, 014 (2010).
69. M. Ackermann *et al.*, *JCAP* **1005**, 025 (2010).
70. A. A. Abdo *et al.*, *Astrophys. J.* **712**, 147 (2010).
71. J. Albert *et al.*, *Astrophys. J.* **679**, 428 (2008); E. Aliu *et al.*, *Astrophys. J.* **697**, 1299 (2009); F. Aharonian *et al.*, arXiv:0809.3894; V. A. Acciari *et al.*, *Astrophys. J.* **720**, 1174 (2010).
72. L. E. Strigari *et al.*, *Phys. Rev. D* **75**, 083526 (2007).
73. L. Bergström, T. Bringmann and J. Edsjö, arXiv:1011.4514 [hep-ph].
74. F. A. Aharonian *et al.*, *Astropart. Phys.* **15**, 335 (2001).
75. Z. Ahmed *et al.*, arXiv:0912.3592; projected sensitivity for SuperCDMS and XENON1t from http://dendera.Berkeley.EDU/plotter/ entryform.html
76. S. Digel, presentation given at the Fermi Symposium 2009; see also http://fermi.gsfc.nasa.gov/ssc/data/access/lat/ BackgroundModels.html
77. F. Aharonian *et al.*, *Phys. Rev. Lett.* **97**, 221102 (2006) [Erratum-ibid. **97**, 249901 (2006)].
78. P. Gondolo *et al.*, *JCAP* **0407**, 008 (2004); www.physto.se/~edsjo/darksusy

# Introducing CLUMPY: a public code for gamma-ray emission from Dark Matter annihilation in the Galaxy

C. COMBET

*Dept of Physics and Astronomy, University of Leicester,*
*Leicester, LE1 7RH, United Kingdom*
*E-mail: celine.combet@astro.le.a.c.uk*

A. CHARBONNIER[†] and D. MAURIN[*]

*LPNHE, CNRS-IN2P3/Universités Paris VI et Paris VII, 4 place Jussieu, Tour 33,*
*75252 Paris Cedex 05, France*
[†]*E-mail: aldee@lpnhe.in2p3.fr*
[*]*E-mail: dmaurin@lpnhe.in2p3.fr*

We present CLUMPY, a public semi-analytic code that calculates the $\gamma-$ray emission from dark matter annihilation in a variety of configurations. CLUMPY is written in C/C++, interfaced with ROOT and is fully documented using the Doxygen package. It can be used in a statistical skymap mode or in a object-specific way. The former usage will be useful, among other things, to test the sensitivity of future $\gamma$-ray experiments, while the latter is dedicated to the study of dwarf spheroidal galaxies.

*Keywords*: Style file; LaTeX; Proceedings; World Scientific Publishing.

## 1. Introduction

Despite the several astrophysical evidences pointing at the existence of dark matter (flat rotation curves of galaxies, gravitational lensing, "bullet cluster", etc.), its nature is still evading us. Indirect detection approaches aim at measuring the end products of dark matter annihilation (e.g., $\gamma$-rays, positrons, anti-protons). In particular, the detection of $\gamma$-rays is one of the most promising channel[1] and has been explored by many authors.[2–6] Such a signal is yet to be measured by the existing $\gamma$-rays observatories (HESS, MAGIC, VERITAS, FERMI) but next-generation instruments, such as HAWC or CTA, may have a better chance. Having a modeling tool that can compute the expected $\gamma$-ray flux from dark matter annihilation in a wide range of astrophysical configurations and making it available to the community should prove very useful. This is the motivation for developing the CLUMPY code, the main features of which are briefly covered in this proceeding.

The $\gamma$-ray flux from DM annihilation in the galactic coordinate direction $(\phi, \theta)$ in the solid angle $\Delta\Omega$ is generically written as the product of a particle physics term times an astrophysical contribution:

$$\frac{d\Phi_\gamma}{dE_\gamma}(E_\gamma, \psi, \theta, \Delta\Omega) = \frac{d\Phi_\gamma^{PP}}{dE_\gamma}(E_\gamma) \times \Phi_\gamma^{astro}(\psi, \theta, \Delta\Omega)\,. \tag{1}$$

In this first release, the CLUMPY code focuses mainly on the calculation of the astrophysical term[a], that is written

$$\Phi^{astro}(\psi, \theta, \Delta\Omega) = \int_0^{2\pi} d\beta \int_0^{\alpha_i nt} \sin\alpha \int_0^{l_{\max}} \rho^2(l, \alpha, \beta; \psi, \theta) dl d\alpha d\beta\,, \tag{2}$$

where $d\Omega = \sin\alpha\, d\alpha\, d\beta$, $(l, \alpha, \beta)$ are spherical coordinates defined in the $(\psi, \theta)$ direction and $\rho$ is the local dark matter density. If formally simple, this integral is actually hard to evaluate as the Galactic halo is composed of a smooth DM distribution and of a population of smaller substructures (clumps) distributed throughout the Galactic halo over a wide range of masses, and because all these entities may have very steep DM profiles.

## 2. Code content

CLUMPY can be operated in two modes:

- *The statistical skymap mode* allows the user the generate realistic skymaps (of $\Phi^{astro}$) for any direction, field of view, resolution and DM profiles (smooth and/or clumps). As the code is semi-analytical (to be opposed to the full numerical N-body approach), one realisation takes only $\sim$ hours which really opens the way to statistical studies that are still beyond the reach of N-body approaches.
- *The dwarf spheroidal mode* has been developed with the specific purpose of studying the $\gamma$-ray emission from these small satellite galaxies of the Milky Way. These are thought to be favoured targets for DM indirect detection via the $\gamma$-ray channel as they do not have any background that would pollute the signal.

[a]Given the plethora of DM particle physics model, only few standard parametrisations of the particle spectrum have been included in the code, but it is really left to the user to input their favorite particle physics model.

Both modes of operation share the same core library/functions. A short description of the main ones is given hereafter.

### 2.1. *DM profiles* – `profiles.h`

In CLUMPY, the DM halos (smooth or clumps) are assumed spherical and the code includes the most standard DM profiles found in the literature, e.g. the NFW,[7] Moore,[8] Einasto,[9] Diemand et al.,[10] Navarro et al.,[11] and isothermal profiles. This list is supplemented with a generic $\alpha, \beta, \gamma$ profile,

$$\rho(r) = \frac{\rho_s}{(r/r_s)^\gamma \left[1 + (r/r_s)^\alpha\right]^{(\beta-\gamma)/\alpha}}\,, \tag{3}$$

aimed at being user defined. These profiles can all be used, in any combination, to describe the smooth Galactic component, the clumps inner profiles and the clumps global spatial distribution through the Milky Way halo (see §2.2).

### 2.2. *Clump treatment* – `clumps.h`

CLUMPY makes use the results of N-body simulations to treat the clumps in a semi-analytical way. The clump distribution is given by the product of a spatial distribution $dP_V/dV$ times a mass distribution $dP_M/dM$. The two distributions are assumed independent form one another and the spatial distribution can follow any of the profile mentioned above.

N-body simulations show that the mass distribution is well modeled by a power law,[3]

$$\frac{dP_M}{dM} \propto M^{-\alpha_M} \;\; \text{with} \;\; \alpha_M \in [1.7, 2],$$

and generally present $\sim 100$ clumps between $10^8$ and $10^{10}$ $M_\odot$. Extrapolating the mass function up to $10^{-6}$ $M_\odot$, one infers $\sim 10^{14}$ clumps in the whole Galactic halo. This is too large a number to be statistically drawn. To circumvent this difficulty, CLUMPY only draws the clumps below a critical distance $l_{\rm crit}(M)$ that is determined for each mass decade on statistical arguments. Above that distance, no clumps are drawn and only their average contribution is computed; below $l_{\rm crit}$, the clumps are drawn following their mass and spatial distributions and their flux is computed individually.

### 2.3. *Integration along the line of sight* – `libintegr_los.h`

This is the core of the CLUMPY code and numerically computes eq.(2) at a given user-defined precision. This triple integral is numerically challenging as some of the DM profiles becomes very steep toward the centre of the Galactic halo/clumps and a lot of care is necessary for the integration not to explode or take a prohibitive amount of time to be evaluated. The details of the strategy to perform this integration can be found in the Doxygen documentation. For illustration, the main trick consists in locating, along the line of integration, the position where the DM density is the highest and splitting the integration into two parts, using on both side the same adaptative log-step Simpson integrator.

## 3. Examples

### 3.1. *Skymap mode*

Figure 1 is an example of a $5 \times 5$ deg$^2$ skymap, along the direction $(\psi, \theta) = (-120, 20)$ produced by CLUMPY in the statistical skymap mode. The smooth halo, clump inner DM profile and clump global spatial distribution are all assumed to follow the Einasto prescription. The mass distribution index is $\alpha_M = 1.9$ and the minimum mass of the clumps chosen to be $M_{\rm min}^{\rm cl} = 10^{-6}\ M_\odot$. The integration angle was decreased from 0.1 deg to 0.01 deg and the same seed of the random number generator chosen is both case.

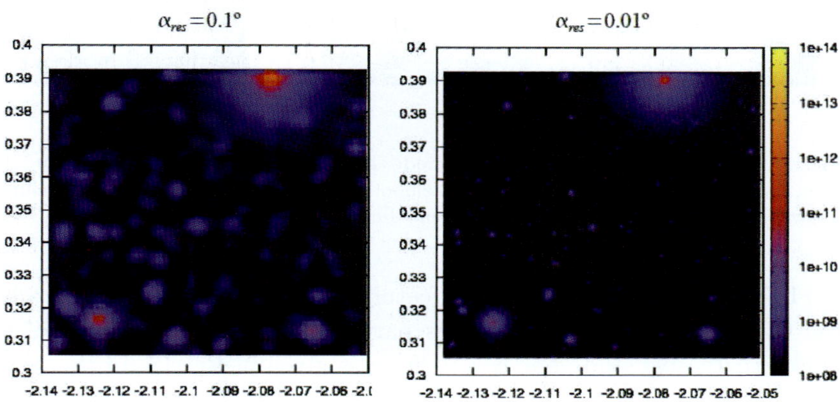


Fig. 1. Output of CLUMPY in the skymap mode. See text for details.

Such maps, once multiplied by the particle physics term could then be convolved with an instrumental response and be used as template for analysis for the future $\gamma$-ray observatories. The seed of the random number generator could also be systematically changed to allow statistical studies.

### 3.2. *Dwarf Spheroidal mode*

When used in the dwarf spheroidal mode, the code computes the astrophysical term of the gamma-ray flux for one individual clump, the density profile of which is user-defined. In this mode, the possibility exists to explicitly draw a distribution of subclumps within the DM halo considered, a feature that is, in the first release of the code, only averaged over in the statistical skymap mode. Figure 2 is an example of a hypothetical dSph galaxy, with a mass $M = 10^7\ M_{\odot}$, located at 100 kpc. The three snapshots correspond to three integration angles. As the resolution increases, the value of the $\gamma$-ray flux decreases but the sub-clumps become more and more apparent.

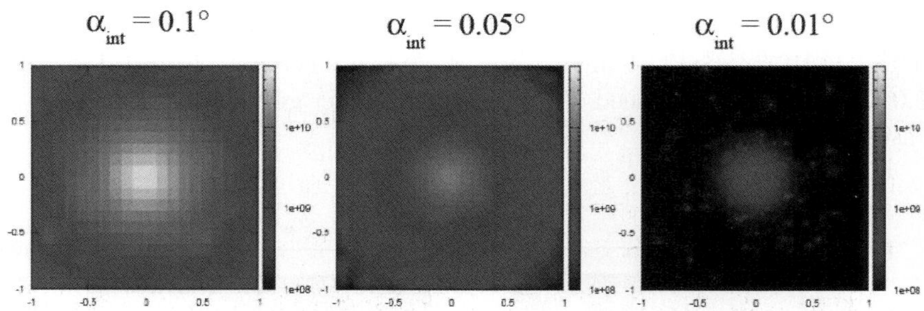


Fig. 2. CLUMPY in dwarf spheroidal mode. In this example, the dwarf is a $10^7\ M_{\odot}$ halo, located at 100 kpc and has a core smooth profile. The population of subclumps follows the smooth distribution, but the subclumps have NFW inner DM profiles.

This figure only illustrates one feature among the many available when the CLUMPY code is used in this particular mode. An example of the kind of extensive study one can perform is given in Ref 12. In the latter, we have re-explored the detectability of the eight classical dSph galaxies by existing and future $\gamma$-ray telescopes. This was done using the latest DM profiles obtained for these galaxies from velocity measurements coupled with the MCMC analysis of Ref. 13. The dwarf spheroidal mode of CLUMPY provides many routines to use results from MCMC (or other statistical methods) runs as an input and calculate the astrophysical factor along with its confidence levels. The reader is referred to that paper for a more ample description.

## 4. Conclusion

CLUMPY is a new public code for the calculation of the $\gamma$-ray flux from DM annihilation in the Galaxy. In its first release, it includes the contribution from the smooth DM halo and that of a population of substructures, randomly drawn from user-defined distributions. It may be used to produce skymaps at any resolution or in a more specific way for the study of dwarf spheroidal galaxies. CLUMPY is fully documented with the Doxygen package and interfaced with the ROOT CERN libraries for its statistical and graphical facilities. Following releases of the CLUMPY should include other DM annhilation tracers, such as anti-protons and positrons, the secondary $\gamma$-ray emission from leptons along with the extragalactic contribution.

## References

1. J. E. Gunn, B. W. Lee, I. Lerche, D. N. Schramm and G. Steigman, *ApJ* **223**, 1015(August 1978).
2. L. E. Strigari, S. M. Koushiappas, J. S. Bullock and M. Kaplinghat, *Phys. Rev. D* **75**, 083526(April 2007).
3. J. Diemand, M. Kuhlen and P. Madau, *ApJ* **657**, 262(March 2007).
4. L. Pieri, A. Pizzella, E. M. Corsini, E. Dalla Bontà and F. Bertola, *A&A* **496**, 351(March 2009).
5. G. D. Martinez, J. S. Bullock, M. Kaplinghat, L. E. Strigari and R. Trotta, *J. Cosmology Astropart. Phys.* **6**, 14(June 2009).
6. M. Kuhlen, *Advances in Astronomy* **2010** (2010).
7. J. F. Navarro, C. S. Frenk and S. D. M. White, *ApJ* **490**, 493(December 1997).
8. B. Moore, F. Governato, T. Quinn, J. Stadel and G. Lake, *ApJ* **499**, L5+(May 1998).
9. D. Merritt, A. W. Graham, B. Moore, J. Diemand and B. Terzić, *AJ* **132**, 2685(December 2006).
10. J. Diemand, B. Moore and J. Stadel, *MNRAS* **353**, 624(September 2004).
11. J. F. Navarro, E. Hayashi, C. Power, A. R. Jenkins, C. S. Frenk, S. D. M. White, V. Springel, J. Stadel and T. R. Quinn, *MNRAS* **349**, 1039(April 2004).
12. A. Charbonnier, C. Combet, M. Daniel, S. Funk, J. Hinton, D. Maurin, C. Power, J. Read, S. Sarkar, M. Walker and M. Wilkinson, *submitted soon* (2010).
13. M. G. Walker, M. Mateo, E. W. Olszewski, O. Y. Gnedin, X. Wang, B. Sen and M. Woodroofe, *ApJ* **667**, L53(September 2007).

# Cosmic rays and dark matter indirect detection

Timur DELAHAYE

*Instituto de Física Teórica UAM/CSIC*
*Universidad Autónoma de Madrid ,*
*28049 Cantoblanco, Madrid, Spain.*
*E-mail: timur.delahaye@uam.es*

Though no firm evidence of dark matter detection can be claimed from the recent data, it is possible nevertheless to derive some interesting constraints already. I show how difficult it is to explain the current positron data only with the point of view of dark matter and also that the anti-proton data start to be constraining for dark matter models.

*Keywords*: Cosmic rays; dark matter; positrons; anti-protons.

## 1. Introduction

Recent cosmic ray data from *PAMELA*[1,2] , *Fermi*[3,4] but also *ATIC*[5] and *HESS*[6] have risen a lot of interest for dark matter indirect detection in the past few years.

Indeed, following the original idea of Silk and Srednicki[7] , looking for dark matter through its annihilation or decay products in cosmic rays is a promising detection method.

Cosmological and astrophysical hints strongly support the hypothesis of the existence of dark matter. In the common understanding of what dark matter is, our galaxy, should be immersed in a very large dark matter halo. In this halo, dark matter particles, depending on their properties, may annihilate or decay[a] into high energy Standard Model particles. Many beyond the Standard Model particle physics models predict a dark matter particle with a mass of the order of 100 GeV in agreement with constraints from cosmology (relic density). This means that the annihilation/decay products are expected to be high energetic and to contribute to the cosmic

[a]Here I will not enter into details concerning dark matter decay, the phenomenology of which is not so different from the annihilating case.

ray budget.

The background being lower for anti-protons than for protons and for positrons than for electrons, it seems natural to focus on these species. However, some experiments are not designed for charge recognition or are affected by systematics that prevent access to accurate measurements of absolute fluxes of these two species. This is why we often have to deal with the sum of electron and positron fluxes (*Fermi* , *ATIC* and *HESS*), the positron fraction (*PAMELA*) that the positron flux divided by the flux of electrons plus positrons or the anti-proton to proton ratio (*PAMELA*). Fortunately, very recently the *PAMELA* collaboration has managed to publish a absolute anti-proton flux up to 200 GeV with an unprecedented precision.

In what follows, I will describe the present situation concerning leptons (electrons and positrons) and anti-protons. Each time I will focus on our understanding of the background and the expected exotic dark matter originating signal.

## 2. Leptons

Cosmic rays studies have given extremely satisfactory agreement between theory and observations for various species since a long time ago. However recent experiment results have changed this bright situation by confirming a discrepancy between theoretical expectations and observations for high energy ($\gtrsim$ 10 GeV) electrons and positrons that had been previously suspected thanks to the *HEAT*[8] experiment.

Up to then, positrons were considered from secondary origin *i.e.* coming from the interaction of cosmic ray protons and $\alpha$ with interstellar hydrogen and helium. Many theoretical uncertainties affect the prediction of this secondary positron flux. A full description of these uncertainties and of the estimation of secondary positrons can be found in Delahaye et al.[9] . However, even taking these uncertainties into account, it is not possible to explain the positron fraction data with secondaries only.

Various astrophysical explanations have been proposed so far: pulsars[10,11] , secondary production in supernova remnants[12,13] , etc. However it is impossible to say that dark matter does not contribute partially or totally to the positron excess.

To compute the positron flux expected from dark matter annihilation, one needs to take into account for the propagation. A detailed analytical computation can be found at reference[14] . Though the propagation parameters are constrained by various secondary to primary ratios[15] , some freedom remains which translate into quite large uncertainties in the expected flux. Thanks to fast analytical methods this uncertainty is quite under control and future secondary to primary ratio data should help diminish this uncertainty.

Computing the expected positron flux, also requires knowing the source term $Q(E,x)$ which is usually written in the following form:

$$Q(E,x) = \eta\langle\sigma v\rangle \left(\frac{\rho(x)}{m_\chi}\right)^2 f(E), \tag{1}$$

where $\eta$ is a statistical factor depending on the spin of the dark matter particle, $\langle\sigma v\rangle$ is the dark matter annihilation cross-section, $\rho(x)$ is the dark matter mass density spatial distribution, $m_\chi$ is the mass of the dark particle and finally, $f(E)$ is the positron spectrum after the annihilation.

A very nice feature is that the positron flux at the Earth, though distorted by propagation effects, should be quite different depending on what is the main dark matter annihilation channel. Hence, it should be possible to distinguish this main annihilation channel, giving access to dark matter properties far more difficult to reach with direct detection method or even with *LHC*.

Another positive aspect of dark matter indirect detection through positrons is that the expected flux is not very sensitive to our precise knowledge of the dark matter distribution in the Galactic halo $\rho(x)$. Indeed this quantity despite many efforts remains affected by large uncertainties, especially around the Galactic center, because N-body simulations and dynamical studies tend to give quite different results. Yet, because positrons loose a lot of energy through inverse Compton scattering and synchrotron radiation when propagating in the Galactic diffusive halo, they cannot propagate very far which means that most the positrons we detect at the Earth have been created in the nearby two kiloparsecs. In this region, the dark matter distribution is expected to be rather similar whatever the method used to estimate it.

However, the dark matter annihilation cross-section $\langle\sigma v\rangle$ is quite constrained by cosmology (relic density computation in $\Lambda$CDM model) and is expected to be rather small. So small that the expected positron flux coming from dark matter annihilation is expected to be quite smaller than the measured spectrum. Various suggestions have been made to *boost* the exotic signal, either by increasing the annihilation cross-section circumventing the cosmological constraints or by invoking the dark matter inhomogeneities predicted by N-body simulation. Indeed, these simulations predict many over-densities (*clumps*) where the annihilation rate should be higher. Actually, because positrons are so local, only one local clump should be enough to boost sufficiently the signal.

Unfortunately, it has been shown[16] that, according to the *Via Lactea II* simulation[17] , the clump required to explain the *PAMELA* excess is very unlikely to exist in the most optimistic cases and is completely ruled out in most cases. Indeed, as one can see in figure 1, even considering a light (100 GeV) dark matter particle annihilating exclusively into electron-positron pair the required clump happen to exist only in less than 1% of the realisation of the N-body simulation.

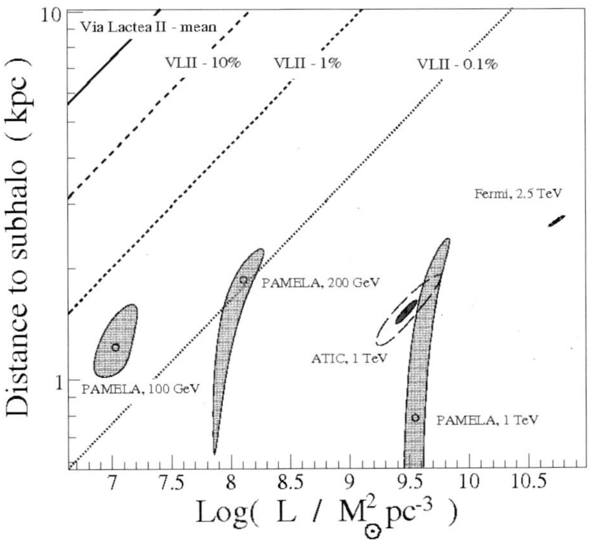


Fig. 1. Best fit results in a clump luminosity-distance plane for different configurations, together with probabilities inferred from *Via Lactea II* results. Each region corresponds to a fit to the experiment the name of which is written next to the mass of the dark matter particle considered. In all the cases only annihilation into electron-positron pairs has been considered.

The other idea to boost the signal, increasing the annihilation cross-section, as already mentioned earlier, can be invoked when considering dark matter annihilating exclusively into electron-positron or muon pairs. However, as soon as another annihilation channel is concerned, it implies a production of anti-protons along with the positrons hence it is necessary to check both species at once.

## 3. Anti-protons

As one can see from figure 2, the most recent data from the *PAMELA* satellite are in very good agreement with the prediction of the secondary and tertiary fluxes[b] (in green) which are of standard astrophysical origin.

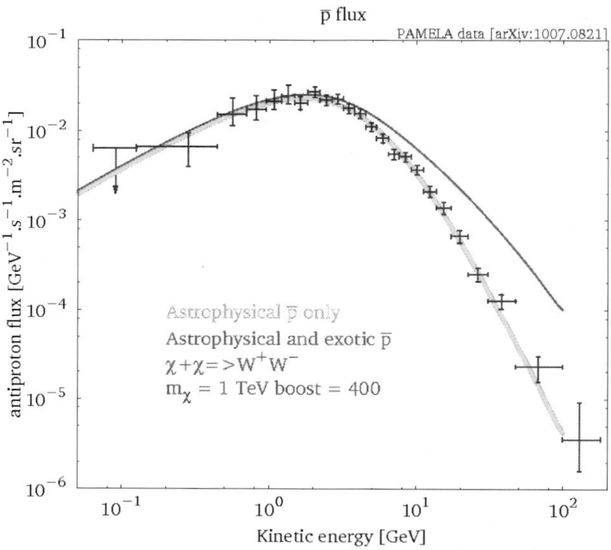


Fig. 2. Recent PAMELA data and expected anti-proton fluxes. The green line correspond to a prediction for secondary and tertiary anti-protons. The red line correspond to the previous one plus the primary contribution expected from a dark matter particle with a mass of 1 TeV annihilating into $W^+W^-$ with a cross-section boosted by a factor of 400. This is an updated version of reference.[18]

In the case where dark matter would only annihilate into electron-positron pairs, the absence of excess of cosmic ray anti-protons is not surprising ; though this possibility cannot be dismissed, it would be quite unexpected from the point of view of particle physics. Most beyond the Standard Model particle physics theories predict a dark matter species which would annihilate into gauge bosons or quarks rather than into electron-positron pair. From this perspective, the interpretation of the positron fraction excess in term of dark matter, is difficult to reconcile with the anti-proton

[b]Tertiary anti-protons are the result of inelastic interaction of secondary cosmic ray anti-protons with interstellar hydrogen and helium.

data. For instance, if one considers a 1 TeV dark matter particle annihilating into $W^+W^-$, an annihilation cross-section 400 times higher than the standard one is required to roughly fit the positron fraction ; as one can see from the red curve of figure 2, this same dark matter would also produce anti-protons and in such an amount that the flux at the Earth should be much higher than the one experimentally observed.

However, the fact that there is no signal in the anti-proton channel is not necessarily sad news, as stressed earlier[19] , this can actually be of some help in putting upper bounds to the couplings of the dark matter particle to quarks and gauge bosons.

## 4. Conclusion

Dark matter indirect detection remains a wonderful way to access some properties of dark matter which cannot be reached otherwise. However the poor knowledge of the involved backgrounds makes the task very difficult and only a multi-messenger analysis can lead to the correct interpretations. Indeed, as I have shown here interpreting the positron fraction data in term of dark matter without taking into account the absence of excess of anti-proton would lead to wrong conclusion. Because the topic of this conference is cosmic ray, I have not spoken about the other messengers that carry information about dark matter annihilation (or decay), yet one should also keep in mind that a thorough analysis requires also to look at neutrinos and, more important, to photons, either due to prompt emission during the annihilation or by the interaction of cosmic rays with the interstellar medium (Bremsstrahlung), with the magnetic field (synchrotron emission) or with the interstellar radiation field (inverse Compton). In the category of cosmic rays, one should also mention anti-deuterium ; however large theoretical uncertainties prevail in this field and no recent data has been published.

## Acknowledgments

This work was supported by the Spanish MICINNs Consolider-Ingenio 2010 Programme under grant CPAN CSD2007-00042. We also thank the support of the MICINN under grant FPA2009-08958, the Community of Madrid under grant HEPHACOS S2009/ESP-1473, and the European Union under the Marie Curie-ITN program PITN-GA-2009-237920.

## References

1. O. Adriani et al., *Nature* **458**, 607(April 2009).
2. PAMELA Collaboration, O. Adriani, et al., *ArXiv e-prints* (July 2010).
3. A. A. Abdo, et al., *Physical Review Letters* **102**, 181101(May 2009).
4. Fermi LAT collaboration, *ArXiv e-prints* (August 2010).
5. J. Chang, et al., *Nature* **456**, 362(November 2008).
6. F. Aharonian, et al., *Physical Review Letters* **101**, 261104(December 2008).
7. J. Silk and M. Srednicki, *Phys. Rev. Lett.* **53**, 624(Aug 1984).
8. S. W. Barwick, and The HEAT Collaboration, *ApJ* **482**, L191+(June 1997).
9. T. Delahaye, R. Lineros, F. Donato, N. Fornengo, J. Lavalle, P. Salati and R. Taillet, *A&A* **501**, 821(July 2009).
10. D. Hooper, P. Blasi and P. Dario Serpico, *J. Cosmology Astropart. Phys.* **1**, 25(January 2009).
11. T. Delahaye, J. Lavalle, R. Lineros, F. Donato and N. Fornengo, *ArXiv e-prints* (February 2010).
12. P. Blasi, *Physical Review Letters* **103**, 051104(July 2009).
13. M. Ahlers, P. Mertsch and S. Sarkar, *Phys. Rev. D* **80**, 123017(December 2009).
14. T. Delahaye, R. Lineros, F. Donato, N. Fornengo and P. Salati, *Phys. Rev. D* **77**, 063527(March 2008).
15. D. Maurin, F. Donato, R. Taillet and P. Salati, *ApJ* **555**, 585(July 2001).
16. P. Brun, T. Delahaye, J. Diemand, S. Profumo and P. Salati, *Phys. Rev. D* **80**, 035023(August 2009).
17. J. Diemand, M. Kuhlen, P. Madau, M. Zemp, B. Moore, D. Potter and J. Stadel, *Nature* **454**, 735(August 2008).
18. F. Donato, D. Maurin, P. Brun, T. Delahaye and P. Salati, *Physical Review Letters* **102**, 071301(February 2009).
19. C. Bœhm, T. Delahaye, P. Salati, F. Staub and R. K. Singh, *J. Cosmology Astropart. Phys.* **6**, 13(June 2010).

# NEUTRINOS FROM DARK MATTER

M. H. RENO*

*Department of Physics and Astronomy, University of Iowa,*
*Iowa City, Iowa 52242, USA*
**E-mail: mary-hall-reno@uiowa.edu*

Dark matter annihilations or decays to neutrinos, or particles that decay to neutrinos, can lead to signals in underground and underwater detectors. Theoretical considerations for neutrino signals and backgrounds from accumulations in the Earth, the Sun and around the galactic center, as well as cosmic diffuse neutrino signals, are reviewed, together with some current limits from neutrino telescopes.

*Keywords*: dark matter; neutrinos; indirect detection.

## 1. Introduction

Neutrinos have been prominent in discussions of dark matter since the understanding that dark matter accounts for significantly more matter than the luminous matter in the universe.[1] Several decades ago, neutrinos were discussed as candidate dark matter particles. Early constraints on the neutrino mass came from the fact that massive neutrino dark matter cannot contribute more than the critical density of matter in the Universe. These first cosmological limits on neutrino masses required the sum of the neutrino masses to be less than tens of electron volts, depending on the Hubble constant and the value of $\Omega$.[2]

The sum of neutrino masses is constrained by the cosmic microwave background temperature fluctuations to be less than a few eV (or less) and lower than 1 eV from structure formation considerations in the Universe.[3] Neutrino masses are expected to be even lower based on neutrino oscillation experiments which measure the differences in squared neutrino masses of the three neutrino species.[4] The current understanding is that massive neutrinos can make up only a small fraction of the dark matter in the universe. Nevertheless, neutrinos can play an important role in determining the particle nature of the dark matter.

The direct detection of dark matter interactions through nuclear recoil in detectors is one approach to constraining the dark matter particle content.[5] Here, indirect detection with neutrinos is the focus.[6] The role of leptons in general has come to the fore with the apparent positron excess reported in the PAMELA[7] and the Fermi Large Area Telescope[8] results for electrons and positrons.

There is a vast literature on lepton signals from dark matter, including many on neutrino signals, either produced directly or indirectly. Dark matter models often considered include supersymmetric models where the lightest supersymmetric particle is the dark matter.[9,10] In some models, the lightest supersymmetric particle is the gravitino (see, e.g., Ref.[11]). Leptophilic models introduce new particles and interactions that favor DM annihilation or decays to leptons (see, e.g., Refs.[12–19]). Another example explored in the literature is with the lightest Kaluza-Klein particle as the dark matter.[20] A broader view of the a dark matter model is also discussed in K. Zurek's contribution to these proceedings.[21]

Indirect detection of dark matter annihilation or decay from neutrino signals is the topic discussed here. Dark matter accumulates in the cores of the Earth or Sun,[9,10,22–25] as well as at the Galactic center. These regions of dark matter accumulation may be regions of enhanced neutrino emission. The neutrinos from dark matter annihilation or decay may be produced directly or via intermediate state weak decays, depending on the DM model. In large underground detectors like Amanda, IceCube, Antares and the proposed KM3net,[26] muon neutrino production of muons inside the detector ("contained"), production of muons outside the detector which travel through the detector ("upward") and production of showers inside the detector ("showers") are the signals most accessible.

For underground neutrino detectors, the irreducible background is the atmospheric neutrino flux. The atmospheric flux generally has a steeper spectrum than the signal neutrino spectrum, and it has a predicted angular dependence where the horizontal flux is larger than the vertical flux.[27,28] An additional background comes from the downward atmospheric muon flux. On the order of $10^6$ muons arrive at the IceCube detector for every $\nu_\mu \to \mu$ conversion. Typically, the DM neutrino signal is therefore an signal from neutrinos arriving from below the detector to eliminate the atmospheric muons. The DeepCore subdetector of IceCube, which would use the larger IceCube detector as a downward muon veto, is another potential avenue for detecting DM neutrinos.[29]

## 2. Essential ingredients

Whether the dark matter annihilation occurs in the Earth's core, in the Sun or in the galactic halo, there are some common ingredients to the evaluation of the resulting signal in detectors on Earth. The density of dark matter plays a role along with the dark matter capture rate with nucleons for the signals from the core, as does the dark matter annihilation or decay rate. The specific model for the final state particles will determine the neutrino energy distribution. For the Sun, there are propagation effects once the neutrino is produced,[22–24] however for the Earth at the energies above about 100 GeV, attenuation is not important until a neutrino energy of about 40 TeV.[30] Muon propagation effects, however, are important in the evaluation of muon energy spectra for a given model.[19,25]

The neutrino flux from the Earth or Solar core is[9]

$$\left(\frac{d\phi_\nu}{dE_\nu}\right)_i = \frac{\Gamma_A}{4\pi R^2}\sum_F B_F \left(\frac{dN_\nu}{dE_\nu}\right)_{F,i}, \tag{1}$$

where $\Gamma_A$ is the annihilation rate, $R$ is the distance from the core to the detector, $B_F$ is the branching fraction for final state $F$ and $(dN_\nu/dE_\nu)_{F,i}$ is the energy distribution of the neutrino for the channel with final state $F$ for neutrino flavor $i$.

The energy distribution depends on the maximum energy available. For direct DM ($\chi$) annihilation to neutrinos for a DM mass of $m_\chi$, $E_\nu = m_\chi$ with a delta function distribution. For $\chi\chi \to \tau^+\tau^-$, the $E_{\nu_\mu}$ distribution in terms of $x = E_\nu/m_\chi$ is soft, dominated at by the low $x$ part. A harder neutrino energy spectrum appears in $\chi\chi \to W^+W^-$, followed by $W \to \mu\nu_\mu$. These three distributions are shown graphically in Fig. 1. The delta function distribution at $E_\nu = m_\chi$ makes the direct annihilation channel the most promising ("golden") for neutrino detection, but the most difficult to constrain by other means such as by electron, positron and photon detection. Neutralino dark matter, however, does not have direct annihilations to $\nu\bar{\nu}$.

With specific model inputs into Eq. (1) through the annihilation rates, branching fractions and annihilation channels, one can evaluate the rate for upward muons, contained muons and shower rates. The upward muon rate depends on the muon range $R(E_\mu^i, E_\mu)$ following from the approximate $dE_\mu/dX = -(\alpha+\beta E_\mu)$ energy loss formula. The neutrino cross section with nucleons enters via

$$\frac{dP_{CC}}{dz dE_\mu^i} = N_A \frac{d\sigma_{\nu N}}{dE_\mu^i} \tag{2}$$

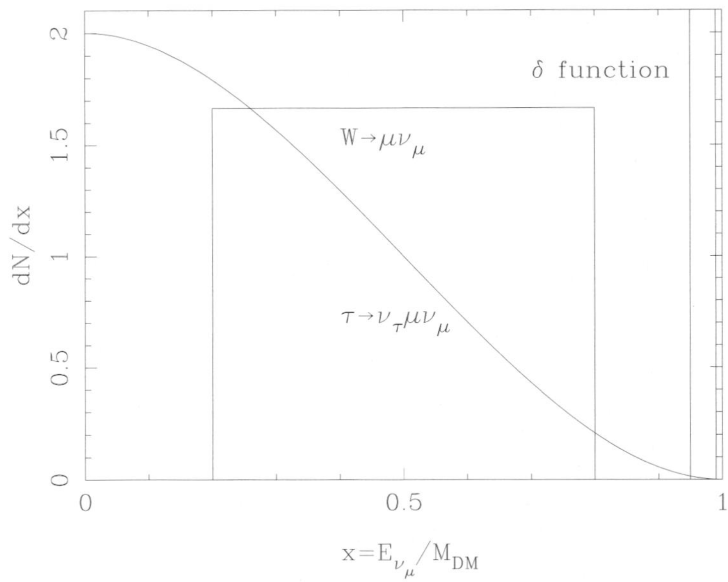


Fig. 1. The neutrino spectra as a function of $x = E_\nu/m_\chi$ for $\chi\chi \to \nu\bar{\nu}$ (delta function), and for $\chi\chi \to W^+W^-$ and $\chi\chi \to \tau^+\tau^-$ followed by $W$ or $\tau$ decay to muon neutrino. The curves are normalized to integrate to one.

where $N_A$ is Avogadro's number and $\sigma_{\nu N}$ is the neutrino nucleon charged current cross section, which at neutrino energies of hundreds of GeV, is often approximated by[31] $d\sigma/dE \propto (a + bE_\mu^2/E_\nu^2)$. At these energies, the muon survival probability is one.

The upward muon flux is then

$$\frac{d\phi_\mu}{dE_\mu} = \int_0^{R_\mu(E_\mu^i, E_\mu)} dz \int_{E_\mu^i}^{E_{max}} dE_\nu \frac{d\phi_\nu}{dE_\nu} \frac{dP_{CC}}{dz dE_\mu^i} \frac{dE_\mu^i}{dE_\mu} , \tag{3}$$

where $E_\mu^i = E_\mu \exp(\beta z) + (\exp(\beta z) - 1)\alpha/\beta$ for column depth $z$. The contained muon spectrum has a similar spectrum, where now it is the size of the detector $\ell$ rather than the muon range that is relevant. This can be written as

$$\frac{d\phi_\mu}{dE_\mu} = \int_0^{\ell} dz \rho \int_{E_\mu}^{E_{max}} dE_\nu \frac{d\phi_\nu}{dE_\nu} \frac{dP_{CC}}{dz dE_\mu} . \tag{4}$$

The initial muon energy is the energy in the detector for contained muons ($E_\mu^i = E_\mu$). The shower flux has the same form as the contained muon flux with $E_\mu \to E_{shr}$.

For the $\chi\chi \to \nu\bar{\nu}$ annihilation channel, the spectrum of the induced muons is changed fairly significantly when going from contained to upward muons. This comes from the muon energy loss. The change in the shape of the spectrum between contained and upward neutrino induced muons from neutrinos from $\tau$, $c$ or $b$ decays is smaller. This is shown for $\chi\chi \to \nu\bar{\nu}$ and $\chi\chi \to \tau\bar{\tau}$ as exhibited in Ref.[25]

For a detector like SuperKamiokande, the characteristic length of the detector is 40 m, so it is the upward muon rate that is most relevant given the scale of the muon range at high energies. The SuperKamiokande collaboration has constrained the WIMP induced upward muon flux from the core of the Earth in the context of neutralino dark matter.[32] The neutralinos annihilate primarily to $b\bar{b}$ (80%) and $c\bar{c}+\tau^+\tau^-$. The neutrino spectrum from these quark semi-leptonic decays is similar, but a little softer than the $\tau^+\tau^-$ spectrum in Fig. 1. The SuperKamiokande constraints put an upper limit on the flux of muons from neutralino annihilation in the range of $\sim 10^{-14} - 2 \times 10^{-15}$ muons per cm$^2$ per sec for neutralino masses between 20 GeV and 10 TeV. There are additional constraints from, for example AMANDA, on the muon flux from neutralino annihilation.[33] For annihilation in the Earth's core, these indirect constraints are weaker than limits from direct searches[5] and probe fluxes higher than what are predicted by SUSY models, for example, as evaluated using the DARKSUSY computer program.[10]

Dark matter annihilation in the Sun is probed by these indirect neutrino signals with neutrino induced muon flux limits in a similar range. For DM annihilations in the solar core, in contrast to annihilations in the Earth, the SuperKamiokande and larger neutrino telescope limits probe deeper into the parameter space constrained by direct DM measurements and limits. The neutrino limits start to constrain SUSY models. The cross section most relevant for the Sun is the spin-dependent wimp-nucleon scattering cross section. SuperKamionkande[32] and other underground experiments have put limits on the spin-dependent cross section, and recently, the IceCube-22 plus AMANDA data was analysed to put constraints on the cross section as low as a few times $10^{-40}$ cm$^2$ for neutralino masses of order a few hundred GeV for the "hard" spectrum ($W^+W^-$).[34] (he limits from a softer spectrum $b\bar{b}$ are considerably weaker for $m_\chi$ in the same mass range. The SuperKamiokande limit on the spin-dependent cross section[32] is better than this IceCube-22 plus AMANDA limit below $m_\chi \sim 100$ GeV, where SuperKamiokande's limit is a few times $10^{-38}$ cm$^2$. The combined full IceCube detector with the DeepCore inner detector has the potential to probe well into the allowed model region.[35] The IceCube collaboration also has flux constraints on the annihilation of the lightest Kaluza-Klein particles[36] in a model by Hooper and Kribs.[20] This model has the primary annihilation into charged lepton channels, with a small branching fraction into $\nu\bar{\nu}$.

## 3. Galactic center and halo

For the neutrino signals from neutrino annihilation at the Galactic center or halo, the dark matter density distribution is required. A common choice is the Navarro, Frenk and White (NFW) density.[37] The density comes in via the integral over the line of sight with a cone half-angle $\theta$ of either the density squared ($n = 2$) for annihilation or the density ($n = 1$) itself for annihilation. The quantity $\langle J_n \rangle_\Omega$ is defined by

$$\langle J_n \rangle_\Omega = \int \frac{d\Omega}{\Delta\Omega} \int_{l.o.s} \frac{dl(\theta)}{R_o} \left( \frac{\rho(l)}{\rho_o} \right)^n , \tag{5}$$

for $l(\theta)$ distance along the line of sight. This is normalized by the local dark matter density $\rho_0 = 0.3$ GeV/cm$^3$. The neutrino flux at Earth, a distance of $R_o = 8.5$ kpc from the galactic center, is[38]

$$\frac{d\phi_\nu}{dE_\nu} = R_o \rho_o^2 B \frac{\langle \sigma v \rangle_0}{8\pi m_\chi^2} \left( \sum_F B_F \frac{dN_\nu^F}{dE_\nu} \right) \langle J_2 \rangle_\Omega \Delta\Omega \quad n = 2 , \tag{6}$$

$$\frac{d\phi_\nu}{dE_\nu} = R_o \rho_o \frac{1}{4\pi m_\chi \tau} \left( \sum_F B_F \frac{dN_\nu^F}{dE_\nu} \right) \langle J_1 \rangle_\Omega \Delta\Omega \quad n = 1. \tag{7}$$

In eq. (6), the typical thermal relic value is $\langle \sigma v \rangle_0 = 3 \times 10^{-26}$ cm$^3$/s. This value is set by the Hubble constant and $\Omega$, and its value is too low to describe the PAMELA and Fermi LAT excesses.[15] A "boost" factor to increase the cross section, $\langle \sigma v \rangle = B \langle \sigma v \rangle_0$, is introduced to increase the annihilation rate by of order 100 or more. This kind of boost factor arises theoretically via a Sommerfeld enhancement of the cross section at velocities much lower than the velocity at freeze-out[39] and/or from density inhomogeneities.[15] For decaying dark matter, the lifetime of the DM is written as $\tau = B_\tau \times 10^{26}$ s.

There have been a number of evaluations of the limits neutrino telescopes can set on the annihilation cross section. For example, Beacom, Bell and Mack[40] integrate over the full sky and over all halos to find limits on $B$ as a function of energy, with $B \simeq 1000$ or more. By focusing on the Milky Way halo, the constraints on $\chi\chi \to \nu\bar{\nu}$ boost factors can be strengthened.[41]

A number of theoretical analyses of neutrino signals from annihilation or decay of dark matter involving intermediate states have appeared in the literature. See, for example, Refs.[13,14,18,19] The intermediate states include $W^+W^-$, muons and $b$ quarks. One theoretical proposal to satisfy the observational lepton excesses, yet avoid constraints from overproduction of photons, is through leptophilic models, for example in Ref.[12] In these models, the DM coupling to standard model particles is only through leptons. Barger et al.[24] have found constraints for $\chi\chi \to \mu\bar{\mu}$ on the boost factor $B$ and the DM mass, and the related condition for the decaying dark matter $\chi \to \mu\bar{\mu}$ with $B_\tau$ and $m_\chi$. With this kind of constraint, kilometer cubed scale underground detectors have the potential to rule out some of these theoretical predictions.

Integrating over the full sky, above or below the detector, reduces uncertainties associated with the model for the dark matter distribution around the galactic center. It is not always optimal to use the full sky to pick out the dark matter signal because of the atmospheric background.[19] This is shown with a few specific models in Fig. 2. The parameters in these models, for a decaying gravitino[11] or leptophilic dark matter particle decay or annihilation[24] are chosen to account for the observed lepton excesses. Also included in the figure is an example of the lightest Kaluza-Klein particle annihilation into multiple channels.[12]

In Fig. 2 for decays, the dotted and upper dot-dashed lines have $m_\chi =$ 400 GeV and $B_\tau = 2.3$, and the lower dot-dashed line is for $m_\chi = 2$ TeV and $B_\tau = 2.9$. For annihilation, the solid line has $m_\chi = 800$ GeV and $B = 200$, while the dashed line is for $m_\chi = 1$ TeV and $B = 400$. While there is some dependence on the channel in $\chi\chi$ annihilation, accounting only for the atmospheric neutrino background,[19] for upward or contained muons the optimal cone with half-angle $\theta_{max}$ centered on the galactic center is of order a few degrees to 10°. Decaying DM signals are optimized for $\theta_{max} \sim 50°$ or greater. For showers, the optimal angle is of order $\theta_{max} \sim 10°$ for annihilations.

Limits on the neutralino induced upward muon flux from SuperKamiokande data in the direction of the galactic center range from $\sim 10^{-14} - 2 \times 10^{-15}$ cm$^{-2}$s$^{-1}$ for $m_\chi$ in the range of $20 - 10^4$ GeV for $\theta_{\rm max}$ between 5° and 30°. IceCube-22 data have put some limits on $\langle \sigma v \rangle$ for several different annihilation channels.[35] The $\nu\bar{\nu}$ channel yields the strongest limit, as expected by the $\delta$-function neutrino spectrum. For the $\nu\bar{\nu}$ channel, $B$ is less than $\sim$ a few times $10^4$ to a few times $10^3$ for the range of $m_\chi \sim 200$ GeV to 10 TeV. Their weakest constraint is for $\chi\chi \to b\bar{b}$, where the limit for $B$ is two orders of magnitude or more larger than the $\nu\bar{\nu}$ limit for the same mass range.

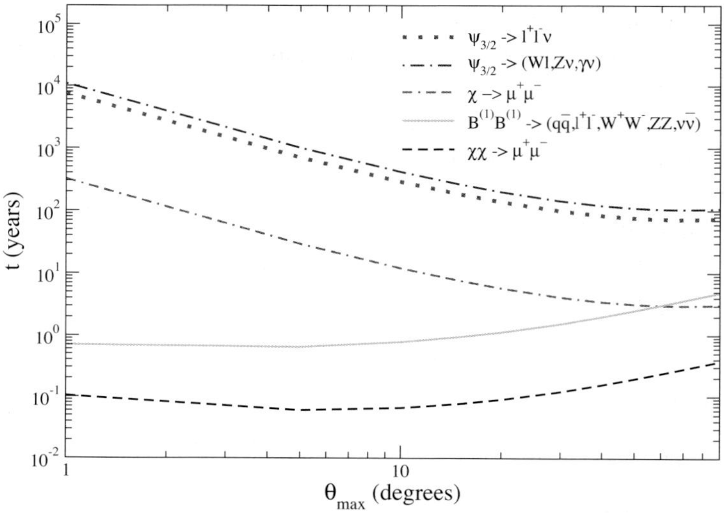


Fig. 2. For a few models (see text for mass and $B$ or $B_\tau$), the time for a kilometer-cubed scale detector to measure a signal over square root of signal plus background greater than two, as a function of cone half-angle $\theta_{max}$ around the galactic center.[19,42]

## 4. Final remarks

Neutrino signals from dark matter have not yet been observed, but as a kilometer-cubed scale neutrino telescope becomes operational, there is the potential to probe some regions of parameter space in particle physics models of dark matter. The $\chi\chi \to \nu\bar{\nu}$ process will be the easiest channel to observe experimentally because of the initial neutrino energy.

The dark matter accumulation in the Earth's core will be difficult to observe experimentally. The predicted neutrino signals are generally quite low compared to the sensitivity of detectors. On the other hand, already some models for DM annihilations in the Sun have been excluded. With the IceCube neutrino telescope, there is the potential to exclude a much larger range of model parameter space.

The DM accumulation in the galactic halo offers the opportunity to observe neutrino signals of DM annihilation or decay. If DM annihilation or decay explains the positron excess in the PAMELA data and the electron positron excess of the Fermi LAT, there are restrictions on the set of cross sections and DM masses. Again, kilometer-cubed scale neutrino telescopes can constrain part of this restricted parameter set.

Neutrino signals provide an important element of the the characterization of the particle nature of dark matter. Ultimately, one needs results from multiple messengers: neutrinos, photons, positrons and antiprotons. Direct DM detection and results from collider experiments will help to distinguish between a range of particle physics models.

## Acknowledgments

This work was supported in part by US Department of Energy contracts DE-FG02-91ER40664. I thank A. E. Erkoca and I. Sarcevic for discussions and collaboration on dark matter topics discussed here.

## References

1. See G. Raffelt, AIP Conf. Proceedings 721, 130 (2004) for a review of neutrinos in cosmology.
2. S. S. Gershtein and Y. B. Zeldovich, JETP Lett. **4**, 120 (1966) [Pisma Zh. Eksp. Teor. Fiz. **4**, 174 (1966)]. See also R. Cowsik and J. McClelland, Phys. Rev. Lett. **29**, 669 (1972).
3. E. Komatsu *et al.*, arXiv:1001.4538 [astro-ph.CO].
4. K. Nakamura [Particle Data Group], J. Phys. G **37**, 075021 (2010).
5. R. J. Gaitskell, Ann. Rev. of Nuc. Part. Sci. **54** 315 (2004); E. Aprile, PoS E **PS-HEP2009**, 009 (2009).
6. See, e.g., J. Feng, arXiv:1003.0904 [astro-ph.CO] for a review.
7. M. Ricci [PAMELA Collaboration], AIP Conf. Proc. **1200**, 968 (2010); O. Adriani *et al.* [PAMELA Collaboration], Nature **458**, 607 (2009).
8. A. A. Abdo *et al.* [The Fermi LAT Collaboration], Phys. Rev. Lett. **102**, 181101 (2009).
9. G. Jungman, M. Kamionkowski and K. Griest, Phys. Rept. **267**, 195 (1996).
10. P. Gondolo, J. Edsj, P. Ullio, L. Bergstrm, M. Schelke and E.A. Baltz, JCAP 07 (2004) 008. See also www.darksusy.org.
11. B. Bajc, T. Enkhbat, D. K. Ghosh, G. Senjanovic and Y. Zhang, JHEP **1005**, 048 (2010).
12. P. J. Fox and E. Poppitz, Phys. Rev. D **79**, 083528 (2009).
13. L. Covi, M. Grefe, A. Ibarra and D. Tran, JCAP **1004**, 017 (2010).
14. A. Ibarra, A. Ringwald, D. Tran and C. Weniger, JCAP **0908**, 017 (2009).
15. I. Cholis et al., Phys. Rev. D **80**, 123511 (2009);
16. M. Cirelli et al., Nucl. Phys. B **813**, 1 (2009).
17. V. Barger, Y. Gao, W. Y. Keung, D. Marfatia and G. Shaughnessy, Phys. Lett. B **678**, 283 (2009).

18. S. K. Mandal, M. R. Buckley, K. Freese, D. Spolyar and H. Murayama, Phys. Rev. D **81**, 043508 (2010).
19. A. E. Erkoca, M. H. Reno and I. Sarcevic, arXiv:1009.2068 [hep-ph]; A. E. Erkoca, G. Gelmini, M. H. Reno and I. Sarcevic, Phys. Rev. D **81**, 096007 (2010).
20. D. Hooper and G. D. Kribs, Phys. Rev. D **67**, 055003 (2003); See also, S. Arrenberg, L. Baudis, K. Kong, K. T. Matchev and J. Yoo, Phys. Rev. D **78**, 056002 (2008).
21. See K. Zurek's contribution to these proceedings.
22. M. Blennow, J. Edsj and T. Ohlsson, JCAP 0801, 021 (2008).
23. M. Cirelli, N. Fornengo, T. Montaruli, I. A. Sokalski, A. Strumia and F. Vissani, Nucl. Phys. B **727**, 99 (2005) [Erratum-ibid. B **790**, 338 (2008)].
24. V. Barger, W. Y. Keung, G. Shaughnessy and A. Tregre, Phys. Rev. D **76**, 095008 (2007).
25. A. E. Erkoca, M. H. Reno and I. Sarcevic, Phys. Rev. D **80**, 043514 (2009).
26. A. Kappes [IceCube Collaboration], Nucl. Phys. A **827**, 567C (2009); R. Abbasi [IceCube Collaboration], Int. J. Mod. Phys. D **19**, 1041 (2010); M. Bou-Cabo and J. A. Martinez-Mora, Nucl. Instrum. Meth. A **617**, 505 (2010); A. Kappes et al, KM3NeT Collaboration, PoS EPS-HEP2009, 102 (2009).
27. T. K. Gaisser and M. Honda, Ann. Rev. Nucl. Part. Sci. **52**, 153 (2002).
28. R. Abbasi et al., Astropart. Phys. **34**, 48 (2010).
29. See D. Grant's contribution to these proceedings.
30. R. Gandhi, C. Quigg, M. H. Reno and I. Sarcevic, Phys. Rev. D **58**, 093009 (1998); R. Gandhi, C. Quigg, M. H. Reno and I. Sarcevic, Astropart. Phys. **5**, 81 (1996).
31. A. Strumia and F. Vissani, arXiv:hep-ph/0606054. For higher energies, see also A. Bulmahn and M. H. Reno, Phys. Rev. D **81**, 053003 (2010).
32. S. Desai *et al.* [Super-Kamiokande Collaboration], Phys. Rev. D **70**, 083523 (2004) [Erratum-ibid. D **70**, 109901 (2004)].
33. A. Achterberg *et al.* [AMANDA Collaboration], Astropart. Phys. **26**, 129 (2006).
34. R. Abbasi *et al.* [ICECUBE Collaboration], Phys. Rev. Lett. **102**, 201302 (2009); M. Ackermann *et al.* [AMANDA Collaboration], Astropart. Phys. **24**, 459 (2006).
35. P. Desiati and f. t. I. Collaboration, arXiv:1007.2621 [astro-ph.IM].
36. R. Abbasi *et al.* [The IceCube collaboration], Phys. Rev. D **81**, 057101 (2010).
37. J. F. Navarro, C. S. Frenk and S. D. M. White, Astrophys. J. **462**, 563 (1996).
38. P. Ullio, L. Bergstrom, J. Edsjo and C. G. Lacey, Phys. Rev. D **66**, 123502 (2002); L. Bergstrom, J. Edsjo and P. Ullio, Phys. Rev. Lett. **87**, 251301 (2001).
39. N. Arkani-Hamed, D. P. Finkbeiner, T. R. Slatyer and N. Weiner, Phys. Rev. D **79**, 015014 (2009); M. Pospelov and A. Ritz, Phys. Lett. B **671**, 391 (2009).
40. J. F. Beacom, N. F. Bell and G. D. Mack, Phys. Rev. Lett. **99**, 231301 (2007).
41. H. Yuksel, S. Horiuchi, J. F. Beacom and S. Ando, Phys. Rev. D **76**, 123506 (2007).
42. A. E. Erkoca, following Ref. 19.

# CHARGED COSMIC RAYS FROM DARK MATTER

P. SALATI*

*LAPTH, Université de Savoie & CNRS*
*9 Chemin de Bellevue, BP110,*
*F-74941 Annecy-le-Vieux Cedex, France*
**E-mail: salati@lapp.in2p3.fr*
*http://lapth.in2p3.fr/ salati/*

There may be a natural connection between galactic cosmic rays and the astronomical dark matter. The nature of the latter is still unresolved. We know that it cannot be made of ordinary atoms. The most favoured candidate for this important and yet enigmatic component of the universe is a massive and weakly interacting particle dubbed WIMP. Should this species exists, it would be the essential constituent of the Milky Way halo and would continuously annihilate, producing in particular charged antimatter cosmic rays. This would lead to a distortion in the spectra of antiprotons and positrons and hence to the possibility of indirectly detecting the dark matter particles. After having insisted on how uncertain is our modeling of the propagation of charged galactic cosmic rays, I review the current status of the searches for the antiprotons and positrons generated by dark matter. Some attention will be paid to the positron excess which the PAMELA satellite has observed above 10 GeV. The WIMP explanation is very contrived if not already excluded by radio and gamma-ray observations.

*Keywords*: Dark Matter; WIMP Indirect Searches; Antimatter Cosmic Rays

A significant portion of the universe is made of a massive and non-relativistic component which is also detected at the galactic scale through the rotation curves of spirals, and inside clusters of galaxies through their X-ray emissions and their gravitational lensing effects on background sources. The astronomical dark matter (DM) is not made of ordinary atoms. Its nature is still unknown, hence the exciting possibility that a new species, yet to be discovered, is the building brick of this enigmatic component. Among the many different candidates which have been so far proposed, the best motived possibility is a weakly interacting and massive particle or WIMP – hereafter denoted by $\chi$ – with a mass in the range from 10 GeV up to a few TeV. Should WIMPs pervade the halo of the Milky Way, their mutual annihilations would yield several indirect signatures. These are potentially detectable on earth under the form of spectral distortions appearing in various cosmic radiations

$$\chi + \chi \rightarrow q\bar{q}, W^{+}W^{-}, \ldots \rightarrow \bar{p}, \bar{D}, e^{+}\, \gamma \,\&\, \nu' s \ . \tag{1}$$

Detection of the DM annihilation products has motivated the spectacular development of several new experimental techniques. Searches for antiprotons, positrons and soon antideuterons are performed by balloon and satellite-borne devices. The AMS 02 mission will even be installed on the International Space Station next February.

## 1. Cosmic ray propagation – uncertainties in the modeling

Whatever the mechanism responsible for their production, charged cosmic rays (CR) subsequently propagate through the galactic magnetic field and are deflected by its irregularities : the Alfvén waves. In the regime where the magnetic turbulence is strong – which is the case for the Milky Way – CR transport needs to be investigated numerically. Monte Carlo simulations[1] indicate that it is similar to space diffusion with a coefficient

$$K(E) = K_0 \, \beta \, (\mathscr{R}/1\,\mathrm{GV})^{\delta} \ , \tag{2}$$

which increases as a power law with the rigidity $\mathscr{R} = p/q$ of the particle. In addition, because the scattering centers drift inside the Milky Way with a velocity $V_a \sim$ 20 to 100 km s$^{-1}$, a second order Fermi mechanism is responsible for some mild diffusive reacceleration. Its coefficient $K_{EE}$ depends on the particle velocity $\beta$ and total energy $E$ and is approximately related to the space diffusion coefficient $K(E)$ through

$$K_{EE} = \frac{2}{9} \, V_a^2 \, \frac{E^2 \beta^4}{K(E)} \ . \tag{3}$$

In the case of positrons, diffusive reacceleration is completely dominated by energy losses above 10 GeV. Finally, galactic convection wipes cosmic rays away from the disc with a velocity $V_C \sim$ 5 to 15 km s$^{-1}$.

In general, steady state is assumed for the various populations of particles. This is a good approximation for nuclei – including antiprotons and antideuterons – but may fail for high-energy positrons. The master equation fulfilled by the space and energy distribution function $\psi = dn/dE$ may be expressed as

$$\partial_z \left( V_C \, \psi \right) - K \, \Delta\psi + \partial_E \left\{ b^{\mathrm{loss}}(E) \, \psi - K_{EE}(E) \, \partial_E \psi \right\} = q\left(\mathbf{x}, E\right) \ . \tag{4}$$

This equation applies to any species – protons, antiprotons or positrons – as long as the rates for CR production $q$ and energy loss $b^{\mathrm{loss}}(E)$ are properly accounted for. It can be solved within the framework of the semi-analytical two-zone model which has been extensively discussed in previous works[2,3] and whose salient features we briefly recall now. According to this approach, steady state is assumed and the region of the Galaxy inside which cosmic rays diffuse – the so-called diffusive halo or DH – is pictured as a thick disc which matches the circular structure of the Milk Way. This picture is well supported by observations of radio continuum haloes around several galaxies such as NGC 4631[4]. The galactic disc of stars and gas, where primary cosmic rays are accelerated, lies in the middle. It extends radially 20 kpc from the center and has a half-thickness $h$ of 100 pc. Confinement layers, where cosmic rays are trapped by diffusion, lie above and beneath the thin disc of gas. The intergalactic medium starts at the vertical boundaries $z = \pm L$ as well as beyond a radius of $r = R \equiv 20$ kpc. Notice that the half-thickness $L$ of the DH is not known as discussed below. The diffusion coefficient $K$ is assumed to be the same everywhere, whereas the convective velocity is exclusively vertical with component $V_C(z) = V_C \, \mathrm{sign}(z)$. This galactic wind, which is produced by the bulk of the disc stars like the sun, drifts away from its progenitors along the vertical directions, hence the particular form assumed here for $V_C$. Notice also that the normalization coefficient $K_0$, the index $\delta$, the galactic drift velocity $V_C$ and the Alfvén velocity $V_a$ are all unknown quantities. This situation may be remedied with the help of the boron to carbon ratio (B/C) which is quite sensitive to CR transport and which may be used as a constraint. The three propagation models featured in table 1 have been drawn from ref. 5. The MED configuration provides the best fit to the B/C measurements whereas the MIN and MAX models lead respectively to the minimal and maximal allowed CR antiproton fluxes which can be produced by WIMP annihilations. These three sets of parameters belong to an ensemble of $\sim 1,600$ different CR propagation models found[2] to pass successfully the B/C test.

Table 1. Typical combinations of diffusion parameters that are compatible with the B/C analysis.[2] As shown in ref. 5, these propagation models correspond respectively to minimal, medium and maximal primary antiproton fluxes.

| CR model | $\delta$ | $K_0$ [kpc$^2$/Myr] | $L$ [kpc] | $V_C$ [km/s] | $V_a$ [km/s] |
|---|---|---|---|---|---|
| MIN | 0.85 | 0.0016 | 1 | 13.5 | 22.4 |
| MED | 0.70 | 0.0112 | 4 | 12 | 52.9 |
| MAX | 0.46 | 0.0765 | 15 | 5 | 117.6 |

The master equation (4) can be also solved numerically without the need to expand $\psi$ as a series of radial Bessel functions. The well known public package GALPROP[6,7] integrates directly that equation with the Crank Nicolson method. The new numerical code DRAGON[8] has been recently developed to account more specifically for a space and direction dependent diffusion. DRAGON allows the comparison between the semi-analytic and numerical approaches. An excellent discussion can be found for instance in DiBernardo *et al.*[9] where the discrepancies between the two approaches are explained. The semi-analytic two-zone model has led to the package USINE, soon to be released publicly. The CR propagation parameters are now derived from the B/C ratio with the help of a Markov chain Monte Carlo analysis[10,11] which allows for a much faster exploration of the parameter space. Notice that the half-thickness $L$ of the DH is not determined accurately, with a value extending from 1 to 16 kpc. Better measurements of the light nuclei abundances are mandatory to improve the determination of the CR propagation parameters.

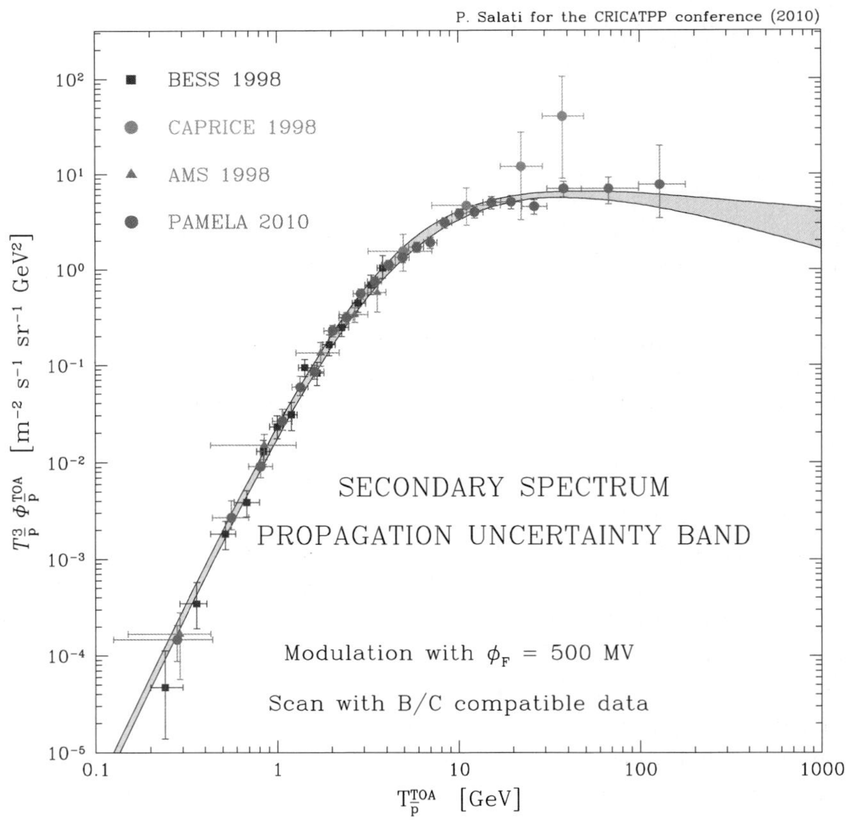


Fig. 1. Theoretical uncertainties in the secondary flux of antiprotons, taking into account the whole range of propagation parameters that is allowed by the existing B/C ratio, featured together with the existing top of the atmosphere data.[12–15] The recent PAMELA measurements (red dots) fall closely to the uncertainty band. Notice the near $T_{\bar{p}}^{-3}$ scaling of the flux at high energies as well as the broadening in the predictions owing to the unknown value of $\delta$

## 2. Antiprotons – a robust probe for indirect searches

The propagation of cosmic ray antiprotons is dominated by diffusion. Energy losses as well as diffusive reacceleration do not play any major role. A very crude approximation for the antiproton Green function is obtained by neglecting galactic convection and solving the resulting Poisson equation in infinite space. This yields the antiproton propagator

$$G_{\bar{\text{p}}}\left(\mathbf{x}_{\odot} \leftarrow \mathbf{x}_S\right) \equiv \frac{1}{4\pi K(E)} \frac{1}{r_{\oplus}} , \tag{5}$$

where $r_{\oplus}$ denotes the distance between the earth and the source. The sole merit of this expression is to exhibit the importance of remote sources, in clear contrast with high energy positrons and electrons which cannot come to the earth from far away. The finite thickness of the DH is nevertheless a limiting factor since cosmic rays may escape through the vertical boundaries as they wander towards the earth. As a consequence, the size of the antiproton production sphere cannot be much larger than the DH half-thickness $L$. Galactic convection comes also into play. If the wind velocity $V_C$ is large, cosmic rays are efficiently blown outside the Milky Way. This process limits further the reach of the antiproton sphere at low energy, below a few GeV. The solution of the master equation (4) has been thoroughly investigated and several different techniques [16,17] lead essentially to the same fluxes at the earth. As shown in Fig. 1, the secondary antiproton flux predicted in the framework of diffusion-reacceleration models is in remarkable agreement with the antiproton data [12–14] and in particular with the recent PAMELA observations. [15] There is no distortion in the energy distribution and hence no signal of the presence of DM in the antiproton cosmic radiation. The latter is of great interest in the search for DM indirect signatures because of the small uncertainty on the flux of secondary antiprotons – produced by the interactions of primary CR nuclei on the interstellar gas. Antiprotons should be therefore considered as a valuable probe for DM. Notice finally that the USINE agreement between its semi-analytic predictions[3] and the antiproton data – illustrated in Fig. 1 – is not quite as good in the case of the GALPROP package. The latter tends to favour low values of the index $\delta$ and requires a strong diffusive reacceleration in order to match the B/C ratio. A simple power law in energy or momentum for the proton injection spectrum does not allow the fit of the proton flux measured at the earth because a bump appears in the calculations at the Gev scale. To erase it, a break in the CR nuclei source spectra is assumed below 9 GV with a harder spectrum at low energy. The antiproton spectrum is then deficient and underestimates [18,19] the observations. This explanation has been proposed by the DRAGON collaboration [9] who, like USINE, fits successfully the B/C ratio and the antiproton data at the same time.

## 3. The positron excess – a recently hot topic

### 3.1. *Evidence for primary positrons*

The ATIC,[20] HESS[21,22] and Fermi[23] collaborations have recently reported measurements of the total cosmic ray (CR) electron and positron flux at high energy. The Fermi-LAT instrument in particular has collected high precision data between 20 GeV and 1 TeV which can be fitted by the simple power-law spectrum[24]

$$\Phi_{e^{\pm}} = \left\{175.40 \pm 6.09 \text{ GeV}^{-1}\,\text{m}^{-2}\,\text{s}^{-1}\,\text{sr}^{-1}\right\}\,\varepsilon^{-(3.045\pm0.008)}\,, \qquad (6)$$

where $\varepsilon$ denotes the electron or positron energy expressed in units of GeV. Primary CR electrons are believed to originate from the interstellar medium – like most of the primary CR nuclei – and are accelerated by supernovae driven shock waves which inject them with a spectral index $\alpha \sim 2.2 \pm 0.1$ inside the Milky Way disc. In the GeV-TeV regime, they also lose energy through synchrotron radiation as they spiral in the magnetic fields and by inverse Compton scattering (ICS) on the CMB and on galactic stellar light. A detailed investigation[25] leads to a CR electron flux at the earth of the form

$$\Phi_{e^-} \propto \varepsilon^{-\alpha - 0.5 - \delta/2}\,, \qquad (7)$$

with a spectral index of order $3.05 \pm 0.15$, in reasonable agreement with the Fermi-LAT measurement. The value of $\delta = 0.7$ used here may seem too large to some colleagues.[7] It suffices then to increase $\alpha$ in order to match the Fermi-LAT spectrum. The latter also contains a small admixture of secondary electrons and positrons which are produced by CR primary protons and helium nuclei impinging on the interstellar gas. These spallation reactions generate many charged pions which eventually decay into positrons and in a lesser extent into electrons. The source spectral index is set by the CR proton and helium nuclei fluxes. With a value $\alpha = 2.7$, we expect a spectral index of $3.55 \pm 0.05$ for that secondary component which is swamped in the total flux.

The PAMELA collaboration has measured[26] the positron fraction $e^+/e^- + e^+$ on a large energy range. From the above mentioned arguments, we would naively expect that fraction to decrease with energy as $\varepsilon^{-1/2}$. A more refined analysis[25] confirms that a CR electron spectral index as hard as 3.045 is associated to a decrease of the positron fraction above 10 GeV. The PAMELA measurements establish on the contrary that the positron fraction increases at high energy. This very important observation is in contradiction with a pure secondary origin for CR positrons. Primary sources which directly inject these particles in the interstellar medium are necessary.

### 3.2. *DM species with quite special properties*

The announcement of a positron excess by the PAMELA collaboration has triggered an intense activity and a lot of excitement. This excess was actually considered as the first hint of the presence of DM species in the Milky Way halo. Should WIMPs exist, their annihilations would be a source of primary positrons[27] with a production rate given by

$$q_{e^+} = \frac{1}{2} \langle \sigma_{\rm an} v \rangle \left\{ \frac{\rho_\chi}{m_\chi} \right\}^2 f(\varepsilon) \; . \tag{8}$$

The positron excess lies above 10 GeV. It can be generated by a particle with a mass in the range 100 GeV to a few TeV, in good agreement with theoretical expectations of the WIMP mass. If the DM species are thermally produced during the big bang, their relic abundance matches the WMAP value[28] of $\Omega_\chi h^2 = 0.1131 \pm 0.0034$ provided that their annihilation cross section at the time of decoupling is equal to $\langle \sigma_{\rm an} v \rangle \sim 3 \times 10^{-26}$ cm$^3$ s$^{-1}$. The energy distribution $f(\varepsilon)$ of the positrons produced in a single annihilation is set by the WIMP type. High energy positrons (and electrons) cannot diffuse on long distances and those detected at the earth must have been produced locally, hence a value of $\rho_\odot = 0.3$ GeV cm$^{-3}$ for the WIMP density $\rho_\chi$. Baring in mind these bench mark values, one can estimate $q_{e^+}$ and eventually the DM positron flux $\Phi_{e^+}$. Alas, the values derived for a large choice of DM candidates and for various DM galactic distributions are way too small to account for the observed excess. For a WIMP mass $m_\chi$ of 1 TeV, the DM positron production rate $q_{e^+}$ needs to be enhanced by a factor of $\sim 10^3$ in order to match the PAMELA measurements, as illustrated in the left panel of figure 2.

To save the WIMP explanation of the PAMELA positron anomaly, a first remedy is the possibility that $\langle \sigma_{\rm an} v \rangle$ is much larger than what is currently assumed. Three directions at least have been explored so far. To commence, WIMP decoupling in the early universe takes place conventionally during a period of radiation domination. If the universe at that time is dominated by another component, like a scalar field rolling down its potential,[29,30] the expansion rate is significantly increased and the same WIMP relic abundance $\Omega_\chi$ requires now a much larger annihilation cross section. Another possibility is that WIMPs do not decouple from the primordial plasma. In this class of less natural scenarios, a non-thermal production takes place through the decay of heavier species like gravitinos or moduli fields.[31] The WIMP relic abundance $\Omega_\chi$ is no longer determined by $\langle \sigma_{\rm an} v \rangle$. It depends on the production rate and lifetime of these unstable heavier states. A final solution relies on the so-called Sommerfeld effect.[32] If the DM species are massive enough and interact through the exchange of a light particle $\phi$, they are attracted towards each other at sufficiently low velocities. Their wave function is focalized at the interaction point and the annihilation cross section can be significantly enhanced. Scenarios based on that effect have been recently proposed,[33–35] reviving the idea of secluded dark matter.[36] WIMPs are still thermally produced in the early universe. Their annihilation cross section at that time is equal to the conventional value of $3 \times 10^{-26}$ cm$^3$ s$^{-1}$. As the universe expands, the average WIMP velocity $\beta$ decreases. The Sommerfeld effect sets in as soon as the WIMP kinetic energy is dominated by the interaction energy, i.e. when $\beta$ no longer exceeds the effective WIMP coupling constant $\alpha'$. The cross section increases and may be today large enough to account for the PAMELA positron excess. The enhancement saturates when the WIMP de Broglie wavelength $\sim 1/m_\chi \beta$ exceeds the range $\sim 1/m_\phi$ of the interaction. This occurs for $\beta \leq m_\phi / m_\chi$. The annihilation cross section is resonantly increased whenever $\alpha' m_\phi \sim \alpha'^2 m_\chi / n^2$ where $n$ is an integer.[37]

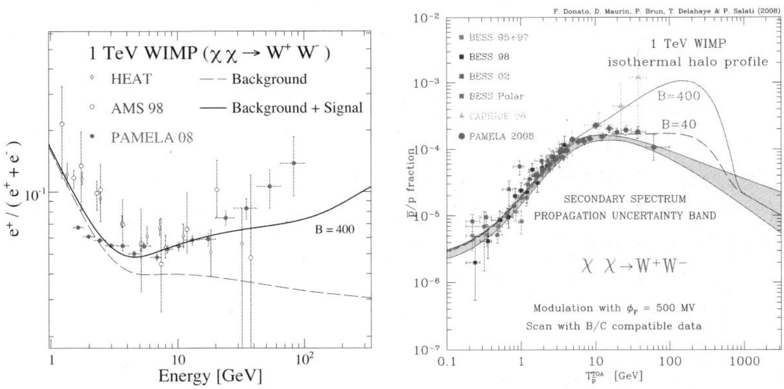


Fig. 2. The pedagogical case of a 1 TeV WIMP annihilating into a $W^+W^-$ pair is presented. In the left panel, the positron signal which this DM species yields has been increased by a factor of 400, hence the solid curve and a marginal agreement with the PAMELA data. Positron fraction data are from HEAT,[38] AMS-01[39,40] and PAMELA.[26] If the so-called Sommerfeld effect[32] is invoked to explain such a large enhancement of the annihilation cross section, the same boost applies to antiprotons and leads to an unacceptable distortion of their spectrum as indicated by the red solid line of the right panel. Figures and caption are borrowed from[41]

Irrespective of the mechanism enhancing $\langle \sigma_{\mathrm{an}} v \rangle$, WIMPs should not overproduce antiprotons.[41,42] The annihilation channels through which these antimatter species are produced must be significantly suppressed. The antiproton to proton ratio measured by the PAMELA collaboration[43] does not show actually any excess and is consistent with a pure secondary origin as featured in the right panel of figure 2. Therefore, on top of an abnormally large annihilation cross section, WIMPs preferentially annihilate into charged leptons, a feature which is unusual in supersymmetry but more typical of Kaluza-Klein theories. From a phenomenological point of view, WIMP annihilations must proceed either directly to lepton pairs or, in some models, through the production of the above mentioned light mediator species $\phi$ which subsequently decay into leptons via

$$\chi + \chi \rightarrow \phi + \phi \rightarrow l^+ l^- l^+ l^- \ . \tag{9}$$

The possibility to explain the PAMELA positron excess by leptophilic DM particles with enhanced annihilation cross section has been extensively investigated. Depending on its magnitude, the injection of high energy positrons and electrons is associated to clear astrophysical signatures.[44] As they spiral along the galactic magnetic field lines, these particles produce a radio emission through synchrotron radiation. They also inverse Compton scatter photons from the CMB and stellar light. Finally, the charged leptons released by the WIMP annihilations can produce final state radiation photons. After an extensive scan[45–47] of the various possibilities, the region of the $(m_\chi, \langle \sigma_{\rm an} v \rangle)$ plane compatible with the PAMELA and Fermi-LAT measurements is found to be excluded for most of the leptophilic DM candidates. The most stringent limit arises from the synchrotron radio emission produced at the galactic center, a region where the DM distribution may be cuspy and where magnetic fields should be strong. That limit weakens considerably though as soon as a cored isothermal DM distribution is preferred to the usual Einasto or NFW profile. In that case, leptophilic WIMPs are not excluded and could be a viable explanation of the positron anomaly. A more reliable tool is provided by the ICS of the positrons and electrons produced by WIMP annihilation at high galactic latitude, a region which is less subject to uncertainties than the center of the Milky Way. A careful investigation,[48] performed only in the case where the WIMPs annihilate directly into lepton pairs, excludes values of $m_\chi$ larger than 1 TeV. Another model independent analysis implies the injection in the intergalactic medium at redshift $\sim 1000$ of the secondary particles produced by WIMP annihilation. The recombination process is affected and an imprint is left on the CMB, which may be searched for in the high precision WMAP data.[49] Most of the PAMELA compatible leptophilic DM candidates still survive that test[50] but will be probed by the Planck mission.

### 3.3. *The effect of clumpiness on DM annihilation*

Another way to increase the positron production rate (8) relies on the existence of clumps inside the smooth DM galactic distribution. Because $\langle \rho_\chi^2 \rangle$ is always larger than $\langle \rho_\chi \rangle^2$, inhomogeneities tend to enhance $q_{e^+}$, a quantity proportional to the square of the DM density. How large is that enhancement depends on the position of the substructures with respect to the DM smooth component of the Milky Way. The inner structure of clumps as well as their mass distribution are also key factors. The lower mass cut-off and the concentration-to-mass relation of the DM substructures come finally into play. A statistical analysis is mandatory since we live inside one particular realization of the clump distribution to be taken out of an infinite set of similar realizations. Using the tools specifically forged for computing the odds of that galactic lottery,[51] a comprehensive analysis[52] indicates that the boost factor to be applied to $q_{e^+}$ in the case of a ΛCDM universe cannot exceed at most a factor of 20 at high energy. The correct value is presumably much smaller, i.e. of order unity.[53]

Although a clumpy DM halo does not lead on average to an increase of the positron flux at the earth, the statistical variance of the boost factor is large at high energy, hence the existence of configurations where $q_{e^+}$ could be significantly enhanced. The possibility that a DM clump lies for instance in the vicinity of the solar system has been suggested[54,55] as an explanation for the PAMELA positron anomaly. Such a local substructure would actually outshine the rest of the galactic DM distribution. Alas, the probability that such a situation occurs in the DM halo of the Milky Way is vanishingly small as demonstrated in a recent analysis[56] based on the results of the cosmological N-body simulation Via Lactea II.[57] In the favorable case of a 100 GeV DM species annihilating into $e^{\pm}$ pairs, the PAMELA positron excess is best fitted by a subhalo with annihilation volume

$$\xi = \int_{\text{clump}} \{\rho_c(\mathbf{x})/\rho_\odot\}^2 \, d^3\mathbf{x} \sim 114 \text{ kpc}^3 \;, \tag{10}$$

located at 1.22 kpc from the earth. That configuration has a probability of only 0.37 percent to occur. Other arrangements are even less probable.

### 3.4. *Decaying dark matter*

Finally, the possibility of unstable DM candidates with very large lifetimes has also been explored.[58,59] Irrespective of their production mechanism in the early universe, unstable species could in principle generate the PAMELA signal. The production rate (8) may be written now as

$$q_{e^+} = \tau_{\text{dec}}^{-1} \left\{ \frac{\rho_\chi}{m_\chi} \right\} f(\varepsilon) \;, \tag{11}$$

where the quantity $1/2 \times \langle \sigma_{\text{an}} v \rangle \times \{\rho_\chi/m_\chi\}$ has been replaced by the decay rate $\tau_{\text{dec}}^{-1}$. For a 1 TeV WIMP, a lifetime of order $2 \times 10^{26}$ sec is required to reproduce the positron excess. The synchrotron radio emission from the galactic center is no longer a problem even in the case of a NFW or Einasto profile since the positron production rate from decaying DM is proportional to $\rho_\chi$ and not to $\rho_\chi^2$. A recent analysis[48] excludes however decaying DM from the isotropic emission inferred from the Fermi data, unless it decays predominantly into muon pairs. The lifetime needs to be fine tuned though. Higher dimension operators must be invoked to explain the large value of $\tau_{\text{dec}}$ required by PAMELA. If a four-fermion point-like coupling is responsible for the WIMP instability, we anticipate[60] a decay rate $\propto m_\chi^5/M^4$ where $M$ denotes the typical scale of the underlying high energy theory. If factors of 2 and $\pi$ are neglected, a GUT mass $M \sim 2.3 \times 10^{16}$ GeV could lead to the required lifetime. Notice that the DM species must also be leptophilic otherwise antiprotons are also overproduced.

## 4. Conclusions and perspectives

Galactic CR propagation needs to be better understood. Several models are now available and public. But the actual values of the height $L$ of the DH as well as of the index $\delta$ and the normalisation $K_0$ of the space diffusion coefficient are not known precisely. If a strong diffusive reacceleration is assumed, a small value of $\sim$ 0.3 is derived for $\delta$ with a moderate extension of the diffusive halo. The secondary antiproton spectrum is in that case under predicted and the pion contribution to the galactic gamma-ray diffuse emission is in conflict with the Fermi measurements. Agreement with the observations is recovered for $L \sim 10$ kpc, small values of the Alfvén velocity and $\delta \sim 0.5$. Forthcoming measurements of the B/C ratio and of the abundances of unstable secondary nuclei such as $^{10}$Be will help discriminate among those possibilities. We may actually be surprised and it is essential then to explore other routes such as the production of secondaries inside the shocks which accelerate them.

Several groups find a remarkable agreement between the secondary antiproton spectrum which they predict and the data collected so far, including the recent PAMELA observations. The uncertainty arising from CR propagation being small, the antiproton background is well known and the antiproton signal provides a robust probe for the indirect DM searches. If an excess was found, its shape would help discriminate between DM and spallation at the sources.

The CR anomalies reported two years ago – and the PAMELA positron excess in particular – have triggered a boiling activity in the field of particle astrophysics. Alas, the possibility that these signals are produced by annihilating WIMPs in the Milky Way halo is fading away. The DM explanation is quite contrived – at least definitely far from the main stream – and the positrons which the DM species are supposed to produce generate in turn radiations which are not seen. As William of Ockham would have pointed out, a more plausible explanation has to be found in pulsars and/or supernova remnants as shown by a recent and very comprehensive[61] study

Finally, secondary antideuterons should be detected by the forthcoming AMS & GAPS missions. A sub-GeV excess would be suggestive of DM. Moreover, the strong correlation[62] between the momenta of the antineutron and antiproton produced by a WIMP annihilation or decay could also lead to an antideuteron excess at high energy. Detecting galactic antideuterons is a real challenge which may happen to be unexpectedly rewarding.

## Acknowledgments

I would like to thank the organizers to have given me the opportunity to take an active part in this exciting and inspiring conference. This work could not have been performed without the help of the Institut universitaire de France.

## References

1. F. Casse, M. Lemoine and G. Pelletier, *Phys. Rev.* **D65**, p. 023002 (2002).
2. D. Maurin, F. Donato, R. Taillet and P. Salati, *Astrophys. J.* **555**, 585 (2001).
3. F. Donato *et al.*, *Astrophys. J.* **563**, 172 (2001).
4. R. D. Ekers and R. Sancisi, *Astron. & Astrophys.* **54**, 973 (1977).
5. F. Donato, N. Fornengo, D. Maurin and P. Salati, *Phys. Rev.* **D69**, p. 063501 (2004).
6. A. W. Strong and I. V. Moskalenko, *Astrophys. J.* **509**, 212 (1998).
7. A. W. Strong, I. V. Moskalenko and O. Reimer, *Astrophys. J.* **613**, 962 (2004).
8. C. Evoli, D. Gaggero, D. Grasso and L. Maccione, *JCAP* **0810**, p. 018 (2008).
9. G. Di Bernardo, C. Evoli, D. Gaggero, D. Grasso and L. Maccione, *Astropart. Phys.* **34**, 274 (2010).
10. A. Putze, L. Derome and D. Maurin, *Astron. Astrophys.* **516**, p. A66 (2010).
11. D. Maurin, A. Putze and L. Derome, *Astron. Astrophys.* **516**, p. A67 (2010).
12. T. Maeno *et al.*, *Astropart. Phys.* **16**, 121 (2001).
13. M. Boezio *et al.*, *Astrophys. J.* **561**, 787 (2001).
14. M. Aguilar *et al.*, *Phys. Rept.* **366**, 331 (2002).
15. O. Adriani *et al.*, *Phys. Rev. Lett.* **105**, p. 121101 (2010).
16. T. Bringmann and P. Salati, *Phys. Rev.* **D75**, p. 083006 (2007).
17. D. Maurin, R. Taillet and C. Combet (2006).
18. I. V. Moskalenko, A. W. Strong, J. F. Ormes and M. S. Potgieter, *Astrophys. J.* **565**, 280 (2002).
19. I. V. Moskalenko, A. W. Strong, S. G. Mashnik and J. F. Ormes, *Astrophys. J.* **586**, 1050 (2003).
20. J. Chang *et al.*, *Nature* **456**, 362 (2008).
21. F. Aharonian *et al.*, *Phys. Rev. Lett.* **101**, p. 261104 (2008).
22. H. E. S. S. C. F. Aharonian (2009).
23. A. A. Abdo *et al.*, *Phys. Rev. Lett.* **102**, p. 181101 (2009).
24. D. Grasso *et al.*, *Astropart. Phys.* **32**, 140 (2009).

25. T. Delahaye *et al.*, *Astron. Astrophys.* **501**, 821 (2009).
26. O. Adriani *et al.*, *Nature* **458**, 607 (2009).
27. T. Delahaye, R. Lineros, F. Donato, N. Fornengo and P. Salati, *Phys. Rev.* **D77**, p. 063527 (2008).
28. G. Hinshaw *et al.*, *Astrophys. J. Suppl.* **180**, 225 (2009).
29. P. Salati, *Phys. Lett.* **B571**, 121 (2003).
30. S. Profumo and P. Ullio, *JCAP* **0311**, p. 006 (2003).
31. G. Kane, R. Lu and S. Watson, *Phys. Lett.* **B681**, 151 (2009).
32. J. Hisano, S. Matsumoto and M. M. Nojiri, *Phys. Rev. Lett.* **92**, p. 031303 (2004).
33. M. Cirelli, R. Franceschini and A. Strumia, *Nucl. Phys.* **B800**, 204 (2008).
34. N. Arkani-Hamed, D. P. Finkbeiner, T. R. Slatyer and N. Weiner, *Phys. Rev.* **D79**, p. 015014 (2009).
35. M. Pospelov and A. Ritz, *Phys. Lett.* **B671**, 391 (2009).
36. M. Pospelov, A. Ritz and M. B. Voloshin, *Phys. Lett.* **B662**, 53 (2008).
37. M. Lattanzi and J. I. Silk, *Phys. Rev.* **D79**, p. 083523 (2009).
38. S. W. Barwick *et al.*, *Astrophys. J. Letters* **482**, L191+ (1997).
39. M. Aguilar *et al.*, *Phys. Lett.* **B646**, 145 (2007).
40. J. Alcaraz *et al.*, *Phys. Lett.* **B484**, 10 (2000).
41. F. Donato, D. Maurin, P. Brun, T. Delahaye and P. Salati, *Phys. Rev. Lett.* **102**, p. 071301 (2009).
42. M. Cirelli, M. Kadastik, M. Raidal and A. Strumia, *Nucl. Phys.* **B813**, 1 (2009).
43. O. Adriani *et al.*, *Phys. Rev. Lett.* **102**, p. 051101 (2009).
44. E. Borriello, A. Cuoco and G. Miele, *Astrophys. J.* **699**, L59 (2009).
45. G. Bertone, M. Cirelli, A. Strumia and M. Taoso, *JCAP* **0903**, p. 009 (2009).
46. L. Bergstrom, G. Bertone, T. Bringmann, J. Edsjo and M. Taoso, *Phys. Rev.* **D79**, p. 081303 (2009).
47. P. Meade, M. Papucci, A. Strumia and T. Volansky (2009).
48. M. Cirelli, P. Panci and P. D. Serpico, *Nucl. Phys.* **B840**, 284 (2010).
49. S. Galli, F. Iocco, G. Bertone and A. Melchiorri, *Phys. Rev.* **D80**, p. 023505 (2009).
50. T. R. Slatyer, N. Padmanabhan and D. P. Finkbeiner, *Phys. Rev.* **D80**, p. 043526 (2009).
51. J. Lavalle, J. Pochon, P. Salati and R. Taillet, *Astron. & Astrophys.* **462**, 827 (2007).
52. J. Lavalle, Q. Yuan, D. Maurin and X. Bi, *Astron. & Astrophys.* **479**, 427 (2008).
53. J. Diemand, (2009), private communication.
54. D. T. Cumberbatch and J. Silk, *Mon. Not. Roy. Astron. Soc.* **374**, 455 (2007).
55. D. Hooper, A. Stebbins and K. M. Zurek, *Phys. Rev.* **D79**, p. 103513 (2009).
56. P. Brun, T. Delahaye, J. Diemand, S. Profumo and P. Salati, *Phys. Rev.* **D80**, p. 035023 (2009).
57. J. Diemand *et al.*, *Nature* **454**, 735 (2008).
58. A. Ibarra and D. Tran, *JCAP* **0902**, p. 021 (2009).
59. A. Ibarra, D. Tran and C. Weniger (2009).
60. C.-R. Chen, M. M. Nojiri, F. Takahashi and T. T. Yanagida, *Prog. Theor. Phys.* **122**, 553 (2009).
61. T. Delahaye, J. Lavalle, R. Lineros, F. Donato and N. Fornengo, *Astron. Astroph.* **524**, p. A51 (2010).
62. M. Kadastik, M. Raidal and A. Strumia, *Phys. Lett.* **B683**, 248 (2010).

# Gamma-Ray and Neutrino Signatures of Unstable Dark Matter

David Tran*

*Physik-Department T30d, Technische Universität München,*
*85748 Garching, Germany*
**E-mail: david.tran@ph.tum.de*

We investigate indirect neutrino and gamma-ray signatures of dark matter in scenarios where the dark matter is weakly unstable, but decays with a cosmological lifetime. In the light of recently observed anomalous results in measurements of high-energy cosmic-ray electrons and positrons, we examine constraints imposed by the upcoming generation of neutrino observatories. We also discuss the generation of gamma-ray lines induced by radiative decay of leptophilic dark matter at the quantum level and present resulting constraints.

*Keywords*: dark matter; cosmic rays; neutrinos; gamma rays.

## 1. Introduction

While there is overwhelming evidence for the existence of copious amounts of non-luminous dark matter in the Universe, little is known about the physical nature of this dark matter. Among its unknown properties is the lifetime of the dark matter. While it is certainly plausible that the dark matter is perfectly stable, the only immediate lower bound on the dark matter lifetime is set by its presence in the Universe today, $\tau_{\mathrm{DM}} \gtrsim 10^{17}$ sec. If the dark matter is in fact unstable, it may be possible to detect its decay products as exotic contributions to the fluxes of photons, neutrinos or charged cosmic rays.

Indeed, over the last couple of years, cosmic-ray telescopes have made a number of unexpected observations that might indicate the presence of an exotic primary cosmic-ray component. Namely, the PAMELA telescope observed a steep rise in the positron fraction extending up to at least 100 GeV,[1] in conflict with expectations from standard astrophysical processes. Furthermore, the Fermi Gamma-ray Space Telescope observed an $e^+e^-$ flux[2] that is harder than expected, also pointing to the possible presence of additional primary component. At the same time, no hadronic excess of antiprotons was observed.[3] Many attempts have been made to interpret the anomalous behaviors of the leptonic cosmic rays in terms of dark matter, as well as more conventional astrophysical sources such as pulsars.

Here we entertain the possibility that the dark matter is indeed composed of weakly unstable particles which decay mostly into charged leptons, thus constituting the source of the cosmic-ray anomalies while avoiding the overproduction of antiprotons. This interpretation suggests a scale for the dark matter mass of a few TeV and a lifetime of the order $10^{26}$ sec.[4] Although this lifetime exceeds the age of the Universe by several orders of magnitude, the resulting fluxes of charged cosmic rays, gamma rays and neutrinos can be in the observable range. We discuss two complementary methods of probing the region of the parameter space suggested by the cosmic-ray anomalies in the following.

## 2. Neutrino Signatures of Unstable Dark Matter

Neutrinos constitute an important probe to constrain decaying dark matter, since their production is almost inevitable in scenarios involving charged leptons among the decay products of the dark matter. Namely, neutrinos can be produced directly in the decay of dark matter particles, e.g. $\psi_{\rm DM} \to \ell^+\ell^-\nu$ for fermionic dark matter. Alternatively, secondary neutrinos can be produced in the decay of charged leptons originating from dark matter, or when the dark matter decays into hadronic final states. Since they do not diffuse or lose energy, neutrinos can therefore provide some important and relatively direct information about the possible decay of dark matter.

A major problem in the detection of a dark matter signal in neutrinos is the presence of large atmospheric backgrounds. The task is made easier, however, by the fact that for the relatively large dark matter masses under consideration here, the corresponding neutrino energies are in the range of hundreds of GeV to a few TeV, where atmospheric backgrounds decrease to a point where the hypothetical dark matter signal becomes comparable to the backgrounds. Nevertheless, the detection of a small contribution remains challenging. In the case of decaying dark matter, the best statistical significance of a dark matter-induced neutrino signal is achieved for full-sky observations, as opposed to observations focused on regions with a high dark matter density such as the center of the Galaxy. Due to the large propagation distances, neutrino flavor information is essentially erased, implying that the same signal is present in all flavors. In the energy range of interest here, muon neutrinos from dark matter decay scatter deep-inelastically, triggering a hadronic shower and a muon track in the detector, which pro-

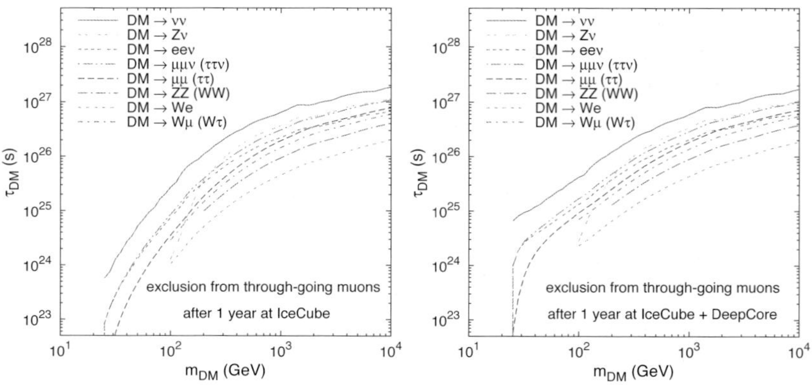


Fig. 1. Projected IceCube exclusion limits from the non-observation of a signifcant excess of neutrino-induced through-going muon events after one year of operation. *Left*: Exlusion limits for IceCube. *Right*: Exclusion imits for IceCube with the DeepCore subdetector.

vides the clearest observational signature.

The non-observation of a neutrino-induced excess of muon events can be used to constrain the parameter space of dark matter mass vs. lifetime, which is essentially unconstrained by existing data. However, it turns out that the upcoming generation of neutrino telescopes of $km^3$ dimensions should be able to constrain the parameter space at a level relevant to the cosmic-ray anomalies within a few years of observations.[5] We show the projected limits we find for IceCube in various dark matter decay modes in fig. 1. If the dark matter interpretation of the cosmic-ray anomalies is correct, a corresponding signal in neutrinos should be detectable within the near future. Should a signal be observed, however, the discrimination between different dark matter decay channels seems almost impossible from neutrino observations, as different decay modes yield very similar signatures at neutrino observatories. To this end, complementary information from other channels like gamma-ray observations are needed.

## 3. Gamma-Ray Lines from Radiative Dark Matter Decay

We now discuss constraints arising from the (non-)observations of monochromatic lines in the gamma-ray spectrum observed by satellite observatories or ground-based Cherenkov telescopes. Due to the absence of an excess of antiprotons, the dark matter must be coupled predominantly or exlusively to charged leptons in order to constitute a viable explanation of the cosmic-ray anomalies. If we assume that the dark matter indeed decays 'leptophilically' into charged leptons at tree-level, then radiative two-body decays are generically induced at the quantum level.[6] Although these higher-order decays are suppressed by powers of the couplings and loop factors, decays into monoenergetic photons are important because of the absence of astrophysical sources of gamma-ray lines. Therefore, the observation of a line in the diffuse gamma-ray flux or in extragalactic sources would be a compelling signature of an underlying particle-physics process.

If one assumes that the dominant decay mode of the dark matter is into charged leptons, where the simplest possibilities are $\phi_{\rm DM} \to \ell^+\ell^-$ for scalar dark matter and $\psi_{\rm DM} \to \ell^+\ell^- N$ for spin-1/2 dark matter, where $N$ is a neutral fermion. One can then combine the external charged lepton lines from the tree-level diagrams into a loop, where photons can be radiated off the loop at next-to-leading order. Examples of Feynman diagrams contributing to the tree-level decay of fermionic dark matter into charged leptons and the induced radiative one-loop decay $\psi_{\rm DM} \to \gamma N$ are shown in fig. 2. We explicitly examined the cases of both scalar and spin-1/2 dark

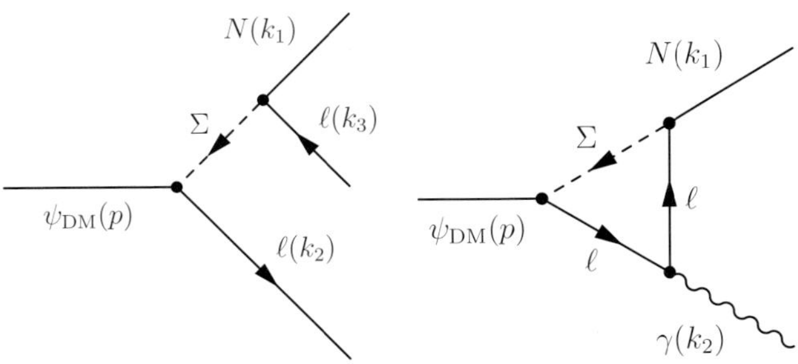


Fig. 2. *Left*: Tree-level diagram contributing to the decay $\psi_{\rm DM} \to \ell^+\ell^- N$ via a virtual charged scalar $\Sigma$. *Right*: One-loop diagram contributing to the radiative decay $\psi_{\rm DM} \to \gamma N$.

matter particles in a simple leptophilic toy model. In the case of scalar dark matter, the helicity suppression in the radiative decay $\phi_{\rm DM} \to \gamma\gamma$ generally leads to a very strong suppression of this decay mode, making the observation of the corresponding gamma-ray line practically impossible. However, for fermionic dark matter the suppression is relatively modest, typically leading to a ratio between two- and three-body decay widths that can be expressed schematically as

$$\frac{\Gamma(\psi_{\rm DM} \to \gamma N)}{\Gamma(\psi_{\rm DM} \to \ell^+\ell^- N)} = \frac{3\alpha_{\rm em}}{8\pi} R \times S \tag{1}$$

The numerical prefactor has the value $3\alpha_{\rm em}/(8\pi) \simeq 10^{-3}$, whereas the factors $R$ and $S$ contain the kinematics as well as the details of the couplings between dark matter and leptons.[a] These factors are typically of order one, so the radiative decay mode amounts to about a permil of the tree-level decay in many cases.

Interestingly, at gamma-ray energies of 100 – 400 GeV, this corresponds to partial lifetimes for decays into monochromatic photons which can be probed by Fermi LAT observations of the diffuse gamma-ray flux. At higher energies, Cherenkov telescopes provide information. At present, constraints do not reach the region of interest, but this is expected to change in the near future. In particular, the proposed Cherenkov Telescope Array (CTA) is expected to improve upon the sensitivity of current Imaging Air Cherenkov Telescopes by about an order of magnitude. This would allow one to probe significant parts of the parameter space relevant to dark matter interpretations of the cosmic-ray anomalies. We show the constraints on the partial

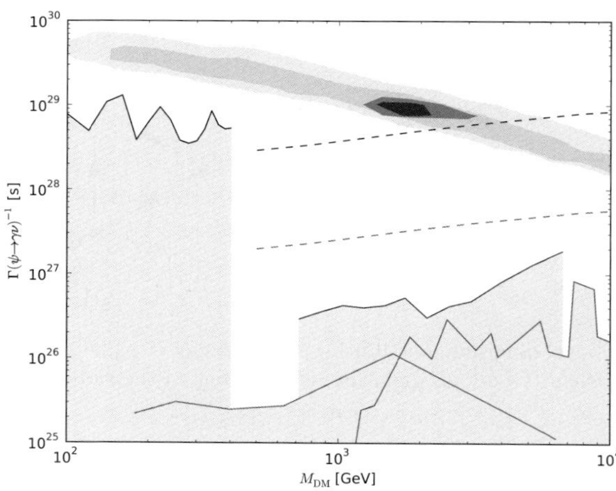


Fig. 3. Bounds on the partial lifetime for the radiative decay $\psi_{\rm DM} \to \gamma\nu$, as induced by the leptophilic decay $\psi_{\rm DM} \to \ell^+\ell^-\nu$. We assume a branching ratio $\mathrm{BR}(\psi_{\rm DM} \to \gamma\nu) = 3\times10^{-3}$ here, which is the maximum if the decay is mediated by a virtual scalar particle. Projected limits for the CTA are shown in dashed lines.

lifetime for $\psi_{\rm DM} \to \gamma\nu$ induced by the leading-order decay $\psi_{\rm DM} \to \ell^+\ell^-\nu$

[a]The dependences on kinematics and couplings factorize when the couplings between dark matter and leptons are chiral, which is the case in many models.

in fig. 3. Existing constraints are derived from line searches in the diffuse flux by Fermi LAT,[7] as well as in the flux from M31 by HEGRA.[8] Further constraints can be derived from measurements of the gamma-ray flux from the Perseus cluster by MAGIC[9] and the diffuse electron flux observed by H.E.S.S.[10] In the figure, we also indicate the region of the parameter space that provides a good agreement with the PAMELA and Fermi observations of charged leptons from the leading-order decay. The contours corresponding to a $3\sigma$ and $5\sigma$ confidence level around the best-fit points for the PAMELA positron fraction and the Fermi LAT electron flux are shown in orange and gray, respectively. Interestingly enough, in some cases the non-observation of gamma-ray lines can actually provide stronger constraints than measurements of charged cosmic rays, even though the lines are only generated at the quantum level.

## Acknowledgements

I would like to thank Laura Covi, Mathias Garny, Michael Grefe, Alejandro Ibarra and Christoph Weniger for pleasant collaborations. Support from the DFG cluster of excellence "Origin and Structure of the Universe" and the DFG Graduiertenkolleg "Particle Physics at the Energy Frontier of New Phenomena" is acknowleged.

## References

1. O. Adriani *et al.* [PAMELA Collaboration], Nature **458**, 607 (2009) [arXiv:0810.4995 [astro-ph]].
2. A. A. Abdo *et al.* [The Fermi LAT Collaboration], Phys. Rev. Lett. **102**, 181101 (2009) [arXiv:0905.0025 [astro-ph.HE]].
3. O. Adriani *et al.*, Phys. Rev. Lett. **102**, 051101 (2009) [arXiv:0810.4994 [astro-ph]].
4. A. Ibarra, D. Tran, C. Weniger, JCAP **1001**, 009 (2010). [arXiv:0906.1571 [hep-ph]].
5. L. Covi, M. Grefe, A. Ibarra and D. Tran, JCAP **1004**, 017 (2010) [arXiv:0912.3521 [hep-ph]].
6. M. Garny, A. Ibarra, D. Tran and C. Weniger, [arXiv:1011.3786 [hep-ph]].
7. A. A. Abdo *et al.*, Phys. Rev. Lett. **104**, 091302 (2010) [arXiv:1001.4836 [astro-ph.HE]].
8. F. A. Aharonian *et al.* [HEGRA Collaboration], Astron. Astrophys. **400**, 153 (2003) [arXiv:astro-ph/0302347].
9. J. Aleksic *et al.* [The MAGIC Collaboration], Astrophys. J. **710**, 634 (2010) [arXiv:0909.3267 [astro-ph.HE]].
10. F. Aharonian *et al.* [H.E.S.S. Collaboration], Phys. Rev. Lett. **101**, 261104 (2008) [arXiv:0811.3894 [astro-ph]].

# Gamma-ray anisotropies from decaying dark matter

C. Weniger

*Max-Planck-Institute for Physics,*
*80805 Munich, Germany*
*E-mail: weniger@mppmu.mpg.de*

Many theoretically well-motivated models predict the decay of dark matter with lifetimes much longer than the age of the universe. This can potentially lead to peculiar features in the observed cosmic-ray fluxes. In particular the gamma-ray channel, which carries directional information, is promising for dark matter searches, since it is sensitive to the underlying source distribution. We will discuss the peculiar dipole-like anisotropy of prompt gamma-ray radiation from dark matter decay inside our galaxy, as well as the angular power-spectrum of gamma-rays from dark matter decay outside of our galaxy. We will comment on the experimental detectability of such anisotropies.

## 1. Introduction

Despite the overwhelming gravitational evidence for the existence of non-baryonic dark matter, very little is known about its particle nature.[1,2] Many theoretically motivated dark matter models suggest that the dark matter particle might actually decay with very long lifetimes.[3–8] If the decays occur at a sufficiently large rate, the decay products could be observable as an exotic contribution to the high energy cosmic ray fluxes of gamma rays, electrons, positrons, antiprotons, neutrinos or antideuterons.[9–13] In particular the electron and positron fluxes attracted a lot of attention recently, due to observations by PAMELA, Fermi LAT and H.E.S.S., which are pointing to an excess of positrons and electrons at energies above 10 GeV.[14–17] Also with this respect, the gamma-ray channel is one of the most interesting to study, due to upcoming data from Fermi LAT, and due to its potential to discriminate between different astrophysical and exotic sources via the angular profile of the observed signal.

The decay of dark matter particles at galactic and extra-galactic distances produces a prompt gamma-ray flux at earth that is given by

$$\begin{aligned}\frac{dJ}{dE_\gamma}(\vec{\Omega}) &= \frac{\Omega_{\rm dm}}{4\pi m_{\rm dm}\tau_{\rm dm}}\times \\ &\times \int_0^\infty dz \frac{\rho_{\rm dm}(r(z),\vec{\Omega},z)}{H(z)}\frac{dN_\gamma}{dE_\gamma}[(z+1)E_\gamma]\; e^{-\tau(E_\gamma,z)}\,,\end{aligned} \tag{1}$$

where $H(z) = H_0\sqrt{\Omega_\Lambda + \Omega_{\rm m}(z+1)^3}$ is the Hubble expansion rate as a function of redshift $z$, $dN_\gamma/dE_\gamma$ is the energy spectrum of gamma rays produced in the decay of a dark matter particle, and $m_{\rm dm}$ and $\tau_{\rm dm}$ denote the dark matter mass and lifetime, respectively. Attenuation effects due to scattering of gamma-rays with the inter-galactic background light are taken into account via the optical depth $\tau(E_\gamma, z)$.[18] The dark matter distribution $\rho_{\rm dm}(r,\vec{\Omega},z)$ depends both on position, with $r = 0$ denoting the position of the sun, and on redshift. It can be split into an extra-galactic and into a galactic halo part, $\rho_{\rm dm} = \rho_{\rm dm}^{\rm halo} + \rho_{\rm dm}^{\rm eg}$. Correspondingly, the total gamma-ray flux from dark matter decay splits into two parts,

$$\frac{dJ}{dE_\gamma}(\vec{\Omega}) = \frac{dJ_{\rm eg}}{dE_\gamma}(\vec{\Omega}) + \frac{dJ_{\rm halo}}{dE_\gamma}(\vec{\Omega})\,. \tag{2}$$

In this note, we will summarize some aspects of the angular anisotropies of gamma rays from dark matter decay on large and small angular scales. We will firstly discuss the gamma-ray anisotropies in the extra-galactic flux associated with the large scale structures of the universe, and secondly the dipole-like anisotropy associated with the Milky Way halo. For a more detailed discussion see Ref. 19.

## 2. Angular Power Spectrum

For the angular power spectrum $C_\ell$ of gamma rays from dark matter decay (for the annihilation case, see *e.g.* Ref. 20–27) spatial fluctuations of the dark matter density on small scales are relevant. These fluctuations are related to subhalos of our own galactic halo, and to the large-scale structure of the dark matter distribution in the nearby universe. We will concentrate on the predictions related to the large-scale structures, since the mass fraction in subhalos is expected to be small and negligible for our purposes.

The fluctuations of the dark matter gamma-ray flux in a certain energy band can be expanded in spherical harmonics $Y_{\ell m}$ with coefficients $a_{\ell m}$ according to

$$\delta J(\vec{n}) \equiv J(\vec{n}) - \langle J\rangle = \langle J\rangle \sum_{\ell m} a_{\ell m} Y_{\ell m}(\vec{n})\,, \tag{3}$$

where $\langle J \rangle$ denotes the mean gamma-ray flux, averaged over the whole sky. Our goal is to calculate the angular power spectrum $C_\ell \equiv \langle |a_{\ell m}|^2 \rangle$, which can be split up into an extra-galactic and a halo part according to

$$C_\ell = f_{\rm eg}^2 C_\ell^{\rm eg} + (1 - f_{\rm eg})^2 C_\ell^{\rm halo} \,. \quad (4)$$

Here, $f_{\rm eg}$ denotes the fraction of dark matter induced photons with extra-galactic origin.

The angular power spectrum of the halo component, $C_\ell^{\rm halo}$, can be simply calculated from the angular profile of the halo flux, $J_{\rm halo}(\psi)$, where $\psi$ denotes the angle to the galactic center. On the other hand, following the Limber approximation, the extra-galactic component can be determined by[21]

$$C_\ell^{\rm eg} = \langle J_{\rm eg} \rangle^{-2} \int_0^\infty \frac{dr}{r^2} W(z)^2 P\left(k = \frac{\ell}{r}, z\right) \,. \quad (5)$$

Here, $P(k, z)$ denotes the power spectrum of the dark matter distribution as a function of redshift $z$, and the comoving coordinate $r$ is related to $z$ by $dr = dz/H(z)$. Furthermore, $W(z)$ is the so-called window-function, which incorporates attenuation effects and information about the dark matter decay spectrum, see Eq. (1). The region of the universe that contributes to the observable gamma-ray fluxes is large for lower gamma-ray energies. At *e.g.* $E_\gamma = 1\,{\rm TeV}$, around 90% of the observable photons come from comoving distances $r \lesssim 400\,{\rm Mpc}$, whereas at lower gamma-ray energies $E_\gamma = 10\,{\rm GeV}$ photons with distances up to 8 Gpc are received. Lastly, the dark matter power spectrum $P(k)$ at $z = 0$ can be derived from $N$-body simulations or analytically in the Halo model.

The total angular power spectrum of gamma rays from decaying dark matter is shown in Fig. 1, including the halo and the extra-galactic component, and assuming that the decay of the dark matter particle only produces mono-energetic photons with energies $E_\gamma = 10\,{\rm GeV}$, 100 GeV or 1 TeV. As apparent from the plot, the halo component of $C_\ell$ increases slightly with increasing $E_\gamma$, which is related to a decrease of the extra-galactic contributions, namely $f_{\rm eg}$ in Eq. (4). The energy dependence of $C_\ell^{\rm eg}$ is more complicated, due to its dependence on the window function and attenuation effects. In the whole considered energy regime, the halo component dominates at large angles as long as $\ell \lesssim 100$, whereas the extra-galactic component takes over at smaller angles with $\ell \gtrsim 100$.

To estimate the observational prospects for Fermi LAT, we will consider a simple situation where the total extra-galactic gamma-ray background

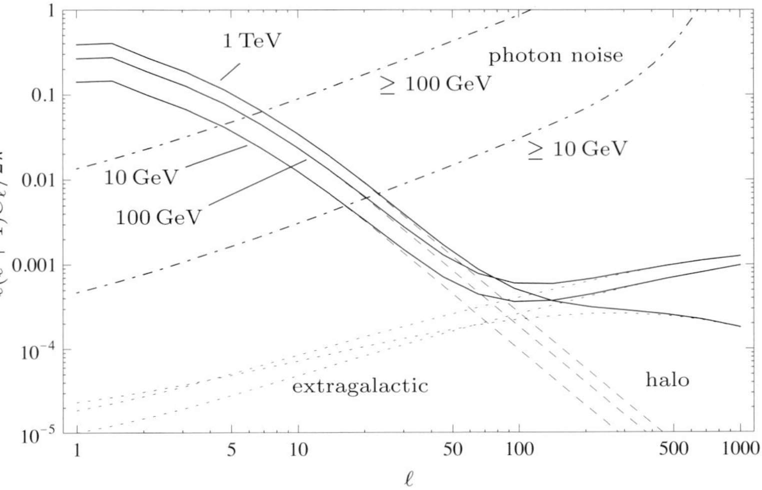


Fig. 1. Angular power spectrum of gamma-ray signal from dark matter decaying via $\psi_{\rm dm} \to \gamma\nu$. The gamma-ray energy is $E_\gamma = 10\,{\rm GeV}$, $100\,{\rm GeV}$ or $1\,{\rm TeV}$. Shown are the contribution from the halo component (*dashed lines*), the contribution from the extragalactic radiation (*dotted lines*), and their sum (*solid lines*). The *dash-dotted* lines show our predictions of the photon noise level of Fermi LAT after five years of observation, for photon energies above 10 GeV and above 100 GeV. Details can be found in Ref. 19.

(EGBG) above a certain energy is assumed to be due to dark matter decay. Then the $1\sigma$ deviation of the power spectrum, averaged over $\ell$-bins with size $\Delta\ell$, is given by[21]

$$\delta C_\ell = \sqrt{\frac{2}{(2\ell+1)\Delta\ell f_{\rm sky}}}\left(C_\ell + \frac{C_N}{W_\ell^2}\right), \tag{6}$$

where $C_N = 4\pi f_{\rm sky} N_\gamma^{\rm tot}/(N_\gamma^{\rm s})^2$ is the photon noise, $N_\gamma^{\rm s}$ and $N_\gamma^{\rm tot}$ denote the number of measured dark matter photons and number of total photons, respectively, $f_{\rm sky}$ is the observed fraction of the sky and $W_\ell = \exp(-\sigma_b^2\ell^2/2)$ is the window function of a Gaussian point-spread function. For the Fermi LAT we take as beam size $\sigma_b = 0.1^\circ$, we set $\Delta\ell = 0.5\,\ell$, and we only consider latitudes with $|b| \geq 20^\circ$, which implies that $f_{\rm sky} = 0.66$. In this part of the sky the ratio between the total gamma-ray fluxes and the EGBG flux is a factor of a few, and we will set $N_\gamma^{\rm tot}/N_\gamma^{\rm s} = 3$ for definiteness. As EGBG we will take the one determined by Fermi LAT.[28]

The resulting photon noise (second term of $\delta C_\ell$ in Eq. (6)) for five-years Fermi LAT observation is shown in Fig. 1 by the two dash-dotted lines, taking into account only gamma rays with energies $E_\gamma \geq 10\,{\rm GeV}$

or $\geq 100\,\text{GeV}$, respectively. As apparent from these lines, the small-scale fluctuations are completely hidden under photon noise in the energy range we consider, and hence they are not directly accessible by current instruments like Fermi LAT. However, since the small-scale angular fluctuations of gamma-rays from annihilating dark matter or astrophysical sources are typically larger then the ones from decaying dark matter and potentially observable, the non-observation of angular fluctuations might finally help to distinguish a decaying dark matter signal from other sources.

## 3. Dipole-like halo anisotropy

The offset between sun and galactic center causes a peculiar angular dependence in the gamma-ray signal from dark matter decaying[29] or annihilating[30,31] in the Milky Way halo, even if the halo profile itself is isotropic. The halo signal is largest in direction of the galactic center and smallest in direction of the galactic anticenter. The observation of an anisotropy in the EGBG that is aligned in this way would be a striking hint for a contribution from dark matter decay.

It is convenient to define the anisotropy parameter

$$A = \frac{\bar{J}_{\text{GC}} - \bar{J}_{\text{GAC}}}{\bar{J}_{\text{GC}} + \bar{J}_{\text{GAC}}}\,, \tag{7}$$

where $\bar{J}_{\text{GC}}$ and $\bar{J}_{\text{GAC}}$ in general denote the total diffuse gamma-ray flux (from dark matter and from astrophysical sources) integrated over $E_\gamma$ in some energy range, and averaged over the hemisphere in direction of the galactic center (GC) and anticenter (GAC), respectively. Sky regions with small galactic latitudes, $|b| < 10^\circ$, are excluded from the average.

As an example, we calculate the anisotropy parameter $A$ for the decay $\psi \to \gamma\nu$ of a fermionic dark matter particle $\psi$, after subtracting astrophysical sources (the galactic foreground and the extra-galactic gamma-ray background), in order to compare the anisotropy expected *purely* from dark matter decay to the anisotropy expected from the galactic models. In this case, the energy spectrum of gamma rays has two components: a monoenergetic line from the decay of dark matter particles in the halo and a redshifted line from decays at cosmological distances. It turns out that the anisotropy of the "pure" dark matter signal ranges between 0.20 and 0.36, depending on the gamma-ray energy, but with only little dependence on the profile of the dark matter halo.

To illustrate the effect of the dark matter induced anisotropy on the *total* fluxes, we show in the left panels of Fig. 2 the anisotropy that would be

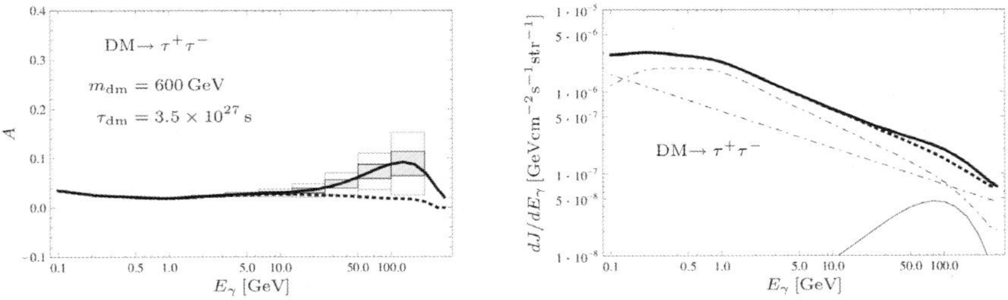


Fig. 2. *Left panel:* Anisotropy of the gamma-ray flux from dark matter decay into $\tau^+\tau^-$ pairs as a function of energy. The *dotted line* shows the background anisotropy as expected from the galactic foreground, while the *solid line* shows the anisotropy of signal + background. The boxes show estimates of the statistical errors for one-year and five-year Fermi LAT observations. *Right panel:* Gamma-ray fluxes averaged over all galactic longitudes as a function of energy. The *thin solid line* shows the gamma rays from dark matter decay (including also inverse Compton radiation from electrons and positrons produced in the dark matter decay). The two *dash-dotted lines* show the adopted astrophysical EGBG and the galactic foreground separately. The *thick solid line* shows the sum of all contributions, whereas the *thick dotted line* shows the sum without contributions from dark matter. For details see Ref. 19.

measurable if the dark matter particles decay exclusively into $\tau^+\tau^-$ pairs. As galactic foreground we take predictions of the conventional GALPROP-model as presented in Ref. 32 (model 44_500180), and as astrophysical EGBG we will use a simple power-law in agreement with the Fermi LAT measurements.[28] The dark matter mass is taken to be $m_{\rm dm} = 600\,{\rm GeV}$, and we set the lifetime to $\tau_{\rm dm} \simeq 3.5 \times 10^{27}\,{\rm s}$. The energy spectra of the photons produced in the decay are calculated with the event generator PYTHIA 6.4.[33] The lifetime is chosen such that the gamma-ray fluxes are below and compatible with the EGBG, as shown in the right panel of Fig. 2. Most strikingly, for the adopted choice of parameters and foreground model, an anisotropy is predicted that is significantly different to the one expected from the diffuse galactic emission alone. Namely, the contribution from dark matter decay shows up as a spectral feature in the anisotropy parameter of the overall flux. Such an anisotropy could be indeed observable by Fermi LAT, as illustrated by the boxes in the figure, which correspond to our estimates of the one-year and five-year statistical errors of Fermi LAT.

As an important application of the anisotropy parameter, one can consider the decay channels that are compatible with the dark matter interpretation of the PAMELA/Fermi positron excess. If dark matter decay is the origin of the excesses, the predicted anisotropies in the gamma-ray flux can

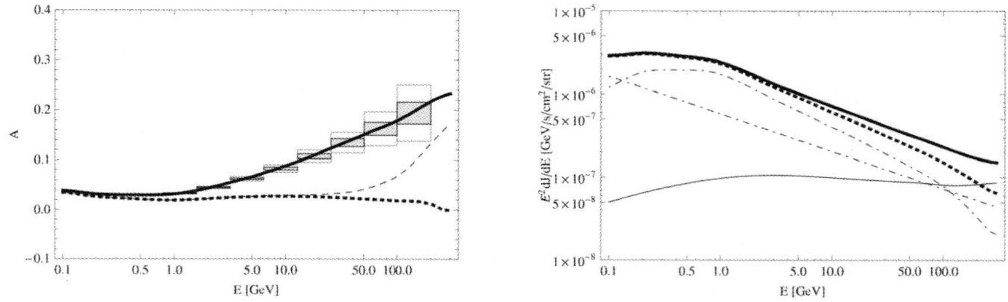


Fig. 3. Like Fig. 2, but for the decay channel $\psi_{\rm DM} \to \mu^+\mu^-\nu$ with $m_{\rm dm} = 2500\,{\rm GeV}$ and $\tau_{\rm dm} = 2.3 \times 10^{26}\,{\rm s}$, which was found to fit the positron excess as observed by PAMELA/Fermi. In the left panel, the *thin dashed line* shows additionally the anisotropy of the overall flux when including the dark matter signal, but neglecting the inverse Compton radiation coming from electrons and positrons produced in the dark matter decay.

be quite large. In Fig. 3 we show our results for the anisotropy, $A$, which is expected to be observed by the Fermi LAT if the dark matter particle decays via one exemplary decay channel that well fits the positron and electron data.[34] Even although the shown channel is compatible with the data of the EGBG, a sizeable anisotropy, around $A \simeq 0.2$, is predicted at high energies. This is due to the prompt gamma rays, coming directly from dark matter decay, as well as due to the gamma-rays stemming from inverse Compton scattering of the electrons and positrons that are produced in the decay. The observation of such an anisotropy would be a strong hint in favor of the dark matter interpretation of the positron excess. The non-observation of such an anisotropy on the other hand could finally be useful to exclude such interpretations, although this requires a robust understanding of the galactic foreground.

Finally, the dipole-like anisotropy described above can be also relevant for cosmic-ray experiments with large isotropic foregrounds, like atmospheric Cherenkov telescopes or observations of cosmic-ray neutrinos.

## 4. Conclusions

We have calculated the angular power spectrum of the gamma-ray signal from dark matter decay, which exhibits imprints of the large scale structure of the universe. We have shown, that for gamma-ray energies above a few GeV these small scale effects are too weak to be directly observed with present instruments. However, the non-observations of these fluctuations could finally help to distinguish decaying from annihilating dark matter or astrophysical sources.

We have discussed the dipole-like anisotropy of the gamma-ray flux that can come from dark matter decaying inside our galactic halo. This anisotropy is due to the offset between Earth and the center of the dark matter halo. We have shown that a contribution from decaying dark matter could show up as a spectral feature in the anisotropy of the overall observed gamma-ray fluxes. Furthermore, we have shown that if the interpretation of the PAMELA/Fermi positron excess in terms of decaying dark matter is correct, a rise in the anisotropy of the overall fluxes at high energies should be observed. The non-observation of such an anisotropy could finally help to exclude this interpretation, which requires however a robust understanding of the galactic foreground.

## Acknowledgments

The author likes to thank the organizers of the *ICATPP Conference on Cosmic Rays for Particle and Astroparticle Physics, Como, Italy 7–8 Oct 2010* for an interesting conference, and Alejandro Ibarra and David Tran for fruitful collaboration.

## References

1. J. Dunkley *et al.*, *Astrophys. J. Suppl.* **180**, 306 (2009).
2. G. Bertone, D. Hooper and J. Silk, *Phys. Rept.* **405**, 279 (2005).
3. W. Buchmüller, L. Covi, K. Hamaguchi, A. Ibarra and T. Yanagida, *JHEP* **03**, p. 037 (2007).
4. A. Ibarra, A. Ringwald and C. Weniger, *JCAP* **0901**, p. 003 (2008).
5. A. Ibarra, A. Ringwald, D. Tran and C. Weniger, *JCAP* **0908**, p. 017 (2009).
6. C.-R. Chen, F. Takahashi and T. T. Yanagida, *Phys. Lett.* **B671**, 71 (2009).
7. K. Kohri, A. Mazumdar, N. Sahu and P. Stephens, *Phys. Rev.* **D80**, p. 061302 (2009).
8. C. Arina, T. Hambye, A. Ibarra and C. Weniger, *JCAP* **1003**, p. 024 (2010).
9. M. R. Buckley, K. Freese, D. Hooper, D. Spolyar and H. Murayama, *Phys. Rev.* **D81**, p. 016006 (2010).
10. A. Ibarra and D. Tran, *JCAP* **0906**, p. 004 (2009).
11. M. Kadastik, M. Raidal and A. Strumia, *Phys. Lett.* **B683**, 248 (2010).
12. K. Ishiwata, S. Matsumoto and T. Moroi, *Phys. Lett.* **B675**, 446 (2009).
13. A. Ibarra and D. Tran, *JCAP* **0902**, p. 021 (2009).
14. O. Adriani *et al.*, *Nature* **458**, 607 (2009).
15. A. A. Abdo *et al.*, *Phys. Rev. Lett.* **102**, p. 181101 (2009).
16. F. Aharonian *et al.*, *Astron. & Astrophys.* **508**, 561 (2009).
17. D. Grasso *et al.*, *Astropart. Phys.* **32**, 140 (2009).
18. F. W. Stecker, M. A. Malkan and S. T. Scully, *Astrophys. J.* **648**, p. 774 (2006).

19. A. Ibarra, D. Tran and C. Weniger, *Phys. Rev.* **D81**, p. 023529 (2010).
20. S. Ando and E. Komatsu, *Phys. Rev.* **D73**, p. 023521 (2006).
21. S. Ando, E. Komatsu, T. Narumoto and T. Totani, *Phys. Rev.* **D75**, p. 063519 (2007).
22. J. M. Siegal-Gaskins, *JCAP* **0810**, p. 040 (2008).
23. J. M. Siegal-Gaskins and V. Pavlidou, *Phys. Rev. Lett.* **102**, p. 241301 (2009).
24. S. Ando, *Phys. Rev.* **D80**, p. 023520 (2009).
25. M. Fornasa, L. Pieri, G. Bertone and E. Branchini, *Phys. Rev.* **D80**, p. 023518 (2009).
26. L. Zhang, F. Miniati and G. Sigl (2010).
27. A. Cuoco, A. Sellerholm, J. Conrad and S. Hannestad (2010).
28. A. A. Abdo *et al.*, *Phys. Rev. Lett.* **104**, p. 101101 (2010).
29. G. Bertone, W. Buchmüller, L. Covi and A. Ibarra, *JCAP* **0711**, p. 003 (2007).
30. C. Calcaneo-Roldan and B. Moore, *Phys. Rev.* **D62**, p. 123005 (2000).
31. D. Hooper and P. D. Serpico, *JCAP* **0706**, p. 013 (2007).
32. A. W. Strong, I. V. Moskalenko and O. Reimer, *Astrophys. J.* **613**, 962 (2004).
33. T. Sjostrand, S. Mrenna and P. Skands, *JHEP* **05**, p. 026 (2006).
34. A. Ibarra, D. Tran and C. Weniger, *JCAP* **1001**, p. 009 (2010).

# LIST OF PARTICIPANTS AND CONFERENCE ORGANIZERS

| | | | |
|---|---|---|---|
| ALAVERDIAN | Artur | P.N. Lebedev Physical Institute | Russia |
| ANTONELLI | Vito | Dipartimento di Fisica and INFN Milano | Italy |
| BANKS | Thomas | University of California, Berkeley | USA |
| BATTISTON | Roberto | INFN Perugia | Italy |
| BARRAU | Aurélien | Laboratory for Subatomic Physics and Cosmology Grenoble | France |
| BELLI | Pierluigi | INFN - Roma Tor Vergata | Italy |
| BLASI | Pasquale | Osservatorio Astrofisico di Arcetri | Italy |
| BLUEMER | Johannes | Karlsruhe Institute of Technology | Germany |
| BOBIK | Pavol | Institute of Experimental Physics of the Slovak Academy of Sciences | Slovak Republic |
| BOLDYREV | Alexey | P.N. Lebedev Physical Institute | Russia |
| BONAMENTE | Emanuele | INFN sezione di Perugia - c/o Dip. Di Fisica | Italy |
| BORISOV | Sergey | P.N. Lebedev Physical Institute | Russia |
| BRIGIDA | Monica | Dipartimento Interateneo di Fisica dellUniversità di Bari | Italy |
| BRINGMANN | Torsten | II. Institute for Theoretical Physics, University of Hamburg | Germany |
| BUCIK | Radoslav | Max Planck Institute for Solar System Research | Slovak Republic |
| BUESCHING | Ingo | Theoretical Physics, Ruhr-University | Germany |
| CAMARRI | Paolo | Universitàdi Roma Tor Vergata and INFN Roma Tor Vergata | Italy |
| CAPRIOLI | Damiano | INAF/Osservatorio Astrofisico di Arcetri - Firenze | Italy |
| CERNUDA | Ignacio | CIEMAT | Spain |
| CIRELLI | Marco | CENR and CNRS CEA Saclay | France |
| CODINO | Luca | INFN and university of Perugia | Italy |

| | | | |
|---|---|---|---|
| COMBET | Céline | Dept. of Physics and Astronomy - University of Leicester | U.K. |
| COMISSO | Luca | INFN Milano Bicocca | Italy |
| CONSOLANDI | Cristina | INFN Milano Bicocca | Italy |
| COVINO | Stefano | INAF/Brera Astronomical Observatory | Italy |
| DEGRANGE | Bernard | CNRS/IN2P3 | France |
| DELAHAYE | Timur | IFT | France |
| DELLA TORRE | Stefani | INFN Milano Bicocca | Italy |
| DI FELICE | Valeria | University of Rome Tor Vergata | Italy |
| DONATO | Florenza | Dipartimento di Fisica Teorica, Un. Torino | Italy |
| DORMODY | Michael | SCIPP | USA |
| DRURY | Luke | Dublin Institute for Advanced Studies | Ireland |
| FANFANI | Alessandra | University and INFN of Bologna | Italy |
| FIGUEROA-FELICIANO | Enectali | Massachusetts Institute of Technology | USA |
| FRUCK | Christian | Max Planck Institut fuer Physik Muenchen | Germany |
| GABICI | Stefano | CNRS - APC | France |
| GERVASI | Massimo | INFN/University of Milano Bicocca | Italy |
| GIANI | Simone | CERN | Switzerland |
| GIROLETTI | Elio | University of Pavia and INFN of Pavia | Italy |
| GOSLING | Christopher | FermilaB MS 226 | USA |
| GRANDI | Davide | INFN Milano Bicocca | Italy |
| GRANT | Darren | University of Alberta | Canada |
| GRASSO | Dario | INFN Pisa | Italy |
| HAINO | Sadakazu | INFN Perugia | Italy |
| HALLIN | Aksel | University of Alberta | Canada |
| HANASZ | Michal | Centre for Astronomy, Nicolaus Copernicus University | Poland |
| HASEBE | Nobuyuki | Waseda University | Japan |
| HAUNGS | Andreas | Karlsruhe Institute of Technology Campus North – IK | Germany |
| IBARRA | Alejandro | Technical University of Munich | Germany |
| JANSEN | Frank | DLR Institute of Space Systems | Germany |
| JORDAN | Thomas | University of Florida/Fermi National Accelerator Laboratory | USA |
| KACHELRIESS | Michael | Department of Physics, NTNU | Norway |

| | | | |
|---|---|---|---|
| KARELIN | Alexander | NRNU MEPhI | Russia |
| KLIMOV | Pavel | Lomonosov Moscow State University Skobeltsyn Institute of Nuclear Physics | Russia |
| KUDELA | Karel | IEP SAS Kosice | Slovak Republic |
| LAVALLE | Julien | Torino University and INFN | Italy |
| LINEROS | Rodriguez | INFN - Univ. Degli Studi di Torino | Italy |
| LEROY | Claude | Université de Montréal | Canada |
| MAILYAN | Bagrat | Alikhanyan National Laboratory | Armenia |
| MALTEZOS | Stavos | National Technical University of Athens NTUA | Greece |
| MARTIN | Pierrick | MPE Munich | Germany |
| MAURIN | David | LPNHE - IN2P3/CNRS/Universites Paris VI et Paris VII | France |
| MEMOLA | Elisabetta | INFN Milano Bicocca | Italy |
| MERTSCH | Philipp | Rudolf Peierls Centre for Theoretical Physics | UK |
| MONTE | Claudia | University and INFN Bari | Italy |
| MONTINI | Paolo | Università degli studi Roma Tre | Italy |
| MORLINO | Giovanni | INAF/Arcetri | Italy |
| ORIO | Filippo | Universit degli studi di Roma La Sapienza | Italy |
| PATO | Miguel | ITP Zurich / IAP Paris / Università degli Studi di Padova | Italy |
| PARKER | Becky | Langton Star Centre | UK |
| PESCE-ROLLINS | Melissa | INFN Pisa | Italy |
| PIERI | Lidia | University and INFN of Trento | Italy |
| PINFOLD | James L. | University of Alberta | Canada |
| PIRO | Marie-Cécile | Université de Montréal | Canada |
| POCHON | Jonathan | Instituto Astrofisica de Canarias | Spain |
| POTGIETER | Marius | Centre for Space Physics | South Africa |
| PUTZE | Antje | Oskar Klein Centre (SU / KTH) | Sweden |
| QUINN | John | University College Dublinn | Ireland |
| RANCOITA | Pier-Giorgio | INFN MIlano-Bicocca | Italy |
| RENO | Mary Hall | University of Iowa | USA |

| | | | |
|---|---|---|---|
| RICCIARINI | Sergio | INFN Structure of Florence | Italy |
| ROBERTS | Jonathan | Center for Cosmology and Particle Physics, New York | USA |
| RUCHTI | Randal | University of Notre Dame | USA |
| SAKO | Takashi | Solar-Terrestrial Environment Laboratory, Nagoya University | Japan |
| SALATI | Pierre | LAPTH and Université de Savoie | France |
| SANTANGELO | Andrea | Kepler Center IAAT | Germany |
| SARKAR | Subir | University of Oxford | UK |
| SCALZOTTO | Villi | INFN Trieste | Italy |
| SCHUESSLER | Fabian | CEA / Irfu | France |
| SEO | Eun-Suk | University of Maryland | USA |
| SERPICO | Pasquale | CERN and LAPTH | France |
| SHUSTOVA | Olga | Moscow State University | Russia |
| SINITSYNA | Vera G. | P.N. Lebedev Physical Institute | Russia |
| SINITSYNA | VERA Y. | P.N. Lebedev Physical Institute | Russia |
| SIRONI | Lorenzo | Princeton University | USA |
| SMOLEK | Karel | Czech Technical University in Prague | Czech Republic |
| SNOW | Gregory | University of Nebraska | USA |
| SOLA | Valentina | Torino University and INFN | Italy |
| SPARVOLI | Roberta | University of Rome Tor Vergata and INFN | Italy |
| SPRINGER | Wayne | University of Utah | USA |
| STAWARZ | Lukasz | ISAS/JAXA | Japan |
| STRELNIKOVA | Olga | Scobeltsyn Institute of Nuclear Physics | Russia |
| STRONG | Andrew | MPE | Germany |
| SUVOROV | Yury | LNGS-INFN | Italy |
| TACCONI | Mauro | INFN Milano Bicocca | Italy |
| TAKAI | Helio | Brookhaven National Laboratory | USA |
| TARTARI | Andrea | Università degli Studi di Milano-Bicocca | Italy |
| TAYLOR | Andrew | ISDC | France |
| THOMSON | Gordon | University of Utah | USA |
| TIMMERMANS | Charles | Nikhef/Radboud University | The Netherlands |
| TOMAS | Ricard | Universitaet Hamburg | Germany |
| TOMASSETTI | Nicola | Perugia University and INFN | Italy |
| TRAN | David | Technical University of Munich | Germany |

| | | | |
|---|---|---|---|
| WALDENMAIER | Tilo | Humboldt-University Berlin | Germany |
| WALTER | Roland | ISDC, University of Geneva | Switzerland |
| WENIGER | Christoph | DESY, Hamburg | Germany |
| WILLIAMS | Crispin | INFN Bologna | Italy |
| WOTTON | Stephen | University of Cambridge | UK |
| ZAYMIDOROGA | Oleg | Joint Institute for Nuclear Research, Dubna | Russia |
| ZUREK | Kathryn | University of Michigan | USA |